Dictionary of Literary Biography® • Volume One Hundred Sixty-Nine

American Poets Since World War II
Fifth Series

Dictionary of Literary Biography® • Volume One Hundred Sixty-Nine

American Poets Since World War II
Fifth Series

Edited by
Joseph Conte
State University of New York at Buffalo

A Bruccoli Clark Layman Book
Gale Research
Detroit, Washington, D.C., London

Advisory Board for
DICTIONARY OF LITERARY BIOGRAPHY

John Baker
William Cagle
Patrick O'Connor
George Garrett
Trudier Harris

Matthew J. Bruccoli and Richard Layman, Editorial Directors
C. E. Frazer Clark Jr., Managing Editor
Karen Rood, Senior Editor

Printed in the United States of America

The paper used in this publication meets the minimum requirements
of American National Standard for Information Sciences–Permanence
Paper for Printed Library Materials, ANSI Z39.48-1984. ∞™

This publication is a creative work fully protected by all applicable copyright laws, as well as by misappropriation, trade secret, unfair competition, and other applicable laws. The authors and editors of this work have added value to the underlying factual material herein through one or more of the following: unique and original selection, coordination, expression, arrangement, and classification of the information.

All rights to this publication will be vigorously defended.

Copyright © 1996 by Gale Research
835 Penobscot Building
Detroit, MI 48226

All rights reserved including the right of reproduction in
whole or in part in any form.

Library of Congress Cataloging-in-Publication Data

American Poets since World War II. Fifth series / edited by Joseph Conte.
　　p. cm. – (Dictionary of literary biography; v. 169)
"A Bruccoli Clark Layman book."
Includes bibliographical references and index.
ISBN 0-8103-9932-6 (alk. paper)
1. American poetry – 20th century – Bio-bibliography – Dictionaries. 2. Poets, American – 20th century – Biography – Dictionaries. 3. American poetry – 20th century – History and criticism.
I. Conte, Joseph Mark, 1960–   . II. Series.
PS323.5.A5   1996
811'.5409'03 – dc20　　　　　　　　　　　　　　　　　　　　　　96-31650
　[B]　　　　　　　　　　　　　　　　　　　　　　　　　　　　　　　CIP

Per la famiglia Conte

Contents

Plan of the Series ix
Introduction xi

David Antin (1932–) 3
  Kenneth Sherwood

Charles Bernstein (1950–) 13
  Loss Pequeño Glazier

Ted Berrigan (1934–1983) 29
  Edward Halsey Foster

Elizabeth Bishop (1911–1979) 35
  Brett C. Millier

Louise Bogan (1897–1970) 54
  Brett C. Millier

Charles Bukowski (1920–1994) 63
  Michael Basinski

Robert Creeley (1926–) 78
  Loss Pequeño Glazier

Irving Feldman (1928–) 98
  Willard Spiegelman

Kathleen Fraser (1935–) 106
  Peter Quartermain

Allen Ginsberg (1926–) 116
  Laszlo K. Géfin

Anthony Hecht (1923–) 137
  Ashley Brown

Ronald Johnson (1935–) 146
  Eric Murphy Selinger

Carolyn Kizer (1925–) 157
  Elizabeth B. House

Robert Lowell (1917–1977) 165
  Ashley Brown

Nathaniel Mackey (1947–) 179
  Mark Scroggins

W. S. Merwin (1927–) 192
  James McCorkle

Michael Palmer (1943–) 215
  Keith Tuma

Ishmael Reed (1938–) 224
  Robert S. Friedman

James Schuyler (1923–1991) 235
  Joseph M. Conte

Anne Sexton (1928–1974) 244
  Diane Wood Middlebrook

Ron Silliman (1946–) 253
  T. C. Marshall

Cathy Song (1955–) 267
  Susan M. Schultz

James Tate (1943–) 275
  Chris Stroffolino

Rosmarie Waldrop (1935–) 284
  Steven R. Evans

Richard Wilbur (1921–) 297
  Richard J. Calhoun

James Wright (1927–1980) 312
  Andrew Elkins

Checklist of Further Readings 327
Contributors 334
Cumulative index 337

Plan of the Series

> ... *Almost the most prodigious asset of a country, and perhaps its most precious possession, is its native literary product — when that product is fine and noble and enduring.*
>
> Mark Twain*

The advisory board, the editors, and the publisher of the *Dictionary of Literary Biography* are joined in endorsing Mark Twain's declaration. The literature of a nation provides an inexhaustible resource of permanent worth. We intend to make literature and its creators better understood and more accessible to students and the reading public, while satisfying the standards of teachers and scholars.

To meet these requirements, *literary biography* has been construed in terms of the author's achievement. The most important thing about a writer is his writing. Accordingly, the entries in *DLB* are career biographies, tracing the development of the author's canon and the evolution of his reputation.

The purpose of *DLB* is not only to provide reliable information in a convenient format but also to place the figures in the larger perspective of literary history and to offer appraisals of their accomplishments by qualified scholars.

The publication plan for *DLB* resulted from two years of preparation. The project was proposed to Bruccoli Clark by Frederick G. Ruffner, president of the Gale Research Company, in November 1975. After specimen entries were prepared and typeset, an advisory board was formed to refine the entry format and develop the series rationale. In meetings held during 1976, the publisher, series editors, and advisory board approved the scheme for a comprehensive biographical dictionary of persons who contributed to North American literature. Editorial work on the first volume began in January 1977, and it was published in 1978. In order to make *DLB* more than a reference tool and to compile volumes that individually have claim to status as literary history, it was decided to organize volumes by topic, period, or genre. Each of these freestanding volumes provides a biographical-bibliographical guide and overview for a particular area of literature. We are convinced that this organization — as opposed to a single alphabet method — constitutes a valuable innovation in the presentation of reference material. The volume plan necessarily requires many decisions for the placement and treatment of authors who might properly be included in two or three volumes. In some instances a major figure will be included in separate volumes, but with different entries emphasizing the aspect of his career appropriate to each volume. Ernest Hemingway, for example, is represented in *American Writers in Paris, 1920-1939* by an entry focusing on his expatriate apprenticeship; he is also in *American Novelists, 1910-1945* with an entry surveying his entire career. Each volume includes a cumulative index of the subject authors and articles. Comprehensive indexes to the entire series are planned.

With volume ten in 1982 it was decided to enlarge the scope of *DLB*. By the end of 1986 twenty-one volumes treating British literature had been published, and volumes for Commonwealth and Modern European literature were in progress. The series has been further augmented by the *DLB Yearbooks* (since 1981) which update published entries and add new entries to keep the *DLB* current with contemporary activity. There have also been *DLB Documentary Series* volumes which provide biographical and critical source materials for figures whose work is judged to have particular interest for students. One of these companion volumes is entirely devoted to Tennessee Williams.

We define literature as the *intellectual commerce of a nation*: not merely as belles lettres but as that ample and complex process by which ideas are generated, shaped, and transmitted. *DLB* entries are not limited to "creative writers" but extend to other figures who in their time and in their way influenced the mind of a people. Thus the series encompasses historians, journalists, publishers, and screenwriters. By this means readers of *DLB* may be aided to perceive literature not as cult scripture in the keeping of intellectual high priests but firmly positioned at the center of a nation's life.

---

*From an unpublished section of Mark Twain's autobiography, copyright by the Mark Twain Company

Plan of the Series

DLB includes the major writers appropriate to each volume and those standing in the ranks immediately behind them. Scholarly and critical counsel has been sought in deciding which minor figures to include and how full their entries should be. Wherever possible, useful references are made to figures who do not warrant separate entries.

Each *DLB* volume has a volume editor responsible for planning the volume, selecting the figures for inclusion, and assigning the entries. Volume editors are also responsible for preparing, where appropriate, appendices surveying the major periodicals and literary and intellectual movements for their volumes, as well as lists of further readings. Work on the series as a whole is coordinated at the Bruccoli Clark Layman editorial center in Columbia, South Carolina, where the editorial staff is responsible for accuracy of the published volumes.

One feature that distinguishes *DLB* is the illustration policy — its concern with the iconography of literature. Just as an author is influenced by his surroundings, so is the reader's understanding of the author enhanced by a knowledge of his environment. Therefore *DLB* volumes include not only drawings, paintings, and photographs of authors, often depicting them at various stages in their careers, but also illustrations of their families and places where they lived. Title pages are regularly reproduced in facsimile along with dust jackets for modern authors. The dust jackets are a special feature of *DLB* because they often document better than anything else the way in which an author's work was perceived in its own time. Specimens of the writers' manuscripts are included when feasible.

Samuel Johnson rightly decreed that "The chief glory of every people arises from its authors." The purpose of the *Dictionary of Literary Biography* is to compile literary history in the surest way available to us — by accurate and comprehensive treatment of the lives and work of those who contributed to it.

The *DLB* Advisory Board

Introduction

Dictionary of Literary Biography 169: American Poets Since World War II, Fifth Series, as well as the sixth series to be published later, is devoted to American poets who have made a significant contribution to their art after 1945. Such a criterion allows for the inclusion of poets of different age groups, diverse styles, and competing poetic principles. Some of the poets presented here had already established their careers by the close of World War II; others have only recently begun to attract — or provoke — the critical attention that their talents deserve. The presence of poets accomplished in traditional forms and familiar genres alongside those practiced in a resolutely avant-garde approach is not an accident of this volume but its intention. Readers of DLB volumes naturally can choose to read according to their inclinations, but the juxtapositions of poetic careers that either reinforce or contend with one another is perhaps the chief advantage of a book with such variety of subjects and contributors. In the estimation of this editor, the poets included all have a strong claim for their importance in the period; however, the thorough critical appraisal, biographical information, and bibliographical support provided by the contributors will enable readers to judge each case for themselves.

The selection of entries for this volume gives special consideration to its relation to those prior volumes in the DLB series that address American poetry after 1945. *DLB 169* extends the selection of contemporary poets made by this editor in *DLB 165: American Poets Since World War II, Fourth Series* (1996) and supplements entries found in *DLB 120: American Poets Since World War II, Third Series* (1992) and *DLB 105: American Poets Since World War II, Second Series* (1991). A particular concern was to review the treatment of poets included in the two-volume *DLB 5: American Poets Since World War II, First Series* (1980). In the fifteen years since the publication of *DLB 5,* several of the poets included there have produced major new works or have since been the beneficiaries of extensive critical studies. Such changes called for an entirely fresh appraisal. In other instances, it has been deemed sufficient to update and revise the treatment originally afforded the poet to account for less dramatic shifts in a career. The passage of even fifteen years has naturally seen the decline of interest in certain midcentury authors, but it also brings a demand for a reassessment of postmodern poetry with entries devoted to previously overlooked — but by no means minor — poets. And finally, this volume presents several poets whose work was still in a gestational stage in 1980 and who thus appear in the DLB for the first time. Such poets may well introduce the poetics that will carry into the next century.

From this vantage point near the close of the twentieth century, the history of postmodern poetry can be described broadly in three generational clusters. The first generation of postmodern poets are those born shortly after the turn of the century. The milestones of high modernism, such as the publication of T. S. Eliot's *The Waste Land* (1922), were written while they were still quite young. In a literal manner they regard the modern poets of their day — Wallace Stevens, Ezra Pound, Eliot, Robert Frost, Gertrude Stein, Marianne Moore, and William Carlos Williams — as both their elders (by some twenty or more years) and as their mentors. With these mentors they carry on an extensive personal and epistolary contact. Their early mature work appears between 1929 and 1944, marked by the grinding demands of the Depression and often inflected by leftist politics and economic theory. They are well aware that their writing follows upon the dramatic breach with the genteel aesthetic of late-nineteenth-century poetry as found in Eliot's "The Love Song of J. Alfred Prufrock" (1917), Pound's first gathering of his epic in *A Draft of XXX Cantos* (1930), or Stein's cubist poetry in *Tender Buttons* (1914). In one sense they are late-modernists; they regard their work as an extension of the modernist campaign to "make it new." But they also struggle "to witness / and adjust" (adapting Williams's phrase) to the modern condition, interjecting within the revolutionary aesthetic the social, political, and personal concerns of their generation.

Louis Zukofsky (1904–1978) is a representative type of the first postmodernist, embarking upon his career under the close and sought-after tutelage of Ezra Pound (chronicled in their extensive correspondence between 1927 and 1963). The first half of Zukofsky's own eight-hundred-page epic poem, *"A,"* resembles the canto structure of Pound's "poem including history." After 1951, however, Zukofsky's work becomes increasingly hermetic, procedural in form, and language-oriented. In short, the latter part of his career represents a departure from his modernist mentor and his hortatory manner and culminates in such distinctively postmodern achievements as *80 Flowers* (1978). However different in style and biography, other members of this first generation follow a similar pattern that realizes a slow sundering of the mentor/disciple relationship. Among poets in this volume, Louise Bogan (1897-1970) immersed herself in the New York avant-garde of the 1920s that included Williams, Lola Ridge, Mina Loy, and Edmund Wilson before crafting her own formal lyric style. The famous meeting of Elizabeth Bishop (1911-1979) and Marianne Moore on the steps of the New York Public Library in 1934 initiated yet another lasting friendship and tutelage of a literary career, though Bishop's poetry assumed a more personal idiom than Moore's. The presence of these poets in this volume is not only justified by the substantial work they completed after 1945 but by their continued influence on contemporary poets. Their careers serve as instructive graphs of the transition between modern and postmodern poetics. In some cases the first postmoderns worked in relative obscurity until discovered, published, and in turn adopted as mentors by members of the postwar generation. Essays on several important poets born in the first decade of this century, including Charles Reznikoff, Edward Dahlberg, Laura Riding Jackson, Langston Hughes, and Stanley Kunitz, can be found in *DLB 45: American Poets, 1880-1945, First Series* (1986) and *DLB 48: American Poets, 1880-1945, Second Series* (1986).

Several recent studies, among them Albert Gelpi's *Coherent Splendor* (1987) and Cary Nelson's *Repression and Recovery* (1989), have shown that modern American poetry was hardly the monolithic program that the New Critical canon had suggested. Fissures arising from differing relationships with romanticism, in the test of domestic or expatriate affiliations, and in the conflicting political allegiances of the 1930s and 1940s suggest that the spate of movements and groupings that were identified or actively promoted after 1950 were an inevitable result of such disagreements.

The postwar generation of postmodern poets are those born in the second quarter of the century, with a particularly distinguished class of 1926-1927 that includes A. R. Ammons, John Ashbery, Robert Bly, Robert Creeley, Allen Ginsberg, James Merrill, W. S. Merwin, Frank O'Hara, and James Wright. Their notable early works appear between 1955 and 1970. It may be judicious to describe the attitude of postwar poets to their modernist forebears as ambivalent; that is, capable of registering strongly positive or negative responses in separate cases. Creeley, for example, rejects the symbolist mode of Eliot for the immediate contact with the real he finds in Williams. This shift from a poetics of transcendence to one of immanence becomes a defining characteristic of midcentury poetry, as Charles Altieri has argued in *Self and Sensibility in Contemporary American Poetry* (1984). Charles Olson, who was the first poet to identify himself as "post-modern" in "The Present is Prologue" (1950), espouses a posthumanism and posthistoricism that runs counter to the romantic ideology found in Stevens. But Ammons, Ashbery, and James Schuyler have retained a close affinity for the romantic imagination expressed in Stevens's "poem of the act of the mind." Robert Lowell's first three volumes emulated the dense symbolic language and impersonal registers of Eliot, only to follow Ginsberg's *Howl* (1956) into a confessional mode in *Life Studies* (1959) that made a virtue of traumatic personal revelation. As M. L. Rosenthal and others have noted, the shift from an impersonal to a personal register and the disclosure of the self in the poem are distinguishing aspects of postwar poetry. In the end, the postwar poets neither denied the continuing relevance of the modernist writers, nor pledged absolute fealty to their principles.

An additional source of generational tension stems from the longevity of the modernist poets whose major late works appear simultaneously with the important early works of the postwar poets. Thus one finds Creeley's landmark *For Love* published in the same year as Williams's remarkable *Pictures from Brueghel* (1962). Ashbery's *Some Trees* (1956) closely follows Stevens's *Collected Poems,* issued on his seventy-fifth birthday in 1954. Bishop, who was slow to publish and whose reputation accrued gradually, arrives at her *Selected Poems* in the same year as Moore's meticulous edition of her *Complete Poems* (1967).

One finds in the work of these postwar poets many acts of homage to their still-productive modernist predecessors. But these homages are accompanied by profuse statements on poetics that suggest their project was not merely to extend modernism but to advance from it. Olson's "Projective Verse," Denise Levertov's "Some Notes on Organic Form," Ginsberg's "First thought best thought," Robert Kelly's "Notes on the Poetry of the Deep Image," O'Hara's "Personism: A Manifesto," Lowell's autobiographical essay "91 Revere Street," and Adrienne Rich's politically charged "When We Dead Awaken" are among the landmark documents that challenge modernist precepts, add new concepts to the encyclopedia of poetics, and found postmodern schools of poetry. The proliferation of movements and manifestos at midcentury can be read as an attempt by second-generation postmodernists to accentuate their differences from and exploit the fissures within the dominant mode of modernism. Each school challenges the New Critical emphasis on an ironic and distant voice, the mastery of a well-wrought form, and the limiting of poetic language and subject matter to the decorous. The proposition of an organic process in poetry by the members of the Black Mountain School, the chatty casualness of the New York School, the reliance upon intuition among the Beats, and the confessional provocations of a psyche-in-distress contribute to the expansive modulations of postwar poetry.

If the poets of the 1950s and 1960s suffered somewhat from the anxiety of influence, they also enjoyed the largest and broadest readership for poetry in this century: unofficial tabulation from royalty statements, course enrollments, political rallies in public parks, and the burgeoning popularity of poetry readings in academic lecture halls and bohemian clubs suggest that the cachet of poetry reached its peak in this countercultural era. Books by Bly, Creeley, Sylvia Plath, Anne Sexton, Gary Snyder, and Ginsberg sold tens of thousands of copies. Like many other aspects of American life at midcentury, poetry enjoyed a period of unbridled expansion.

Third-generation postmodern poets are most likely the contemporaries of many readers of this volume. Born during the twenty years after World War II, they count themselves among the Baby Boomers. Their numbers are legion; their achievements are still under evaluation. They are college graduates for whom "Modern Poetry" was a three-credit course in their major; most hold graduate degrees in literature or creative writing; many teach their craft in university writing programs. They have had the opportunity of sitting at a seminar table with acclaimed members of the postwar movements who have been the holders of distinguished professorships and whose papers fetch large sums from special collections. But contemporary poets do not identify themselves with postwar movements beyond an acknowledgment of their historical importance. They often feel only slight indebtedness to the prior generation for the battles fought – mostly between the academics and the bohemians, the "cooked" and the "raw" poetries – that are now largely resolved. Few marvel that the renegade poets of the 1960s are now ensconced in the university or in textbook anthologies or that "open" and "closed" forms cohabit in little magazines or in the latest volume by Ashbery.

The field of contemporary poetry cannot be solely described in terms of movements and schools. Instead one finds a four-sided mandala of less tightly bonded interest groups, or PACs (Poetry Action Committees). In one quarter reside the traditionalists who have assumed the mantle of the academic poets of the postwar generation, though in a more discursive and less intricate language. Aided by recent collections – such as *The Direction of Poetry: An Anthology of Rhymed and Metered Verse Written in the English Language Since 1975* (1988), *Expansive Poetry: Essays on the New Narrative and the New Formalism* (1989), and *A Formal Feeling Comes: Poems in Form by Contemporary Women* (1994) – an alliance of New Formalists calls for a return to traditional rhyme and meter (to recoup what they view as the slack practice of the 1960s and 1970s) and a renewed emphasis on narrative to foster a general, educated audience for poetry (which the diffuse poetries of the counterculture had supposedly lost). Alfred Corn (b. 1943), Timothy Steele (b. 1948), Vikram Seth (b. 1952), Brad Leithauser (b. 1953), and Gjertrud Schnackenberg (b. 1953) revive many stanzaic forms and metrical devices fallen into disuse. Though some among their group combine neoconservatism with a return to traditional forms, one should note that the lesbian formalist Marilyn Hacker (b. 1942) addresses the milieu of homosexual life in her sonnets, and businessman-poet Dana Gioia (b. 1950) has successfully resisted both an academic appointment and the republicanism of the management class. Entries on all of the New Formalists mentioned here can be found in *DLB 120: American Poets Since World War II, Third Series;* a fresh assessment of the career of an important mentor and sponsor of the formalists, Anthony Hecht (b. 1923), can be found in the present volume.

In the opposite quarter reside the experimentalists whose antiestablishment convictions preclude participation in the "professional verse culture." They are widely published (in small-press books, financially precarious but daring magazines, and on always-warm laser-jet printers), but poorly distributed. They form an underground wholly ignored by the Associated Writing Programs and by the prize-selection committees that announce the winners of the latest competition in their newsletters. The Language poets are the most identifiable of this group, attacking the conventions of the personal voice and transparent, "absorptive" language in the lyric. Important poets of the Language movement included in this volume are Charles Bernstein (b. 1950), Kathleen Fraser (b. 1935), Michael Palmer (b. 1943), and Ron Silliman (b. 1946); others appearing in recent DLB volumes are Lyn Hejinian (b. 1941), Susan Howe (b. 1937), and Bernadette Mayer (b. 1945). These poets muster an array of antiabsorptive techniques that provoke self-consciousness about the reading process. Blatant artifice, syntactical disruptions, phonetic play, typographical anomalies, impermeability, and splicing and co-opting of itinerant texts and popular iterations combine to assert what Bernstein in *A Poetics* (1992) calls the "skepticism, doubt, noise, [and] resistance" of postmodern culture. The Language poets' elaboration of poststructuralist theories of language and their recourse to marxist attacks on a publishing industry that commodifies referential language have heightened their appeal to a theoretically aware university readership. A few Language poets, such as Bernstein, Howe, and Bob Perelman (b. 1947), have in fact entered the professoriat. In contrast to the foregrounding of textuality in Language writing, performance poetry – such as the shamanistic songs and Yiddish vaudeville of Jerome Rothenberg (b. 1931), the parodic lectures of the scholar-translator in Armand Schwerner's (b. 1927) *The Tablets,* and the free-form "talk poems" of David Antin (b. 1932) in this volume – stresses the improvisation of an oral performance incorporating autobiographical, ethnographic, musical, or other nonliterary sources.

In a third quarter reside poets for whom identity politics are a prominent consideration. Issues of race, gender, ethnicity, and sexual preference constitute the subject matter of their work, which ranges from the most intimate personal revelation to broad public pronouncements. They are less concerned than either traditionalists or experimentalists by debates over poetic form, though they pursue styles that differentiate their work from the speech and experience of white middle America. The espousal of an opaque or overly literary language runs counter to the political statements they feel compelled to make. Their charge is to give voice to previously repressed segments of American society, and therefore to introduce a pluralist concept of community. They speak first to an audience with whom they share their experience of alterity and marginality, but they consequently seek the understanding of a larger readership. In the gender and hemispheric politics of Carolyn Forché (b. 1950), in the pan-African rituals and the tonalities of jazz as recorded by Nathaniel Mackey (b. 1947) and the ever-combative Ishmael Reed (b. 1938), in the reconciliation of Asian tradition and modern Hawaiian life in Cathy Song (b. 1955) and Indonesian-born Li-Young Lee (b. 1957), in the bilingual culture of Simon Ortiz (b. 1941), American poetry overcomes its monochromatic and monotonal historical origins.

The fourth quarter is occupied by the practitioners of the most pervasive mode in American poetry today, the personal (or postconfessional) lyric. These poets extend the confessional mode of Lowell, Plath, W. D. Snodgrass, John Berryman, and Sexton, though they are less strident in their attack on decorum and perhaps no longer able to shock through autobiographical revelation in the era of tabloid journalism. Despite the moderation in tone, these poets continue to explore the psyche and emotions in poems that test the propositions of the self against the experience of the world. They exhibit a general disregard for formalist techniques (including meter) that might impede immediate expression; and they refuse to distract from the presentation of the self by calling attention to the language-as-object. Poets such as James Wright (1927–1980) and James Tate (b. 1943), who are discussed in this volume, as well as Louise Glück (b. 1943), Robert Hass (b. 1941), Jonathan Holden (b. 1941), and Sharon Olds (b. 1942) have pursued the family drama and childhood's traumatic incidents, psychic distress and substance abuse, sexual adventuring and marital strife as their common subjects. Beyond the immediate relation to confessionalism, these poets share an exploration of subjectivity that is the legacy of the romantic lyric. Like the odes of William Wordsworth or John Keats – in a language only slightly heightened from the American vernacular and soothing to the contemporary ear – these poems call upon remembrance within a dramatic setting and often reveal the poet's sensibility and identity. Despite Olson's midcentury warning against "the lyrical interference of the individual as ego" in "Projective Verse," the self in the postconfessional lyric once again assumes the role of arbiter of meaningful experience. These are the poems that dominate such verse magazines as *American Poetry Review* (with its author photographs accompanying poems), *Shenandoah, Prairie Schooner,* and the *Denver Quarterly.* And these poems represent the majority of those discussed in writing workshops where the dynamic of group therapy now reigns. The personal lyricists are the most likely to be directors of writing programs and to have published award-winning volumes in university press poetry series.

The four-part mandala of contemporary poetry here described is not intended to locate poets permanently or exclusively in particular quarters. Adrienne Rich (b. 1929), for example, began her career as a formalist in the school of W. H. Auden and underwent a conversion to become one of the foremost exponents (in poetry and prose) of feminist politics. One observes among recent writers a more flexible alliance not permitted by the close-knit movements of midcentury. Poets are often not responsible for the labels attached to their work, and contemporary poets frequently resist the scholar's restrictive and ultimately reductive labels. While Amiri Baraka (LeRoi Jones, b. 1934) abandoned his association with the New York bohemian schools to lead the Black Arts movement, Reed, Lorenzo Thomas (b. 1944), and Mackey now move more easily between an African American poetics and an experimental mode. Similarly, female poets such as Levertov (b. 1923), Fraser, and Carolyn Forché have worked to correlate the terms of lyric poetry and political activism.

The third generation of postmodern poets may also be characterized by outright resistance to the aesthetic and political program of modernism. In this regard the poets are in alignment with much that has transpired in literary criticism and theory in the past twenty years. The poets of identity politics have been especially critical of the Eurocentric and masculinist bias that permeated modernism. The pluralism that espouses the equal validity and aesthetic worth of disparate cultural experience directly challenges the elitist, sexist, and discriminatory attitudes implicit in modernism. The postmodern lyric's renewed emphasis on personal expression serves as repudiation of the impersonal and objectivist slant of modern verse and reinstates the individual to authority over totalizing systems. Whereas the modernists held the word as sacred and symbolic *Logos,* with intrinsic and incantatory meaning, the postmodern avant-garde regards language as a plastic medium that can be reshaped without lingering impressions. Rather than aspire to a pure, refined art, the postmodern poet appropriates theoretical jargon and demotic speech, the embedded phraseology of commerce and the free signifier. Lastly, there exists a reactionary postmodernism that dubs the modernist revolt against nineteenth-century aesthetics a failure because of its disregard for the general audience and urges a return to traditional forms, public statement, and coherence of narration and setting. On all fronts there is little doubt that the contemporary poet now disdains many of the precepts of modernism. A new literary period has begun.

Contemporary poetry has been even more agitated by debates about the canon than other fields of literary endeavor. Popular acclaim and satisfied booksellers temper judgments as Susan Sontag displaces E. B. White in essay collections, or as Toni Morrison captures the Nobel Prize that eluded Vladimir Nabokov. But because poetry continues to operate within a limited economy and an eroding readership outside the university — only a few volumes are published by trade houses, and reviews of poetry in major newspapers and magazines have virtually vanished — arguments as to who the important figures are, how they are identified, and why they should be taught preoccupy the field. The literary canon demands selectivity. It should guide the reader toward works that are significant or rewarding. In the past decade especially, pluralists have argued that such "guides to reading" are neither benign nor impartial. The canon creates a hierarchy of writers, and it traditionally reinforces the dominant culture at the expense of the marginal or disenfranchised. Literary history has a greater obligation to inclusivity, in an attempt to establish a thorough cultural record and to recover unjustly neglected or repressed works. Thus the canon-reinforcing process of evaluation sometimes clashes with the politics of inclusion and efforts at suitable representation of the diversity of American voices.

In 1929 Pound asserted in his own effort at canon formation, "How to Read," that poetry should be chosen for an anthology "because it contained an invention, a definite contribution to the art of verbal expression." He argues that one should not "sub-divide the elements in literature according to some non-literary categoric division. You do not divide physics or chemistry according to racial or religious categories" (*Literary Essays*). Although Pound deplored the conservatism of the anthologist, he nevertheless equates the terms of literary selection with the supposedly impersonal and universal truths of science, which is precisely the object of complaint among today's pluralists: the universal category too often turned out to be male and white.

In their efforts to expand the canon, pluralists have introduced a remarkable number of special-interest anthologies that identify poets by gender, race, ethnicity, sexual preference, and nationality, or some combination thereof. Among these are *Breaking Silence: An Anthology of Contemporary Asian American Poets* (1983), *Harper's Anthology of 20th Century Native American Poetry* (1987), and *Gay and Lesbian Poetry in Our Time* (1988). As Alan Golding observes in "American Poetry Anthologies," an essay in *Canons* (1984), such collections have the notable virtue of preserving a specific tradition and rehistoricizing our understanding of literary heritage. But they are also symptoms of an increasing literary balkanization through which one reader's familiar figures of contemporary poetry escape the notice of another. And as Charles Bernstein points out in his 1992 essay "State of the Art," "Too often, the works selected to represent cultural diversity are those that accept the model of representation assumed by the dominant culture in the first place" (*A Poetics*).

Pound's premise about anthologies should not be considered invalid, but the criteria he posits must evolve as the tradition advances. Inventive writers of all descriptions continue to be neglected by the canon in favor of aesthetically conservative writers. Cultivating those works that, as Pound says, contain "an invention, a definite contribution to the art of verbal expression" is crucial to the survival of poetry in America. The canon of contemporary poetry persists because one must finally discriminate between inventive and stale work. New critiques that persuasively describe daring work to a skeptical readership, in tandem with the poetics of inclusion that represent a panoply of American traditions, make a revised and expanded canon essentially beneficial to American poetry. One recent anthology that proposes a multicultural and international solution to the dilemma of representation is *Poems for the Millennium* (1995). In the first of two volumes, extending "From Fin-de-Siècle to Negritude," editors Rothenberg and Pierre Joris offer a capacious gathering that embraces a wide cultural heritage and delineates the innovative contributions to twentieth-century poetry.

The entries in *DLB 169: American Poets Since World War II, Fifth Series* are substantial enough to supply biographical and literary-historical context in addition to an extensive evaluation of the poet's oeuvre. At the same time the entry length limits the number of poets that can be treated. Although each volume contributes to a thorough understanding of the literary history of the genre and period, it cannot offer a complete representation of the work in the field. One should regard this volume as a companion and supplement to entries in the earlier volumes of the series. Readers will also find entries on important poets writing after World War II in *DLB 16: The Beats: Literary Bohemians in Postwar America* (1983); *DLB 41: Afro-American Poets Since 1955* (1985); *DLB 82: Chicano Writers, First Series* (1989); and *DLB 122: Chicano Writers, Second Series* (1992). As an incremental series these volumes combine critical selection and comprehensive literary history.

The movements and schools that were so prevalent at midcentury partially depended on the personal association of the poets. The New York School poets were fellow students at Harvard before relocating to lower Manhattan's art community. Olson and Creeley kept up a voluminous correspondence before they met at Black Mountain in North Carolina. With the institutionalization of creative writing programs, contemporary poets move to jobs at colleges and universities across the country. Among the results of this distribution of talent are the affiliation of poets by publishing venues and by their practice in certain genres and forms.

While the personal lyric is the most prevalent contemporary type of poem in terms of quantity, the meditative poem appears to be growing in importance. The lyric devotes itself to the physical and the passionate; in its intimate voice, the lyric provokes an emotional response. The meditative poem retreats from the turbulent desires of the ego; it is cognitive rather than sensual, abstract rather than particular. The lyric is hot, the meditative poem cool. The chief modernist predecessor in the meditative mode is Stevens, who sought to define modern poetry as "the act of the mind." The major exponents of the meditative mode are now Ashbery and Ammons. Their excursions of thought find a comfortable rhythm in longer works such as "Self-Portrait in a Convex Mirror" (1975), "A Wave" (1984), and *Flow Chart* (1991) by Ashbery; and *Tape for the Turn of the Year* (1965), *Sphere: The Form of a Motion* (1974), and *Garbage* (1994) by Ammons. Younger poets at work in meditative poetry include Robert Hass and Ann Lauterbach (b. 1942). In the present volume one finds a strong meditative vein in the work of performance poet David Antin. His *Meditations* (1971) represents a formal investigation of the mode, and his later "talk poems" should be considered as improvisational meditations on a given theme. James Schuyler's long impromptus, such as "The Morning of the Poem" (1980) and "A Few Days" (1985), combine the Stevensian "act of the mind" with the casual conversational style that one encounters among other members of the New York School such as O'Hara and Ted Berrigan (1934–1983).

The appeal of the meditative poem resides in the patience with which the mind of the poet deploys, maps, and inscribes itself. As a reaction to the abstract language and indeterminacy of scene in meditative poetry, there have been several recent exponents of a return to narrative verse. With James Merrill's elegant masque, *The Changing Light at Sandover* (1983), now completed, poetic works deploying many characters and eventful linear narratives followed. Vikram Seth chronicles the foibles of Bay Area yuppies in his 307-page novel-in-verse, *The Golden Gate* (1986). Frederick Turner and Frederick Feirstein issue a manifesto on behalf of the New Narrative in *Expansive Poetry* (1989).

The fizzling of several modernist epic poems and a distaste for the hierarchical structures and belief systems that frame them has led many postmodern poets to serial composition. Poems written in many loosely associated parts also signify the impatience of poets with the short, personal lyric demanded by journals. The series is a modular form in which individual sections are both discontinuous and capable of multiple orderings. In contrast to the linear causality of most narrative forms, the serial poem is desultory and polyvalent, accommodating an expanding and heterodox universe. Among the first postmodern examples is George Oppen's *Discrete Series* (1934). Midcentury practice includes the open-ended "Passages" of Robert Duncan (1919–1988), published through several volumes; Robert Creeley's *Pieces* (1969); *The Journals* (1975) of Paul Blackburn (1926–1971); and the later books of Jack Spicer (1925–1965). Among contemporary poets the examples include *Notes for Echo Lake* (1981) and *At Passages* (1995) by Michael Palmer (b. 1943), who is included in this volume, as well as Robin Blaser's (b. 1925) *Pell Mell* (1988) and Robert Kelly's (b. 1935) *flowers of unceasing coincidence* (1988). The serial poem represents postmodern poetry's innovative contribution to the long form.

In contrast to the return to traditional poetic forms espoused by the New Formalists, some postmodern poets have invented their own constricting formal devices. These procedural forms consist of predetermined and arbitrary constraints that are relied upon to generate the context and direction of the poem during composition. Unconvinced by the presence of any grand order in the world, the poet discretely enacts a personal order. Procedural forms present themselves as alternatives to the well-made metaphorical lyric once touted by the New Criticism. Louis Zukofsky composes the densely inscrolled *80 Flowers* in honor of an eightieth birthday he did not live to see; each "flower" comprises eight five-word lines. In such books as *Themes and Variations* (1982) and his Norton "lectures" *I-VI* (1990), John Cage (1912–1992) invents the mesostic, a form of acrostic poem in which he "writes through," or "across," a proper name or aesthetic term centered vertically in the text. Ron Silliman (b. 1946) employs a mathematical sequence known as the Fibonacci series to determine the number of sentences in each paragraph of his book-length prose poem, *Tjanting* (1981). Rosmarie Waldrop (b. 1935) has written an abecedarium based on a text by George Santayana. These poets advocate constraint for its paradoxically liberating and generative effect.

Many critics and poets have lamented the increasing marginalization of poetry in American culture and intellectual life, with the postmortem examination performed in essays such as Joseph Epstein's "Who Killed Poetry?" in *Commentary* (1988). Few were disturbed by Epstein's declaration that there were no longer any great poets who spoke of language as an "exalted thing" and went forth as "a kind of priest." The passage of such romantic postures was not lamented because neither American poets nor their readers were any longer comfortable with the production or consumption of a cultural artifact in an elitist or quasi-religious vein. Few contemporary poets wish to see themselves as so detached from the secular and egalitarian American experience; few readers wish to worship much of anything.

Epstein scored more heavily when he attacked poets where they live, challenging the cultural efficacy of "poetry professionals" who are wholly supported as teachers in creative writing programs and whose publications are largely subvented by grants and foundations. Poetry became irrelevant – or at least marginal – to American life when poets needed only to perform their academic obligations of workshops, readings, and the publication of a quadrennial volume to secure their careers. Epstein's accusations hurt because he pointed to the most prestigious institutions among the society of poets as the culprits of the genre's decline. Responses that prescribed solutions, rather than merely denying that the patient had expired, include Dana Gioia's "Can Poetry Matter?" in *The Atlantic Monthly* (1991), Jonathan Holden's *The Fate of American Poetry* (1991), and Vernon Shetley's *After the Death of Poetry* (1993). All stop short of suggesting that poets resign their tenured positions. These essays contend that the intensive and self-absorbed "difficulty" of poetry – promoted by modernists as a required response to the complexities of their world, and in disdain for the common reader – has increasingly repelled a general audience. Poetry, these commentators argue, must appeal to and engage the intellectual and cultural concerns of the general reader whose attentions have been captured by prose.

While critics argue over the death of poetry, the writing of poetry has never been more democratically practiced. Gioia estimates that writing programs "will produce about 20,000 accredited professional poets over the next decade." The quantity alone is impressive, but these poets – whether professional or freelance laborers – will surely be more diverse in their backgrounds than their predecessors. The result of American pluralism is that there are now many more types of poets and poetry than there were in the homogenized, New Critical 1940s. The absence of a "major" poet may be the price paid for the gradual dissolution of the dominant culture that would have identified and rewarded him. Production has increased with the workforce. As Rochell Ratner observed with a touch of weariness in an essay for *American Book Review* titled "Superfluous?" in 1994, "a recent 'Poetry Showcase' at Poets House in New York City had nearly 1000 books on display, all published in 1993." As readership erodes in the relentless surf of popular broadcast media, one wonders whether the chapbooks and small-press publications have not already outnumbered the people who purchase them. But the almost tenfold growth in noncommercial literary presses between 1965 and 1990 documented by Loss Pequeño Glazier in *Small Press* (1992) suggests something important: even as poetry appears to decline in prestige and in the attention paid to it by major markets, there is a thriving "back channel" of writing and exchange that escapes the notice of culturally conservative institutions. This alternative ferment, separate from the résumé stuffing of some eminent poets and critics, may yet provide the next significant advance in poetics and speak to poetry's role in the next century.

Acknowledgments

This book was produced by Bruccoli Clark Layman, Inc. Karen L. Rood is senior editor for the *Dictionary of Literary Biography* series. George P. Anderson and Kenneth Graham were the in-house editors. They were assisted by Anthony Scotti.

Production manager is Samuel W. Bruce. Photography editors are Julie E. Frick and Margaret Meriwether. Photographic copy work was performed by Joseph M. Bruccoli. Layout and graphics supervisor is Emily Ruth Sharpe. Copyediting supervisor is Jeff Miller. Typesetting supervisor is Kathleen M. Flanagan. Systems manager is Chris Elmore. Laura Pleicones and L. Kay Webster are editorial associates. The production staff includes Phyllis A. Avant, Ann M. Cheschi, Melody W. Clegg, Patricia Coate, Joyce Fowler, Brenda A. Gillie, Stephanie C. Hatchell, Penelope M. Hope, Rebecca Mayo, Kathy Lawler Merlette, Pamela D. Norton, Delores Plastow, William L. Thomas Jr., and Allison Trussell.

Walter W. Ross and Steven Gross did library research. They were assisted by the following librarians at the Thomas Cooper Library of the University of South Carolina: Linda Holderfield and the interlibrary-loan staff; reference-department head Virginia Weathers; reference librarians Marilee Birchfield, Stefanie Buck, Stefanie DuBose, Rebecca Feind, Karen Joseph, Donna Lehman, Charlene Loope, Anthony McKissick, Jean Rhyne, Kwamine Simpson, and Virginia Weathers; circulation-department head Caroline Taylor; and acquisitions-searching supervisor David Haggard.

The editor would like to acknowledge the dean of Arts and Letters, Kerry Grant, and the chair of the Department of English, Kenneth Dauber, at the State University of New York at Buffalo for providing research support that contributed to the timely completion of this volume. The Poetry Collection at the University at Buffalo has again proved an invaluable resource for materials and information related to contemporary poets; the curator, Robert J. Bertholf, and assistant curator, Michael D. Basinski, deserve gratitude for their assistance. I would like to express my deep appreciation to family members for their generous support during the past year – my parents Ralph and Ann Conte, brothers Ralph and Christian, and sisters Mary Louise and Anne Marie. Special thanks are due to the friends who tolerated an editor's irritability in reaching after fact and reason, especially Joy Leighton, Robert Basil, Charles Jones and Kerry Maguire. Several colleagues and contributors served as unofficial consultants to the project, taking time away from their own work to offer invaluable advice, particularly Peter Baker, Ronald Baughman, Elisabeth Frost, Laszlo Géfin, Burt Kimmelman, Patrick Meanor, Diane Middlebrook, Susan Schultz, Willard Spiegelman, Keith Tuma, and Peter Quartermain. Finally, George Anderson and Kenneth Graham at Bruccoli Clark Layman deserve gratitude and praise for their skillful editing of the text and preparation of the materials for the volume.

Dictionary of Literary Biography® • Volume One Hundred Sixty-Nine

American Poets Since World War II
Fifth Series

Dictionary of Literary Biography

David Antin
(1 February 1932 -)

Kenneth Sherwood
State University of New York at Buffalo

BOOKS: *Definitions* (New York: Caterpillar Press, 1967);

Autobiography (New York: Something Else Press, 1967);

Code of Flag Behavior (Los Angeles: Black Sparrow Press, 1968);

Meditations (Los Angeles: Black Sparrow Press, 1971);

Talking (New York: Kulchur Foundation, 1972);

Talking at the Boundaries (New York: New Directions, 1976);

Whos Listening Out There (College Park, Md.: Sun & Moon Press, 1979);

Tuning (New York: New Directions, 1984);

Selected Poems: 1963-1973 (Los Angeles: Sun & Moon Press, 1991);

What It Means To Be Avant-Garde (New York: New Directions, 1993).

RECORDINGS: *The Principle of Fit, 2,* Watershed Tapes, 1981;

The Archaeology of Home; Lemons, High Performance Audio, 1987.

SELECTED PERIODICAL PUBLICATIONS – UNCOLLECTED: "Notes for an Ultimate Prosody," *Stony Brook,* 1 (Fall 1968): 173-178;

"Modernism and Postmodernism: Approaching the Present in American Poetry," *Boundary II,* 1 (Fall 1972): 98-133;

"Some Questions About Modernism," *Occident,* 8 (Spring 1974): 6-38;

"David Antin: A Correspondence with the Editors, William V. Spanos and Robert Kroetsch," *Boundary II,* 3 (Spring 1975): 595-650;

David Antin (photograph by Byron Pepper)

"Talking to Discover," *Alcheringa: Ethnopoetics,* 2, no. 2 (1976): 112-119;

"Radical Coherency," *OARS,* 1 (1981): 175-191;

"The Stranger at the Door," *Genre,* 20 (Fall-Winter 1987): 463-481;

"determination, suspension, diversion, digression, destruction," *Conjunctions,* 19 (1992): 51-78;

"Fine Furs," *Critical Inquiry,* 19 (Autumn 1992): 151-163;

"The Theory & Practice of Postmodernism: A Manifesto," *Conjunctions,* 21 (1993): 335-343;

"The Begger & the King," *Pacific Philogical Review* (November 1994).

David Antin is most famously known as the "talk poet." Since 1972, rather than typing poems in private, he has composed his poems by talking them out in front of audiences, improvising across genre boundaries and combining features of epic poetry, academic lecture, stand-up comedy, and storytelling into compelling, narrative collages. These postmodern talk pieces link Antin with Beat poetry, performance art, and writers such as John Cage, Jackson MacLow, Charles Olson, and Jerome Rothenberg – all of whom have been fellow participants in the informal movement that critic Henry M. Sayre addresses in his essay, "David Antin and the Oral Poetics Movement" (1982). Antin himself suggests that this loose coalition of poets was held together by the "conviction that poetry was made by a man on his feet, talking," a quotation that comes from Antin's 1972 essay, "Modernism and Postmodernism," which critically supports his own practice of "talking" poetry out.

In all his work, not just his lauded talk pieces, Antin's consistent sense of poetry as the motion of thinking, *dianoia*, exceeds that of most Black Mountain poets; the meditative experimentalism of his work anticipates that of a subsequent generation of Language poets. In *In Search of the Primitive* (1986), one of several important books to address Antin's work, Sherman Paul explicitly connects talk poetry with projective verse, writing, with Olson's formula, that "it is 'language as the act of the instant' not 'as the act of thought about the instant.'" In the opening pages of his first complete book of talks, *Talking at the Boundaries* (1976), Antin famously insists:

```
                                    if robert lowell is a
   poet i dont want to be a poet     if robert frost was a
   poet i dont want to be a poet   if socrates was a poet
   ill consider it[.]
```

Antin does not "want to be considered a poet   if a poet is someone   who adds art to talking." The expressed affinity for the dialogic philosopher Socrates is not chosen casually. In claiming for poetry the domains of philosophy, talking, joking, and storytelling, Antin opens up the field of poetry to the full range of language activities, a project shared with many of his contemporaries.

Antin's career has "involve[d] a broadening and inclusive notion of poetry," writes Hank Lazer, considering Antin and Rothenberg within larger cultural contexts in the essay "Thinking Made in the Mouth." According to Lazer, Antin's work "directs us toward sources and disciplines beyond what are conventionally considered to be 'literary' and 'poetic.'" This intellectually deliberate inter- or anti-disciplinarity accounts for the fact that Antin's iconoclastic work has inspired no school of poetry; indeed, it is hardly even taught in school as poetry. However, his seminal involvement in the more formalized, academic interdiscipline, ethnopoetics, is reflected by his representation in its defining anthology, *Symposium of the Whole* (1983). As a contributing editor to the journal *Alcheringa: Ethnopoetics*, Antin interprets ethnopoetics (a term coined by Rothenberg) as the poetics of natural language genres; ethno-poems freely use universal forms of talk such as greetings, introductions, inquiries, invocations, or demands, rather than the more conventional poetic forms such as the sonnet or the ode. Antin himself is less interested in developing a mythopoetics, in the sense of oral-as-primitive, than he is in practicing a linguistically conscious, demythologizing poetry of the present. His work proposes a radical redefinition of poetic forms and, perhaps more important, of their implications. Antin expresses this position best in an interview in a 1975 issue of the magazine *Vort,* dedicated to his and Rothenberg's work.

> I really try to direct my work into the *domain* question . . . the *domain of the human mind* and its relation to the world. . . . I claim all of that for poetry.

Antin was born in New York City on 1 February 1932. He attended City College of New York with poets Rothenberg, Robert Kelly, Armand Schwerner, and the filmmaker and performance artist Eleanor Fineman, whom he married in 1960. A friend since 1948, Rothenberg remembers Antin as a notable talker even in college and recalls, in *Vort,* the two roommates listening to classical and jazz music and "black sermons on the radio on Saturday nights." United with Rothenberg against what both saw as the "academic decorative poetry" of the day – only later did they encounter magazines such as *Origin* and *Black Mountain Review* – Antin sought out foreign poetry, apprenticing himself to the translation of André Breton.

When a registrar waived his remaining degree requirement, Antin found himself inadvertently graduated from the City College of New York in 1955, the pursuit of his fifth or sixth major interrupted. With an eclectic education in literature, linguistics, and engineering, he became a freelance editor and translator of technical books during the next few years. In the years following graduation he wrote mostly prose; his story "Balanced Aquarium" was published in the *Kenyon Review,* and his deep-image-influenced poem, "Death of the Platypus," shared the Longview Award (along with poems by William Carlos Williams and Denise Levertov). While this early work was appreciated by James Wright and Robert Bly, Antin himself became dissatisfied with it by 1960 and has subsequently declined to reprint it.

By 1961 Antin had committed to a linguistic poetry, finding that the symbolic depth of deep image became, he says in *Vort,* an idea he "not only didn't understand but thoroughly disliked." Soon dissociating himself from deep-image poetry, Antin pleasurably recalls the decisive reading of an early linguistic poem, "Definitions for Mendy," to a shocked and disappointed Wright. The apparent lack of sentimentality – the use of definitions from an insurance manual, of technical vocabularies, and so on – made such material "unpoetic" for readers like Wright, according to Antin in his *Vort* interview.

Antin imagined himself as a novelist rather than a poet at this time, and, taking advantage of an editorial consulting arrangement with Dover Press, he retreated upstate to write. By 1964 Antin had abandoned his planned novel, "The Stigmata," returned to New York City, and begun to write and give readings of poetry that disruptively employed clichéd forms of language. Under the influences of the poet MacLow and his own training in linguistics, which culminated in a master's degree in 1966, Antin wrote poems that investigated language as a means of getting at truth "backwards through the lie," as he recalls in *Vort.* His focus on language itself as a poetic problem led to his most interesting experimentation with the conventions of syntax.

Finding establishment literary journals uninterested in such challenging work, Antin initiated and edited a magazine, *Some/Thing,* with Rothenberg in 1964. In its pages appeared some of the poems that would later comprise Antin's first book, *Definitions* (1967). The magazine also published the work of contemporaries such as Rothenberg, MacLow, Paul Blackburn, Larry Eigner, Clayton Eshleman, Allen Ginsberg, David Ignatow, Schwerner, Gary Snyder, and Diane Wakoski. *Some/Thing* expired in 1968 after four issues, one of which sported a cover (designed by Andy Warhol) of perforated "Bomb Hanoi" stamps.

Characteristically, Antin works with deliberately placed obstacles and situational accidents. He sets up formal problems for writing because, according to Toby Olson in *Vort,* he "chooses to look *into* the way we use words rather than *behind* that use." For Antin formal obstacles, like controls in a scientific experiment or compositional constraints in painting, activate the process of discovering through words. His work in science and increasingly – as his essays began appearing in major journals such as *Artforum* – his art criticism informed his experiments in the writing of poetry, motivating him literally rather than merely metaphorically to employ in his own work procedures from linguistic experiment and visual art.

In 1967 Antin became a curator at the Institute of Contemporary Art in Boston, Massachusetts, and saw two of his books published. Designed by Eleanor Antin and printed by Eshleman's Caterpillar Press, *Definitions* had the appearance of a spiral-bound notebook, and its poems were printed over a light blue grid. The reverse cover was marked by a small blob and the caption: "The Caterpillar Glyph is a small napalmed Vietnamese child. Until the end of the war this black caterpillar."

In *Definitions* Antin explores such philosophical themes as the meaning of "loss" and the knowability of pain, themes at once abstract and topical. One poem, "Definitions for Mendy," written for a dying friend, uses deliberately antielegiac language; in fact, its typically cool first line is drawn from an insurance book: "loss is an unintentional decline in or disappearance of a value arising from a contingency." The seeming insufficiency of this definition – particularly in the context of his friend's situation – reflects Antin's sense that any such definition in language is bound to be inadequate, unaccountable to the circumstances of real loss and felt pain. Rothenberg, in a review reprinted in *Vort,* remarks about this poem that "what begins to distinguish it is that Antin won't let himself drift into a 'poetic' softness: a refusal that grows fiercer with each successive work." This aversion to softness, like his earlier noted suspicion of depth, has a political context. Amid the upheaval of the Vietnam conflict and the civil rights movement, many felt the need for a poetry rigorous enough to engage the duplicities of governmental and military rhetoric. "The Black Plague," the other long poem of this volume, is dedicated to three murdered civil rights workers:

a man cries out he is in pain   we want to say we know
what that means   what does that mean[.]

Partially collaged from Ludwig Wittgenstein's *Philosophical Investigations* (1953), Antin again implies the

Dust jacket for Antin's 1976 book, his pioneering collection of improvisational oral poetry

insufficiency of language. He tries to respect the horror of his subject by refusing to infuse his poetry with the *feeling* language of emotion, adopting instead a sterile diction.

Employing somewhat arbitrary constraints — phrases from philosophy and insurance texts — Antin reveals the varying beauty and barrenness of language. Like a documentary photographer, he lays bare the set and exposes his subjects without makeup, wanting neither to brush aside nor get beyond the limits of words. He writes in the "Black Plague" that "we can imagine a language in which a person could write down or give vocal expressions to his inner experiences and feelings[.]" The poem continues with the qualification that, although one might imagine such a language, "it is not our language."

Antin explains in the *Vort* interview that he wrote *Autobiography* (1967) because he "was probably the least personal poet anybody could imagine." This purportedly tell-all poem reveals so little that

Antin jokingly appends a third-person note of "Biographical Information." The apocryphal information includes the useful facts that the author "said 'suppositories' before he was two" and that he wrote his first poem "during a lull in an English class and was promptly expelled for showing it to the boy in back of him." The main text consists of the narrative experiences of Antin or friends or relatives or acquaintances, seemingly fragmented or truncated in what appears arbitrary ways because Antin only left in what interested him, believing that anybody's experience that formed part of his thinking was part of his autobiography. Antin explains his compositional procedure in the preface to *Selected Poems: 1963–1973* (1991), within which the poem is reprinted:

> I began to work with prefabricated and readymade materials, recycling texts and fragments of texts, enclosing valuable and used up talk and thought and feeling, hoping to save what was worth saving, liberate it and throw the rest away.... *Autobiography* had all of the personal history — mine and the people I knew — that I wanted to think about and keep.

The subject of these disjointed mininarratives, often referred to in third person, is not David Antin; rather than telling the story of his own life, *Autobiography* presents, as a narrative, the stories encountered in the course of one person's living.

Like *Definitions*, *Autobiography* turns the structure of a particular genre into an obstacle, dissolving the narrative continuity and the stable identity of characters. While each stanza seems to have a subject, exactly who the pronouns refer to remains ambiguous, shifting with each superseding stanza:

> He kept eating carrots and jumping off chairs. They ended by accepting him. He flew 24 missions and married a girl who looked like she'd been coated with colorless nail polish.

Perfectly sensible as sentences, the combinations lay bare the holes in the story, the absent context. In postmodern style these minimalist stories enact the fragmented forms that the act of remembering often takes. The lost details and continuities correspond to the absences and faults of memory that a conventional autobiography would smooth over; Antin makes his poem by exposing the gaps an autobiography normally would omit or repair.

An elaborate, red, white, and blue cloth-covered book, *Code of Flag Behavior* (1968) contains shorter poems from these same years. In the serial "Novel Poem," Antin continues the juxtaposition of story fragments, nominally achieving his novelistic ambition. Professing a love/hate relationship with novels, Antin dislikes novelistic filler but wants to recover narrative for his own work. "If you flip a novel very fast it reduces to what's interesting about it very quickly," Antin explains in *Vort.* Following this compositional method exactly, he browses through novels by Ernest Hemingway and Ayn Rand and notes points of interest, composing the first derivatives of these works through an improvisatory selection that anticipates his later talk pieces.

Several of the poems in the long series, "Novel Poem," are subtitled "songs" yet seem hardly song-like. The whole sixth section of "10 songs" reads:

> will you do it Tom
> will i do what
> will you leave[.]

Despite the flatness of the language, it is not difficult to imagine this as a conceptual song, marked by the repetition, line breaks, or perhaps the trite eulogizing of lost love. Simply in calling these unsonglike poems songs, Antin expands the notion of what song might be, in a gesture similar to John Cage's expansion of the conventional parameters of music. Antin's feelings about song — especially important in his oral poetry, since other "oral poets" draw on musical notions of performance — are as conflicting as his feelings about the novel. He makes his complaint clear in *Vort:* "When poets say they're writing a song, they trivialize the prosodic resources of language and reduce them to the trivial prosody of the tradition — from the great music of language to the trivial music of the profession." Antin was testing the prosody of his language in performance long before he began giving talks. He read one of his more explicitly political poems, "Code of Flag Behavior," at a 1965 antiwar reading he helped organize called "Three Penny Poets Reading Against the War." Composed entirely of flag-handling rules such as "the flag should never be displayed with the union down except as a sign of distress" and "the flag should never be used to cover a ceiling," the poem, by the listing of these prohibitions, recontextualizes the flag; the traditional object-of-respect is transfigured into an object-of-repression. The last line, "preferably by burning," turns the code against itself in an invitation to destruction.

Another powerful list poem, "Delusions of the Insane / What They Are Afraid of," performs a similarly ironic transformation. Antin manages to turn the psychological assessment back on American society by recontextualizing a list of supposedly delusory fears such as:

> the police
> being poisoned
> being killed
> being alone[.]

As the poem concludes with images of burning children evocative of the political violence of the time, the fears of the delusional begin to seem quite rational, even justifiable, while the calm rationality of the sane — what Rothenberg calls the "legislated reality," referring to this book in *Vort* — is called into question.

"I hate those poems that realize other times. They're a sort of imperialist drag," Antin proclaims in *Vort.* Coming across a phrase in a footnote — "the places where you are now" — he realizes that the phrase inadvertently helps him crystallize the compositional method of *Meditations* (1971) and the preliminary conception of his talk method: "What I was going to do was to begin at the places where I was then . . . as my beginnings." His ubiquitous interest in the present place, joined with his contempt for nostalgia, emerges as a primary principle. Often this "place" is situated among deliberate obstacles, as in "the first hundred" and "the second hundred," which are structured by proceeding alphabetically through the words in a high-school spelling book. Here is one precise attempt to inhabit the moment in "The Second Hundred: for Sid Luft," section 13:

> to a man on a tightrope the Falls at Niagara are the truth of a
> river   and living to ninety appears like a sudden nomination[.]

The words "Niagara," "ninety," and "nomination" must be written to satisfy the constraint of using the vocabulary in alphabetical order, while the poem is formulated in dealing with this contingent difficulty. The dangerous, sudden presence of the "nomination" is here a linguistic act, to name.

The most striking of the long poems in this book is "Stanzas," a series of mostly three-line stanzas that are modeled on the form of the logical syllogism but attempt to deflate it. The poem contains silly and false proofs such as:

> no ducks dance
> no officers refuse to dance
> no officers are poultry[.]

There are nonsequiturs such as:

forty percent of the population have savings
many are in dire want
Descartes was certain of his own existence[.]

Antin's own description in *Vort* of the difficulty of composing with words from pregiven lists illuminates the method that invigorates many of his poems, that of proceeding via obstacles:

> They were like obstacles, pebbles in my mouth that gave shape to what I had to say. So you could say I was meditating on each of these little meaningful obstacles while I was going on. So the meditative act . . . was to respond to these givens. . . . You encounter somebody in the street – it's a given. You write a poem and put in somebody you bump into in the street – it's a given. Why not with a word you encounter on a page?

The complex, formal obstacle courses typical of Antin's early poems assume a different, increasingly improvisational direction in his book *Talking* (1972). Begun in 1968, the year of his move to San Diego to become director of the University Art Gallery, these poems needed to be "arbitrary enough to represent our American fate," he later recollects (in the preface to their reprinting in *Selected Poems*). For example, "The London March" is contingent upon a card game. In this piece Antin and his wife, Eleanor, discuss the potential success of an antiwar rally while she deals out cards, performing a tarotlike solitaire. Together the two improvise another poem in this volume, "In Place of a Lecture," which is recorded and then transcribed. This piece is built over a text from a book on experimental design, which Eleanor reads until she is interrupted by Antin with questions or comments that she answers or comments upon in return. The commentaries are free improvisations – the text a kind of "ground base" to which they continually return.

While in earlier poems formal constraints were often secondary, they become an obvious dimension of the poems in *Talking*. In one of the last poems that Antin will "write," the "November Exercises," the constraint of writing quickly in noisy situations is marked by exactly inscribing the day, hour, and minute of composition in the margin during the course of a month in 1971. As his mock editorial, "Silence/Noise," in the first issue of *Some/Thing* maintains, silence is possible only by an act of deliberate insulation from the world; his meditative poems attempt to think without divorcing themselves from the world and are written under conditions of noise. The most pertinent moment in this piece is asserted manifesto style, in capital letters: "by DEFINITION LANGUAGE IS PUBLIC – THERE IS NO SUCH THING AS A PRIVATE LANGUAGE." As an art critic and professor of art at the University of California, San Diego, Antin regularly gave public lectures and gallery talks. But it was not until 1972, after Antin had been improvising privately and given several talks without notes, that he publicly improvised a poem.

Listening with him to the recording of a talk given at Pomona College on 12 April 1972, Eleanor Antin, his wife, dramatically observed, "my God, it's a poem" – so Antin reports in *Vort*. It is the transcription of "Talking at Pomona" that closes *Talking*, a book that began with the epigraph, "if someone came up and started talking a poem at you how would you know it was a poem?" In *Vort* Antin further explains the origins of the Pomona talk, which was born out of a sense that talk should be adapted to present needs, rather than being prepared in advance for an unknown, unanticipated audience:

> I really didn't know what it was those students needed to hear. . . . All I needed to know was who I was going to be talking to. Because, for my purposes, what I wanted in talking was this sense of address.

In the following year alone he improvised eight talks at universities and museums. Feature issues in the journals *Boundary II* and *Vort* followed in 1974, although a complete book of talks had not yet been published.

With the publication of *Talking at the Boundaries* in 1976, the maturity of an innovative poetic form became evident; recognition followed with reviews in the *Nation* (11 December 1976) and *The New York Times Book Review* (28 November 1976). The talks typically begin with the formalized introduction of a broad question such as "what am i doing here?" or "is this the right place?" or a condensed account of how he came to doing talk poems, acknowledging the implicit questions of the audience. These opening frames resemble the formulaic framing a traditional oral poet or stand-up comic might employ, but they are particularly tuned to Antin's estimation of each audience. "Talking at the Boundaries" begins by explaining the motivation for talk poetry and the invention of the particular talk's title:

>   when i agreed to come here   to indiana   barry
> alpert didnt   have a title for what i was going to talk
>   about   i think maybe he   forgot to ask me
>   which was   i suppose   just as well
>   because he had an idea   i had given it to him
> of the kind of thing i was now doing   and i had an
>   idea too of going places to talk to people   i was

> seeking an occasion   for the kind of talking
> i want to do   which would of course modify the
>   kind of talking i wanted to do   and how i was no
>   longer   so clearly a poet   a linguist   an art
>   critic   all of which i had so clearly been[.]

The talks then proceed through various narrative and discursive forms in a kind of weaving or knotting motion, one which Marjorie Perloff discusses, along with John Cage's work, in *The Poetics of Indeterminacy* (1981): "The talk poem incorporates as many different threads as will allow it to retain its improvisatory quality, yet those threads are all relational." Borrowing a term from Northrop Frye, Perloff identifies an "associative rhythm" – William Spanos opts to call it a "structural rhythm" or a "rhythm of exploration" in Antin's "Correspondence" – to describe the musical correspondence of themes and ideas that gives coherence to an Antin piece. This collage is organized by a "sense of address," as Antin puts it in "Correspondence." The audience, subject, and the potential for discovery are like targets he talks toward, so that the talks are as concerned with the nature and forms of talking as with anything else, a self-reflexive proclivity typical of postmodern art.

The literate, postmodern poet, "our foremost mythmaker" – as Stephen Fredman dubs him in *Vort* – cannot afford simply to talk his poems; Antin transcribed selected performances to appear in book form. The talks' double dynamic of being made for the present and for possible future publication is often signified by an acknowledgment of the tape recorder:

> im looking at this to see if im still on tape because
>   though its a private occasion in a public place its
>   eventually going to have to become something else
>   because i dont intend   to let it disappear into
>   this occasion and i want it on tape somehow this
>   anxiety that she had[.]

With his literate, technological consciousness, Antin would find Fredman's comparison of his work with mythic poetry strange unless it were meant in Levi Strauss's sense of myth as an art particular to the occasion and audience. "My interest is essentially in engaging an audience . . . in some social relationship," he explains in "Talking and Thinking," a 1993 interview with Hazel Smith and Roger Dean. Antin's own sense about myth is quite ambivalent. In his first book of talks Antin says, "the word 'myth' is the name given to the lies told by little brown men to men in white suits with binocular cases." Thus, his interest in myth is a qualified one, differing substantially from Roth-

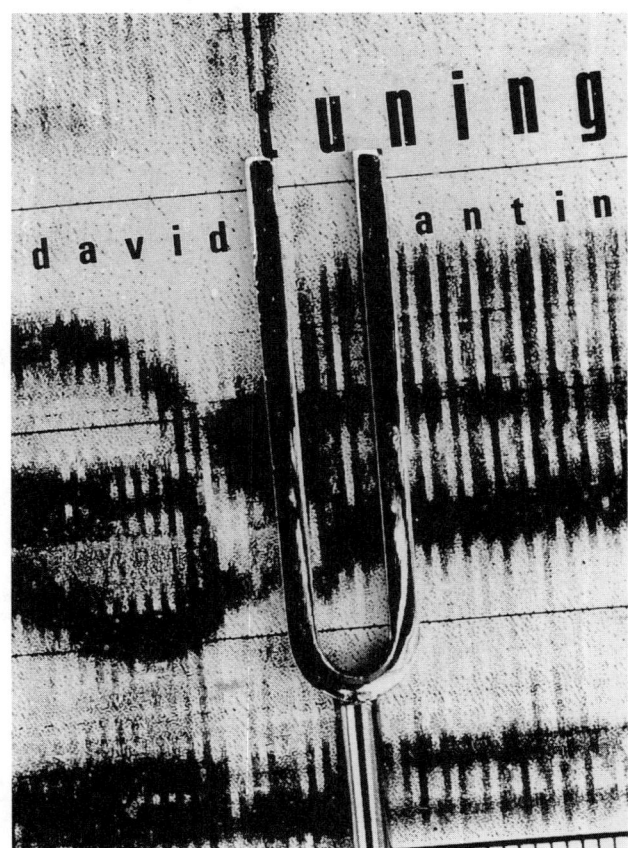

Cover for Antin's 1984 book, in which he seeks "a common space" with his audience

enberg's. In "Correspondence" a distinction between the two poets is offered by Spanos, who proposes that Rothenberg "is oriented towards the timeless formal patterns . . . of collective ritual expression of mythic societies, whereas . . . [Antin] is oriented towards the temporal (time-ridden) openness of individual or existential speech."

Transcribing the talks presents Antin with the problem of retaining the rhythmical, flowing quality of his talk. The elimination of standard punctuation, ragged "right and left margins," lower-case typography, and rambling, paratactic sentences mark his books as something other than standard prose or poetry. Their sense of the present occasion – traditional oral poems often have an appropriate time or season for telling – while aligning him with traditional poetry is also part of a twentieth-century tradition of attention to the written poem as score, a tradition that includes Rothenberg, Charles Olson, and Robert Creeley. Like traditional oral poetry, Antin's pieces are neither formulaic and prefabricated nor composed without preparation or preconceived ideas of performance. He prepares, reads, and thinks about what he will do, and then the poems are composed in performance, sometimes as reworkings of favorite tropes.

In book form the transcribed talks are regularly prefaced and often concluded with additional contextual information that links them together, giving the chronological effect of performance in one continuous talking tour, despite the fact that sometimes several years separate two contiguous talks in a given book. Antin's poems locate themselves at the boundaries between discourses and direct their talk at the margins – the fully justified edges of disciplinary knowledge – synthesizing art and criticism with sociology, politics, and folklore. When he speaks about the division between public and private in "a private occasion in a public place," his talk ranges from the private occasion of marriage, to the privacy of a schizophrenic's self, to his own situation in the room as a private person who is talking publicly:

> who speaks for me when i speak? do i have a
> quorum? im here in this public place talking in
> the context of some old friends some new ones and
> people who have come here with expectations
> maybe of what a poet is you know? what a poet
> will do what hell talk about and i come here
> with my private thing so to speak and i tell you
> im here to define myself and im telling you who i am
> and what im doing it for and you can believe it as
> much as you can believe me or any poet or as
> much as you can believe your wife or your child
> or yourself[.]

Involved here are almost all of Antin's classic concerns about occasion, audience, dialogue, place, truth, privacy, and thinking. Antin rethinks some of these questions of self in another piece called "a more private place," where he wonders about the notion that the self is interior.

> its a funny idea i
> mean the idea that the self is inside is itself a funny
> notion whats inside? and inside of what? i
> mean
> im not even inside of myself if i come here and talk to
> myself or if i sit in my own room and talk to myself
> what do i see as my self? the part that comes out
> see i cant get out of saying out it keeps coming
> "out" as i say it it comes out and i suddenly see it
> out there and seeing it out there i hear what i
> say and then i know what i mean[.]

Here the relation of self-knowledge to the publicly performed self becomes an occasion of discovery.

For Antin this active discovery can only be accomplished by present thinking, another of his methods. Antin performs present thinking in *Tuning* (1984); when he says "you" or "this place here" the reference is not to textual people or places of a distant time or location, but to the people in the particular room with whom at that moment Antin proposes a relationship. As a native New Yorker speaking to an audience in New Jersey, Antin establishes an identification with his audience:

> so now
> you have my images of the common places we have
> between us atlantic city toms river
> barnegat though we may be coming from them at
> different angles to this place here where for a
> short time weve come more or less together[.]

What he tries to accomplish quite concretely in this talk, as in many others, is a bringing together of people into the same place.

The striking development of his book *Tuning* is its primary metaphor – talking as a tuning of speakers to one another – through which people are not only brought into the same physical space but into a shared place in thought. Here Antin finds the most apt formulation for the experience of dialogue his form allows:

> so we are more precisely considered within
> range of each others senses within walking and
> talking range which still does not automatically put
> us in step or tune with each other or with the place
> now the point of doing these pieces for me is that
> it gives me a chance by a kind of subtle but
> ordinary human concentration to get a
> sense of where youre coming from and how
> and to allow that sense to put some pressure on
> my own way of moving to bring me into somewhat
> closer range of you close enough to compare
> our ways of moving our sense with each
> other and in this situation to find a fundamental
> human act this negotiation in a common space[.]

The possibility of practicing an art tuned to the present occasion continues to occupy Antin in his most recent book, *What It Means To Be Avant-Garde* (1993). Except for the growing prominence of personal stories in it, the form of his improvisations and their commitment to the occasion remain largely unchanged. The present in many of these pieces is an aging or passing one, tuned perhaps to Antin's own increasing years. For instance, "the river" concludes with the story of his friend, an "unfamous painter" who suffers from a problem because "a 90 year old painter has to be famous because he cant be promising any more." There are stories of his mother's memory loss and of her friend's expulsion from a retirement home. Another piece, "the structuralist," narrates the reading of a linguist friend's poem and the poem's disappearance after his sudden death:

> so this extraordinary epic poem   written in no
> known language by my structuralist friend never saw
>   the light of day   and the truth of the divinity of
>   language and its universal constituents
> revealed   through the speech of this man as evoked
>   by the beauty and sympathy of that woman
>   will probably   remain obscure to us forever[.]

Perhaps the most thematically unified of his books, *What It Means To Be Avant-Garde* proceeds to lament various absences, the passing of friends and family, and particularly the loss of the stories they tell about their lives. In this vein "the fringe" provides a witty, slightly peripheral, almost retrospective assessment of Antin's own work. This opening piece recounts a search through the newspaper's calendar pages to see if the present talk had been noted there, only to find the troubling category of "the fringe." Antin finds his placement in the fringe like being both mentioned and not mentioned at once:

>   i knew they didnt think it was
> really a good thing to be on the fringe   but they
>   wanted to cheer us up if we were on it   because
> it was a los angeles fringe and we were trying hard
>   they seemed to suggest that we might be here
>   and there could be an agent out there who
> might see us
>   is anybody here an agent?   no agents out there[.]

Displaying a comic disdain toward his own potential for notoriety, Antin complicates the sympathetic treatment of cessation in the other talks. He wants to celebrate the fringe and the pleasurability of his accustomed place there; his is not an avant-garde striving to lead his troops into the "white light." Continuing the serious stand-up routine, Antin alludes to the reception of his work and his own strange career as the poet who gave up writing poems:

>   so i realized i didnt have this
>   right   and i asked myself how do i get this
>   right   i dont want to be a success   and ive
>   succeeded   theres an ambiguity about not wanting
>   to be a success
>   i wanted to be a
>   poet   when you say you want to be a poet you
> already know that you dont want to be a
>   success   the model for poetry is not to be a
>   success   that's the first model you learn
>   the first ones went blind   later they died of

Cover for Antin's 1993 book, in which he speaks about the position of the poet in U.S. culture

> tuberculosis or committed suicide or drowned[.]

Antin satirically recalls a familiar literary formula — write to gain posthumous fame — that runs counter to his method of presentness, cognizant that his art will not become immortal through books.

He reiterates his peculiar take on success throughout "the fringe" and in other talks, recalling jobs and lovers refused. The reception and intermittent recognition of his work satisfies him. Antin is content to continue escaping to and talking in the fringe of the present. But as he almost ruefully observes:

>   in the
> united states you cant escape that
>   somebody thinks
> you should not really be in the shade but out there
> in the white light[.]

In 1976 Antin endured the moderately white light of a Guggenheim Fellowship, and in 1983 he endured it again when he won a National Endowment for the Arts Fellowship. In the late 1980s, besides performing talks, Antin conducted several public art events, including a series of fleetingly present skywriting poems described in his "Fine Furs" (1992). With the publication of his *Selected Poems* in 1991 by the Language poetry press, Sun and Moon, Antin's long-out-of-print early work will surely influence a new generation of readers. At the Nina Freudenheim Gallery in Buffalo, New York, on Thursday, 8 October 1992, Antin began talking through the question of whether to accept an offer of early retirement from his "white light" position as chair of the art department at the University of California at San Diego.

He subsequently decided to retire from academia, but, pondering a conversation in the conclusion to "what it means to be avant-garde," Antin illustrates the goal of presentness his poetry continues talking toward:

>   nothing within the horizon
> of my discourse had prepared me for that moment with
>   my aunt fanny   who had just lost the
> husband she'd lived with for over forty years and was
>   now on the telephone
>   and now it seems to me that if you cant respond
> to that youre not in the avant-garde[.]

Letters:

"An Exchange," *Vort: Twenty-first Century Pre-views: David Antin – Jerome Rothenberg*, 7 (1975): 72–80;

"A Response to Ellen Zweig," in *The Poetry Reading: A Contemporary Compendium on Language and Performance*, edited by Stephen Vincent and Ellen Zweig (San Francisco: Momo's Press, 1981), pp. 187–191;

"from David Antin," in *In Search of the Primitive: Rereading David Antin, Jerome Rothenberg, and Gary Snyder*, by Sherman Paul (Baton Rouge: Louisiana State University Press, 1986), pp. 62–70.

Interviews:

Barry Alpert, "David Antin – An Interview," *Vort: Twenty-first Century Pre-views: David Antin – Jerome Rothenberg*, 7 (1975): 3–33;

Hazel Smith and Roger Dean, "Talking and Thinking," *Postmodern Culture*, 3 (May 1993).

References:

Barry Alpert, "Post-Modern Oral Poetry: Buckminster Fuller, John Cage, and David Antin," *Boundary II*, 3 (Spring 1975): 665–681;

Charles Altieri, "The Postmodernism of David Antin's *Tuning*," *College English*, 48 (January 1986): 9–26;

Stephen Fredman, *Poet's Prose: The Crisis in American Verse* (Cambridge: Cambridge University Press, 1983);

Frederick Garber, *Repositionings: Readings Contemporary Poetry, Photography, and Performance Art* (University Park: Pennsylvania State University Press, 1995);

Charles O. Hartman, *Jazz Text: Voice and Improvisation in Poetry, Jazz and Song* (Princeton: Princeton University Press, 1991);

Lita Hornick, *David Antin: Debunker of the "Real"* (Putnam Valley, N.Y.: Swollen Magpie Press, 1979);

Hank Lazer, "Thinking Made in the Mouth: The Cultural Poetics of David Antin and Jerome Rothenberg," in *Picturing Cultural Values in Postmodern America*, edited by William G. Doty (Tuscaloosa: University of Alabama Press, 1995), pp. 101–139; also included with an additional essay devoted to Antin in *Opposing Poetries: The Cultural Politics of Avant-Garde American Poetry* (Chicago: Northwestern University Press, 1995);

Sherman Paul, *In Search of the Primitive: Rereading David Antin, Jerome Rothenberg, and Gary Snyder* (Baton Rouge: Louisiana State University Press, 1986);

Marjorie Perloff, *The Poetics of Indeterminacy: Rimbaud to Cage* (Princeton, N.J.: Princeton University Press, 1981);

Henry M. Sayre, "David Antin and the Oral Poetics Movement," *Contemporary Literature*, 23 (Fall 1982): 428–450;

Vort: Twenty-first Century Pre-views: David Antin – Jerome Rothenberg, 7 (1975): 2–91;

Ellen Zweig, "Where Is the Piece: An Account of a Talk by David Antin," in *The Poetry Reading: A Contemporary Compendium on Language and Performance*, edited by Stephen Vincent and Zweig (San Francisco: Momo's Press, 1981), pp. 174–186.

Charles Bernstein
(4 April 1950 –)

Loss Pequeño Glazier
State University of New York at Buffalo

BOOKS: *Asylums* (New York: Asylum's Press, 1975);
Parsing (New York: Asylum's Press, 1976);
Shade (College Park, Md.: Sun & Moon Press, 1978);
Disfrutes (Needham, Mass.: Ganick, 1979);
Poetic Justice (Baltimore: Pod Books, 1979);
Senses of Responsibility (Berkeley: Tuumba Press, 1979);
Legend, by Bernstein, Bruce Andrews, Steve McCaffery, Ron Silliman, and Ray DiPalma (New York: L=A=N=G=U=A=G=E/Segue, 1980);
Controlling Interests (New York: Roof Books, 1980);
The Occurrence of Tune (New York: Segue, 1981);
Stigma (Barrytown, N.Y.: Station Hill Press, 1981);
Islets/Irritations (New York: Davies, 1983);
Resistance (Windsor, Vt.: Awede Press, 1983);
Content's Dream: Essays 1975–1984 (Los Angeles: Sun & Moon Press, 1986);
Artifice of Absorption (Philadelphia: Singing Horse Press/Paper Air, 1987);
The Sophist (Los Angeles: Sun & Moon Press, 1987);
Veil (Madison, Wis.: Xexoxial Editions, 1987);
Four Poems (Tucson: Chax Press, 1988);
The Nude Formalism, by Bernstein and Susan Bee (Los Angeles: Sun & Moon Press, 1989);
The Absent Father in Dumbo (La Laguna, Islas Canarias (Spain): Zasterle Press, 1990);
Fool's Gold, by Bernstein and Bee (Tucson: Chax Press, 1991);
Rough Trades (Los Angeles: Sun & Moon Press, 1991);
A Poetics (Cambridge, Mass.: Harvard University Press, 1992);
Dark City (Los Angles: Sun & Moon Press, 1994);
The Subject (Buffalo: Meow Press, 1995);
Technology/Art: 20 Brief Proposals for the National Academy of Engineering, by Bernstein, Bruce Andrews, and James Sherry (Buffalo: Meow Press, 1995).

RECORDINGS: *Class,* Tapeworks from Widemouth Tapes, 1976;

Charles Bernstein (photograph by Susan Bee)

Guess Language, Audio Muzixa Qet, 1986;
Live at the Ear, edited by Bernstein, Elemenope/Oracular Recordings, 1993.

OTHER: *L=A=N=G=U=A=G=E,* edited by Bernstein and Bruce Andrews, 1–3, no. 13 (1978–1981); volume 4 also published as *Open Letter,* 5, no. 1 (1982);
"Language Sampler," edited by Bernstein, *Paris Review,* no. 86 (1982), pp. 75–125;
Claude Royet-Journoud, *The Maternal Drape,* translated by Bernstein (Windsor, Vt.: Awede Press, 1984);

The L=A=N=G=U=A=G=E Book, edited by Bernstein and Andrews (Carbondale: Southern Illinois University Press, 1984);

"43 Poets (1984)," edited by Bernstein, *Boundary 2,* 14 (Fall 1985/Winter 1986): 1–113 (1987);

"L=A=N=G=U=A=G=E Lines," edited by Bernstein and Andrews, in *The Line in Postmodern Poetry,* edited by Robert Frank and Henry Sayre (Urbana: University of Illinois Press, 1988), pp. 175–216;

Olivier Cadiot, *Red, Green, and Black,* translated by Bernstein (Hartford: Potes & Poets, 1990);

"Inventing Wordness: Gertrude Stein's Philosophical Investigations," in *Gertrude Stein Advanced: An Anthology Of Criticism,* edited by Richard Kostelanetz (Jefferson, N.C.: McFarland, 1990), pp. 57–62;

Patterns / Contexts / Time: A Symposium On Contemporary Poetry, edited by Bernstein and Phillip Foss (Santa Fe, N. Mex.: Recursos de Santa Fe, 1990);

The Politics of Poetic Form: Poetry and Public Policy, edited by Bernstein (New York: Roof, 1990);

"13 North American Poets," edited by Bernstein and Susan Howe, *TXT,* 31 (1993): 2–32;

Poetics, E-mail discussion group moderated by Bernstein (http://wings.buffalo.edu/epc/poetics);

Bernstein Web-site (http://wings.buffalo.edu/epc/authors/bernstein).

Charles Bernstein is mostly widely known for his early influence in Language poetry. A loose constellation of experimental writers in New York, San Francisco, and Toronto, Language poetry emerged in the late 1970s and early 1980s and promoted new forms of experimental writing that paid attention to language itself. Much as, in the art world, movements such as Abstract Expressionism called attention to the materiality of their artistic medium, Language poetry engaged a similar task for writing. Exploring some of the overlooked breakthroughs of literary and philosophical predecessors such as Gertrude Stein, the Objectivist poets, and Wittgenstein, Language poetry at times is seen as developing and extending some of the experimentalism of the earlier Black Mountain and New York Schools, particularly as those schools had begun to explore the place of the "I" in the poem. At the center of the emerging school was *L=A=N=G=U=A=G=E* magazine, edited by Bernstein and Bruce Andrews. *L=A=N=G=U=A=G=E* (from which the movement also got its name) allowed for the circulation of a rich web of essays and literary writings. These writings explored the literary influences, theories, and modes of writing that formed a loose nucleus around which many of these writers conducted their investigations. Bernstein and Andrews edited *The L=A=N=G=U=A=G=E Book,* a volume that allowed for the much wider circulation of many of the key essays and writings from the magazine. *The L=A=N=G=U=A=G=E Book* was instrumental in laying a foundation for a new sense of writing and vividly articulated the multidisciplinary and polytextual sweep of this writing's core investigations.

Beyond his highly visible activities at the beginning of Language writing, Bernstein's presence in the world of experimental writing has been consistent and extraordinary. His work has been published in nearly every major experimental magazine over the past two decades; he has been a key voice in many significant conferences. Besides the voluminous amount of poetry he has written, he has been tireless in his devotion to the work of writing. His reviews, essays, and other writings have been influential in championing new voices in literature, in calling attention to overlooked writers from the past, and in exposing the complacent lethargy of what Bernstein calls "official verse culture," that body of conventional opinion that often seems at odds with new possibilities. Now the David Gray Professor of Poetry and Letters at the State University of New York at Buffalo, Bernstein remains a strong literary influence on many through his continued writing and teaching, the many publications published through Buffalo's Poetics program, the Poetics discussion list on the internet (which Bernstein moderates), and the Electronic Poetry Center (for which Bernstein serves as chief advisor). Bernstein has been acclaimed by such important critics as Marjorie Perloff, Jerome McGann, and John Shoptaw; he is a frequent keynote speaker at major literary events, and there are panels about his writing at gatherings such as the Modern Language Association and the Twentieth Century Literature conferences. He continues to be active as both a poet and critic.

Charles Kegel Bernstein was born on 4 April 1950 in New York City to Herman and Sherry Bernstein. His father worked in the garment industry, mainly as a dress manufacturer. The youngest of three children, Bernstein grew up near Central Park in Manhattan. He attended the Bronx High School of Science and edited the school newspaper. In 1968 he met his future wife, the artist Susan Bee (born Susan Bee Laufer). From 1968 to 1972 Bernstein attended Harvard

University and was active in the antiwar movement. He concentrated in philosophy, studying with Stanley Cavell and Rogers Albritton. His senior thesis, "Three Compositions on Philosophy and Literature," was a reading of Gertrude Stein's *Making of Americans* (1925) through Ludwig Wittgenstein's *Philosophical Investigations* (1953). Some of his thesis has subsequently been published in Richard Kostelanetz's *Gertrude Stein Advanced: An Anthology of Criticism* (1990). Bernstein edited Harvard's freshman literary magazine and published *Writing,* a photocopy magazine. He was also actively involved in radical theater, directing several productions. He graduated Phi Beta Kappa in 1972.

In early 1973 Bernstein won a William Lyon MacKenzie King fellowship, enabling him to move, with Bee, to the Vancouver area. At Simon Fraser University, Bernstein took a seminar on Emily Dickinson with Robin Blaser, an experience that proved influential for him. Following nine months in Vancouver, Bernstein and Bee moved to Santa Barbara where Bernstein worked part-time as a health education coordinator in a community free clinic; his writing from that time would be later published as *Asylums* (1975) and *Disfrutes* (1979). Bernstein and Bee moved back to New York in early 1975. They married in New Hampshire in 1977; their children, Emma and Felix, were born in 1985 and 1992, respectively.

Although he had met poets in Vancouver, Santa Barbara, and San Francisco, where he had traveled to meet Ron Silliman, it was not until his return to New York that Bernstein became fully engaged in poetry as public practice. He met Bruce Andrews shortly after his return, spent much time going to St. Mark's Poetry Project and other New York readings, and in 1978 cofounded with Ted Greenwald the acclaimed Ear Inn series, a significant venue for emerging writers. A compact disk of some of the readings was produced by Bernstein in 1993.

While Bernstein devoted himself to poetry and poetics, his income was not related to literary activity. For nearly twenty years his living would come from his work as a medical and healthcare editor and writer. In 1977 and 1978, for example, he wrote medical abstracts full-time for the Canadian edition of *Modern Medicine*. In an unpublished August 1995 interview he commented that "this immersion in commercial writing and editing — as a social space but more in the technical sense of learning the standardized compositional rules and forms at the most detailed, and numbingly boring, level — was informing in every way." He also noted that he

Cover for Bernstein's 1978 collection of poems, the first book published by the influential avant-garde press Sun & Moon

and Bee designed the *Health Manpower Consortia Newsletter* in the same format that would be later used for *L=A=N=G=U=A=G=E* magazine. Bernstein has adapted medical vocabulary for poetic purposes and has, like Stein, Wittgenstein, and the Objectivist poets before him, continued to explore "rules and forms."

Brought out by his and Bee's Asylum's Press in New York, Bernstein's first two books, *Asylums* and *Parsing* (1976), were produced in the manner of such influential publications as *Adventures In Poetry, C* and *The World* as side-stapled photocopies of manually typed pages. The poem "Asylum," which Bernstein chose to reprint in *Islets/Irritations* (1983), begins:

> rooms, suites of rooms, buildings, plants
>   in line. Their encompassing or total
>   character
>   intercourse with the outside and to departure
> such as locked doors, high walls, barbed
>   wire, cliffs, water, forests, moors
>   conflicts, discreditings, failures
>   of assimilation. If cultural change
>   the outside. Thus, if the inmates stay
>   victory. They create and sustain
> a particular kind of tension
>   dangers to it, with the welfare

> jails, penitentiaries, P.O.W.
> camps, concentration camps[.]

This poem is striking in several ways. First, the *field* of the poem, the way the lines look on the page, is constantly shifting, constantly modified. Not only does this repeatedly modified field suggest chaos and a clash of multiple voices and tones, it marks a direct refusal to accept the concept of poetic "order." This poem will not assume tidy stanzas or give the reader the comfort of predictable line lengths. Second, there is no fixed narrative element to this poem. The first line of the poem seems to be the middle of a sentence, and the poem is built from a series of objects and their sounds, which are placed "in line." Third, this poem is dense in its diction; it is hard to read aloud and the words at time seem to clash, butt up against each other, and grate. Last, the poem is about institutions. (The title is a reference to one type of institution; poetry itself is another type.) The reader can feel the artificial pressure and stark reality of the institution throughout this poem. It has "locked doors, high walls, barbed / wire"; the poem "create[s] and sustain[s] / a particular kind of tension." It warns of danger and also speaks of a victory that apparently comes as long as the literal surface tension of the poem can be maintained.

Critic John Shoptaw describes "Asylum" as "a disturbing poem composed of broken lines unjustly arranged upon the page." Shoptaw asserts that in the poem "Bernstein confronts the objectifying capacity of the list." *Asylums* also includes an early version of "Out of this Inside," a poem Bernstein later rewrote for *Poetic Justice* (1979). *Parsing* contains two sections – the first, "Sentences," giving way to the second, "Parsing," as sentences break up into phrases. Bernstein's concern for order is evident in "Roseland": "this axis this / the human order / more or less / you have a map."

In 1978 Bernstein and Bruce Andrews founded *L=A=N=G=U=A=G=E* magazine. In "An Interview with Tom Beckett," included in *Content's Dream: Essays 1975–1984* (1986), Bernstein recalls that the motivation to publish the magazine grew from the exchanges he had in correspondence with Andrews and Silliman. "The impulse for the magazine was to make that kind of exchange . . . more public, to share the thinking and conceptualizing with as large a group of people as we could interest." As to the magazine's prose format, Bernstein comments that the project as conceived would have included poetry, reviews, and art, but this was not affordable. The scope of the original vision was eventually realized through a small group of periodicals. "*L=A=N=G=U=A=G=E* was one piece of this larger project," Bernstein suggests, "which to some degree did come into being in a modular way – if you think of *This*, or *Roof*, or *A Hundred Posters* or *Tottel's* as other parts."

Produced as photocopied, coverless stapled booklets, *L=A=N=G=U=A=G=E* was published until 1981. This run encompassed thirteen regular issues, three supplements, an index, and a final number copublished as an issue of the journal *Open Letter*. Despite its modest format and limited circulation, *L=A=N=G=U=A=G=E* has proven to be one of the most influential little magazines of the period. It introduced many writers who have since enjoyed significant recognition as poets and scholars. *L=A=N=G=U=A=G=E* also became the namesake for a constellation of literary concerns now common in academic and poetic discourse.

Although contributors to *L=A=N=G=U=A=G=E* shared a common interest, the magazine was never intended to serve as the center of a literary school or movement. According to Steve McCaffery, "Though many contributors conceived the practice of writing to be primarily a social fact and saw the production of meaning as occupying, with a certain inevitability, a sociopolitical position within the politics of representation, there was never a suggestion of a unitary group of movement." In *Content's Dream* Bernstein's assertions show him to be uncomfortable with the notion of a *L=A=N=G=U=A=G=E* school:

> Certainly, I do see the magazine, *L=A=N=G=U=A=G=E*, and my own work, as expressing certain shared views about reading and about the constituting power of language, about seeing language itself as the medium of the work and foregrounding that medium. And yet this is not a movement in the traditional art sense, since the value of giving an aesthetic line such profile seems counterproductive to the inherent value of the work.

He expresses agreement with Frederic Jameson's statement from *The Prison House of Language* (1972) that when a literary writing "marks my affiliation with a given social group, it signifies the exclusion of all the others also – in a world of classes and violence, even the most innocuous group-affiliation carries the negative value of aggression with it."

L=A=N=G=U=A=G=E advanced poetic thought by foregrounding specific interests that stood in opposition to the prevailing academic tastes. "Language writers have used structuralist

Cover for Bernstein's 1981 collection, which includes photographic montages by his wife, Susan Bee

and poststructuralist theory at times, to furnish ad hoc support for assertions about the problematic status of description, self, and narrative in poetry," Bob Perelman asserts in "Write the Power," "But these positions have come out of writing practices closely informed by the modernists, especially Gertrude Stein, and the Objectivists, especially Louis Zukofsky, and to some extent by the New American poets and the New York school. Thus — to be schematic about it — language writing occupies a middle territory bounded on the one side by poetry as currently instituted in the academy and on the other by theory." As to its position between "the workshop poem's commitment to voice and immediate experience" and the ambiguities of texts generated by "theoretical regimes," Language writing is different in its "foregrounding [of] the activity of writing."

Bernstein's next three books were published during the L=A=N=G=U=A=G=E years: *Shade* (1978), *Poetic Justice,* and *Senses Of Responsibility* (1979). These books began to establish his work at the vanguard of experimental writing. *Shade,* a typescript publication in the style of Bernstein's first two books, enjoys the distinction of being the first book publication of Sun & Moon Press, now a major publisher of avant-garde writing. It includes sixteen poems constructed of short lines that challenge the reader to leap from thought to thought. In "Ballet Russe," for example, Bernstein writes: "I love motion & dancing. / I did not understand God. / I have made mistakes." The jaggedness in the progression of phrases becomes even more emphasized when lines are shorter, as in "The Bean Field": "Yet I / gelatinous mildewy tether / hissing of urn / screech-owl or / this vast. Range / too."

In contrast to the predominant white space of *Shade,* Bernstein's *Poetic Justice* exhibits a twisting landscape of intermingled prose utterances stretching from margin to margin, as is shown in the following line from "Lo Disfruto": "One a problem with a fragment sitting. Wave I stare as well at that only as if this all and not form letting it but is it." Also present in this collection are works such as "eLecTrIc" and "AZOOT D'PUUND," a poem in which Bernstein experiments with dislocated typography in the spirit of Jackson Mac Low's change permutations:

iz wurry ray aZoOt de pound in reducey ap crrRisLe
ehk nugkinj sJuxYY senshl. Ig si heh hahpae uvd r
fahbeh at si gidrid. ImpOg qwbk tuUg. jr'ghtpihqw.

In this work, it is clear that Bernstein is exploring the permeability of language. What does one see when one can not be lulled into reverie by what the words say? What is the emotional result when capitals and other typographic markers are radically recast? What does the reader get from such typography – is it sound, a visual work, an assault on conservative poetic value? A final and very important point is the question that such texts raise about where a poetic work comes from. Maybe the above characters are intentional. Maybe they occur as a result of errors by the typist. How could one decide?

One of Bernstein's often cited poems is called "Lift Off," which from its opening challenges the reader:

HH/   ie,s   obVrsxr;atjrn   dugh   seineocpcy   I iibalfmgmMw
er,,me"ius ieigorcy¢jeuvine+pee.)a/nat" ihl"n,s

What does one "see" in such a poem? Pieces of words here leap out (*HH*, a diminutive of *HRH*; *seineocpcy*, the Seine River; *me"ius*, a parody of Latin; *jeuvine+pee*, juvenile plus urination) in a hilarious cacophony and raucous visual clutter. If one of the issues that experimental writing explores is to remove the authorial "I," then this poem is an excursion to a terrain that is purely experimental. If one of the issues is where a poem comes from, this poem not only begs the question but laughs at one for asking since it is a transcription of the correction tape from a self-correcting typewriter. As such it is a primarily poetic work constructed from the literal typographic detritus of other works, both literary and nonliterary. The emphasis on typographical elements in these poems creates texts where, as Shoptaw writes, "the characters themselves sometimes make the music."

Originally published by Lyn Hejinian's Tuumba Press, *Senses Of Responsibility* was reprinted in 1989 by Paradigm Press. This small collection contains six poems made up mostly of short lines, though its sonorous sentences often continue over several lines. Shoptaw has pointed to this collection as showing that Bernstein has "occasionally stopped and hovered over [John] Ashbery's fluid measures." *Senses Of Responsibility* shows not only Bernstein's consideration of Ashbery's poetics but also his command and translation of them into his own form.

In 1980 Bernstein received a National Endowment for the Arts Creative Writing Fellowship. Also that year he led a workshop at the Poetry Project at St. Mark's Church in New York, and his works *Legend* and *Controlling Interests* were published. In *Legend* Berntein collaborated with Andrews, Ray DiPalma, McCaffery, and Silliman to explore a wide range of textual presentations. *Controlling Interests* was a milestone volume in Bernstein's career, the first of his books to present poems in heterogeneous formats. The formal variance of its seventeen poems, ranging from single stanza poems to mixtures of prose and verse in fields of words, foreground an intentional unevenness.

"Disrupting chronology," Silliman notes in his essay in *Controlling Interests*, "is a defense against the reduction of poetry to 'mere' autobiography, particularly when the formation of a subject is taken as the persistent content of the work." Bernstein's use of a temporally fragmented authorial subject, Silliman argues, allows him to focus on the "constitutive aspect of language, rather than treating it as a self-forming *object*." In the "discontinuous units" that comprise this collection, language is foregrounded as subject. Barrett Watten asserts that "The politics of the work are in [Bernstein's] internalization of 'radical structural means'; oppression, seen as an act of language, will be increasingly revealed. 'To push things into further nature' is the impulse; *Controlling Interests* intends a further statement conceived entirely on the ability to act." *Controlling Interests* contains such crucial Bernstein poems as "Matters of Policy," "Sentences My Father Used," and "Standing Target." Writing of "Sentences My Father Used," Perelman maintains: "Language here is investigating itself, proclaiming its opacity, revealing words as code or husk. The bonds of grammar are loosened . . . Statement and image half appear and then fade into something other."

The last issue of *L=A=N=G=U=A=G=E* was published in 1981 as were three of Bernstein's books, *Disfrutes, The Occurrence of Tune,* and *Stigma*. *Disfrutes* was written in 1974, published privately in 1979, and then reprinted that same year by Potes & Poets Press. It is an unpaginated, chapbook-sized volume that evokes the alphabet with its twenty-six untitled poems (or sections of a poem). Highly abstract, each poem/page is a single column of text containing one to several words. Phrases, parts of phrases, grammatical parts such as "of it / on / the it / she / on the / it she" make the reader aware of the white space surrounding the words, the silence surrounding language's content. The regularity of the white space or silence is occasionally blurred or broken as certain lines malfunction:

( ) unched
th... b...rb...n th...mb...l...n...[.]

The poems thus suggest an "alphabet" that refuses language's capacity for communication. Shoptaw notes that *disfrutes* in Spanish means "enjoyment." Alternatively, the poems might be read as the disaffiliated "fruits" of language.

Bernstein collaborated with Susan Bee, who provided photographs, in the creation of *The Occurrence of Tune*. The third line of the text declares that there is "no presence, no things," a statement that in the context of the volume asserts the limitations of concrete reality (what words describe, what images depict) to imply meaning. The altered photographs by Bee contain recognizable images whose familiar meanings are either overpowered or undermined by the manner of their presentation. In Bee's cover for the book, for example, American icons are overbearing despite their physical decay: a run-down fast food restaurant advertising "Frankfurters, Table Service, Fountain, Hamburgers" hovers over a pattern that suggests a highway landscape. Likewise, the text establishes patterns of words that are at odds with "normal" reality by juxtaposing snatches of conversation in quotes, aphoristic declarations, and blocks of prose showing "much flurry, little *regard*" for standard syntax.

Consisting of eleven single stanza poems, each headed with a title in bold type, *Stigma* presents a more conservative textual format than *The Occurrence of Tune*. It includes such traditional titles as "New Year's," "December," and "April." This presentation, however, merely serves as a conventional frame from which Bernstein immediately departs. "April," for example, begins:

Webbed space
akin to almost ash
gathered at entrance
a sadness
basically projected
all this
haecceity but not
one thing discerned from another[.]

The flow of juxtaposed words and phrases in the "webbed space" of the poem is neither logical nor predictable. Even as "projected," it breaks down until "haecceity" is invoked. The title word *stigma* carries the senses of both a mark and a stain. In this collection, it is as if the words themselves stain the text; they are not words but faults that "convene, argue plans, yet point / At any loss, so much, erasing / Our undoing, greatest wildness. Continuous / Focus – shift, blur, become transparent, persists."

The most important event in the reception of Bernstein's early work was the 1982 publication of the "Charles Bernstein Issue" of *The Difficulties,* edited by Tom Beckett. The issue included Beckett's interview with Bernstein that was reprinted in *Content's Dream* (a continuation of this interview was published as "Censers of the Unknown-Margins, Dissent, and the Poetic Horizon" in *A Poetics* in 1992) as well as critical essays, homages, and poems. The contributors include Silliman, Robert Creeley, Perelman, Jackson Mac Low, Watten, Alan Davies, and others. James Sherry commented on the poet's attempt to reshape thought in language: "sentences reflect the constant intrusion of the world on the mind and the impossibility of meditation/mediation.... Bernstein is not concerned with grammar and sentences as correct... [but with] ... patterns of thought in the patterns of [his] phrases rather than given standards of the shape of the thought as in the grammatically oriented notion of the sentence." Craig Watson maintains that in Bernstein's work "the apprehension of meaning is not conditioned by substance or continuity alone, but by the constitution of relations in a network of possibility; the object is mediated and named by our whole experience with it. Cognition is made up of a vast circuitry within which language is gestural and continually subjective. Through these circuits we conduct and create our selves, a reality." Perelman writes that "Bernstein's ... language centered writing never fails to be centered around the person, and the pressures of culture and history that thwart language's power."

The issue also contains twenty-five pages of poems by Bernstein, concluding with "Substance Abuse" in which the poem's speaker relates, "I feel (felt) stripped by these / changes, I / don't know what I received and what / I was shut up with" and "I feel like a very nervous man." These sentiments are relevant to the occasion of the special issue, and the poem in this context seems a comment on Bernstein's new level of public prominence: "So these sorrows pronounce themselves / in rhymes before my eyes" and "I make this point because your gazing / at a so projected grouping 'at a / distance' clouds your view – ." Bernstein astutely displays the dynamics of what such a recognition does to the speaker: "One / guise disguises itself within myself, / the other within my text." The sounds of these lines echo the sentiments of the speaker: "guise" is within "disguise" as the guise is "within myself." Secondly, the guise "within myself" is repeated by the sound of "within my text." "Myself"

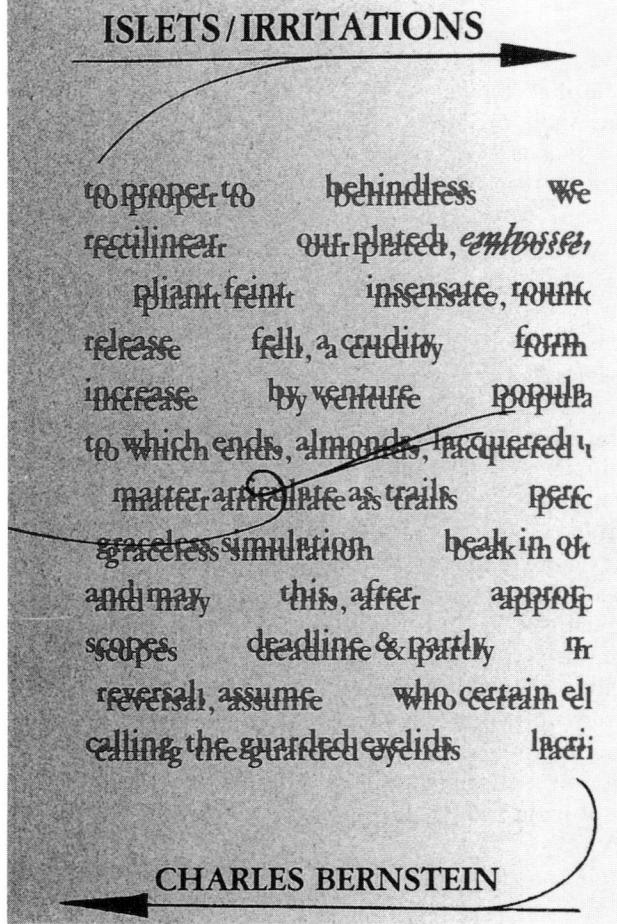

Cover for Bernstein's 1983 collection that includes "The Klupzy Girl," a poem beginning with the lines "Poetry is like a swoon, with this difference: it brings you to your senses"

and "my text" are parallel elements and reveal their difference only in the disparity between "self" and "text." Near the end of the poem, the speaker says, "Anxious and waiting for something, but not / definable-amorphous. What pans out? / I'm afraid to set it down, to contend with / the medium at hand." In "Substance Abuse," Bernstein purposefully dramatizes the poet's public persona.

After editing a selection of Language poetry for the *Paris Review* in 1982, Bernstein the next year brought out two collections – *Resistance,* a gathering of eighteen hard-edged, densely packed poems, and *Islets/Irritations*. The poems in *Islets/Irritations* – including "Sprocket Damage," "Gradation," "Asylum," and one of Bernstein's best-known poems, "The Klupzy Girl" – are particularly interesting because they display a wide range of poetic form. The title poem, the first in the collection, is an example of a modified field format – the terrain Berstein can establish through his densely packed, often ironic weave of words. Its first two lines indicate how Bernstein uses regular spacing at irregular intervals to form his field: "to proper to behindless weigh in a rotating, / rectilinear our plated, *embosserie des petits cochons*." He uses a completely different form in the poem "Contradiction Turns To Rivalry" by juxtaposing colorful sentence-length paragraphs:

On-the-street subjects render fragmented versions; a two-way mirror provides some unexpected "reflections"; a pair of outdoor phonebooths and two muddled conversations befuddle a man.

A backstage view is interwoven with a tragic story.

A detective is captured by a mobster who plans to hook him on heroin and then deny him a fix until he reveals the whereabouts of the jealous hood's former girlfriend.

A retarded young man witnesses a murder but is not articulate enough to tell his story to the police.

In his 149-line single-stanza poem, "The Klupzy Girl," Bernstein engages the reader from its opening lines: "Poetry is like a swoon, with this difference: / it brings you to your senses." Early on Bernstein delares, "Not clairvoyance, / predictions, deciphering – enacting," a line that indicates the energy he strives to attain in his dense poem. He juxtaposes diverse voices (including Walter Benjamin's "'There is no document of civilization / that is not at the same time a / document of barbarism'"), lines propelled by clipped grammar ("It has / more to me than please to note acquits / defiant spawn"), and various *forms* of address:

> That's why I'm perplexed
> at your startlement, though obviously
> it's startling to see contexts changed on you
> to have that done to you and
> delivered unbeknownst. The Ideal
> swoops, and reascends.

"The Klupzy Girl" operates most of all through its sustained sense of music, the sound of its enactment maintaining the poem's velocity throughout.

Bernstein's next project was to edit *The L=A=N=G=U=A=G=E Book* (1984), in which he included a selection of work published in *L=A=N=G=U=A=G=E* magazine. The same year he also published some of his own translations of Claude Royet-Journoud's work from *The Maternal Drape* (1984). The nearly three-hundred-page *The L=A=N=G=U=A=G=E Book* reprints material from the first three volumes of *L=A=N=G=U=A=G=E*, about half of the material published in the original magazine. Awarded a John Simon Guggenheim Memorial Fellowship in 1985, Bernstein attended the New Poetics Colloquium in Vancouver that year. During this period he wrote the essay "Being A Statement On Poetics," later published in *Rough Trades* (1991). He makes a strong argument for a new poetics, declaring that "A poem should not be but become" and humorously insisting on the poem's social materiality: "The problem / is not the bathwater but the baby. I want / a poem as real as an Orange Julius."

Bernstein's first essay collection, *Content's Dream: Essays 1975–1984,* contains work that originally appeared in *L=A=N=G=U=A=G=E, Open Letter, Sagetrieb, Boundary, Code of Signals, Poetics Journal, The Paris Review, Sulfur, The Poetry Project Newsletter, The Difficulties, Credences, What Is A Poet?,* and *Writing/Talks*. Titles include such crucial works as "Three or Four Things I Know about Him," "Thought's Measure," "Writing and Method," "Characterization," "Words and Pictures," "The Objects of Meaning: Reading Cavell Reading Wittgenstein," and "Blood on the Cutting Room Floor." As Jerome McGann points out,

> The problem of poetry's relation to prose (and vice versa) is a recurrent preoccupation of Bernstein's work. *Content's Dream,* for example, is subheaded 'Essays 1975–1984' but the collection itself calls into question the distinction between poetry and prose. 'Three or Four Things I Know About Him' (which opens the collection); 'A Particular Thing'; 'G--'; are these texts poetry or are they prose? It is not easy to say. Similarly, his books of 'poetry' all print texts which do not, for various different reasons, 'look like' poetry at all.

Bernstein investigates writing as a practice that is not subordinate to a particular genre. In "Writing and Method" he asserts that "One vision of a constructive writing practice I have, and it can be approached in both poetry and philosophy, is of a multi-discourse text, a work that would involve many different types and styles and modes of language in the same 'hyperspace.' Such a textual practice would have a dialogic or polylogic rather than monologic method." In "Thought's Measure" Bernstein argues that "Language is the material of both thinking and writing. We think and write in language, which sets up an intrinsic connection between the two." *Content's Dream* allows the reader to trace the interweavings of writing and philosophy that inform the poetry that Bernstein writes and admires. He believes that "language is measure. And it is with this that we make our music – by ourselves, privately (if so that the measure's heard) – a private act, a revelation of the public. So that writing that had seemed to distance itself from us by its solitude – opaque, obscure, difficult – now seems by its distance more public, its distance the measure of its music. A privacy in which the self itself disappears and leaves us the world."

Bernstein also puts forth a precise description of the forces that are at odds with the poetry that attracts him. He refers to the "nexus of poets and critics who enforce norms of poetic value based on transparent language" as "official verse culture." In the essay "The Academy in Peril: William Carlos Williams Meets the MLA" he explains his definition:

> Let me be specific as to what I mean by "official verse culture" – I am referring to the poetry publishing and reviewing practices of *The New York Times, The Nation, American Poetry Review, The New York Review of Books, The*

New Yorker, Poetry (Chicago), *Antaeus, Parnassus,* Atheneum Press, all the major trade publishers, ... [and most] major university presses.... Add to this the ideologically motivated selection of the vast majority of poets teaching in university writing and literature programs ... as well as the interlocking accreditation of these selections through prizes and awards judged by these same individuals.

Bernstein's objection is to "biased, narrowly focussed and frequently shrill and contentious accounts of American poetry, [which claim,] like all disinformation propaganda, to be giving historical or nonpartisan views."

Bernstein's struggle in the arena of literary politics is a major thrust of his work. His premise is that writing is the practice of ideology, not an occupation that uses a transparent or neutral medium. As McGann comments in *Black Riders: The Visible Language of Modernism* (1993), "Bernstein manipulates the visible and audible features of his work because those features give material shape to the writing's social and intellectual commitments. The writing means to declare its ideological goals." In describing the concerns of Bernstein's writing, McGann uses the phrase "the poet's office," an expression that well fits Bernstein's conception of writing as a political act. Bernstein has edited a volume titled *The Politics of Poetic Form: Poetry and Public Policy* (1990). In the essay "Optimism and Critical Excess (Process)" in *Content's Dream* he argues that "Poetics is the continuation of poetry by other means. Just as poetry is the continuation of politics by other means," echoing Silliman's assertion that "Poetry, like war, is the pursuit of politics by other means." In his consideration of Bernstein's *A Poetics* (1992) in "Write the Power," Perelman argues that "Bernstein frames the book with gestures that make poetry a counter-State or non-State." Such a view of poetry may account for the title of Bernstein's forthcoming collection of poems, "Republics of Reality."

In 1986 Bernstein received the University of Auckland Foundation Fellowship and became a visiting lecturer in the English department of the New Zealand school, an appointment that enhanced his international reputation and brought him into contact with active New Zealand poets. His sojourn in New Zealand was followed by the publication of several works, including *Veil* (1987), a work he had written in 1976. The collection is prefaced by a quotation from Nathaniel Hawthorne's story "The Minister's Black Veil": "There is an hour to come ... when all of us shall cast aside our veils. Take it not amiss ... if I wear this piece of crape till then." The "veil" of this book is manifested by unreadable print a text of superimposed and blurred characters. In effect *Veil* is a collection of visual poems. According to Perelman, "its wordless pages, light gray with an overlay of letters, furnishes a textbook illustration of [how] ... dissonances, foregrounding their own incompleteness [can be] considered in some sense normative." Bernstein also published *Artifice of Absorption* and the collection of poems *The Sophist* in 1987. The next year he brought out *Four Poems,* a fine press edition of short poems on a single large sheet. He also edited an issue of the journal *Boundary 2* titled "43 Poets (1984)" and a collection of statements about the line, "L=A=N=G=U=A=G=E Lines," which was included in *The Line in Postmodern Poetry* (1988).

The poems in *The Sophist* have sparked considerable critical interest. Literary critics such as Marjorie Perloff, Jerome McGann, and John Shoptaw have hailed such poems as "Safe Methods of Business," "The Simply," "The Year As Swatches," and "Amblyopia" as important to contemporary poetry. "Dysraphism" is a poem from the collection that charts new poetic ground:

> Did a wind come just as you got up or were
> you protecting me from it? I felt the abridgment
> of imperatives, the wave of detours, the sabre-
> rattling of inversion. *All lit up and no*
> *place to go.* Blinded by avenue and filled with
> adjacency. Arch or arched at. So there becomes bottles,
> hushed conductors, illustrated proclivities for puffed-
> up benchmarks. Morose or comotose. "Life is what
> you find, existence is what you repudiate." A good example

A major poem in his oeuvre, "Dysraphism" brings together many of the approaches Bernstein made in earlier work: his humor, the auditory density of his words, and the dramatic compression of a multitude of voices into a continuous hiss or apparent scrawl. The impermeability of texts such as "Shade" and the accidentalism of "Lift Off" collide here. The qualities of typographic compression and density are present despite the readability of the text. Further, the accidental qualities of the voices and phrases seem completely natural and effortless though the poem unwaveringly drives home its powerful presence, establishing an argument and an enactment of Bernstein's poetics.

In a footnote to this poem, Bernstein makes clear that he is using a medical term to describe a method of constructing poems. *Raph* means seam, Bernstein writes in a note to the poem, "so for me disraphism is mis-seaming — a prosodic device." Perloff remarks in *The Dance of the Intellect:*

Studies in the Poetry of the Pound Tradition (1985) that " 'Dysraphism' playfully exploits such rhetorical figures as pun, anaphora, epiphora, metathesis, epigram, anagram, and neologism to create a seamless web of reconstituted words" and that "sensitivity to etymologies and latent meanings is reflected in the poem itself, which is an elaborate 'dysfunctional fusion of embryonic parts' a 'disturbance of stress, pitch, and rhythm of speech' in the interest of a new kind of urban 'rhapsody.'" The sound of "Dysraphism" almost achieves the status of a rhapsody or a song while in it Bernstein achieves moments of brilliant poetic speech. For example, "blinded by avenue and filled with adjacency" demonstrates exactly how this "seaming" that Bernstein proposes would work. The monolithic, ordered path ("avenue") is blinding; that is the traditional form of poeisis. What intrigues the reader is the idea of "adjacency," the materials at hand, next to us, next to each other. These are the materials that will be dysraphically combined. Even as Bernstein makes this assertion, his next words are "arch or arched at," words that are "adjacent" because of their sound.

Bernstein's *Artifice of Absorption,* published as a special issue of *Paper Air* in 1987, is a further articulation of his poetics. Called "the best introduction to Language poetry likely to be written" by Shoptaw, this essay in poem format immediately foregrounds the question of genre by its own form. In his introduction to *A Poetics* (where an emended version of this text was later republished), Bernstein comments, "If there's a temptation to read the long essay-in-verse . . . as prose, I hope there will be an equally strong temptation to read the succeeding prose as if it were poetry." He makes a similar point later in *Rough Trades,* "Nowadays, you can often spot a work / of poetry by whether it's in lines / or no; if it's in prose, there's a good chance / it's a poem." Shoptaw points out that genre-blurring occurs in Bernstein's "versified narrative summaries" where he "blurs the lines between poetry, fiction, and prose."

In *Artifice of Absorption* Bernstein writes an essay that looks, on the page, like a poem. It is through this positioning that this text is able to so clearly set the stage for the description of the frame of Language poetry. The operative vocabulary of this description consists of two main terms, "absorption" and "artifice." A poem, Bernstein asserts, is writing "designed to absorb." Artifice is "a measure of the poem's intractability to being read as the sum of its devices and subject matter." Perelman notes that "even though its line breaks foreground its artifice, the poem-essay contains much conventional

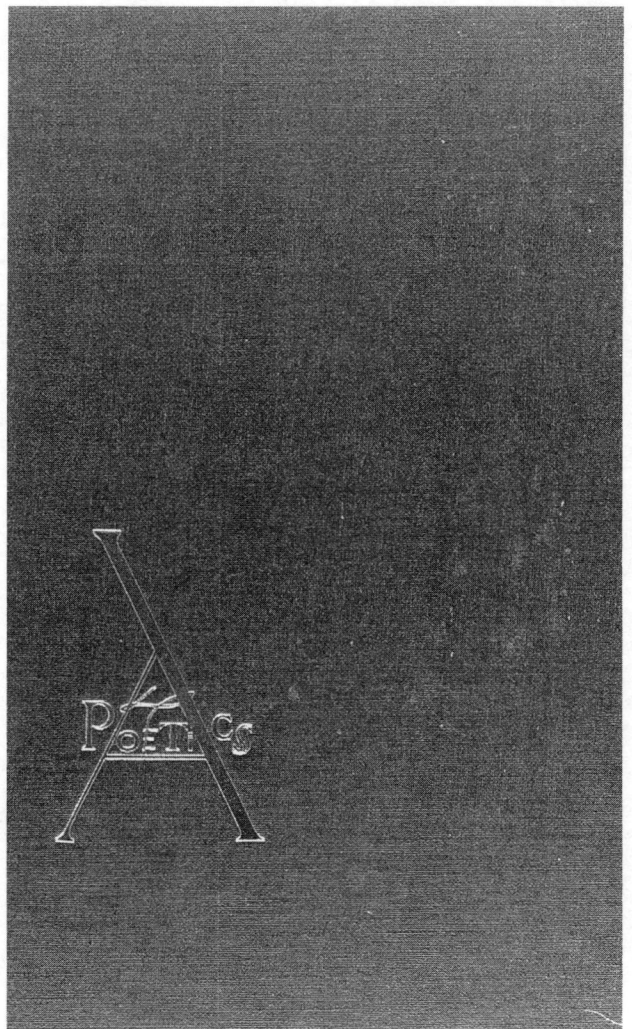

Cover for Bernstein's 1992 book of essays, in which he argues, "Poetics is the continuation of poetry by other means. Just as poetry is the continuation of politics by other means."

wiring. A large part of it is devoted to an omnibus review of contemporary poetry, providing capsule summaries of a number of language writers," including McCaffery, Susan Howe, Perelman, Nicole Brossard, David Antin, and Lyn Hejinian. The essay is an excursion, concisely stated, into poetry's terms, and gives, through references to other crucial texts in the terrain, specific descriptions of the textual particulars that constitute such writing specifically, what alternatives there are to a view of the text as a "transparent carrier of 'meaning.' " Bernstein writes that meaning, in such poetry, "is not absent or / deferred but self-embodied as the poem / in a way that is not transferable to another code / or rhetoric." Thus, the significance of this essay is that it does not surrender the ground of the essay, but redefines it – and in a way that is consistent with Bernstein's poetic achievements.

"Most of [Bernstein's] poetry," McGann writes in *Black Riders,* "does not surrender any of the territory of writing, any of the means to meaning. As a consequence, his writing extends across an unusual semiotic range – from the most minimal textual units, pre-morphemic, to the most complex rhetorical and semantic structures; and it carries out this 'opening of the field' not as an exercise in, or display of, imaginative mastery, but as an enactment – literally – of the world of writing." Bernstein's achievement in *Artifice of Absorption* is that he goes beyond simply mapping this new territory to claiming it through his own parody of artifice.

Bernstein had several appointments as a visiting professor during the late 1980s at such schools as the University of California at San Diego, the New School for Social Research, the State University of New York at Buffalo, and at Princeton University. In 1989 Bernstein and Bee published the lavishly illustrated *The Nude Formalism,* a book of short parodic poems, described by McGann as "parodies of the emblem tradition in [a] witty collaborative collection." Also interested in musical theater, Bernstein worked with composer Ben Yarmolinsky and wrote three librettos, *Blind Witness News, The Subject* (1995), and *The Lenny Paschen Show.* Bernstein also wrote *Cafe Buffe,* a libretto commissioned for composer Dean Drummond's New Band.

A milestone in Bernstein's career was his appointment in 1990 as the David Gray Professor of Poetry and Letters at the State University of New York at Buffalo, where he is also an associate member of the Department of Comparative Literature. After his appointment, Bernstein cofounded the Poetics Program, through which he has supported the New Coast and other conferences as well as publications such as *Chain, Poetic Briefs,* and *Situation* and resources such as Leave Books, Meow Press, and the Electronic Poetry Center. He coordinates the "Wednesdays At Four Plus," a reading series that has brought in important literary figures to read and discuss their work, and also directs the Poetics discussion list, an international online discussion group. Bernstein received a New York Foundation for the Arts Fellowship in 1990.

Among Bernstein's publications in the early 1990s was *Rough Trades,* a large collection that presented poems in three sections: "The Riddle of the Fat Faced Man," "Rough Trades," and "The Persistence of Persistence." Notable for its display of Bernstein's sense of comedy, the collection opens with a clever poem whose title evokes a tropical geography:

THE KIWI BIRD IN THE KIWI TREE

I want no paradise only to be
drenched in the downpour of words, fecund
with tropicality. Fundament be-
yond relation, less 'real' than made, as arms
surround a baby's gurgling: encir-
cling mesh pronounces its promise (not bars
that pinion, notes that ply). The tailor tells
of other tolls, the seam that binds, the trim,
the waste. & having spelled these names, move on
to toys or talcums, skates & scores. Only
the imaginary is real – not trumps
beclouding the mind's acrobatic versions. The first fact is the social body,
one from another, nor needs no other.

The reference to the "downpour of words" suggests not only a tropical climate but also a profusion of tropes and literary devices. Bernstein uses the poem to call for heterogeneous writing, not the "paradise" of a rigidly hegemonic model in which all is subordinated to a central image or insight. This fourteen-line poem makes notable use of enjambment in its decasyllabic lines. The hyphenated words *beyond, encircling,* and *versions* add a stammering element to the sound of the poem but also relate to its meaning: *be-* emphasizes being; *encir-cling* breaks the very circle the word means to circumscribe while clinging to the attempt; and *vers-ions* suggests both *verse* and *vers,* which means toward or against in Latin. The sense of direction in "vers-" is reinforced by the word's second half's being in a different direction, on the far left of the next line. Thus, the word *version* itself becomes a contortionist, dramatizing its own adjective, *acrobatic.* In this poem and others Bernstein calls for writing described in "Reading the Tree: 1," that is "turned into creamed figures, like constant / commotion, [is] repeatedly connoting."

A significant theme of *Rough Trades* – one that recurs in Bernstein's work – is suggested by its title. For Bernstein poetry not only is a vocation but also a business – and in either sense, a rough one. This theme is more than literary; it is true to Bernstein's generous and tireless involvement with poetry. As McGann notes in *Black Riders,* "The meaning of the work of poets like Lyn Hejinian and Bernstein is partly a function of their involvement with the social and material production of texts," and "like Morris and Pound, both have been actively involved in every aspect of poetry's production –

Draft for a poem to be included in a collection titled Republics of Reality *(courtesy of the author)*

from writing to book design to editing to distribution." While the difficulty of Bernstein's poetry limits his audience, he believes in the poet's social role: "The first fact is the social body, / one from another, nor needs no other."

Bernstein's "How I Painted Certain of My Pictures" was included in *Best American Poetry, 1992,* but the more significant event of the year was the publication of Bernstein's second book of essays, *A Poetics.* Significant essays, including "State of the Art," "Play It Again, Pac-Man," "Optimism and Critical Excess (Process)," "The Second War and Postmodern Memory" and the revised "Artifice of Absorption," explore poetics, philosophy, and the social dimensions of the text. Perelman comments, "Bernstein's continual humor – certainly a pleasure in the critical landscape – and the finesse with which he traverses institutional contexts (contemporary photography and painting, TV news and ads, video games, poststructuralism) mask the intransigence that ultimately informs the book where poetry is concerned."

Everywhere in his writing Bernstein's direction is toward the plural. In "State of the Art" he asserts, "*Poetry is the aversion of conformity.*" "Occasionally," Perelman points out, "Bernstein envisions a capacious unity, as in his call for 'a poetry and a poetics that do not edit out so much as edit in....'" This is difficult ground, however. As Perelman observes, "Such unities of poetry and poetics, modern and postmodern – e pluribus unum – flip into their opposite. Instead of an enlarged identity... writing becomes the site of an intimate demonstration of the impossibility of identity."

An engaging piece in this collection is "Optimism and Critical Excess (Process)," the recast transcription of a talk given in Buffalo in 1988, which begins with the statement, "This is not a transcription." This essay is particularly valuable because of Bernstein's consciousness of it as a performance and his working through his main points with multiple voices – the literary historian, the comic, the truly poetic in a manner that crosses multiple disciplines. Poetics itself, in Bernstein's view, is the site of a confluence of possibilities. It is exploratory and declines reductive conclusions; it is "an activity that is ongoing, that moves in different directions at the same time, and that tries to disrupt or problematicize any formulation that seems too final or preemptively restrictive."

Furthering his efforts to give experimental poetry a platform, Bernstein edited two works in 1993. He chose and edited recordings for the compact disk *Live At the Ear,* which made available selections from the reading series conducted from 1984 to 1993 at Ear Inn, a restaurant-bar in lower Manhattan. The poets who can be heard include Bernstein as well as Susan Howe, Ron Silliman, Bruce Andrews, and Rosmarie Waldrop. Bernstein also worked with Howe to edit "13 North American Poets" for *TXT* magazine in France.

Bernstein's thirteen-poem collection *Dark City* shows his continuing mastery of multiple approaches to experimental poetry. Its major poems include "The Lives of the Toll Takers," "How I Painted Certain of My Pictures," "Emotions of Normal People," "The Influence of Kinship Patterns upon Perception of an Ambiguous Stimulus," and its title poem. According to Hank Lazer, "The Lives of the Toll Takers," the first poem in *Dark City,* "establishes a consideration of the state of poetry today as one ... recurring concern for Bernstein." Lazer also suggests that this poem contains "some seemingly simple and straightforward axioms or conditions for poetry." Early on in the poem Bernstein rejects an alliance with theories of empty signifiers:

>   Difference or
> *différance*: it's
> the distinction between hauling junk and
> removing rubbish, while
> I, needless not to say, take
>   out the garbage
>   (pragmatism)[.]

As Bernstein has consistently argued, no single theoretical approach can ever be as fruitful as a framework that allows multiple possibilities for writing. "The Lives of the Toll Takers" exhibits Bernstein's humor ("What? No approach / too gross if it gets a laugh"), his use of intrusions from the external world ("Fatal Error F27: Disk directory full"), and playful typography:

> She can slip and she can slide, she's every
>   parent's j
>   oy & j
>   i
>   b
>   e
>   (guide)
> In dreams begin a lot of bad
> poetry.

A diverse collection, *Dark City,* as Lazer explains, is "built on a principle of difference – i.e., each poem different from those which surround it and a book of poems which offers conscious resistance to signature and the cults of personal voice [and] personality." Lazer points to "Emotions of Normal People," a catalogue of seemingly found language in the form of "an extended collage-poem," as one of the most important in the collection and praises Bernstein's "compositional arsenal," which includes "a wide-ranging vocabulary. . ., the recurrence of a peculiarly clotted sound-effect, a kind of line and sound that is deliberately but interestingly difficult to say, [and] a kind of anti-mellifluousness." The collection has the difficulty, rewards, and delight ("as difficult as / Keeping a hat in a hurricane / Or an appointment with an erasure," Bernstein writes in "The Influence of Kinship Patterns") that one would expect from a major poet.

Bernstein's work has commanded the attention of leading national scholars and been widely read. Lazer reports that his writing has been published in Argentina, China, Spain, Australia, New Zealand, France, Switzerland, Germany, Italy, the Netherlands, Portugal, England, Canada, Mexico, Finland, Yugoslavia, and Japan. A correspondent for *Sulfur*, Bernstein also serves on the editorial boards of *M/E/A/N/I/N/G* (New York), the Segue Foundation (New York), and many other organizations. His readership and reputation as a poet will probably rise when his next collection, *Republics of Reality: Poems 1975–1995*, is published by Sun & Moon Press. It will include Bernstein's earlier, nearly unobtainable books – *Parsing, Shade, Poetic Justice, Resistance, Stigma, Senses of Responsibility, The Occurrence of Tune*, and *The Absent Father in Dumbo* (1990) – and will feature a new sequence of poems titled "Residual Rubbernecking."

Bernstein has been called a "difficult" poet, but such a label surely in part indicates his integrity as an artist. The body of Bernstein's writing is evidence of his tireless investigation not only of but also inside language. It is a record that cannot be collected in a single volume because the disparate nature of each investigation must, of necessity, modify the form it undertakes. As Bernstein remarks in his essay "Optimism and Critical Excess (Process)," "We don't know what Art is or does but we are forever finding out." McGann puts the case well in his essay in *Contemporary Poetry Meets Modern Theory*:

> For Bernstein, poetic "meaning" is never a product, and hence cannot be coded or decoded. It is a process of writing through which [in Shelley's words] "the before unapprehended relations of things" have to be *attended to* (in both senses of that phrase). Among the most important of those unapprehended relations are the ideological formations – the constellated sets of different social opinions and understandings – which define (sometimes even dominate) "the way we live now." The poet's office, for Bernstein, is to put those constellations at the reader's disposal.

"Like Shelley," McGann continues, "Bernstein pursues this revelation not for its own sake, but to break the spell of ideology by dislocating its forms of representation."

Much of the difficulty of Bernstein's work is part and parcel of his aim as a writer. "The function" of such writing, according to Silliman, is "to make the reader aware of the role of projection *as a response to form* in the constitution of the reader as subject." McGann in *Black Riders* also recognizes the centrality of the reader in this process: "Poetry pursues its truth-functions by revealing agencies of meaning and by implicating the reader in the processes of revelation." Being implicated in this "process of revelation" places a demand on the reader. In *Radical Artifice* Perloff suggests that a text is called difficult because it does not meet the expectations of readers: "It is not because meaning won't reveal itself to a receptive reader, but because the culture has preconceptions of how images should be articulated and connected. The stumbling block, that is to say, is not so much obscurity as convention."

Discussing the kinds of texts that have been called difficult in his essay "In the Middle" from *A Poetics*, Bernstein suggests that one must first differentiate between types of textual fragmentation: "To do this, one has to be able to distinguish between, on the one hand, a fragmentation that attempts to valorize the concept of a free-floating signifier unbounded to social significance . . . and, on the other, a fragmentation that reflects a conception of meaning as prevented by conventional narration and so uses disjunction as a method of tapping into other possibilities available within language." In "Writing" from *Content's Dream* he argues that the latter kind of text is not so much difficult as incomplete, in the sense that it must be completed by the reader: "In contrast to the predetermined interpretations of a text based on the primacy of the self or of logic, it is the formal autonomy of the text as model that elicits a response, an interpolation." The response that such a text demands makes the text inclusive rather than exclusive, since the reader's interpolation enters the fabric of the text. Ultimately, such a text enters a new realm of truth because "its autonomy is not of the self or logic but of nature, the world. Its truth is not assumed but made." What Bernstein attempts to make with his reader is a truth far greater than the tradition of author-centered texts could ever achieve. Bernstein's work – beyond the issue of its so-called difficulty – positions the text, its reader, and its meanings into a constellation of activities that open new possibilities made real in writing.

Letters:

Steve McCaffery, "Steve McCaffery, Ron Silliman & Charles Bernstein, Correspondence May 1976–December 1977," *Line 5* (Spring 1985): 59–89.

Interviews:

Bernstein and Tom Beckett, "An Interview with Tom Beckett," in *Content's Dream: Essays 1975–1984* (Los Angeles: Sun & Moon Press, 1986), pp. 385–410;

Bernstein and Beckett, "Censers of the Unknown-Margins, Dissent, and the Poetic Horizon," in *A Poetics* (Cambridge, Mass.: Harvard University Press, 1992), pp. 179–192.

Bibliography:

"Charles Bernstein: A Bibliography" in special issue on Bernstein, *The Difficulties,* 2 (Fall 1982): 115.

References:

The Difficulties, special issue on Bernstein, edited by Tom Beckett, 2 (Fall 1982);

Hank Lazer, "Charles Bernstein's Dark City: Polis, Policy, and the Policing of Poetry," *American Poetry Review,* 24 (September/October 1995): 35–44;

Jerome McGann, *Black Riders: The Visible Language of Modernism* (Princeton, N.J.: Princeton University Press, 1993);

McGann, "Charles Bernstein's 'The Simply'" in *Contemporary Poetry Meets Modern Theory,* edited by Anthony Easthope and John Thompson (Toronto: University of Toronto Press, 1991), pp. 34–39;

McGann [as Anne Mack, J. J. Rome, and Georg Mannejc], "Private Enigmas and Critical Functions, with Particular Reference to the Writing of Ch. Bernstein," *New Literary History,* 22 (Spring 1991): 441–464;

Bob Perelman, "Write the Power," *American Literary History,* 6 (Summer 1994): 306–324;

Marjorie Perloff, *The Dance of the Intellect: Studies in the Poetry of the Pound Tradition* (Cambridge & New York: Cambridge University Press, 1985);

Perloff, *Poetic License: Essays on Modernist and Postmodernist Lyric* (Evanston, Ill.: Northwestern University Press, 1990);

Perloff, *Radical Artifice: Writing Poetry in the Age of Media* (Chicago: University of Chicago Press, 1991);

John Shoptaw, "The Music of Construction: Measure and Polyphony in Ashbery and Bernstein," in *The Tribe of John: John Ashbery and Contemporary Poetry,* edited by Susan Schultz (Tuscaloosa: University of Alabama Press, 1994), pp. 211–257;

Ron Silliman, "Controlling Interests," in his *The New Sentence* (New York: Roof Books, 1985), pp. 171–184.

Ted Berrigan
(15 November 1934 - 4 July 1983)

Edward Halsey Foster
Stevens Institute of Technology

See also the Berrigan entry in *DLB 5: American Poets Since World War II, First Series.*

BOOKS: *A Lily for My Love* (Providence, R.I.: Privately published, 1959);
The Sonnets (New York: Lorenz & Ellen Gude, 1964; second edition, New York: Grove, 1966; enlarged, New York: United Artists, 1982);
Seventeen, by Berrigan and Ron Padgett (New York: Privately published, 1964);
Living with Chris (New York: Boke Press, 1965);
Bean Spasms, by Berrigan and Padgett (New York: Kulchur Press, 1967);
Many Happy Returns (New York: Corinth Books, 1969);
Doubletalk, by Berrigan and Anselm Hollo (Iowa City, Iowa: Privately published, 1969);
In the Early Morning Rain (London: Cape Goliard, 1970);
Memorial Day, by Berrigan and Anne Waldman (New York: Poetry Project, 1971);
Train Ride (New York: Vehicle Editions, 1971);
Back in Boston Again, by Berrigan, Tom Clark, and Padgett (New York: Telegraph Books, 1972);
The Drunken Boat (New York: Adventures in Poetry, 1974);
A Feeling for Leaving (New York: Frontward Books, 1975);
Red Wagon (Chicago: Yellow Press, 1976);
Clear the Range (New York: Adventures in Poetry/Coach House South, 1977);
Yo-Yo's with Money, by Berrigan and Harris Schiff (Henniker, N.H.: United Artist Books, 1979);
So Going Around Cities: New and Selected Poems 1958-1979 (Berkeley, Cal.: Blue Wind Press, 1980);
In a Blue River (New York: Little Light, 1981);
A Certain Slant of Sunlight (Oakland, Cal.: O Books, 1988);
Selected Poems, edited by Aram Saroyan (New York: Viking/Penguin, 1994).

Ted Berrigan in 1981

SELECTED PERIODICAL PUBLICATION – UNCOLLECTED: "Poems from *500 American Postcards,*" *Talisman: A Journal of Contemporary Poetry and Poetics,* 3 (Fall 1989): 82-97.

Ted Berrigan – Edmund Joseph Michael Berrigan Jr. – was born in Providence, Rhode Island, the oldest of three children of Margaret Dugan and Edmund Berrigan, the chief engineer at Ward's Baking Company. On both sides the family was Irish Catholic. Berrigan attended local schools and entered Providence College, a local Catholic school, but left after a year and enlisted in the army.

Berrigan was sent to Korea in 1954 but never saw action. Sixteen months later he returned to the United States and was based in Tulsa, Oklahoma. He enrolled in the University of Tulsa under the GI bill, receiving his B.A. in 1959 and his M.A. in 1962, after writing his master's thesis on Bernard Shaw. Berrigan returned the diploma for his master's degree to the university with a note saying that he was "the master of no art"; he was, he would tell friends, a poet because he wrote poetry, not because he had mastered poetics. To say that one could "master" an art was to imply that it was a matter of learning lessons and following rules.

Berrigan considered himself a "late Beat," and, like Allen Ginsberg and Jack Kerouac, he traced his lineage as a writer to the American Expressionist tradition. The first writer to excite him was Thomas Wolfe, who had also been important to Kerouac. Berrigan had very high regard as well for Kerouac, whom he interviewed for the *Paris Review*, and for another writer at the core of the Expressionist tradition, William Saroyan.

American Expressionism, which grounds literary authority in the personality of the writer rather than, say, a political creed or traditional aesthetics, can be traced to the work of mid-nineteenth-century writers such as Ralph Waldo Emerson, Henry David Thoreau, and Walt Whitman. Its most influential twentieth-century practitioner was Gertrude Stein, whose disciples included Sherwood Anderson and William Faulkner, as well as Saroyan and Wolfe. Kerouac stated his own Expressionist position in "Essentials of Spontaneous Prose," where he said that a writer should pursue "not 'selectivity' of expression but following free deviation (association) of mind into limitless blow-on-subject seas of thought, swimming in a sea of English with no discipline other than rhythms of exhalation and expostulated statement."

Berrigan argued that the world as presented in his poetry was a projection of his self. In an interview with Anne Waldman and Jim Cohn, Berrigan noted that:

> one of my principal desires is to make my poems be like my life ... I can't see myself the way that you can see me, but I can see everything else around me. If I can make everything around me be the way that is, presumably I can create the shape of the self inside the poem, because there is a person inside almost all of the poems.

Expressionist aesthetics were anathema to many academic and professional literary critics in the 1940s and 1950s, particularly the New Critics, and they largely excluded Wolfe, Saroyan, Henry Miller, and others from most textbooks and courses in American literature. Faulkner was admitted because of his use of myth and various traditional literary elements in his writing; his roots in American Expressionism were ignored.

At the same time, Kerouac and other Beats extended the Expressionist tradition and were in turn widely ridiculed by writers and critics such as Norman Podhoretz, who felt that Kerouac's "conception of feeling was one that only a solipsist could believe in." Nonetheless, through the Beats, Expressionism again became a dominant trend in American writing, one with great consequence for Berrigan as well as Clark Coolidge, Anne Waldman, and other poets with whom Berrigan was associated.

Berrigan's principal precursor is usually said to be Frank O'Hara, rather than Kerouac, and it is true that Berrigan owes much to O'Hara's "I do this, I do that" style. O'Hara's phrasing emerges in early Berrigan works as well, although — as in "Personal Poem #9" — it is sometimes parodied, a fact that critics often overlook. In her book on O'Hara, for example, Marjorie Perloff wrote that "Personal Poem #9" was "merely derivative," catching "the O'Hara manner without the substance," but Berrigan (a great admirer of Lewis Carroll) was in fact simply burlesquing the O'Hara manner in lines like these:

> It's 8:54 in Brooklyn it's the 26th of July
> and it's probably 8:54 in Manhattan but I'm
> in Brooklyn I'm eating English muffins and drinking
> Pepsi and I'm thinking of how Brooklyn is
> York City too how odd[.]

There are characteristic O'Hara phrasings in many Berrigan poems, particularly in the early work, but the sensibility they express is unmistakably Berrigan's. He borrowed procedures and even lines from other poets but fused them with a persona that was distinctly his own. His ambition was always to absorb whatever he borrowed, so well that it would become wholly his own. His control in doing this was in fact so good that he was able to write an "interview" with John Cage, whom in fact he had never met, which was selected by an unwary George Plimpton as the best interview of the year for *National Literary Anthology*.

Berrigan's closest poetic associate and collaborator was Ron Padgett, but their respective poems are radically different from each other. While much of Berrigan's poetry draws on intimate or private matters, Padgett's is rarely personal, at least on the surface. His poetry is self-conscious, and the tone is sophisticated and urbane; he has been particularly successful in creating a characteristically French wit in his poems and so becoming, it has been said, America's greatest French poet. Berrigan, on the other hand, projects a sensibility that is confiding, sad, graceful, affectionate, and indistinguishable from the sensibility he projected in person. His poetry can also be — although never in a pretentious manner — cerebral and erudite; he was enormously well read, much better read, perhaps, than most poets of his generation.

In spite of differences in age as well as sensibility (Padgett is eight years younger than Berrigan), they became close friends shortly after meeting at the bookstore in Tulsa where Padgett worked in 1959. Padgett was still in high school but was already known to avant-garde poets throughout the United States for a mimeographed magazine, *The White Dove Review*, which he edited with his boyhood friends, the poet Dick Gallup and the artist and writer Joe Brainard.

Padgett entered Columbia College at Columbia University in 1960, and Berrigan, Brainard, and Gallup soon followed him to New York. Berrigan lived for several months with Brainard and then with Padgett before finding his own apartment on the Lower East Side. Aside from years spent teaching at universities outside New York, he spent the rest of his life there.

Berrigan, Padgett, Gallup, and Brainard shared great enthusiasm for poets who made up the so-called New York School, particularly O'Hara, John Ashbery, and Kenneth Koch. Critics in turn saw the four, together with a few other poets, such as Waldman and Coolidge, as a second generation of the "school"; many of these poets objected to being classified in this manner, but not Berrigan, who greatly valued any sense of community, especially among poets.

Berrigan initially supported himself in New York by writing papers for Columbia students (papers that consistently earned an A or better), then tried assorted odd jobs, and finally taught at various universities. He insisted that any work he did be somehow related to writing. As his wife Alice Notley said, he "famously believed that being a poet was a 24-hour-a-day job — you did it in your sleep too, in your dreams when you gave in to sleep. It was full-time also in the sense that it was worthy of all one's attention, and a poet shouldn't have to have another job as well."

Berrigan was much respected as a teacher of poetry, in part, no doubt, because of his uncompromised dedication to his work, but also, it may be, because of his democratic ability to appreciate poetry and poetics radically different from his own. At a time when American poets were divided between academic, traditional poets and those, like Berrigan, whose work was more innovative or experimental, he spoke highly of comparatively conventional poets such as Conrad Aiken. (Aiken was less generous, insisting that Berrigan submit himself to a regimen of Dantean and Virgilian poetics.)

Portrait of Berrigan by Alex Katz

Berrigan advised his students and other poets who approached him for advice to find their own voices and technical means, rather than imitate his. Aram Saroyan remembered Berrigan's criticisms as "tactful, sweet–natured and full of a real generosity of spirit." That attitude earned him a strong following in his classes at schools as various as Yale, the University of Iowa, the Stevens Institute of Technology, and the City College of New York.

In 1962 Berrigan married his first wife, Sandra Alper, with whom he had two children, David and Kate. The following year, he wrote *The Sonnets* (1964), perhaps his most famous book, and founded *"C"* magazine. In 1964 he began publishing *"C"* books and quickly won a modest fame in what was soon to be called the counterculture. He also wrote essays for *Art News* on Jane Freilicher, Red Grooms, and Alice Neal. In 1966 the Poetry Project was founded at St. Mark's Church-in-the-Bowery, near Berrigan's home on the Lower East Side, and he became one of the project's principal organizers, taught workshops there, and was closely associated with it until his death. In 1971 he married the poet Alice Notley, and they had two sons, Anselm and Edmund.

By this time Berrigan occupied a place at the center of a community of avant-garde New York artists and writers, including such painters as Alex Katz, Philip Guston, Jasper Johns, and George Schneeman. The publication in 1969 of *An Anthology*

Cover for Berrigan's 1970 book of poems, demonstrating his interest in a minimalist aesthetic

of New York Poets, edited by Ron Padgett and David Shapiro, helped to define the second generation of New York poets, while insisting that "most poets of any interest these days . . . automatically reject, in their lives and work, the unhealthy idea of being part of a literary movement."

Healthy or not, poetic movements and communities, at least to the degree that they defined a circle of like-minded individuals, were important to Berrigan, and his own community included many who were identified as composing the second generation of New York poets: Padgett, Gallup, Coolidge, Waldman, Bernadette Mayer, Lewis Warsh, Simon Pettet, and Eileen Myles, among many others. He also made alliances with poets from strikingly different backgrounds, such as Robert Creeley, Tom Clark, Aram Saroyan, and Philip Whalen. Two of his closest companions were the Finnish American poet Anselm Hollo, whom he met when teaching in the Writers' Workshop at the University of Iowa in 1968–1969, and the British poet Douglas Oliver, whom he met in 1973–1974 when teaching at Essex University.

Berrigan's books and chapbooks of poetry were published regularly, sometimes two or three a year. *Clear the Range*, a novel created by substituting words in a popular Western, appeared in 1977. A volume of selected poems, *So Going Around Cities*, was published in 1980. A second volume of selected poems appeared in 1994.

Much of Berrigan's reputation throughout his career depended on *The Sonnets,* privately published in 1964 but reissued by Grove Press in 1966. According to Notley, at times he dismissed the volume "as being too reflective of a poetic education," while at other times he spoke "with awe of the exalted place" from which it had been written. Certainly the book does reflect "a poetic education." One of the more obvious influences, and one that Berrigan freely acknowledged, is T. S. Eliot's *The Waste Land* (1922), which he had read with great excitement as an undergraduate. In particular, the disjunctive structure and the mixture of cadences and voices in *The Sonnets* is modeled on Eliot's example. What did not interest Berrigan was Eliot's wearied tone and sensibility, and in its place one finds Berrigan's distinctive Irish American temper, his own blend of humor and grace. Instead of Eliot's allusions to, and quotations from, canonical works of high culture, Berrigan employed a great range of references from popular as well as serious works.

Berrigan was also influenced by the playfulness and fractured texts of Marcel Duchamp and John Cage and the cutups of William Burroughs. *The Sonnets* were, however, far more sympathetic to traditional form than one generally finds in the works of experimental writers; the sonnet, Berrigan demonstrated, was not fourteen lines of iambic pentameter, but rather a form essentially grounded in certain intricate relationships of cadence and meaning. Duchamp, Cage, and Burroughs disrupted traditional form; Berrigan, meanwhile, argued that form operated on far more subtle levels than critics and poets generally realized or acknowledged.

Berrigan was particularly attentive in his work to the delicate shifts in connotation and sound, and it is not surprising that he was enthusiastic about Wallace Stevens and the precise modulations in poems such as "The Idea of Order at Key West." Berrigan's "Things To Do in Providence" (1970), another of his highly regarded works, is a fine instance of his ability to create subtly shifting evocations of meaning and tone. Its surface is as disjunctive as *The Sonnets,* but the breaks from one tone of feeling to another are never casual and map a precise personal trajectory from emotional exhaustion to joy.

Like his friend Robert Creeley in the late 1960s and early 1970s, Berrigan tried stripping his poems of all but the most essential words. These minimalist works incensed critics such as David Lehman, who savagely reviewed *In the Early Morning Rain* (1970) for *Poetry*, singling out one poem as "moronic drivel." However, Berrigan's minimalism was very important in further refining his capacity for presenting exact nuance in tone and mood.

That precision is evident in Berrigan's greatest works, such as "Red Shift" and "A Certain Slant of Sunlight," written toward the end of his life. These poems are notable for their strong emotions; their intricate, subtle modulations in sensibility and cadence; and their occasionally baroque grammar. Here, for example, are the concluding lines to "Red Shift":

I'm only pronouns, & I am all of them, & I didn't ask for this
You did
I came into your life to change it & it did so & now nothing
will ever change
That, & that's that.
Alone & crowded, unhappy fate, nevertheless
I slip softly into the air
The world's furious song flows through my costume.

Occasionally Berrigan's works were reviewed in mainstream publications, but it was not there that he had, or has, his principal reputation. There have been a few critical studies by academics such as Marisa Januzzi, but most of the work on Berrigan has been done by other poets. The most important resource is the collection of memoirs, photographs, and homages edited by Waldman in *Nice To See You: Homage to Ted Berrigan* (1991), but there are individual memoirs by Tom Clark, Ron Padgett, and Edward Halsey Foste, as well as extended reminiscences and commentary in books (listed in the bibliography) by Aram Saroyan, Eileen Myles, and Lita Hornick. Notley wrote a brief reminiscence as an introduction to Berrigan's *Selected Poems* (1994), and her life with Berrigan is also reflected in various works, including her play *Anne's White Glove* (1987) and her poem, "At Night The States" (1987). Other memoirs, as yet uncollected in book form, have appeared, and there will undoubtedly be more, because Berrigan was a presence of great importance to many poets. Some might disagree with his poetics or find him at times tendentious and argumentative, but he had few strong enemies. Padgett calls his inflexibly honest memoir "disloyal," but even when the less congenial aspects of Berrigan's character are acknowledged (notably his great mood swings that perhaps were caused by the heavy use of amphetamines), he seems one of his generation's most respected and well-liked poets. Above all, his absolute dedication to poetry in a culture that has only a marginal interest in that art was virtually heroic.

Because Berrigan was in poor health during his last years, other poets were more likely to visit his apartment at 101 St. Mark's Place than he was to visit them. He spent much of his time in bed, smoking Chesterfields and drinking Pepsis, surrounded by established and younger writers who listened to his elaborate disquisitions on poetry and poetics. Notley wrote that she and Berrigan "never had any money," but if an individual were willing to make "a small loan," he adopted that person as a student.

Berrigan died 4 July 1983, at the age of forty-eight. He had contracted hepatitis in 1975, and his liver was severely damaged. As Notley wrote, "He was, in a way, always sick during [his last] years. The illness went untreated, because there was no treatment really; we couldn't afford doctors anyway, he didn't want to change his lifestyle *that much*, and he didn't want his illness named and charted by doctors."

Berrigan's funeral and burial were held at Calverton National, a military cemetery on Long Island. The ceremony was perfunctory, the priest doing no more than the occasion required, but the casket was surrounded by a multitude of poets. Cemetery regulations required that a person's full name appear on the grave marker, but in this case officials made an exception, and the stone reads simply "Ted Berrigan." Had Berrigan's wishes been carried out, however, the stone would also have read "Nice To See You."

Interview:

Stephen Ratcliffe and Leslie Scalapino, *Talking in Tranquility: Interviews with Ted Berrigan* (Bolinas, Cal.: Avenue B / Oakland, Cal.: O Books, 1991).

Bibliography:

Anne Waldman, "Ted Berrigan's Published Work: A Checklist," in *Nice to See You: Homage to Ted Berrigan,* edited by Waldman (Minneapolis, Minn.: Coffee House Press, 1991), pp. 242–244.

Biographies:

Tom Clark, *Late Returns: A Memoir of Ted Berrigan* (Bolinas, Cal.: Tombouctou, 1985);

Lita Hornick, *The Green Fuse* (New York: Giorno Poetry Systems, 1989);

Anne Waldman, ed., *Nice To See You: Homage to Ted Berrigan* (Minneapolis, Minn.: Coffee House Press, 1991);

Aram Saroyan, *Friends in the World: The Education of a Writer* (Minneapolis, Minn.: Coffee House Press, 1992);

Ron Padgett, *Ted: A Personal Memoir of Ted Berrigan* (Great Barrington, Mass.: The Figures, 1993);

Edward Foster, *Code of the West: A Memoir of Ted Berrigan* (Boulder, Colo.: Rodent Press, 1994);

Eileen Myles, *The Chelsea Girls* (Santa Barbara, Cal.: Black Sparrow Press, 1994);

Alice Notley, "Introduction" to Ted Berrigan's *Selected Poems* (New York: Viking/Penguin, 1994).

References:

Marisa Januzzi, "Ted Berrigan," *Talisman: A Journal of Contemporary Poetry and Poetics,* 12 (Spring 1994): 211-215;

Tod Thilleman, "The Berrigan Case," *Talisman: A Journal of Contemporary Poetry and Poetics,* 12 (Spring 1994): 209-210.

Papers:

A substantial collection of Berrigan manuscripts is housed in Special Collections at Columbia University.

Elizabeth Bishop
(8 February 1911 - 6 October 1979)

Brett C. Millier
Middlebury College

See also the Bishop entry in *DLB 5: American Poets Since World War II, First Series.*

BOOKS: *North & South* (Boston: Houghton Mifflin, 1946);

Poems: North & South – A Cold Spring (Boston: Houghton Mifflin, 1955; London: Chatto & Windus, 1956);

Brazil, by Bishop and the editors of *Life* (New York: Time/*Life* World Library, 1962; London: Sunday *Times,* 1963);

Questions of Travel (New York: Farrar, Straus & Giroux, 1965);

Selected Poems (London: Chatto & Windus, 1967);

The Ballad of the Burglar of Babylon (New York: Farrar, Straus & Giroux, 1968);

The Complete Poems (New York: Farrar, Straus & Giroux, 1969; London: Chatto & Windus, 1970);

Geography III (New York: Farrar, Straus & Giroux, 1976);

The Complete Poems, 1927–1979 (New York: Farrar, Straus & Giroux, 1983);

The Collected Prose (New York: Farrar, Straus & Giroux, 1984).

OTHER: "It All Depends," in *Mid-Century American Poets,* edited by John Ciardi (New York: Twayne, 1950), p. 267;

Alice Dayrell Brant, *The Diary of Helena Morley,* translated by Bishop (New York: Farrar, Straus & Giroux, 1957);

An Anthology of Twentieth-Century Brazilian Poetry, edited by Bishop and Emmanuel Brasil (Middletown, Conn.: Wesleyan University Press, 1972).

SELECTED PERIODICAL PUBLICATIONS – UNCOLLECTED: "Gerard Manley Hopkins: Notes on Timing in His Poetry," *Vassar Review,* 23 (February 1934): 5-7;

Elizabeth Bishop, 1954 (photograph © Rollie McKenna)

"What the Young Man Said to the Psalmist," *Poetry,* 79 (January 1952): 213;

"'I Was But Just Awake,'" review of *Come Hither: A Collection of Rhymes and Poems for the Young of All Ages,* by Walter de la Mare, *Poetry,* 93 (October 1958): 50-54;

"On the Railroad Named Delight," *New York Times Magazine,* 7 March 1965, pp. 30-31, 84-86.

Elizabeth Bishop's critical reputation has grown steadily since her death in 1979. Always a respected poet honored by her peers, Bishop was not well known outside the poetry circles of New York and Boston during her lifetime. Today her poems are widely read and taught in classrooms, and a steady stream of essays and books about her life and work has followed since the publication of *The Complete Poems, 1927–1979* in 1983. One can account for this extraordinary attention in several ways: the rise of feminist scholarship and the attendant search for neglected female writers; the continuing respect for Bishop's poems shown by teachers and other poets; and most important, the simultaneous profundity and accessibility of the poems and the remarkable appeal of the distinctive and personal voice in which they are written.

Elizabeth Bishop was born the only child of William Thomas Bishop and Gertrude May Boomer Bishop on 8 February 1911 in Worcester, Massachusetts. The wealthy and comfortable circumstances of her birth were disrupted eight months later when William Bishop died of Bright's disease, and the child and her mother were left to the care of relatives. Gertrude Bishop was deeply disoriented by her husband's death. For the next five years she was in and out of hospitals and mental institutions and bounced between Boston, Worcester, and her native Great Village, Nova Scotia. As Elizabeth Bishop wrote of her mother in her prose poem/memoir "In the Village": "First, she had come home, with her child. Then she had gone away again, alone, and left the child. Then she had come home. Then she had gone away again, with her sister; and now she was home again." In 1916 Gertrude Bishop was diagnosed as permanently insane and institutionalized. She lived for eighteen more years, but her daughter never saw her again.

By 1916 Bishop had managed, with the uncanny adaptability of a child, to construct for herself a secure and apparently timeless world in the home of her maternal grandparents in Nova Scotia. But her father's wealthy family, worried that their then only grandchild would grow up backward (and Canadian) among the rural poor, uprooted her a year later. She now began what would be a lifetime of living as a guest in other people's homes. Her stay with the Bishop grandparents was miserable from the start; there in the Worcester mansion, among her aunts and uncles and the Swedish household staff, she reacted by developing a debilitating depression that she would never really overcome. In addition she developed a severe case of asthma and several allergies.

The Bishops gave up this experiment after nine nearly fatal months and arranged for Elizabeth to live with her mother's sister Maud, near Boston. They also allowed her to spend two months of each summer with her Nova Scotia grandparents. Because of her illnesses, Bishop had little formal schooling until she enrolled at Saugus (Massachusetts) High School in 1925. Her aunt Maud introduced her to poetry – Alfred Tennyson, Robert Browning, Ralph Waldo Emerson, Thomas Carlyle, and other Victorians. In bed with asthma or bronchitis, Elizabeth memorized so many of these poems that she said later that she felt they must be "an unconscious part" of her. She started with a fascination for hymns and fairy tales – those of Hans Christian Andersen and Jacob and Wilhelm Grimm – and discovered Walt Whitman at age thirteen. At summer camp later that same year she became familiar with Harriet Monroe's famous anthology of modern poets and developed a fondness for Gerard Manley Hopkins. She reported also going through "a Shelley phase, a Browning phase, and a brief Swinburne phase." In 1926 Elizabeth attended the North Shore Country Day School in Swampscott, where her first published poems appeared in the school literary magazine. At the Walnut Hill School for Girls in Natick the following year she became a literary star by editing and contributing most of the contents of *The Blue Pencil*. She also starred in many skits, plays, and musical performances. In the fall of 1930 she started classes at Vassar College.

It was at Vassar that the terms of Elizabeth Bishop's adult life were set, for better and for worse. For better, she found herself among an extraordinary group of gifted female writers – Mary McCarthy, Muriel Rukeyser, and Eleanor Clark – and she began, tentatively, to make her own way as a writer in the larger world. She received a first-rate education in English literature and the classics; she found enduring poetic models in Hopkins, George Herbert, and other seventeenth-century poets; and she received much encouragement from her professors and friends. For worse, she was lonely and had no home outside of college; she began to drink heavily, and the shame and embarrassment she felt over it fed the guilt she had concerning her mother's illness. She became secretive and withdrawn, and her academic performance was erratic. Most alarming of all, she fell desperately, hopelessly in love with her roommate, and she had to confront for perhaps the first time the idea that she would never have a "normal" life. As a result her sense of shame increased. She graduated from college in 1935 with the following conditions well established: she would suffer from depression, asthma, and alcoholism; she would wander the world in search of a place to call home; the passion in her life would be for women; and she would hide these things from almost everyone she knew. She would also be a

Bishop with her maternal grandparents, William and Elizabeth Boomer, in 1911 (courtesy of Phyllis Sutherland)

poet, a poet whose themes would remain remarkably consistent throughout her writing life. Bishop believed in accuracy and discretion, and in both unconscious inspiration and craft. Her poetry shows a preoccupation with the losses inherent in the passage of time; with problematic and alienated perspective; with the gap between the observer/poet and what she sees, and with the vexed morality of appropriation that such distance involves.

In the spring of her senior year at Vassar, Elizabeth Bishop found the courage to contact the poet Marianne Moore, whose work she had admired for several years. After a somewhat comical meeting "on the bench at the right of the door leading to the reading room of the New York Public Library," the two became friends. Bishop told the story of the relationship in the essay "Efforts of Affection," which was posthumously published in *The Collected Prose* (1984), and in the poem "Invitation to Miss Marianne Moore" (1948). It is impossible to overstate the importance of the friendship with Moore in the development of Bishop's sense of herself as a poet. Though it took her four months to bring herself to admit to Moore that she had literary ambitions, from that time forward until the fall of 1940, Moore influenced Bishop's poetic career both by example and by direct assistance. By nature a meticulous writer and a shy person, Bishop had no inclination toward the public life. She learned from Moore during this period that it was possible to be a poet without subjecting oneself to the stresses of reading, teaching, or editing (to which Moore returned after the death of her mother in 1947). Moore maintained that the focus should be on morality, on the responsibility of the poet toward what she sees and what she writes. Bishop wrote in "Efforts of Affection" that she never left the apartment that Moore shared with her mother at 260 Cumberland Street in Brooklyn "without feeling happier: uplifted, even inspired, determined to be good, to work harder, not to worry about what other people thought, never to try to publish anything until I thought I'd done my best with it, no matter how many years it took – or never to publish at all."

What really changed for Bishop when she began reading Moore and listening to her was the idea of what, among the objects and emotions in the world, was suitable subject matter for poetry. "The Map," a poem in *North & South* (1946), Bishop's first book, shows dramatically the difference that Moore's influence made in Bishop's thinking about her work. First published alongside "Three Valentines" and "The Reprimand" in an anthology called *Trial Balances* in 1935, it presents a contrast to those mannered, imitative college poems. The meaning generated through contemplation of a single object, the familiar object reseen, the commitment to accuracy, the reach of simile, the wisdom of tone, the naturalness of diction — these are all gifts that Marianne Moore gave to Bishop, along with infinite subjects less hackneyed than "sorrow" and "love." Moore also instilled in Bishop the confidence to trust her own instincts in handling those subjects. "The Map" is the first "Elizabeth Bishop" poem.

"The Map" arose from Bishop's contemplation of a glass-covered map of the North Atlantic — including the Maritime Provinces of Canada and northern Europe. It proceeds, as many later Bishop poems do, with semirhetorical questions of perspective: "Shadows, or are they shallows," "Or does the land lean down to lift the sea from under, . . . / is the land tugging at the sea from under?" This is the first manifestation of her early conviction that poetry should portray the mind thinking rather than in repose, as well as the beginning of a lifelong concern with shifting perspective and scale. The Scandinavian countries resemble the fingers of "women feeling for the smoothness of yard-goods" (women she had grown up among, perhaps). Bishop's poem asks questions that the map cannot answer. It is successful in part because it simultaneously holds forth both the map and the actual geography (the physical world not drawn to scale) as realities. (See "Large Bad Picture" (1946), "Brazil, January 1, 1502" (1960), and "Poem" (1978) for later examples of this technique.) The poem's famous last line, "More delicate than the historians' are the mapmakers' colors," expresses Bishop's fascination with the artist's "delicate" creations and her awareness of the limitations of art.

Bishop left college with the desire to be a writer — she actually thought of herself as a novelist at that point — but with a great deal of anxiety about how to accomplish that. The purpose of art in the Depression, which made such "frivolous" activity seem self-indulgent, also troubled her. Skeptical of the leftist commitments of many writers of the period, she worked out her crises of artistic faith in the poems themselves. Several critics have noted that the conflict expressed in "The Map" is common in Bishop's earliest adult work. Several poems question the adequacy or usefulness of the intensely introspective vision they employ. This irony can be seen in works such as "The Imaginary Iceberg" (1935), "The Gentleman of Shalott" (1936), and "The Weed" (1937). Anxieties about the purpose of art can be seen in "The Monument" (1939) and "Large Bad Picture." In this regard she conducted an implicit quarrel with Wallace Stevens, whose 1934 book *Harmonium* she claimed to know "by heart." Most critics agree that in the "south" portion of *North & South,* Bishop began to realize that the world of things outside the self — of other people, of nature — would be her true subject. Poems such as "Florida" (1939), "The Fish" (1940), and "Roosters" (1941) demonstrate this change of focus. Bishop satirized the artist's isolation and self-indulgence in two prose fables from 1936–1937 as well, "The Sea and Its Shore" and "In Prison." (Both are reprinted in *The Collected Prose.*)

After college Bishop was painfully on her own. With a small income from her father's estate, she was free of the necessity to work but troubled by such a lifestyle in the Depression. She tried to live in New York, though the pressures of the city and of the competitive literary world made her miserable. She traveled in Europe in 1935 and again in 1937; in between, she discovered Key West, Florida, where she would live for a part of each year until 1945. This travel was for pleasure and education, but at the same time it exacerbated Bishop's desire for a true home. She suffered from debilitating asthma — the only treatment for which was adrenaline injections — and depression. She also struggled to keep her drinking under control.

Although her poems appeared regularly in magazines after 1937, Bishop had a long wait to see her first book published. Starting in 1939 she circulated the manuscript of what would be *North & South,* but it was not until she won the Houghton Mifflin Poetry Prize Fellowship in 1945 that the book would appear. The disappointment of so many rejections cast a pall of uncertainty over Bishop's sense of her gifts and ambitions that not even the fellowship and the book's excellent reviews could dispel.

The reviewers identified within the poems a moral vision, a firm sense of right and wrong, of control, that Bishop hardly recognized in herself. For example, the critic Randall Jarrell praised her poems in specifically moral terms: "Instead of crying, with justice, 'This is a world in which no one can get along,' Miss Bishop's poems show that it is barely but perfectly possible — has been, that is, for her. . . . [Her poems tell us] that morality, for the individual, is usually a small, personal, statistical, but heartbreaking or heartwarming affair of omissions and commissions the greatest of which will seem infinitesimal, ludicrously beneath notice to those who govern, rationalize, and deplore." This praise may have been sweet to Bishop but must also have seemed ironic, so much was her life at this time a slippery moral struggle to "do well" — with alcohol; in her relationships; in the eternal dilemma of

Bishop (third row, center) in tenth grade at the Walnut Hill School (courtesy of Archives, Walnut Hill School)

"work." That she had managed to project, at least for Jarrell and several other reviewers of her poems, a relatively firm artistic and moral voice must have struck her as a thing for laughter or for tears.

When *North & South* appeared in the summer of 1946, Bishop was traveling through Nova Scotia after a sixteen-year absence. This trip, which was something of an accident (she'd headed north for Keene, New Hampshire, and just kept going), turned out to be, in poetic terms, one of the most important she would ever make. In Nova Scotia for the first time since her mother's death, she found herself drawn to the difficult memories of her childhood, of her aunts and grandparents, and to wondering how those people and circumstances had combined to make her who she was. The trip was both disturbing and significant to Bishop in ways that it would take her years to articulate. In her notebook from this trip are the beginnings of her poems "At the Fishhouses" (1948); "The Prodigal" (1950), her only poem about alcoholism; and "The Moose" (1972).

"At the Fishhouses" is typical of the development in Bishop's style that took place between the publication of *North & South* in 1946 and of *Poems: North & South – A Cold Spring* in 1955. Her ambivalence toward introspection and overdependence on imagination yielded her a frank first-person speaker contemplating her own circumstances. While the facts behind those circumstances may not be explicitly revealed, the poems examine psychological truth. "At the Fishhouses" records an actual visit by the poet with an old fisherman outside the fishhouses of a small Nova Scotia port. After nearly fifty lines of luminous description of the physical scene and the "heavy surface of the sea," the poet is tempted to philosophical speculation but retreats from it:

Cold dark deep and absolutely clear,
element bearable to no mortal,
to fish and to seals. . . . One seal particularly
I have seen here evening after evening.

She goes on to describe her humorous encounters with the seal but is drawn again to the meaning of the scene to herself, only again to retreat – this time to contemplate the "dignified tall firs . . . a million Christmas trees stand / waiting for Christmas." And then, unable to avoid the pull of the water any longer, she mentally plunges in:

> I have seen it over and over, the same sea, the same,
> slightly, indifferently swinging above the stones,
> icily free above the stones,
> above the stones and then the world.
> If you should dip your hand in,
> your wrist would ache immediately,
> your bones would begin to ache and your hand would burn
> as if the water were a transmutation of fire
> that feeds on stones and burns with a dark gray flame.
> If you tasted it, it would first taste bitter,
> then briny, then surely burn your tongue.
> It is like what we imagine knowledge to be:
> dark, salt, clear, moving, utterly free,
> drawn from the cold hard mouth
> of the world, derived from the rocky breasts
> forever, flowing and drawn, and since
> our knowledge is historical, flowing, and flown.

The water so cold it burns is first, of course, a physical description of the icy cold water of the North Atlantic. But at the same time, in a poem in which Bishop is considering her origins, the cold water reflects the absence of maternal warmth in her memories of her childhood. Contemplating the sum of her inheritance from her mother, what she "derived" from that troubled relationship – her "knowledge" of herself and her Nova Scotia past – Bishop saw that such knowledge was shifting, "flowing and drawn," and hopelessly temporal and irremediable, "historical, flowing, and flown."

Other poems from *A Cold Spring* (which was published in combination with a reissuing of *North & South* in order to compensate for the book's thinness) show a similar focus on the individual's complex negotiations with the contingent world of objects and other people. It also shows Bishop's mastery of the technique of investing personal emotion and significance in the details of her descriptive poems. "The Bight" (1949) is a fine example of a Bishop speaker viewing a scene, telling "what really happened" (the poem is based on her observations of excavations underway at Garrison Bight in Key West in 1948), and at the same time demonstrating her individuality, even personality. The poem avoids all explicit reference to the viewer of the excavation except in its subtitle "(On my birthday)" and one brief first-person reference. The "objective" description of the animals and machines in the poem is charged by Bishop's personal perspective. The poem's central lines speak of its metaphoric method:

> Some of the little white boats are still piled up
> against each other, or lie on their sides, stove in,
> and not yet salvaged, if they ever will be, from the last bad storm,
> like torn-open, unanswered letters.
> The bight is littered with old correspondences.

This is a pun, in part: in a letter Bishop compared the scene to her messy desk; but surely these are also Charles Baudelaire's *correspondances;* indeed, she mentions him in the poem: "If one were Baudelaire / one could probably hear [the water] turning to marimba music." But Bishop's version of Baudelaire's "profound unity" in the world of symbols is simply the organizing force of her own perspective. The poem describes the activity of the bight as "awful" before "cheerful," and activity here is not just of dredges and bulldozers but of strangely mechanized, razorlike animals as well: crashing pelicans "like pickaxes" and "man-of-war birds" with tails like scissors. The sense that the whole scene might at any moment explode into a fireball is supported by the juxtaposition of "pilings dry as matches" and water that one can smell "turning to gas." Things that ought to have power are strangely ineffective: the "water ... doesn't wet anything" while "the frowsy sponge boats" are stove in and unsalvageable. The poem has many metaphors, and they are expressed in terms of potential violence. In one of these, Bishop gives us the only first-person reference in the poem – the pelicans crash unnecessarily hard, "it seems to me, like pickaxes." Along with the subtitle, this reference tells us that the poem is about disorder outside as well as inside oneself. About this birthday Bishop also remarked, "Thirty seven / and far from heaven."

The years 1947 to 1951 were almost unremittingly miserable for Bishop as she continued to suffer from asthma and depression, continued to struggle with alcoholism, and was involved in several unhappy relationships. But her poetic career took a significant positive turn when, in January 1947, she was introduced at a party in New York to the poet Robert Lowell, who had also recently published his first book. Although six years younger than Bishop, Lowell already had more access to the inner circles of poetry than she did, and he soon became her confidant and link to the grants and awards that would support her work. She benefited from his rapid rise in the poetry world right from the start. Their friendship, which verged from time to time on romance, would last until the end of Lowell's life in 1977.

In April 1947 Bishop was awarded a Guggenheim Fellowship, the first of many such awards she would receive in her career. Two years later, through Lowell's intervention, she was named poetry consultant to the Library of Congress, and also at his suggestion she spent two long visits at Yaddo, the prestigious artists' retreat in Saratoga, New York, where she established important literary and artistic friendships. But despite this professional success, Bishop was unable to find a way to live comfortably, to stop drinking, or to rise above the depression that dogged her every attempt to gain control of her life.

The poems from these years reflect her difficulties. "The Prodigal" is a double sonnet about the moral struggle of an alcoholic. Bishop distanced herself from the poem's events by calling it an imaginative re-creation of what must have been going on, when, as she said in a letter to Joseph Summers, "one of my aunt's stepsons offered me a drink of rum, in the pig styes at about 9 in the morning" on her trip to Nova Scotia in the summer of 1946. But the poem is also a self-conscious presentation of her own solitary struggle with alcohol in 1949–1950. Its protagonist is self-exiled in the pigsties of his family's farm, "hid[ing] the pints behind a two-by-four." The poem's final lines locate responsibility for this exile in the protagonist himself:

> Carrying a bucket along a slimy board,
> he felt the bats' uncertain staggering flight,
> his shuddering insights, beyond his control,
> touching him. But it took him a long time
> finally to make his mind up to go home.

What links this poem most to Bishop's own self-described experience of alcoholism is its presentation of a person exiled from normal human relationships, from "home," by his own bestial behavior. Living among pigs, he is at times inured to the smell, the mud, the dung, but Bishop insists that his exile is at least in part a matter of will – the prodigal, the wastrel, not yet having decided to return to be forgiven, need only decide, and he may return. Bishop's prodigal must "make his mind up to go home," and he is at times aware, "in shuddering insights, beyond his control," of what is required of him. The vaguely hopeful ending, suggesting that although he took a long time to make his mind up he eventually went home, leaves the emphasis on willpower, echoing Bishop's belief that the solution to her own drinking problem lay in self-discipline. While Bishop saw no clear "home" from which she was exiled, the picture of her relative, removed from the home of her favorite aunt out to the pigsty behind the barn, is a fitting analogy to her own experience of shame and isolation.

Bishop in Key West, 1940 (photograph by James Laughlin)

Bishop also completed a poem about asthma in 1950. "O Breath" is metered and punctuated to reflect the lung capacity and speaking pace of an asthma sufferer:

> Beneath that loved   and celebrated breast,
> silent, bored really   blindly veined,
> grieves, maybe   lives and lets
> live, passes   bets,
> something moving   but invisibly,
> and with what clamor   why restrained
> I cannot fathom   even a ripple.

"O Breath" was the last of a group of four ambivalent love lyrics, titled "Four Poems," all of which express frustration at the possibilities for communication between people. In this context, the negotiations demanded by breathing itself come to stand for what Bishop saw as the inevitable compromises and losses of love. Other poems in *A Cold Spring*, such as "Argument" (1947), about the quarrel between "days" and "distance" on the "gentle battleground" of love, "neither proving you less wanted nor less dear"; "Insomnia" (1951), with its projection of a mirror-world, "where left is always right . . . and you love me"; and "Varick Street" (1947), with its famous refrain, *"And I shall sell you sell you / sell you of course, my dear, and you'll sell me,"* all reflect a dark view. In poems less explicitly about love, a similar sense of lost opportunity or dead-endedness prevails. In "Cape Breton" (1950), "Whatever the landscape had of meaning appears to have been abandoned." In "Over 2000 Illustrations and a Complete Concordance" (1948), an account of her 1935–1937 travels in Europe and Africa, Bishop laments the apparent aimlessness of those travels: "Everything only connected by 'and' and 'and.' " "Why couldn't we have seen this old Nativity while we were at it?" In "View of the Capitol from the Library of Congress" (1951), Bishop's only poem about her experience in Washington, D.C., the Air Force band plays on the east steps of the Capitol Building, "but – queer – / the music doesn't quite come through." And in "Faustina, or Rock Roses" (1947), the questions raised by the relationship between the privileged but dying white woman and her much more powerful black caretaker proliferate "helplessly": "There is no way of telling. / The eyes say only either." Even the celebrations in the book are muted and qualified. The title poem, "A Cold Spring" (1952), a tribute to Bishop's friend Jane Dewey, whose farm near Havre de Grace, Maryland, was an important refuge for her for many years, sees the ambiguousness of early spring with ambivalence. "The violet is flawed on the lawn"; a calf is born "in a chill white blast of sunshine"; its mother "took a long time eating the after-birth, / a wretched flag." As spring progresses, the landscape softens and the sense of unease abates; the fireflies rise "exactly like the bubbles in champagne." The final lines of "The Bight" reflect this attitude as well: "All the untidy activity continues, / awful but cheerful."

By the fall of 1951 Bishop had run out of energy for finishing *A Cold Spring* as well as for managing her literary and personal life. She had traveled restlessly that summer, including another trip to Nova Scotia, but she was unable to decide on where or how to live. After several months of living with friends and in hotels, she finally made up her mind, at her doctor's insistence. As soon as possible, she would take a freighter to Tierra del Fuego and the Strait of Magellan and a trip around the world. Her first stop on this ambitious and solitary journey would be Rio de Janeiro, where she would visit friends. On 10 November, almost literally at her wit's end, she boarded the S.S. *Bowplate* in New York harbor and headed south.

Even before the freighter reached the coast of Brazil, Bishop saw that the trip would change her life. A few days in, she wrote in her notebook:

> This trip is a "shake-down" trip for me, all right. I know I am feeling, thinking, looking, sleeping, dreaming, eating & drinking better than in a long long time, & when I read something like "The question about time is how change is related to the changeless" – & look around – it doesn't seem so hard or so far off. The nearer clouds seem to be moving quite rapidly; those in back of them are motionless – Watching the ship's wake we seem to be going fast, but watching the sky or the horizon, we are just living here with the engines pulsing, forever.

That she would glimpse freedom from her anxiety about time, expressed in such poems as "Paris, 7 A.M." (1939), "The Colder the Air" (1936), and "Anaphora" (1945), was the first sign of how deeply a life in Brazil would agree with her.

Bishop disembarked at Santos, Brazil, and traveled immediately to Rio de Janeiro in order to visit Lota de Macedo Soares and Mary Morse, two women she had met in New York in 1942. Soares and Morse owned an apartment in Rio and were building a house on Soares's estate, Samambaia, near Petrópolis. Bishop spent three weeks touring the countryside in Soares's Jaguar and writing in the Rio apartment. She was scheduled to rejoin her freighter on 26 January. However, sometime in mid December Bishop sampled the exotic fruit of the Brazilian cashew tree and fell violently ill with an allergic reaction – her face and hands swelled so that she could neither see nor write. Three weeks later she was still bedridden but was nonetheless astounded by the love and care shown her by Soares and Morse as well as by the household servants and their children. By mid January she had decided to stay and recuperate for at least another month. From her sickbed she wrote to her doctor, "Aside from my swelled head and the asthma I feel fine & although it is tempting Providence to say so I suppose, happier than I have felt in ten years." By 10 February she admitted that the idea of continuing her trip, or of going back to the United States, was further and further from her mind. For her birthday a Samambaia neighbor gave her a toucan, and together they settled into Soares's half-finished house, high on a mountainside amid "fantastic" scenery. In April, Bishop, accompanied by Soares, returned to New York to arrange for the shipping of her possessions to Brazil.

The life that Bishop left behind in New York and the United States held little nostalgia for her. She said a few good-byes, but really there was nothing about the place that she would miss. The six years (1945–1951) that she spent in New York with its racing, competitive literary scene had offered her little personal sustenance. Her life in Brazil, having an opposite emphasis, was already taking shape. Lota Soares offered to build Bishop a writing studio on the hillside behind the new house, next to the stream, and it must have seemed to Bishop that all of the practical difficulties of her life were being remedied. She would not have to live alone, or get a job; she would be out of reach of the dissipating influences and competitiveness of New York; and in the "timeless" Brazilian world she would be free at last from the pace of city life, which had seemed to her a dizzying plunge toward loss and death. The very impracticality and inefficiency of the Brazilian way of doing things charmed her thoroughly and seemed to indicate that here, anyway, one could control the rate of one's decline. When she returned to Rio on 7 June, she was glad to be "home."

During those first few months back in Brazil Bishop continued her recovery and in the process fell in love with Lota Soares. Meanwhile she planned to complete *A Cold Spring*. It would be three years before the book appeared, and she would be unable to finish any of the older poems she hoped to include. But right away she began "Arrival at Santos" (1952) and by the end of August had finished "The Shampoo" (1955). The two Brazilian poems would give *A Cold Spring* a strange, forward-looking lift at the end; it is apparent to a reader that somewhere near the book's finish the poet's life had changed.

"Arrival at Santos" celebrates the life-changing voyage south. By using real people ("Miss Breen," "the elderly Negro pilot") and actual events from the S.S. *Bowplate's* arrival in Brazil, the poem describes the uneasy sensation of disembarking in a strange land. "Oh, tourist, / is this how this country is going to answer you / and your immodest demands for a different world, / and a better life . . . ?" This early "question of travel" is answered in the poem's luminous last lines, suggesting an agenda for Bishop's future: "We leave Santos at once; / we are driving to the interior."

What Bishop found in the interior is in some sense explained in "The Shampoo," the most hopeful love poem she would ever publish. Its scene, the pool in the stream behind Soares's house; its ostensible subject, the two symmetrical streaks of bright gray in Soares's long, straight, black hair (an earlier draft was titled "Gray Hairs"); its real subject, Bishop's gentle truce with time ("Time is / nothing if not amenable"); and the easy intimacy of the final lines were all newly acquired pleasures that came to her with Brazil:

The shooting stars in your black hair
in bright formation
are flocking where,
so straight, so soon?
– Come, let me wash it in this big tin basin,
battered and shiny like the moon.

By the fall of 1952 Bishop said that she had accomplished "more work in this last stretch than in several years." To work was to write, and suddenly she was writing about her childhood. She found it odd that she should have such vivid recollections of Nova Scotia in its geographical mirror image, Brazil. In fact, she had trouble writing about Brazil. After "Arrival at Santos" she managed only "The Mountain" as a poetic exploration of the country in her first few years there. But she was writing frankly autobiographical prose "like mad." In rapid succession she completed "Gwendolyn" (1953), her marvelously evocative account of her childhood fascination with frail, dying Gwendolyn Appletree; and "In the Village" (1953), the story of Gertrude Bishop's last breakdown, something that she had been trying to write since college. The story describes several weeks in the life of a five-year-old girl who lives with her grandparents on a farm in Nova Scotia. During this time her mother returns to the farm from a stay in a mental hospital, suffers a relapse, and goes back to the hospital with a troubling air of finality. In the central event the mother screams in fear as her dressmaker fits the first colorful dress she'll wear after five years of mourning black, terrifying her imaginative daughter. The story traces the scream's echoes through the child's days and nights in the weeks following, indicating its results in her fragmentary perception. Bishop thought of this as the first in a series of autobiographical short stories, which would include "Gwendolyn," "Memories of Uncle Neddy" (1977), "Primer Class" (about the one-room school in Great Village), and "A Country Mouse" (about her unhappy "kidnapping" from Nova Scotia to Worcester). The last two were published for the first time in *The Collected Prose* in 1984. A draft of "Mrs. Sullivan Downstairs," about her time with her Aunt Maud in Revere, Massachusetts, remains among Bishop's papers.

Though she remained unable to write poems about Brazil for several years, Bishop found a way of expressing her feelings and of continuing her explorations of childhood by translating a beloved Brazilian classic, *Minha Vida de Menina* (My Life as a

Bishop's writing studio on the Samambaia estate, near Petrópolis, Brazil (courtesy of Vassar College Library)

Young Girl), by "Helena Morley" (Alice Dayrell Brant). The book is the diary of a girl between the ages of thirteen and fifteen in the remote provincial village of Diamantina in the mid 1890s. Bishop began the project as a way of practicing her Portuguese, but as she began to read she was taken by Morley's remarkable voice. Life in Diamantina in 1893 had an uncanny resemblance to life "In the Village" in Nova Scotia in 1916. For five years and to the exclusion of almost all other work Bishop devoted her literary energy to the translation. Her introduction to the volume, nearly her longest piece of published prose, provided her the medium she sought finally to write about Brazil. In it she articulates her astonishment in finding that this book by a stranger, written in a different genre and language, could accurately express her own literary and personal values: "*it really happened;* everything did take place, day by day, minute by minute, once and only once, just the way Helena says it did."

Because Bishop was unable to complete more poems for *A Cold Spring,* her publishers decided to reissue *North & South* at the same time in a double volume. The compromise turned out happily when *Poems: North & South – A Cold Spring* won the 1956 Pulitzer Prize. Putting the book behind her was an important milestone for Bishop, and the prize renewed her confidence in her poetic gift. Still unable to write convincing poems about Brazil, she turned again to her childhood for subject matter. "Manners" (1955) was to be the first in a book of children's poems, which she never completed, though in her review of Walter de la Mare's *Come Hither: A Collection of Rhymes and Poems for the Young of All Ages* in 1958, she said she found the values she would like to have included in her project. In his introduction, de la Mare instructed his young readers: "Learn the common names of everything you see ... and especially those that please you most to remember: then give them names also of your own making and choosing – if you can." "He loves 'little articles,'" Bishop went on, "home-made objects whose value increases with age, Robinson Crusoe's lists of his belongings, homely employments, charms, herbs." Bishop's "Sestina," the other childhood poem of this period, takes objects from her Nova Scotia past – child, grandmother, house, stove, tears, almanac – and through the repetitions of the sestina form richly expresses the mystery, loss, sorrow, and love of Bishop's early life. The first poem that Bishop was able to write about Brazil as home (rather than as tourist destination) was also about childhood, "Squatter's Children" (1956). This group of poems would eventually be joined by "First Death in Nova Scotia" (1962), the only published poem in which Bishop presented her mother directly.

"Questions of Travel" (1956), the title poem of Bishop's third volume, is her richest exploration of the traveler's dilemma, the relationship between home and destination, between the observer of a culture and its natives, between freedom and fate. It is more successful than Bishop's other attempts to write about her adopted home because of its modest perspective; the poem is about the limitations of one's knowledge and understanding of a foreign culture. In that sense it opened the way for the Brazil poems that were to follow, the best of which continue to ask their own questions.

"Questions of Travel" opens with an observation about the Brazilian climate that Bishop had made often in letters: "There are too many waterfalls here; the crowded streams / hurry too rapidly down to the sea." After this somewhat recalcitrant opening, the poem poses its questions: "Should we have stayed at home and thought of here?" "Where should we be today?" "Oh must we dream our dreams / and have them, too?" Its tentative answers come from a wholly aesthetic, tourist's appreciation of the landscape and the people:

> But surely it would have been a pity
> not to have seen the trees along this road,
> really exaggerated in their beauty, . . .
> – Not to have had to stop for gas and heard
> the sad, two-noted, wooden tune
> of disparate wooden clogs
> carelessly clacking over
> a grease-stained filling-station floor.
> (In another country the clogs would all be tested.
> Each pair there would have identical pitch.)
> – A pity not to have heard
> the other, less primitive music of the fat brown bird
> who sings above the broken gasoline pump
> in a bamboo church of Jesuit baroque:
> three towers, five silver crosses.

In the end, however, the traveler-poet becomes exasperated by the question of "Where should we be today?" and writes in her "notebook":

> *"Continent, city, country, society:*
> *the choice is never wide and never free.*

> *And here, or there . . . No. Should we have stayed at home,*
> *wherever that may be?"*

Though she saw "Manuelzinho" (1956) as her first attempt to speak seriously about Brazilian culture, Bishop covered her anxiety about her outsider's perspective by writing the poem in Lota Soares's voice. A portrait of the endearingly unreliable gardener at Soares's estate, the poem delineates Brazilian social distinctions carefully. We are told in the first two lines that Manuelzinho and his family are tenants and that they are white. Lota Soares is their keeper in the paternalistic (in this case, maternalistic) social system. The family's long tenancy ("the steep paths you have made – / or your father and grandfather made – / all over my property") gets them the landowner's tolerance, but little else. The poem does not deeply question the almost feudal system and borders on condescension toward its subject. This is what Bishop had feared in her Brazilian poems, and in the next one she published she returned to her questioning mode.

"The Armadillo" (1957) explores the ambiguous phenomenon of illegal fire balloons launched on St. John's Day (24 June) on the mountain behind Lota Soares's house. The poet's aesthetic appreciation of the light show is tempered by the fire danger and the terror of the small animals – owls, rabbits, the armadillo – frightened by the flames. In the end the poet scolds herself for exploitative metaphor making: "*Too pretty, dreamlike mimicry!*" She reminds us of the poet's responsibility to observe and describe accurately and not to appropriate the pain and terror of others for merely artistic ends.

Her happiness in Brazil – her pleasure in Lota Soares's love, the successful treatment of her asthma, her general sobriety, the rich and fruitful correspondences she carried on with distant friends – took many pressures off Bishop and allowed her to write more freely. In "Exchanging Hats" (1956) she explores the complexities of gender identification in an *Alice in Wonderland* rubric, and in "Visits to St. Elizabeths" (1957) she puts to rest the figure of Ezra Pound and the fanatic ambition his career represented. (She had visited Pound in his hospital room during her year in Washington, D.C., from 1949 to 1950.) She had made herself a poet on her own terms – away from New York, writing only what she chose – thus, she could afford to pity "the wretched man / that lies in the house of Bedlam."

In the second group of Brazil poems Bishop was more comfortable with her outsider's voice. "Brazil, January 1, 1502," concerns the rape of the Brazilian landscape by sixteenth-century Portu-

PINK DOG

The sun is blazing and the sky is blue.
Umbrellas clothe the beach in every hue.
Naked, you trot across the avenue.

Oh, never have I seen a dog so bare!
Naked and pink, without a single hair...
Startled, the passersby draw back and stare.

Of course they're mortally afraid of rabies.
But you're not mad. You have a case of scabies,
~~a terminal one, I'd say. Where are your babies?~~

~~In what slum have you hidden them, poor bitch?~~
A nursing mother, by those hanging teats,
(in what slum have you hidden them, poor bitch?

Didn't you know? It's been in all the papers,
to solve this problem, how they deal with beggars?
They take ~~them out~~ and throw them in the tidal rivers.

Yes, idiots, paralytics, parasites,
go bbbbing in the ebbing sewage, nights,
out in the ƥsuburbs, where there are no lights.

If they do this to anyone who begs,
~~drunk, drugged, or sober,~~ with or without legs,   drugged, drunk or sober.
what might they do to bare, four-legged dogs?

In the cafés and on the   corners
the joke is going round that all the beggars
who can afford them, now wear life-preservers.

In your condition you would not be able
to ~~get away~~   ~~even to dog-paddle.~~   even to float, much less to dog-paddle
                    this time of year.
Ash Wednesday's coming; Carnival is here.
~~What~~ samba will you dance? What will you wear?

The Carnival's ~~of course~~ degenerating
—radios, Americans, or something,
have ruined it completely. They're just talking.

Carnival is always wonderful!
A depilated dog will not look well.
Get dressed! Get dressed and dance at Carnival!

Revised typescript for one of Bishop's poems inspired by her experiences in Brazil (courtesy of Alice Methfessel)

guese explorers and the country's complicity in its own demise. "The Riverman" (1960), a "fairy tale" in dramatic monologue, is about a man who studies to become a *sacaca,* or witch doctor, who works with the Amazon River spirits. "Electrical Storm" (1960) and "Song for the Rainy Season" (1960) are both about insects and animals at Samambaia. All these poems succeed in making Brazilian settings and details into vehicles for Bishop's lifelong poetic and personal concerns: the losses inherent in the passage of time, the shock of the human encounter with nature, and the experiences that lie on the fringes between sleep and consciousness.

In 1961 the Brazilian life of Elizabeth Bishop and Lota de Macedo Soares changed dramatically. Soares took a job in Rio de Janeiro as the director of the project that would eventually produce the enormous and beautiful Flamengo Park. As a result she entered the maelstrom of Brazilian politics more than full-time. The job meant transferring their lives from the cool, spread-out, rural Samambaia to the hot, crowded, and somewhat dysfunctional city of Rio. Soares's work left Bishop alone (with servants) in the apartment most days and evenings, and the happy and peaceful life they had made began to come undone.

Seeing firsthand the massive poverty and corruption of Brazilian life in the city, Bishop again found herself unable to express what she saw in poems. Coincidentally, in June 1961, Time Books asked her to write the *Life* World Library volume on Brazil, and despite her hatred of deadlines and her distrust of journalists, Bishop agreed to take on the task. Seeing it as a chance to make some money, collaborate with Soares, and to say about Brazil what she had been unable to say in poems, she approached it with some enthusiasm. But from the start she quarreled with the editors over content and contested every deadline. The finished book was a great disappointment to her, and the copies she delivered to friends she "corrected" in green ink. Nevertheless, much of the book is recognizably Bishop's, and it stands as the only extensive consideration of Brazil that she managed to complete.

In 1964 Bishop completed the last two poems for *Questions of Travel* (1965). "Twelfth Morning; or What You Will" considers the beach at Cabo Frio, an as yet undeveloped resort town north of Rio, where she and Soares spent several memorable vacations. The poem expresses Bishop's ambivalence about the place's poverty and beauty and its potential for exploitation. Its tone is typical of all of Bishop's later thinking about Brazil:

> Like a first coat of whitewash when it's wet,
> the thin gray mist lets everything show through:
> the black boy Balthazár, a fence, a horse,
> a foundered house,
>
> – cement and rafters sticking from a dune.
> (The Company passes off these white but shopworn
> dunes as lawns.) "Shipwreck," we say; perhaps
>   this is a housewreck.

This bleak view is redeemed, as almost all such views are redeemed in Bishop's poems, by the insistent contentment of the child Balthazár himself:

> But the four-gallon can
> approaching on the head of Balthazár
> keeps flashing that the world's a pearl, *and I,*
>   *I am*
> *its highlight!* You can hear the water now,
> inside, slap-slapping. Balthazár is singing.
> "Today's my Anniversary," he sings,
>   "the Day of Kings."

Despite numerous attempts to write poems specifically about the deteriorating political and economic situation in Brazil, Bishop found that she could not. Instead, it is in poems such as "Twelfth Morning" and the only other Brazil poem she was able to complete for *Questions of Travel,* "The Burglar of Babylon" (1964), that such things get said. The latter poem is a *faux naif* ballad about a real-life manhunt conducted on the impoverished and *favela*-covered hillsides of Rio, in full view of the balconies of the wealthy. The poem's rigid structure and childish rhyme scheme gave Bishop license to deal in political generalizations she could not make convincing in her own voice. Its bare facts and its imaginative entry into the consciousness of Micuçú, an individual product of the unjust Brazilian economic system, seemed to her the best way to express her political message. She had learned this lesson in the use of popular forms from the Rio Samba schools, which each year made songs of current events to be sung at Carnival.

Questions of Travel was a thin volume, and the publishers, Farrar, Straus and Giroux, decided to solve that problem by including the whole of the prose poem/memoir "In the Village" between the "Brazil" and "Elsewhere" sections of the book. This turned the book in a sharply autobiographical direction, but the inclusion of a prose memoir in a book of poems had precedent in Robert Lowell's *Life Studies* (1959). Bishop had been thinking about autobiography in the year before the book appeared as well; the poet Anne Stevenson was writing a book about her for the Twayne U.S. Authors Series.

Quite surprisingly and for the first time, Bishop was sharing the facts of her childhood and her thinking about poetry in a systematic way. The letters she wrote to Stevenson in 1964–1965 are extraordinary autobiographical documents, the closest Bishop came to a full consideration of how the circumstances of her life affected her poems.

When the reviews of *Questions of Travel* began to appear in the spring of 1966, Bishop was in the United States, having taken on the first teaching job of her life. At the University of Washington in Seattle she taught poetry and creative writing. Bishop had left Brazil badly in need of a break from the constant struggles of life in Rio and from the terrifying intensity with which Lota Soares worked at her job. Also before her departure, Bishop had purchased an eighteenth-century house in Ouro Prêto, in the state of Minas Gerais, and she was suddenly in need of money to fund its renovation.

The job at the University of Washington terrified her; she had never worked for a living; she had been out of the country for fifteen years, and all of the transformations of the 1960s were new to her. But she found that she was able to do it, especially with the help of good friends. Early in her stay in Seattle, Bishop fell in love with a young woman, and with that relationship her Brazilian life began to fall apart.

When she did return to Brazil in June 1966, Bishop found that her worst fears about Lota Soares's health had come true. Within months Soares suffered a serious mental breakdown, and her circle of Brazilian family and friends began to close around her and to shut Bishop out. Soares and Bishop were both hospitalized in separate facilities for illness and mental exhaustion until March 1967, when they moved back up to Samambaia to try to remake their peaceful rural life. This experiment lasted until early July, when it became clear to Bishop that Soares would be better off without her until Soares was better.

Bishop flew to New York on 3 July; Soares was to join her when she was well enough to travel. Bishop spent a lonely summer in the empty New York apartment of her friends the painter Loren MacIver and her husband, Lloyd Frankenberg, before Soares, against her doctors' advice, joined her on 19 September. The two spent a peaceful afternoon and evening together, but sometime during the night, Soares got up and took an overdose of tranquilizers. By the time Bishop awoke at six in the morning she was nearly comatose. Bishop had Soares rushed to the hospital, and it seemed at first that she would survive. Doctors found only Valium in her system (though she had been clutching an empty bottle of Nembutal when Bishop found her), and overdoses of Valium are rarely fatal. But Soares had arteriosclerosis and a history of heart trouble. Moreover, no one knew if she had had any alcohol to drink before taking the pills. After five days in an unbroken coma, she died at the age of fifty-seven. It took Elizabeth Bishop a long time to recover from this loss.

After going back to Brazil and arranging to have her belongings shipped to the United States, Bishop moved for a time to San Francisco. There she lived with the young woman she had met in Seattle. After seventeen months Bishop's longing for her Brazilian life convinced her to move her household – including the woman and her child – to the beautiful house in Ouro Prêto, still being restored. The stay in Brazil was an unrelieved nightmare; without Lota Soares to intercede for her, Bishop lost her infatuation with Brazilian inefficiency. The house took months to complete. At the same time, amid the memories of her losses, Bishop lost control of her drinking. The young female friend evidently cracked under the strain as well; she was hospitalized at Bishop's direction, then flown back to the United States with her child. Bishop was once again rescued by Robert Lowell, who arranged for her to fill his teaching post at Harvard while he took a sabbatical. She arrived in Cambridge, Massachusetts, in September 1970 desperately in need of a new life.

During this unhappy period, Bishop's literary reputation continued to rise. She had published five poems since *Questions of Travel*: "House Guest" (1968), about a young visitor to Samambaia in 1956; "Trouvée" (1968), which deals with Lota Soares's suicide in a contemplation of the adage about chickens and roads, in this case a red-and-white hen run over on West 4th Street in Greenwich Village; the prose poems "Rainy Season; Sub-Tropics" (1967), which explore the troubled perspectives of three oversized, alienated Brazilian animals; and two very effective poems about Brazilian life. The first, "Under the Window: Ouro Prêto" (1966), is a "slice-of-life" observation of travelers who stopped to drink at a rustic fountain across from Bishop's Ouro Prêto house. Its voices are entirely convincing:

> "She's been in labor now two days." "Transistors
> cost much too much." "For lunch we took advantage
> of the poor duck the dog decapitated."
>
> The seven ages of man are talkative
> and soiled and thirsty.
>   Oil has seeped into

the margins of the ditch of standing water

and flashes or looks upward brokenly,
like bits of mirror – no, more blue than that:
like tatters of the *Morpho* butterfly.

The other Brazilian poem, "Going to the Bakery" (1968), was begun in 1960 and finished in the spring of 1967. Its dark view of life in Rio de Janeiro, a life of rationed electricity, "adulterated" bread, and sickly citizens, troubled Bishop, and she revised the poem numerous times before it was finally published. By 1967 she felt the criticism of life in Rio was justified.

In 1969 Farrar, Straus and Giroux brought out Bishop's *Complete Poems,* and she mustered enough courage to do several readings in conjunction with the book's publication, thus effectively launching her reading career. In March 1970, while Bishop was still in Ouro Prêto, it won the National Book Award for poetry.

Bishop's life improved from the moment she arrived at Harvard. Though she sometimes bristled at the conservatism and sexism of the institution, she found a circle of writer/friends – including not only Lowell, but Seamus Heaney, Octavio Paz, John Malcolm Brinnin, Frank Bidart, Lloyd Schwartz, James Merrill, Robert Fitzgerald, and her editor, Robert Giroux. She also found a devoted friend and caretaker in Alice Methfessel, the young administrative assistant at Harvard's Kirkland House, where Bishop first lived in Cambridge. Bishop continued to struggle in the last years of her life with depression, asthma, and alcoholism and their attendant health concerns; but she lived well. She taught her courses, socialized, renovated an apartment on Boston's Lewis Wharf to live in, spent summers traveling with Methfessel – including trips back to Ouro Prêto, the Galápagos Islands, Scandinavia, and Greece – and, after 1974, spent a part of each summer on North Haven Island, Maine, which became her spiritual home. She wrote only six new poems in her last nine years – "Night City" (1972), "The End of March" (1975), "Five Flights Up" (1975), "One Art" (1976), "North Haven" (1978), and "Sonnet" (1979). These and the older poems she was able to finish – "Crusoe in England" (1971), "In the Waiting Room" (1971), "The Moose" (1972), "Poem" (1972), "12 O'Clock News" (1973), "Santarém" (1978), and "Pink Dog" (1979) – extended her range and reputation dramatically. She also published several translations, notably of Octavio Paz's "*Primero Enero*" ("January First," 1975) and "Objetos y Apariciones" ("Objects and Apparitions," 1974).

Bishop with Robert Lowell at the beach in Rio de Janeiro, July 1962 (courtesy of Special Collections, Vassar College Libraries)

Bishop's later poems differ from her earlier work primarily in the freedom with which they address her personal concerns. "In the Waiting Room," for example, recounts a real moment of existential awareness from Bishop's childhood – "In Worcester, Massachusetts, / I went with Aunt Consuelo / to keep her dentist's appointment" – and at that moment of terrifying realization, Bishop names herself:

I said to myself: three days
and you'll be seven years old.
I was saying it to stop
the sensation of falling off
the round, turning world
into cold, blue-black space.
But I felt: you are an *I*,
you are an *Elizabeth*,
you are one of *them*.
Why should you be one, too?
I scarcely dared to look
to see what it was I was.

In "Crusoe in England" Bishop revises the story of Daniel Defoe's Robinson Crusoe so that his concerns become her own. The details of his island come from her own travels; his faith in the Bible becomes her agnosticism ("The books / I read were full of blanks"); his dreams of salvation become her fear of the observer's responsibility ("I'd have nightmares of islands / . . . knowing that I had to live / on each and every one, eventually, / for ages, registering their flora, / their fauna, their geography"); and his companionship with Friday becomes the lost love of Bishop's life ("And Friday, my dear Friday, died of measles / seventeen years ago come March.")

"The Moose" is Bishop's most complete poetic consideration of her Nova Scotia past. The poem had been with her since her 1946 bus trip to Boston from Nova Scotia; she finished it in June 1972, just in time to read it as her Phi Beta Kappa address at Harvard. The poem recounts, in quite realistic detail, the journey of the bus from Great Village to New Brunswick. The geography of Nova Scotia, the falling light on typical plants and people, the familiar smells of salt and hay, are all recounted in the long, narrow (trimeter lines) poem, with its elaborately rhymed stanzas. The memory of the landscape prompts a more personal recollection:

"Grandparents' voices"

uninterruptedly
talking, in Eternity:
names being mentioned,
things cleared up finally;
what he said, what she said,
who got pensioned...

"Yes..." that peculiar
affirmative. "Yes..."
A sharp, indrawn breath,
half groan, half acceptance,
that means "Life's like that.
We know *it* (also death)."

Talking the way they talked
in the old featherbed,
peacefully, on and on,
dim lamplight in the hall,
down in the kitchen, the dog
tucked in her shawl.

Now, it's all right now
even to fall asleep
just as on all those nights.

A moose stumbles out of the woods into the highway to interrupt this reverie, and slowly, sleepily, the passengers domesticate this visitor from the other world: "Why, why do we feel / (we all feel) this sweet / sensation of joy?"

"Poem" also revisits Nova Scotia, and the great-uncle/painter whose work Bishop had disparaged in the early "Large Bad Picture." Examining a small painting given her by her aunt, the poet finds to her surprise that she recognizes the scene, "I can almost remember the farmer's name." She and the uncle she never knew apparently had this very spot in common:

Our visions coincided — "visions" is
too serious a word — our looks, two looks:
art "copying from life" and life itself,
life and the memory of it so compressed
they've turned into each other. Which is which?

In "The End of March" Bishop revisits her lifelong fantasy of complete retirement from the world in her vision of the "proto . . . crypto dream house" on the cold, gray beach at Duxbury, Massachusetts, where she walks with two friends. Ultimately she rejects the fantasy of withdrawal and accepts such revelation as the early spring sun can offer:

On the way back our faces froze on the other side.
The sun came out for just a minute.
For just a minute, set in their bezels of sand,
the drab, damp, scattered stones
were multi-colored,
and all those high enough threw out long shadows,
individual shadows, then pulled them in again.

Despite these tentative approaches to the confessional mode, Bishop retained her belief in discretion as a poetic method. Too much direct revelation of personal grief would limit a poem's significance to others. The most personal poem among Bishop's last work, and the last to be finished in time to be included in *Geography III* (1976), is the villanelle "One Art."

The drafts of "One Art" remain among Bishop's papers, and from them one can see the poem's genesis. It arose from a crisis in Bishop's personal life. In the fall 1975 she feared losing Alice Methfessel, the woman she called her "saving grace." The first draft of the poem is a painful catalogue of the losses in her life, written almost in prose. " — This is by way of introduction. I really / want to introduce myself — I am . . . / fantastic lly good at losing things / I think everyone shd. profit from my experiences." Moreover: "You may find it hard to believe, but I have actually lost / I mean lost, and forever, two whole houses," along with "one peninsula and one island," "a small-sized town . . . and many smaller bits of geography or scenery" — indeed a continent, "the whole damned thing!" She concludes:

Bishop in her later years (courtesy of Alice Methfessel)

> One might think this would have prepared me
> for losing one average-sized not ... exceptionally
> beautiful or dazzlingly intelligent person
> except for blue eyes) (only the eyes <u>were</u> exceptionally
> beautiful and the hands <u>looked</u>
> intelligent) the fine hands
> But it doesn't seem to have, at all.

By the second draft the poem took shape as a villanelle, and through fifteen more drafts Bishop whittled the catalogue of losses into a discreet and resonant form. At the same time, she attempted to come to terms with her loss. Midway through the seventeen drafts she allows for the possibility that maybe she will survive. As the final version of the poem expresses it:

> — Even losing you (the joking voice, a gesture
> I love) I shan't have lied. It's evident
> the art of losing's not too hard to master
> though it may look like (*Write* it!) like disaster.

One way to read this tentative resolution is to see that in the writing of such a demanding poem lies the mastery of the loss. Working through each of her losses — from the bold, painful catalogue of the first draft to the finely honed and privately meaningful final version — is the way to overcome them. It is all, perhaps, "one art" — writing elegies, mastering loss and grief, self-mastery. The losses in the poem are lifelong and real, as seen with time, in the form of the "hour badly spent," and more tellingly for the orphaned Bishop, "my mother's watch"; the lost houses in Key West, Petrópolis, and the one then still in doubt, Ouro Prêto. The city of Rio de Janeiro and the whole South American continent were lost to her with Lota Soares's suicide. And currently, in the fall of 1975, she seemed to have lost her dearest friend and lover, she of the blue eyes and fine hands. But each version of the poem distanced the pain a little more, depersonalized it, moved it away from the tawdry self-pity and "confession" that Bishop disliked in many of her contemporaries. In the development of this poem, discretion is a poetic method, as well as part of a process of self-understanding, the seeing of a pattern in her own life.

Elizabeth Bishop wrote a few poems after the publication of *Geography III,* notably the lovely elegy for her friend Robert Lowell, who died in 1977. "North Haven" (1978) takes the timeless elements of the Maine coast, which Bishop and Lowell had in common, as a figure for the stillness of death:

> You left North Haven, anchored in its rock,
> afloat in mystic blue ... And now – you've left
> for good. You can't derange, or re-arrange,
> your poems again. (But the Sparrows can their song.)
> The words won't change again. Sad friend, you cannot
>   change.

And she finished, after eighteen years, the poem she had hoped to write about her 1960 trip down the Amazon River from Manaus to Belém. "Santarém" is another poem of lifelong concerns – the wish for retreat or withdrawal; the struggle with the passage of time; the complexity of choice, represented by the two rivers, one blue and one brown, that meet at the village; the alienation inherent in the poet's perspective. The poem has a valedictory air, which Bishop developed in its later drafts.

Elizabeth Bishop died suddenly of a cerebral aneurysm on Saturday, 6 October 1979, at the age of sixty-eight. She had written letters that morning, and was dressing for dinner at the critic Helen Vendler's house when she died. She had been scheduled to read at Harvard's Sanders Theater the following evening, and, in her absence, her friends gathered and read her poems for her.

Bishop was remembered in numerous obituaries and at many memorial services in terms she would have liked: the greatness of her poems and the modesty of her self-presentation. She was not "famous" when she died, except in the Boston and New York poetry circles she had feared so much as a young poet, but her work was widely respected. James Merrill remarked on her "instinctive, modest, life-long impersonations of an ordinary woman." Lloyd Schwartz remembered that her favorite example of iambic pentameter had been "I hate to see that evening sun go down."

She has become famous since her death, defying the more typical inverted arc of the American poet's reputation by refusing to disappear immediately after dying, only to reappear twenty or thirty years later. As critical judgments of her poetry have developed, the early comments on her "objectivity" and "impersonality" have yielded to the gentle insistence of the personal voice in her poems, as her readers have come to see that she, like most other poets, told the story of her life in her work. She told it with sorrow, with humor, and with an almost perfect understanding of her own strengths and failures. "Awful, but cheerful," she asked Alice Methfessel to inscribe on her tombstone in the Bishop family plot in Worcester.

Letters:

One Art: Elizabeth Bishop, edited by Robert Giroux (New York: Farrar, Straus & Giroux, 1994).

Interviews:

Ashley Brown, "An Inteview with Elizabeth Bishop," *Shenandoah,* 17 (Winter 1966): 3-19;

George Starbuck, "'The Work!': A Conversation with Elizabeth Bishop," *Ploughshares,* 3, nos. 3 & 4 (1977): 11-29;

Alexandra Johnson, "Artists and Their Inspiration: Poet Elizabeth Bishop: Geography of the Imagination," *Christian Science Monitor,* 23 March 1977, pp. 20-21;

Elizabeth Spires, "The Art of Poetry XXVII," *Paris Review,* 23 (Summer 1981): 57-83.

Bibliographies:

Candace McMahon, *Elizabeth Bishop: A Bibliography, 1927-1979* (Charlottesville: University Press of Virginia, 1980);

Diane E. Wyllie, *Elizabeth Bishop and Howard Nemerov: A Reference Guide* (Boston: G. K. Hall, 1983).

Biographies:

Brett C. Millier, *Elizabeth Bishop: Life and the Memory of It* (Berkeley: University of California Press, 1993);

Gary Fountain, *Remembering Elizabeth Bishop: An Oral Biography* (Amherst: University of Massachusetts Press, 1994).

References:

Harold Bloom, ed., *Elizabeth Bishop* (New York: Chelsea House, 1988);

Bonnie Costello, *Elizabeth Bishop: Questions of Mastery* (Cambridge, Mass.: Harvard University Press, 1991);

Carole Doreski, *Elizabeth Bishop: The Restraints of Language* (New York: Oxford University Press, 1993);

Lorrie Goldensohn, *Elizabeth Bishop: The Biography of a Poetry* (New York: Columbia University Press, 1991);

Victoria Harrison, *Elizabeth Bishop: The Poetics of Intimacy* (New York: Cambridge University Press, 1993);

David Kalstone, *Becoming a Poet: Elizabeth Bishop with Marianne Moore and Robert Lowell* (New York: Farrar, Straus & Giroux, 1989);

Kalstone, *Five Temperaments* (New York: Oxford University Press, 1977);

Lynn Keller, "'Words Worth a Thousand Postcards': The Bishop/Moore Correspondence," *American Literature,* 55 (October 1983): 405–429;

Marilyn May Lombardi, ed., *Elizabeth Bishop: The Geography of Gender* (Charlottesville: University Press of Virginia, 1993);

Jeredith Merrin, *An Enabling Humility: Marianne Moore, Elizabeth Bishop, and the Uses of Tradition* (New Brunswick, N.J.: Rutgers University Press, 1990);

Barbara Page, "Shifting Islands: Elizabeth Bishop's Manuscripts," *Shenandoah,* 33, no. 1 (1981–1982): 51–62;

Robert Dale Parker, *The Unbeliever* (Urbana: University of Illinois Press, 1988);

Adrienne Rich, "The Eye of the Outsider: The Poetry of Elizabeth Bishop," *Boston Review,* 8 (April 1983): 15–17;

Lloyd Schwartz and Sybil Estess, eds., *Elizabeth Bishop and Her Art* (Ann Arbor: University of Michigan Press, 1983);

Anne Stevenson, *Elizabeth Bishop* (New York: Twayne, 1966);

Thomas Travisano, *Elizabeth Bishop: Her Artistic Development* (Charlottesville: University Press of Virginia, 1988).

Papers:

The major collection of Elizabeth Bishop's papers, notebooks, letters, unpublished works, and drafts is at the Vassar College Library in Poughkeepsie, New York. Quotations from unpublished material in this essay can be found in the Vassar collection. Bishop's letters to Marianne Moore are in the Marianne Moore Papers at the Rosenbach Museum and Library, Philadelphia. Her letters to Robert Lowell are at the Houghton Library, Harvard University, and at the Harry Ransom Humanities Research Center, University of Texas at Austin. Important correspondence is also stored at the Washington University Library, Saint Louis; at the Princeton University Library; at the University of Delaware Library; and in the Henry W. and Albert A. Berg Collection of the New York Public Library.

Louise Bogan
(11 August 1897 – 4 February 1970)

Brett C. Millier
Middlebury College

See also the Bogan entry in *DLB 45: American Poets, 1880-1945, First Series.*

BOOKS: *Body of This Death* (New York: McBride, 1923);

Dark Summer (New York: Scribners, 1929);

The Sleeping Fury (New York: Scribners, 1937);

Poems and New Poems (New York: Scribners, 1941);

Achievement in American Poetry, 1900-1950 (Chicago: Regnery, 1951);

Collected Poems, 1923-1953 (New York: Noonday Press, 1954);

Selected Criticism: Prose, Poetry (New York: Noonday Press, 1955);

The Blue Estuaries: Poems 1923-1968 (New York: Farrar, Straus & Giroux, 1968);

A Poet's Alphabet: Reflections on the Literary Art and Vocation, edited by Robert Phelps and Ruth Limmer (New York: McGraw-Hill, 1970);

Journey Around My Room: The Autobiography of Louise Bogan, edited by Ruth Limmer (New York: Viking, 1980).

OTHER: *Works in the Humanities in Great Britain 1939-1946,* compiled by Bogan (Washington, D.C.: Library of Congress, 1950);

Yvan Goll, *Elegy of Ihpetonga* and *Masks of Ashes,* translated by Bogan (New York: Noonday Press, 1954);

Emily Dickinson: Three Views, by Bogan, Archibald MacLeish, and Richard Wilbur (Amherst, Mass: Amherst College Press, 1960);

Ernst Juenger, *The Glass Bees,* translated by Bogan and Elizabeth Mayer (New York: Noonday Press, 1961);

Goll, *The Myth of the Pierced Rock,* translated by Bogan (New York: Allen Press, 1962);

Johann Wolfgang von Goethe, *Elective Affinities,* translated by Bogan and Mayer (Chicago: Regnery, 1963);

Louise Bogan in 1922

The Journal of Jules Renard, translated by Bogan and Elizabeth Roget (New York: Braziller, 1964);

The Golden Journey: Poems for Young People, edited by Bogan and William Jay Smith (New York: Reilly & Lee, 1965);

Virginia Woolf, *A Writer's Diary, Being Extracts From the Diary of Virginia Woolf,* afterword by Bogan (New York: New American Library, 1968);

Goethe, *The Sorrows of Young Werther* and *Novella*, translated by Bogan and Mayer (New York: Random House, 1971).

The critic Malcolm Cowley remarked in a review of Louise Bogan's slim volume *Poems and New Poems* (1941) that she had "done something that has been achieved by very few of her contemporaries: she has added a dozen or more to our small stock of memorable lyrics. She has added nothing whatever to our inexhaustible store of trash." Bogan's reputation as a poet is secure on exactly that scale. She is remembered and studied as one of the finest lyric poets America has produced, though the fact that she was a woman and that she defended formal, lyric poetry in an age of expansive experimentation made evaluation of her work, until quite recently, somewhat condescending. Her achievement in poetry has also been overshadowed by her extensive critical writings; for thirty-eight years she was the poetry critic for *The New Yorker* magazine, the arbiter of taste in such matters for a literate and influential audience.

Louise Bogan was born of the unhappy marriage of Daniel and May Shields Bogan in Livermore Falls, Maine, a bustling mill town on the Androscoggin River. In 1897, the year of Louise's birth, Daniel Bogan was superintendent in a pulp mill in the town, the first of many such relatively white-collar mill jobs he would hold during her childhood. Louise was their third child; a son, Charles, had been born in 1884, and a second son, Edward, had died in infancy. Bogan grew up in the Irish communities of deepest New England, moving often with her family to a variety of hotels and boardinghouses and other temporary dwellings: to Milton, New Hampshire, in 1901; to Ballardvale, Massachusetts, in 1904; to Roxbury, near Boston, in 1909. These moves were prompted both by economics and by the family's unhappiness. May Shields Bogan was a beautiful and unstable woman prone to flaunting her many extramarital affairs (on at least one occasion witnessed by her daughter) and to mysterious and lengthy disappearances.

Despite these disruptions, Bogan was quite well educated, in a New Hampshire convent (1906–1908) and at Boston's excellent Girls' Latin School (1910–1915), where she received a classical education in Latin, Greek, French, mathematics, history, science, and the arts. Having fallen under the poetic spell of A. C. Swinburne and the French Symbolists, she was a constant contributor to Latin's literary magazine, *The Jabberwock*, until she was told by the headmaster to trim her ambitions: "No Irish girl could be editor of the school magazine." Such prejudice was prevalent in Boston at the time, and Bogan never ceased to resent it. She transcended these limitations, however, and continued to publish her high school poems, including four in the *Boston Evening Transcript*, and was named class poet. She was a wide and constant reader who followed her own tastes and developed early and intense literary ambitions.

The difficulties and instabilities of her childhood produced in Bogan a preoccupation with betrayal and a distrust of others, a highly romantic nature, and a preference for the arrangements of art over grim, workaday reality. She would suffer for most of her life from serious depression, which resulted in three lengthy hospital stays for treatment. She would drink heavily, and her work would suffer from what Elizabeth Frank in *Louise Bogan: A Portrait* (1985) has called "a principle of arrest": "Something stopped Louise Bogan dead in her tracks, not once, but many times." But her fine early education would form the foundation of her poetry and criticism, as would, in some sense, her unhappiness.

Sent by her parents to Boston University in 1916, Bogan did extremely well and earned a scholarship to Radcliffe for her sophomore year. She turned it down, however, for the chance to leave home in the company of a husband, Curt Alexander, a soldier of German origin nine years her senior. She moved with her husband to New York City, and then, when war was declared in 1917, to Panama, where she gave birth to their daughter, Mathilde (Maidie) Alexander. Miserable in the role of military wife and in the heat and humidity of Panama, Bogan wrote poems about her condition, including "Betrothed" (which appeared in her first collection) and "The Young Wife" (which she never collected) and schemed to get back to New England. She left Panama with her child in May 1918 and moved in with her parents in Massachusetts. At the end of the war she was briefly reconciled with her husband, and they lived for a time on army bases near Portland, Maine, and near Hoboken, New Jersey. In the summer of 1919 she left Alexander for good, delivering her daughter to her parents and finding herself an apartment in New York from which to launch her career as a woman of letters. Alexander died in 1920, and his army widow's pension enabled Bogan to stay in the city.

From her temporary job at Brentano's bookstore in New York, Bogan quickly became involved in the city's active literary community. Her earliest friendships included Lola Ridge, Malcolm Cowley,

William Carlos Williams, Mina Loy, Maxwell Bodenheim, John Reed, Louise Bryant, Conrad Aiken, and, most important, Edmund Wilson. At the fringes of, but deeply skeptical about, the leftist politics common in the group, Bogan nonetheless found herself in an intense love affair with a young radical named John Coffey, who would shoplift (his speciality was furs) and then plead the cause of the poor in his courtroom appearances. Bogan drove the getaway car on one of these escapades, a fact that embarrassed her from time to time for the rest of her life. When Coffey finally succeeded in making his motives clear to a judge, he was committed to a hospital for the insane.

Bogan set about educating herself and honing her writing skills with great seriousness and dedication. Ever conscious of her educational deficiencies (she carried a lifelong resentment toward people with advanced degrees), she sought to make up for them by reading. In this period she discovered the poetry of William Butler Yeats, who, like Rainer Maria Rilke and W. H. Auden later, would become a poetic touchstone and an important influence on her work. Bogan's poems appeared in the best journals of her day, almost from the beginning of her stay in New York. She published often in Harriet Monroe's influential *Poetry: A Magazine of Verse* and was involved from the start with *The Measure*, a "little magazine" devoted to the formal lyric. She became acquainted with the work of the most important female poets of the day, including Elinor Wylie and Edna St. Vincent Millay, as well as the poetry of Edward Arlington Robinson, perhaps her most direct American influence. Also through her work on *The Measure* (and a brief stint as a card filer in the office of the anthropologist William Fielding Ogburn) she met and became friendly with Margaret Mead and the poet Léonie Adams. In 1922 Bogan spent six months alone in Vienna, absorbing European culture and writing, and by the fall of 1923 she had secured a publisher, Robert M. McBride, for her first book of poems.

Body of This Death (1923) contains several of Bogan's most memorable poems and in general reveals its author's preoccupations and tastes. Betrayal, particularly sexual betrayal, is a constant theme, though the poems are in no way "confessional." In private writing included in *Journey Around My Room: The Autobiography of Louise Bogan* (1980), Bogan echoes Emerson's charge that the poet tell his life story in "cipher": "The poet represses the outright narrative of his life. He absorbs it, along with life itself. The repressed becomes the poem. Actually, I have written down my experience in the closest detail. But the rough and vulgar facts are not there."

Like Emily Dickinson's "gift of screws," Bogan's poems are made of meticulously distilled experience, distanced from the source by objective language. Often presented by their titles as songs or chants or arias, her poems call attention to themselves as rhetorical acts in a common language. Such commitment to public discourse did not protect Bogan from occasional obscurity — the distillation sometimes reduced emotion to indecipherable symbolism. But Bogan saw herself in the tradition of sixteenth- and seventeenth-century English lyric poetry, and she disciplined her poetic emotion to the formal rhyme and meter she instinctively preferred. Her well-known poem "The Alchemist" speaks to the method:

> I burned my life, that I might find
> A passion wholly of the mind,
> Thought divorced from eye and bone,
> Ecstasy come to breath alone.
> I broke my life, to seek relief
> From the flawed light of love and grief.

The poem's concluding second stanza admits the necessity of "unmysterious flesh" after all.

Several of the poems in *Body of This Death* address specifically female concerns and point to Bogan's ambivalent relationship with the tradition of female lyric poets. Her poems are by no means dogmatically feminist; Bogan held a deep distrust for all ideological commitment. In fact, she has been castigated somewhat unfairly by contemporary feminists for the dry pronouncements of her much-anthologized lyric "Women": "Women have no wilderness in them, / They are provident instead, / Content in the tight hot cell of their hearts / To eat dusty bread." Missing the ironic self-criticism in the poem ("As like as not, when they take life over their door-sills / They should let it go by"), feminist critics have read it as general condemnation of women and their ways of viewing experience. While the situation of women in Bogan's poems is rarely preferred to the situation of men, she is capable of wise and penetrating insight. "Betrothed," "Portrait," "My Voice Not Being Proud," "Medusa," "The Crows," "The Changed Woman," "Chanson Un Peu Naïve," and "Fifteenth Farewell" are all strong poems with female speakers or subjects. She saw herself and her work as arising from that definite tradition of female lyricists, represented in the generation just older than Bogan herself by the strong figures of Sara Teasdale, Millay, and Wylie. For Marianne Moore and Hilda Doolittle, the more typically modernist women poets of the period, she felt less affinity.

Early in 1924 Bogan's close friend Edmund Wilson suggested that she try her hand at criticism. That spring she published her first book review, of D. H. Lawrence's *Birds, Beasts and Flowers* (1923), in *The New Republic*. She would continue to write poetry reviews for the rest of her life, and as poems came to her less and less frequently as she grew older, Bogan became better known as a critic than as a poet. In 1931 she wrote her first review for *The New Yorker,* and twice a year until 1969 she presented the season's new poetry books to that magazine's discriminating audience while also continuing to write for *The New Republic* and the *Nation*.

Bogan's predilections in her criticism are similar to those in her poems: she showed a marked preference for crafted eloquence over free-verse expansiveness; she directed her readers away from contemporary fashions and toward what she called in the February 1925 issue of *The Measure* "the heft and swing of English poetry in the tradition"; and she would tolerate no slackness in thought or expression. She thought of herself as educating her audiences and shared with them the enthusiasms of her own reading, particularly William Butler Yeats, Rainer Maria Rilke, and W. H. Auden. She was critical when she saw the need to be, regardless of her relationship to the writer, and she lost friends in the process. In 1932 she reviewed her friend Allen Tate so harshly that he wrote to protest. Her reply, in a private letter dated 1 April, defended her objective view: "I was reviewing a book of poetry which aroused in me respect and irritation in about equal measure. If you objected to the tone of my review, I objected, straight down to a core beyond detachment, to the tone of some of the poems."

In 1925 Bogan married Raymond Holden, a sometime poet and novelist who had been a friend of Robert Frost. She retrieved her daughter from her parents and moved with Holden to Boston. Although Holden came from a wealthy family, he was in financial straits by the time he married Bogan. From the start their marriage suffered from economic strain, but for a time the relationship was relatively happy. They moved in a social and literary circle that included Rolfe Humphries and Adams, both of whom would be lifelong friends to Bogan. In 1926 they moved back to New York but spent the winter in Sante Fe, New Mexico, for Holden's health. In 1928 they bought a farmhouse in Hillsdale, New York, and amid the chaos of renovations Bogan found a measure of happiness and new poems. But in December 1929 the house burned to

Bogan with her brother Charles in Boston, 1911

the ground (including almost all of Bogan's books, letters, and manuscripts). While the insurance money enabled the couple to set up a new life in New York, the happiest period in Bogan's life had clearly come to an end.

Dark Summer (1929), her second volume of poetry, marks Bogan's first work with her most helpful editor, John Hall Wheelock of Scribners. In what would become a pattern in Bogan's publishing life, the volume includes a selection of poems from *Body of This Death* as well as new work, which included the only two long poems she ever published: "The Flume," an autobiographical narrative based on the many waterways of her mill-town childhood (which she never again included in a collection); and "Summer Wish," a moving argument between two voices concerning the possibility of spring's renewal and the necessity for acceptance in fall. "Summer Wish" reflects the contemplative happiness of Bogan's stay in the house in Hillsdale and as such was almost anachronistic by the time it saw print.

The shorter lyrics of *Dark Summer* again show Bogan's mastery of observation, diction, meter, and rhyme in poems that generally emphasize acceptance and fulfillment rather than the disappointment and betrayal of her earlier work. "Cassandra" captures the mythical figure's sorrowing mood: "To me, one silly task is like another. / I bare the shambling tricks of lust and pride. / This flesh will never give a child its mother." "Winter Swan" and "The Cupola" show the poet's descriptive powers, reminiscent of those of Moore. Other poems such as "The Crossed Apple" recall the language and tone of Robert Frost:

> This apple's from a tree yet unbeholden,
> Where two kinds meet, –
> . . . .
> Eat it; and you will taste more than the fruit:
> The blossom, too,
> The sun, the air, the darkness at the root,
> The rain, the dew,
>
> The earth we came to, and the time we flee,
> The fire and the breast.
> I claim the white part, maiden, that's for me.
> You take the rest.

The seasons and the passage of time are the subject of several of the strongest lyrics in the book, including "Division," "Girl's Song," "Feuer-Nacht," "Fiend's Weather," and "Come, Break with Time." The lovely first stanza of "Simple Autumnal" illustrates Bogan's preoccupation:

> The measured blood beats out the year's delay.
> The tearless eyes and heart, forbidden grief,
> Watch the burned, restless, but abiding leaf,
> The brighter branches arming the bright day.

In the year following the publication of *Dark Summer*, the marriage between Bogan and Holden began to fail, and Bogan fell ill with severe depression. In the spring of 1931 she checked herself into New York's Neurological Institute in hopes of finding a cure. "I refused to fall apart," she wrote to Wheelock, "so I have been taken apart, like a watch." In the mood of self-reflection following her release from the hospital, she wrote the autobiographical essay "Journey Around My Room," which would become the basis of the "autobiography" edited by her friend Ruth Limmer in 1980.

In 1932 Bogan was awarded a Guggenheim Fellowship "for creative writing abroad" and set sail alone for Italy in April 1933. While she was away in Italy, France, and Austria, struggling to write and often depressed, her marriage fell apart completely. When she returned home several months early to remake her life, the enterprise was not immediately successful. In November 1933 she checked herself into New York Hospital's Westchester Division, this time admitting a "bad nervous crack-up."

She stayed at the hospital for nearly seven months and returned home a good deal healthier than when she had left. She divorced her husband, gathered her good friends – Edmund Wilson and Morton Dauwen Zabel in particular – around, and set about to do her work. Though she was able to write only a few poems, she took up her critical prose with enthusiasm and began to write short stories and autobiographical prose, which she hoped to make into a novel tentatively titled "Laura Dailey's Story." An excellent prose stylist and storyteller, Bogan eventually published thirteen stories in *The New Yorker,* but she did not complete her novel. She gave up writing both fiction and autobiography after 1936.

The years between 1935 and 1941 or so were some of the most fulfilling in Bogan's life, despite financial troubles (she was evicted from her apartment in September 1935 and had to retrieve her possessions from the street). She continued to write critical prose and began to prepare her third book of poems, *The Sleeping Fury* (1937). In June 1935 she began a happy love affair with the young poet Theodore Roethke, a dozen years her junior. She wrote to Wilson of her enthusiasm:

> I, myself, have been made to bloom like a Persian rosebush, by the enormous love-making of a cross between a Brandenburger and a Pomeranian, one Theodore Roethke by name. He is very, very large (6 ft. 2 and weighing 218 lbs) and he writes very very small lyrics. 26 years old and a frightful tank. We have poured rivers of liquor down our throats, these last three days, and, in between, have indulged in such bearish and St. Bernard-ish antics as I have never before experienced . . . Well! Such goings on! A woman of my age!

The affair lasted several months, and the two remained friends. Bogan had much to teach Roethke about lyric poetry, and she quickly assumed that role in his life. Several poems came to Bogan during this relationship, perhaps the last such spell of extended creativity she would experience. The new poems enabled her to publish *The Sleeping Fury,* which was then generally regarded as her strongest volume.

The lyrics of *The Sleeping Fury* reflect the hard-won wisdom of Bogan's psychological recovery as well as her renewed health and vitality. Its reviewers remarked the collection's "sparseness" but praised its integrity. Her friend Zabel noted in the 5 May 1937 *New Republic* the book's freedom from the fashionable ideologies of the day and defended its "old fashioned" values:

> It is because they show so firmly what this depth can yield that these poems bring the finest vitality of the lyric tradition to bear on the confusions that threaten the poets who, by satire or prophecy, indignation or reform, have reacted against that tradition and cast it into contempt.... Her work, instinctive with self-criticism and emotional severity, speaks with one voice only; her rewards and those of her readers have a common source in the discipline to which the clarity of her music and her unsophistic craftsmanship are a testimony. It should be a model for poets in any decade or of any ambition.

The book opens and closes with "songs" and in between contains a handful of Bogan's finest lyrics. The opening of "Roman Fountain" reflects her memories of Italy and, perhaps, some of the sexual vitality of her affair with Roethke:

> Up from the bronze, I saw
> Water without a flaw
> Rush to its rest in air,
> Reach to its rest, and fall.
>
> Bronze of the blackest shade,
> An element man-made,
> Shaping upright the bare
> Clear gouts of water in air.

Several poems offer advice to an imagined reader who has suffered what Bogan has. "Henceforth, from the Mind" counsels acceptance of the diminished emotional intensity of a healthy adult life. "Exhortation" repudiates that resolution through painful irony: "Give over seeking bastard joy / Nor cast for fortune's side-long look. / Indifference can be your toy; / The bitter heart can be your book."

In her title poem, "The Sleeping Fury," Bogan looks mental torment in the face:

> Your hair fallen on your cheek, no longer in the semblance of serpents,
> Lifted in the gale; your mouth, that shrieked so, silent.
> You, my scourge, my sister, lie asleep, like a child,
> Who, after rage, for an hour quiet, sleeps out its tears.
> ....
> And now I may look upon you,
> Having once met your eyes. You lie in sleep and forget me.
> Alone and strong in my peace, I look upon you in yours.

"Kept" shows a mature denial of sentimentalized memories of childhood and youth:

> Time for the wood, the clay,
> The trumpery dolls, the toys
> Now to be put away:
> We are not girls and boys.
> ....
> Time for the pretty clay,
> Time for the straw, the wood.
> The playthings of the young
> Get broken in the play,
> Get broken, as they should.

Bogan also looks hard at alcohol, the friend and sometime nemesis of her life. In "To Wine" she ironically exhorts the "Cup, ignorant and cruel," to

> Take from the mind its loss:
> The lipless dead that lie
> Face upward in the earth,
> Strong hand and slender thigh;
> Return to the vein
> All that is worth
> Grief. Give that beat again.

In 1937 Bogan applied for and was granted the remainder of her Guggenheim Fellowship, which she had been unable to complete in 1932. In April she sailed for Ireland. From the start the trip was a struggle for Bogan, who was frequently depressed and anxious. In the country of her ancestors she was unable for find a place for herself, and she sailed home several weeks later in a state of near collapse. A man on the boat train to Southampton came to her rescue and cared for her throughout the voyage home. After a week of recovery Bogan, who had remarked in her notebook three years before that "There can be no new love at 37, in a woman," began a relationship with the man, an electrician from the Bronx, that would last eight years. She kept it largely a secret from her friends, but by her own account the relationship was as happy and fulfilling as any she would ever have.

In the spring of 1938 Bogan moved into the apartment on West 169th Street where she would live for the rest of her life, engaging in a lively and energetic career as a literary critic and a woman of letters, squabbling with her fellow poets and critics, and publishing many incisive and insightful reviews and essays. She championed the cause of W. H. Auden as he arrived in the United States in 1939 and did a great deal in bringing public attention to his work. But poems came to her only occasionally. She struggled to produce enough work for a new book, and *Poems and New Poems* included old work as well as what Bogan called her "light verse," clever occasional poems on contemporary topics. In this category is the memorable couplet titled "Solitary

Bogan at the home of John Dos Passos in Provincetown, Massachusetts, in 1936

Observation Brought Back from a Sojourn in Hell": "At midnight tears / Run into your ears."

The group of new poems in the volume opens with "Several Voices Out of a Cloud," a sharp attack on the ideological hacks Bogan saw dominating the world of poetry. The poem is uncharacteristically contemporary, in a manner reminiscent of Auden: "Come, drunks and drug-takers; come, perverts unnerved!" she invites and concludes by naming the pretending poets:

> Parochial punks, trimmers, nice people, joiners
>   true-blue,
> Get the hell out of the way of the laurel. It is deathless
>   And it isn't for you.

Other new poems have a variety of subjects: "Animal, Vegetable and Mineral" is a contemplation of the glass flowers exhibit at Harvard's Museum of Natural History; "To Be Sung on the Water" is a tender and playful love lyric; and "Zone" captures with icy accuracy the disquieting ambiguity of New England in the month of March: "Now we hear / What we heard last year, / And bear the wind's rude touch / And its ugly sound / Equally with so much / We have learned how to bear."

Bogan wrote no poems between the publication of *Poems and New Poems* in 1941 and 1948. The horror of World War II discouraged her about the power of poetry against such hatred, and she was troubled by what she saw as the obscurity of her own position. In hopes of finding a publisher that would promote her work more forcefully, Bogan left Scribners and Wheelock, with unfortunate results. She would not find a new publisher until 1954, and the sense of being on her own made writing poems more difficult. She once remarked to Wheelock that "A woman writes poetry with her ovaries." As she entered middle age Bogan began to feel that her time had past.

Bogan's essays and reviews did much to keep her name before the public in the war years and their wake. She began new, lasting friendships among young admirers of her work, including William Maxwell, a novelist and *New Yorker* staffer when he met Bogan in 1938; May Sarton, an established poet whom Bogan invited to her apartment in 1953; and Ruth Limmer, an English professor whom Bogan met in 1956 when she received an honorary doctorate from the Western College for Women (Limmer would become her literary executor). With the winding down of the war, literature could once again command attention, and Bogan began to be asked to serve on various poetry-prize juries. In 1944 she gave the Hopwood lecture at the University of Michigan. She also began to read her poems in public and accepted the position as Consultant in Poetry to the Library of Congress (1945–1946). She was in the Library Fellows group that, amid controversy, awarded the first Bollingen Prize to Ezra Pound, then incarcerated in a mental institution in Washington, D.C., having been judged incompetent to stand trial for treason at the end of the war. She accepted a teaching position at the University of Washington and went on to teach at the University of Chicago and New York University, among other places. Bogan also began her work as a translator, working with Elizabeth Mayer on works by Johann Wolfgang von Goethe and Ernst Juenger and with Elizabeth Roget on works by Jules Renard. In 1951 she published her critical study, *Achievement in American Poetry, 1900–1950,* which included an anthology with her selection of worthy poems from the period.

Bogan found a publisher in the new Noonday Press in the early 1950s and set about preparing her

Bogan at work, 1937

Collected Poems 1923–1953 (1954). She had only three new poems to add to the whole: "After the Persian," her contemplation of Persian art at New York's Metropolitan Museum; the light poem "Train Tune"; and, most remarkably, "Song for the Last Act." This poem is built around a refrain, varied to marvelous effect in each of its three stanzas: "Now that I have your face by heart, I look"; "Now that I have your voice by heart, I read"; "Now that I have your heart by heart, I see." The poem links desire and memory in a tone unmistakably valedictory, "O not departure, but a voyage done!" A year later Bogan published a volume of *Selected Criticism: Prose, Poetry* (1955). Both volumes were respectfully reviewed, and she shared the 1955 Bollingen Prize with Léonie Adams.

Bogan's last years were a combination of honors, continued hard work, dark depression, and alcoholism. Between 1957 and 1964 she went annually to the MacDowell Colony in Peterborough, New Hampshire, where she found the time and peace to write poems as well as critical prose. Her collection *The Blue Estuaries: Poems 1923–1968* (1968) includes a dozen new poems, most of which had been begun much earlier. Most of her last poems are in free verse, as Bogan grew more willing to accept poems in the forms in which they came to her. Notable among these last works is "The Dragonfly," written on commission from the Corning Glass Company, which had a Steuben glass dragonfly carved to illustrate it. The poem "Night" provided the title for the collection. It recalls in both setting and sound poems by Elizabeth Bishop, Bogan's somewhat younger contemporary:

> The cold remote islands
> And the blue estuaries
> Where what breathes, breathes
> The restless wind of the inlets,
> And what drinks, drinks
> The incoming tide;
> ....
> — O remember
> In your narrowing dark hours
> That more things move
> Than blood in the heart.

The Blue Estuaries received strong reviews. William Meredith in the 13 October 1968 *New York Times Book Review* called Bogan "one of the best woman poets alive" and wondered at how her "reputation has lagged behind a career of stubborn, individual excellence." Hayden Carruth in the August 1969 issue of *Poetry* praised the poems despite their small number: "this book's best pages make it fundamentally irreducible." That assessment seems accurate; perhaps the forces or frailties that prevented Bogan from writing more afforded her marvelous control over her art.

Louise Bogan died at her apartment of a coronary occlusion on 4 February 1970. A memorial service, arranged by her friend William Jay Smith, was held at the Academy of Arts and Letters on 11 March, attended by 120 of her friends and admirers. At the service W. H. Auden noted Bogan's personal strength. "Aside from their technical excellence," he said of her poems, "what is most impressive . . . is the unflinching courage with which she faced her problems, and her determination never to surrender to self-pity, but to wrest beauty and joy out of dark places."

Letters:

What the Woman Lived: Selected Letters of Louise Bogan, 1920-1970, edited by Ruth Limmer (New York: Harcourt Brace Jovanavich, 1973).

Bibliographies:

William Jay Smith, *Louise Bogan: A Woman's Words* (Washington, D.C.: Library of Congress, 1971);

Jane Couchman, *Louise Bogan: A Bibliography of Primary and Secondary Materials, 1915-1975,* part 1, *Bulletin of Bibliography,* 33 (February-March 1976): 73-77; part 2, *Bulletin of Bibliography,* 33 (April-June 1976): 111-126, 147;

Claire E. Knox, *Louise Bogan: A Reference Source* (Metuchen, N.J.: Scarecrow, 1990);

Robert A. Wilson, "Louise Bogan: A Bibliographical Checklist," *American Book Collector,* 7 (September 1986): 31-36.

Biography:

Elizabeth Frank, *Louise Bogan: A Portrait* (New York: Knopf, 1985).

References:

Gloria Bowles, *Louise Bogan's Aesthetic of Limitation* (Bloomington: Indiana University Press, 1987);

Martha Collins, ed., *Critical Essays on Louise Bogan* (Boston: G. K. Hall, 1984);

Mary DeShazer, "My Scourge, My Sister: Louise Bogan's Muse," in *Coming to Light: American Women Poets in the Twentieth Century,* edited by Diane Wood Middlebrook and Marilyn Yalom (Ann Arbor: University of Michigan Press, 1985), pp. 92-104;

Elizabeth C. Dodd, *The Veiled Mirror and the Woman Poet: H.D., Louise Bogan, Elizabeth Bishop, and Louise Glück* (Columbia: University of Missouri Press, 1992);

Jeanne Larsen, "Lowell, Teasdale, Wylie, Millay, and Bogan," in *The Columbia History of American Poetry,* edited by Jay Parini and Brett Millier (New York: Columbia University Press, 1993), pp. 203-232;

Michael Paul Novak, "Love and Influence: Louise Bogan, Rolfe Humphries, and Theodore Roethke," *Kenyon Review,* 7 (Summer 1985): 9-20;

Douglas Peterson, "The Poetry of Louise Bogan," *Southern Review,* 19 (Winter 1983): 73-87;

Jacqueline Ridgeway, *Louise Bogan* (Boston: Twayne, 1984).

Papers:

The major collection of Louise Bogan's papers is at the Amherst College Library. The Library of Congress also has significant Bogan holdings.

Charles Bukowski

(16 August 1920 – 9 March 1994)

Michael Basinski
State University of New York at Buffalo

See also the Bukowski entry in *DLB 5: American Poets Since World War II, First Series* and *DLB 130: American Short-Story Writers Since World War II.*

SELECTED BOOKS: *Flower, Fist and Bestial Wail* (Eureka, Cal.: Hearse Press, 1960);

Longshot Pomes for Broke Players (New York: 7 Poets Press, 1962);

Run with the Hunted (Chicago: Midwest Poetry Chapbooks, 1962);

It Catches My Heart in Its Hands (New Orleans: Loujon Press, 1963);

Grip the Walls (Storrs, Conn.: Wormwood Review Press, 1964);

Crucifix in a Death-Hand (New York: Loujon Press/Lyle Stuart, 1965);

Cold Dogs in the Courtyard (Chicago: Literary Times/Cyfoeth Publications, 1965);

Confessions of a Man Insane Enough to Live with Beasts (Bensenville, Ill.: Mimeo Press, 1965);

The Genius of the Crowd (Cleveland: 7 Flowers Press, 1966);

All the Assholes in the World and Mine (Bensenville, Ill.: Open Skull Press, 1966);

Night's Work (Storrs, Conn.: Wormwood Review Press, 1966);

2 Poems (Los Angeles: Black Sparrow Press, 1967);

The Curtains Are Waving and People Walk through the Afternoon Here and in Berlin and in New York City and in Mexico (Los Angeles: Black Sparrow Press, 1967);

Poems Written Before Jumping Out of an 8 Story Window (Glendale, Cal.: Poetry X/Change, 1968; Gersthofen, Germany: Maro, 1974; enlarged edition, Salt Lake City: Litmus, 1975);

At Terror Street and Agony Way (Los Angeles: Black Sparrow Press, 1968);

A Bukowski Sampler (Madison, Wis.: Quixote Press, 1969);

Notes of a Dirty Old Man (North Hollywood, Cal.: Essex House, 1969);

Charles Bukowski (photograph by Michael Montfort)

If We Take (Los Angeles: Black Sparrow Press, 1969);

The Days Run Away Like Wild Horses Over the Hills (Los Angeles: Black Sparrow Press, 1969);

Another Academy (Los Angeles: Black Sparrow Press, 1969);

Fire Station (Santa Barbara, Cal.: Capricorn Press, 1970);

Post Office: A Novel (Los Angeles: Black Sparrow Press, 1971);

Mockingbird Wish Me Luck (Los Angeles: Black Sparrow Press, 1972);

Erections, Ejaculations, Exhibitions and Tales of Ordinary Madness (San Francisco: City Lights Books, 1972); abridged and republished as *Life and Death in the Charity Ward* (London: London

Magazine Editions, 1974); reissued as two books: *The Most Beautiful Woman in the World* (San Francisco: City Lights Books, 1983) and *Tales of Ordinary Madness* (San Francisco: City Lights Books, 1983);

While the Music Played (Los Angeles: Black Sparrow Press, 1973);

South of No North (Los Angeles: Black Sparrow Press, 1973);

Burning in Water, Drowning in Flame (Los Angeles: Black Sparrow Press, 1974);

Africa, Paris, Greece (Los Angeles: Black Sparrow Press, 1974);

Factotum (Los Angeles: Black Sparrow Press, 1975);

Scarlet (Santa Barbara, Cal.: Black Sparrow Press, 1976);

Tough Company (Santa Barbara, Cal.: Black Sparrow Press, 1976);

Maybe Tomorrow (Santa Barbara, Cal.: Black Sparrow Press, 1977);

Art (Santa Barbara, Cal.: Black Sparrow Press, 1977);

Love Is a Dog from Hell: Poems 1974–1977 (Santa Barbara, Cal.: Black Sparrow Press, 1977);

Women (Santa Barbara, Cal.: Black Sparrow Press, 1978);

Play the Piano Drunk Like a Percussion Instrument Until the Fingers Begin to Bleed a Bit (Santa Barbara, Cal.: Black Sparrow Press, 1979);

Shakespeare Never Did This (San Francisco: City Lights Books, 1979);

Dangling in the Tournefortia (Santa Barbara, Cal.: Black Sparrow Press, 1981);

Ham on Rye (Santa Barbara, Cal.: Black Sparrow Press, 1982);

Bring Me Your Love (Santa Barbara, Cal.: Black Sparrow Press, 1983);

Hot Water Music (Santa Barbara, Cal.: Black Sparrow Press, 1983);

There's No Business (Santa Barbara, Cal.: Black Sparrow Press, 1984);

War All the Time: Poems 1981–1984 (Santa Barbara, Cal.: Black Sparrow Press, 1984);

You Get So Alone at Times That It Just Makes Sense (Santa Rosa, Cal.: Black Sparrow Press, 1986);

The Movie, "Barfly": an original screenplay by Charles Bukowski for a film by Barbet Schroeder (Santa Rosa, Cal.: Black Sparrow Press, 1987);

The Roominghouse Madrigals: Early Selected Poems 1946–1966 (Santa Rosa, Cal.: Black Sparrow Press, 1988);

Hollywood: A Novel (Santa Rosa, Cal.: Black Sparrow Press, 1989);

Septuagenarian Stew: Stories & Poems (Santa Rosa, Cal.: Black Sparrow Press, 1990);

In the Shadow of the Rose (Santa Rosa, Cal.: Black Sparrow Press, 1991);

The Last Night of the Earth Poems (Santa Rosa, Cal.: Black Sparrow Press, 1992);

Run With the Hunted: A Charles Bukowski Reader, edited by John Martin (New York: HarperCollins, 1993);

Pulp (Santa Rosa, Cal.: Black Sparrow Press, 1993);

Confession of a Coward (Santa Rosa, Cal.: Black Sparrow Press, 1995);

Heat Wave (Santa Rosa, Cal.: Black Sparrow Graphic Arts, 1995);

Betting on the Muse: Poems and Stories (Santa Rosa, Cal.: Black Sparrow Press, 1996).

Charles Bukowski is recognized as an international literary figure. His first poem, "Hello!," celebrating the black driver of a horse-drawn trash wagon, was published in a mimeograph magazine called *Matrix* in 1946, when Bukowski was twenty-six years old. He published little poetry or prose in the 1940s and did not publish with any frequency until after he turned thirty-five. When he began to write and publish in the late 1950s, it was with a vengeance and a fury. From 1960, when he published his first book of poetry, *Flower, Fist and Bestial Wail,* until 1994, when he died of leukemia, his name appeared frequently in little literary magazines and newspapers around the world. In his long career he published more than sixty books of poetry and prose, including both short-story collections and novels. Many of these books have been translated into more than a dozen languages. Bukowski also wrote and published a screenplay, *Barfly* (1987), which became a major motion picture starring Faye Dunaway and Mickey Rourke.

Bukowski's rise to fame was not meteoric; his wide-ranging popularity was gained slowly and by great effort. A self-motivated poet from the blue-collar class and a literary eccentric, Bukowski was the product of the small press, little literary magazine, and underground alternative journal. This publishing world – outrageous, raucous, volatile, and generally unstable – was one that Bukowski fit perfectly, and as the independent publishing ventures of the 1960s gained notoriety, Bukowski's reputation also grew. To the end of his life Bukowski continued to publish poetry in little literary magazines. His poetry motivated scores of editors and poets to write and publish independently and with harsh, sometimes crude vigor. His obstinate stance against convention and the conservative in his own

Bukowski with his parents, 1947

life, society, and the world of literature endeared him to his many fans. Until his death he remained active and prolific, a dominant literary figure in the little magazine world and in underground literary circles; he also became one of the most imitated of American poets. Black Sparrow Press, one of the most successful and respected independent publishers, published his books for more than twenty-five years. All Bukowski's major Black Sparrow editions passed through numerous printings and remain in print. Although his death ended a prolific career, collections of his prose and poetry, as well as his letters, reviews, and essays, will continue to appear for many years. A biography by Neeli Cherkovski, *Hank: The Life of Charles Bukowski,* was published in 1991.

Bukowski's methodical rise to international literary prominence was filled with endless obstacles. He was forty years old in 1960 when he published his first slim volume of poetry, *Flower, Fist and Bestial Wail,* a tiny chapbook containing only fourteen poems. E. V. Griffith published the small book under his Hearse imprint. The publishing project took nearly two years to complete and began more than a decade after Bukowski's poetry started appearing in magazines. During the mid 1940s (he did not participate in World War II) through the mid 1950s, Bukowski lived the life of a wandering hobo and skid-row alcoholic. An exile from the mainstream of American culture, he survived on the lowest rungs of society, moving from city to city – New York, Philadelphia, Atlanta – and working as a day laborer at menial tasks that provided him with enough money to buy alcohol (a lifelong obsession) and pay for his small rooms in skid-row hotels and rooming houses. Eventually his wandering led him back to Los Angeles. After a decade of heavy drinking he was hospitalized with a severe bleeding ulcer that threatened his life. Making a remarkable recovery, he acquired a job with the U.S. Postal Service as a mail carrier, and, after nearly a decade of almost complete literary silence, began to write poetry again. He continued to write poetry without pause for nearly the next forty years.

Bukowski was only loosely connected to the poetic movements of the late 1950s and early 1960s.

Neither a part of the Beat generation nor a Black Mountain poet, he was also not a confessional poet or an academic author. His poetry, nevertheless, secured him in these years a loyal, albeit small, audience. In the early 1960s he published a series of chapbooks in quick succession, among them *Longshot Pomes for Broke Players* (1962), *Run with the Hunted* (1962), and *It Catches My Heart in Its Hands* (1963). As a leading figure in the so-called mimeo revolution (magazines and books printed cheaply and quickly with a mimeograph machine), he wrote poetry that appeared in many of the important — and infamous — underground magazines of the period, such as *The Wormwood Review, Olé, Mica,* and *Renaissance*. During this time he became friends with John Byrum, a young radical publisher who edited the Los Angeles–based newspaper *Open City*. Byrum convinced Bukowski to write a weekly prose column for his paper; the column, appropriately titled "Notes of a Dirty Old Man," eventually led to the publication of a book of short fiction by the same name in 1969. Although Bukowski's column brought him wide recognition, it was two publishers, John Edgar Webb of Loujon Press and John Martin of Black Sparrow Press, who were principally responsible for publishing and, more important, distributing Bukowski's poetry to an ever-expanding and seemingly insatiable audience. This loyal readership remains the envy of almost all other American poets.

Henry Charles Bukowski Jr. was born in Andernach, Germany, on 16 August 1920. The only child of an American soldier, Henry Charles Bukowski Sr., and a German mother, Katherine Fett Bukowski, he immigrated with his family to Los Angeles in 1922. The young Bukowski (who dropped the "Henry" when he began to publish, because he felt it was not a literary name) grew up in the middle-income, working-class neighborhoods of Los Angeles. Bukowski later recalled that one of his earliest memories was being ridiculed for his German birth, an experience that was only the beginning of the relentless ridicule that the young Bukowski endured. His mother was a docile and subservient housewife, and his father was a milkman. This almost typical American family was ruled by the strict, harsh, disdainful senior Bukowski, a cruel master who imprisoned his family in a regimented set of unbreakable and often unbearable rules designed to maintain the facade of middle-class respectability. In his poem "My Non-ambitious Ambition," in *You Get So Alone at Times That It Just Makes Sense* (1986), Bukowski writes: "it seemed to me that I had never met / another person on earth / as discouraging to my happiness / as my father." The elder Bukowski imposed his ideals with the strap, which he used liberally on his son. The young Charles hated his father and what his father represented — the economic and social success that came from strict adherence to the American Dream, the pursuit of which translated into hard work and patriotism that were supposed to reward conformity.

In the poem "My Old Man" (1977) Bukowski refers to some of his father's actions:

16 years old
during the depression
I'd come home drunk
and all my clothing –
shorts, shirts, stockings –
suitcase, and pages of
short stories
would be thrown out on the
front lawn and about the
street.

my mother would be
waiting behind a tree:
"Henry, Henry, don't
go in . . . he'll
kill you . . .[.]"

Disdain for conformity and any hypocrisy, particularly that of middle-class America, became major themes in Bukowski's poetry and appeared from his early poems through those he wrote in the 1990s. Again in the poem "My Non-ambitious Ambition" Bukowski attributes his adult lifestyle to his reaction against his father's ideals: "and I thought, if being a bum is to be the / opposite of what this son-of-a-bitch / is, then that's what I'm going to / be." Bukowski's cynical, biting, and satiric poetic themes can also be traced to his unhappy childhood, well documented in his prose, particularly in the novel *Ham on Rye* (1982). The specific, day-to-day, miserable occurrences Bukowski endured as a child and teenager are not recorded in his poetry so much as the attitudes and philosophical perspectives that he formed as psychological defenses against his abusive home life.

The Depression of the 1930s, which both strengthened the senior Bukowski's resolves to attain middle-class respectability and made those standards more ironic and repellent to his son, began as the younger Bukowski entered his teenage years and high school. In the poem "John Dillinger Marches On" (1984) Bukowski wrote about the Depression years as a "good training ground. / people learned to live with adversity as a common everyday thing." Economic hard times, however, fueled the abusive nature of Bukowski's father. To add to the misery, Bukowski, then a teenager, was victimized by a severe case of acne that forced him to miss half his sophomore year and endure endless hours of painful treatment. His doctors noted that his was the worst case of acne vulgaris that they had ever treated. Bukowski was left with extensive facial scars that matched the psychological scars already in place. The rage and pain of these years were released in the poetry he wrote in the 1950s. Because of his skin condition and his heinous home life, Bukowski became a forlorn teenager who took delight in his toughness and ability to withstand pain.

Regardless of his public literary persona, Bukowski was always an introspective person who found solace and companionship in reading. Early favorite books were Upton Sinclair's *The Jungle* (1906) and Sinclair Lewis's *Main Street* (1920); he also admired the early prose of Ernest Hemingway. Among Bukowski's other early influences were D. H. Lawrence, James Thurber, Theodore Dreiser, John Dos Passos, and Sherwood Anderson. He later added John Fante, Franz Kafka, and Fyodor Dostoyevsky to his list of favorites. Late in his career Bukowski wrote poems of homage and admiration to many of his early literary heroes: for example, to Lawrence, "I Liked Him" (1979), "One For Sherwood Anderson" (1981), and, to E. E. Cummings, "What a Writer" (1992). The poets and the writers of the 1930s always held a romantic allure for Bukowski, and his own poems refer to such figures as Gertrude Stein, Ezra Pound, and H. D. (Hilda Doolittle). This entire literary period is praised in his poem "The Last Generation" (1984). Recalling when and where he first encountered many books and authors, he wrote in "Days like Razors, Nights Full of Rats" (1992) that "as a very young man I divided an equal amount of time between / the bars and the libraries." And recalling his youthful reading habits, he wrote, after the Los Angeles Public Library burned down, a poem titled "The Burning of the Dream" (1990), in which he observed that the shelves of that library had "tremendous grace."

In tone and philosophy Bukowski's early poetry resembles that of Robinson Jeffers. Jeffers admired the swift, violent strength of hawks and the enduring strength of rocks. Bukowski, too, admired toughness and particularly the will to endure physical as well as emotional pain. Like Jeffers, Bukowski is identified with violent and base human emotions, and his poetry often features violent confrontations between men and women. The tenacity and strength of will that Bukowski welcomes allowed him to endure his many battles with women and scourging literary critics. Bukowski also took supreme pleasure in basic, primal human drives and compulsions such as defecation, intoxication, copulation, and creation. Bukowski's own early book titles, for example *Run with the Hunted* and *Crucifix in a Death-Hand* (1965), resemble Jeffers's titles, such as *Be Angry at the Sun* (1941) and *Give Your Heart to the Hawks* (1941). One of Bukowski's first poems of literary homage, "He Wrote in Lonely Blood" (1972), honored Jeffers and mentioned *Be Angry at the Sun*. Other poems salute Jeffers, among them a poem simply titled "Jeffers" (1992), in which Bukowski wrote, "his gigantic crushed earth / bellows / against dumb / time." Bukowski's first major collection of poetry, *It Catches My Heart in Its Hands,* draws its title from Jeffers's poem "The Hellenistics" (1937). Bukowski's own proselike line structure also shows the influence of Jeffers; however, Bukowski's poetry was never as bleak and pessimistic as Jeffers's was. Bukowski believed too much in luck and was far too idealistic and romantic to be a Jeffers clone. Bukowski modified Jeffers's poetic line to include his own personal cadence and inflection. This spoken or speaking poetic voice is the device that links Bukowski with the postwar experimental, or progressive, poetry that claims by way of William Carlos Williams to engage the cadences of spoken language. Rather than capturing cadence out of the mouths of working-class Polish mothers, as Williams suggested, Bukowski relied upon his own, personal voice and ear (many of the poems are composed in part or totally of dialogue) to guide the reader into, through, and out of the poem. Bukowski, in a letter in 1962, called this "bar-talk." The ease with which he is able to lead the reader through the poem marks the simple, effortless beauty of his hard-edged poetic narratives.

Bukowski was also influenced by Hemingway's quick, clear, unadorned reportage. Mentioned more than any other writer in Bukowski's poetry, Hemingway had a special place in Bukowski's list of literary influences. Bukowski especially appreciated and emulated Hemingway's early prose, admiring its clarity, precision, crispness, and particularly its (sentence) line form. Like Hemingway, Bukowski admired simplicity in art; like Hemingway, he, too, was fascinated by bullfights, matadors, and boxing. Beyond these artistic attractions, Bukowski was interested in Hemingway's popular image, his exaggerated masculinity, and his suicide, which Bukowski regarded as an

Bukowski in his Los Angeles apartment, 1969 (photograph by Samuel Cherry)

honorable and pugnacious way to confront the inevitability of defeat and death.

Bukowski later added Albert Camus's *The Stranger* (1942) and the novels of Louis-Ferdinand Céline to the list of his important influences. In his poem "The Word," in *The Last Night of the Earth Poems* (1992), Bukowski called Céline's *Journey to the End of the Night* (1932) the greatest book ever written. From his literary heroes, including Hemingway, Jeffers, and others, Bukowski reveals that his literary influences were formed by classic American and world literature. All his influences predated the postmodern poetics and literature of the 1950s and 1960s. His own ideas about literature were formed in the 1930s. Nearly fifty years old before he gained wide recognition, Bukowski was, therefore, not attuned to the poetic movements of his literary peers. This position he did not wish to amend; in fact, it gave him a degree of pleasure. His unbending ideology (social and literary), formed by years of survival on skid row, dictated that he would be a rebel poet. His revolutionary, anarchistic, and obstinate attitude fit the pop and youth-culture philosophy of the 1960s, when he began to write wildly and publish widely. Although critics have not been able to fit him into any of the literary or poetic schools associated with the era, Bukowski exemplified, more than Allen Ginsberg, the anarchistic, anti-middle-class attitude of the decade.

The first third of Bukowski's life was filled with brutality and taught him the cruel lessons of survival in a hostile world. It is, therefore, not surprising that, when he began to write poetry, his metaphors were sarcastic and violent. Also, when he addressed his fellow poets, those, so to speak, of his generation, or at least those publishing alongside him in magazines, it is not surprising that he bitterly resented them, usually pointing out their effete natures and their confused notions about poetry. Many of his poems comment on his peers and on the writing scene. In "An Unkind Poem" (*Love Is a Dog from Hell*, 1977), Bukowski writes:

> they go on writing
> pumping out poems –
> young boys and college professors
> wives who drink wine all afternoon
> they go on writing
> the same names in the same magazines
> everybody writing a little worse each year[.]

Later in the same poem he writes:

> most of them live in North Beach or New York City,
> and their faces are like their poems:

alike,
and they know each other and
gather and hate and admire and choose and discard
and keep pumping out more poems . . .
tap tap tap, tap tap, tap tap tap, tap tap . . . [.]

Early in his publishing career Bukowski wrote several reviews and statements in which he presented an alternative to rigid poetics and what he considered the boring world of current poetry. In 1960, writing in *Nomad,* a progressive literary magazine, Bukowski, in a sarcastically academic literary style, wrote "Manifesto: A Call for Our Own Critics." Here he noted that "The insurgency of criticism from a nosography on poetics to a censorious dictum by certain university groups who write the laws of poetry, and spawn, with sumptuous grace and style, their own puppeteers – these, and their half-brethen and their purlieu, form a most deadly and snobbish poetic fixation. They create, record, and argue their own history, charmed with the largesse of their chosen circumference." In direct opposition to "them" in a review of Louis Zukofsky's "A Test of Poetry," published in *Olé* in 1967, Bukowski wrote about the type of poetry he imagined: "We will no longer accept dry and safe bread. Poetry is going into the streets, into the whorehouses, into the sky, into the picnic basket, into the whiskey bottle. the fraud is over – certain people will not be allowed to live while others die. at least not from this typewriter." Following his own lead, Bukowski, as he said in an interview in 1974 for *Poetry Now,* set out to "loosen and to simplify poetry, "make it more humane," make "it easier for others to follow," and write poetry that would "even be entertaining."

Bukowski published only four poems in the 1940s, and those were in *Matrix,* a Philadelphia/New York City–based magazine that also published his short stories. The four poems were "Hello!," "Voice in the Subway," "Soft and Fat Like Summer Roses," and "Object Lesson." These poems foreshadow the tone and temperament of Bukowski's poetic fury, and the foreshadowing is particularly evident in "Object Lesson." Written in short, variable stanzas, it addresses the reader directly, and the philosophical narrator is concerned with the nature of existence. The question the poem asks is how to kill or destroy (metaphorically perhaps) another person. Various ways to achieve this end are suggested, including stuffing an old bathing cap down the throat, or blinding and strangling a victim. But "Not so. Not so," writes the cynical sage (Bukowski). "Tell him to seek the stars and he will kill himself with climbing." In this poem Bukowski reveals his attitude toward success in America. He finds that obtaining success can blind a person or compel a person to sell his soul or compromise, thereby betraying the primal self, the self that celebrates its independence in an unstructured, uncomplicated lifestyle. Contrary to and contrasting with this poem is "Hello!," in which Bukowski imagines that a trash wagon driven early in the morning by a black driver is actually a chariot, and that the blood in the horse and in the driver are vibrating and full of life and sound. With his vision of this working man, whom he does not know, he feels an intense unity and bond. His imagery speaks of wombs, stars, roses, wounds, love, and of a thousand poems to the poet Federico García Lorca dying in the rain. The poem concludes with a salute to the trash wagon driver: "the top of the morning to you!"

Even before his ten-year self-exile into the underworld of society, Bukowski chose the way of the meek and downtrodden to find his literary revelry. He would celebrate the common man. He would write about the day-to-day, usually painful, sometimes pleasurable lives of outsiders – those outside privilege and power. He would celebrate himself. In Bukowski's poetic world there is always a balance between life and death and pleasure and pain, the yin-yang balance of existence itself, where the various parts make up a whole. In Bukowski's view, too much wealth, too much greed, too much success, and too much of the excess offers only blindness, a lifeless soul, conformity. Human beings lose their originality and become obvious, dull, the living dead. Primitive emotions, originality, and spontaneity bring life into true perspective. Without these there is only death, despair, and destruction. To correct and balance the artistic world of poetry, which was drawing the newer poetry into the dull, dim, proper sphere of middle-class conformity and stale academia, Bukowski wrote in a prose piece "Portions from a Wine-Stained Notebook (1960), that "what we need is a single hero to uphold the defeated, A Quijote of the windmills, here and now." Bukowski imagined himself that hero, and to that particular image he remained virtually faithful throughout his personal and professional life.

In the late 1950s E. V. Griffith, a rebellious California small-press editor and publisher with liberal poetic taste, agreed to publish a small chapbook of Bukowski's poetry, *Flower, Fist and Bestial Wail.* Griffith's Hearse Press and magazine specialized in toughly worded narrative poems. There was nothing polite in what he published, which included poetry by writers such as Judson Crews and Bukowski. Continuing to affront academic, mundane, and artistic poetry, Griffith later edited magazines such as *Gallows* and *Poetry Now*, both of which featured Bukowski and aggressive, often unsettling poetry. *Flower, Fist and Bestial Wail* was published in an edition of two hundred in November 1960. The poems in this volume were written during the period 1957–1960 and are representative of the poems Bukowski wrote between 1955 and 1965. The collection begins with a poem titled "10 Lions and the End of the World." Here Bukowski writes about a pride of lions stationing themselves in a village, blocking traffic on the main road. The poem suggests that one should do the same as the lions do before one's life runs out (or before Bukowski's life does). So begins Bukowski's poetic fixation with death and his endless, relentless striving for spiritual and physical and unbridled revelry, which in Bukowski's view is the essence of life. In this poem Bukowski, in his no-nonsense, talk/telling fashion, plops himself into the midst of American poetry. He also contrasts life and death in the poem "All-Yellow Flowers," in which a casket is covered by all-yellow flowers. The casket and the corpse within it are an image of his own body, in which his soul is confined, a soul that wishes with deep desire to be free of its human flesh. In a prose poem in the collection, "His Wife, The Painter," is the sentence "I can paint – a flower eaten by a snake," which sums up a poetic stance that guided Bukowski through the hundreds of poems he wrote after *Flower, Fist and Bestial Wail*. The sentiment is echoed in a short prologue to *It Catches My Heart in Its Hands,* where Bukowski writes that "small birds who go / the way of cat sing / on inside my head." These short passages reveal the life-and-death balance and the contrast between the beautiful and the ugly that are so prevalent in Bukowski's poetry.

During the 1960s Bukowski published ten books of poetry. He also dealt with ten different publishers. Each of these collections of poetry is distinct, but, taken together, they present poems written while Bukowski was still a postal employee (a position he left in January 1970 to become a full-time author). It was also the period before he wrote his first novel, *Post Office* (1971), and during which he was writing his column, "Notes from a Dirty Old Man." Among his smaller collections were *Cold Dogs in the Courtyard* (1965), *Poems Written Before Jumping Out of an 8 Story Window* (1968), and *A Bukowski Sampler* (1969). To satisfy the demands of editors for books of poetry, many poems were collected and recollected throughout these early books. *It Catches My Heart in Its Hands, At Terror Street and Agony Way* (1968), as well as *The Days Run Away Like Wild Horses Over the Hills* (1969), are all collections that draw from Bukowski's chapbooks. With each collection Bukowski kept in print the poems he considered his strongest old work as well as his best new work. This tradition was maintained by Black Sparrow when the press issued *The Roominghouse Madrigals: Early Selected Poems 1946–1966* (1988). Like the other books, this collection brings together previously uncollected poems as well as poems from Bukowski's early, out-of-print chapbooks.

Publishing as many books with as many publishers as Bukowski did in the 1960s manifests a facet of his personality not revealed by his literary persona. The recently published *Screams from the Balcony: Selected Letters 1960–1970* (1993) shows that Bukowski was a prolific letter writer and worked diligently at his career as a poet. His poems were constantly circulating, and he maintained close, friendly, and frequent contact with his editors.

His volume *It Catches My Heart in Its Hands* is a good example of an early book of poetry. It was published by John Edgar Webb and Louise Webb after Bukowski had become friends with Webb, a former convict who was an exceptionally fine printer with a passion for obstinate poetry. Bukowski's association with Webb solidified and tremendously aided his literary career. Not only did Webb edit the beautifully printed journal *The Outsider,* which featured the most innovative writing of the 1960s, from Charles Olson to Ginsberg and Gregory Corso to Kenneth Patchen and Bukowski, but Webb himself had a taste for Bukowski's maverick personality. Webb printed and distributed Bukowski's *It Catches My Heart in Its Hands,* complete with John William Corrington's introduction, "Charles Bukowski at Midflight," in which Corrington notes that Bukowski's poetry is "free of literary pretense." Two years later the Webbs published Bukowski's *Crucifix in a Death-Hand*. *It Catches My Heart in Its Hands* opens with a poem, "The Tragedy of the Leaves," that reveals much about Bukowski's perspective in these years. The tragedy of the leaves is that they live and they die. Bukowski equates this birth-and-death cycle with his own life, remembering when he was young and supposed to be an author with genius. But he aged. The romance of life had cheated him. In the poem he meets his raving, aged landlady, a hag who wants her rent, in the hall of his rooming house. The world and life had also failed her. The romantic imagination that conceives leaves to be more than they are and that supposes that life is without pain is itself a tragedy. Yet in this poem Bukowski claims that the sun is, nevertheless, good, thereby moderating the bleakness of the poem with a note of optimism.

The poems collected in *It Catches My Heart in Its Hands* and *Crucifix in a Death-Hand* are filled with loss, despair, pain, and cruelty but also contain an undercurrent of joy. It was during the 1960s that Bukowski met John Martin, who began Black Sparrow Press in 1966. Black Sparrow's first publication was a broadside published in April 1966, featuring Bukowski's poem "True Story." Four more Bukowski broadsides were published in 1966. Martin then published Bukowski's *At Terror Street and Agony Way* in 1968 and subsequently most of Bukowski's major books. Through Martin's impetus and financial support Bukowski left his post-office job and devoted himself full-time to his art. Martin's gamble paid off. Bukowski became a prolific novelist and an even more prolific poet.

In 1972 Bukowski published *Mockingbird Wish Me Luck,* a collection that featured new poems written between 1969 and 1970, poems that showed several shifts in Bukowski's work. The poetry in this collection features the enjambed line break, which became more prevalent in his later poems. The line breaks make the poems more staccato and allow for ironic and sarcastic shifts in tone and temperament, characteristics particularly evident when the poems are read aloud. It was also during this period that Bukowski began to read his poems in public. He comments on his style in his poem "Style," in which he notes, "style is the difference, / a way of doing / a way of being done." The poems in *Mockingbird Wish Me Luck* are narrative in structure. Subverted in the poetry is the intense pondering on the inevitable pain, rage, and suffering of life. Dominating the poems are the day-to-day experiences of the raucous Bukowski. The poetry, therefore, is on one level more personal, but also more egotistical. Throughout this collection the poet refers to himself as Bukowski, rather than Henry Chinaski, his fictional persona. Chinaski is much less complex than the poet Bukowski; however, the Bukowski in the poetry is also a veiled character.

Nevertheless, Bukowski's poetry is somehow more real or legitimate than his realistic fiction; it is more personal, immediate, direct. Bukowski said in his *New York Quarterly* craft interview in 1985 that "I am 93 percent the person I present in my poems; the other 7 percent is where art improves upon life, call it background music." Bukowski's increasing popularity and fame provided subject matter for his poetry, as he lived less and less the life of a cynical, monkish, introverted philosopher. The poetry of this period grew out of his immediate response to

Cover for Bukowski's 1981 book, which includes poems about gambling at the racetrack

his life as he experienced it and not from the well of his memory or from his intense emotions. The shift to the narrative in Bukowski's poetry also followed his increased prose production. His novel *Post Office* found great success, as did *Notes of a Dirty Old Man*. *South of No North* (1973), a collection of Bukowski's short stories, was also published. Bukowski was writing, at the time, a second novel, *Factotum* (1975). The reading public's favorable response to his fiction influenced his poetry, and he began to relate his uninhibited existence in a narrative fashion. He discovered that his unsavory alcoholic life, filled with pain and suffering, excited his readers. Bukowski's fictional, veiled life became exotic, and literary tourists flocked to the forlorn and dark Los Angeles of his underground existence.

To further intoxicate his readers, Bukowski, in *Mockingbird Wish Me Luck,* began writing poems that were more deviant than his previous poetry and filled with rape and murder. They were also more expressionistic than introspective. Mystifying his public, Bukowski transformed himself into an image of a mythic, Dionysian, madman poet, becoming more savage, at least in his poetry and his public appearances, as he became more successful. His fans thrived on his vile habits. As a defense against his own sensitivity, he constructed an antisocial image of himself. The narrative, confessional mode, however, was a type of poem that fit Bukowski's new material well. He was a poet who relied upon ironic shifts of tone and double meanings rather than any deep poetic mystery. Shifting from metaphoric incidents, such as that in which lions are sitting in the middle of a road, to writing about himself sitting in his bathtub reading Céline (to introduce a commonsense bit of philosophical musing) was not difficult.

In general, however, while Bukowski's poetic form was changing, his basic philosophy and attitude were not. In the poems "4th of July" and "Those Sons of Bitches," he describes again the dullness he perceived in most people. In "4th of July" he writes, "It is amazing / the number of people who can't feel / pain." In "Those Sons of Bitches," writing about humanity, he notes, "the dead are drunk on New Year's Eve / satisfied at Christmas / thankful on Thanksgiving / bored on the 4th of July," and finishing the poem, he says, "one tombstone for the mess, / I say: / humanity, you never had it / from the beginning." Frustrated by having to deal with society because of his increased popularity, Bukowski, in his poetic persona in the poem "Ants," attempts to drown and scald the ants that he finds in the drains of his washbasin and bathtub, but to no avail. The ants seem indestructible. His fame cannot keep pain and frustration out of his life, because life is filled with pain as well as success and pleasure.

Bukowski's next collection of poetry, *Love Is a Dog from Hell,* features poems written between 1974 and 1977 and includes many poems about the endless, sad, violent encounters Bukowski had with his many girlfriends during this period. His fame and his unencumbered lifestyle allowed him to engage in many frantic affairs, and the book's major themes are love and love's failure. These poems were composed at the time he was also writing *Women* (1978), his novel that deals with much of the same material that is in *Love Is a Dog from Hell.* For Bukowski love and sex are desperate acts that are nevertheless deeply interconnected. Sex is the primeval, uncontrollable biological (beastly) drive that causes people to blunder relentlessly into relationships that end in emotional disaster. His fascination with whores grew from his attempt to rid sexuality of love. In his poem "The Girls at the Green Hotel" he writes, "they are whores / they are whores without / souls / and they are magic / because they lie / about nothing." However, he knows he cannot really separate love from sex because he is basically a romantic who searches for love even among whores. As if to prove himself a jilted lover, correct about his deadly assumption, he fails horribly in relationships. The whores that he loves are as dishonest as everyone else, female or male. Bukowski recognizes the frailty of relationships and knows that he himself is not immune to love's corruption. Yet his sensitivity makes him feel victimized by the women he encounters. Recognizing the complexity on both a real and an abstract level, he writes in his poem "Trapped" that "she is a / child / and a mannequin / and death." Engaged totally in the cycle of life through his relationships with women, relationships that he could not escape, separate from, or resolve, Bukowski kept forming new relationships, repeating over and over his same mistakes. Again, for Bukowski, existence is filled with both pleasure and inevitable pain. In "Another Bed" the gentleness of a love affair leads to eventual sadness, and, accepting this melancholy fact, the poet begins to contemplate his next fling. This idea is reinforced by his equating love with a spider's web in "The Escape," where he writes, "she's so good that I almost miss my death, / but not quite; / I've escaped. I view the other / webs." "It's only human," Bukowski writes about the sexual act, in a poem bluntly titled "Fuck"; he then explains that sex is "like a party – / two trapped / idiots." To equate love and violence and pain is in Bukowski's world to understand the larger pattern of animal life, a pattern echoed in the title *Flower, Fist and Bestial Wail.* Yet, in a poem such as "A Flower in the Rain," love is sexual and sensual, and sometimes the war between men and women that is so evident in Bukowski's poems ends in a peace that brings happiness to both sides.

Following *Love Is a Dog from Hell* is another book of poetry, *Play the Piano Drunk Like a Percussion Instrument Until the Fingers Begin to Bleed a Bit* (1979). This book incorporates poems that Bukowski published in his chapbook *Fire Station* (1970) and those published in *While the Music Played* (1973) and in *Africa, Paris, Greece* (1974), both of which were part of Black Sparrow's Sparrow series. The selection also collects new poems and is dedicated to Linda Lee Beighle, who became Bukowski's wife. A poem titled "Mermaid," which describes her bathing, is a love poem in which Bukowski receives an unnameable something from this woman's essence. This type of unnameable pleasurable essence, perhaps

Bukowski with his wife, Linda, at the horse races (photograph by Michael Montfort)

love, proved a great comfort to Bukowski, and Beighle provided him with an emotional anchor. His emotional life solidified, he was able to spend more time writing. Among other newly collected poems in *Play the Piano* are those with an emphatic, enjambed, one- or two-word line that became a structural mainstay in Bukowski's late poetry. This line structure appears in the poem "Leaning on Wood":

> there are 4 or 5 guys at the
> racetrack bar.
> there is a
> mirror behind the
> bar.
> the reflections are not
> kind.
> of the 4 or 5 guys at the
> racetrack bar.

The structure allowed Bukowski to break his line in unpredictable places and causes the reader to fall into a cadence and to hear, if the piece is read aloud, both emphatic breath stops and the trailing off of lines.

Bukowski is almost never overtly political. He does not directly comment upon social class in society. In fact, his barbs are usually pointed at the masses or mass culture. While his sympathies (when he has them for humanity) are with the lower-class working people, miscreants, whores, and alcoholics, it is usually through ironic circumstances that he comments about their plight. Two poems in *Play the Piano,* "Face of a Political Candidate on a Street Billboard" and "The Drunk Tank Judge," are politically direct. "Face of a Political Candidate on a Street Billboard" comments on the type of perfect person that is usually elected to political office in America, one with "7 pairs of shoes / a son in college / a car one year old / insurance policies / a very green lawn." "The Drunk Tank Judge" comments ironically upon the political system. The judge is "young / well-fed / educated / spoiled" and has "smooth / delicate / skin" and "fat / jowls." Those the judge is judging are Mexicans and poor whites whose cars "look worse than / the ones in the / junkyards." Bukowski's final comment upon this class division: "justice."

Published in 1981, *Dangling in the Tournefortia* contains poems written during the period between late 1977 and 1980. The collection is dedicated to John Fante, a California novelist whose prose style and material are similar to Bukowski's. The tournefortia is a tree particularly suited to the southern California climate with, Bukowski notes at the beginning of the book, small, delicate flowers and a kind of fleshy fruit. Bukowski considers himself a piece of tournefortia fruit, swinging in the southern California wind and indigenous to life in southern California. The collection begins with a series of memory poems that deal with periods in Bukowski's past; for example, "The Lady in Red" is about the years of the Depression, and "The Immortals" is about black jazz musicians in New Orleans. The latter poem's tone is similar to that of Bukowski's first published poem, "Hello!," in the sense that the narrator identifies with and feels an admiration for the neglected black jazz artists, just as he felt sympathy and admiration for the black driver of the trash wagon in his earlier poem. But in "Immortals" there is also a comparison and a communion. The jazz greats are old men forced to perform for the tourists. Bukowski, old and an artist, perceives himself in a similar situation. He recalls how the black jazz greats and he (when he was young and living the life of a bum) used the same toilet. Since all people use the toilet, it is absurd that the poor, magnificent creative artists were forced to entertain the ignorant, economically comfortable tourists.

Poems about Bukowski's youth, his teenage years during the Depression, and his life as a bum became more prevalent in the poetry he wrote in the 1980s. The poems in *Dangling in the Tornefortia* were written by the successful poet who, content and secure, obviously had much free time. There are poems about Bukowski's driving his black BMW and about the beautiful setting of his house overlooking the San Pedro harbor. Bukowski now had little to do but write and bet at the racetrack, and the book contains many poems about the races. The racetrack, like life, is a gamble where most lose. It becomes a microcosm for the world, and, like a social scientist, Bukowski studies it and comments upon it. There are also poems that deal with the tension between men and women, but these are fewer than in the previous collections. However, Bukowski's general vision of male-female relationships is the same. In his poem "We Both Knew Him" he wrote, "discard or be discarded / it was endless." But he also reveals his innate sentimentality about women and relationships. "I get attached," he writes in his poem "Pretty Boy," a piece that goes far in explaining why Bukowski continued to search for love in the face of love's obviously fleeting nature.

There is a basic belief in Bukowski's poetry that eventually he would become lucky in love. Meeting Linda Lee Beighle proved to be the fortunate circumstance he anticipated. He met Linda Lee, owner of the Dewdrop Inn restaurant, in 1976, and they married in 1985. His friend and publisher John Martin was best man at their wedding. The couple remained together the rest of Bukowski's life. Linda Bukowski prompted Bukowski to pay attention to his health, and for the first time in his life he had a somewhat stable personal life. Without the unsettling day-to-day traumas of his earlier days, he began writing poems that seem generally complacent. To compensate for the comfort and security he enjoyed, Bukowski began intensifying and magnifying his "bad" behavior. He writes about his sexual fantasies with young girls and makes frequent references to urinating or vomiting on "everything" as he maneuvers through his suburban life. He records a car accident, a dead cat, a domestic argument. In semiretirement in the suburbs, Bukowski sensationalized everything and used shock techniques to maintain his hard-boiled, tough reputation. In his poem "Suckerfish" he wrote, "I'm *Buke*, like in puke." To his detriment he sometimes parodies his own writing. His age and situation therefore become artistically problematic because his poems mainly draw their power from his expressionistic and eccentric encounters with life. His more mundane existence resulted in poetry that is less pathological; it even becomes reflective, as Bukowski meditates about his daily encounters rather than actively involving himself in their narratives. Becoming defensive, he wrote a poem, "I Can't Stop," directed at his critics. In it he claims that "I Can't Stop" is in fact just another boring poem.

Barricaded in his San Pedro writing room, Bukowski wrote poems that are often like daybook entries, recording simple occurrences upon which he reflects. In "Fear and Madness" he writes, "barricaded here on the 2nd floor / I am in a small room again." He slips in these poems into his sage persona, philosophizing as he writes. The poems become longer, more meditative, less spontaneous. In *Dangling in the Tournefortia* Bukowski tries to come to grips with being sixty years old and famous. There are several references to his trip to Europe, where he met with a welcome reception. The tone of the poetry is not bitter, and fame, contemplation, and reflection suit Bukowski rather well. He simply creates poetry out of whatever comes his way. In his poem "War" he writes about a near-miss accident as he was on his way to purchase stamps. Life for Bukowski becomes rather uncomplicated, and this book leads the way toward the poetry he composed during the rest of his life. A summation of this period in his life occurs in the poem "Guava Tree," in which he writes about his former life in factories and skid-row hotels and how it related to his present success. Lounging in the sun, he writes that he is like a "wolf who got out of the trap without gnawing a / leg away." Many poems in the book are simply ironic views of what he sees happening as he travels through his normal days, meeting this person or that, or just reading the paper or having coffee at an oceanside restaurant. His former relentless sexual encounters give way to watching television. But Bukowski, from the beginning of his career, was always suspicious of success. His early poem "Object Lesson" warned that success brought destruction.

Bukowski was aware of the fact. An executed man or a man who has committed suicide often dangles (from a tournefortia) at the end of a rope.

During the 1980s Bukowski continued to write poetry and prose at a rapid pace. His major poetry collections, *War All the Time* (1984), *You Get So Alone at Times That It Just Makes Sense,* and *Septuagenarian Stew: Stories & Poems* (1990), account for more than one thousand pages of poetry. The narratives of the poetry deal with Bukowski's daily visits to the racetrack, his short errands around San Pedro, and encounters with his suburban neighbors and with members of the medical profession. This poetry of the 1980s, however, is principally reflective. In poems such as "Sparks," "My Friend," "Streetcars," "Rags, Bottles, Sacks," and "Education," he recalls the glories of his long life. A salvager, he resurrected his past, reinterpreting and reinventing the meanings of his memories. In his poem "My Buddy," in *You Get So Alone at Times That It Just Makes Sense,* he recalls an old man that he once befriended in New Orleans, when he was twenty-one. Bukowski recalls the old man's adages and notes that "he was a / sage." In these poems Bukowski became the ancient pessimistic sage and referred to himself as "Chinaski," an alter ego that gave him further distance from his material. The raging lunatic who inhabited the poems of the late 1960s and 1970s was left behind in Los Angeles when Bukowski moved to San Pedro. Honoring his heroes and paying old debts, there are many tributes in these collections, including poems to John Edgar Webb, John Martin, Hemingway, Jeffers, Céline, and Marlon Brando.

The collections of the 1980s are clearly the work of an aging poet adjusting to his fame and wealth. In his poem "Here I Am," in *War All the Time,* Bukowski, writing into the night, comments upon the state of his health: "liver gone / kidneys going / pancreas pooped / top-floor blood pressure." He was on his own deathwatch, waiting and watching, and many poems repeat this scenario. However, in his poem "Pernicious Anemia," in *Septuagenarian Stew,* he writes that in the face of death, "I must mount a / comeback. / I must crawl / inch by inch / back in- / to the sun of / creation." His tenacious personality remained intact. In his poem "For the Concerned," in *You Get So Alone at Times That It Just Makes Sense,* Bukowski confronts those who think his poetry has lost its power since he no longer writes about or lives the life of a skid-row bum. He writes, "please have some / cheer: agony sometimes changes / form / but / it never ceases for / anybody." And in his poem "Concerned," in *You Get So Alone at Times That It Just Makes Sense* he writes, "it has been a beautiful / fight / still / is." Bukowski's philosophical outlook remained intact. His opinion of the masses became less ironic and metaphoric and more emphatic, and in his poem "The Masses," in *Septuagenarian Stew,* he condemns people for being "totally unappetizing, dutifully unoriginal, they are / cowardly and placid, sunk in self-pity." He continued to develop as an artist, and the poems that he wrote in the 1980s continue to challenge the poetic line and the form of the poem. Throughout these poems he avoids the obvious; the forms the poems take are never repetitive or predictable. In his poem "The Rape of the Holy Mother" (1990), still calling for an all-out poetic revolution, he writes, "we insist that there are / other voices / other ways of creating."

Bukowski's last book of poems published in his lifetime was *The Last Night of the Earth Poems,* his longest collection. Like his other collections, this book is rich in sarcasm, anti-authoritarian triads, Diogenistic cynicism, madness, satire, and death. Death had always been a facet of Bukowski's poetry, but in this collection it is not the death that stalked Bukowski through forty years of poetry. This is not death as a result of alcohol abuse, suicide, depravity, cultural repression, misunderstanding, or a broken heart. It is the end, death as the eventual climax of a long-lived life. Bukowski reveals that he is and has been involved in the great seasonal cycles of life. There are birth, death, and rebirth, and there are pain, sorrow, and love.

In "Jam," the opening poem of the collection, death is the jam in which all humanity is trapped. The metaphor is that of a traffic jam, which is also likened to a dinosaur "crawling feebly home somewhere, somehow maybe / to / die." Bukowski returns to the dinosaur image in the poem "Dinosauria We," a condemnation of human weaknesses that lead to human extinction. Yet after humankind has destroyed itself, something peculiar occurs in the final lines of the poem, where Bukowski writes: "and there will be the most beautiful silence never heard / born out of that / the sun still hidden there / awaiting the next chapter." After the complete destruction of society on doomsday, the sun rises, and the sun (son) awaits the next chapter of life on earth, or the next poem, or the next novel. It is resurrection.

Writing was always Bukowski's salvation. Throughout his life his identity was anchored in the word. In his poem "The Creative Act" Bukowski reveals that writing is in fact the sole purpose in life "for the 5th of July / for the fish in the tank / for the

Title page for Bukowski's 1992 book, in which he writes, "I just wanted to get the word / down; / fame, money, didn't matter"

old man in room 9 / for the cat on the fence – for yourself." In "Death is smoking my Cigars" he writes, "I just wanted to get the word / down; / fame, money, didn't matter." And again, in "Only One Cervantes" Bukowski observes that "writing had been my fountain / of youth / my whore / my love, / my gamble." And all this writing goes on in the face of death, as it does, for example, in the poem "Are You Drinking?"

> so here I am
> propped against my pillows
> again.
>
> just an old guy
> just an old writer
> with a yellow
> notebook.
>
> something is
> walking across the
> floor. toward
> me.

In the poem "The Creative Act" he writes of "this life dancing in front of / Mrs. Death." Here death is female, femininity being symbolic for both life and death. It is in the face of Mrs. Death that Bukowski's creativity (birth and rebirth) becomes balanced and is part of the eternal, mythic realm. In the poem "No More No Less" Bukowski writes about this contrast and the eternal balance of the universe: "each day is still a / hammer, / a flower." Bukowski's poetry is always in the midst of both life and death, pleasure and pain, and always part of both.

Bukowski's poems are always full of surprises, revelations, and disclosures, and the poems in *The Last Night of the Earth Poems* are no exception. In "The Bluebird" Bukowski writes,

> There's a bluebird in my heart that
> wants to get out
> but I'm too clever, I only let him out
> at night sometime
> when everybody's asleep.

Perhaps even more revealing is the poem "The Confession," which again discloses a personal sensitivity well veiled in Bukowski's early poetry. "The Confession" is one of many obvious love poems in the collection, written most certainly for Linda Lee Bukowski. In the poem the sentimental Bukowski writes in the last two lines, "I love / you."

Bukowski was a prolific writer his entire life, and in his waning years his productivity increased. He wrote and submitted dozens of new poems each month to magazines throughout the world. Shortly before his death he finished his sixth novel, *Pulp* (1993). In 1993 the first volume of his selected letters, *Screams from the Balcony: Selected Letters 1960–1970,* was published. There is a Bukowski newsletter, titled *Sure.* Critics have begun to praise the poet's tenacity and frankness. Much of the current praise echoes Corrington's preface to Bukowski's 1963 collection, *It Catches My Heart in Its Hands,* in which Corrington characterizes Bukowski's writing as "the spoken voice nailed to paper." The first biography of Bukowski, *Hank,* reveals a sensitive and imaginative man who lived through many of the capers he wrote about. It also reveals a more complex and rounded Bukowski than the mythic hard guy of his prose and poetry.

Few scholarly articles have been written about Bukowski's work. The problem is that Bukowski was always a maverick who was perceived by many academics to be hostile and antipoetic. His frank approach to life and writing is still too often considered simplistic. This perspective, however, is changing. *Screams from the Balcony* shows a hardworking, oftentimes naive, sensitive, and tortured poet who was filled with self-doubt but deeply committed to his art. Additional volumes of letters will eventually be published, and the first major critical study of Bukowski, *Against the American Dream: Essays on Charles Bukowski,* by Russell Harrison, was published in 1994. While Bukowski's literary achievements are still widely unrecognized and critically undervalued in the United States (in Europe he is already considered a classic American author), the publication of Harrison's study and Seamus Cooney's editorial work on Bukowski's letters seems to mark the beginning of new era, in which Bukowski's poetry and prose will be taken more seriously by American scholars and critics. In the coming years Bukowski's uncollected material — poems, reviews, essays, as well as letters and ephemeral writings — will also be published, helping to solidify Bukowski's proper place of importance in late-twentieth-century American poetry.

Interviews:

William Childress, "Interview with Charles Bukowski," *Poetry Now,* 1, no. 6 (1974): 1, 19, 21, 32;

William Packard, "Craft Interview with Charles Bukowski," *New York Quarterly,* no. 27 (Summer 1985): 19–25.

Letters:

The Bukowski/Purdy Letters, edited by Seamus Cooney (Sutton West, Ontario & Santa Barbara, Cal.: Paget, 1983);

Bukowski: Friendship, Fame & Bestial Myth, edited by Sherman Jory (Augusta, Ga.: Blue Horse, 1988);

Screams from the Balcony: Selected Letters 1960–1970, edited by Cooney (Santa Rosa, Cal.: Black Sparrow Press, 1993);

Living On Luck: Selected Letters, 1960s–1970s, edited by Cooney (Santa Rosa, Cal.: Black Sparrow Press, 1995).

Bibliographies:

Sanford Dorbin, *A Bibliography of Charles Bukowski* (Los Angeles: Black Sparrow Press, 1969);

Hugh Fox, *Charles Bukowski: A Critical & Bibliographical Study* (Sommerville, Mass.: Abyss, 1969).

Biography:

Neeli Cherkovski, *Hank: The Life of Charles Bukowski* (New York: Random House, 1991).

References:

Review of Contemporary Fiction, special issue on Bukowski and Michel Butor, 3 (Fall 1983);

Second Coming, special issue on Bukowski, 2, no. 3 (1974);

Neeli Cherkovski, *Whitman's Wild Children* (Venice, Cal.: Lapis Press, 1988);

Glenn Easterly, "The Pock-Marked Poetry of Charles Bukowski," *Rolling Stone,* no. 215 (17 June 1976): 28–36;

Al Fogel, *Under the Influence: A Collection of Works by Charles Bukowski* (Sudbury, Mass.: Jeffery H. Weinberg Books, 1984);

Loss Pequeño Glazier, *All's Normal Here: A Charles Bukowski Primer* (Fremont, Cal.: Ruddy Duck Press, 1985);

Russell Harrison, *Against the American Dream: Essays on Charles Bukowski* (Santa Rosa, Cal.: Black Sparrow Press, 1994);

Jack Matthews, "The Search for Bukowski," in *Critical Survey of Poetry,* edited by Frank N. Magill (Englewood Cliffs, N.J.: Salem Press, 1973), pp. 354–365;

Norman Moser, "Charles Bukowski," in *Contemporary Poets,* third edition, edited by James Vinson (London & Basingstoke: Macmillian, 1980), pp. 205–208;

"The Charles Bukowski Issue," *The Outsider,* special issue on Bukowski, edited by J. E. Webb, 1, no. 3 (1963).

Papers:

Bukowski's papers are held at the University of California, Santa Barbara; in the American Literature Collection of the University of Southern California; and in the special collections of Temple University in Philadelphia, Pennsylvania.

Robert Creeley
(21 May 1926 -)

Loss Pequeño Glazier
State University of New York at Buffalo

See also the Creeley entries in *DLB 5: American Poets Since World War II, First Series* and *DLB 16: The Beats: Literary Bohemians in Postwar America.*

BOOKS: *Le Fou* (Columbus, Ohio: Golden Goose Press, 1952);
The Kind of Act Of (Palma, Spain: Divers Press, 1953);
The Immoral Proposition (Karlsruhe-Durlach & Baden, Germany: Jonathan Williams, 1953);
The Gold Diggers (Palma, Spain: Divers Press, 1954); enlarged as *The Gold Diggers and Other Stories* (London: Calder, 1965; New York: Scribners, 1965);
A Snarling Garland of Xmas Verses, anonymous (Palma, Spain: Divers Press, 1954);
All That Is Lovely in Men (Asheville, N.C.: Jonathan Williams, 1955);
If You (San Francisco: Porpoise Bookshop, 1956);
The Whip (Worcester, U.K.: Migrant Books, 1957);
A Form of Women (New York: Jargon Books/Corinth Books, 1959; Fontwell, U.K.: Centaur, 1960);
For Love: Poems 1950-1960 (New York: Scribners, 1962);
The Island (New York: Scribners, 1963; London: Calder, 1964);
Words (Rochester, Mich.: Perishable Press, 1965; enlarged edition, New York: Scribners, 1967);
Poems 1950-1965 (London: Calder & Boyars, 1966);
The Charm: Early and Uncollected Poems (Mount Horeb, Wis.: Perishable Press, 1967; enlarged edition, San Francisco: Four Seasons Foundation, 1969; London: Calder & Boyars, 1971);
A Sight (London: Cape Goliard Press, 1967);
The Finger (Los Angeles: Black Sparrow Press, 1968); enlarged as *The Finger: Poems 1966-1969* (London: Calder & Boyars, 1970);
5 Numbers (New York: Poets Press, 1968);
Numbers (Stuttgart: Edition Domberger/Düsseldorf: Galerie Schmela, 1968);
Divisions & Other Early Poems (Mount Horeb, Wis.: Perishable Press, 1968);

Robert Creeley, 1988 (photograph by Elsa Dorfman)

Pieces (Los Angeles: Black Sparrow Press, 1968; enlarged edition, New York: Scribners, 1969);
Mazatlan: Sea (San Francisco: Poets Press, 1969);
In London (Bolinas, Cal.: Angel Hair Books, 1970);
A Quick Graph: Collected Notes and Essays, edited by Donald Allen (San Francisco: Four Seasons Foundation, 1970);
1-2-3-4-5-6-7-8-9-0 (Berkeley: Shambala/San Francisco: Mudra, 1971);
St. Martin's (Los Angeles: Black Sparrow Press, 1971);
A Day Book (Berlin: Graphis, 1972); expanded edition, including "In London" (New York: Scribners, 1972);
Listen (Los Angeles: Black Sparrow Press, 1972);
Notebook (New York: Bouwerie Editions, 1972);

A Sense of Measure (London: Calder & Boyars, 1972);
The Class of '47, by Creeley and Joe Brainard (New York: Bouwerie Editions, 1973);
The Creative (Los Angeles: Black Sparrow Press, 1973);
For My Mother: Genevieve Jules Creeley, 8 April 1887–7 October 1972 (Rushden, U.K.: Sceptre Press, 1973);
His Idea (Toronto: Coach House Press, 1973);
Inside Out (Los Angeles: Black Sparrow Press, 1973);
Thirty Things (Los Angeles: Black Sparrow Press, 1974);
Backwards (Knotting, U.K.: Sceptre Press, 1975);
Away (Santa Barbara, Cal.: Black Sparrow Press, 1976);
Hello (Christchurch, N.Z.: Hawk Press, 1976);
Mabel: A Story and Other Prose (London: Boyars, 1976);
Presences: A Text for Marisol (New York: Scribners, 1976);
Selected Poems (New York: Scribners, 1976);
Was That a Real Poem or Did You Just Make It Up Yourself (Santa Barbara, Cal.: Black Sparrow Press, 1976);
Myself (Knotting, U.K.: Sceptre Press, 1977);
Thanks (Old Deerfield, Mass.: Deerfield Press / Dublin, Ireland: Gallery Press, 1977);
Desultory Days (Knotting, U.K.: Sceptre Press, 1978);
Hello: A Journal, February 29–May 3, 1976 (New York: New Directions, 1978; London: Boyars, 1978);
Later: A Poem (West Branch, Iowa: Toothpaste Press, 1978);
Was That a Real Poem and Other Essays, edited by Allen, with a chronology by Mary Novik (Bolinas, Cal.: Four Seasons Foundation, 1979);
Later (New York: New Directions, 1979; London: Boyars, 1980);
Corn Close (Knotting, U.K.: Sceptre Press, 1980);
Mother's Voice (Santa Barbara, Cal.: Am Here Books/Immediate Editions, 1981);
The Collected Poems of Robert Creeley, 1945–1975 (Berkeley: University of California Press, 1982);
Echoes (West Branch, Iowa: Toothpaste Press, 1982);
A Calendar: 1984 (West Branch, Iowa: Toothpaste Press, 1983);
Mirrors (New York: New Directions, 1983);
The Collected Prose of Robert Creeley (New York: Boyars, 1984; corrected edition, Berkeley: University of California Press, 1987);
Memory Gardens (New York: New Directions, 1986);
The Company (Providence, R.I.: Burning Deck, 1988);
Window (Buffalo: Poetry/Rare Books Collection, SUNY at Buffalo, 1988);
7 & 6 (Albuquerque, N.M.: Hoshour Gallery, 1988);
The Collected Essays of Robert Creeley (Berkeley, Los Angeles & London: University of California Press, 1989);
Dreams (New York: Periphery/Salient Seedling Press, 1989);
It (Zurich: Bruno Bischofberger, 1989);
Have a Heart (Boise, Idaho: Limberlost Press, 1990);
Places (Buffalo: Shuffaloff Press, 1990);
Windows (New York: New Directions, 1990);
Autobiography (Madras & New York: Hanuman, 1991);
Gnomic Verses (La Laguna, Canary Islands: Zasterle Press, 1991);
The Old Days (Tarzana, Cal.: Ambrosia Press, 1991);
Selected Poems (Berkeley: University of California Press, 1991);
Life & Death (New York: Gagosian Gallery, 1993);
Echoes (New York: New Directions, 1994);
Loops: Ten Poems (Kripplebush, New York: Nadja, 1995).

RADIO SCRIPT: *Listen,* London, 1972.

OTHER: Charles Olson, *Mayan Letters,* edited, with a preface, by Creeley (Palma, Spain: Divers Press, 1953; London: Cape, 1968);
Black Mountain Review, edited by Creeley, nos. 1–7 (Spring 1954–Autumn 1957); republished, with an introduction by Creeley, 3 volumes (New York: AMS Press, 1969);
New American Story, edited by Creeley and Donald Allen (New York: Grove, 1965; Harmondsworth, U.K.: Penguin, 1971);
Selected Writings of Charles Olson, edited, with an introduction, by Creeley (New York: New Directions, 1966);
The New Writing in the U.S.A., edited by Creeley and Allen, with an introduction by Creeley (Harmondsworth, U.K.: Penguin, 1967);
Whitman: Selected Poems, edited, with an introduction, by Creeley (Harmondsworth, U.K.: Penguin, 1973);
Creeley, "Thinking of You," *Review of Contemporary Fiction,* 8 (Fall 1988): 82–85;
The Essential Burns, edited, with an introduction, by Creeley (New York: Ecco, 1989);

Olson, *Selected Poems,* edited, with an introduction, by Creeley (Berkeley: University of California Press, 1993).

For the second half of this century, Robert Creeley's work as an innovative poet has occupied a singular place in postwar American letters. The contributions of Creeley, with Charles Olson, were instrumental, through Projective Verse, to the definition of emerging senses of poetic form in the 1950s. The *Black Mountain Review* (1954–1957), which Creeley edited, was a landmark literary journal of the period. Creeley's work, along with that of other poets connected with Black Mountain College, has also been a benchmark against which subsequent poetry has been measured. Perhaps most important, Creeley's career has placed him as immediately active in the record of the contemporary American "new poetries." Though known primarily as a poet, Creeley's poetry is intricately tied to other genres of writing, prose the most important among them. He has also made substantial contributions in essays, letters, editing, autobiography, collaborations with artists, the interview, and various forms of miscellaneous writing. The fact of literary engagement is crucial to Creeley's work; his sense of writing as a multifaceted engagement may well stand as a definition of the contemporary literary endeavor.

Known for the dictum, "form is never more than an extension of content," Creeley has largely been connected with Black Mountain College and with Projective Verse. The latter, developed primarily in the early 1950s, was not meant as a defining style for any poet but as a means of breaking from poets associated with the New Criticism and their insistence on form as extrinsic to the poem, dominant at the time. Creeley also had considerable interchange with other poetic alternatives, including the Beat poets. Further, he coedited the anthologies the *New American Story* (1965) and *The New Writing in the U.S.A.* (1967), which presented the work of divergent groups of writers. To date, Creeley has been mentor to or advocate of many American poetry movements including the Beats, Objectivists (retrospectively), the San Francisco Renaissance, poets of the New College of California and the Naropa Institute, writers associated with a multitude of small presses, and most recently, Language poets. Creeley's involvement with this last school includes his inclusion in *This 1* (1971), credited by Ron Silliman as being a harbinger of language-oriented poetry. In addition, two of Robert Grenier's five critical texts in the issue were about Creeley; one of these included the assertion, as related by Silliman in his introduction to *In the American Tree* (1986), that "Projective Verse is *Pieces* On," an acknowledgment of the importance of Projective Verse and Creeley's collection of poems *Pieces* (1968) to emerging poetries. Creeley's literary involvement and lasting contribution, more than with any single poetic idiom however, has been with the possibilities of new writing – and he has been a tireless campaigner for efforts exploring these engagements.

Though popularly considered a love poet, Creeley's poems are given to explore the incongruities of such relations, rather than allowing for their easy statement. Creeley has, especially in his early work, presented the persona of an "unsure egoist"; many of these poems carry a veneer of accessibility yet often befuddle the reader with senses of conflicting identity, shortfalls in the possibilities of interpersonal communication, and a persistent characterization of the self as isolated. He is also known for his formal investigations in the late 1960s; here his texts are multiform, stanzaic verse giving way to brief epiphanies, observations, sketches, prose interruptions in the form of journal entries, personal reflections, and rhythmic notation reminiscent of the improvised jazz that was formative to his early creative development. This period was to prove the most rankling to literary critics representing conservative American poetic ideologies. Though the poetics of his recent work take a turn as yet unexplored to any depth by contemporary critics, Creeley's entire oeuvre stands as a monumental contribution to the realignment of poetic form in the second half of the twentieth century, and his poetic achievement renders, as Robert Hass suggests in *Twentieth-Century Pleasures* (1984), "what the mind must, slowly, in love and fear, perform to locate itself against, previous to any discourse."

Robert White Creeley was born in Arlington, Massachusetts, on 21 May 1926. His parents, Oscar Slate and Genevieve Jules Creeley, soon moved the household – Robert, his older sister, Helen, and Teresa Turner, the family housekeeper – to a farm in West Acton. In 1928 Robert Creeley suffered an injury that resulted in the loss of his left eye three years later. Oscar Creeley, a successful physician, died in 1930, and Robert Creeley found himself growing up in a household of women. He enjoyed a closeness to nature and in later years reflected on his youth in West Acton as essentially a small-town experience despite their geographic closeness to Boston. He considers himself to have been raised in "the New England manner, compact of puritanically deprived senses of speech and sensuality," an envi-

The Creeleys' house in New Hampshire, circa 1949

ronment in which one did not feel encouraged to linger over words and in which the sentiment was that "life was real and life was earnest, and one had best get on with it."

At age fourteen he entered Holderness School, a boarding school for boys in Plymouth, New Hampshire. His early interest in writing was evidenced by the articles and stories he published in the Holderness School literary magazine, the *Dial*. In 1943 Creeley entered Harvard University, but his attendance was interrupted by service as an ambulance driver for the American Field Service in the India-Burma theater during World War II. In winter 1945 Creeley returned to Harvard and married Ann McKinnon the following spring. With Ann he would have three children, David, Thomas, and Charlotte, born in 1948, 1950, and 1952, respectively. He assisted in editing the special E. E. Cummings issue of the Harvard *Wake*, which included contributors such as William Carlos Williams, Marianne Moore, Allen Tate, Karl Shapiro, Wallace Stevens, and Mark Van Doren. The *Wake* also included Creeley's first published poem, "Return."

With its street ending in "darkened doors" (a "darkness" that recurs in subsequent work) and the disaffiliated sense of home evoked in the closing lines — "enough for now to be here, and / to know my door is one of these" — "Return" makes clear that Creeley's homecoming from war, as for many, was less than satisfying. As Creeley has noted, a new social perspective had to be found: "The value of one's life as a progression toward some attention was gone because the war demonstrated that no matter how much you tried, as Morganthau said: *facts have their own dynamic.*"

The Creeleys lived in Truro, Massachusetts, with Robert Creeley commuting to Harvard until the summer of 1947, when in his last semester he withdrew from the university. The Creeleys then moved to Rock Pool Farm near Littleton, New Hampshire, and practiced subsistence farming while living on Ann Creeley's income from a trust fund. At Rock Pool Farm Creeley also bred pigeons and chickens, which he showed in Boston. During these years Creeley began to listen to jazz, especially Charlie Parker, Thelonious Monk, Max Roach, Jacky Byard, and Dick Twardzik. He was "fascinated with what these people did with *time*. Not to impose this kind of intellectual term upon it ... this was where I was hearing 'things said' in terms of rhythmic and sound possibilities." Creeley credits Henry Miller, Kenneth Patchen, D. H. Lawrence, and Hart Crane with making clear to him the possibility that a writer has "access to your feelings and can really use them as a demonstration of your own reality." However, jazz took him further: besides the acoustic and improvisatory elements these musicians employed, Creeley credits them with extending the possibilities of his writing by making clear "how *subtle* and how sophisticated ... how *refined* that expression might be."

In December 1949 Creeley heard Cid Corman's Boston radio program *This Is Poetry* and began a correspondence with Corman, which lasted for six years. In 1950 Creeley gave his first public poetry reading on Corman's program and became American editor for Rainer Gerhardt's German magazine *Fragmente*. During this period Creeley and Jacob Leed collected manuscripts for a proposed (but never published) experimental magazine, the "Lititz Review."

Meanwhile a friend had sent to Creeley two of Charles Olson's poems for the magazine. Creeley responded to Olson with some criticism, thus beginning an extraordinary correspondence that lasted until Olson's death in 1970. These literary exchanges produced more than a thousand pieces of correspondence; the "range and articulation" of the letters, Creeley has written, "took me into terms of writing and many other areas indeed which I otherwise might never have entered." The correspondence was vital to the developing interests of both writers and formed, as George F. Butterick noted, "a critical document for understanding the emerging poetics of a generation, as well, perhaps, of the poetries yet to come." One of the principles of these "emerging poetics" is Projective Verse, usually circumscribed by Creeley's declaration that "form is never more than an extension of content" and Olson's instruction in his essay "Projective Verse" (1950) that poems proceed from "the HEAD, by way of the EAR, to the SYLLABLE" and from "the HEART, by way of the BREATH, to the LINE." One of the motivations behind Projective Verse was to provide a means of breaking from poets associated with the New Criticism. As Creeley has subsequently explained, "The forties were a hostile time.... The colleges and universities were dominant in their insistence upon an *idea* of form extrinsic to the given instance." What was objectionable was the "assumption of a *mold,* of a means that could be gained beyond the literal fact of the writing *here and now,* that had authority." Projectivism was not meant as a defining style for any single poet but recognized "that writing could be an intensely specific revelation of one's own content, and of the world the fact of any life must engage." Creeley later reflected on Olson's "Projective Verse" essay and its subsequent "curiously mystic effect upon people." According to Creeley: "I took it to be a series of observations akin to Pound's rules of thumb, 'a few don'ts' for example, and used it in that sense. [Olson] showed me how the line might be organized in terms of the breath involved in it. [This approach] is a fairly practical application of emotion in terms of a given series of beats and is not at all a mystique."

In 1951, with children to support, the Creeleys moved to France, hoping to survive more economically on Ann Creeley's income. Much of the material gathered for the "Lititz Review" went into Corman's *Origin* I (1951), the Olson issue, in which Creeley's poem "Hart Crane" appeared. *Origin* II (1951) featured a special section of Creeley's work. Because of the inflation in France the Creeley family moved to Majorca, Spain. Creeley was briefly editor at Martin Seymour-Smith's Roebuck Press at Majorca, an experience that – along with the increasingly bitter tensions in his marriage – played a key role in his novel *The Island* (1963). In 1952 *Le Fou,* his first book of poems, was published by Golden Goose Press. On Majorca Creeley and his wife started the Divers Press, an independent press that focused on publishing experimental writers and was founded to create "a place defined by our own activity and accomplished altogether by ourselves – a *place* wherein we might make evident what we, as writers ... hoped our writing might enter." The press published Corman's *Origin* VIII (1953) as well as books by Olson, Paul Blackburn, Irving Layton, Douglas Woolf, Larry Eigner, Robert Duncan, and Katue Kitasono. Divers Press also published Creeley's second book of poems, *The Kind of Act Of* (1953), and a collection of Creeley's stories, *The Gold Diggers* (1954).

Through Ezra Pound, Creeley had met René Laubiès. Illustrations by Laubiès, with their simple yet bold brush strokes and emphasis on materiality, were published on the covers of Creeley's *The Kind of Act Of, The Gold Diggers,* and his later book *The Whip* (1957). *The Immoral Proposition* (1953), published by Jonathan Williams as *Jargon* 8 and printed on long, folded pages loosely bound by string, comprises poems by Creeley, each accompanied by a drawing by Laubiès.

As a poet Creeley was intensely conscious of "elders" such as Pound, Williams, and Louis Zukofsky. Nevertheless he commented that "initially I had thought that my work as a writer would be primarily in prose." This is evidenced by his manifesto in *Origin* II, which discussed Projectivism and a possible new approach to prose. Any consideration of Creeley's poetry must include a reading of his prose. Visual art also opened up new avenues

Creeley with his son Dave on Majorca, Spain, circa 1953

for Creeley. In 1953 at a gallery in Paris he was struck for the first time by Jackson Pollock's work, discovering in Abstract Expressionism specific approaches to writing.

A decisive opportunity arose when Olson proposed that Creeley edit the *Black Mountain Review*. Intended as a promotional vehicle for the college, the Black Mountain magazine had under Creeley's editorship some affinities to *Origin*. Creeley hoped to extend these interests and to explore "the *ground* that an active, ranging critical section might effect" by including "critical writing that would break down habits of 'subject' and gain a new experience of context generally," a vision that has stayed with Creeley to this day. In March 1954 the first issue of the *Black Mountain Review* came out just before Creeley left Majorca to teach at Black Mountain College, meeting Olson for the first time. At Black Mountain he was introduced to artists and writers with whom he would maintain lifelong contact and began his first, somewhat awkward, teaching experience. Black Mountain was an artistic community that included visual artists, musicians, and dancers as well as writers; it offered students a chance to develop their own approaches to creative work. Because of the journal the label "Black Mountain School" soon came into use to describe contemporary poets such as Olson, Duncan, Denise Levertov, Edward Dorn, and Blackburn – writers that, though sharing specific interests, vary in their writing styles.

In July Creeley returned to Majorca and unsuccessfully tried to repair his troubled marriage. He continued to edit the *Black Mountain Review*, publishing issues in the summer, fall, and winter of 1954. (Only three more issues of the magazine appeared, in 1955, 1956, and 1957 – each with cover art by Laubiès.) Creeley's last Divers Press book was an anonymous pamphlet, *A Snarling Garland of Xmas Verses* (1954), "a wallet pocket-book" of five poems fashioned from one folded pullout strip of long paper. Distraught about the breakup of his marriage, he returned to Black Mountain College and resumed teaching.

At Black Mountain there was a great deal of interaction with painters known as Abstract Expressionists. Creeley formed close relationships with Pollock, Philip Guston, Jack Tworkov, Willem de Kooning, Esteban Vicente, John Chamberlain, Dan Rice, and John Altoon. Later, at the Cedar Bar in New York, Creeley came to know Franz Kline. With Guston and another painter, Ashley Bryant, Creeley "gradually ... began to come into the relationship to painters that does become decisive."

(Bryant had contributed the frontispiece drawing to *Le Fou*.) Creeley considers Altoon extremely important because "the things he drew, made manifest in his work, were images in my own reality." Most important was the Abstract Expressionists' approach to art: "Their ways of experiencing activity, energy — that whole process, like Pollock's 'when I am in my painting' — that the whole condition of their way of moving and acting and being in this activity was so manifestly the thing we were trying to get with Olson's 'Projective Verse,' the open field." Painting, in the mid 1950s, Creeley felt, "was far more fresh as imagination of possibility than what was the case in writing, where everything was still argued with traditional or inherited attitudes and forms." The affinities between Projectionism and Abstract Expressionism are evident in Tworkov's statement in 1957: "My hope is to confront the picture without a ready technique or a prepared attitude, a condition which is nevertheless never completely attainable; to have no program and, necessarily then, no preconceived style." Creeley's *All That Is Lovely in Men* (1955), published by Jonathan Williams, includes drawings by Rice. These illustrations had "disposition toward, for, and of, SPACE," which Creeley liked tremendously. He insisted that "there is no matter more urgent, now, than how we occupy our space."

In 1956 Creeley resigned from Black Mountain College and headed for San Francisco "to see the Pacific Ocean, if nothing else." Creeley spent a remarkable three months in San Francisco, where he met Beat poets Allen Ginsberg, Jack Kerouac, Philip Whalen, Michael McClure, Gary Snyder, and other writers. San Francisco at the time, writes Creeley, was "an intensive meeting ground" full of "blasts of sound, and talk of Pollock, *energy*" and "packed with things *happening*." Creeley published poems by some of these writers in the seventh and final issue of *Black Mountain Review*. Through vital participants in the new poetries of the West Coast, Creeley also made lasting friendships.

The next few years would see several important events in Creeley's life. He received a B.A. from Black Mountain College in 1955 and married Bobbie Louise Hoeck two years later. Children in the Creeley household at the time included Kristen (stepdaughter), Leslie (stepdaughter), Sarah (born 1957), and Katherine (born 1959). In 1960 he completed an M.A. from the University of New Mexico; he lectured there periodically until 1966. His publications during this time included *If You* (1956), *The Whip,* and *A Form of Women* (1959). He was awarded the Levinson Prize for ten poems in *Poetry* and a D. H. Lawrence Fellowship, both in 1960. In addition he was included in Donald Allen's anthology *The New American Poetry: 1945-1960* (1960), the defining anthology of the decade.

For Love: Poems 1950-1960, Creeley's first widely circulated book of selected poems, was published by Scribners in 1962. The collection was widely reviewed and nominated for a National Book Award, selling more than forty-seven thousand copies. Of Creeley's popularity at the time, Robert Hass remembers going into college lounges "jammed with people sitting on the floor, nodding their heads in profound sympathy and agreement with some [Creeley] poem they had heard only once."

Consisting of three chronological sections covering 1950 to 1960, *For Love* is best known for its presentation of the love lyric, in which, in Olson's words, "the intimate / is an exactitude." In his prefatory note to *For Love,* Creeley comments: "Insofar as these poems are such places, always they were ones stumbled into: warmth for a night perhaps, the misdirected intention come right; and too, a sudden instance of love, and the being loved, wherewith a man also contrives a world (of his own mind)." Though popularly considered a love poet, Creeley typically explored the incongruities of such relations. Especially in this early work, Creeley presents, as Charles Altieri has written, the persona of an "unsure egoist." Many of the poems have a veneer of accessibility yet befuddle the reader with varied senses of conflicting identity, shortfalls in the possibilities of interpersonal communication, and a persistent characterization of the self as isolated. Creeley has since commented that he had not expected a collection concerned with "marital confusion, loneliness, and isolation" to be so popular; that it struck such a chord in readers at the time was a significant statement about modern culture itself.

The poems in *For Love* tend to be compressed and urgent, focusing on a single event or fact of observation. One of the most often discussed of these poems is "The Warning," in which

For love — I would
split open your head and put
a candle in
behind the eyes.

Love is dead in us
if we forget

> the virtues of an amulet
> and quick surprise.

The phrase "for love" echoes the title of the collection. An immediate "quick surprise" for the reader is that "The Warning" *is* a warning; it is not about love's pleasantries but its limitations. The act proposed in the first stanza is immediate, ritualistic, and violent. (Creeley has suggested that this proposed act is rather "a true measure of an ability to love," a defining boundary in the possibility of the relation.) The poem clearly conveys to the reader pent-up energy and violent pressure, as illustrated by the phrase "split open your head."

A similar pressure is expressed in "The Whip." (Creeley has written about the idea of "the whip" as having to do with the conflict between the mind and the body: "It's a weird tension and the torque that's created by that systematization of experience is just awful . . . something was really, you know . . . slashing and cutting me.") In this poem the narrator is in bed. Two women are present: one "on / the roof," a troubling, dreamlike presence who is a "woman I / also loved"; and the other tangibly next to him and, by contrast, inert. His reaction to this crisis of values, "lonely," is to cry out. The woman next to him puts her hand on his back, and, paradoxically, this tender act "whips" the narrator. The message for the marital relationship then is doubt and misperception.

The narrative of this poem also has a parallel in Creeley's story "The Musicians." Here, there is also a woman on the roof, dressed in a housecoat, seemingly deranged because of a love gone wrong. The persistent presence of three characters – of whom one is always detached – typifies the relations of both the poem and the story. The title also proceeds from the jazz sense of "whip it" or "whip that thing," which serves as an exhortation for a musician to play an instrument. (This phrase can also refer to sexual activity.) There is music playing on a phonograph, and musical instruments are present, though unused. What the characters in the story "play" however, are proposed narratives, that is, other stories; hence the title might suggest "tell that story." "The Whip" also operates rhythmically. Observing the terminal junctures at the end of each line, one gets a clear sense of a rhythm that communicates "the bleak confusion from which [the poem] moves emotionally" in bursts of jazzlike phrasing. Creeley states that it is jazz "that informs the poem's manner in large part. Not that it's jazzy, or about jazz – rather, it's trying to use a rhythmic base much as jazz of this time would. . . . That is,

the beat is used to delay, detail, prompt, define the content of the statement or, more aptly, the emotional field of the statement. It's trying to do this while moving in time to a set periodicity – durational units call them." The third sense of the title is "whipped up," meaning exhaustion of the characters. In this sense the narrator is recounting the poem's activity "wrongly," since the exacerbation of interpersonal situations really leads to emotional constriction.

Creeley taught at the University of British Columbia in 1962 and 1963, then returned to New Mexico as a lecturer until 1966. During this time he completed *The Island,* set on Majorca in the early 1950s. This novel, "a process of discovery," on one level discovers that "no wife – indeed no other person – can reify one's existence, that the 'love' that demands such reification is really a form of infantile dependency." But *The Island* involves discovery in other ways and is crucial to Creeley's poetry. Resonant in breadth to Olson's *Maximus Poems* (1953-1975), Duncan's *Passages* (1968-1987), Ginsberg's *Howl* (1956) and *Kaddish* (1960), Williams's *Paterson* (1946-1963), and Pound's *Cantos* (1917-1969), *The Island* was Creeley's "first 'long poem'; it was the first piece of serial writing that went on for many days, weeks, and so forth." As such, it opened up, as Creeley relates, "a great deal of possibility for me. . . . I think it permits poems like 'The Finger' to be written." In terms of structure, *The Island* follows a numeric procedure: "Each chapter is an economy of five pages in length, with five chapters to each of the four parts. And five times four is twenty, which is the number of chapters in the book." *The Island,* then, should be viewed as a formal investigation. In this way it shares more with poetry than fiction of the period. Creeley is insistent that *The Island* had no preexisting narrative plan or outline: "I did *not* 'work out the novel in my mind. . . .' I 'worked it out' *literally* as I wrote it."

In 1963 Creeley read and lectured with Olson, Duncan, Ginsberg, Levertov, and others at the Vancouver Poetry Festival, another defining literary occasion. Creeley was a Guggenheim Fellow in 1964, the same year he won the Oscar Blumenthal Prize for thirteen poems in *Poetry* magazine. He also received a Rockefeller Foundation grant in 1965. That year the Scribners edition of *The Gold Diggers and Other Stories* was published, as was the *New American Story,* which he coedited with Donald Allen. In addition Creeley participated in the Berkeley Poetry Conference, another significant gathering of new poets. In 1966 he was visiting professor at the State University of New York at Buffalo. (Although he

Creeley (right) and artist Dan Rice at Black Mountain College, 1955 (photograph by Jonathan Williams)

has had concurrent appointments, his affiliation with Buffalo has been continuous since he became professor of English there in 1967; he was appointed David Gray Professor of Poetry and Letters in 1978 and Samuel P. Capen Professor of Poetry and the Humanities in 1989.) He was the subject of the National Educational Television film *Poetry USA: Robert Creeley* (1966), and a British edition of collected poems, *Poems 1950–1965* (1966), was published. Creeley also edited *Selected Writings of Charles Olson* (1966). This volume, introduced by Creeley, includes the *Mayan Letters* (1953), an extraordinary group of letters written to Creeley while Olson was in Mexico; a selection of Olson's essays and poems; and Olson's defining "Projective Verse" essay. The following year Creeley received the *Poetry* Magazine Union League Civic and Arts Foundation Prize, participated in the World Poetry Conference in Montreal, and, with Donald Allen, edited *The New Writing in the U.S.A.*

At a time when Creeley could have perpetuated the well-received forms of writing in *For Love*, he pushed forward, extending the writing process and his investigation of its materiality, moving from a "generative" to a "conjectural" mode. *Words* (1965) was an important advancement in Creeley's poetic project. In this work Creeley explores several aspects of the possibility of the poem. Adhering consistently to his idea of "measure," the poem emerges in Creeley's work as an active engagement with its own language. As he commented in Berkeley in 1965, "measure" is for Creeley "the actual measure of the speech, the way the words are going . . . the topography or actual ground, in no metaphoric sense, of where it is one is moving. . . . In other words, how does one gain a use of that place where he or she *is*, in no sentimental or enlarging way? How do you get to ground?" Words themselves are experienced as physical in relation to the poet's presence. In *Words* Creeley's work explores a literal physicality of subject matter. This is prefigured by the epigraph (a quote from Williams), which evokes "a counter stress, / born of the sexual shock, / which survives it." In addition to the clearly intended sense of the text as the body, poems such as "The Woman" ("you have left me / with, wetness, pools /

of it, my skin / drips"), "The Dream," and "Distance" also graphically explore elements of the heterosexual experience. This is the element of Creeley's work to which M. L. Rosenthal reacted so vehemently in his well-known essay "Problems of Robert Creeley." (Sherman Paul's often cited response to Rosenthal's attack appeared in a 1975 issue of *Boundary 2*. These two essays provide opposing arguments in the reception of Creeley's work of this period.) The heterosexual experience is reiterated by the recurring image of "the hole," as illustrated in this section of "The Language":

> I heard words
> and words full
>
> of holes
> aching. Speech
> is a mouth.

The image of "the hole" refers to both the incapacities of language and to sexuality from a male viewpoint. In these poems "the hole" is a site of incredible tension. The gendered view of sexuality in these poems may be the most difficult element of Creeley's work to reconcile. In this regard, Charles Bernstein remarks in *Contents Dream* (1986) that "Creeley's work attests to the experience of maleness as a social condition, replete with the troubling and problematic values that are so central a part of that role." Certainly "troubling" values are evoked in these poems. As the tension of gender escalates, *Words* leads to explicit gender-based violence. Wendy Brabner alludes to this mechanism as a tendency by the poet to "confront [his] fears and conquer them through some form of violent action." This brutality emerges most strikingly in "Enough" ("Your body is a garbage can") and "Hello" (where "he" caught "the edge of // her eye and / it tore, down, / ripping.")

Violence can also be seen as a statement of Creeley's poetics, "ripping" language to pieces. Part of this process involves discarding logical relations and embracing verbal positioning resembling Abstract Expressionism in image, representation, and arrangement of compositional elements, thus working to eschew traditional methods of knowing. In his essay "A Sense of Measure" Creeley rejects the assumption that order "can be either acknowledged or gained by intellectual assertion, or will, or some like intention to shape language to a purpose which the literal act of writing does not in itself discover." Early into *Words* logical associations begin to disintegrate as Creeley's writing pursues this course. In "I Keep to Myself Such Measures" the narrator asserts that "there is nothing / but what thinking makes / it less tangible." As the poem "The Window" declares, "I can / feel my eye breaking" or the proposal of "Intervals" that "*who / am I – / identity / singing.*"

Toward the end of *Words* short poems stand out, illustrating modes of abstraction in this collection and also substantially prefiguring work to be collected in *Pieces*. Among these are the four-line poem "The Farm" ("Tips of celery, / clouds of // grass – one / day I'll go away"), "Joy," "A Piece," and "The Box." The last is dedicated to the artist John Chamberlain and uses words as a material to create a work of sculpture:

> Three sides,
> four
> windows. Four
>
> doors, three
> hands.

Like a work of sculpture, the poem has defined, solid parameters. Its construction is less than typical, a fact that the reader (who would expect four sides of a box) observes with the first line. As Creeley has noted about painting, "If no one sees a painter, or, rather, what he is doing – finally, not 'doing' – doesn't he still have *things?*" Creeley's poetics insist here on words as real and concrete components of the poem operating, as Joseph M. Conte has suggested, as "paradigmatic forms whose elements continually evaluate *their own* affinity and dissimilarity." The poem has "things" substance: it attempts to constitute an object with mass – one constructed of common objects – but pieced together the way an Abstract Expressionist sculpture might be. This constellation follows Creeley's account of Pound's instruction that "poetry is a form cut in time as sculpture is a form cut in space." Through this construction, Creeley shows his engagement with "making the world / tacit description / of what's taken / from it."

Creeley collaborated with R. B. Kitaj on *A Sight* in 1967, the same year *Robert Creeley Reads,* a recording of Creeley reading from *Words,* was released. It was followed by *The Finger* (1968), which was illustrated with collages by Bobbie Creeley. *Numbers* (1968), written at the suggestion of Robert Indiana, is a "sequence of poems involved with experiences of numbers," accompanied by ten strikingly colored folio serigraphs by Indiana. The subsequent publication of the sequence in *Pieces,* though textually accurate, cannot convey the visual dimen-

Title page for Creeley's 1962 collection of poems about "marital confusion, loneliness, and isolation"

sion and graphic richness of this earlier presentation of the text.

Creeley's maturation as a poet is closely linked to *The Island*. According to Creeley, writing the novel "led me to feel through things in a more various way.... I'm more at ease with myself; I have much more very literal confidence." As a result Creeley has described the poems collected in *Pieces* as approaching "a far freer context of statement." Preceding the better-known 1969 Scribners collection *Pieces* was the 1968 Black Sparrow Press book of the same title, which has thirteen pages of poems with eight collages by Bobbie Creeley. These full-page collages are composed of fractured images that cohere through their dynamics on the page and are a significant element in the original edition of *Pieces*. The epigraph to *Pieces* is Ginsberg's statement "I always wanted, / to return / to the body / where I was born." Though one critic has contended that this epigraph evidences saturation "by a discourse on masculinity," Creeley has argued for a sense of the body as "the [poetic] 'field' and ... equally the experience of it." As Creeley insists: "It is, then, to 'return' not to oneself as some egocentric center, but to experience oneself as *in* the world, thus, through this agency or fact we call, variously, 'poetry.'" *Pieces* explores this "field" through its engagement with form.

Creeley is direct about this project, opening the collection with an untitled poem:

> As REAL as thinking
> wonders created
> by the possibility —
>
> forms. A period
> at the end of a sentence
> which
>
> began *it was*
> into a present,
> a presence
>
> saying
> something
> as it goes.

Creeley's emphasis on form argues that the tangible presence of discrete words and phrases, freed from conventional syntactic relations, makes the text "real." According to Williams this freeing of words is comparable to what Gertrude Stein had accomplished in *Tender Buttons* (1914). Stein had "completely unlinked [words] ... from their former relationships in the sentence." The goal of such unlinking is to present words as "things" that occupy relational space. These "things," Creeley has written, "are large or small objects, having the fact of space in whatever dimension becomes them." Thus *Pieces* is composed of many fragments of text, "pieces" of poems. "Sometime in the mid-sixties," Creeley later wrote, "I grew inexorably bored with the tidy containment of clusters of words on single pieces of paper called 'poems.'... My own life, I felt increasingly, was a *continuance* ... and here were these quite small *things* I was tossing out from time to time, in the hope that they might survive my own being hauled on toward terminus." Breaks in poetic flow, uncertainty about exactly where a poem begins or ends, and the appearance of seemingly incongruent forms, such as journal entries and fragments of letters, contribute to the interruptive quality of *Pieces*. Some of these "pieces" have conventional titles, but more frequently the poems begin with a phrase in all capital letters, exist within a textual flow set off by bullets, or achieve a status of being curiously attached to titled poems, reflecting the chronological accumulation of this text. *Pieces* emphasizes the "fact of process" of composition in an effort to "trust writing." The proposal is to move the poem into the present; its "realness" will be qualified by "presence" – process being the route to the poem's realization. As to method, in "Here" it is proposed:

> My plan is
> these little boxes
> make sequences . . .[.]

With the "plan" of the journal-like progression of these poems, the following consideration arises:

> Lift me
> from such I
> makes such declaration.

"Such I" alludes to a primary dilemma in *Pieces*. The narrator of these poems, as he tries to be more and more immediate, suffers a consequent split in identity. Creeley has described this dilemma by writing that "As soon as / I speak, I / speaks." In "They" Creeley writes of the mind following what is "true" then adds "and *I* also," indicating the contrast between what is thought and the "I" of the poem. However, this is more than a contrast; it is a constant friction, "*a* 'poet' of such impossibilities 'I' makes up," as Creeley writes in "Echo." The instability of the "I" stands out in *Pieces* for the intensity of conflict; through his later poems and autobiography, Creeley will continue to explore the dimensions of the "I" positioning, in Creeley's words, "the sense of 'I' into poet Louis Zukofsky's 'eye' – a locus of experience, not a presumption of expected value."

Creeley was a visiting professor at the University of New Mexico when *Words* and *Pieces* were published. The retrospective collection *The Charm* (1969) was also published by the Four Seasons Foundation at this time, bringing back into print works that were long unavailable. These include "Return," poems from *The Kind of Act Of,* and parts of *Le Fou* not included in *For Love*. In 1970 Creeley participated in the International Poetry Festival at the University of Texas and the Neuvième Biennale Internationale de Poèsie at Knokke-le-Zoute, Belgium. His books, *A Quick Graph: Collected Notes and Essays* and *The Finger: Poems 1966–1969* were published in the same year. He moved his family to Bolinas, California, in order to serve as visiting professor at San Francisco State College in 1970–1971. This move provided an opportunity to meet various writers, including Joanne Kyger, Tom Clark, Aram Saroyan, Bill Berkson, Clark Coolidge, David Meltzer, and Philip Whalen, among others. Also in the community were musicians such as the Jefferson Airplane, the Rowan Brothers, and Steve Swallow (who would later compose music based on Creeley's poems) and artists such as Kitaj and Arthur Okamura (who Creeley knew from Majorca). Some of Creeley's poems and prose of this period directly draw on Bolinas experiences and persons. In 1972 his radio play, *Listen,* was produced in London. *Listen, A Sense of Measure* (essays and an interview), and *A Day Book* were all published the same year. *A Day Book* is composed of a series of journal entries and "In London," a selection of poems later printed in Creeley's *Collected Poems, 1945–1975* (1982). "In London" contains a substantial number of poems, similarly performing an act of "direct recording." Fragmentary poetic "pieces" dominate the first part of the collection with many references to "place." The original publication of *A Day Book* was a lavishly produced collaboration with Kitaj.

In 1973 Creeley's *Whitman: Selected Poems,* a tribute to Whitman's "instruction that one speak for oneself," was published. This same year he established residence in Buffalo, where he and his family maintain a permanent home. His next few collections of poetry, *His Idea* (1973), *Thirty Things* (1974), *Backwards* (1975), and *Away* (1976), are the only sections in the *Collected Poems* which appear as originally printed by small presses. These books record the "factual life" of the development of Creeley's poetics during this period and, as such, exemplify an insistence on the practice of writing. *His Idea* seems to stand as a continuation and denouement of his work in *Pieces,* a single series of individual fragmentary poetic pieces each beginning with a phrase or word in capital letters, in which "Days go by / uncounted." *Thirty Things* consists of thirty, mostly occasional, individually titled short poems maintaining, in Duncan's words, "a tension between resignation and resolution." Some of these poems are directed to individuals, sometimes evoking Bolinas by direct reference or facts of geography; other poems express a consciousness of time's passing such as "Still" ("Still the same / day? / Tomorrow") and "One Day" ("One day after another – / perfect. / They all fit"). *Backwards* (the title suggesting direction, though regressive), is a short collection, similarly composed of discrete, titled poems showing a like discontent with time standing still, as in the two-line title poem of the collection: "Nowhere before you / any of this." *Away* consists of poems in a variety of formats, individually titled poems as well as sequences of poetic phrases separated by

Cover for Creeley's 1972 book of journal entries and poems

bullets. "For My Mother: Genevieve Jules Creeley" stands as an excursion into thematic material that emerges strongly in his later works. In this collection the idea of abstraction in poetry is again questioned. In "Berlin: First Night & Early Morning," an interlocutor interrupts a flow of poetic phrases to say:

You: "too abstract,
try it,
all,
over again . . .[.]"

As Creeley moves forward in his career, it is evident that he will leave behind "exacerbated tension" and explore, as in the Williams epigraph to *In London,* "But what to do? and / What to do next?"

Creeley's prose works are also important in understanding his literary accomplishments. As Stephen Fredman relates, "Creeley [says] that not until toward the end of writing *Words* . . . did he begin to loosen his sense of formal necessity, to free ideas and expressions from the imperative of reaching a rested conclusion. He began then to write serial compositions, starting with *Pieces* . . . and continuing in the three prose books collected in 'Mabel.'" The three texts in "Mabel: A Story" as published in the *Collected Prose* (1984), are "A Day Book," "Presences," and "Mabel: A Story." *A Day Book,* writes Creeley, is "precisely what it says it is, thirty single-spaced pages of writing in thirty similarly spaced days of living." As such, it resonates with *Thirty Things* and its accompanying prose works. Written for a collaboration with Kitaj, "A Day Book" is direct, exploratory, and improvisatory; it is principally an investigation of the possibilities of form, particularly of writing within specified form. (As originally published, *A Day Book* is physically an experiment in form. In an oversized format, Creeley's writing appears "solid" because of its large font and is juxtaposed with striking large plates by Kitaj.) *Presences,* proposed because Marisol had seen *Numbers,* Creeley's collaboration with Robert Indiana, was originally published in the United States by Scribners as *Presences: A Text for Marisol* (1976). It was designed with large type and little gutter area and was accompanied by photographs of sculptures by Marisol. The text of *Presences* consists of "a series of improvisations upon Marisol's images, both sculptural and personal." Fredman has noted that in *Presences* "there is much for Creeley to identify with in Marisol's art: the isolation, the immensity of the heart, the repetition of frontality, and the constant use of the self." The text contains five sections of six single-spaced pages of text in permutations of 1-2-3, beginning and ending with 1. Of this numeric arrangement, Creeley writes, "I wanted a focus, or frame, with which to work, and *one, two, three* seemed an interesting periodicity of phrasing." "Mabel," "begun as an imagination of women" for a collaboration with Jim Dine, contains five sections of six single-spaced pages of text in permutations of the same 1-2-3 sequence.

Creeley's essay *Was That a Real Poem or Did You Just Make It Up Yourself* (1976) is an extraordinary statement of poetics; it has also served as an exemplar for the craft of the essay, influencing contemporary writers. A distinctly individual blend of autobiography and poetics, *Was That a Real Poem* presents to the reader an account of Creeley's development as a poet. Acknowledging that, "my thinking about poetry may or may not have anything actively to do with my actual work as a poet," he describes his work based on personal feelings, his past, and a resilient sense of poetry qua poetry. He believes that "what we call *poems* are an intrinsic fact in the human world whether or not there be poets at this moment capable of their creation." The matter of poetic creation is beyond him, and although at times a source of despair, it is somehow comfortingly *there* nonetheless. In a sense, this manifesto allows for personal involvement in writing while letting the process stand by itself, beyond the personal. Moreover, this statement is a definitive accomplishment, providing a locus of activity for the contemporary poem, "a *place,* in short, one has come to, where words dance truly in an information of one another, drawing in the attention, provoking feelings to participate."

Creeley divorced his second wife, Bobbie, in 1976 and, in the spring of the same year, embarked on an extensive reading tour of New Zealand, Australia, and Asia. In 1977 Creeley married Penelope Highton, whom he met in New Zealand. He would have two children with Penelope, William and Hanah, born in 1981 and 1983, respectively. The next year he accepted the position of David Gray Professor of Poetry and Letters at Buffalo and found a new publisher, New Directions. Creeley's first New Directions book was *Hello: A Journal, February 29–May 3, 1976* (1978), poems written in the form of journal entries from tours and readings Creeley had given in Fiji, New Zealand, Australia, Singapore, the Philippines, Malaysia, Hong Kong, Japan, and Korea. Creeley abandons the use of heterogeneous forms, structuring the poems as discrete units, often in stanzaic form or as poetic fragments under the aegis of a specific geographic location. The result is a reflective and consistent tone of voice in which a sense of hope emerges as a direct result of observations. For example, in "Soup (Palmerston North, New Zealand)":

Bye-bye, kid says,
girl, about five –
peering look,
digs my one eye.

A change of direction is indicated in "Out Here." The narrator of the poem sits in the corner of an airport bar. He relates how "a few minutes ago" he was thinking in a sexual mode characteristic of his earlier writing, but then retracts the thought, explaining that such an approach applies "not any more," because "it's later." The speaker finds himself "spooked, tired, and approaching / my fiftieth birthday" as he relates his in-transit status, surroundings, and thoughts. Then, finally:

I'll be a long way away
when you read this – and I won't
remember what I said.

Correspondingly, the text itself is "not any more, it's later," a direct invocation of the title of his following collection of poems. *Robert Creeley: A Gathering* (1978) seems to have marked a change in the direction of the author. This collection, a 570-page issue of *Boundary 2,* sought to present "critical orientations and disclosures ... as diverse as the occasion of a first gathering demands," effectively constituting a forum on Creeley's innovations and assembling many important statements about his work. The following year, *Was That a Real Poem and Other Essays* was issued by the Four Seasons Foundation, making available, along with the title essay, some previously uncollected poetic statements, including his introduction to *Whitman: Selected Poems* and "The Creative," "On the Road," and "A Sense of Measure."

Even as these collections were published, Creeley was entering a completely different arena of poetic investigation, abandoning formal fragmentation. It might have been a disillusionment with such experiment, as he writes in "Still Too Young," "all I'd hoped for / is going up in abstract smoke" and "I'm too old to do it again / and still too young to die" or because simply that formal discontinuity does not suit the project at hand. As indicated in the poem "Age," the poet has accepted the altered situation or at least does not indicate he wishes to change it: "There's no surprise now, / not the unexpected / as it had been. He's agreed / to being more settled." The facts of Creeley's changed circumstances shift the "I" into a retrospective position, leading to reflection. The titles of Creeley's five later collections of poetry all consider various forms of reflection, whether relating to memory, in *Later* (1979) and *Memory Gardens* (1986), optics, in *Mirrors* (1983) and *Windows* (1990), or acoustics, in *Echoes* (1982).

Concomitant with this exploration, Creeley increasingly worked with different surface modes of the poem; the subject matter of his poems shifts to include natural phenomena, a more accepting gesture toward the marital relationship, a willed, though not always positive, sense of human endurance, and, as Fred Moramarco writes in an article collected in *Robert Creeley's Life and Work* (1988), content "characterized by a greater emphasis on memory, a new sense of life's discrete phases, and an intense preoccupation with aging." These facts of aging, significantly, are facts of human life and, like childhood ("When I was a kid, I / thought like a kid – // I *was* a kid, / you dig it"), are presented in an "angle of incidence" to the form of the poem. Age has many faces in these later poems, from simple facts of observation ("now my hands are // wrinkled and my hair / goes grey" or "I am no longer / one man – // but an old one") to a sense of despair, such as in "Prayer to Hermes," where "these days / of physical change" bring to the poet "a weakness, / a tormenting, relieving weakness." Here this reference is tied to the advent of winter, the season likened to the poet's body. Writing, as in earlier works, continues to be simultaneous with the body. Creeley insists on a "*physical* sentence." It is a sentence on the page; it is a "life" sentence (physical aging). And as a writer, his sentences also become "older." The poem's activity must similarly accommodate this "sentence," and the structure of the poem must respond to and feel "the meat contract, // or stretch, upon bones" with a sense of compacted attention and tightened poetics. With the contraction of subject matter in these poems comes a greater potency of the activity within the circumscribed area. His insistent return to the idea of "echoes" delineates a poetic form of his own making. (The idea is introduced early in his writing, evoked significantly in *Words* and *Pieces,* and brought to focus by the fact that an entire volume is titled *Echoes*.) The "echo," as investigated by Creeley, is expressed as something ultimately beyond grasp; what becomes foregrounded then is the act of *reaching* for the echo. In "My Own Stuff" Creeley expresses this idea: "It is a / flotsam I could / neither touch quite / nor get hold of . . . yet / insistent to touch / it . . . [I] kept poking, trying / with my stiffened / fingers to get hold of" and "its substance I had / even made to be / there its only / reality my own." Of course, grasping for echoes can only leave the hand empty. The place at which the poet has arrived seems an empty one because "there is / no one here but words, / nothing but echoes." Yet these poems are not empty. Creeley's later work is true to his longstanding commitment to the poem as direct observation and to textual elements of the poem constituting real objects.

The poem "Helsinki Window" in *Windows* is an example of a poem where words work in a circumscribed area. This poem is dedicated to the Finnish-born writer Anselm Hollo, whom Creeley has elsewhere described as "a man who lives daily, humanly, in the physical event of so-called existence." In "Helsinki Window" direct observation clings to Creeley's compositional urge, "daily" and "humanly" bounded in nine poems of roughly twelve lines each. Here, "at the edge of this / reflective echo," each rush of phrasing issues from what is immediate: the "same roof," "old sky," "windows now lit," "a bicycle / across the way," a "spare pool / of light." The elements of the poem are drawn from what may almost exclusively appear through a single window. The window's view is full of *things;* yet the things cohere only within the compacted structure of their frame. The window, as a frame, does not create a "picture," rather an area for the relations of "things." Creeley has pointed out that Abstract Expressionism "regains the canvas as surface – or literally imposes as significant surface anything on which the painting occurs" thereby causing painting to lose "its historical sense of *picture,* insofar as our sense of a picture seems to imply something which is referential." The same focus on surface applies to Creeley's poems and the result is "classic emptiness." Yet it is an emptiness which is full, the poem itself – "it / is *there* here *here*" – and it is concrete and immovable – "no / other thing can for a / moment distract it be / beyond its simple space."

"Sonnets," in *Echoes,* consisting of six twelve-line and one fourteen-line poem, also witnesses form compacting the information of the poem. The "window" of observation in this case is memory. Memory pushes "a twisting / away tormented unless" into the "presence" of writing. The rhythm of these poems arises from the regularity that the enclosure of form exerts on the text. This text can be highly rhythmic at times:

> teeth wearing hands wearing
> feet wearing head wearing
> clothes I put on take now
> off and sleep or not or sit . . .[.]

Of greatest importance is the density of the writing. This density, Creeley has stated, is "a situation in which each word becomes not so much singular as though its meaning were to be abstracted from its companies, but each word is a possible pivot or shift or relocation of what it is that I seem to be . . . trying to get said." In "Sonnets" there is the feature of past events (including literary precedent, as the use of the sonnet form suggests); there is the action of the poet's "I," bridging from these events to the present; but equally, there is the active sense of writing as an immediate and central "thing" operating to "get said" what is necessary. Despite the element of recollection, the focus of this work is not on what is reflected but on reflection itself, as in the collection's epigraph, "echo or mirror seeking of itself, / . . . makes a toy of Thought." Since thought has itself become a compositional element, "*I* is not / the simple / question // after all, / nor *you* / an interesting answer." What emerges in this collection is the interrogation of reflection, its circumscription by thought processes, and its condensation within

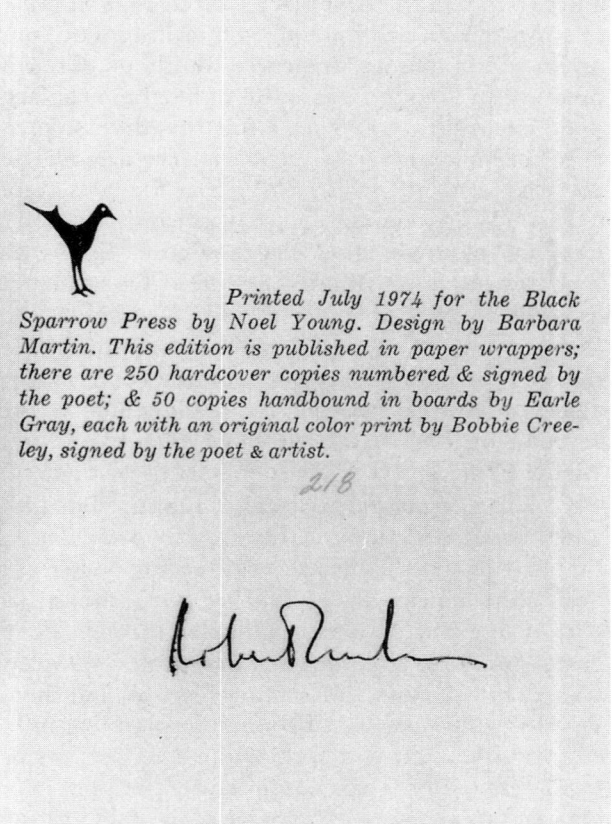

Title page and signed colophon for Creeley's 1974 book, which was illustrated by his wife

formal frames. Thought itself becomes "consequential, / itself an act, a // walking round rim / to see what's within."

The first two volumes of *Charles Olson and Robert Creeley: The Complete Correspondence* were published by Black Sparrow Press in 1980, marking the first availability of this copious correspondence. Two major volumes of collected writings also appeared in the early 1980s. *The Collected Poems of Robert Creeley, 1945–1975,* including a section of "Uncollected Poems," works printed in magazines that never appeared in books, and *The Collected Prose of Robert Creeley,* reprinting Creeley's diverse excursions into the genre, including *The Gold Diggers, The Island, Listen,* and "Mabel: A Story." These were followed in 1989 by *The Collected Essays of Robert Creeley,* bringing together a diversely published body of critical writing that would be nearly impossible to assemble otherwise. The influences for these essays include Lawrence, Edward Dahlberg, Williams, and Olson, though Pound is also clearly in Creeley's mind. These writings manifest Creeley's agreement with Pound that critical writing should come "from those defined in the arts relating" and follow creative works as "the two feet of one biped" follow each other.

The same year Creeley's autobiography not only explored his personal past but extended the form of the autobiography itself. Much as Creeley's *Was That a Real Poem* offered an alternative vision of the literary essay, his autobiography, first published in the Contemporary Authors Autobiography Series (1989), reconsiders the narrative. (It was published separately as *Autobiography* in 1991 and republished in Tom Clark's *Robert Creeley and the Genius of the American Common Place* in 1993.) Though asserting that "there is an awful, self-consciously recognized limit to what may be called my sincerity," Creeley is perhaps unparalleled in making his feelings appear as facts. Thus, though this autobiography has facts necessary to its genre, Creeley's insistence that "we believe a world or have none" creates a narrative of a life lived inseparable from a writer's poetics. "It is the pleasure and authority of writing that it invents a life to live in the first place," Creeley maintains. His prose is an example of such writing. Focusing on his youth, family, and early adulthood, Creeley presents an account that is full of biographical facts and also argues for "a scale for [the] diverse presence" of humanness. As such, this very personal account presents the facts of his life as the facts argue a life. "What cannot be objectified is oneself," Creeley acknowledges, "Yet the fiction, finally for real, is attractive." In this way, the self is presented as only one possible version of lived selves. As Bernstein notes, "Amidst the onslaught of events, the self provides only an *apparent* centering or agency, always subject to readjustments and recentering." What the autobiography manifests is "fictions" made fact in the literal act of writing.

The Essential Burns was also published in 1989 and offered selections, in Creeley's words, among his "first delights in hearing and reading poetry as a boy." The appearance of Creeley's correspondence with Irving Layton made available documents covering, "perhaps the most seminal period in recent Canadian and American literary history" and showing the rich transnational communication that influenced the literature of both countries. Creeley's *Selected Poems* (1991) was compiled by the poet from his vast output. Creeley's choice of the contents was guided by his feelings about the place of poetry: "Why poetry? Its materials are so constant, simple, elusive, specific. It costs so little and so much. It preoccupies a life, yet can only find one in living. It is a music, a playful construct of feeling, a last word and communion." Standing by his earlier work, he drew heavily on *The Charm* and *For Love,* with an even but sparser inclusion of his subsequent writings. The last poem in the book, "Body," is set off as if making a statement, concentrating on the presence of the body and the physical circularity of human experience, suggesting themes from *Pieces,* though this poem is from *Echoes.* In 1993 Olson's *Selected Poems* was edited by Creeley, and it offered him the opportunity of presenting his personal selection of the works of the poet who was "the first practical influence upon me of a contemporary." Accordingly, Creeley acknowledges the selections as "unavoidably [his] own" and proposes that they be read keeping in mind that their relationship provided "a measure, an unabashed response to what either might write or say." Creeley points to the precedent of their collected correspondence to suggest that this collection stands as Creeley reading Olson, certainly true to their writing relationship.

Creeley has unarguably attained and has accepted the status of an elder statesman of poetry. His recognition during the 1980s and 1990s has been impressive. His works have been translated and published in the Netherlands, Germany, Austria, Spain, Mexico, Italy, Denmark, Norway, Czechoslovakia, and France, among other countries. His many awards have included the Shelley Memorial Award from the Poetry Society of America (1981), a National Endowment for the Arts grant (1982), two DAAD Fellowships in Berlin (1983 and 1987), the Leone d'Oro Premio Speziale, Venice (1985), the Poetry Society of America Frost Medal (1987), a Distinguished Fulbright Award as Bicentennial Chair in American Studies, Helsinki University, Finland (1988), distinguished professor rank at the State University of New York (1989), the Walt Whitman Citation, which included service as New York State Poet (1989–1991), and an honorary doctor of letters granted by the University of New Mexico (1993). He also received the Horst Bienek Lyrikpreis from the Bavarian Academy of Fine Arts in Munich (1993), Germany's highest honor to a living foreign poet and an award that few other American authors have received. Perhaps most significantly Creeley enjoys the distinction of having been elected to the American Academy of Arts and Letters.

Though the physical presentation of Creeley's poems has changed considerably through his career, Creeley's attention has remained focused on writing as determined by the facts of experience, rather than by preconceived forms. This attention has advanced a sense of writing that, in addition to having roots in what is intensely familiar (the "common place" of his later lectures), also has no determined limit of genre. Some of his most notable later literary accomplishments include his autobiography, the occasional prose essay, interviews, orally presented poetry, and lectures. Aside from teaching he makes literary tours of the United States and the world. In addition to these genres of "writing," Creeley's commitment to discourse between the arts and technology have witnessed his poems being set to jazz (Steve Swallow's *Home,* appearing in 1980, and Steve Lacy in the 1985 release *Futurities*) and his participation in films (*Creeley,* produced by Documentary Research in 1988 and *Robert Creeley,* released as part of the Lannan Literary Series in 1990), CD-ROM (*Poetry in Motion*), and recently, the Internet. He has, however, expressed caution about this latest medium, suggesting that one must be conscious of "what happens to knowledge when its traditional relation to ordering and retrieval (memory)

Poem and monoprint from Creeley's Thirty Things

Of great significance have been his contributions to gallery and public art (such as his eight poems engraved on the bollards at Seventh and Figueroa Streets in Los Angeles) and his collaborations with visual artists. Creeley's engagement with the visual arts has been a consistent factor in his writing. His recent collaborators include Susan Barnes, John Chamberlain, Francesco Clemente, Cletus Johnson, Susan Rothenberg, Robert Therrien, and Martha Visser't Hooft. His collaborations with Johnson, in which Creeley provided texts for "visual transformation" by Johnson, were exhibited at galleries in Buffalo and in New York. Creeley's work with Clemente has been significant. *It* (1989), published by Bruno Bischofberger in Zurich, is a substantial volume containing sixty-four pastels by Clemente and twelve poems by Creeley. In *7 & 6* (1988) Creeley provides poems for seven of Therrien's works. (The collaboration also includes prose by Michel Butor.) Creeley's responses to the works are factual, sometimes humorous, and address the image (presented on the facing page) with a "literalism [that] twists into pleasurable knottiness." Creeley's collaborations are "wholly in keeping with Therrien's own no-nonsense approach." *Life & Death* (1993) is particularly illustrative of the intensity with which Creeley approaches such collaborations. Creeley viewed Clemente's series *The Black Paintings* (1991–1993) on a bright Sunday morning in the artist's studio. He relates that the light made the paintings "tangible" to him and that "for me they tell a story, a very old one, of how humanly we live both as one and as many, in a world particular to our lives but also far vaster and more communal than such personal limits can ever acknowledge." Creeley wrote a poem for each image but returning home realized he wished to say more and so wrote a second half to each poem. True to the spirit of this collaboration, the paintings are reproduced in *Life & Death* with the poems in the order Creeley viewed them and with bullets separating the pair of textual creations for each painting.

Creeley remains a popular poet, though recent critical attention to his work has varied in frequency. Two recent critical "gatherings" in *Poetics Journal* (1991) and *Sagetrieb* (1991) include several articles on Creeley's work. A special issue of *The Review of Contemporary Fiction* recently appeared, focusing on Creeley's prose and offering contemporary evaluations of *The Gold Diggers, The Island,* Creeley's autobiography, and his experimental texts. Two new books have appeared: *Tales Out of School: Selected Interviews* (1993), which has five lengthy interviews (four from *Contexts of Poetry*), and *Robert Creeley and the Genius of the American Common Place* by Clark. In this hybrid critical and biographical work Clark focuses on Creeley's view of the "common-place" through interviews with Creeley, a transcript of "Some Senses of the Commonplace" (a talk Creeley gave at the New College of California in 1991), and Creeley's "Autobiography." The most important form of recognition, however, has been the very real and lasting changes Creeley's work has effected in postmodern poetry. Clark, in *The Poetry Beat,* credits Creeley with "creating a mass-democratic cottage verse that ushered postwar American poetry outside the halls of fading ivy, and in turn made possible a whole new generation of 'post modern' academicism." Thus, the visibility of the tradition for which Creeley speaks, as Bernstein points out, is substantially greater than it was thirty years ago, and "while this tradition and its current manifestations are hardly in the mainstream of official culture, the magnitude of related subterranean, ground level, and occasionally aboveground anthologies, presses, magazines, and public readings is incomparably greater, and far more entrenched – a circumstance that Creeley has significantly helped to bring about."

Letters:

Charles Olson & Robert Creeley: The Complete Correspondence, volumes 1–8 edited by George F. Butterick, volume 9 edited by Richard Blevins (Santa Barbara, Cal.: Black Sparrow Press, 1980–);

Irving Layton & Robert Creeley: The Complete Correspondence, 1953–1978, edited by Ekbert Faas and Sabrina Reed (Montreal: McGill-Queen's University Press, 1990).

Interviews:

Richard Eberhardt, interview with Creeley [audiotape] (Washington: Library of Congress, 1 June 1961);

Donald Allen, ed., *Contexts of Poetry: Interviews 1961–1971* (Bolinas, Cal.: Four Seasons Foundation, 1973);

Ekbert Faas, *Towards a New American Poetics: Essays and Interviews* (Santa Barbara, Cal.: Black Sparrow Press, 1979), pp. 165–198;

Ernesto Livon Grosman, interview with Creeley [radio broadcast] (New York: Radio Reading Project, 1992);

Tales Out of School: Selected Interviews (Ann Arbor: University of Michigan Press, 1993).

Bibliographies:

Mary Novik, *Robert Creeley: An Inventory, 1945–1970* (Kent, Ohio: Kent State University Press, 1973);

Willard Fox, *Robert Creeley, Edward Dorn, and Robert Duncan: A Reference Guide* (Boston: G. K. Hall, 1989), pp. 1–170;

Vincent Prestianni, "Robert Creeley: An Analytical Bibliography of Bibliographies," *Sagetrieb,* 10, nos. 1–2 (1991): 209–213.

References:

Charles Altieri, "Robert Creeley's Poetics of Conjecture," in his *Self and Sensibility in Contemporary American Poetry* (Cambridge: Cambridge University Press, 1984), pp. 101–131;

Athanor, special issue on Creeley, edited by Douglas Calhoun, 4 (Spring 1973);

Charles Bernstein, "Creeley's Eye and the Fiction of the Self," *Review of Contemporary Fiction,* 15 (Fall 1995): 137–140;

Bernstein, "Hearing 'Here': Robert Creeley's Poetics of Duration," in his *Contents Dream: Essays 1975–1984* (Los Angeles: Sun & Moon Press, 1986), pp. 292–304;

Boundary 2, special issue on Creeley, edited by William V. Spanos, 6 (1978);

Tom Clark, *The Poetry Beat: Reviewing the Eighties* (Ann Arbor: University of Michigan Press, 1990);

Clark, *Robert Creeley and the Genius of the American Common Place* (New York: New Directions, 1993);

Joseph M. Conte, "One Thing Finding Its Place with Another: Robert Creeley's *Pieces,*" in his *Unending Design: The Forms of Postmodern Poetry* (Ithaca: Cornell University Press, 1991), pp. 87–104;

Robert Duncan, "A Reading of *Thirty Things*," *Boundary 2*, 6 (Spring/Fall 1978): 293-299;

Cynthia Edelberg, *Robert Creeley's Poetry: A Critical Introduction* (Albuquerque: University of New Mexico Press, 1978);

Ekbert Faas, "Robert Creeley," in *Towards a New American Poetics: Essays and Interviews* (Santa Barbara: Black Sparrow Press, 1979), pp. 147-164;

Arthur L. Ford, *Robert Creeley* (Boston: Twayne, 1978);

Edward Halsey Foster, "Robert Creeley, Poetics of Solitude," in his *Understanding the Black Mountain Poets* (Columbia: University of South Carolina Press, 1994), pp. 81-121;

Stephen Fredman, "'A Life Tracking Itself': Robert Creeley's *Presences: A Text for Marisol*," in his *Poet's Prose: The Crisis in American Verse*, (Cambridge: Cambridge University Press, 1990), pp. 57-100;

Robert Hass, "Creeley: His Metric," in his *Twentieth-Century Pleasures: Prose on Poetry* (New York: Ecco, 1984), pp. 150-160;

Anselm Hollo, *Sojourner Microcosms* (Berkeley, Cal.: Blue Wind Press, 1977);

Bruce Jackson and Diane Christian, *Creeley* [filmstrip] (Buffalo: Documentary Research, 1988);

Lewis MacAdam and John Dorr, *Robert Creeley* [videocassette] (Los Angeles: Lannan Foundation, 1990);

Ann Mandel, *Measures: Robert Creeley's Poetry* (Toronto: Coach House Press, 1974);

"Multiples & Objects & Books," *Print Collector's Newsletter*, 24 (January/February 1994): 227-228;

Charles Olson, "For R. C.," *Olson: the Journal of the Charles Olson Archives*, 6 (Fall 1976);

Olson, *Selected Writings of Charles Olson* (New York: New Directions, 1966);

Sherman Paul, "A Letter on Rosenthal's 'Problems of Robert Creeley,'" *Boundary 2: A Journal of Postmodern Literature*, 3 (Spring 1975): 747-760;

Paul, *The Lost America of Love: Rereading Robert Creeley, Edward Dorn, and Robert Duncan* (Baton Rouge: Louisiana State University Press, 1981), pp. 1-73;

Ted Pearson, "Robert Creeley and the Politics of the Person," *Poetics Journal*, 9 (June 1991): 159-164;

Marjorie Perloff, "Four Times Five: Robert Creeley's *The Island*," *Boundary 2*, 6 (Spring/Fall 1978): 491-507;

M. L. Rosenthal, "Problems of Robert Creeley," *Parnassus*, 2 (Fall/Winter 1973): 205-214;

Ron Silliman, "Language, Realism, Poetry," in his *In the American Tree* (Orono: National Poetry Foundation, University of Maine at Orono, 1986), pp. xv-xxiii;

Warren Tallman, *Three Essays on Creeley* (Toronto: Coach House Press, 1973);

Carroll F. Terrell, ed., *Robert Creeley: the Poet's Workshop* (Orono: National Poetry Foundation, University of Maine at Orono, 1984);

Jack Tworkov, *The New American Painting As Shown in Eight European Countries, 1958-1959* (New York: Museum of Modern Art, 1959);

Robert Von Hallberg, "Robert Creeley and John Ashbery: Systems," *American Poetry and Culture, 1945-1980* (Cambridge, Mass.: Harvard University Press, 1985), pp. 36-61;

John Wilson, ed., *Robert Creeley's Life and Work: A Sense of Increment* (Ann Arbor: University of Michigan Press, 1988);

William Carlos Williams, *Imaginations,* edited by Webster Schott (New York: New Directions, 1970).

Papers:

The major collection of Creeley manuscripts and correspondence is housed in Special Collections, Stanford University; other collections include those at the Beinecke Rare Book and Manuscript Library, Yale University (correspondence with William Carlos Williams); the Harry Ransom Humanities Research Center, University of Texas, Austin (correspondence with Ezra Pound and Louis Zukofsky); the John M. Olin Library, Washington University (manuscripts and correspondence predating 1965); the Lilly Library, Indiana University (manuscripts and correspondance with Cid Corman); the Simon Fraser University Library (correspondence with Richard Emerson); and the University of Connecticut Library, Storrs (correspondence with Charles Olson).

Irving Feldman
(22 September 1928 –)

Willard Spiegelman
Southern Methodist University

BOOKS: *Works and Days and Other Poems* (Boston: Little, Brown / London: Deutsch, 1961);
The Pripet Marshes and Other Poems (New York: Viking, 1965);
Magic Papers and Other Poems (New York: Harper & Row, 1970);
Lost Originals (New York: Viking, 1972);
Leaping Clear and Other Poems (New York: Viking, 1976);
New and Selected Poems (New York: Viking, 1979);
Teach Me, Dear Sister and Other Poems (New York: Viking, 1983);
All Of Us Here and Other Poems (New York: Viking, 1986);
The Life and Letters (Chicago: University of Chicago Press, 1994).

Irving Feldman is one of the rare American poets who have made a successful crossing from high modernism to what is generally labeled the postmodernist era. He has continuously revised his style and reimagined his poetic personae, while at the same time remaining true to a sensibility that combines Jewish earnestness and moral intensity with vigorous humor and a skepticism tempered by hope. His work has become, during the past three and a half decades, more distinctive and confident. In retrospect he may turn out to have been one of the few poets to have reflected, or perhaps even to have defined with prescience, American culture and sensibility at the end of the twentieth century.

Irving Mordecai Feldman was born on 22 September 1928 in the Coney Island section of Brooklyn, New York. His parents were Russian Jews who came to America in 1906. Like many children of immigrant parents, Feldman was educated at the City College of New York (B.S., 1950); he then went on to Columbia University for graduate work (M.A., 1953). Most of his life has revolved around universities. From 1954 to 1956 he taught at the University of Puerto Rico, where he met the sculptor Carmen Alvarez del Olmo. Married in 1955, they have

Irving Feldman (courtesy of the Lilly Library, Indiana University)

one son, Fernando. Between 1956 and 1958 he lived in Paris and Lyons as a Fulbright Fellow; he has also lived at various times in Palma.

After six years at Kenyon College (1958–1964), Feldman moved to Buffalo to join the faculty of the State University of New York, where, since 1990, he has been the SUNY Distinguished Professor of English. Feldman's reputation, modest at first, has grown steadily as he has come to speak more authoritatively for the society he represents and addresses. During the course of his career he has accumulated important grants and honors: fellowships from the Ingram-Merrill and the Guggenheim Foundations, the Academy of American Poets, and the National Endowment for the Arts were capped by a coveted five-year prize from the MacArthur Foundation in 1992. Both *The Pripet Marshes and Other Poems* (1965) and *All Of Us Here and Other Poems* (1986) were nominated for the National Book Award.

Feldman has carved out for himself a unique place within the literary landscape. Perhaps because he was city born and bred, he has never been a nature poet. In *The Poetry of Irving Feldman* (1992), edited by Harold Schweizer, critic David Huerta notes, "the poetic space of Irving Feldman is the human condition in modern cities." Feldman's strongest precursors include Charles Baudelaire and Franz Kafka. Description of the external world, especially one devoid of human beings, figures less prominently in his work than in the work of most other major American poets. At the same time, he is often down-to-earth and colloquial in his diction, but he is seldom overtly confessional or conventionally autobiographical. The reluctance to parade his own feelings may have resulted from his own coming of age during the waning of high modernism. He has always worked by indirection and evasion, obliquely and wryly.

Feldman seems to have taken to heart many of the dictates and principles first suggested by T. S. Eliot and Ezra Pound, although his poetic style does not resemble theirs. As was true for Wallace Stevens, Feldman gives only occasional glimpses of his own family and life. He fits John Keats's famous formula: "A man's life of any worth is a continual allegory." Keats's observations on William Shakespeare are also appropriate to Feldman: "[he] led a life of allegory: his works are the comments on it." Feldman's poems comment upon but do not inevitably reproduce his life and lack Robert Lowell's autobiographical glamour.

Although he shares similarities with other contemporary poets, Feldman has developed his own inimitable poetic voice and style. Like Ben Belitt and Theodore Weiss, two older poets with a comparable Jewish moral intensity, Feldman has stood by difficulty as a means of achieving aesthetic ends. While Feldman, like Robert Pinsky and the later Howard Nemerov, exemplifies a Jewish interest in ethical and philosophical questions, he often veers away from the contemplative Horatian "middle" voice into a far more experimental register. From the beginning of his career he has not been afraid to reassert magniloquent diction and an almost baroque syntax in his poetry. Furthermore, unlike such contemporaries as Anthony Hecht, James Merrill, and Richard Wilbur, who have all stayed closer to formalist principles, Feldman has ventured into starker and freer stylistic territory. Reviewing *Leaping Clear and Other Poems* (1976) in *Poetry,* the poet-critic J. D. McClatchy identified in 1977 two of Feldman's salient qualities: an ability to handle raw social and political issues and a "kinetic voice which, volume by volume . . . grows more mercurial, authoritative, and interesting."

Although Feldman's poetry has a modernist impersonality, it also often possesses – especially in the years following *Teach Me, Dear Sister and Other Poems* (1983) – an idiosyncratic brand of postmodernist irony. Feldman, like John Ashbery, often writes poems that lack an autobiographical center, though Feldman's work is in most ways the opposite of Ashbery's sprawling, discontinuous poetry. Feldman addresses or describes nameless people, as Walt Whitman did in "Crossing Brooklyn Ferry," and treats some people as though they were art objects. Often he seems to prefer to evoke rather than to record straightforwardly, to approach obliquely rather than to tackle his subjects head-on.

It is a significant happenstance that the first word in Feldman's first book, *Works and Days and Other Poems* (1961), is *God* and the last word in the mordant, whimsical last poem in *The Life and Letters* (1994) is *dark*. At the beginning of his career Feldman, like many poets of his generation, was a slavish imitator of William Butler Yeats, hewing strictly to the styles and lessons of high modernist poetics. From a mythic, semireligious outlook on the world, Feldman moved steadily toward a psychological, secular understanding of its bleakness.

The poems in his first volume (the title of which he borrowed from Hesiod) examine such themes as pain, wonder, and burdens unasked for though endured (as in the case of "The Saint" and "Prometheus") through personae that are often historic or divine agents of doom. The speaker in "The Ark" announces, "I learned all history's a *pogrom*." The volume combines the two elements Matthew Arnold in *Culture and Anarchy* (1869) labeled the major components of Western civilization, the Hellenic spontaneity of conscience and the Hebraic strictness of conscience, or moral rigor. The first book also contains a six-poem sequence on Francisco José de Goya that anticipates what many regard as Feldman's most original and exciting work, the title sequence in *All Of Us Here.*

As in his first volume, Feldman in *The Pripet Marshes* poses many of the post-Holocaust questions that exercised most of the poets who came of age during World War II. The title poem, a powerful elegy for the Holocaust victims, is typical of Feldman's seriousness and of his handling of dark themes. Written in prose with some Old Testament cadences, it begins:

> Often I think of my Jewish friends and seize them as they are and transport them in my mind to the *shtetlach* and ghettos,
>
> And set them walking the streets, visiting, praying in *shul*, feasting and dancing. The men I set to arguing, because I love dialectic and song — my ears tingle when I hear their voices — and the girls and women I set to promenading or to cooking in the kitchens, for the sake of their tiny feet and clever hands.
>
> And put kerchiefs and long dresses on them, and some of the men I dress in black and reward with beards. And all of them I set among the mists of the Pripet Marshes, which I have never seen, among wooden buildings that loom up suddenly one at a time, because I have only heard of them in stories, and that long ago.

Opening as an act of recovery and recuperation, describing Jews in ordinary village lives, the poem reaches a midpoint that introduces the dangers of history:

> But in a moment the Germans will come.

The poet, like Zeus in a Homeric epic, then imagines saving his people — "I snatch them all back, For, when I want to, I can be a God. . . . one by one I cover them in mist" — only to realize the inevitable failure of such acts of restitution and of both literal and figurative recovery:

> But I can't hold out any longer. My mind clouds over. I sink down as though drugged or beaten.

Haunted by the nightmare of contemporary history, Feldman in another poem, "To The Six Million," poses the eternal religious questions about the identity of a God who will allow such horrors to occur:

> If there is a god,
> He descends from the power.

Title page for Feldman's first book, which includes a sequence of poems on the Spanish painter Francisco José de Goya

> But who is the god rising from death?

The anguish in the volume produces no definitive answers, nor does the poet achieve a final religious or philosophical understanding of the problem of evil.

In *The Pripet Marshes* as well as several of his subsequent collections, Feldman chooses to conclude on a light or hopeful note. He ends this dark second collection with the delicate eleven-line poem "Song" (originally titled "Orpheus' Song"), which balances its horrors with a statement of muted hope and acceptance:

> So you are
>
> Stone, stone or star,
> Flower, seed,
> Standing reed,

> River going far
>
> So you are
>
> Shy bear or boar,
> Huntsman, death,
> Arising breath,
> Stone, stone or star
>
> So you are.

Like a clear dawn after the dark night of the soul, this lyric purges fear and pity and prepares both the poet and his readers for a fresh beginning. Similarly, at the end of *Leaping Clear,* Feldman concludes with "The Gift of Life," a tender elegy for Lionel Trilling, the great critic-teacher. The last poem in *New and Selected Poems* (1979), "The Tortoise," is a lovely prose allegory modeled on the fable of the tortoise and the hare. The last poem in *All Of Us Here,* following the long title sequence and the Kafkaesque fable "The Flight from the City," is a little joke titled "Art of the Haiku," which dramatizes what it explains:

> His finger then, now yours
> here, where master stopped, went back,
> counted syllables.

Since the Haiku form requires three lines of five, seven, and five syllables, the poem needs correction because the first line contains an extra syllable. By offering such poems at the end of his collections, Feldman leavens the deeper anxieties expressed in his work.

In his third collection, *Magic Papers and Other Poems* (1970), Feldman moves in a new direction, one that shows he had been heeding the postmodernist style of Ashbery. Opening the volume, the title poem, "Magic Papers," announces:

> Before we came with our radiance
> and swords, our simulacra of ourselves,
> our injurious destinies
> and portable exiles,
>   women were here
> amid incredible light that seemed
> to have no source, that seemed suspended.

Who these women are or might be – ancestors, angels, role models, dreams – is never explained. They move through "our" lives,

>   pausing to converse,
> turning to where we were not yet,
> saying, Here we live the victory
> of the senses over the senses.

The tone and diction sound Hebraic and religious, but the word *simulacra* (with its overtones of the work of the French philosopher Jean Baudrillard) suggests the contemporary idea that no true identity or reality exists apart from imitation, repetition, and the strangeness of familiarity. Poems such as this (or "Poet at Thirteen" and "Four Passages," both chiseled and unyielding) combine a cold opacity with an undervoiced autobiographical component. Although Feldman later in the poem creates a first-person speaker, that speaker's relationship to the poem's subject – women? childbirth? loneliness? – seems quasi-mythic and remains unclear.

In his fifth collection, *Leaping Clear,* Feldman achieves a new level of control of subject, tone, and rhythm. An avid squash player since 1966, Feldman admitted in a 1995 letter that he would like to believe that he "learned a vital lesson in style from playing squash: to keep my sentences on their toes, balanced (not leaning this way or that), ready to move swiftly in any direction, to keep the sentences, phrasing, etc. from being predictable." "A Player's Notes" in this volume proceeds as a numbered series of prose epigrams in the manner of William Blake, which explains the games of squash and poetry writing equally. The performance, in the sense used by the American critic Richard Poirier in *The Performing Self* (1971), is paramount. One is reborn during the act, but one must not be caught daydreaming; one must have readiness, delicacy, and a willingness not to think. The poem concludes after an out-of-breath speaker asks, "Is it possible to survive here without being reborn?":

> 12 Rebirth – if that is the issue – comes from renewed contact with the eternal, as in its momentary flight on the court.
>
>   Therefore, these are forbidden, these are enjoined.
>
> 13 Forbidden: hairsplitting, pessimism, fantasy.
>
> 14 Enjoined: knowledge, good humor, the exchanging of gifts.

Feldman's rhythmic control, vital to sport as well as poetry, is exemplified by the beginning of the volume's opening poem, "The Handball Players at Brighton Beach":

> And then the blue world daring onward
> discovers them, the indigenes, aging,
> oiled, and bronzing sons of immigrants,

the handball players of the new world
on Brooklyn's bright eroding shore
who yawp, who quarrel, who shove,
who shout themselves hoarse, don't
get out of the way, grab for odds,
hustle a handicap, all crust,
all bluster, all con and gusto all
on show, tumultuous, blaring,
grunting as they lunge.

The energy of the sentence – with its opening in medias res, its vocabulary mingling the Latinate (*indigenes* and *tumultuous*) and the street lingo of Walt Whitman (*yawp, Hustle,* and *bluster*), and especially its rhythm that starts and stops like the game itself, nervously moving from shorter to longer phrases (as in the enjambed lines "all con and gusto all / on show") – epitomizes the vigor that Feldman has been able to control and then release in the poems of his middle age.

The whole of *Leaping Clear* registers the political and social turbulence of its decade; it is more overtly public than Feldman's previous collections. Its title poem celebrates the New York cityscape in the tradition of Whitman, Herman Melville, and Hart Crane. In "The City and Its Own" Feldman writes:

The city is the realm of selves in rut
and delirium of ownership, is property,
objects made marvelous by prohibition
whereby mere things of earth become ideas,
thinkable beings in a thought-of world
possessed by men themselves possessed by gods.

The poem shows Feldman's objectification of an urban landscape as well as his tendency to animate objects and translate them into an aesthetic sequence. Objects become ideas, ideas become persons, the world becomes the possession of men themselves controlled by gods or at least by the idea of gods. All is reticulated, interconnected. Throughout this volume and elsewhere in Feldman's work, the person whom Whitman might label the "me myself" is hard to pin down, just as a walker in the city is hard to dissociate from the thousands of walkers beside him. Feldman deals with the anonymity of city life powerfully in "Millions of Strange Shadows," in which he echoes the "polite meaningless words" of Yeats's "Easter 1916" and the society chatter of the women in Eliot's "The Love Song of J. Alfred Prufrock." The collection also contains some chillingly comic poems. "The Golden Schlemiel" is part Arabian fable, part Yiddish joke. Taken together the poems register the collective human condition of middle age: fear, melancholy, regret, and hostility accompanied by gratitude and some pleasurable emotions.

The fact of Feldman's middle age may explain the looser, funnier, more autobiographical poems that appear in *New and Selected Poems* and *Teach Me, Dear Sister*. In many of these poems Feldman becomes a psychologist of process, examining how people become what they are and asking whether there is a "self" beyond the masks that are taken up and worn. It is hardly possible to get an accurate or adequate view of Feldman the person even from the poems that seem most overtly autobiographical. In the title poem, "Teach Me, Dear Sister," Feldman presents an updated version of Johann Wolfgang von Goethe's *Ewig-Weibliche* (Eternal Feminine), who assumes the different forms of "teacher, muse, sybil." The process of teaching and learning melts into the poet's memories of "cruising snatch," gawking at naked goddesses in art museums ("large jubilant babes in a marble playpen"), as the sexual impulse merges with artistic aspirations in an adolescent boy. Is this the young Feldman? It scarcely matters. In "The Biographies of Solitude" Feldman writes that "solitude has no biographers." Later he notes "how America is immense and filled with solitudes." The individual often stands in for the type in Feldman's work, and his personae are anonymous precisely because their words speak for many, not just himself.

Feldman's work has moved from strength to strength, each book building on and progressing beyond the previous one. *All Of Us Here,* considered his strongest collection by many, is divided into three major parts: the title sequence of ekphrastic poems that are responses to an exhibit of realistic plaster cast sculptures by George Segal; a fable titled "The Flight from the City" that is inspired equally by Kafka, George Orwell, Jorge Luis Borges, and John Hollander's *Reflections on Espionage* (1976); and, between them, a middle series of lengthy narrative, satiric, sometimes autobiographical pieces.

The title sequence provides a complex lesson in art appreciation that combines satire and a serious contemplation of art's power to provoke, annoy, and sustain. Dealing with representational art leads Feldman to contemplate some basic aesthetic questions: Whom do these statues represent? How do they make one feel? Are they any good? In other words, he views the sculptures mimetically, responsively, sociologically, and evaluatively. He also examines the sculptures historically, since in "Surely they're just so large" he imagines the statues as revenants from the plaster-cast collections that universities used for instructional purposes throughout the nineteenth century (such casts were mostly destroyed or relegated to dusty basements once slide reproductions of original works became widely available). The pieces seem to have "come racing in place from all the way back / to stand in no time at all – all of them, all of us / together here." The wonderful duplicity in the phrase "in no time at all" suggests both the eternity of art and the immediacy of the statue-creatures's hasty exit from down below. The plural first-person *we* of the poem speaks for the human observers of the animated statue-creatures.

Whom do these statues resemble? People we know? Human types? Ourselves? From one perspective they look alive and seem to admonish their observers: "We yield this place to them." We also "acknowledge that Being (however poor) precedes / *becoming* (though swift, exuberant, fascinating, gorgeous)." But from another viewpoint these statue-creatures are a "mirror of literalness"; because they are too easy to identify with they consequently disappoint us, like John Keats's Grecian urn that speaks volumes but remains eternally silent. Perhaps they remind observers, unpleasantly, of their own mortality, since the ghostly pallor of the statues portends the common fate of the grave, or perhaps they are like the corpses of those people at Pompeii who were caught unawares by the volcanic ash and buried in place before they ever knew what hit them.

One feels squeamish in the presence of such death-in-life. The statues remind us "of our own obtrusive, / too tedious, too obvious, too mortal selves." Such mimetic art is both an occasion for self-congratulation and a reproach to our self-delusions, as unsettling as it is gratifying. The narcissistic pleasure of seeing any human representation is quickly replaced by discomfort and annoyance, as weariness overcomes eagerness (who has not been overwhelmed by a long stretch in a museum or gallery?), and disappointment supplants hope.

Even such frustrations, however, indicate the human hunger for something greater in life, something that only art can give. Consolations, however fleeting, are possible: "And so it's oddly flattering to feel welcomed / by being ignored, as if we belong here like stones / and trees, are features natural to the scene." Who is looking at whom? One plaster couple looks like a heightened and improved version of a human couple:

Cover for Feldman's 1979 book, which includes poems from his first five collections

> This couple strolling here beside the BRICK WALL
> look the way we might look if it were us
> in raincoats walking slowly in the city dusk
> – but how much better at it these figures are!

Does art imitate nature, or is it the other way around? It hardly matters, since Feldman slyly reminds his readers of the implicit metaphor in all representation and consequently in all knowledge: "About these figures we don't ask, 'Who are they?' / We ask, 'Who, who is it they remind us of?'" In other words, they lack a genuine identity or integrity; they are merely *like* something else. Feldman returns his readers to the domain of the simulacrum.

The poetic sequence combines general reflections on aesthetics and aesthetic experience with deliberations on this particular Segal exhibit. One might label this multisectioned poem an allegory or a dramatic rendering of the entire spectrum of feelings, thoughts, and questions that an individual –

or many individuals — might have in reference to a specific artistic provocation. The lumpish but strikingly realistic sculptures, possessing what Feldman calls a "radiant density," like all art are both less and greater than actual life. The last poem in the sequence, "They Say To Us," is nominally concerned with family snapshots but may as well be discussing the sculptures. The represented figures, whether sculpted or photographed, "appreciate under our gaze," the speaker announces. A sense of satisfaction comes through at last as steady looking produces results: the figures gain in value as they are appreciated. Feldman's final response is one of contentment, purging all fear, pity, and even questioning:

> There they are — "little William,"
> "her brother," "the neighbor's girl" —
> all of them pressed close
> in a single pack in dark
> and radiant density.
> All of us here
> are deeply satisfied.

One wants to linger on these poems, reviewing and rereading them in the same way one might return to an art exhibit to get several takes on a single subject.

The miscellaneous poems in the middle section of the volume belong to the comic vein that Feldman has been increasingly mining. "An Atlantiad" is a literary mock epic, about a composite rival figure, part Walt Whitman, part John Ashbery, and part the Atlantic itself. "The Judgment of Diana" updates and translates the fable of Diana and Actaeon to an academic setting and places a new female poet amid a bemused bunch of male colleagues.

In the final section in the volume, the long poem "The Flight from the City," Feldman shows his increasing interest in doublings and simulacra as well as his preoccupation with the anonymity and threats of contemporary urban life. This grim parable allegorizes self, landscape, and literature as it debases personal identity and linguistic creation. "Nothing there seemed my own experience," says Kleinwort, the central figure who like Kafka's Joseph K. in the novel *The Trial* (1930) is accused of an unnameable crime. "The Flight from the City" shows that one cannot recover an original in a world where everything is a copy.

The Life and Letters is a diverse collection. Some of the poems are wildly funny; some plot the lives of born losers and are desperately sad. The book compiles extended character studies and also extended narratives, perhaps showing that Feldman is trying to develop the resources of prose in his poetry. The critic Charles Altieri has called Feldman "suspicious of unmodified traditional lyricism": here his lyricism is complemented by satire, outrage, and political fervor, all of which one finds more often in narrative prose than in lyrical verse.

In the title poem, "The Life and Letters," Feldman continues to explore the use of doubles in the creation of character. Feldman's main character writes lying letters in a very bad hand to his mother, who then copies them down in her own perfect script and mails them back to him. The letters *make* his life, rather than merely reporting it, and the poem ends ingeniously with his copying back a letter he finds in his pocket from his mother. Echoing those words of hers that echo words of his makes him happy. All is repetition; all is failure.

As opposed to his earlier work in which he explicitly or implicitly raised religious questions, Feldman in this collection has turned his attention almost wholly to the concerns of the secular world. In many poems he takes as his main theme the nature of obsession. When, in "Warm Enough," an elevator man in an apartment building asks the clichéd question, "Warm enough for you today?," the question inspires a series of literary and mythic responses. Feldman also indulges his penchant for satire more than he has in previous volumes. In the poem titled "In Theme Park America" he suggests that the entire country has become an imitation of itself, where violence asserts itself so ubiquitously that Americans are largely immune to it except as part of the overall spectacle: "nothing has changed / and everything is different." Feldman's use of the word *and* instead of the expected *but* makes all the difference in this desperate comedy, proving that appearances may not reflect reality.

One of Feldman's most accomplished poems is "The Little Children of Hamelin." He employs the power of incantation in his examination of the themes of loss, crossed wires, and missed opportunities. In an early stanza one of the doomed children recalls the Pied Piper's charm:

> Like a reed, hollow green tender heaven-high,
> he played, and the world sang out of his mouth
> in flowers; crocus and columbine and daisy
> and rose the iris the lilac the lily blew
> and scattered and flew, arrows we chased after
> in the light in the wind into the world
> until we had no names left to call them only
> our shouts and cries that burst from his mouth
>   and bloomed.
> The stones in the road clattered and clay laughed.

The impelling, nonstop lines mimic the excitement and the insinuating temptation of the piper's seductive music, while the stanza's last line, set apart as a single sentence, has an ominous finality that brings readers back to their apprehension of what will happen to his victims.

Most of the poems in *The Life and Letters* are either dramatic monologues or other kinds of performances; they all demand to be read aloud and heard. In the language of French literary theorist Roland Barthes, they are *lisible* rather than *scriptible*, because they deliver a quasi-theatrical revelation through the voice. Feldman develops tension in his verse through his abrupt changes in rhythm, syntax, and imagery. Even when the mood is elegiac the verse never plods or sinks. In the autobiographical poem "Adventures in the Postmodern Era," he correctly distinguishes between himself and his contemporaries: he possesses, or so he claims, subjects for his poems, however volatile or mercurial his tone, whereas they take as theme only the banal formula "Omnipotence in the realm of words." Of these poets and this idea he takes a dim view: "No fact or feeling, circumstance or being / ever affronted their far reconnoitering."

In a review of *Lost Originals* (1972), Feldman's fourth collection, the poet Richard Howard complained in *Poetry* that Feldman in his early poetry demonstrated "the tug between wit and wisdom." It is clear that wisdom and wit are still present in Feldman's work, sometimes at odds with and sometimes complementing each other. In "Fragment," one of the last poems in *The Life and Letters,* Feldman announces what Homer and all of the major narrative poets throughout history would have instinctively known: "the language isn't saved by style / but by a tale worth telling." With modernism played out, Feldman does not want "to purify the old words / but to bring new speech into / the lexicon of the tribe." Such speech includes everything from slang to literary language. Like Whitman, Feldman is democratic in his habits of speech, even when his themes reflect the rarefied interests of a hieratic intellectual caste. His latest work braves nonsense and whimsy at the same time that it tells poignant human tales. Moving away from his early fascination with myth-making, Feldman has made his poetry more powerful by making its base more human.

Reference:

Harold Schweizer, ed., *The Poetry of Irving Feldman* (Lewisburg, Pa.: Bucknell University Press, 1992).

Kathleen Fraser
(22 March 1935 –   )

Peter Quartermain
University of British Columbia

BOOKS: *Change of Address & Other Poems* (San Francisco: Kayak, 1966);

Stilts, Somersaults, and Headstands: Game Poems Based on a Painting by Peter Breughel (New York: Atheneum, 1968);

In Defiance of the Rains (San Francisco: Kayak, 1969);

Little Notes to You, from Lucas Street: Poems 1970–1971 (Urbana, Ill.: Penumbra, 1972);

What I Want (New York: Harper & Row, 1974);

Magritte Series (Willits, Cal.: Tuumba, 1977);

New Shoes (New York: Harper & Row, 1978);

Each Next: Narratives (Berkeley, Cal.: The Figures, 1980);

Something (even human voices) in the foreground, a lake (Berkeley, Cal.: Kelsey Street, 1984);

Notes Preceding Trust (Santa Monica, Cal.: Lapis, 1987);

Boundayr, with original acquatints by Sam Francis (Santa Monica, Cal.: Lapis, 1988);

From a text . . ., with original paintings by Mary Ann Haydn (N.p., 1993);

When New Time Folds Up (Minneapolis, Minn.: Chax, 1993);

Wing (Mill Valley, Cal.: Em Press, 1995).

VIDEO: *Women Working in Literature,* video anthology written and narrated by Fraser (San Francisco: Poetry Center/American Poetry Archives, 1992).

OTHER: *Feminist Poetics,* edited by Fraser (San Francisco: San Francisco State University, 1984);

"Line. On the Line. Lining up. Lined with. Between the lines. Bottom line," in *The Line in Postmodern Poetry,* edited by Robert Frank and Henry Sayre (Urbana: University of Illinois Press, 1988), pp. 152–174;

"One Hundred and Three Chapters of Little Times: Collapsed and Transfigured Moments in the Fiction of Barbara Guest," in *Breaking the Sequence: Women's Experimental Fiction,* edited by Ellen G. Friedman and Miriam Fuchs (Princeton:

Kathleen Fraser, 1978 (photograph © the Estate of Thomas Victor; courtesy of Kathleen Fraser)

Princeton University Press, 1989), pp. 240–249;

"The Tradition of Marginality," in *Where We Stand: Women Poets on Literary Tradition,* edited by Sharon Bryan (New York: Norton, 1993), pp. 52–66.

SELECTED PERIODICAL PUBLICATIONS – UNCOLLECTED: "How Did Emma Slide?," *Trellis,* 3 (1979);

"Why HOW(ever)?," *HOW(ever),* 1 (May 1983): 1;
"Things that do not exist without words," *Talisman,* 9 (Fall 1992): 144–149; republished in *Wallace Stevens Journal,* 17 (Spring 1993): 38–43;
"This Phrasing Unreliable Except As Here," *Talisman,* 1 (Fall 1994–Winter 1995).

As director of the Poetry Center at San Francisco State University from 1973 through 1975, founder of the American Poetry Archives (possibly the largest collection of audio- and videotape recordings by contemporary poets in North America), and founding editor of *HOW(ever)* (1983–1991), a much-imitated radical journal of women's innovative poetry, Kathleen Fraser has had an important influence on American poetry and poetics for a quarter century. That her work nevertheless remains little recognized is at least in part a direct result of her own decision, after the publication of *New Shoes* by Harper and Row in 1978, to withhold all her future work from major New York trade publishing companies in favor of little magazines and small private presses. Her earliest work appeared in such well-established and prestigious journals as *Poetry* (Chicago), *The New Yorker, The Hudson Review, The Nation,* and *Mademoiselle.* Since the mid 1970s, however, Fraser has chosen to publish almost exclusively in little magazines such as *Temblor, Hambone, Conjunctions, Sulfur,* and *Avec.*

Nevertheless, Fraser's poetic output shows remarkable consistency. Though she seems in the early years to have concentrated on and excelled at writing the "well-made" expressionist lyrical poem, grounded in clearly identifiable personal experience, and in her later years to show clear and even compelling affinities with the work of the language writers, deliberately foregrounding language as material object, criticizing habits of meaning by defamiliarizing customary language patterns, her work has, throughout its shift from a publishing career to a writing career, been marked by a delight in rhetorical forms and strategies and what one reviewer, Peter Scheldahl, has called "an eager and uncomplicated impulse toward love and friendship." Fraser's work is notable for the immediacy and directness of both the sensual and the emotional, while avoiding the pitfalls of display to which confessional poetry is so prone. At heart, it is a poetry of exploration and of zest. It is remarkably accessible yet by no means conventional.

The eldest of four children (one boy, three girls), Kathleen Joy Fraser was born in Tulsa, Oklahoma, on 22 March 1935 to Marjorie Joy (Axtell) Fraser and James Ian Fraser II. One year after beginning his practice as an architect, her father – a graduate of the University of Tulsa School of Architecture – reached the decision to change professions and entered the Union Theological Seminary in Chicago to prepare for the Presbyterian ministry, a calling which he followed until his death from a head-on car crash in 1966. In several essays Fraser recalls with affection her father's inveterate habit of reciting nonsense verse and singing silly songs – a practice that, coupled with his highly vocal love for the rhetorical orotundity of the King James Bible, would have a lasting and in some respects problematic effect on her poetry. After a childhood in which she spent three years (from grades six through nine) in the high mountains of Glenwood Springs, Colorado, and finally settled in Covina, California, where she lived until graduating from high school, Fraser knew by heart "great chunks" of the Bible, as well as a great deal of English nonsense verse. She also acquired from her father an irrepressible delight in the sounds of words.

Such a heritage was, however, a mixed blessing. As Fraser herself records in her remarkable 1992 essay on Wallace Stevens, "Things that do not exist without words," in the seventh grade in Glenwood Springs, Colorado,

> all doors to playfulness slammed shut. In the seventh grade Miss Elsie Foote began to teach poetry lessons. What had been joyful became flattened and restrictive. *Do Not Sit On The Grass.* Acts of terrorism reigned: four large pages of the slow, sleepy prelude to Longfellow's "Evangeline" would be committed to memory and recited by each student in front of the class.
>
> Poetry's instruction soon undermined that early appetite for words; there was, instead of pleasure, the considerable question of getting it "right."

In that narrative, perhaps, one can see the seeds of Fraser's lifelong struggle – a struggle she came to feel was shared by many women, whether writers or not – against her education. By the time she reached high school Fraser found herself essentially well trained to fear poetry. Alienated by its refusal to yield meanings other than those few handed down by her teachers, Fraser had, like countless others, learned not to take classes that foregrounded poetry ("I'd found myself to be an idiot in that department, never understanding what these things 'meant'").

Much later, teaching creative writing at San Francisco State in the 1970s, Fraser came to believe that this particular problem of meaning is gender related. As she observed in the introduction to her 1984 anthology of student writing, *Feminist Poetics:*

"there is an expectation in [women students themselves] of failure, of not doing it right and never being quite sure they understand what right means, when they've been told that the materials, feelings and structures of many of their poems are inappropriate to the professional world of poetry." In the case of her own development as a poet, Fraser found that the persistence of the rhythms and sounds and attitudes, the mindset encapsulated and embodied in the grand English tradition, was a serious inhibition in the development of her own poetic, imposing as it did a mellifluous sonorous continuity on a life that she felt, as a woman, to be essentially discontinuous, fragmented, marginalized, and multiple. Her need to escape a grand tradition to which she felt strongly attracted came to constitute in complex ways a form of almost self-inflicted intimidation, which influenced her decision to major in philosophy when she entered Occidental College in Los Angeles as a sophomore in 1954.

Yet in her senior year Fraser switched to an English literature major, thereby postponing graduation until January 1959. This change was partly a result of what she learned in a humanities/classics two-year, required course, where she read Herman Melville and Walt Whitman as well as standard English classics, but it was mainly the result of browsing in the library or bookstores – where she discovered works such as Virginia Woolf's *The Waves* (1931). Through intense college friendships she was introduced to the work of writers such as E. E. Cummings, James Joyce, and William Carlos Williams, and she began scribbling poetry ("along with my girl-friends, among whom writing poetry was a badge of 'sensitivity' "). For her birthday her college friends joined together to give her T. S. Eliot's *Collected Poems* (1936), Williams's *Journey to Love* (1955), and Cummings's *i: six nonlectures* (1953) – works that, along with the writings of Dylan Thomas, became models for her early verse. Cummings was especially attractive because his radical breaking of normative syntax and grammar spoke directly, as she would put it in a letter written some years later, to the "split between my resolve to be an attractive and acceptable female student, and my stubborn resistance to all rules – including those of prosody, which did not appear to describe my hesitant and multiple ways of perceiving and forming thought." This struggle within and against her education came to inform Fraser's whole career as a writer, a writing life perpetually venturing into what she called, in *How(ever)* in 1982, "the tentative regions of the untried."

On graduating from college in 1959 Fraser left an increasingly incompatible southern California for New York City, where she began work writing copy for the fashion magazine *Mademoiselle*. By this time committed to a life of writing poetry, she felt she knew virtually nothing of the contemporary writing scene: at college she had never in class been introduced to the work of any living writers (other than perhaps William Faulkner or Ernest Hemingway), though she had attended a reading at Occidental by Robert Lowell. She had never read or even come across such major women modernists as H. D. (Hilda Doolittle), Marianne Moore, or Gertrude Stein. Hungry for news, she found herself excited by everything she read – Pablo Neruda, César Vallejo, Paul Célan, Giuseppe Ungaretti, Eugenio Montale – but with no basis for comparison. She found herself perpetually learning, unlearning, and then learning to do things differently. In "The Tradition of Marginality" (written in 1985), Fraser recalled, "Already I was aware that I carried a number of clamoring voices in me, arguing, protesting, obsessively repeating themselves . . . my mind was polyphonic and fragmenting, as I heard it." It was some years before she could evolve a poetic philosophy and technique commensurate with her experience, but her hunger for writing news was such that almost immediately on her arrival in New York she took a poetry course offered by poet Stanley Kunitz at the Ninety-second Street Young Men's Hebrew Association (YMHA), enrolling in his workshop in the fall of 1959 and again in the following semester. Fraser was "thrilled," she would later say, by Kunitz's "Yeatsian language and passionate metaphysical vision," and she learned through him to admire the work of Elizabeth Bishop. Kunitz, a generous and sympathetic teacher, was at this time still strongly traditionalist in his own verse, which was strictly formal in structure and lofty in theme. (It was only in the 1960s, for instance, that Kunitz began some of his lines with lowercase letters.) "A high style," he said, "wants to be fed exclusively on high sentiments." Such an approach to poetry could not satisfy for long someone of Fraser's immediacy and passion. For by instructing Fraser to cast herself and her experience as representative – and to think of her condition as writer as both universal and transcendent, unaffected by the world, free of such quotidian distractions as race or gender – Kunitz made an icon of the lyric poetic self, elevating it to a position superior to that of the reader by installing its own power as seer, transformer, and possessor of meaning. In thus subordinating the reader to the role of witness seeking to "under-

Fraser reading her poetry, July 1993 (courtesy of Kathleen Fraser)

stand" the poem, rather than participant in the construction of meaning, it perpetuated the very condition Fraser had found so crippling as a student in school.

In Kunitz's workshop Fraser met the young poet Jack Marshall. In 1960 he became her first husband. Their conversations opened up for Fraser the world of poetry and painting: he took her to see the work of Willem de Kooning, Franz Kline, Jackson Pollock, and Sam Francis. "I was turned-on," she says of these New York School painters, "by the collective humor and the freedom to invent work in new shapes, on new terms . . . Their painting further stimulated a very strong visual component in my nature." She began to meet other New York writers – having been drawn almost from her arrival in New York to Greenwich Village and to the "downtown" poets, including Robert Kelly, Paul Blackburn, Jerome Rothenberg, Armand Schwerner, Carol Berge, and Diane Wakoski; to Black Mountain writers such as Robert Creeley, Robert Duncan, Denise Levertov, and Charles Olson; and to the New York School poets Frank O'Hara, Kenward Elmslie, Edwin Denby, James Schuyler, and especially Barbara Guest. Such avant-garde writers on the margin were, whether they knew it or not, Fraser's teachers. In her mind they made up her literary family.

During this same period she came across the work of Wallace Stevens, hearing two young men recite his work at a party in Greenwich Village. As she recounts the story in "Things that do not exist without words," she was transported into an "untranslatable elation." The next day she bought his poems, which she read, "transfixed," every day in the office at lunch hour. For Fraser, Stevens was the great "unloosener" who, freeing the reader from ordinary habits of reading, quickening her through his music and through "particulars that do not, cannot have existed, without his words." Stevens thus became the first of several figures who drew Fraser away from poetry as she had learned it in school, who drew her back toward the poetry of her childhood.

Fraser's reading fed into her increasing dissatisfaction with her own writing situation and thus drew her in the summer of 1964 to enroll in Daisy Alden's two-week course at Wagner College. Because Alden was ill, Kenneth Koch taught the course instead. Koch, famously imaginative and strong-minded as a teacher, gave silly nonsense-writing assignments that restored to Fraser the playful attitude toward poetic language she had acquired from her father in her childhood. Koch's hostility in class toward any sign of high seriousness or emotional vulnerability, whether in the writing or in the individual, and his disdain for sentimental poetic retreads were crucial for Fraser at this stage in her career, when she felt in need of liberation from older forms and inhibitions. Soon after, through Frank O'Hara, Fraser met Barbara Guest, whose intensely disciplined attention to the accuracy of her language, whose "linguistic mysteries . . . composed and collaged from the precise fragments of her painterly witness" (the words are Fraser's) would have a pervasive and lasting effect. Guest's sense of language as a prison, her sense that language is inadequate to the writing event itself, had its kinship with the work of the New York painters about whom she wrote so brilliantly.

During these years Stevens and Guest became Fraser's great exemplars and inspirations; finally, hearing George Oppen read his work early in 1967 firmly secured her sense of her own difference from the mainstream. Striking Fraser as wholly without posture, modest yet severe in its unflinching attention to detail and nuance, Oppen's work appealed to her as a new kind of attentiveness, speaking to some neglected level of gravity. Oppen joined Stevens and Guest as prime constituents of Fraser's writing life. Fraser's personal life, meanwhile, had not been without its troubles. In 1965 her father was killed in a head-on car crash. Later, in 1969, her sister Mary died. A mezzo-soprano, she too appears in Fraser's writings. Christmas of 1966 saw the birth of her son, David, on 26 December. The New York neighborhood in which Fraser lived was becoming increasingly the site of drug trafficking and consumption; in September 1967 she and Marshall returned home from walking their son to find their apartment completely trashed and everything portable (including typewriters, stereo, and tape recorder) stolen. Within the week they left for San Francisco, with one month's rent and few prospects.

The move to San Francisco proved extremely fortunate. George and Mary Oppen became close friends, George reading (and advising her about) her poems. As a writer Fraser found the atmosphere of the Bay Area congenial, contrasting sharply with the flash and dazzle of writing performance characteristic of so many New York poets. Fraser's first book, *Change of Address & Other Poems* (1966), had already been published in San Francisco by George Hitchcock's Kayak Press; in 1968 the prestigious publishing house Atheneum brought out *Stilts, Somersaults, and Headstands,* a book of poems for children. Much of the energy for this book no doubt came from Fraser's work caring for her son, but the childlike directness of the language and the simplicity of the sound patterns so prominent in these verses for children carry over into Fraser's subsequent work, most directly in the poems gathered in *In Defiance of the Rains* (1969), again published by Hitchcock.

The title of Fraser's first book, *Change of Address,* suggests the extent to which her poetry is drawn directly from her immediate daily experience in a physical world. The poems themselves, often centering on an "I" or a readily imaginable "you," investigate different relations in and of speech and exploit some of the puns implicit in the title. The poems in *In Defiance of the Rains* explore line breaks, punctuation, and sentence pattern in their construction of sound. Thus, the title poem is reminiscent of Stein:

Solidly, her. (the essence)
Except that to conceal like a little cup
of sanity, she is careful.
Domesticity, how white!

Some of the poems play with context and reader expectation, at the same time punning the alphabet, as in the epistolary "Letters: to him," "to her," and "to Barbara," and the title of the collection itself can be read as a punning and coded resistance to another's rule. These two Kayak books are Fraser's initial steps toward exploring the physicality of language as experience: while they clearly and emphatically deal with the apparent trivia of daily life, especially in its domesticity (a focus in later years to be closely identified with feminist writing), they certainly do not regard language as a clear glass through which to regard the world. These poems, later gathered in *What I Want,* are intensely personal and intimate, paying astonishingly close attention to the physical, the immediate materiality of experience.

On a visit to San Francisco in late spring 1969 George Starbuck, director of the Iowa Writers' Workshop, offered Fraser and Marshall teaching posts at the workshop, where Fraser taught from 1969 to 1971. The following year she was writer-in-residence at Reed College in Portland, Oregon.

(The couple was divorced in 1970.) Teaching turned out to be Fraser's vocation, and at Reed her latent feminism began to emerge into informal classes on Stein and H. D. held at her house. At that time the notion that the modernist women writers had any relevance to feminism was unfashionable in feminist politics, where a more commonly spoken language was virtually de rigueur for any woman who wanted acceptance in the women's writing community.

From Portland, Fraser moved to San Francisco, where she directed the San Francisco Poetry Center (1973–1976) and taught creative writing at San Francisco State University. In 1985 she began to spend up to five months of each year working and living in Rome, where, that same year, she married the philosopher A. K. Bierman, whom she had met during her first year at San Francisco State. Her experience in Italy afforded radical linguistic and cultural challenges that significantly informed and colored the poems gathered in *Notes Preceding Trust* (1987) and later collections. She retired from San Francisco State as a full professor in 1993, having converted her position from full- to part-time a few years earlier. Concurrent with her first years in San Francisco, Fraser got to know women writers whose work seemed generatively close to her own. As she puts it in "The Tradition of Marginality," their work offered "a new kind of attentiveness: it wasn't the witty polish or posturing of 'great lines,' but a listening attitude, an attending to unconscious connections, a backing off of the performing ego to allow the mysteries of language to come forward and resonate more fully." In 1974 Harper and Row published *What I Want*, which consists mainly of work gathered from her previous collections, with some new poems that show an intensifying commitment to syntactical experiment and unconventional form.

The first major indication, however, of Fraser's increasing alignment with experimental and even avant-garde writing is *Magritte Series* (1977), published by Lyn Hejinian as number six in her Tuumba series of chapbooks. (Other writers appearing in that "First Series" include Dick Higgins, Susan Howe, and Kenneth Irby, as well as Hejinian herself.) In *Magritte Series* – to an even greater extent than in "The History of My Feeling" and "Six Uneasy Songs" at the close of *What I Want* – it becomes clear that the writing is creating the situation to which it refers, a mode no conventional reader can comfortably accept, since reference is at a minimum. The poems of *Magritte Series* wittily and disturbingly play familiar ordinary syntax with the grotesque, thereby constituting a stylistic equivalent to the paintings of René Magritte, to which the poems seek to be companions.

These poems were later collected in *New Shoes* (1978), Fraser's last book with a major New York trade publisher. The wit and the cultivation of the bizarre function as controlling devices to keep the reader distant from – but at the same time intensely aware of – the controlled but never hidden high emotional charge of these intensely personal poems. "One of the Chapters," for instance, tempers the poet's sheer outrage at the preposterous difficulties of living in a university town (where the men make up the universe) through a carefully controlled comic ironic tone, coupled with the important news that the poem draws on someone else's text. Yet Peter Schjeldahl, reviewing this book in *The New York Times Book Review* (13 August 1978), praised Fraser's "delight in rhetorical forms and . . . sense of what words mean" but nevertheless concluded – perhaps with the deliberate grotesques of the *Magritte Series* specifically in mind – that, "lushly synesthetic" and "full of appeal to the senses," the work is at times "self-absorbed to (and sometimes over) the brink of solipsism and incoherence." *What I Want* and *New Shoes* reveal the extent to which Fraser has learned to trust rather than bully the reader – no mean feat when the poems are full of scorn or anguish. The poems are remarkably skillful, with Fraser firmly in control of the meaning, which is transmitted with great emotional impact to the reader.

By the late 1970s Fraser found herself increasingly reluctant to submit her work to male editors and no longer found it possible to accept their well-meaning but patronizing "corrections" of her work. Some of her difficulties and a great deal of her passion as a writer came in her eyes to have an increasingly gendered origin, a theme she would explore with remarkable and indeed devastating effectiveness in *Each Next: Narratives* (1980). This book marks the great turning point in Fraser's career.

Since the title declines the label *fiction*, for example, there is no means to tell whether these largely prose works are fictions or not; they bear the stamp of direct and immediate autobiography. A passionate defense of nontraditional writing by women, the narratives were written exactly at the time when Fraser felt isolated as a writer in San Francisco, unable to "submit" her work to male editors and finding precious few if any feminist journals prepared to publish stylistically innovative heterosexual work. The book was also published exactly at the time when Fraser, Beverly Dahlen, and Frances Jaffer were embarking on a series of conver-

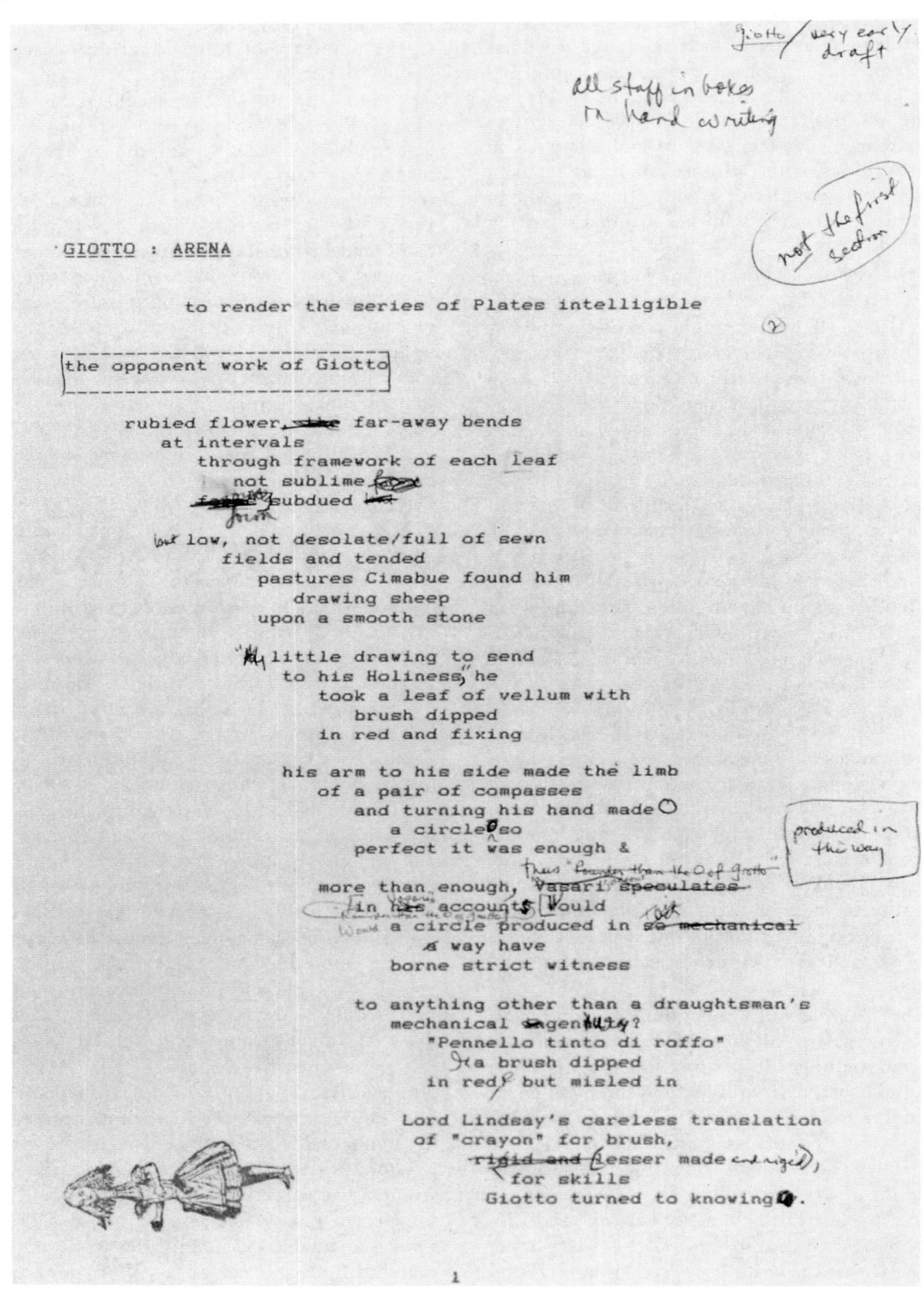

Draft for the first two pages of a poem included in When New Time Folds Up *(courtesy of Kathleen Fraser)*

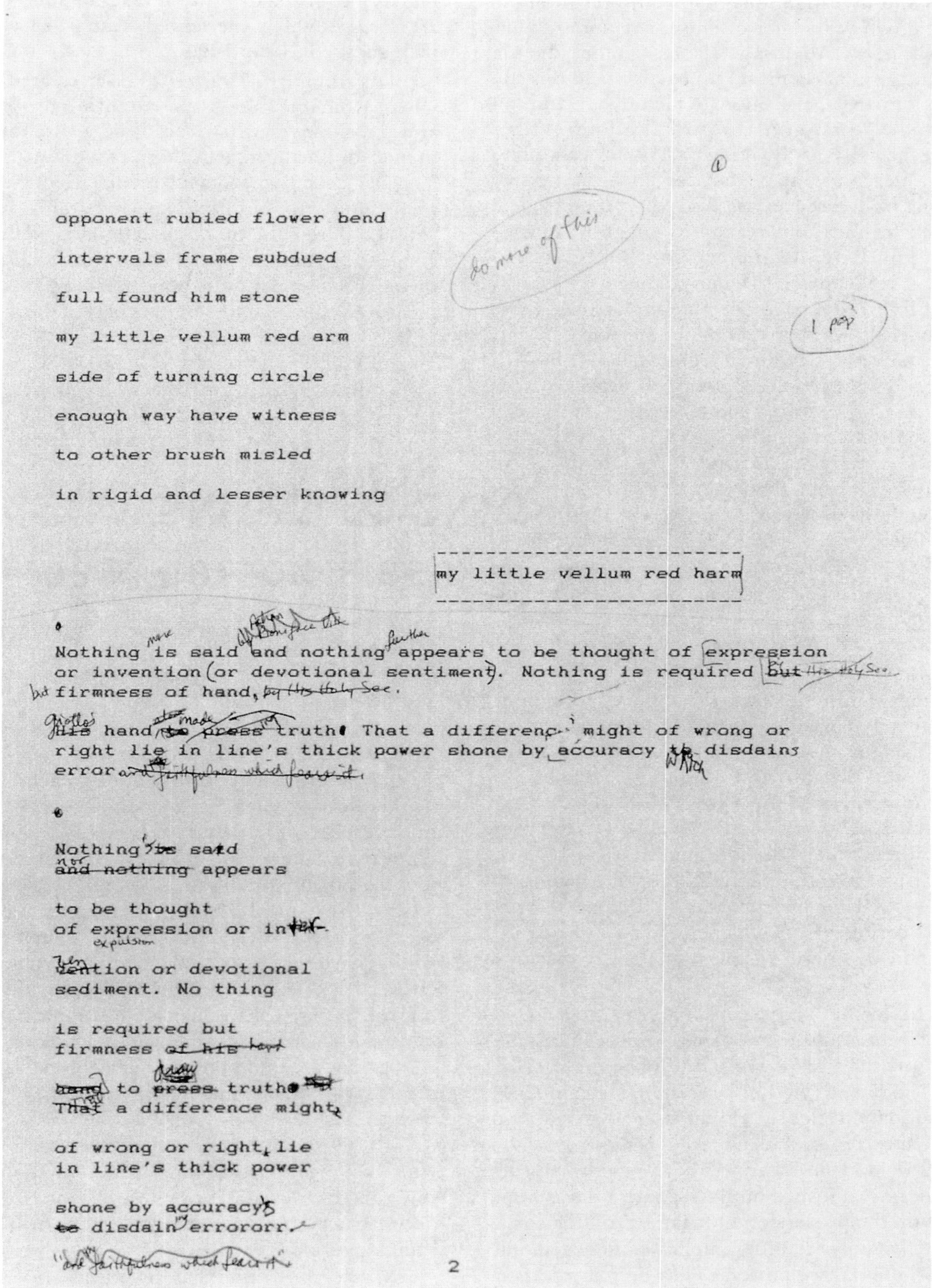

opponent rubied flower bend

intervals frame subdued

full found him stone

my little vellum red arm

side of turning circle

enough way have witness

to other brush misled

in rigid and lesser knowing

my little vellum red harm

Nothing is said and nothing appears to be thought of expression or invention (or devotional sentiment). Nothing is required but firmness of hand, ~~by His Holy See~~.

~~His~~ hand ~~to press~~ truth. That a difference might of wrong or right lie in line's thick power shone by accuracy to disdains error ~~and faithfulness which fears it~~.

Nothing ~~is~~ said
~~and nothing~~ appears

to be thought
of expression or in~~ver~~-

~~vention~~ or devotional
sediment. No thing

is required but
firmness ~~of his~~

~~hand~~ to ~~press~~ truth.
~~That~~ a difference might

of wrong or right lie
in line's thick power

shone by accuracy
~~to~~ disdain errororr.

"and faithfulness which fears it"

2

sations and investigations that would result in the founding two years later of Fraser's important journal *HOW(ever)*. Adapting Olson's famous dictum that "one perception must immediately and directly lead to a further perception" — already a feature of some of the poems in *New Shoes* and before — Fraser finds, especially in "Talking to Myself Talking to You," a fierce narrative drive to push the writing headlong from discovery to discovery; multiple and complex feelings and responses suggest the fragmenting of daily life and self so alien to the discourse of male power, at times fragmenting syntax, continuity, and image. At the same time, these prose poems erase conventional narrative concepts such as point of view in an ironic melange of fragments, and they dissolve generic boundaries. All the poems in *Each Next* are "about" writing (and necessarily then about reading) as a woman, and the reader, forced by the linguistic play into the active construction rather than the passive reception of meaning, finds procedural and even methodological clues. The narratives of *Each Next* pave the way to discovery by participating in it and moving the reader into such participation: "She wanted a 'flow' she thought, but an error was made in transcription, displacing *o* and substituting *a*. She could give herself to an accident."

Such writing is marked by overt risk-taking, by an acutely painful honesty of revelation and detail of dialogue and response, of thought and desire, of anger and delight. These are powerful poems of desire, interrogating the beloved, interrogating the very nature of "other," and — by means of the comments others reportedly make in these stories — interrogating the self, or selves: "I begin working to piece together the fragments, a skill I have learned to take seriously." The poems also interrogate, then, their own writing and invite the reader into the act.

The formal and especially thematic achievement of *Each Next* is to dissolve generic contrasts between "prose" and "verse" and the boundaries between "fact" and "fiction," rendering such distinctions not only irrelevant but intrusive. Fraser's next books, *Something (even human voices) in the foreground, a lake* (1984) and *Boundayr,* written some three years later but not published until 1988, are a devastating assault on the possession of meaning, on the social and intellectual certainties implicated in hegemonic and institutionalized powers. They do so by destabilizing the text and by flattening out the voice. Some of the poems in *Something* abandon referentiality almost completely; those in *Boundayr* cultivate error as compositional principle. These writings are intimately connected with Fraser's founding of *HOW(ever)*, which she edited from 1983 to 1991, with guest editors in 1990.

As editor of *HOW(ever)*, Fraser created a place where women could "focus attention on language and . . . discover what [could] be written in other than traditional syntactical or prosodic structures." The magazine gave women an immensely important opportunity to publish experimental work that would call into question conventional models of language usage and the social institutions that enforce those conventions. A groundbreaking enterprise, the project shared many of the goals of the language writers, who deliberately foreground language as material object and seek to criticize habits of meaning by defamiliarizing customary language patterns. Its most important features were its openness to new writers of whatever persuasion and its refusal to adopt a partisan feminist position: "If we are politically engaged as poets," Fraser asked in the twelfth issue of *HOW(ever)*, "are we not called upon to investigate our experience through the re-structuring of it?" At the same time the magazine undertook the important work of retrieving work by forgotten women modernists such as Mary Butts, Mina Loy, Lorine Niedecker, and others, and it became the model for *f(lip)* magazine in Vancouver and *6ix* magazine in Philadelphia.

The great work of editing *HOW(ever)*, so closely linked as it was with Fraser's own writing, set the pattern for her future books, each of which deliberately pursues a course of discovery implicit in her earlier work — and each of which is formally innovative. Thus, the work in *Boundayr* freed Fraser to write poems such as *Giotto: Arena,* first published as an entire issue of *Abacus* (15 November 1991) and collected in *When New Time Folds Up* (1993). As Meredith Quartermain observed in an important review (*West Coast Line,* Winter 1994–1995), the title poem affords the reader a remarkable complex of manifold relations and voices, formally invoking and interrogating the tradition of which it declares itself a part. Carol Muske, in a judicious but nevertheless enthusiastic review in *The New York Times Book Review* (6 February 1994), called Fraser a maverick who "belongs to no school" and commented that through her "voracious desire" to enter and deconstruct language Fraser "demonstrates how thinking evolves, how we think what we see and vice versa," remarking that the poems are to a great extent the harvest of her rich early work (influenced by poets as diverse as Frank O'Hara and George Oppen) and her subsequent language experiments. Rachel Blau DuPlessis, writing in *Sulfur* (Spring

1994) on "the rich tinctures of multifarious webbings" in these poems, pointed to Fraser's use of H. D. "for the pensive, illuminating reading and rereading on signs (a play with repetition and recirculation most striking in the final, and title poem)" and of Virginia Woolf for "the delicate, determined 'deliberate burdens through the temporal.' "

Despite its extraordinary technical sophistication and capacity to disturb the reader, the work in *When New Time Folds Up* and Fraser's more recent books is completely unintimidating. Drawing on a great range of resources, including graphics – and in *Wing* (1995) playing with great charm and indeed passion with the visual shape of the poem – Fraser's work has become more and more playful and less and less dogmatic, while at the same time cultivating a meditative and cogitative stance and habit that continually cultivate and exploit the random and the accidental. The projected publication of "Il Cuore: The Heart. Selected Poetry 1964–1994" in 1997 will make available a great deal of work long out of print and make possible a comprehensive overview of her entire writing career. Reading through Fraser's work in chronological order makes one see clearly that her whole career has been a move away from certainty and into discovery. Overall, it reflects a great generosity of spirit, a necessary corollary to a deep and abiding curiosity. The chief characteristic of her work, persisting through its abiding lyrical intensity and condensation and love of color and the sheer body-ness of the language and the vision, is its stubborn and courageous refusal to rest satisfied with any sort of status quo, a determination to find a language adequate to the writing occasion of which it is witness: the perturbability of the writer. "How is it," she asked in 1993, that the scrubbed and well-brushed historic formulas of the known so often begin as our internalized judges in both the sciences and the arts – apart from any intention of the poem or the cell's functioning to represent any final paradigm? Does that static resting place, often regarded as "the perfect solution," actually function as the carrot dangling from the stick, the lure urging us forward with its possibility of temporary sustenance so that we may go on to risk our own idiosyncratic depictions and commit our own perfection-resistant "errors"?

The essay from which these words are taken, "This Phrasing Unreliable Except As Here" (published in *Talisman,* 1995), could well serve as a motto for the collected work, obedient as each poem is to the writing occasion itself.

Interview:

Robin Tremblay-McGaw, "How Error Pops Up: An Interview with Kathleen Fraser," *Poetry Flash,* 246 (September–October 1993): 1, 6, 8–10.

References:

Rachel Blau DuPlessis, "Noticings," *Sulfur,* 34 (Spring 1994): 191–194;

Linda A. Kinnahan, *Poetics of the Feminine: Authority and Literary Tradition in William Carlos Williams, Mina Loy, Denise Levertov and Kathleen Fraser* (Cambridge & New York: Cambridge University Press, 1994);

Carol Muske, "Outside the Frame: Three Renegade Stylists," *New York Times Book Review* (6 February 1994): 32;

Meredith Quartermain, "Word Tunnels," *West Coast Line,* 15 (Winter 1994–1995): 134–139;

Peter Scheldahl, "Love and Rage and Loss," *New York Times Book Review* (13 August 1978): 15.

Allen Ginsberg
(3 June 1926 -)

Laszlo K. Géfin
Concordia University

See also the Ginsberg entry in *DLB 5: American Poets Since World War II, First Series* and *DLB 16: The Beats: Literary Bohemians in Postwar America.*

BOOKS: *Howl and Other Poems* (San Francisco: City Lights Pocket Bookshop, 1956);

Kaddish and Other Poems: 1958-1960 (San Francisco: City Lights Books, 1961);

Empty Mirror: Early Poems (New York: Totem Press/Corinth Books, 1961);

Reality Sandwiches: 1953-60 (San Francisco: City Lights Books, 1963);

The Yage Letters, by Ginsberg and William Burroughs (San Francisco: City Lights Books, 1963);

T.V. Baby Poems (London: Cape Goliard, 1967; New York: Grossman, 1968);

Airplane Dreams: Compositions From Journals (Toronto: House of Anansi Press, 1968; San Francisco: City Lights Books, 1969);

Ankor Wat (London: Fulcrum Press, 1968);

Planet News: 1961-1967 (San Francisco: City Lights Books, 1968; London: Villiers, 1968);

Notes After an Evening With William Carlos Williams (New York: Samuel Charters, 1970);

Indian Journals: March 1962-May 1963 (San Francisco: David Haselwood/City Lights Books, 1970);

Ginsberg's Improvised Poetics, edited by Mark Robinson (Buffalo: Anonym Press, 1971);

The Fall of America: Poems of These States 1965-1971 (San Francisco: City Lights Books, 1972);

Iron Horse (Toronto: Coach House, 1972; San Francisco: City Lights Books, 1974);

The Gates of Wrath: Rhymed Poems, 1948-1952 (Bolinas, Cal.: Grey Fox, 1972);

The Visions of the Great Rememberer (Amherst, Mass.: Mulch Press, 1974);

Allen Verbatim: Lectures on Poetry, Politics, Consciousness, edited by Gordon Ball (New York: McGraw-Hill, 1974);

Allen Ginsberg (photograph by Lisa Law)

Chicago Trial Testimony (San Francisco: City Lights Books, 1975);

First Blues: Rags, Ballads & Harmonium Songs 1971-74 (New York: Full Court Press, 1975);

Journals: Early Fifties Early Sixties, edited by Ball (New York: Grove, 1977);

Mind Breaths: Poems 1972-1977 (San Francisco: City Lights Books, 1978);

Poems All Over the Place: Mostly 'Seventies (Cherry Valley, N.Y.: Cherry Valley Editions, 1978);

Composed on the Tongue, edited by Donald Allen (Bolinas, Cal.: Grey Fox, 1980);

Straight Hearts' Delight: Love Poems and Selected Letters, by Ginsberg and Peter Orlovsky, edited by Winston Leyland (San Francisco: Gay Sunshine Press, 1980);

Plutonian Ode: Poems 1977–1980 (San Francisco: City Lights Books, 1982);

White Shroud: Poems 1980–1985 (New York: Harper & Row, 1986);

Your Reason and Blake's System (New York: Hanuman Books, 1988);

Allen Ginsberg: Photographs (Altadena, Cal.: Twelvetrees, 1990);

Kaddish for Naomi Ginsberg, 1894–1956 (San Francisco: Arion, 1992);

Honorable Courtship: From the Author's Journals, January 1–15, 1955, edited by Ball (San Francisco: Coffee House, 1993);

Snapshot Poetics: A Photographic Memoir of the Beat Era (San Francisco: Chronicle Books, 1993);

Cosmopolitan Greetings: Poems 1986–1992 (New York: HarperCollins, 1994);

Journals Mid-Fifties, 1954–1958, edited by Ball (New York: HarperCollins, 1995);

Illuminated Poems (New York: Four Walls Eight Windows, 1996).

Collections & Editions: *Collected Poems 1947–1980* (New York: Harper & Row, 1984; Harmondsworth, U.K.: Viking, 1985);

Howl: Original Draft Facsimile, Transcript & Variant Versions, Fully Annotated by Author, edited by Barry Miles (New York: Harper & Row, 1986).

SELECTED RECORDINGS: *Allen Ginsberg Reads Howl and Other Poems,* Fantasy-Galaxy Records, 1959;

Allen Ginsberg Reads Kaddish, A Twentieth Century Ecstatic Narrative Poem, Atlantic Verbum Series, 1966;

Allen Ginsberg/William Blake: Songs of Innocence and Experience, M-G-M Records, 1969;

Birdbrain!, by Ginsberg and the Gluons, Alekos Records/Wax Trax Records, 1981;

First Blues: Rags, Ballads and Harmonium Songs, Folkways Records, 1981;

The Lion For Real, Great Jones/Island Records, 1990;

Cosmopolitan Greetings, words by Ginsberg, music by George Gruntz, Migros-Genossenschafts-Bund Musikszene, Scheitz, 1993;

Hydrogen Jukebox, libretto by Ginsberg, music by Philip Glass, Electra Nonesuch, 1993;

Howls, Raps & Roars: Recordings from the San Francisco Poetry Renaissance, Fantasy Records, 1993;

Allen Ginsberg: Holy Soul Jelly Roll – Songs and Poems (1949–1993), Rhino Records, 1994.

Allen Ginsberg's career as writer and performer of poetry now spans nearly half a century. In an interview aired on British television on 9 January 1995 the critic Jeremy Isaacs called him "America's best known poet," a title Ginsberg conceded with characteristic modesty and generosity to Bob Dylan. Asked "How would you like us to remember you?" he replied without hesitation, "I think 'Father Death Blues,'" which he then proceeded to sing. Ginsberg's response would probably surprise the many readers who think of him primarily as the renowned author of "Howl" and "Kaddish," two poems he wrote in the 1950s. Although his rendition of "Father Death Blues," a poem written in memory of Ginsberg's father, has a poignancy and condensed wisdom, as a song it falls well short of Dylan's best pieces. As a poem, one of seven in a poetic sequence titled "Don't Grow Old" that was collected in *Mind Breaths: Poems 1972–1977* (1978), "Father Death Blues" is less pithy and exact than the segment preceding it.

For Ginsberg to choose to be remembered by such a song instead of one of his far more complex and memorable poems raises the question of the nature of his artistry as it has evolved during the last two decades of his career. One might argue, of course, that his choice was whimsical or that it reflected a lapse of aesthetic judgment – an instance of the oft-proven truth that poets are no more reliable judges of their own productions than any other reader. But on a deeper level Ginsberg's choosing to sing rather than recite is an indication that he has come to see himself increasingly as a singer rather than *just* a poet. He has become a successful public performer – reading, chanting, and singing his own works, those of William Blake he "tunes" himself, or Hindu mantras, accompanied by rock bands or by himself on his ubiquitous harmonium.

Ginsberg invariably attracts large audiences in the United States and elsewhere; in March 1994 at Concordia University in Montreal, for example, he filled an auditorium of seven hundred to capacity, with over five hundred disappointed fans having to be turned away. In the same year a boxed set of four compact disks of his poems and songs was issued, and Ginsberg participated in the world premiere of Lee Hyla's "Howl" at Carnegie Hall, chanting the poem to the music played by the Kronos Quartet. These examples illustrate the vibrancy

Ginsberg (right) with singer-songwriter Bob Dylan, 1975 (photograph by Elsa Dorfman)

of Ginsberg's vocation. While pursuing his long-held desire to return poetry to orality, Ginsberg continues to play the role of a post-Whitmanian national expresser. Even at a time when conservative forces are once again on the rise, he disseminates his libertarian, counterculture ideals to new generations in the language of the pop media as distinct from modes prevalent in an older print culture that still characterize much of American poetry. Instead of opting for the latter's often abstruse language games intended for the select few, Ginsberg the pop-star poet sings out with the assured relevance of ancient bards made new for the electronic age.

Such a view of Ginsberg in the 1990s, while correct in many ways, would still be incomplete or distorted since too great an emphasis on his successes as a performing artist singing blues and rock lyrics obscures other significant facets of his career. After all, when confronted with the question of whether he is a writer or a performer, Ginsberg yet maintains that he is "primarily a writer." Ginsberg is also a teacher of poetry at the Jack Kerouac School of Disembodied Poetics of the Naropa Institute at Boulder, Colorado, and at Brooklyn College (where he is the Distinguished Professor of Poetry). More important, he is the creator of a poetics that underlies his creative practice. His poetics makes it clear that Ginsberg the analytical theorist is not incompatible with Ginsberg the inspired singer. His ability to reconcile these seemingly opposite aspects of his personality makes him the important figure he is in American and world literature.

Ginsberg's poetics can best be understood as a synthesis of the divergent strains in American poetry that Roy Harvey Pearce labeled *Adamic* and *mythic* in *The Continuity of American Poetry* (1961) – the former relying for poetic authorization on the self, exemplified by Walt Whitman's *Leaves of Grass,* and the latter on myth and history, as in Ezra Pound's *Cantos*. Although these terms of archetypal criticism indicate the syncretism that is central to his career, Ginsberg transcends even these conceptual boundaries. Ginsberg praises Whitman not only as the poet of the self but also of history, whose aim had been, as he asserted in a 1975 interview, to "alter the consciousness of a nation"; and he has lauded the *Cantos* in *Allen Verbatim: Lectures on Poetry, Politics, Consciousness* (1974), not as Pound often proclaimed "the tale of the tribe," but in Adamic terms, as a personal document in which Pound had constructed "a model of his consciousness over a fifty-year time span." The Ginsbergian convergence of the two principal traditions has been supplemented by material from various other aesthetic, philosophical, and religious systems, including those of William Blake, William Carlos Williams, the French Surrealists, gnosticism, Buddhism, Hinduism, Sufism, shamanism, and the Hebrew prophets.

As represented by his *Collected Poems: 1947–1980* (1984), Ginsberg stands not as a "major minor poet," as Roger Rosenblatt put it in the 4 March 1985 *New Republic,* nor as "a modern Whitman," in James Atlas's more complimentary phrase in the December 1984 *Atlantic,* but as a major poet *tout court*. Even Helen Vendler's astute observation in the 13 January 1986 *New Yorker* that Ginsberg is "responsible for loosening the breath of American poetry at mid-century" only partially can describe Ginsberg's contribution to literature. If Pound's definition of the two types of major poet – the "inventors" and "masters" – is correct, then Ginsberg is a combination of the two: he has invented and mastered a process of fusing orality and writing that represents an advance not only on the print-bound poetry of his time but also on the practices of earlier poets who set their poetry to music.

Ginsberg the innovator has also brought about a genuine blending of Western and Eastern modes of poetic composition unequaled in the entirety of the Euro-American literary tradition. While Ginsberg is a serious practicing Buddhist, his is a Buddhism intermixed with a quasi-gnostic feeling of cosmic alienation and a large dose of American pragmatism. For him the art of poetry is both a religious act and a social practice, and in this he carries on the Transcendentalist poetics of Ralph Waldo Emerson and Henry David Thoreau. His continued faith in the poet as inspired visionary is due in no small part to the native American tradition, whose predilection for intuitive utterance has also subtly influenced his penchant for spontaneous writing, which Ginsberg incorporated in his poetic methods by following the advice of Kerouac. His motto "First thought, best thought" has been the same for decades and has been reconfirmed by his deeper immersion in Hindu and Tibetan Buddhist poetic practices. This aspect of his poetics is also partly responsible for his expansion of the language of poetry and his treatment of taboo subjects such as sexuality. His is a voice whose unprecedented candor has broken through the puritanical hypocrisy that has tended to suppress for centuries the truthful expression of the deepest of human passions.

Since the late 1970s Ginsberg's work increasingly has been getting serious recognition. He has progressed from the young outsider of the Beat Generation, published in the samizdat-like little volumes of City Lights Books, to the canonized "elder statesman" of the more-than-eight-hundred-page *Collected Poems*. Part I of "Howl," uncensored, has been included in the *Norton Anthology of American Literature* since 1979, and an annotated edition of the poem was published in 1986. Important work on Ginsberg includes *On the Poetry of Allen Ginsberg* (1984), a varied compendium of studies by more than fifty writers; two massive biographies, *Ginsberg: A Biography* (1989) and *Dharma Lion: A Critical Biography of Allen Ginsberg* (1992); and *The Works of Allen Ginsberg 1941–1994: A Descriptive Bibliography* (1995). In addition, his renown as a talented photographer has been growing as his work has appeared in dozens of exhibitions in America and Europe. And he has collaborated with composers Philip Glass and George Gruntz on operas. In occupying a realm encompassing orality and visuality, music and text, spontaneity and craft, humor and utmost seriousness, sublime expression and everyday speech patterns, Ginsberg's overall achievement is certainly a major one.

The development of Ginsberg's career is more easily traced thanks to his chronological arrangement of his poetry in *Collected Poems 1947–1980,* where the sequence of poems is often rearranged from the way they appeared in the individual volumes. As reading the poems makes clear, Ginsberg's poetics, though springing from the romanticism of the Beat movement, certainly outgrew what can now be seen as a historical moment. Like the monumental works *Leaves of Grass* and the *Cantos,* the *Collected Poems* deserve to be read as a whole, as opposed to the practice of picking out certain "poetically intense" parts and discarding what is deemed to be prosaic or journalistic. In the "Author's Preface, Reader's Manual" Ginsberg writes that the work "may be read as a lifelong poem including history." While he points out that the poems comprise "a panorama of valleys and plateaus with peaks of inspiration every few years," every piece included in the collection is relevant to the total "image" of his epic autobiography.

Born on 3 June 1926 in Newark, New Jersey, Allen Ginsberg grew up in Paterson in a family with political, intellectual, and literary interests. His mother, Naomi Ginsberg, was born in Russia and grew up speaking Yiddish. Influenced by her family's sympathy with the anticzarist revolutionary movement, she became a Communist in the United States. Louis Ginsberg, though born in Newark, also came from a Russian-immigrant, Jewish

Ginsberg in 1940 with his mother, Naomi, the subject of his important poem "Kaddish" (Collection of Allen Ginsberg)

background. A socialist, he taught English at Central High School in Paterson and became a well-known poet, whose verse appeared in the literary magazines of the day, including *The New Masses* and *The New York Times Magazine*. Ginsberg's home life as a child was marked by tension between his parents that was exacerbated by Naomi Ginsberg's mental illness. The family struggled emotionally with her mental breakdowns and financially with the expense of her treatment.

Always a voracious reader, a habit encouraged by his teacher-poet father, the young Ginsberg read widely in Metaphysical and Romantic poetry. He read the poetry of Edgar Allan Poe, Emily Dickinson, and other American poets during his high school years in Paterson, New Jersey. At Columbia University, which Ginsberg entered in 1943, the list widened to include Continental, mainly French, poets. Although he enjoyed the classes he had with Lionel Trilling, Mark Van Doren, and Raymond Weaver (a biographer of Herman Melville), Ginsberg later came to consider his formal education there largely a waste of time. As he recalled to Jane Kramer in 1967, Whitman was hardly taught while John Crowe Ransom and Allen Tate were "the supreme literary touchstones."

While attending Columbia, Ginsberg received a supplementary education from William Seward Burroughs, who had an apartment near the school that became a meeting place for would-be writers, Ginsberg and Kerouac among them, beginning in 1944. Burroughs lived surrounded by a library of books wholly unfamiliar to Ginsberg, and he lent or suggested books by Franz Kafka, Louis-Ferdinand Céline, Arthur Rimbaud, Jean Cocteau, and others. Ginsberg began attempts at writing serious poetry in early 1945, producing a long quasi-symbolist work titled "The Last Voyage," which was an amalgam of his heterogeneous reading, mainly of Rimbaud and Charles Baudelaire but also of Poe; in content the poem echoes Whitman in the young poet's aspiring to some form of "cosmic consciousness." In the ensuing five years there occurred two events that decisively influenced Ginsberg's growth as poet and thinker: his Blake vision and his meeting with William Carlos Williams.

Collected Poems contains only one poem written earlier than 1948, the watershed year of Ginsberg's poetic and spiritual education. In the summer of 1948 he experienced what he has variously described as "an auditory illumination" or a "cosmic vibration breakthrough": while reading Blake's "Ah, Sunflower" and "The Sick Rose" and occasionally glancing out the window, he suddenly heard a grave, deep voice reciting the poems. At different times he has identified the voice as his own, or as Blake's, or as that of "the Ancient of Days." The voice was accompanied by a sensation of bliss and a knowledge of having penetrated the mystery of the universe. In subsequent years Ginsberg tried to capture the experience in writing. He also tried to repeat and to expand on the experience to attain even higher states of visionary awareness. This mystical hunger prompted him to experiment with a wide assortment of drugs, from hashish through LSD to laughing gas and magic mushrooms; like Burroughs, he even embarked on a search for the mysterious yage root of South American Indians.

The Blake vision and the taking of hallucinogenics had at least two lasting consequences for Ginsberg's poetic development. First, they strengthened his sense of alienation from mainstream notions of selfhood and being, initiating an inquiry into esoteric and non-Western religious systems and their concomitant modes of expression in voice and writing. He always encountered a harmonious balance between orderly mental processes and spontaneous verbal eloquence; his slogan "If mind is shapely, art will be shapely" proceeds from his immersion in Hindu mantras, the great sutras of Buddhist teaching, and the oral poetry of the Baul singers of Bengal or of the Aborigine songmen of Australia. Second, writing under the influence of drugs helped loosen the bonds of customary language usage, not only in scrambling grammar and getting rid of what the poet termed "syntactical sawdust" but also in encouraging, as in the Surrealists' automatic writing, his free association and juxtaposition of disparate words and phrases. The practice was later theoretically expounded by Ginsberg as ellipsis, which has given a new charge to the Whitmanian long line and revitalized Pound's ideogramic method.

In 1948 and 1949 most of Ginsberg's attempts to capture the Blake vision in poetry show the obvious influence of Blake ("On Reading William Blake's 'The Sick Rose'"), but "Vision 1948" shows some evidence of his struggle toward his own voice in its oxymoronic and synesthetic juxtapositions: "blind vision," "[d]umb roar of the white trance," and "I / Wake in the deep light / And hear a vast machinery / Descending without sound." (Ginsberg brought these last few phrases to full poetic realization in the famous line "angelheaded hipsters burning for the ancient heavenly connection to the starry dynamo in the machinery of night" in "Howl," Part I.) The lines are clearly intended to approximate in language the mystical experience that by its nature is ineffable; but instead of striving for a metaphoric or even metonymic substitution the poet presents the unpresentable. During the ensuing years Ginsberg made several attempts to write about his Blake experience, but it was only ten years after this spiritual turning point that he captured it definitively in "The Lion for Real."

In the spring of 1950, Ginsberg for the first time heard William Carlos Williams read his poetry, at the Guggenheim Museum. Although initially baffled by Williams' rhythms, he wrote later, "I suddenly realized he [Williams] was hearing with raw ears, [hearing] the pure sound and rhythm as it was spoken around him," rather than reproducing "literary rhythms he would hear in a place inside his head from having read other writings." This was almost exactly Williams' own view of what he wanted to achieve with his "variable rhythm," an arrangement of lines in which, as he put it, "the rhythmic pace was the pace of speech." After this first encounter Ginsberg sent a handful of rhymed, formal poems to Williams that failed to impress him. Before and after the meeting, however, Ginsberg was already making attempts to write in a Williamsian manner in such poems as "Cézanne's Ports" and "Two Boys Went into a Dream Diner"; yet, as he wrote his mentor, "I may need a new measure myself." In the midst of trying to take seriously Williams's dictum "No ideas but in things" – a practice in which everyday objects and events are viewed as the proper subjects for poetry and are rendered in a straightforward manner that resemble ordinary speech – Ginsberg was dissatisfied with the short, jagged line and stanza pattern favored by Williams. His own manner was too elastic, the pressure of his influences too heterogeneous, leading him in directions diametrically opposed to Williams's unpretentious, at times threadbare, objectivist practice.

One important influence on Ginsberg during this period was Surrealism, to which he was introduced by Burroughs and again later by Carl Solomon, whom he met at the Columbia Psychiatric Institute in 1949 during an eight-month period of hospitalization. On Solomon's urging he began reading writers such as Antonin Artaud and Henri Michaux. Artaud's mental problems and his ideas on the writer's necessity of being "cruel" and Michaux's experiments with mescaline and other drugs attracted Ginsberg's visionary, esoteric temperament that Williams's sobriety was unable to touch.

By the early 1950s Ginsberg was not simply a novice in search of his personal style but an experimental poet, taking on and discarding poetic masks in a quasi-Poundian manner. His experiments included imitations of Metaphysical poets such as Andrew Marvel; Williamsian imagistic shorthand; long-line stanzas modeled on Whitman and Christopher Smart; post-Dadaist, quasi-absurdist songs such as "Pull My Daisy," a playful collaboration between himself, Kerouac, and Neal Cassady, and "Bop Lyrics"; as well as poems in which he transcribes journal jottings and strives toward an uncensored expression of sexual and other emotional subject matter. Ginsberg's formal and thematic diversity, though at times less pronounced, has remained a constant in his writing. From the beginning his writing contains the seeds of the great syncretistic achievement of his maturity.

Such variety does not, of course, preclude Ginsberg's having a voice of his own. Louis Simpson correctly noted in Lewis Hyde's *On the Poetry of Allen Ginsberg* (1984) that of all the early verse collected in *Empty Mirror* (1961), the poem "Paterson" is "absolutely Ginsberg's." It unites in an indissoluble whole fragments of actual reality and the ecstatic surreal intuitions of the visionary, all shot through with a unique pathos mixed with bemused self-irony. Asking in the first line "What do I want in these rooms papered with visions of money?" Ginsberg later begins to describe his preference:

> I would rather go mad, gone down the dark road to Mexico, heroin dripping in my veins,
> eyes and ears full of marijuana,
> eating the god Peyote on the floor of a mudhut on the border
> or laying in a hotel room over the body of some suffering man or woman;
> rather jar my body down the road, crying by a diner in the Western sun;
> rather crawl on my naked belly over the tincans of Cincinnati[.]

"Paterson" uses the fixed base of Whitman's catalogues, but the pace is quickened, and the juxtapositions of shocking data, the unabashed yet justified use of taboo words and themes, carry the poem far beyond the limits that circumscribe the poetry of both Whitman and Williams. The grotesque imagery prefigures the agony, alienation, and black humor of "Howl" and much of Ginsberg's best work. Even in such a poem as "Bop Lyrics," so different in form and mood from the large-breathed "Paterson," Ginsberg states something otherwise inexpressible. His poetry is from the other side of language, a little mad, going freely wherever the flow of signifiers would take the speaker, all the magic wordplay coming down to "I'm so lucky to be nutty."

Poems such as "A Dream" and "The Shrouded Stranger" do not make *Empty Mirror* a "very morbid book," as Paul Christensen has suggested; rather, they show a rare instance in American poetry of the influence of Poe, absorbed by Ginsberg either directly or via a detour of Poe's own influence on French writers from Baudelaire through Stéphane Mallarmé to the Surrealists. It is as if Ginsberg had borrowed some of Poe's rhetorical and thematic figures (graveyards, haunted houses, ghosts, and, more important, ghoulish symbolic strangers) and deliberately disfigured them, so that the morbidity becomes suffused with irony and parody, the inevitable corollaries of intertextual adaptation. At times he transcends the ever-present ironic mode, as in "The Terms in Which I Think of Reality," where an extended simile of man compared to a River Street prostitute is filled with a simultaneous post-Baudelairian longing and a gnostic intuition of estrangement. The underlying theme is the recognition of all humans' birthright to realize themselves and evade the alienating forces of social control.

The two poems written before "Howl" that stand out from the other pieces of the early 1950s are "The Green Automobile" and "Siesta in Xbalba." The former, broken up into stanzas of four staggered lines that when read aloud quite remarkably resemble the breath units of "Howl," is written about, and at times to, Cassady, with whom Ginsberg had a brief love affair that grew into a strong emotional and intellectual relationship lasting until Cassady's death in 1968. If Herbert Huncke may be seen as one extreme aspect of "beatness" — furtive, cool, cynical, stationary, apathetic, depressed — then Cassady is its other side: exuberant, enthusiastic, full of life and inexhaustible manic energy.

In "The Green Automobile" the Ginsbergian persona assumes a wild joyousness as he imagines a trip to Denver to visit Cassady, driving a green car so that they would then together go "on the road" to "deal with each other in princely gentleness once more." The poem finally celebrates their attachment to one another as a memorial that withstands the ravages of time. In a deflection from the Horatian or Donnean aesthetic object, the speaker seeks as "an ageless monument to love / in the imagination: / memorial built out of our own bodies / consumed by the invisible poem," as if to suggest that the art into which life had been fused must be as transparent as possible so not to obscure the physical and metaphysical passion that once united the lovers. Ginsberg echoes Whitman in defining a friendship that on the highest level is both private and public, not denying but including reality:

> Neal, we'll be real heroes now
>   in a war between our cocks and time:
>   let's be the angels of the world's desire
>   and take the world to bed with us before we die.

The image evoked here will be recast in an even more effective manner in the third section of "Howl" to become the vortex not only of that poem but, arguably, of the entirety of Ginsberg's poetic undertaking.

Although stylistically uniform, "The Green Automobile" contains disjunctive devices such as evocatively juxtaposed word-pictures ("poolhall flophouse jazzjoint jail") and surreal telescopings ("The windshield's full of tears, / rain wets our naked breasts"). In contrast, "Siesta in Xbalba" is quite openly an example of projectivist "composition by field" as set out by Charles Olson in the

Maximus Poems (1960): the poem is a collage of different formal modes and various line arrangements corresponding to events and states of mind. In depicting a stay at the cocoa plantation in Chiapas, Mexico, of his friend Karena Shields, Ginsberg self-consciously deromanticizes the romantic topos of presenting a self divided against itself.

As Ginsberg described "Siesta in Xbalba" to Cassady, the "old forms" of *Empty Mirror* have become synthesized so that "the actual fragments of thought utilized for short poems before are now linked together in a natural train of thought or images" corresponding to his mind's meditative state. The poem contains clinically dispassionate passages in which the narrating "I" turns its cold gaze on the more naive narrated self as enmeshed in its social setting. The poetry produced by the naive self is observed during these moments of reverie and distancing in unflattering terms as "crude night imaginings" and "primitive illumination." The speaker is caught up in the spirit of the place, amidst the ruins of past civilizations, awakening in him a nostalgia for Europe, "the ancient continent," and a craving to embark on a pilgrimage with no known destination, yet in search of a "future, unimaginable God." A Whitmanian "enough!" puts an end to the deeply private reflections, yet the concerns of the social critic are also heard.

Although some scholars have argued that Ginsberg's metaphysical search dominated his thought and poetic interests from the Blake vision until he reoriented his pursuits in 1963 following his trip to India (as is reflected in "The Change"), his mystical yearnings were never able to extinguish his fierce indignation at the sight of suffering and injustice. Writing poetry for him meant not only an inner journey to uncover the layers of the self but also the taking on of responsibility to face up to the evils perpetrated not by cosmic archons but by men. That a difficult balance between the subjective and the larger social realms is essential is encapsulated in the poem "On Burroughs' Work," which should be read as a manifesto both of poetics and politics:

> The method must be purest meat
>   and no symbolic dressing,
> actual visions & actual prisons
>   as seen then and now.
>
> Prisons and visions presented
>   with rare descriptions
> corresponding exactly to those
>   of Alcatraz and Rose.
>
> A naked lunch is natural to us,
>   we eat reality sandwiches.
> But allegories are so much lettuce.
>   Don't hide the madness.

The speaker clearly does not deny the symbolic and allegoric aspects of writing nor its rhetorical structure; the poem is rich in figurative language. But in a post-Poundian insistence on exact presentation ("rare description") he also maintains that "sandwiched" between the figures the poem's linguistic signs must bear a referential relationship to reality. While aware of a phenomenal realm (for example, prisons) that the mind has not made, the poet is a visionary to the extent that his insights are capable of penetrating the veneer of reality posing as sanity, necessity, and common sense. The double meaning of "naked lunch," at once referring to the title of Burroughs's satiric antinovel and the truth about reality such a work represents, ironically complements the poem's running metaphor of consumption, accenting the differential attitude of the counterculture to the mainstream's materialism and blindness. The task of the writing is thus to uncover the underlying repression and exploitation, the deranged, destructive aggressiveness stemming from human greed and fear.

It is a hallmark of Ginsberg's poetic journey to wed, as Ihab Hassan remarked, language to the flesh, fusing the uncensored vision of the individual to an accurate vision of the world. Nowhere in Ginsberg's early poetry, nor in the work of his contemporaries, does such a convergence of private and public, mystical and realistic, Adamic and mythic poetic modes take place as in the signal poem "Howl." In this extraordinary verbal convulsion Ginsberg brought to a focus all his previous experiments in form and technique and at once changed the topography of American and world literature. Four years later he was able to follow it up with "Kaddish," the long poem that many critics have judged to be his greatest literary accomplishment.

After some four months in early 1954 at Shields's plantation, Ginsberg traveled to stay with Cassady, who lived near San Francisco. During the next few years Ginsberg became part of a literary milieu that favored unconventional poetic approaches and experimentation with new forms. He frequented the literary salon of Kenneth Rexroth — anarchist, poet, editor, and a generous elder brother to younger aspiring writers such as Ginsberg, Robert Duncan, and Philip Lamantia. It is a sign of the

A portion of the first draft for "I Beg You Come Back & Be Cheerful," written in Ginsberg's notebook on 25 September 1959, with commentary (lines 3–7) by Jack Kerouac (by permission of Allen Ginsberg). The poem was published in Reality Sandwiches *(1963).*

times that "Howl," written in the summer of 1955, was not originally intended for publication. "I thought I wouldn't write a *poem,*" Ginsberg remarked in 1959, "but just write what I wanted to without fear, let my imagination go, open secrecy, and scribble magic lines from my real mind – sum up my life – something I wouldn't be able to show anybody, writ for my own soul's ear and a few other golden ears." One in Ginsberg's golden audience was Peter Orlovsky, whom he had met six months before. The two men formed a deep attachment that was to last for decades. Thus, Ginsberg's maturity of style and form, enabling him to open the floodgates of the poem's "huge sad comedy of wild phrasing," was complemented by the final acceptance of his sexual orientation.

The emotional complexity of "Howl," one of the central literary documents of the Beat Generation, has often been misunderstood. Many of its early critics disparaged the poem for its graphic language and flouting of mid 1950s standards of decorum, yet its complexity may have been missed by even some its most astute and sympathetic interpreters. Only Helen Vendler clearly articulated the central insight about the poem embodying "Ginsberg's comic rage, or raging comedy." It is not surprising that when on the twenty-fifth anniversary of the publication of the poem Ginsberg gave a reading of it at his alma mater, Columbia University, the reporter for the 7 December issue of *Time* was disappointed: he found Ginsberg's tone mocking and satiric, out of line with what he perceived to be the "real" import of the poem, to give voice to "nobly expended pain." He saw the poet not as protesting the viciousness of the times but as "chatting, singing, wearing a necktie and making his howl a thigh-slapping hoot."

Although it hardly seems likely that "Howl" could ever become a "hoot," the reporter's reaction raises an important point, that despite its pathos and catalogue of suffering the poem generically belongs to comedy rather than tragedy. In this light, Ginsberg was not parodying himself and his poem in the commemorative reading; on the contrary, the parody, travesty, and other forms of burlesque and black humor were there from the beginning. Ginsberg succeeded in writing a comic, travestied "Song of Myself" and not simply to copy or modernize Whitman. Ginsberg calls attention to the poem's comic tone in Part III when in apostrophizing Carl Solomon he writes: "I'm with you in Rockland / where you laugh at this invisible humor." Springing from the poem's nature as a satiric protest poem, the humor is fed by Ginsberg's use of ellipsis as well as the descriptive passages of his worldview, lifestyle, and literary influences. The humor is the easily overlooked veil in which the poem's all-too-serious matter is shrouded; as in Yeats's "Lapis Lazuli," it is the invisible gaiety that transfigures all the dread.

"Howl" is the testimony of a witness to an outrage, although more violently and explicitly it tells a story similar to T. S. Eliot's *The Waste Land* (1922) in that it registers the disaffection of a generation facing a morally bankrupt, hypocritical, oppressive social milieu. As Pound had been for an earlier generation, Ginsberg was the catalyst for the Beats. His first public reading of "Howl" at the 6 Gallery on 7 October 1955 in the company of Gary Snyder, Michael McClure, Lamantia, Philip Whalen, Rexroth, and Kerouac launched the San Francisco Poetry Renaissance.

In "Howl," as in Burroughs's *Naked Lunch* (1959) and earlier in the work of Jonathan Swift and Juvenal, rage is the motivating emotion; but while Beat satire shares the *saeva indignatio* (fierce indignation) of its classical and neoclassical forebears, it characteristically lacks their sense of moral superiority and is more closely linked to André Breton's "révolte supérieure de l'esprit" (supreme revolt of the mind), which gave rise to the "humour noir" of the Surrealists. Of the Beat writers it was Ginsberg, along with Burroughs and to a lesser extent Gregory Corso, who mastered satire and burlesque, the forms most appropriate to protest and defiance. But Ginsberg's rage also has something uniquely American about it that he shares not only with some of his Beat comrades but also with prose writers such as Norman Mailer, Joseph Heller, Ken Kesey, James Purdy, and John Barth. Theirs is a rage, as Douglas Davis noted in *The World of Black Humor* (1967), "that erupts, scarring everyone near it. Rage that holds itself in, behind deadpan, rage that rarely reveals itself. Rage as well that loves what it hates, mixing with arrogance a strange humility, foreign to Europe."

The rage, energy, and humor in "Howl" come across to the reader or listener largely through Ginsberg's technique of ellipses, as he called the paratactic juxtaposition of "disparate thinks put down together." Ginsberg's juxtaposed elements, which in journal notes he also called "shorthand notations of visual imagery" or "abstract haikus," are the micro-building blocks of his long Whitmanian lines. His reference to haikus is especially significant, for shortly before writing "Howl" he had been reading H. R. Blyth's collection of haikus and writing his own. His experiments with haiku, the form he called the primary mode of the ellipsis, he thought to be "useful for advancement of practice of western metaphor." Humor as a particular linguistic gesture is a figural device that defies logic yet has concrete emotional validity; Ginsberg's incongruous juxtaposition of elements often leads to a comic insight, an invisible link that provides the bridge over the ellipses.

"Howl" begins on a note of high rhetoric, with lines that are as canonical as Whitman's "I sing myself and celebrate myself" from *Leaves of Grass* or Eliot's "April is the cruellest month" from *The Waste Land* or Pound's "And then went down to the ship" from the *Cantos*:

> I saw the best minds of my generation destroyed by madness, starving hysterical naked,
> dragging themselves through the negro streets at dawn looking for an angry fix[.]

Although the original version has *angry streets* and *negro fix,* the adjective-noun phrase works both ways to express the seething anger of Harlem and the presence of heroin addicts. The ellipsis *angry fix* ironically presents the "best minds" in search of the drug. The humor begins to make itself felt in the exaggerated and hyperbolic descriptions of the various activities of the Beats.

The reader gets a constantly shifting, nightmarish, yet wildly comic series of images of these "angelheaded hipsters" cowering in "unshaven rooms in underwear," eating "fire in paint hotels" and drinking "turpentine in Paradise Alley," as they "purgatoried their torsos night after night / with dreams, with drugs, with waking nightmares, alcohol and cock and endless balls." Ginsberg uses ellipses to perform various surreal acts of transference, as when individual states are projected on the environment (for example, *unshaven rooms*); but on the whole he creates tropes of excess bordering on caricature. The hipsters are presented as grotesque giants in various stages of intoxication, performing prodigious acts such as "leaping towards poles in Canada & Paterson" (as if the two were geographically equivalent), flying through Brooklyn and the Battery to the Bronx where they are finally brought down "all drained of brilliance in the drear light of Zoo." It turns out that the flight was actually a subway ride, and their feverish activity mainly talk. The narrating "I" is both inside and outside the "lost battalion of platonic conversationalists" who went

> yacketayakking screaming vomiting whispering facts and memories and anecdotes and eyeball kicks and shocks of hospitals and jails and wars,

this time vanishing in "nowhere Zen New Jersey." This oscillation between figures of inflation and deflation parallels the fantastic lifestyle of frenzied activity and exhaustion as presented by the speaker.

Ginsberg goes on to depict the surreal travels of the Beats, where the ellipses reflect the frantic efforts of these bumbling Brobdingnagians to experience visions in the most unlikely places — in Kansas, Baltimore, or seeing "visionary indian angels" in Idaho. In parodying his own experience with Blake, Ginsberg is implying that dream visions and hallucinations are countered by social reality; hence these moments of enlightenment, as he writes in Part II, are all "gone down the American river." The speaker's silent humor is turned inward and toward his comrades who are constantly wondering "where to go," crisscrossing the continent and seeking (in Houston, for instance) "jazz or sex or soup" — apparently the staples of their existence and perhaps in that order. If there is self-mythologizing going on here and elsewhere in "Howl," as some critics have argued, then such tactics are always undercut by the unsentimental comic gaze of the narrating "I."

In the sad, comic catalogue of political and sexual activities that follow, the absurdity heightens as readers are confronted with the heroes, "in beards and shorts with big pacifist eyes," investigating the FBI no less. Protesting the "narcotic tobacco haze of capitalism," they distribute not just communist but "Supercommunist pamphlets," while "weeping and undressing" in public. Looming colossal one moment and reduced to Lilliputian insignificance the next, the Beat antiheroes are allegorically not only Paul Bunyan but also Charlie Chaplin's Little Tramp. Their dominant identity is the schlemiel of Jewish literature, jokes, and anecdotes, a character who uses self-mocking humor to turn his weakness into an invisible weapon. The schlemiel uses his comic stance, as described by Ruth Wisse in *The Schlemiel as Modern Hero* (1971), "as a stage from which to challenge the political and philosophical status quo." All the Beats, Jews and Gentiles alike, are in part schlemiels in "Howl." When arrested for distributing their leaflets, they counter by "bit[ing] detectives in the neck and shriek[ing] with delight" in policecars, for they, like their counterparts in Jewish humor, are basically innocent (or guilty only, as the speaker admits, of their "wild cooking pederasty and intoxication").

Their innocence is one of two main factors alienating the Beats from mainstream society, the other being their sexual openness, whether heterosexual or homosexual. The speaker's account of their sexual activities, without precedent in American poetry for its graphic descriptiveness, is yet always tinged with absurd humor, the surest antidote to reading the passages as in any sense pornographic:

> who copulated ecstatic and insatiate with a bottle of beer a sweetheart a package of cigarettes a candle and fell off the bed, and continued along the floor and down the hall and ended fainting on the wall with a vision of ultimate cunt and come eluding the last gyzym of consciousness.

Even "N.C." (Neal Cassady), the "secret hero of these poems, cocksman and Adonis of Denver," presented as a combined Casanova–Don Juan figure of limitless sexual prowess, comes down a peg or two, as when he is caught by the speaker's quick verbal snapshots with "gaunt waitresses" in the washrooms of roadside diners and gas stations. Like N.C. in his choice of sex partners, the Beats are indiscriminate in sexual and other matters because of a sense of cosmic alienation; as Ginsberg wrote in his poem "Laughing Gas" three years later, from the Beat vantage point the whole universe is but "a funny horrible / dirty joke."

Ginsberg's juxtapositions become especially caustic as the speaker outlines the hypocrisy and counterfeit, not of "square" society but of the Beat lifestyle. These antiheroes go from parties straight to the unemployment office; if not lost in complete inertia, they "scribbled all night," only to find in the morning their "lofty incantations" were nothing but "stanzas of gibberish." Although professing to be Buddhists and dreaming of "the pure vegetable kingdom," they eat the meat of "rotten animals." They seem even more pathetic and ridiculous as they attempt suicide "successively unsuccessfully," or when "demanding instantaneous lobotomy" in mental hospitals. Instead of lobotomy they receive "the concrete void of insulin Metrasol electricity hydrotherapy psychotherapy occupational therapy pingpong & amnesia." The black wit in this catalogue turns outward; not only does the speaker satirize the various treatments as ineffective, but by equating the mismatched components of the list, he shows up the absurdity of the whole exercise. Ginsberg's humor is mordant, but it never effaces the fact that the Beats do indeed suffer; yet even in some of the most painful scenes the narrator manages to inject a dose of irony or bathos or both. The

Ginsberg with Neal Cassady (Collection of Allen Ginsberg)

image of one his comrades "returning years later truly bald except for a wig of blood" is both appalling and ludicrous.

Before Part I ends on a hyperbolic scene of mock crucifixion and apocalypse, the speaker in a thoroughly postmodernist gesture articulates the poem's poetics in remarks that are significant for Ginsberg's later poetry as well. The speaker describes running "through icy streets obsessed with a sudden flash of the alchemy of the use of the ellipse." In the next line he speaks of making "incarnate gaps in Time & Space through images juxtaposed." Ginsberg offers a more direct explanation of the rhetorical practice of word pairings such as *angry fix, hydrogen jukebox, heterosexual dollar, stale beer afternoon* in his comments on "abstract haikus": "objective images written down outside mind the result is inevitable mind sensation of relations. Never try to write of relations themselves, just the images which are all that can be written down on a subject." The injunction is an echo of Chinese translator Ernest Fenollosa's axiom – "Relations are more real and more important than the things which they relate" – that was an influential theoretical point in Pound's turn to ideogramic composition.

Ginsberg's thinking on the juxtaposed images of the haiku was reinforced not by an interest in Chinese translation but by his study of the paintings of Paul Cézanne, whose formal arrangements fascinated him. Describing the effect of Cézanne's paintings, Ginsberg said in 1967, "it's just juxtaposition of one color against another color," so that "by the unexplainable, unexplained non-perspective line, that is juxtaposition of one *word* against another, a *gap* between two words" would be created "which the mind would fill in with the sensation of experience." The juxtaposed words, unrelated either empirically or logically, transfer a portion of their intellectual-emotive-historical content onto each other. In "hydrogen jukebox," for example, the noun *jukebox* replaces the conventional word *bomb,* yet retains some of its power, even if only figuratively, lending a quasi-apocalyptic quality to the otherwise harmless record player by the act of defamiliarization. This principle of juxtaposition is the basis of the collage and in writing also the source of Burroughs's cut-up method.

Part II of "Howl," written some weeks after Part I under the influence of peyote, is the most ecstatic and frenzied in the poem. The humor is all in the excess, in the catachrestic incongruity of every juxtaposition. Each sentence, even each phrase, is an exclamation pointing an accusing finger at modern civilization for conspiring, as Emerson had written, against the manhood of every one of its members. Moloch, the god of the ancient Ammonites to whom children were sacrificed (the name was also invoked to stand for the vast machinery in Fritz Lang's 1926 film *Metropolis*), is superimposed over the particulars of industrial civilization rendered in violent images. After cataloguing individual suffering, Ginsberg escalates his rhetoric to an unexpected crescendo:

> Moloch whose love is endless oil and stone! Moloch whose soul is electricity and banks! Moloch whose poverty is the specter of genius! Moloch whose fate is a cloud of sexless hydrogen! Moloch whose name is the Mind!

In subsequent self-exegeses Ginsberg has tended to make the last exclamation in this passage into the central fact about the nature of Moloch; in the 1995 BBC interview he called it the "ultimate accusation," asserting that "the all-devouring god" is not "out there, it's our own imagination." In adding that this is "a piece of wisdom teaching" that he gleaned from Blake, Ginsberg likely was alluding to the "mind-forg'd manacles" of the poem "London," which captured the misery brought on the urban population by industrial capitalism.

Similar to Part I, the speaker's indignation is directed both at the harm and devastation worked by the mind's capacity for evil and the forces variously attempting to resist or escape its destructive power:

> Visions! omens! hallucinations! miracles! ecstasies! gone down the American river!
> Dreams! adorations! illuminations! religions! the whole boatload of sensitive bullshit!

In the last line of Part II the birth of the entire counterculture is sketched: "Real holy laughter in the river! They saw it all! the wild eyes! the holy yells! They bade farewell!" The implication is that it was precisely those who saw most clearly the dangerous deformities of post–World War II America that turned their back on society and dropped out. Their turning inward historically relates them to the American Transcendentalists and also to the Romantics, who similarly sought solace in inner experience and in select companionship as a result of their dashed expectations in the aftermath of the French Revolution.

Part III stands ostensibly as a testimony to a new sensibility that celebrates private relationships, yet the absurdist language inescapably calls attention to itself. In the litany-like sequence the speaker addresses his friend Solomon, who is in the madhouse, assuring him of his solidarity in the repeated line "I'm with you in Rockland." The images grow progressively more exaggerated, as in the lines "I'm with you in Rockland / where you've murdered your twelve secretaries." Ginsberg's black humor cuts at least two ways: the controllers at Rockland treat Solomon, this somewhat paranoid post-Dadaist schlemiel, as if he were a mass murderer; but also, Solomon himself may feel as if he has committed such an enormous crime against society by being unfit to be a "normal" part of it. Throughout the series of absurd projections about Solomon's condition, Ginsberg's irony mitigates and transposes the horrors of the asylum onto the plane of rhetoric, safeguarding against the dangers of hero-worship and kitsch, proving Breton's axiom that black humor is the deadly enemy of sentimentality.

The entire poem reaches its supreme moment of pathos and humor near the end of Part III:

> I'm with you in Rockland
>> where we hug and kiss the United States under our bedsheets the United States that coughs all night and won't let us sleep.

The image suggests more than the aspect of American black humor that loves what it hates. The walls of the madhouse, as well as those of inwardness, suddenly crumble. The concept of the whole United States assumes the fragility of a sick child whom the two marginalized characters, now seen as dutiful parents or elder siblings, hug and kiss in the madhouse. Even though the sick child is also an incarnation of Moloch and responsible for the "armed madhouse," the two outcasts do not respond to shock therapy and lobotomy in kind; they try to cure America in the only way they know: through love. Ginsberg's humor is still there in the inflated image of the Beats as parents to a country, but with historical hindsight it may be said that this image was neither a pose nor a hoax. As Williams wrote in his introduction to *Howl and Other Poems* (1956), "the spirit of love survives to ennoble our lives if we have the wit and courage and the faith – and the art! to persist." Ginsberg's wit does not dissolve the rage and the pain, but it transfigures them and transforms them into art.

In art form is what matters. To write without fear is one thing; to attain an achieved concentration of mind so that the resulting art would be shapely, as Ginsberg had intended, is quite another. Ginsberg's combination of Whitmanian long lines, the fixed-based repetitive structure of Smart, and the short Williamsian haikulike phrases clashing in juxtaposition provide the texture of his mature work. While there is no reason to doubt his own account of the poem's birth, according to which Part I and a sketch of Part III were all "typed out madly in one afternoon," with the publication of *Howl: Original Draft Facsimile, Transcript & Variant Versions, Fully Annotated by Author* (1986) it has become clear that the finished version is markedly different from the first few drafts. Most of the memorable telescoped phrases and prophetic exclamations are present in the first draft; yet detailed study reveals that Ginsberg's revisions not only tightened and intensified the headlong torrent of phrases by excision and rearrangement but also made the poem less wild, less disordered, while leaving intact its emotive and rhetorical core.

"Footnote to Howl," in which everything is pronounced holy, further emphasizes the visionary conjunction of opposites, the negative capability of holding contraries in suspension without, as John Keats wrote, "any irritable reaching after facts and reason," a definition Ginsberg has often quoted. Some of the finest poems Ginsberg wrote in 1955 and 1956 – "America," "A Supermarket in California," "Sunflower Sutra" – strike the difficult balance between private and public spheres, serious social awareness and absurd humor, and acceptance of human limitations along with a larger spiritual dimension. Yet the gnostic-romantic conviction still persists, as stated best in "Sunflower Sutra," that "we're not our skin of grime" but "all golden sunflowers inside." All these poems, and most of what Ginsberg was to write in subsequent years, commonly use varieties of the long line, with cadences and cuts suited to the poet's "inspiration," both the physical breath and the nature of the subject matter. More regular rhythms are reserved only for his songs and blues.

Ginsberg reached a particularly high level of formal sophistication in "Kaddish: For Naomi Ginsberg 1894–1956," which commemorates the mental illness, deterioration, and death of the poet's mother. Several critics have come to consider the poem the pinnacle of Ginsberg's achievement, although it is valued more often for the unusually frank presentation of the harrowing subject matter than for its inventions. A. Alvarez in *On the Poetry of Allen Ginsberg* praises its "poetic accuracy," asserting that Ginsberg "lets the appalling story speak for itself." Clearly, however, if Naomi Ginsberg's story speaks to readers, it does so because of Ginsberg's artistry, because he was able to attain new heights in the art of elliptical juxtaposition, of counterpoint and rhythmic arrangement of words and larger units. If "Howl" is basically a poem of American black humor, then "Kaddish" is an example of the American sublime: through constructing in a style and form of complex intensity the figure of his mother, the poet registers his confrontation with the divine. The "Hebraic-Melvillian bardic breath" Ginsberg noted in connection with "Howl" is more truly present in "Kaddish" than in any other poem he has written.

"Kaddish," written between 1957 and 1959, is composed of six formally distinct parts: proem, narrative, hymmnn & lament, and litany & fugue. In the proem, whose major themes include mutability and the mystery of death, Ginsberg combines memory of his mother with his recording of sensory perception of the present, a convention that suffuses the absent object with the immediacy of subjective insight. Such disjunct coherence is attempted through the poet's retracing the steps of Naomi "where [she] walked 50 years ago" in New York City's Lower

Ginsberg with Ezra Pound in Portofino, Italy, 1967 (photograph © Ettore Sottsass)

East Side – the area where the poet still lives to this day. As he wanders through the streets to find Naomi by experiencing the traces she herself experienced, his memories wind around various approximations of the nature of death. The poem moves in a similarly meandering fashion to find its true place of utterance. Disjunct images, the unchecked flow of juxtaposed bits of memory, create a fragmented collage of the young girl from Russia, who becomes the wife and mother moving inexorably toward paranoia and finally death, for which all life is bound.

Some of the leisurely elegiac lines are suddenly cut up into "naked haiku," as in the remarkable evocation of Naomi's lobotomy:

> No flower like that flower, which knew itself in the garden, and fought the knife – lost
> Cut down by an idiot Snowman's icy – even in the Spring – strange ghost thought – some Death – Sharp icicle in his hand – crowned with old roses – a dog for his eyes – cock of a sweatshop – heart of electric irons.

Such a cluster of seemingly disjunct phrases owes little to surrealistic association; rather, it is constructed out of abstract though affective images, as in the work of the Objectivists (especially Louis Zukofsky and George Oppen) where one component of the image calls forth names of objects – *sweatshop, electric irons* – that define the emotive content of the whole image. The fractured presentation of traces suggests, however, that the bits of observed fact and memory are nothing but patches of a verbal mask that cannot coalesce into a unified whole. Similarly, when the proem ends in a psalmlike adoration of the "Nameless, One Faced, Forever beyond me, beginningless, endless, Father in death," there is no apparent metaphoric intention to fuse all the discordant facets of the divine together.

The proem's long lines are variously organized, though as the first section builds to its conclusion the lines take on a staccato rhythm as they are increasingly composed of short phrases separated by dashes. Ginsberg's use of the dash, which bears a strong resemblance to the French novelist Céline's use of dots in his later work, dominates the narrative section. His dashes both connect and introduce gaps between thoughts, creating a style suited to the construction of Naomi's changeable mental, spiritual, and physical personae. But Naomi's reinvented persona is inextricably fused with that of her son: on the one hand, Ginsberg assembles data for Naomi out of remembered verbal traces of her political obsessions and general paranoia (being pursued by Hitler and Mussolini's agents, by her mother and sister), from transcriptions of old photographs, and from remembered bits of family lore; but equally, the Naomi character is part of Ginsberg's autobiography, the painful story of growing up with a mentally ill mother that results in inextricably tangled feelings of love and revulsion.

The reader is always conscious that the story of the speaker and Naomi is *written* rather than just remembered, *constructed* rather than merely recorded. One of the most disturbing episodes has to do with the twelve-year-old son's leaving his distraught mother in a rest home; the simple phrase "I shouldn't have left her" reverberates with the guilt of betrayal. It is surely not accidental that immediately juxtaposed to this event come lines describing Ginsberg's first intimations of his homosexuality. Throughout Ginsberg pairs the revelations of Naomi and the speaker. At times his writing attains a near-objective naturalism that is as shocking as the subject matter it sets out to present, as in another traumatic event that occurred some five years later:

> One time I thought she was trying to make me come lay her – flirting to herself at sink – lay back on huge bed that filled most of the room, dress up round her hips, big slash of hair, scars of operations, pancreas, belly wounds, abortions, appendix, stitching of incisions pulling down in the fat like hideous thick zippers – ragged long lips between her legs – What, even, smell of asshole? I was cold – later revolted a little, not much – seemed perhaps a good idea to try – know the Monster of the Beginning Womb – Perhaps – that way. Would she care? She needs a lover.
>
> Yisborach, v'yistabach, v'yispoar, v'yisroman, v'yisnaseh, v'yishador, v'yishalleh, v'yishallol, sh'meh d'kudsho, b'rich hu.

The clinical reinvention of Naomi's ruined body analogously joins it to her ruined mind, while the son becomes little more than an exhibitionist-voyeur evincing both fascination and revulsion in confronting a primal scene. What lifts the passage rhetorically is the power of elliptic juxtaposition, for without transition there follow the Hebrew words of the Kaddish, the Prayer for the Dead, specifically the praise of the Name of the Lord. (The notes on "Kaddish" in *Collected Poems* explain that translation of the Hebrew words can be found in the first lines of "Hymmnn," which follows the conclusion of the narrative section: "In the world which He has created according to his will Blessed Praised / Magnified Lauded Exalted the Name of the Holy One Blessed is He!") The image of a ravaged female body, the body of a mother on which her history of sufferings is inscribed, is linked with the mystery of the unspeakable and unpresentable divine.

In the manner of "Footnote to Howl," Ginsberg in "Hymmnn" blesses not just the name of the Lord but all creation, especially Naomi in her failed life and miserable death:

> Blessed be you Naomi in tears! Blessed be you Naomi in fears! Blessed Blessed Blessed in sickness!
> Blessed be you Naomi in Hospitals! Blessed be you Naomi in solitude! Blest be your triumph! Blest be your bars! Blest be your last years' loneliness!
> Blest be your failure! Blest be your stroke! Blest be the close of your eye! Blest be your withered thighs!
> Blessed be Thee Naomi in Death! Blessed be Death! Blessed be Death!

Ginsberg celebrates the whole of experience, extinction as well as creation. Though in a less exalted manner, this acceptance is carried over into the lament section of the poem, where the confused, pathetic lines from Naomi's last letter to her son about "the key" being in the window, quoted at the end of the narrative, are taken up and given deeper symbolic meaning. The vision of the key subverts notions of magnitude and origin: the "slice of light in hand" opens onto a universe the size of Naomi's grave – a synecdoche of the life-breath of "divided creation." In the litany section Ginsberg invests Naomi's physical parts with surreal attributes, ending with a long catalogue of the sufferings and loneliness encompassed by her eyes.

The poem's final section is a fugue in which Ginsberg counterpoints the "caw caw caw" of crows and the speaker's cries of "Lord Lord Lord." This juxtaposition of human and animal voices suggests multiple possible meanings: an affirmation of divine authority, an accusation of the "Grinder of giant Beyonds," or a lamentation of meaningless anguish. The "strange cry of Beings" toward the Almighty may not be heard at all, since, as the speaker says, "my voice" is sounded "in a boundless field in Sheol," the hell of the Bible transposed onto this world. The final line, "Lord Lord Lord caw caw caw Lord Lord Lord caw caw caw Lord," reinforces the similarity and mystery of human and animal cries in the void. It also poetically underlines the power of the voice as a bond that unites humans and other creatures caught in their inexplicable existential predicament. While the cry "Lord" may not

Cover for an issue of The American Poetry Review that includes eight previously unpublished poems by Ginsberg

ultimately be more meaningful than "caw," the two are linked metonymically to signify a knowing-blind desire to establish some tie with the transcendent other. Whether an actual Lord really registers these cries or not is not relevant; the sounds in the abyss have their own validity.

"Kaddish" transposes into language matters that are deeply personal; yet it is much more than a confessional poem concerned only with self-expression, as many critics have maintained. To label the author of the poem as a Jewish poet or even a Jewish mystic, as Allen Grossman suggests in *On the Poetry of Allen Ginsberg*, is also too limiting. The Ginsbergian strain of syncretism, of combining private reverie and Jewish tradition in a wholly new poetic-linguistic-oral creation, is what makes "Kaddish" one of the high-water marks of American literature. Ginsberg in these poems combines historic data and poetry to create a transgeneric mode, blurring the Aristotelian separation of poetry and historiography. The ultimate value of these long texts of Ginsberg's mid period lies, in Vendler's formulation, in their being "the largest attempt since Whitman to encompass the enormous geographical and political reality of the United States."

Resolutely countercultural, Ginsberg complemented and enhanced his political iconoclasm in protesting the Vietnam War by his explicit depictions of homosexual practices, as in "Kansas City to St. Louis." Ginsberg's deliberate use of the language of sexuality serves as a human antidote to the language of propaganda justifying destruction; but it is also an index of his refusal to abide by the code imposed on speech and writing exactly by "leaders" whose aim is to control the minds and emotions of the population, especially the young. The exigencies of political poetry led Ginsberg to adopt a style more settled and linear in syntax and grammar than his earlier work. Since the aim of the poetry is to reach as large an audience as possible – the poet's declared intention, as stated in "Memory Gardens," being "To ease the pain of living" – Ginsberg largely avoids ellipses while retaining his sharp-eyed observation of visual detail. This more traditional language use is also evident in Ginsberg's gradual turn toward openly oral expression and song. Yet pieces such as "Bixby Canyon Ocean Path Word Breeze" from *The Fall of America: Poems of These States 1965–1971* (1972), with its minute attention to names of flora carefully examined and inclusion of illustrations to expand the visual effects of the page, show his continued zest for experimentation.

Ginsberg's attitude toward the erotic is always forthright in his poetry. In "The Change" Ginsberg not only comes to grips with his mystical yearning and turned toward a more activist, public role but also bids farewell to his bisexual past in favor of homosexuality. All of Ginsberg's sexual poetry is unrestrained in its desire to tell all, but the tone varies greatly. "On Neal's Ashes" is brief and tender in its catalogue of the beloved friend's body that is now all returned to ash; "Sweet Boy, Gimme Yr Ass" (1974) celebrates the mutuality and affection that often motivates sexual desire. In poems such as "Please Master," written in 1968, but also in the much later song "Violent Collaborations," written in 1992, the speaker posits a near-masochistic self-image in order to teach himself to renounce machismo by unconditionally surrendering to another. Ginsberg's graphic descriptions of homosexuality also serve to educate the heterosexual majority to tolerate or condone homoerotic emotions and practice. The series of poems in *Cosmopolitan Greetings*, where both the passion and its object are apostrophized as Love, combines the simplicity of seventeenth-century love lyrics and the singable rhythms of popular songs. These poems are also testimonies to the fading of Ginsberg's erotic relationship with Orlovsky, transmuted into a mutually respectful but nonsexual friendship.

The lyrics of "Capitol Air," the concluding song of *Collected Poems* that was originally collected in *Plutonian Ode: Poems 1977–1980* (1982), are simple and deliberately stilted. The words come alive only in Ginsberg's impassioned performance. It is a poem of protest, full of the satirist's righteous anger, yet still containing word and phrase telescopings of naked haiku adapted to regular rhythm and rhyme. Ginsberg's two subsequent collections of poetry, *White Shroud: Poems 1980–1985* (1986) and *Cosmopolitan Greetings: Poems 1986–1992* (1994) are also characterized by a spirit of improvisation and bold experiments. *White Shroud* contains a dazzling variety of work: spontaneous improvisations such as the haikuesque "Porch Scribbles" and "Irritable Vegetable," meditative exercises such as the fixed-based "Why I Meditate" and the more complex "Thoughts Sitting Breathing II," long-line travelogues such as "One Morning I Took a Walk in China" and "Reading Bai Juyi," homages to Whitman and Williams, and erotic poems.

In his poetry of the 1980s and 1990s Ginsberg combines explicit sexuality with a rueful and gently self-ironic recognition of growing old. Yet Eros is still a strong motivating poetic force even for the older Ginsberg. In "Old Love Story," for example, the poet runs down a list of some famous homosexuals of myth, literature, and history in the West, ending the poem with the plea:

Reader, Hearer, this time Understand
How kind it is for man to love a man,
Old love and Present, future love the same
Hear and Read what love is without shame[.]

A much shorter piece, "Sphincter," is one of the most original and successful of his erotic poems, playing off sexual desire against inevitable death and using humor and candor to move the reader toward understanding.

The title poem, "White Shroud," can stand beside "Howl" and "Kaddish" as a work that matches intense inner experience with a tour-de-force formal achievement. As in most of his best work, Ginsberg was inspired to write the poem from a particular experience, in this case a dream vision: while looking for a place to live in New York's Lower East Side the poet happens to encounter an old bag lady who turns out to be his mother. The poem traces a poet-hero's archetypal search for guidance, and as such it gains in emotional richness by being intertextually

Ginsberg in concert (photograph by John Mureo)

linked to Dante's *Inferno,* Virgil's *Aeneid,* and Homer's *Odyssey.* In the dream Naomi reappears as the Wise Old Woman who, even though poor and destitute, is now independent and fiercely self-reliant; when her astonished son asks what she is doing there, she replies with mocking self-assurance, turning the question back on him:

"I'm living alone,
  you all abandoned me, I'm a great woman, I came here

by myself, I wanted to live, now I'm too old to take
    care
of myself, I don't care, what are you doing here?"

The son-wanderer-quest hero realizes in a flash that his years of exile have ended, for he could share the life of his newfound mother and they could mutually take care of each other:

Those years unsettled – were over now, here I could
    live
forever, here have a home, with Naomi, at long last
. . . My breast rejoiced, all my troubles over, she was
content, too old to care or yell her grudge. . . .

The speaker's exclamation "What long-sought peace!" might be taken as a sign of Ginsberg's psychological reconciliation with his mother — a reconciliation that had not taken place while she was alive — but the poem is more than a subjective record of psychic healing. As the surrealist painter Max Ernst had said, it is not the dream that makes the image but the image that makes the dream. Upon waking, the speaker declares, "I returned / from the Land of the Dead to living Poesy," in other words, not to life but to art, the means whereby the dream may be given significance. The other dream poem in the collection, "Black Shroud," can be seen, as Michael Schumacher suggests, as a "mirror poem" to "White Shroud," but on psychological, rather than artistic, grounds. The matricide it recounts in grisly detail has neither the allusive resonance nor the poetic sophistication displayed in the earlier poem.

Ginsberg's most recent book of poetry, *Cosmopolitan Greetings,* appears to be a clear answer to Eliot's *Four Quartets* (1935-1942): Ginsberg's later career shows that he believes "Old men ought to be explorers," for, arguably, no American poet of his generation has stuck so resolutely to a path of experimentation. Whatever the form he chooses, he demonstrates a sureness of hand and ease in execution. Judging from the meticulously recorded dates, at times even the hour, that he gives to the poems in the collection, Ginsberg's preferred mode of composition is spontaneous or automatic writing. The only exceptions concern poems started or conceived years before that were taken back up and brought to completion. After some five decades of writing poetry, Ginsberg is so attuned to poetry — his mind so shapely — that he can sit down anywhere in the world, from China to Boulder, and improvise on the spot, producing without fail poems full of wit, invention, rhythmic interest, humor, and, on occasion, wisdom. The poetic personae of *Cosmopolitan Greetings* have become imbued with the adventurous spirit of Dante's and Tennyson's Ulysses: nearly every poem displays the explorer-experimenter's insatiable curiosity and appetite for new discoveries, the desire to increase sensual awareness of the riches of existence.

Ginsberg has never shown signs of any anxiety about influence; even in his later years, no matter how far he may have gone beyond his precursors, he has never been reluctant to acknowledge them. In "Improvisation in Beijing" he sets side by side and pays homage to the instigators of the two major poetic streams that converged in his poetry and poetics:

I write poetry because Walt Whitman gave world permission to speak with candor.
I write poetry because Walt Whitman opened up poetry's verse-line for unobstructed breath.
I write poetry because Ezra Pound saw an ivory tower, bet on one wrong horse, gave poets permission to write spoken vernacular idiom.
I write poetry because Pound pointed young Western poets to look at Chinese writing word pictures.

Ginsberg freely credits Williams, Snyder, Chögyam Trungpa, various blues singers, and Russian poets, and a host of others as his inspiration.

As a Buddhist and postmodern poet, Ginsberg does not believe in a substantial, permanent self. The protest poet is still alive and well in strong, passionate pieces such as "You Don't Know It," "On the Conduct of the World Seeking Beauty Against Government," "CIA Dope Calypso," and "Get It?" The collection also includes many deeply personal lyrics depicting the frailties and anxieties as well as the joys of old age, notably "Personal Ad" and the serene, yet vibrant "Autumn Leaves":

At 66 just learning how to take care of my body
Wake cheerful 8 A.M. & write in a notebook
rising from bed side naked leaving a naked boy asleep by
                                           the wall
mix miso mushroom leeks & winter squash for breakfast,
Check bloodsugar, clean teeth exactly, brush, toothpick, floss, mouthwash
oil my feet, put on white shirt white pants white sox
sit solitary by the sink
a moment before brushing my hair, happy not yet to be a corpse.

Apart from showing an unpretentiousness about a lifetime of learning, the poem is both a tribute to and example of the Williamsian theory of "No ideas but in things"; the careful cultivation of particulars fosters a union of thought and feeling, of world and self. The arrangement of lines and economy of presentation derive from a poetics of ellipses that the poet has faithfully maintained and adapted to the various themes and topics of his art.

While *White Shroud* and *Cosmopolitan Greetings* — which include more songs with accompanying musical scores than in the whole of *Collected Poems* — indicate Ginsberg's increasing turn to song, it is important to note that his singing cannot be separated from his oral poetics or from his stance as a public figure. As self-proclaimed prophet, high priest, and guru of the counterculture during the height of "flower power," Ginsberg gave many performances where he spoke, chanted, and sang, expending con-

Cover for Ginsberg's 1996 book, illustrated by Eric Drooker and published forty years after Howl and Other Poems

siderable amounts of his enormous energy in trying to bring about a revolution in politics and sexuality. But in 1971 he admitted that "I'm probably better off as a poet ... so what I would like to do is get myself back, get my energy focused there, on language and on poetry, rather than tromping around in the streets." His renewed focus on poetry, however, was to include an emergent strain along the lines Pound had envisaged: the reunification of words and music. His later work in part answers Pound's wistful (and wishful) remark that "verse to be sung is something vastly worth reviving." In tracing the turns of his own career, Ginsberg described the overall efforts and achievement of his close poetic allies as a "return to the oral tradition of actual speech," which then "led to chanting" in his own earlier work and back to the "minstrel tradition" as in Bob Dylan's songs.

The singing Ginsberg of the 1990s is neither a nostalgic throwback to the 1960s nor an imitator of Dylan. Songs such as "Do the Meditation Rock," "Airplane Blues," "Europe Who Knows," and "Put Down Your Cigarette Rag (Don't Smoke)" – the last being, as the poet quipped during a performance, his contribution to the war on drugs – are highly accomplished and powerfully affecting. In addition to showing Ginsberg's ongoing desire and capacity for formal/thematic experimentation, his songs testify to his unswerving libertarian convictions, a counterweight to what he termed in his interview on British television the "neo-conservative theo-political televangelists" in America. Together with poetry written to be read aloud, chanted, or chanted and sung (as in the wholly unique phonic poem "Hum Bom"), the songs complement a lifetime of active involvement with opening up the self to the world through a radical and unprecedented refashioning of the language of poetry. His poetry and poetics of synthesis are those of a spontaneous visionary, whose achievement is already part of the American poetic tradition.

Letters:

Ginsberg and Richard Eberhart, *To Eberhart From Ginsberg: A Letter About Howl 1956* (Lincoln, Mass.: Penmaen Press, 1976);

Ginsberg and Neal Cassady, *As Ever: The Collected Correspondence of Allen Ginsberg & Neal Cassady*, edited by Barry Gifford (Berkeley, Cal.: Creative Arts, 1977).

Interviews:

Tom Clark, "Interview with Allen Ginsberg," in *The Paris Review Interviews,* third series (New York: Viking, 1967);

Paul Carroll, "Interview with Allen Ginsberg," *Playboy* (April 1969);

Alison Colbert, "A Talk with Allen Ginsberg," *Partisan Review* 38, no. 3 (1971);

Allen Young, *Gay Sunshine Interview: Allen Ginsberg* (Bolinas, Cal.: Grey Fox, 1974);

"An Interview with Allen Ginsberg," *Unmuzzled Ox,* 3, no. 2 (1975);

Richard Kostelanetz, "An Interview with Allen Ginsberg," in *American Writing Today,* volume 1 (Washington, D.C.: Voice of America, 1982);

Linda Hamalian, "Allen Ginsberg in the Eighties," *Literary Review,* 29, no. 3;

Robert Stewart and Rebekah Presson, "Sacred Speech: A Conversation with Allen Ginsberg," *New Letters,* 54 (Fall 1987): 72-78;

Suranjan Ganguly, "Allen Ginsberg in India: An Interview," *Ariel,* 24 (October 1993): 21-32.

Bibliographies:

George Dowden, *A Bibliography of Works by Allen Ginsberg* (San Francisco: City Lights Books, 1971);

Michelle P. Kraus, *Allen Ginsberg: An Annotated Bibliography 1961–1977* (Metuchen, N.J.: Scarecrow Press, 1980);

Bill Morgan, *The Works of Allen Ginsberg 1941–1994: A Descriptive Bibliography* (Westport, Conn.: Greenwood Press, 1995);

Morgan, *The Response to Allen Ginsberg, 1926–1994: A Bibliography of Secondary Sources* (Westport, Conn.: Greenwood Press, 1996).

Biographies:

Jane Kramer, *Allen Ginsberg in America* (New York: Random House, 1968);

Barry Miles, *Ginsberg: A Biography* (New York: Simon & Schuster, 1989);

Carolyn Cassady, *Off the Road: My Years with Cassady, Kerouac, and Ginsberg* (New York: Morrow, 1990);

Michael Schumacher, *Dharma Lion: A Critical Biography of Allen Ginsberg* (New York: St. Martin's Press, 1992).

References:

Glen Burns, *Great Poets Howl: A Study of Allen Ginsberg's Poetry, 1943–1955* (New York: Peter Lang, 1983);

Ann Charters, ed., *Scenes Along the Road* (New York: Gotham Book Mart, 1971);

Michael Davidson, *The San Francisco Renaissance and Postmodern Poetics* (Cambridge: Cambridge University Press, 1983);

George Dowden, *Allen Ginsberg: The Man/The Poet on Entering Earth Decade His Seventh* (Montreal: Alpha Beat Press, 1990);

Lewis Hyde, ed., *On the Poetry of Allen Ginsberg* (Ann Arbor: University of Michigan Press, 1984);

David R. Jarraway, " 'Standing by His Word': The Politics of Allen Ginsberg's Vietnam "Vortex," *Journal of American Culture,* 16 (Fall 1993): 81–88;

Thomas F. Merrill, *Allen Ginsberg* (New York: Twayne, 1969; revised edition, 1988);

Bill Morgan and Bob Rosenthal, *Best Minds: A Tribute to Allen Ginsberg* (New York: Lospecchio Press, 1986);

Eric Mottram, *Allen Ginsberg in the 60's* (Brighton, U.K.: Unicorn Bookshop, 1972);

Paul Portuges, *The Visionary Poetics of Allen Ginsberg* (Santa Barbara, Cal.: Ross-Erikson, 1978);

Willard Spiegelman, *The Didactic Muse: Scenes of Instruction in Contemporary American Poetry* (Princeton, N.J.: Princeton University Press, 1989);

John Tytell, *Naked Angels: The Lives and Literature of the Beat Generation* (New York: McGraw-Hill, 1976).

Papers:

Stanford University, Columbia University, and the Humanities Research Center, University of Texas at Austin, have collections of Ginsberg's papers.

Anthony Hecht
(16 January 1923 -)

Ashley Brown
University of South Carolina

BOOKS: *A Summoning of Stones* (New York: Macmillan, 1954);

The Hard Hours (New York: Atheneum, 1967; London: Oxford University Press, 1967);

Millions of Strange Shadows (New York: Atheneum, 1977; Oxford: Oxford University Press, 1977);

The Venetian Vespers (New York: Atheneum, 1979);

Obbligati: Essays in Criticism (New York: Atheneum, 1986);

Collected Earlier Poems (New York: Knopf, 1990);

The Transparent Man (New York: Knopf, 1990);

The Hidden Law: The Poetry of W. H. Auden (Cambridge, Mass.: Harvard University Press, 1993);

On the Laws of the Poetic Art, Bollingen Series XXXV: 41 (Princeton, N.J.: Princeton University Press, 1995).

OTHER: Aeschylus, *Seven Against Thebes,* translated by Hecht and Helen H. Bacon (New York & London: Oxford University Press, 1973).

Anthony Hecht is one of the outstanding poets of his generation, which is the generation that came of age during or immediately after World War II. In 1923, the year of his birth, at least four other poets of distinction were born: Louis Simpson, Denise Levertov, Daniel Hoffman, and James Dickey. Edgar Bowers, James Merrill, and a dozen others soon came along. It is a very successful generation, one that has frequently been awarded prizes, fellowships, and teaching posts at leading universities, and almost without exception these poets have continued to publish work that is remarkably good. Their success has been earned. Perhaps more than some other generations, they have considered poetry a profession, a calling of great dignity.

Anthony Hecht's poetry has the high finish that one finds in the work of some of his contemporaries (Richard Wilbur, Merrill, Bowers, and John Hollander, for instance), and, indeed, at one time he was admired for his elegance of manner as much as anything else. But the real case to be made for him should be based on his profound sense of tragedy. His poetry reverberates with the disasters of the age; in reading it one is frequently reminded, willingly or not, of the chaos into which mankind could descend. At times the reader is brought up against truly horrifying episodes, all the more painful because they are isolated and then heightened by the art of a completely dedicated poet. In reading Hecht one is also reminded, sometimes, of the Oedipus plays and *King Lear*. In his work a noble tradition of tragic sentiment brings the contemporary scene into focus.

Hecht was born in New York City, and, after all his travels, he remains a New Yorker. He took his bachelor's degree at Bard College in 1944 and a master's degree at Columbia University in 1950. But his academic career was broken up by the war: he served in the U.S. Army for three years, first in Europe, then in Japan. This wartime experience has greatly influenced his poetry.

In 1947 he started a series of teaching jobs that took him across the Midwest and into New England and then back to New York. Perhaps the most important of these was his year (1947–1948) at Kenyon College in Ohio, where he was

Anthony Hecht

a student of the famous southern poet John Crowe Ransom. At that time Ransom was also the editor of the *Kenyon Review,* a literary quarterly of great prestige, and that is where Hecht's first poems appeared. These early pieces, which have not been reprinted, register a direct response to the author's wartime experience in Germany. They were followed by a few poems that show the influence of Ransom, a poet who tends to resist imitation. By the late 1940s, however, when he was back in New York, Hecht was studying informally with Allen Tate, another southern poet who has been admired by several men in Hecht's generation. In 1959 Hecht said the most important thing he got from Tate was "the way a poem's total design is modulated and given its energy, not by local ingredients tastefully combined, but by the richness, toughness and density of some sustaining vision of life." (His tribute to Tate appears in the *Sewanee Review,* Autumn 1959.)

In the early 1950s Hecht was thought of as a virtuoso. Louise Bogan, the distinguished poet who was at that time the poetry critic of *The New Yorker,* said of his first book, "If Hecht, often disturbed by disorder and death, is drawn by ideal proportion, elegance, and color ... he is yielding to promptings that at his age only a fanatically serious spectator would deny him." This first book was *A Summoning of Stones* (1954), whose epigraph is taken from George Santayana: it is the poet's duty "to call the stones themselves to their ideal places, and enchant the very substance and skeleton of the world." The quotation is appropriate in more than one sense. Santayana had been living for many years in Rome in a kind of exile of his own choice. After the war, when the Allies liberated the city, the old philosopher-poet received literary Americans, military and civilian, in the convent where he had taken refuge. He is the subject of three memorable poems of rather different character: Wallace Stevens's "To an Old Philosopher in Rome," Robert Lowell's "For George Santayana," and Hecht's "Upon the Death of George Santayana."

When Santayana died in 1952, Hecht had been living in Rome for a year or so after receiving a writing fellowship at the American Academy. His Santayana poem, written somewhat later, imagines the old man, resisting to the last the claims of the Catholic Church where he has taken refuge, descending to a classical underworld:

> Loving the Greeks,
> He whispered to a nun who strove to woo
> His spirit unto God by prayer and fast,
> "Pray that I go to Limbo, if it please
> Heaven to let my soul regard at last
> Democritus, Plato and Socrates."

This poem, though not long, gives Hecht the occasion to introduce a large, serious subject, the death of a noble philosopher, with considerable wit. Santayana identifies himself with Socrates rather than Christ; on the other hand, the convent where he dies, which all his literary visitors will remember, is a reminder of the drama of Christ's life and death as well as the long stretch of history that he introduced. It is typical of Hecht to be able to think, poetically, in these large terms.

The more familiar poems of the 1950s, however, were the virtuoso pieces that Bogan referred to, such as "La Condition Botanique" and "The Gardens of the Villa d'Este." The former, which is actually set in the Brooklyn Botanical Gardens, begins:

> Romans, rheumatic, gouty, came
> To bathe in Ischian springs where water steamed,

Puffed and enlarged their bold imperial thoughts, and which
Later Madame Curie declared to be so rich
  In radioactive content as she deemed
    Should win them everlasting fame.

The poem then proceeds to wind its way through nineteen formal stanzas; the sentences often run to fifteen or twenty lines and are played off against the stanzas. This kind of virtuosity was typical of the 1950s, mainly because of the presence of W. H. Auden's elaborate yet colloquial poetic structures, which, it sometimes seemed, could accommodate anything. (In those days Auden spent part of each year at Ischia, which is a taking-off point for this poem.) "The Gardens of the Villa d'Este" is even more elaborate than "La Condition Botanique" (although it uses the same stanza), and its playful eroticism is a constant delight. A more obvious piece of wit is "Divisions upon a Ground," which is a parody of Wallace Stevens's "Le Monocle de Mon Oncle." When Hecht included it again in his second book, *The Hard Hours* (1967), he called it "Le Masseur de Ma Soeur." Stevens, as much as Auden, was a great influence during the 1950s.

Hecht begins to move closer to his "serious" subject in "Alceste in the Wilderness," which takes Molière's misanthrope to the unpopulated secluded place where he threatens to go at the end of the play, *The Misanthrope*. But the natural goodness that Alceste seeks is betrayed on every hand, and the focal point of the poem is almost repellent:

One day he found, topped with a smutty grin,
The small corpse of a monkey, partly eaten.
Force of the sun had split the bluish skin,
Which, by their questioning and entering in,
A swarm of bees had been concerned to sweeten.

Perhaps even closer to the center of Hecht's work is "Christmas Is Coming," in which an old verse from the past, an echo of a genial spirit of charity, is set against an almost surrealist battlefield: "*Christmas is coming. The goose is getting fat. / Please put a penny in the Old Man's hat.*" Hecht here returns to the scene of his earliest poems, but with a new kind of indirection that is far more effective. This poem is new in another way: it is one of the few things that he wrote in blank verse at that time; he generally used rhymed stanzas, as in "Alceste in the Wilderness."

The Hard Hours includes several of the poems for which Hecht is best known: "Behold the Lilies of the Field," "The Vow," "More Light! More Light!," and "The Dover Bitch." The first and third

Dust jacket for Hecht's first book, which takes its title from George Santayana's assertion that it is the poet's duty "to call the stones themselves to their ideal places"

of these, which tend to set the tone for the book, are shocking, as history can sometimes be. In "Behold the Lilies of the Field" the Roman emperor Valerian (253–260) is executed and humiliated in the most dreadful way. This vignette from ancient history, however, is an image of a disturbing malaise that recurs in our time, and in fact it is a bad dream that takes over the mind of a patient on the couch of a modern psychoanalyst. The poem "More Light! More Light!" – these are the last words of Johann Wolfgang von Goethe, the great poet-humanist who supposed that a benign enlightenment would eventually prevail in history – juxtaposes two incidents: a religious martyrdom of the Reformation, horrible but at least retaining a "pitiful dignity," and a brutal execution at Buchenwald, in a Germany that Goethe could never imagine. The title is typically sardonic.

"The Vow" is an intensely personal poem, yet one that has tragic associations of a large sort. Hecht has twice chosen this poem as his favorite for anthologies: *Poet's Choice,* edited by Paul Engle and Joseph Langland in 1962; and *Preferences,* edited by Richard Howard in 1974. (In the latter case Hecht paired his poem with *King Lear,* act 4, scene 7, when asked to pair one of his own poems with a favorite work from the past.) The subject is a dead unborn child. The poet boldly allows the child to speak:

> And for some nights she whimpered as she dreamed
> The dead thing spoke, saying: "Do not recall
> Pleasure at my conception. I am redeemed
> From pain and sorrow. Mourn rather for all
> Who breathlessly issue from the bone gates,
>   The gates of horn,
> For truly it is best of all the fates
>   Not to be born."

The last lines spoken by the dead child allude to a passage in Sophocles' *Oedipus at Kolonos;* as Hecht remarked, it is an "aged, bitter, Sophoclean wisdom."

In contrast to this poem are two slighter but charming pieces, "Jason" and "Adam," named for the poet's first two sons, who, through their names, are a link with the Judaic-Greek traditions. In these poems the children re-create their mythical roles, but just through random gestures; in the second poem the poet-father says:

> Adam, there will be
> Many hard hours,
> As an old poem says,
> Hours of loneliness.
> I cannot ease them for you;
> They are our common lot.
> During them like as not,
> You will dream of me.

This passage alludes to the title of the book, *The Hard Hours,* which in turn suggests the tragic potentialities of human existence. (The book is dedicated to the children.) At the end of the last poem in the book the father offers this curt prayer:

> And that their sleep be sound
> I say this childermas
> Who could not, at one time,
> Have saved them from the gas.

The Hard Hours is a successful book in various ways; it has gone through six printings since it was first published in 1967. Tate said at that time, "We have gotten into the bad habit of ranking our poets. I refuse to do this. I can only say that whoever else may be at the top, Hecht is there too; for there is nobody better." The book was awarded the Pulitzer Prize in 1968. By then Hecht had been appointed to the faculty of the University of Rochester, where he was the John H. Deane Professor of Poetry.

Like some other poets in his generation, Hecht has done outstanding work in translation, beginning with short poems from the French of Joachim du Bellay, Charles Baudelaire, and Guillaume Apollinaire; he has also translated Voltaire's "Poem Upon the Lisbon Disaster," a small masterpiece that Hecht has rendered in heroic couplets. In 1973 he collaborated with Helen Bacon, a distinguished classicist, in a version of Aeschylus's *Seven Against Thebes* – a major event in its way, because this play has not been dealt with as much as Aeschylus's other surviving works. At one time he was translating *Oedipus at Kolonos* (a chorus was included in *Millions of Strange Shadows,* 1977) and poems by the Russian poet Joseph Brodsky, who figures in one of Hecht's own poems.

Millions of Strange Shadows comprises thirty-one poems, including the translations from Voltaire and Sophocles. The book has an extraordinary range and a new freedom that the poet permits himself. The title of the volume suggests a kind of motif that the author of this entry once described thus: "Youth and age figure prominently in these poems, and it is part of Mr. Hecht's great talent to be able to move with ease among accumulated memories. He does this as though he were riffling through a heap of photographs, deciding what to discard, what to preserve, and actual photographs are the focal points of several poems." In "Exile," which is dedicated to Brodsky, the Russian poet who was living in the United States, Hecht uses a photograph by Walker Evans to illustrate the gritty dead level of the American scene, the "*terra deserta.*" Then he turns round the biblical story, as it were, and says gently:

> This is Egypt, Joseph, the old school of the soul.
> You will recognize the rank smell of a stable
> And the soft patience in a donkey's eyes,
> Telling you you are welcome and at home.

The most remarkable of the "photograph" poems is "A Birthday Poem," written for the poet's second wife, Helen, in 1976. The time is late June, the atmosphere is heavy and thick with midges. Trying to bring the scene into focus, he imagines it as the background to the *Crucifixion* by Andrea Mantegna, then as a group portrait by Hans Holbein. Then, getting a mini-history of artistic perception, the reader is in the age of optics and lenses. The poet takes up a photograph of his wife as a child of four in her red sneakers:

> The picture is black and white, mere light and shade.
>   Even the sneakers' red
> Has washed away in acids. A voice is spent,
>   Echoing down the ages in my head:
> *What is your substance, whereof are you made,*
> *That millions of strange shadows on you tend?*

And the poet calls on William Shakespeare (sonnet 53) to speak for him momentarily: a voice "echoing down the ages" from the greatest moment in English poetry. Elsewhere in this volume he calls on George Herbert, just as in *The Hard Hours* he brought du Bellay and Sophocles into his poems. Hecht, in fact, has a supreme respect for his predecessors, and he perpetuates, more deliberately than some poets, these voices from the past in order to understand the present.

The major piece in the volume is called "Apprehensions," and it is in some ways the boldest. It is a visionary poem in which the poet stakes a good deal on sheer sincerity. This long evocation of his childhood suggests a large enveloping action: the stock market crash, which leads to other disasters, public and private; his German governess with her "special relish for inflicted pain," who somehow anticipates the "Wagnerian twilight of the *Reich*"; and so on. At the center of this unstable world, the child Anthony Hecht, living in an apartment house on Lexington Avenue in New York City, has a primal vision, what Auden (who is actually quoted) calls "The Vision of Dame Kind." It is a sultry, late-summer afternoon:

> The streetcar tracks gleamed like the path of snails.
> And all of this made me superbly happy,
> But most of all a yellow Checker Cab
> Parked at the corner. Something in the light
> Was making this the yellowest thing on earth.
> It was as if Adam, having completed
> Naming the animals, had started in
> On colors, and had found his primary pigment
> Here, in a taxi cab, on Eighty-ninth street.
> It was the absolute, parental yellow.

The vision has "to be put away / With childish things" in the course of time, but it has made itself felt. The poet has relaxed his severity to some extent even though the world's tragic course never stops. A dozen shorter poems in this volume bear out the effect of the primal vision, especially "An Autumnal" and "The Lull."

Hecht was now at the height of his powers. The creative momentum that he built up carried him into larger and possibly more dangerous enterprises. In 1978 he published a long poem in *Poetry* magazine, "The Venetian Vespers," which runs to

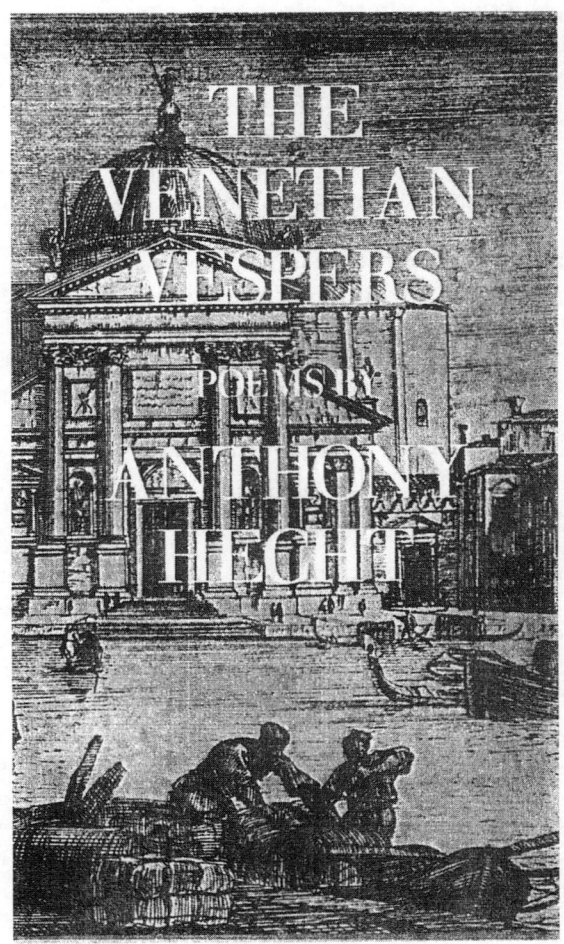

Dust jacket for Hecht's 1979 book. The title poem is a dramatic monologue of an American who seeks refuge in Venice after his experiences during World War II.

approximately a thousand lines and was also the title poem of a new volume. In its versification and construction it looks back to "Green: An Epistle" and "Apprehensions," but it is more daring. It is a dramatic monologue whose main problem is to create a character, and it manages this feat brilliantly.

"The Venetian Vespers" is Hecht's equivalent of T. S. Eliot's "Gerontion." The scale is much larger, but the structure is the same: the verse paragraphs of Eliot's poem have their counterparts in the six sections of Hecht's. Like Eliot in his time, Hecht has studied the blank verse of the Elizabethan dramatists, and the tone for "The Venetian Vespers" is appropriately set by a quotation from *Othello* (lines spoken by Iago in Act 3, scene 3), as "Gerontion" is started off by a quotation from *Measure for Measure*. Hecht's poem, however, is grander in certain passages, being much longer, and it frankly owes something to John Ruskin: a quotation from *The Stones of Venice* (1851-1853) provides the other epigraph. The reader with a well-stocked literary imagination will doubtless find allusions to works by Ben Jonson; Thomas Otway; George Gordon, Lord Byron; Percy Bysshe Shelley; perhaps Henry James; "Baron Corvo" (Frederick William Rolfe); Marcel Proust; and Thomas Mann. Venice has been a favorite setting for corruption, if not death, for the last few centuries.

The literary references are only incidental to the subject, but certainly they allow Hecht to create an unusually rich poetic texture when he wants to. The narrator is a nameless, middle-aged American, the son of immigrant parents who came to Lawrence, Massachusetts, from Latvia. (We learn these facts gradually.) The child grew up in an A & P store that his father and uncle ran, but the father, caught in the ambitions of American life, went west to make his fortune; they never saw him alive again. His mother died when he was six. Years later he hears about the fate of his father in Toledo, Ohio: robbed and beaten and knowing little English, his father was taken to the state mental hospital. The uncle, in a terrible, misguided burst of shame, fearing a scandal, refuses to acknowledge his brother until he dies. The day after the burial the narrator, now a young man during World War II, enlists as an aidman with an infantry company. Refusing to bear arms, he

> Was constrained to bear the wounded and the dead
> From under enemy fire, and to bear witness
> To inconceivable pain, usually shot at
> Though banded with Red Crosses and unarmed.

After two and a half years he is mustered out, "mentally unsound." Now he has come to Venice for an endless death-in-life:

> Lights. I have chosen Venice for its light,
> Its lightness, buoyancy, its calm suspension
> In time and water, its strange quietness.
> I, an expatriate American,
> Living off an annuity, confront
> The lagoon's waters in mid-morning sun.
> Palladio's church floats at its anchored peace
> Across from me, and the great church of Health,
> Voted in gratitude by the Venetians
> For heavenly deliverance from the plague,
> Voluted, levels itself on the canal.

The verse is wonderfully responsive to the shifting moods. In this passage, for instance, the reader moves from the limp rhythms of "I, an expatriate American, / Living off an annuity" (reminiscent of Gerontion's "I an old man, / A dull head among windy spaces") to the muted splendor of the lines that evoke the great church across the lagoon. If the main problem of the poem is the creation of character, the analogous problem is the creation of an action, but this is perhaps impossible. "Gerontion," an archetypal modern poem, comes full circle, but it has only the semblance of an action; and that is the case with "The Venetian Vespers," which ends thus:

> I, who have never earned my way, who am
> No better than a viral parasite,
> Or the lees of the Venetian underworld,
> Foolish and muddled in my later years,
> Who was never even at one time a wise child.

"The Venetian Vespers" is finally a great meditation on the tragic displacements of the time. And it is a kind of monument that comes at the midpoint of Hecht's career.

The poet-critic has been a recurring figure in American literary history, especially in the twentieth century. Ransom, Tate, and Robert Penn Warren (to name only three) in their essays helped to establish the context within which their poetry might be read. By the time that Anthony Hecht came along after World War II, there was a small but eager and intelligent audience for modern poetry. It was no longer necessary for a young poet to engage in polemics or justify his work. But the habit of writing reviews and even extended essays continues, and most of the poets in Hecht's generation have published criticism of a high order. All the same, the criticism follows the poetry. In 1986 Hecht published a volume called *Obbligati: Essays in Criticism*. The title is a sort of pun based on the Latin *obligare*. As Hecht explains in his preface, "My title refers to a complex set of obligations undertaken by these essays — obligations to the works and poets they deal with." *Obbligato* is also a technical term in music, meaning an accompaniment that has its own character but is necessary to the proper performance of a work. Hecht's essays are all celebrations of poets: Shakespeare, Andrew Marvell, John Keats, Emily Dickinson, Auden, Elizabeth Bishop, Lowell, and Wilbur, among others. None of them requires defending or promoting.

By the time this book came out Hecht had moved from the University of Rochester to Washington, D.C., where he was first the consultant in poetry at the Library of Congress (the post that is now called the poet laureate), then university professor in the graduate school of Georgetown University. Rather than teaching creative writing, as so many poets would have expected to do, Hecht turned to Shakespeare as the subject of one of his courses. Hence it is not surprising to find that *Othello* and *The Merchant of Venice* are the topics for two of the essays in *Obbligati*. The essay on *The Merchant of Venice* is in fact the most extended piece in the book, a dazzling account of what the play might have meant to the original audience. The subtitle of the essay is "An Exercise in Hermeneutics," a term originally associated with the interpretation of biblical texts but now (like the term *canon*) frequently used in a secular context. Hecht is in a sense a learned rabbi who in this case uses the biblical texts to elucidate many passages in Shakespeare's play. As he points out toward the end of his essay, Shakespeare "made use of ideas that he could count on his audience instantly recognizing and accepting." It is interesting that Hecht, author of "The Venetian Vespers," so rich in allusions to Venice, should have had a special fondness for *The Merchant of Venice* and *Othello,* plays that are often concerned with sensitive ethical and theological issues, to say nothing of racial factors. But here and elsewhere he confronts these large issues with an unusual knowledge of the traditions of Western culture, both Judaic and Greco-Roman.

In writing about his contemporaries Bishop, Lowell, and Wilbur, Hecht is especially generous in his praise. When he wrote the Lowell essay (originally a lecture at the Library of Congress), it was well known that Lowell's troubled life, marked by recurrent mental illness, often figured in his poetry. Hecht faces this situation and finally speaks of Lowell's "constant moral and artistic endeavor to situate himself in the midst of our representative modern crises, both personal and political." Bishop was still alive when Hecht wrote his piece on her for the London *Times Literary Supplement,* and he concluded for an audience mainly British, "Hers is about the finest product our country can offer the world; we have little by other artists that can match it; and it beats our cars and films and soft drinks hollow."

In 1990 Hecht changed publishers; he moved from Atheneum, where he had been since 1967, to Knopf. His new publisher marked the occasion by bringing out two books simultaneously: a new collection of poems, *The Transparent Man;* and an omnibus *Collected Earlier Poems,* which includes the complete texts of *The Hard Hours, Millions of Strange Shadows, The Venetian Vespers* (1979), and all the poems that he wishes to preserve from *A Summoning of Stones*. Thus, most of Hecht's work was brought to a new generation at once. By this time Hecht was being praised by leading reviewers such as Harold

Dust jacket for Hecht's 1990 book that includes "A Love for Four Voices," a long poem based on Franz Joseph Haydn's last string quartet

Bloom and Christopher Ricks, and indeed during the preceding year, 1989, he was the subject of a book of essays called *The Burdens of Formality,* edited by Sydney Lea. This collection of thirteen pieces is the most important critical presentation that Hecht has yet received. It includes a short tribute by the late Joseph Brodsky, the Nobel laureate from Russia who was on the way to becoming an American poet. Brodsky states: "Anthony Hecht is, without question, the best poet writing in English today." Brodsky reaffirms the tragic nature of Hecht's poetry, and several other critics (some of them poets) in this volume discuss his work in these terms. As for the title of the book, *The Burdens of Formality,* most of the contributors agree that Hecht's technical mastery has paid off handsomely.

The Transparent Man carries forward some of the themes and technical procedures of *The Venetian Vespers*. The two longest poems, which occupy sections of their own in the book, are "See Naples and Die" and "A Love for Four Voices." The latter work is based on Joseph Haydn's String Quartet Op. 77, No. 2, the last that the composer finished. In a radio program on a Washington station a few years ago, Hecht chose this quartet as one of his favorite works to be played. He is somewhat unusual among poets in being equally responsive to music and painting. (Eliot and Auden, for instance, loved music and sometimes worked with musicians; William Butler Yeats and Stevens were closer to painting and the other visual arts.) It is possible that listening to a recording of the Haydn quartet might enhance one's enjoyment of Hecht's poem, but the poem stands by itself. Hecht probably thinks that the idea of several voices blending in various combinations is sufficient as an idea for the structure of the poem; the four movements are described exactly as the composer names them. As for the four voices, they are those of the lovers (Hermia, Helena, Lysander, Demetrius) in *A Midsummer Night's Dream.* Perhaps rereading the play, like listening to the quartet, might add something to one's enjoyment of the poem, but Hecht has taken the characters out of Shakespeare, as it were, and let them act on their own. This is not a pastiche of Elizabethan idiom. The characters are contemporary; they speak our language in a great variety of poetic forms. This is one of those works that, like Auden's *The Sea and the Mirror* (1944), operates through a superior kind of virtuosity and the pleasure that formality can bring to art, but the general subject – the varying nature of human love – is much in evidence. There is many a double entendre awaiting the reader who likes wit along with romantic love – as Shakespeare obviously did.

"A Love for Four Voices" is a celebration of love; it follows "See Naples and Die," which is concerned with love's failure, that is, the breakup of a marriage. The narrator, long after the event, re-creates a spring holiday in Naples with his wife, Martha. Nothing momentous occurs between them; the husband admits that he is sometimes foolish or insensitive; gradually small events lead to a kind of mutual annoyance, and toward the end of his recollection he says:

> Marriages come to grief in many ways.
> Our own was, I suppose, a common one,
> Without dramatics, a slow stiffening
> Of all the little signs of tenderness,
> Significant silences, self-conscious efforts
> To be civil even when we were alone.
> The cause may be too deep ever to find,
> And I have long since ceased all inquiry.

> It seems to me in fact that Martha and I
> Were somehow victims of a nameless blight
> And dark interior illness.

This erosion of affection is played out against a spectacular setting: "The greatest amphitheater in the world: / Naples and its Bay." It promises much, and in a way the narrative is a series of false starts and misunderstandings, with a scene near the end at the Elysian Fields that yields only "The bewildered tourists, acres of desolation." But Naples itself is vibrant with physical details, and Hecht in this poem has done for the city what he did for Venice in "The Venetian Vespers." As Brodsky says, "Anthony Hecht's ability for visualization is absolutely extraordinary."

The two long poems are the center of the book. They are flanked by a variety of shorter pieces, including the chorus from Sophocles' *Oedipus at Colonos* that Hecht refers to in "The Vow": "Not to be born is, past all yearning, best." The Sophoclean passage thus links Hecht's most recent book with *The Hard Hours* and *Millions of Strange Shadows,* which includes another chorus from *Oedipus at Kolonos* (spelled with a *K* in the earlier text). Other notable short poems include "Devotions of a Painter," a beautiful monologue in blank verse that exhibits Hecht's visual side, in contrast to "Meditation," which is concerned with the lingering effect of music. Toward the end are three fine elegies for friends whom Hecht has outlived; these are conducted with wit as well as affection and a certain delicacy; and in their way they attest as much as anything else to the range of Hecht's poetic sensibility.

Hecht has recently retired from his academic post at Georgetown, but he is more active than ever. One of his recent books is a long critical study of Auden's poetry called *The Hidden Law* (1993). This book is the result of a lifetime's admiration, because Hecht has read Auden since he was eighteen. At one period, during the early 1950s, both poets lived temporarily at Ischia, the island near Naples. The book is at once scholarly and personal. The reader who already knows Auden's poetry will find much to augment his or her knowledge, and at least one chapter (on *For the Time Being,* Auden's Christmas oratorio of 1944) might be considered a definitive scholarly treatment of the subject. The book is personal not only in its great admiration for Auden's poetry but in its freely ranging speculations about many literary and theological issues that the poems suggest to the critic. In the end it may tell the reader as much about Hecht as it does Auden.

This is surely the best kind of criticism, when one distinguished poet writes about another.

Now established on the Washington scene, Hecht gave the 1992 Mellon Lectures at the National Gallery of Art, an honor that is usually extended to a famous art historian such as Kenneth Clark. The topic of his six lectures was "On the Laws of the Poetic Art." The first lecture, appropriately concerned with poetry and painting, concluded with an extended account of Stevens's "Sea Surface Full of Clouds" and the Impressionist paintings that influenced it. The subsequent lectures were concerned with poetry and music, public and private art, and other issues. They were published in the famous Bollingen Series by the Princeton University Press. Hecht is the first American poet to have given the Mellon Lectures; Herbert Read, Stephen Spender, and Kathleen Raine, poets from England, have preceded him long before now. Hecht is part of an artistic intelligentsia that in some respects is as much European as American. He has never hesitated to draw on his learning and his international experience, nor has he adopted the populist role that some American poets have taken up. He regards the literary art as a serious public matter as well as a source of private enjoyment. The fact that he has been honored in various ways is perhaps an indication that our culture is more sophisticated than it used to be.

References:

Ashley Brown, "The Poetry of Anthony Hecht," *Ploughshares,* 4, no. 3 (1978): 9–24;

Richard Howard, *Alone with America: Essays on the Art of Poetry in the United States Since 1950* (New York: Atheneum, 1969), pp. 164–173;

Sydney Lea, ed., *The Burdens of Formality: Essays on the Poetry of Anthony Hecht* (Athens & London: University of Georgia Press, 1989).

Ronald Johnson
(25 November 1935 -   )

Eric Murphy Selinger
DePaul University

BOOKS: *A Line of Poetry, A Row of Trees* (Highlands, N.C.: Jargon Press, 1964);

Io and the Ox-Eye Daisy (Dunsyre, Scotland: Wild Hawthorn Press, 1965);

Assorted Jungles: Rousseau (San Francisco: Auerhahn Press, 1966;)

GORSE/GOOSE/ROSE and Other Poems (Bloomington: Indiana University Press, 1966);

Sun Flowers (Gloucester, England: John Furnival, 1966);

The Book of the Green Man (New York: Norton, 1967);

The Round Earth on Flat Paper (Urbana, Ill.: Finial Press, 1968);

Reading 1 and *Reading 2* (Urbana, Ill.: Finial Press, 1968);

Valley of the Many-Colored Grasses (New York: Norton, 1969);

Balloons for Moonless Nights (limited edition, Urbana, Ill.: Finial Press, 1969);

Songs of the Earth (San Francisco: Grabhorn-Hoyem Press, 1970);

Maze / Mane / Wane (Cambridge, Mass.: Pomegranate Press, 1973);

Eyes & Objects (Highlands, N.C.: Jargon Press, 1976);

RADI OS I–IV (Berkeley, Cal.: Sand Dollar Press, 1977);

ARK: The Foundations 1–33 (San Francisco: North Point Press, 1980);

ARK 50: Spires 34–50 (New York: Dutton, 1984);

ARK (Albuquerque, N.M.: Living Batch Press, 1996).

OTHER: Erik Satie, *Sports & Divertissements,* translated, with elaborations, by Johnson (Dunsyre, Scotland: Wild Hawthorne Press, 1965);

Le *Facteur* Cheval and Raymond Isidore, *The Spirit Walks, The Rocks Will Talk,* translated by Johnson (Highlands, N.C.: Jargon Press, 1969);

Jack Sharpless, *Presences of Mind: The Collected Books of Jack Sharpless,* introduction by Johnson (Frankfort, Ky.: Gnomon Press, 1989).

SELECTED PERIODICAL PUBLICATIONS – UNCOLLECTED: "The Eye" and "The Ear," *Parnassus*, 1 (Fall/Winter 1972): 21–24;

"Charles Ives: Two Eyes, Two Ears," *Parnassus*, 3 (Spring/Summer 1975): 345–349;

"On Looking Up 'The Pyramid is a Pure Crystal' in Webster," review of John Taggart, *The Pyramid is a Pure Crystal,* by Johnson, *Parnassus*, 3 (Spring/Summer 1975): 147–152;

"Two Poems: Wor(l)ds 20, *Jan 1st* and Wor(l)ds 23," *Iowa Review*, 6 (Summer/Fall 1975): 101–103;

"A Fairy Tale for Robert Duncan," *Parnassus*, 5 (Fall/Winter 1976): 260–262;

"The *Italics* are Guy Davenport's," *Vort*, 3, no. 3 (1976): 42–45;

"Persistent Light Upon the Inviolably Forever Other," *Vort*, 3, no. 3 (1976): 38–42;

"Wor(l)ds 25–27," *Vort*, 3, no. 3 (1976): 86–99;

"In Memoriam L.Z.," *L=A=N=G=U=A=G=E,* 1 (August 1978): 2–3;

"Jonathan Williams," in *Dictionary of Literary Biography, Vol. 5: American Poets Since World War II, part 2: L–Z,* edited by Donald J. Greiner (Detroit: Gale, 1980), pp. 406–409;

"A Flag for Bunting," *Conjunctions,* 8 (Fall 1985): 197;

"For R.D.," *Sagetrieb,* 4, nos. 2–3 (1985): 15;

"Nearly Twenty Questions," *Conjunctions,* 7 (Spring 1985): 225–238;

"The Planting of the Rod of Aaron," *Northern Lights Studies in Creativity,* 2 (Winter 1985–1986): 1–13;

"From the Imaginary Menagerie," *ACTS,* 5, no. 1 (1985–1986): 106–111;

"Road Side (Desert to Prairie)," *LVNG,* 5 (Winter 1994): 66–71;

"Postcard," *LVNG,* 6 (Winter 1995–1996): 7;

ARK 66 (Finial for EZ), *Chicago Review*, 42, no. 1 (1996): 20;

"From *Hurrah for Euphony*," *Chicago Review*, 42, no. 1 (1996): 20.

Fred Astaire and Louis Zukofsky, Tin Pan Alley and the Transcendentalists — these are among the muses of Ronald Johnson, whom Guy Davenport has called "America's greatest living poet." The claim, although surprising, bears consideration. In his concrete and shorter poems, in his book-length seasonal poem *The Book of the Green Man* (1967), and in the ninety-nine Beams, Spires, and Arches that make up his long poem *ARK* (1980–1996) (a rewriting of John Milton's *Paradise Lost* by excision forms the hundredth piece, a constellated dome over the whole), Johnson has authored a body of work whose ambition, charm, and intelligence are all but unmatched among his contemporaries. An "Artist of Abundancies," in Robert Duncan's phrase, Johnson is devoted to puns, to rhyme, and to *bricoleur*-mystics such as The *Facteur* Cheval and Simon Rodia, whose *Palais Idéal* in France and Watts Towers in Los Angeles, respectively, are models for the collage architecture of *ARK*. A poet of science, Johnson finds the visionary organicism of William Wordsworth, Ralph Waldo Emerson, and Henry David Thoreau confirmed by contemporary cosmology and physics. He thus may sing, in his own postmodern key, a "spousal verse" that celebrates the fit between the world and the mind that has evolved to behold and re-create it.

Johnson has been publishing since the mid 1960s, a period in many ways more receptive to his work than the more skeptical fin-de-siècle view of the late twentieth century. His poems grow out of what Charles Altieri calls the "immanentist" postmodernism of the 1960s, with its faith in natural orders waiting to be discovered and modeled in the poet's equally "natural" act of writing; they have little in common with either the lyric of domestic epiphany that flourished in the 1970s or, at the opposite extreme, with the studied disjunction and (often) politicized avant-gardism of the so-called Language poets. Although his work subtly incorporates both autobiographical echoes and radical artifice, and although the later, metaphorically "higher" portions of *ARK* seem shadowed by the ravages of AIDS on his beloved San Francisco, Johnson's long poem, as a rule, consciously *excludes* history, both personal and political. If a war-torn poetry of witness claims the late-century moral and aesthetic high ground (poets as different as Michael Palmer and Carolyn Forché aspire to the historical sublime), Johnson's epic looks at once back to a more optimistic period in American poetics and forward to an audience and critical climate as inspired and heartened by Complexity theory as it is shaken by the aftershocks of twentieth-century atrocities. Playful, capacious, artful, eloquent, Johnson is the preeminent contemporary poet of the Beautiful.

Johnson was born in 1935 in Ashland, Kansas, a town of two thousand set on the prairie, parched and driven by Depression-era dust storms. His father, A. T. Johnson, was a carpenter, whose lumberyard was for the young poet "a secret ground of changing piles and smells, with hid hollow cupolas, exits to tree-topped roofs, stored bins and nooks long lost, rooms of whirling saws, sharpened pencils": an early model for the architecture of *ARK*, in which his father sometimes figures as The Carpenter. The poet's mother, Helen (Mayse) Johnson, was trained in dance in her youth: in *ARK* she appears at times as The Dancer, and the poet puns on her family name in the lovely self-portrait poem "Of Circumstance, The Circum Stances": "Mayse, my mother's / family name / & had it crest, Maize / / one would make it: a brown field sprouting / Indian corn, / / of red & yellow / kernels / that various, still / / variegated display of / / / ancestry." (By the end of the poem he supplements this crest with a second, that of the Indians who inhabited the land before whites came. "Their crest was a *human hand* / / & in the palm of it, / an eye" — a juxtaposition central to the poet's later effort to join "the / work of vision" to "the word / at hand," thus shaping "paradise.")

Like most young children, Johnson began making up nonsense jingles at an early age, doting on what he later called "the pure pleasure of rhythm and sound." His adult verse, with its ear-candy wordplay, unabashedly draws on this resource of infant joy. He studied piano, and remains deeply influenced by music, particularly the compositions of Gustav Mahler and Charles Ives. However, as Johnson recalls in an interview, "Probably the most seminal thing in my life was growing up and discovering the Oz books. I was about twelve or thirteen before I finally had to face the fact that there was no way to get to Oz." Much of Johnson's poetry can be profitably read in this autobiographical light, because it repeatedly returns to the effort to, as he himself puts it, "make a special place — a garden of some kind — which was a surrogate for that imaginary land," an Oz "where anything is possible and in which the imagination lives."

Leaving Kansas for college, Johnson received a B.A. in 1960 from Columbia University. While there, he met the poet and publisher Jonathan Williams, about whom he was later to write an entry for the *Dictionary of Literary Biography*. The two were companions for the next eleven years. Williams introduced him not only to the work of Zukofsky, Duncan, Robert Creeley, and Charles Olson, but to the poets themselves; in New York they also met and talked shop with the Abstract Expressionist painters and with the many composers and photographers who made the Cedar Tavern the *Boeuf Sur Le Toit* or *Deux Magots* of its day. Johnson hiked the length of the Appalachian Trail with Williams, and their walking tour of England, recollected in tranquillity, informs his rewriting of the English seasonal poem *The Book of the Green Man*. (Along with a boyhood chafing at dry Kansas flatness, these hikes account for the poet's passion for pastoral.) In the 1970s Johnson was a writer-in-residence at the University of Kentucky, and he held the Roethke Chair for Poetry at the University of Washington before settling in San Francisco. Aside from temporary travels, he has made his home in the Bay Area ever since. In 1991 he led the Wallace Stegner Advanced Workshop at Stanford, and in 1994 he was poet-in-residence at the University of California, Berkeley.

From the start Johnson has written in a variety of poetic modes. He has a long-standing interest in collage and concrete poetry, often using these architectural techniques of composition to revivify such older genres as pastoral and Romantic "spousal verse." If he is a poet of nature, however, it is of a natural world illuminated at once by scientific inquiry and by Transcendental vision – as in Thoreau, the two are not opposed – and Johnson's "nature" also includes the second nature of language and human culture, which he treats as a rich loam from which new poems may grow. He thus often shapes his poems out of found materials: quotations from American and British naturalists, from scientists, from older poets, all "allowed to 'speak for themselves' as objects in the text," as Norman Finkelstein observes in *The Utopian Moment in Contemporary American Poetry* (1993), even as "the shaping subject speaks through them." Johnson draws our attention to the page before us and the world around us, engaging readers in what Davenport calls "the simple but difficult business of seeing the world with eyes cleansed of stupidity and indifference." This effort links him to Ezra Pound – who told Johnson, in their one meeting, "I have only pointed out a few things might else have been forgotten" – and to Olson, to whom Johnson dedicated his first book, *A Line of Poetry, A Row of Trees* (1964).

A Line of Poetry, A Row of Trees takes its title from a buried etymological pun: the Greek root *stich*, it so happens, means both. Like many of Olson's poems – the *Maximus Poems*, "The Kingfishers," and so on – these pieces are "stitched" out of borrowed material, including scraps from Samuel Palmer, William Bartram, Emerson, Thoreau, and a "collage poem from *Six Months in Kansas, by a Lady*, / published 1856." Many of Johnson's pieces have a packed, "projective," Olsonian movement: "Columbus, as the first Western eyes, called it / panic grass – Maize, of a 'quaking' ancestry, i.e., the / attempt, always, at classification" ("Indian Corn"); "That what we know of the world is Physiognomy, 'face'. As / Haida square a bear to its corners, / join profiles / – edge to edge – join its head, trunk & limbs / with eyes . . ." ("Landscape with Bears, for Charles Olson"). But even in this first collection Johnson's distinctive voice and interests are clear. Against the "landscape with bears" of the Olson poem, where wilderness is all, the poem "Shake, Quoth the Dove House" sets the art of poetry in an evergreen topiary grotto:

> 'A laurustine bear in blossom
> with a juniper hunter in berries,
> a lavender pig with sage growing in his belly
> & a pair of maidenheads in fir, in great forwardness'.
>
> > This is the Garden, where all is a poet's
> > topiary. Where even the trees
> > shall have tongues, green aviaries,
> > to rustle at his will.
> >
> > And as I sit here, my pipe
> > alight, coos like a turtle-dove in the wood –
> >
> > its smoke a live-oak, in still air.

In "Lilacs, Portals, Evocations," meanwhile, the poet invokes "Kansas, of / sand plums & muddy rivers": a place that all roads once led away from, but to which the poetry of Zukofsky and Williams and the music of composers Carl Ruggles and Ives – artists at once local and innovative – have at last become "ways homeward."

Between *A Line of Poetry, A Row of Trees* and Johnson's next full-length collection, *The Book of the Green Man* (1967), the poet wrote several shorter books, published in limited small-press editions. *Sports & Divertissements* (1965) offers a translation and elaboration of the comic, surreal "performance notes" of the French composer Erik Satie, whose works Johnson had played as a boy and whose words Johnson spaced and paced to make "a music on the page." Artist John Furnival supplied drawings to accompany the poems, and the two collaborated on Johnson's subsequent concrete poem *Io and the Ox-Eye Daisy* (1965), for which Furnival did the lettering. Both books were published in Scotland by Ian Hamilton Finlay, the Scots concrete poet and formal gardener, with whom Johnson stayed for two weeks during a trip to Scotland. Another book of concrete poems, *GORSE / GOOSE / ROSE, and Other Poems* (1966), is dedicated to Finlay: it is a series of "Scotch Shapes / & landscapes" written day by day during Johnson's visit as an experiment in writing a "narrative concrete poem." (Johnson was never pleased with how the book was printed, and it was not widely distributed.) In 1996 the Auerhahn Press of San Francisco published a beautiful, palm-sized, limited edition of *Assorted Jungles: Rousseau*, an ekphrastic suite (*ekphrastic* refers to a poem based on a work of visual art) later reprinted in *Valley of the Many-Colored Grasses* (1969). Centered on the page, printed in an art-nouveau typeface, the poems are written in a tempting, Stevensian diction. Johnson pieces his language out in slow motion, with each word or phrase as discrete and neatly bordered as one of the painter's leaves: an attempt, as he explains elsewhere, to "use words in the naive and exotic way Rousseau painted his jungles." "The dream is Java & / impenetrable, / / its pomegranates clearly / impossible," one poem thus begins. In this painterly incarnation of the poet's grottoes and gardens – an adult Oz, "where anything is possible, and the imagination lives" – the limits of verisimilitude yield to the lushness of possibility, and "Huge orange / oranges / / & pendulous imaginings / of banana / / proliferate / a veritable / Tanganyika."

On his walking tour of England, especially the Lake Country, Johnson found a landscape that, while less exotic and more pastoral, answered his imaginative needs just as powerfully as the one he saw in Rousseau. Norton published the book-length poem that grew out of those English travels and meditations, *The Book of the Green Man*. Derived from the writings of British naturalists – Robert Francis Kilvert and Gilbert White, among others – as much as from the poet's own explorations, the book is a still-young Kansan's effort to "work endless changes" (in Wordsworth's phrase) on the age-old British seasonal poem: an attempt to transplant his American imagination into English soil, and into the "rich silt of bibliography" that covers it. (The gesture is as old as Washington Irving's learning "The Art of Book-Making" in the British Museum, but Johnson takes his cue more directly from Tho-

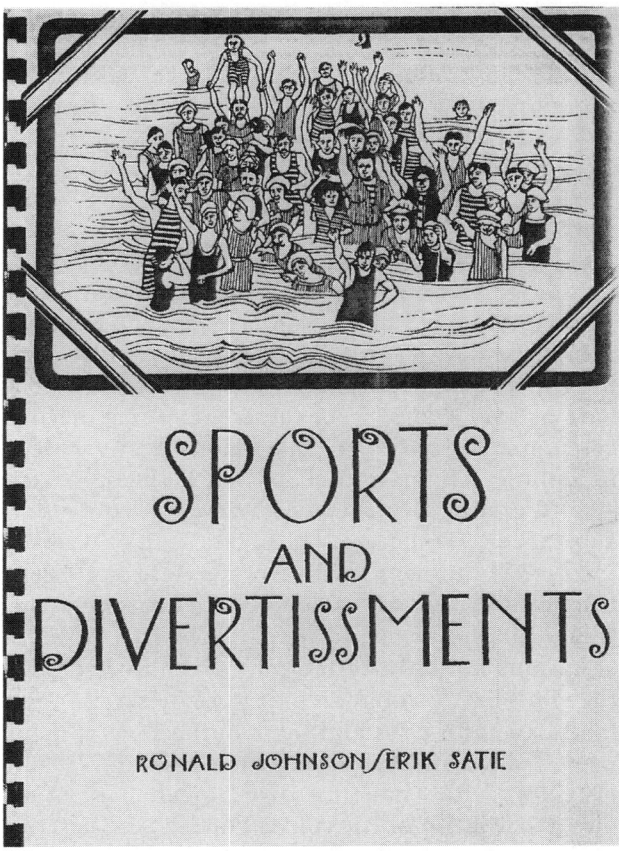

Cover for Johnson's 1965 book, in which he translates and arranges French composer Erik Satie's notes on performances to make "a music on the page"

reau, who noted that "Decayed literature makes the richest of all soils.") In this book the poet's impulse to compose "Of the seasons, / seamless, a garland" bears fruit in a frugal aesthetic where sounds are reused, recombined, rewoven, line by line; more broadly, the poet works to shape the world's "inarticulate / warble / & seething" into "A maze / of sound," to "catch / the labyrinthine wind, / in words – / / syllable, following / on syllable . . ." (Notice that both the wind and the poem are described as mazelike.) This felt symmetry between the world and the poet's words recalls the Transcendentalists and will prove the keel of *ARK*.

The book begins in winter, in Grasmere Churchyard. Wordsworth is its guiding spirit – or, more accurately, both Wordsworths are: the visionary William, "who could not see / daffodils / only / 'huge forms,' *Presences* & earth 'working / like a sea,' " and the more practical Dorothy, who saw the landscape in its sweet and rough particulars of lichen, moss, and water. (Future scholars will no doubt tease out the relationship between this Dorothy and her fictive Kansas cousin in Johnson's pantheon.) As it goes on, *The Book of the Green Man* includes both dream visions, sublime in their leaps of scale, and more strictly empirical, observational passages; and the poet insists on the connections between these realms, which mirror the bonds "between an earth, sentient with moles / & the owl's / radiant eyes – " and those that leap "from earth, to mistletoe, ivy & lichen, to owl's – / wing, to thunder, to lightning, to earth – & back." By spring the reader meets the Green Man in propria persona: the nature genius who came to Sir Gawain as the Green Knight, and who appears in writings of many others here as well. (The "I" of these poems is a bit of a green man himself. As he hikes along the river Wye he quotes Whitman to claim that he incorporates "fruits, grains, esculent roots," his eyes "containing substance / / of the sun, / [his] ears built of beaks & feathers.") Summer is a season of close and scientific scrutiny, full of dissections and anatomies; autumn, too, holds fast to specificities, rich with descriptions not only of nature (a poem that follows the sun's arc by numbered degrees, describing the landscape as it rises and sets), but of several man-made follies and grottoes. One reads of William Stukeley, who "made his own Stonehenge" of an old orchard, and of Alexander Pope, whose description of one garden ("a laurustine bear in blossom," and so on) was quoted earlier in "Shake, Quoth the Dove House." By the end of the poem the landscape of Shoreham is revealed as Albion, a "*Paradys / / Erthely*"; but it is also Johnson's most fully realized Kansas-as-Oz so far: "'*a country / where there is no / night*' / / *but of moons / & with heads of fish / / in the furrow, / / & on each, ear, beneath a husk / of twilight / / were as many suns as / kernels, / / & fields were far / / as the eye / could reach* . . ."

The Book of the Green Man was widely and favorably reviewed. At once "symbolic and magical," Charles Philbrick wrote in *The Saturday Review*, the book seemed a species of "literary mistletoe." (He goes on to note that, unlike Robert Lowell's *Near the Ocean* [1967], "the disappointment of the season," Johnson's work "embodies an epiphany and results in a triumph.") Later critics have explored its use of quotation and bricolage – an aesthetic of "the migratory phrase," as Steve McCaffery puts it – and have dwelled on what Norman Finkelstein calls the poet's "complexly mediated and startling immediate . . . engagement with the Romantic, visionary, and pastoral traditions in English poetry." "Like the green world of rural England," Finkelstein writes, "and like the myth of the Green Man itself, poetry partakes in a potent, celebratory natural impulse; it regenerates itself out of precursor texts that have likewise been initiated into such knowledge." Although in technique the book looks back to Olson and Zukofsky, it also deserves to be read in a broader context of the postmodern poetry of nature, in the company of A. R. Ammons and Gary Snyder. Johnson is as ecological a poet, in many ways, as either, but he offers a third, distinctive poetic: one neither Buddhist ("the Eastern idea of emptiness and the void I find unattractive. Blake didn't believe in it") nor resolutely secular.

Two years later, when Norton published *Valley of the Many-Colored Grasses* – a book that includes most of *A Line of Poetry, A Row of Trees* and the Rousseau poems, along with an extensive collection of ambitious work from 1966 to 1967, all with an afterword by Davenport – it, too, was praised in a range of publications. ("A major book," Margaret Randall called it in *Poetry*, for example, "humble and brilliant at the same time.") The second half of the book, which contains the new poems, is framed by two epigraphs: one from Blake, one from Thoreau, signaling the "Different Musics" the poet heard harmonized in *The Book of the Green Man,* and that he draws on here as well, along with other tutelary figures, notably Whitman and Ives. In the poem "The Different Musics" one thus finds a Thoreau-like attention to the found poetry of etymology and to the unending richness of detail in any patch of nature closely examined, but on the level of form the poem enacts an Ivesian composition in two simultaneous voices, coming together in a chorus and flourish at the end.

Several poems from *Valley* have drawn recent scholarly attention, especially the "Letters to Walt Whitman" and the double-columned "The Unfoldings." In his essay "Whispering Whitman to the Ears of Others" (1992), Ed Folsom calls the "Letters," which open with seed lines from Whitman and then blossom into Johnson's "own" words – though these root down into and draw upon other Whitmanesque passages – "one of the most sustained and suggestive of all the poetic encounters with Whitman." In them Johnson "carries on one of Whitman's deepest concerns," Folsom notes, which is "how the poet can 'indicate' to men and women 'the path between reality and their souls." (In Letter 9 he poses the question through a favorite, recurring motif. Whitman promises *"Landscapes projected masculine, / full-sized and golden."* "Are these landscapes to be imagined," Johnson wonders, or are they, rather, "an actual / Kansas – the central, earthy, prosaic core of us," needing only to be seen anew, following the poet's gesture and exultant gaze?) Davenport has recently argued that the "Letters" respond to another element in Whitman as well: his curiosity about and enthusiasm for science, which seemed to the older poet to describe a harmonic and orderly universe where, as Davenport puts it, "passionate friendship is an example of its harmony." Johnson embodies this harmonic vision in "The Unfoldings," too. The astronomers and writers quoted in the poem's two columns echo and confirm each other's testimony, while the love that draws a Thoreau or a Johannes Kepler or a Leonardo da Vinci to the things of this world — the love that draws Johnson as well — takes shape as the love that moves the sun and other stars. Presumably it also moves the two galaxies whose staticky collision, heard by radio astronomers and broadcast on the BBC, gave Johnson the climax of the book.

As the 1960s came to an end, Johnson published several more concrete poems, again mostly in limited editions. They are witty, spare, lyrical efforts to let the reader "see, not through, but with the letters," the poet explains, lightly reversing Blake's admonition, in "The Everlasting Gospel," that "This life's five windows of the soul / Distorts the Heavens from pole to pole, / And leads you to believe a lie / When you see with, not thro', the eye." Indeed, the poet wryly notes, "one could spend a lifetime writing with just the 26 letters of the alphabet," unfolding their implications: *"Tree: the 't' leaves. An 'r' branches. the 'e's have annual rings . . ."* In *Reading 1* (1968) — a single poem published on a single folded page — the word *book* is printed in solid black lettering with an *r* printed below it, just between the *b* and first *o*. A proofreader's insertion mark, handwritten, signals where the *r* should fit in: where Blake saw books in babbling brooks, Johnson "reads" in the other direction, while the wavery blue pen of his writing restores the fluidity that print masks or denies. (After a moment you see *brooks,* instinctively, at which point Blake's original dictum comes back into play.) In *Songs of the Earth* (1970) Johnson shapes a series of "listenings, as poems must listen and sing simultaneously," inspired by Gustav Mahler's *Das Lied von der Erde* and by Jonathan Williams's *Mahler,* a suite of long poems inspired by and answering to Mahler's symphonies. Johnson's *Songs of the Earth* ranges from the visually simple — at the center of a page, framed by that visual silence, "wood / winds" — to more complex arrays, in which (for example) a spacious square of capital letters spelling *WANE* encloses a lowercase square of *anew,* which rises into italicized capitals a little lower on the page. In *Eyes & Objects,* a longer collection published in 1976, Johnson tried his hand at poems where "every sound is mirrored by some other sound," producing what are, in effect, concrete poems carved out of aural space. "A catalogue for an exhibition which was itself the exhibition," Johnson calls this book; it is a fascinating text, curious in its play of ideas and delicate in music.

As he published these concrete poems, Johnson began work on *Wor(l)ds,* the long poem that occupied him for the next twenty years. (By 1980 he had retitled it *ARK;* it was completed in 1991 and published in its entirety in 1996.) This magnum opus was originally conceived, he has explained, as a "personal epic" in the mode of Wordsworth's *The Prelude* (1805), but there is little explicitly autobiographical material or narrative in the poem as written. Indeed, given his instinct toward architectural structures (the follies and grottos and topiary gardens of his earlier work) and his interest in concrete poetry, Johnson's epic seems in some ways closer in spirit to Wordsworth's unfinished *The Recluse:* that "gothic church" to which the autobiographical *Prelude* was merely the "ante-chapel," as Wordsworth explained, and into which the poet's shorter work might fit as "little cells, oratories, and sepulchral recesses." It is no accident that the first published work pertaining to *Wor(l)ds,* although not designated as part of it, was the slim pamphlet called *The Spirit Walks, the Rocks Will Talk,* in which Johnson gathered his "eccentric translations of two eccentrics": The *Facteur* Cheval and Raymond Isidore, creators, respectively, of the *"Palais Idéal"* in Hauterives, France, and of a mosaic house in the shade of Chartres. Like these men, and like American *bricoleur* folk artists and architects James Hampton and Simon Rodia, Johnson has built a fantastical and visionary monument from scraps and *trouvailles* and glittering fragments, here drawn from sources that range from *Paradise Lost* and the Book of Psalms to Protestant hymns, "The Star-Spangled Banner," Fourth of July celebrations (fireworks included), and Tory Peterson's *A Field Guide to Western Birds* (1969), along with earlier concrete poems by Johnson himself. While this collage model has been used in American long poems from Pound's *Cantos* (1925-1960), Eliot's *The Waste Land* (1922), William Carlos Williams's *Paterson* (1946–1958), Olson's *The Maximus Poems* (1960), and Zukofsky's *"A"* (1940-1978), *ARK* stands apart in the rigor and shapeliness of its architecture and in the poet's professed desire to write, unlike Pound or Williams, Olson or Zukofsky, "a poem *without* history."

ARK is divided into three books of thirty-three sections; *The Foundations* (made up of Beams), *The Spires,* and *The Ramparts* (whose cantos are called

Arches). "Over" these, as a metaphorical dome, rests ARK 100: a rewriting of *Paradise Lost* by excision. There is a temporal sequence to the poems as well, since *The Foundations* begin at sunrise and end at noon, *The Spires* go to sundown, and *The Ramparts* overlook a "midnight of the soul," even as they show the ARK transformed into a metaphorical starship ("all arrowed a rainbow midair, / *ad astra per aspera* / countdown for Lift Off," ARK 99 concludes). ARK 100, the poet has promised, will return us to dawn, looking at once backward in tense and forward in trajectory, with "The world all before." The first portion of ARK to appear in book form, however, was a much earlier section of this overarching dome, published in 1977 as *RADI OS I–IV*.

Inspired by British artist Tom Phillips, who painted and refigured pages from the third-rate Victorian novel *A Human Document* to turn it into his mysterious "treated" text *A Humament* (1980, 1987), and provoked by composer Lukas Foss, whose *Baroque Variations* (1968) presents a piece by George Frideric Handel with pieces cut out ("I composed the holes," Foss explains), Johnson began his revision of *Paradise Lost* rather casually. He soon found, however, that he had entered a grand Romantic tradition of wrestling and rewriting Milton's epic; found, indeed, that he had begun a specifically Blakean "infernal reading" of the prior text, composed by etching away surfaces to display "the infinite which was hid." The first page of *RADI OS* – and the book is composed in pages, each one a visual "text" – thus revises Milton's famous invocation "Of man's first disobedience, and the fruit / Of that forbidden tree whose mortal taste / Brought death into the world" into the following lines:

> O   tree
>
>   into the World,
>
>
>   Man
>
>
>   the chosen
>
> Rose out of Chaos:
>
>
>   song[.]

Johnson has not simply ventilated the dense Miltonic sentence, following the lead of Pound and Edgar Allan Poe. He has erased the divine sentence that stands behind Milton's opening. Disobedience and death have vanished, along with the fact that the Tree was forbidden and along with the "greater" nature of the "Man" (Christ) that Milton invokes. Here "Man / the chosen / Rose out of Chaos," not by the fiat of some creator, but as the rose appears in Pound's steel dust, according to the universe's blessed rage to order, complexity, beauty – and its impulse to create something that will echo that complexity in song. Johnson's cosmos is organized on what physics calls the "anthropic principle"; that is, it assumes a role for the observing human intelligence in the shaping of the creation that shapes our sentience. "After a long time of light, there began to be eyes, and light began looking at itself," Johnson writes in Beam 4 of *ARK: The Foundations*; in Beam 7 we learn that "Matter delights in music, and became Bach."

The four books of *Paradise Lost* that make up *RADI OS* can be read as a covert or implicit narrative, both as a running commentary on Milton and as a story of creation and the human "fall" into a sleepy forgetfulness, which the visionary poet will teach us to wake up from and thereby enable us to ascend to our true stature. (The echoes of Emerson and Thoreau are quite deliberate.) The poem includes sublime passages in which archangels appear – note the pun on *ARK* – and others, more plangent, where Milton's mourning for his blindness is transformed into a lament over the flickering of poetical insight Emerson laments throughout *Nature* (1836). "To find / the more / clear song," Johnson's page unfolds:

> Shine inward, and
>
>   there plant eyes
>
>   that I may see and tell
>
> Of things invisible
>
>   once
>
>   thick as stars
>
> The radiant image
>
>   the only

Garden

On the bare outside of this World[.]

At the close of Book IV of *RADI OS* Johnson returns to this defense of the visionary-cum-utopian perception of everyday life, challenging his readers to test his poetry's claims on the night or sunlit sky, as much as against his Miltonic original. "For proof look up," this book ends, "And read / Where thou art."

For many years Davenport's elegant, perceptive "Afterword" remained the only substantial critical work on *RADI OS*. Unlike Johnson's previous books, this volume was not widely reviewed; even in the Johnson/Davenport issue of *Vort* magazine (1967), an essential source for Johnson scholarship, the poem is not treated in the detail accorded his earlier volumes. (No one, for example, has explored in any detail Johnson's possible debt to Zukofsky's *"A"-14*, which includes another "writing-through" of *Paradise Lost*.) In part this may be due to the unusual nature of the project itself, which has seemed even to some sympathetic readers as quixotic, even maddening. In part, however, it is also due to a shift in sensibility in the American poetry community: a turn away from the immanentist postmodernism of the 1960s toward either the narrower lyric of domestic epiphany or to the politicized avant-gardism of what would soon be called the "Language poets." These poets share Johnson's debt to Zukofsky, but they are, as a rule, quite skeptical of visions, especially when those visions are of organic wholeness, a blissful fit between signifier and signified, language and the world. It is no accident that Ron Silliman, for example, found *RADI OS* to be a less interesting use of prior text than work by Jackson MacLow, William S. Burroughs Jr., or Kathy Acker. MacLow, the aleatoric poet – Johnson is apocalyptic – and Burroughs and Acker, those fashioners of bitter, twisted fictions of the lost, lack Johnson's Blakean apocalyptic vision, as well as his Ariel-like desire to re-create "Paradise, and groves / Elysian" (Wordsworth) in a "verbal earthly paradise" (W. H. Auden). "It's going to take a lot of piss and sweat to balance out these angels (and angles) of light and darkness," Silliman warns revealingly. Such writers also lack Johnson's sense that the physical sciences, especially physics and biology, confirm his optimistic vision – a sense that underwrites *ARK: The Foundations*.

ARK: The Foundations was one of the first three books published by San Francisco's now-defunct North Point Press. This book is a tour de force of harmonies, echoes, balancings, flooded with a light that Johnson hymns in both Neoplatonic and scientific terms. Begun before *RADI OS*, but not published for four years afterward, the book shares with *RADI OS* an impulse to rethink Romanticism in terms of modern physics. As he worked on *RADI OS,* Johnson explains, "I was taken over by Blake, but with my vision of the physical universe and . . . able to try to figure out how we order the universe now. Blake couldn't even look at Newton. I felt if I were to do this I would have to be a Blake who could also look at what we know of modern cosmology."

"The Sun's light when he unfolds it / Depends on the Organ that Beholds it," Blake observed. Beam 1 of *The Foundations* begins with a similar insight, set at sunrise:

Over the rim

body of earth   rays exit sun

rest to full velocity to eastward pinwheeled in a sparrow's

eye

– Jupiter compressed west to the other –

wake waves on wave in wave striped White Throat song

. . . .

as if a several silver

backlit in gust[.]

Photons "rest" at the speed of light, and the sun will fire photons when electrons in its constituent atoms, raised to a higher quantum level by the impact of one photon, "rest" to a lower energy level as the atom sends out another. Some of the sun rays that result are "Pinwheeled" as they enter the pinwheel-shaped iris of a bird's eye, while others, reflected back to Earth from Jupiter, are "compressed" or refracted by the earth's atmosphere, as the other planet sinks below the horizon. (Since a sparrow's eyes are on opposite sides of its head, "able to see

Title page for Johnson's 1969 book that includes a suite of poems on the work of French painter Henri Rousseau

the seed beneath its bill – and at the same instant the hawk descending," as Beam 4 explains, the bird may see both eastward and westward sights at the same time.) The "wake" of those rays, which are themselves waves, or wavicles, wakes an answer in the sound waves of the White Throat song, so that as each bird joins in separately it is as though spots of silver "backlit," or reflected, lit back the sunlight. They do so "in gust," or keen delight, as well as in gusts, or surges, of melody; and human song, poetry, which sings from the electrically sparkling "nervetree," is by extension an equally natural and reflective pleasure, since "out of a stuff of rays, particles, and pulses" comes the poet who notes down and writes up the scene. "(Mid-age. Brought to my knee.)" Johnson calls himself, invoking Dante as he learns himself to be, like any language animal, "the artificer of reality" (Beam 12).

The Beams of *ARK: The Foundations* are a mix of prose and verse, scientific description and etymological inquiry-cum-invention, quotes and concrete poems, even an Orphic writing through of the Book of Psalms: a summa, in short, of Johnson's earlier modes. Added to these are Beams modeled on Christopher Smart's *Jubilate Agno* (1763) and a variety of illustrations. The whole is, to borrow one of Johnson's lithe, blithe phrases, "lucid as Euclid": a world where all parts, on some level, fit and echo one another, so that the handprint that forms Beam 18 displays the whorls of fingerprints, a "rhyme" for earlier references to the whorl of galaxies and to a snail's spiraled shell, while the first stage of a cell-division diagram, found in Beam 25, the "Bicentennial Hymn," looks like the sunrise with which Beam 1 began, and the last stage shows twin cells pressed against each other like two hemispheres of the brain, or like the "balanced dissent" of the United States, of matter and antimatter, and of other divided, procreative pairs mentioned elsewhere in the text. The poem is at once an apotheosis of Johnson's desire to create a garden or Oz – a world "where anything is possible and in which the imagination lives," as he said in the *Vort* interview – and the fulfillment of his hope to find that garden *outside* the act of the mind, grounded in the facts of the physical world. As Paul Lake observed in "Unfolding the Manifold," an admiring essay-review, the poem is "a compelling spiritual document, turning a science that has given us nuclear fission and the principle of indeterminacy into poetry, and turning poetry, in turn, into a medium that can once again engage the full range of human thought."

As their place in Johnson's architecture suggests, the Beams of *ARK: The Foundations* are both dense and spacious. They outline and support the poetry to come, setting out, the poet has explained, "all themes necessary to the work ahead, to have room to turn around in over the years" ("Planting," 3). Four years passed before the second long section of *ARK* was published as *ARK 50: Spires 34–50,* selected for the National Poetry Series by Charles Simic. As a rule, these *Spires* have a delicate motion quite different from the expansive claims, illumination, and revelations of the Foundations. More lines are lean and streamlined, composed of one or two words; and there are frequent references to birds, mountains, stairs, and other appropriately rising phenomena. In the first *Spire,* for example, a lovely elegy for Zukofsky, "Lord Hades" appears, "crackling up / like a wall of prairie fire / in a somersault silver / to climb blank air / around us"; in *Spire 43,* one of three "Lot's Pillar" *Spires,* the poem itself is figured as "one lyric / alptly slapstruck / contraption." Indeed, in the *Spires, ARK* becomes increasingly self-referential, naming as well as enacting its "Strains / legion and ingenious / put to the uses of blessing," in which "s h / a p e / s / abound enobled" (*ARK 46, Fountain I*). The earlier concern with cosmic and evolutionary detail has hardly been abandoned, but now the poet's desire to "Dance / howbeit / about us" and "ply" the " 'nocount / Abyss" often involves calling our attention to the art-skill of the "Poems plain / as Presbyterian pews" he has crafted. Rather than transfiguring Kansas, as the poet did so often in earlier work, Johnson here turns the force of art against the threat of oblivion posed by Lord Hades. All will meet him, these Spires well know; but "to say then head wedded nail and hammer to the / work of vision / of the word / at hand," the Zukofsky *Spire* goes on, "that is paradise ..."

It is tempting to see *The Foundations* of *ARK* as being under the sign of Johnson's mother, the dancer, since they are choreographed to melodies and harmonies taken from "Physics, the music of our time." *The Spires*, in turn, recalls the poet's childhood in the father's lumberyard: they are a structure where "acrobat thought" will "bounce planed planned plank" to "handstand more split images than you could shake a stick at." But in the "reeled world" this poem captures, such distinctions, temporarily helpful, never hold. " 'Step it up a round,' " a voice calls to acrobat thought, and the high-stepping grace of Johnson's artistry bows to its carpenter partner in a sort of square-dance *logopoeia* ("the dance of the mind among syllables," in Pound's phrase). As one leaves the spiral-staircases of *ARK 50* and moves to later *Spires* – that stair a figure for the double helix of DNA, one notes in passing, and also for the twined snakes of Hermes's caduceus – one reaches not only a waltzing Balanchine *Spire* (*ARK 53*), but a luminescent set of Fireworks *Spires*, composed of a "vast smithy spray" from the poet's "foundry" of hammered quotations and lucky finds.

Unlike *ARK: The Foundations*, which was acclaimed in *The Threepenny Review* and *Parnassus*, among other journals, *ARK 50* received only a few reviews. Without a narrative "hook" or thematic fanfare (like the remarkable transcendental physics of the earlier volume), the book rewards close and playful attention but doesn't force or command it; and it seemed out of key with the political poetics championed by a wide variety of poets and critics, experimental and otherwise, in 1984. Since the publication of *ARK 50,* Johnson has finished *ARK,* and most of *The Ramparts* (each one an Arch, as each part of *The Foundation* is a Beam) have been published in literary journals. Like a few of the Beams, and like several of the later *Spires*, almost all the Arches are written in a highly compressed verse whose syntax is either implicit or absent altogether. Grouped in trios of Arches, each written in mosaic tercets, *The Ramparts* thus glint as much as they flow.

While Johnson continues to display his "legerdemain in the Elaboratory," as he puts it in *ARK 72,* there is often a sadder, darker tone to the Arches. The Zukofsky *Spire* was set in sunlit summer, and artifice could stand as paradise; but in the "vigil elegiac" of these later verses a moment comes to "click the ruby slippers" and return to, if not bare Kansas, at least a more wistful mood. Many tutelary dead are mentioned: Duncan, Guillaume Apollinaire, Stéphane Mallarmé, Emily Dickinson, and Whitman, along with a figure named only in initials, who seems at once Henry James ("my true Penelope," Johnson has called him) and the poet's mother, Helen, who is " 'like silver smiting silver' / H. J. on the harp / behind order / Utopia cut figure." And sometimes, between the lines, perhaps a certain sadness about the poet's neglect by readers, the loneliness of his effort to extract "the singing necessities" from his material and leave "not a whit one mightn't want about." Along with Arches taken from Thoreau's *Journals* and from various Protestant hymnals, there is one, "Arches IX," drawn from Vincent Van Gogh's *Letters*. "I have rented a house / yellow outside, whitewashed within / in full sun," the artist writes. "Wishing to see a different light, / exile and stranger / I am dead set on my work."

Johnson's studies in "a different light" indeed set him apart from most contemporary poets, not least in the detailed, researched sense of the world as a Cosmos that underwrites his postmodern poetics of the Beautiful. ("Beauty is difficult," says Aubrey Beardsley to William Butler Yeats at several points in the *Cantos*. "Beauty is easy," Johnson retorts in Beam 14 of *ARK*. "It is the Beast that is the secret.") Insisting on a link between the self-referential construction of the poem and the mathematical coherence of the natural world, Johnson suggests that the stochastic style of "textured information" characteristic of so much late-century American poetry (according to Roger Gilbert, among others) may be at least as much a distortion of the ordering impulse of the human mind as linguistic structure and closure are a human imposition on the buzzing, booming chaos of events.

Johnson's work is all brought back into print with the publication of the complete *ARK*. He may at last be read in the broad context he deserves and is seen not only as a student of Olson and

Zukofsky, more mellifluous than either, and not only as a late-century Transcendentalist, with the intelligence to make his observations and his visions new, but also as an American counterpart to the great Spanish modernist Jorge Guillén, whose *Cántico* (*Canticle,* 1950) is, like *ARK,* at once a hymn to light and to the triumphant pleasures of finding oneself "invented" by a world that calls us to its praise. And he will be seen, perhaps most of all, as the only American poet to transplant the aesthetics of the garden into poetry — an English garden like Pope's at Twickenham, with its mix of formal elegance and wilder Follies and Grottoes, and an American garden like Thoreau's at Walden. Here the gardener decides not only to "make the earth say beans" at his command, but to listen as well to the "cinquefoil, blackberries, johnswort, and the like" that are its native song. In the eyes and hands of The Gardener, as Johnson signs himself in *ARK*'s Beam 30, the buzzing, booming chaos blooms from chaos into complexity, drenched with sunlight, heady with bees.

Interviews:

Barry Alpert, "Ronald Johnson — An Interview," *VORT,* 3, no. 3 (1976): 77–85;

Peter O'Leary, "An Interview with Ronald Johnson," *Chicago Review,* 42, no. 1 (1996): 32–53.

References:

Guy Davenport, "Introduction" to *Valley of the Many-Colored Grasses* and "Afterword" to *RADI OS,* collected in *The Geography of the Imagination* (San Francisco: North Point Press, 1981): 190–204;

Davenport, "Whitman a Century after His Death," *Yale Review,* 80 (October 1992): 1–15;

Ed Folsom, "Whispering Whitman to the Ears of Others: Ronald Johnson's Recipe for *Leaves of Grass,*" in *The Continuing Presence of Walt Whitman: The Life after the Life,* edited by Robert K. Martin (Iowa City: University of Iowa Press, 1992), pp. 82–92;

William Harmon, "The Poetry of a Journal at the End of an Arbor in a Watch," *Parnassus,* 9 (1981): 217–232;

Steve McCaffery, "Synchronicity, Ronald Johnson and the Migratory Phrase," *Vort,* 3, no. 3 (1976): 112–116;

Eric Selinger, " 'I Composed the Holes': Reading Ronald Johnson's *RadiOs,*" *Contemporary Literature,* 33 (Spring 1992): 46–73;

Selinger, "Important Pleasure and Others: Michael Palmer, Ronald Johnson," *Postmodern Culture,* 4 (May 1994).

Papers:

A collection of drafts, typescripts, and corrected proofs of published and unpublished poems, as well as a selection of the poet's correspondence with Guy Davenport, Jonathan Williams, Ian Hamilton Finlay, and others, is housed in the Spencer Library of the University of Kansas.

Carolyn Kizer
(10 December 1925 -   )

Elizabeth B. House
Augusta State University

BOOKS: *The Ungrateful Garden* (Bloomington: Indiana University Press, 1961);

Knock Upon Silence (Garden City, N.Y.: Doubleday, 1965);

Midnight Was My Cry: New and Selected Poems (Garden City, N.Y.: Doubleday, 1971);

Yin: New Poems (Brockport, N.Y.: BOA Editions, 1984);

Mermaids in the Basement: Poems for Women (Port Townsend, Wash.: Copper Canyon Press, 1984);

The Nearness of You (Port Townsend, Wash.: Copper Canyon Press, 1986);

Proses: On Poems & Poets (Port Townsend, Wash.: Copper Canyon Press, 1993);

Picking and Choosing: Prose on Prose (Cheney, Wash.: EWU Press, 1996);

Harping On: Poems 1985-1995 (Port Townsend, Wash.: Copper Canyon Press, 1996).

OTHER: Robert Peterson, *Leaving Taos,* selected for *The National Poetry Series,* and introduced by Kizer (New York: Harper & Row, 1981);

Carrying Over: Poems from the Chinese, Urdu, Macedonian, Yiddish, and French African, translated by Kizer (Port Townsend, Wash.: Copper Canyon Press, 1988);

The Essential John Clare, edited by Kizer (Hopewell, N.J.: Ecco Press, 1992);

100 Great Poems by Women, edited, with an introduction, by Kizer (Hopewell, N.J.: Ecco Press, 1995).

Although Carolyn Kizer has not been a prolific poet, her meticulously crafted work has received high critical praise. She has also made important contributions to the arts as translator, teacher, editor, and critic. In 1985 Kizer won the Pulitzer Prize for *Yin: New Poems* (1984), and in 1988 she won the Theodore Roethke Memorial Foundation Poetry Award. *Proses: On Poems & Poets* was published in 1993 as the inaugural issue in *Writing Re: Writing,* a series whose editors intend "each annual volume to be a major collection of essays on poetry by a leading poet."

In "The Stories of My Life" (the first section of *Proses*), as well as in prose pieces in *Yin* and *The Nearness of You* (1986), Kizer writes of her extraordinary childhood and traces the forces that destined her to be a woman of letters. Kizer was born in Spokane, Washington, the only child of Benjamin Hamilton and Mabel Ashley Kizer. The poet's father was a lawyer who was active in various liberal and civic causes. Before marrying Ben Kizer, Mabel Ashley earned a Ph.D. in biology from Stanford University, studied art and philosophy at Harvard University, organized unions, and ran the first federally sponsored drug clinic in New York. The scholarly Kizer household welcomed such eminent guests as Vachel Lindsay, Percy Grainger, and Lewis Mumford, in addition to intellectuals and statesmen from China and other nations.

After Carolyn's birth Mabel Kizer devoted herself to nurturing the child's abilities. In "A Muse," a prose section of *Yin,* Kizer recounts how as a young girl she randomly committed three letters, *A R T,* to a scrap of paper. Kizer's proud mother quickly pronounced her offspring an artist and spent most waking moments encouraging the poet's development. As a child Kizer resented her mother's overzealous nurturing, and in "A Muse" she explains that only after her mother died did her own "serious life as a poet" begin. Then, Kizer says, "At last I could write, without pressure, without blackmail, without bargains, without the hot breath of her expectations." Yet, acknowledging great love for and indebtedness to her mother, Kizer adds, "I wrote the poems for her. I still do."

Kizer earned a B.A. degree from Sarah Lawrence College in New York in 1945 and later did graduate work at Columbia University (1945-1946) and the University of Washington in Seattle (1946-1947). At Columbia, Kizer had a Chinese Cultural Fellowship in Comparative Literature and for a

Carolyn Kizer (photograph © 1983 the Estate of Thomas Victor)

time interrupted her studies to visit China, where her father was in charge of United Nations relief efforts. During her time at the University of Washington Kizer studied under the poet Theodore Roethke. She married Charles Stimson Bullitt in 1948 but divorced him six years later. Kizer married John Marshall Woodbridge, an architect, in 1975 and now resides in Sonoma, California. Her three children – Ashley Ann, Scott, and Jill Hamilton – are all from her first marriage.

Kizer's literary activities include founding *Poetry Northwest*, a journal she edited from its beginning in 1959 until 1965; serving as a U.S. State Department specialist in Pakistan during 1964 and 1965; and directing literary programs for the National Endowment for the Arts from 1966 to 1970. She has also been poet-in-residence at the University of North Carolina (1970–1974); Hurst Professor at Washington University, Saint Louis, Missouri (1971); acting director of the graduate writing program at Columbia University (1972); a lecturer at Barnard College in New York (1972); McGuffey Lecturer and poet-in-residence at Ohio University (1975); and a professor of poetry in the Iowa Writers Workshop (1976). In 1985 she was a senior fellow in humanities at Princeton University.

At the end of "The Stories of My Life," an autobiographical section of *Proses,* Kizer confides that having experienced "war, love, marriage, separation, loneliness, children, the death of those I loved," she is now "able to say with Chaucer's Criseyde, ' 'I am my owne woman, wel at ese.' " The life journey that has led to this enviable state is intimately tied to Kizer's poetry and its evolving style and subject matter. In her early work Kizer treats abstract, universal topics; typical themes are tensions between humans and nature, civilization and chaos, and the perusal of defenses people raise against pain. In later years her poems have become more personal in theme but at the same time less formal in style.

No matter what her subject, though, Kizer never engages in the pettiness of which women writers have sometimes been accused: "stamping a tiny foot at God, whining in pentameter, dealing with only life's surfaces," as Roethke wrote in "Pro Femina" (part 3), Kizer emphatically rejects "the sad sonneteers, toast-and-teasdales we loved at thirteen . . . / when poetry wasn't a craft but a sickly effluvium" and opts rather for writing poetry which has accurately been labeled "toughminded." In a 1986 interview Kizer noted, "Self-pity is the most useless quality in the world; it destroys any poem that it touches," and, indeed, Kizer faces life's harshest realities without flinching.

Especially in her first book, and to a lesser degree in the second, Kizer seizes upon grotesque images – lice cozily snuggling in a captured bat's wing, carrion birds devouring the last pulp of hellbound bodies – in order to test the ability of poetic form to contain terror. In later volumes she deals with more familiar landscapes, the terrors of human confrontations. In her treatments of all these unsettling topics, Kizer almost invariably maintains a tone of stoic serenity and acceptance, a philosophic stance that she has perhaps gleaned in part from her translations of oriental poetry.

The calm, almost aloof tone characteristic of Kizer's work owes itself in part to the poet's technical skills, especially her ability to use form to distance herself from pain. Particularly in her earlier poems Kizer contains her emotions by imposing on them intricate rhyme schemes or complicated French and oriental verse patterns, and most often her form-giving strategies succeed. Although William Dickey complained in his November 1961 *Poetry* review of *The Ungrateful Garden* (1961) that many of Kizer's poems "are more concerned with the manner of their expression than with the material to be expressed," most other critics have felt that Kizer uses her considerable technical gifts in ways that bring together form and substance. In the July 1961 *Saturday Review*, Robert Spector greeted Kizer as "an important new voice" and especially commended her ability to "cut savagely through all sentimental disguises."

One of the facades that Kizer most frequently explores in her first book is the assumption that humans can still be comforted by things that have traditionally given people solace. The potential sheltering abilities of government, poetry, love, and especially nature become objects of scrutiny in *The Ungrateful Garden*. In the book's title poem, for example, Kizer describes the folly of Midas, who, after having his eyes scorched by reflections from flowers he has turned to gold, decides that nature is evil. Similar false conclusions about nature's indifference as well as humans' supposed bonds with nature are dealt with in other poems.

One of Kizer's best-known pieces, "The Great Blue Heron," depicts nature as a concrete reality that neither cares for nor hates human beings but simply is as it is. In the first few stanzas Kizer evokes a ghostly portrait of the tattered-winged bird who first appears as a "shadow without a shadow" and then flies away on "vast unmoving wings." By the middle of the lyric she has made clear that the heron is a harbinger of death, but she never suggests that the bird is evil. As part of nature he merely reflects the cycle of life and death that time imposes on all living creatures.

Another poem, "The Intruder," emphasizes the distance between humans and nature. Here the persona's mother, feeling "instinctive love," rescues a bat from her cat's jaws. However, when she sees the "lice, pallid, yellow, / Nested within the wingpits," the woman drops her facade of sympathy and flings the wild creature back into the cat's mouth before washing "the pity from her hands."

In "The Suburbans" Kizer laments the substitution of "cardboard-sided suburbs" for the nature that inspired and comforted nineteenth-century poets. Poets now, Kizer suggests, cannot find inspiration in nature so easily. Bound to a culture in which domestic cats are all that remain of "tygers, mystery / Eye-gleam at night," the modern poet must create the animals' "ancient freedom in a cage / of tidy rhyme." In this process, ultimately, poetry rather than nature becomes the writer's final subject and solace. The poet's "limited salvation" is found not in nature but in "the word."

If Kizer feels that nature neither comforts nor inspires twentieth-century people, at least she does not find birds, trees, or even bats actively threatening. Governments, on the other hand, do alarm the poet. The dangerous artificiality of modern governments is reflected in the elaborate, highly stylized villanelle form in which Kizer writes "On a Line from Julian": "I have a number, and my name is dumb. / Such a barbarian have I become!" the persona cries. Kizer shows that such plights occur because governments urge their citizens to relinquish individual identity; in a cardboard society names, the most personal of words, are replaced by mere numbers.

In "The Death of a Public Servant," another uneasy poem about government, Kizer memorializes a Canadian diplomat who committed suicide after a U.S. Senate subcommittee accused him of being a Communist. Ideally, governments should protect their citizens, and words should be used to create form, not destroy it. Yet the Senate's words of accusation end life for the gentle man whom Kizer implores, "Though you escape from words, whom words pursued, / take these to your shade: of rage, of grief, of love."

In contrast to the dearth of comfort Kizer finds in nature and government, poetry and human relationships do provide her with means for dealing with terror. In "Columns and Caryatids" Kizer

"I am a premature Women's Liberationist," writes Carolyn Kizer. "I was writing poems on the subject ten years before it became fashionable, and a great many people, then, didn't understand what the hell the fuss was all about. Now, in many college courses on the subject of women, my poems are used, particularly *Pro Femina*."

Miss Kizer's poems have appeared in two previous collections, *Knock Upon Silence* and *The Ungrateful Garden*. Currently Poet-in-Residence at the University of North Carolina at Chapel Hill, she was in charge of the literary program of the National Council on the Arts during the Johnson Administration.

Back of the dust jacket for Kizer's 1971 collection of poems, Midnight Was My Cry

endows three caryatids (female-shaped, edifice-supporting pillars) with powers to speak; the three roles the pillars assume — wife, mother, lover — are those through which women often form bonds with other people. The mother and lover find satisfaction in their ways of life, but the wife Kizer chooses to epitomize conjugal relationships is Lot's spouse, who, as a pillar of salt, melts away under God's derision. Even in such circumstances Lot's wife is brave, however, and ultimately all three caryatids become emblematic both of women's strengths and of their entrapment in the responsibilities that society metes out to them.

Conflicting facets of human relationships are also found in some of Kizer's love poems. Many writers have found that sexual love creates bonds that protect and shelter, and in "What the Bones Know" Kizer seems to agree. Rejecting Marcel Proust's "wheezy," anemic ideas about love, Kizer decides "that Yeats was right / That lust and love are one." However, the poem's form, paradoxically, both casts doubt on and supports the lyric's message. Apparently Kizer is not completely at ease with sex unless its wildness is tempered with the rigidity of form, for she uses a modified sestina stanza throughout the poem. At the same time, though, the poet uses the constricting scheme of the sestina to help affirm the life-giving properties of sex. In the poem's first six-line stanza the word *death* ends the sixth line while *breath* is the last word in line five. *Breath* remains the final word in each of the three remaining stanzas, but *death* becomes in the second, third, and fourth stanzas the end word for lines four, three, and two, respectively. Thus, as the persona hymns the power of sex, *death* moves farther and farther away from life and *breath*. Kizer expresses a similar ambivalence about the consoling power of sex in "A Widow in Wintertime." Here the persona sees connections between her own desires and those of her philandering cat, but she also counts as part of the discipline that makes her human the fact that she tried "not to dream / Of grappling in the snow, claws plunged in fur."

In *The Ungrateful Garden* the only consolation that Kizer shows to be invariably effective is art. Especially in the section that she labels "In the Japanese Mode," she demonstrates that poetry, unlike nature or governments, can unfailingly combat the chaos of life. One of the poems, "A Poet's Household," for example, is dedicated to Theodore Roethke, Kizer's teacher, and consists of three tankas that extol and are exemplars of the form-giving translation of experience into poetry. Similarly, in "From an Artist's House" Kizer celebrates the immutability of poetry, the fact that in the writer's hands, an old compote full of withered oranges becomes "immortal / On twenty sheets of paper."

Kizer takes the title of her second book, *Knock Upon Silence* (1965), from the *Wen-Fu* of Lu Chi, a third-century Chinese poet; her translation of a key passage is quoted on the title page:

> We wrestle with non-being
> to force it to yield up being;
> we knock upon silence
> for an answering music....

A merging of Eastern and Western cultures, the music in *Knock Upon Silence* is divided into four parts: two long poems, "A Month in Summer" and "Pro Femina"; a section of Chinese imitations; and eighteen translations of Tu Fu, an eighth-century Chinese poet. In the October 1965 *Library Journal* John Willingham asserted that Kizer "clearly ... possesses one of the more impressive talents of our day." Echoing the same view, Richard Moore wrote, "This book is a rare event. One senses in it a fully developed gift in the service of an urgent and unifying perception."

The only part of Kizer's second book that critics have not unanimously praised is "A Month in

Summer," a diary of a collapsing love affair recorded in a mixture of haiku and prose. While William Jay Smith argued in the August 1966 *Harper's* that this poem is the volume's best, other reviewers sided with Richard Moore, who felt the diary to be "the weakest part of the book.... It is moving in places, witty in others; but there is also a tendency to be straggling and repetitive." Some of the prose sections of "A Month in Summer" are bland, but the modified haiku are exquisite, exploring

> The terror of loss:
> Not the grief of a wet branch
> In autumn, but the absolute
> Arctic desolation.

"Pro Femina," perhaps Kizer's best-known work, is a satiric poem about liberated women, especially those who are writers. The piece is written in hexameters, a meter that Kizer has said she derived from Juvenal's satires; the irony in Kizer's selection of meters becomes clear when the poet reminds us of Juvenal's misogyny, that he "set us apart in denouncing our vices / Which had grown, in part, from having been set apart."

In the poem Kizer first notes historical differences between the sexes ("While men have politely debated free will, we have howled for it / Howl still, pacing the centuries"), and then, while cautioning against overconcern with surface appearances, she concludes that women should maintain both their faces and their minds. To succeed in letters, Kizer says, women writers must "struggle abnormally," "submerge our self pity" all the while

> Keeping our heads and our pride while remaining unmarried;
> And if wedded, kill guilt in its tracks when we stack up the dishes
> And defect to the typewriter. And if mothers, believe in the luck of our children,
> Whom we forbid to devour us, whom we shall not devour,
> And the luck of our husbands and lovers, who keep free women.

Midnight Was My Cry: New and Selected Poems (1971) contains thirty-six of the thirty-seven poems published in *The Ungrateful Garden,* eight poems from *Knock Upon Silence,* and sixteen new poems. The book received a warm critical reception. John Willingham in the November 1971 *Library Journal* labeled it a "distinguished volume." Richard Howard in the August 1972 issue of *Poetry* praised the sixteen new poems for being "frank in their response to hope and horror alike" and for showing clear evidence of Kizer's increasing maturity.

The sixteen new poems differ in several ways from those in Kizer's two earlier books. The poet's technical elegance is as apparent in these pieces as it is in her earlier ones, but the ugly images characteristic of *The Ungrateful Garden* and *Knock Upon Silence* are not emphasized, and Kizer's choice of subjects has altered. The concern with nature that the poet displays in *The Ungrateful Garden* is replaced in these later poems by interest in contemporary social and national problems, particularly those of the 1960s. "Poem, Small and Delible," for example, deals with antisegregation sit-ins at Woolworth's; "The First of June Again" describes actions of American marines in Saigon; and "Seasons of Lovers and Assassins" uses blank verse to equate Robert Kennedy's murder and the "caul of vulnerability" that surrounds even those who think they have found safety in love.

Although Kizer uses new subject matter in the sixteen new poems in *Midnight Was My Cry,* she displays in them the same stoic acceptance found in her earlier work. Ultimately, as she notes in the villanelle "On a Line From Sophocles," "Time, time my friend, makes havoc everywhere," both on personal and national levels. Yet even realizing this truth, she does not succumb to panic or terror or whining. Rather she summons courage similar to that which she lauds in "Lines to Accompany Flowers for Eve," a poem dedicated to a woman who has attempted to commit suicide. Life must have value, Kizer says, for the human spirit has surprising resilience even in the face of horror. Most often when people are tempted to "buy peace" by relinquishing their lives, "the spirit rouses . . . / . . . signaling / Self-amazed, its willingness to endure," for with courage it is possible to "live in wonder, / . . . Though once we lay and waited for death."

In 1984, thirteen years after the publication of *Midnight Was My Cry,* Kizer brought out two volumes, *Yin: New Poems* and *Mermaids in the Basement: Poems for Women.* In 1986 *The Nearness of You* appeared; this book, Kizer's companion piece to *Mermaids in the Basement,* features "poems for men." While *Yin* contains only work Kizer had not previously published in book form, *Mermaids in the Basement* and *The Nearness of You* include many poems from the writer's earlier volumes.

In contrast to earlier work, Kizer's poetry from the 1980s is intimately linked to the circumstances of her life. Kizer's mother and father, particularly, as well as other family and friends, become inspirations for and subjects of major new poems.

Even the new offerings that are not openly about Kizer's family and friends are "personal" in the sense that, rather than dealing with abstractions, the poet deals with specific historical/mythical figures as they confront human problems. Also, Kizer is less concerned with formal rhyme schemes and elaborate stanzas than she had been previously.

In Chinese philosophy yin is the female principal, supposedly the more passive, darker side of cosmic forces. In addition to "A Muse," which is the second section of *Yin,* the book has three other parts: "Believing/Unbelieving," "Dreams and Friends," and "Fanny and the Affections." "Believing/Unbelieving" includes "Semele Recycled," a modernized version of the myth of Dionysus's mother. When Zeus appeared to her in all his glory, the original Semele promptly disintegrated; however, in the "recycled" version of the myth, a stronger Semele is reunited with her lover. "The Blessing," also in this section, was written for Kizer's daughter Ashley and lovingly links and celebrates Kizer's strong mother and daughter: "Child and old woman / soothing each other, / sharing the same face / in a span of seventy years, / the same mother wit."

The third section, "Dreams and Friends," includes the poem "Antique Father," in which Kizer describes Ben Kizer during his last illness. The poet says her stern father had been successful in "quelling all queries / of my childhood," and now, ironically, the man wishes "urgently / to communicate / to me" but cannot. Illness has robbed him of speech and imposed on him a "terrible silence," not unlike the formidable quiet he used to separate himself from the poet when she was a child.

The last section in *Yin* features "Fanny," ostensibly a diary of Fanny Osbourne Stevenson, wife of the writer Robert Louis Stevenson. Written in hexameters and later added as a fourth section to "Pro Femina" in *Mermaids,* the piece explores the role of a wife who sacrifices herself to sustain her husband's creativity. Fanny cares for her husband in Samoa, far from the English weather that threatens his health, and in doing so she suffers trials and threats as varied as headhunters and hurricanes. The isolated woman finds her only consolation, her only creative release, in growing magnificent gardens.

In addition to winning the Pulitzer Prize for poetry, *Yin* received a warm critical reception. In *Poetry* (March 1985) Robert Phillips labeled *Yin* a "marvelous book" and noted, "One could never say with certainty what 'a Carolyn Kizer poem' was — until now. *Yin,* her fourth collection, is her most unified, original, and personal. Now we know a Kizer poem is brave, witty, passionate, and not easily forgotten." Patricia Hampl, in the November 1984 *New York Times Book Review,* also praised *Yin* and listed "A Muse" as the volume's "most striking offering." Similarly, Grace Schulman, in *Poetry* (November 1985), cited "Fanny," "Semele Recycled," and "Antique Father" as poems particularly exemplifying Kizer's mature wisdom.

Mermaids in the Basement, Kizer's fifth book, takes its title from Emily Dickinson's lines "I started early — Took my Dog — / And visited the Sea — / The Mermaids in the Basement / Came out to look at me — ." Divided into seven sections — "Mothers and Daughters," "Female Friends," "Pro Femina," "Chinese Love," "Myth: Visions and Revisions," "A Month in Summer," and "Where I've Been All My Life" — the collection examines various roles, personas, and incarnations that women assume.

In the *Hudson Review* (Summer 1985) Mark Jarman rightly judged that the "major service" of the publisher of *Mermaids* was "to reprint many fine [Kizer] poems that were included in now out-of-print editions." "The Intruder," "The Great Blue Heron," The Blessing," "A Widow in Wintertime," "Lines to Accompany Flowers for Eve," "Columns and Caryatids," "Semele Recycled," "A Month in Summer," and "Pro Femina" all reappear in *Mermaids.* The groupings in this volume give form to the poet's vision of yin, the feminine force.

Interspersed with these reprinted pieces are several new lyrics such as "Thrall," a poem that links two scenes; in each, the focus is a room containing "a chair, a table and a father." As a child the poet felt estranged from her stern father, and in "Thrall" she remembers him as a man who "read for years without looking up / Until . . . childhood was over." The poem's second scene pictures the now aged father and his grown daughter, the poet. Again in a sparsely furnished room, she reads aloud to her father and waits "for his eyes to close at last / So [I] . . . may write this poem." In both instances the father and daughter love each other, but their relationship is tempered and muted by a similar devotion to literary pursuits.

In the preface to *The Nearness of You,* Kizer dedicates the book "For the men I love, especially John." The title is taken from a song by Hoagy Carmichael and Ned Washington. Divided into four sections — "Manhood," "Passions," "Father," and "Friends" — *The Nearness of You* deals with the roles of men as viewed from Kizer's feminine perspective.

One of the most revealing pieces in the "Father" section is the narrative "My Good Father."

Kizer composed the essay for a member of Ben Kizer's Study Club who wished to write a paper on her father. In her reply to the member's queries she describes her father as a supremely intelligent person and one devoted to good causes. However, he was also a self-absorbed, aloof thinker who frightened his daughter until she was thirty. Kizer summarizes her father by concluding, "To him people were chiefly important as vehicles by which he could express his passion for abstractions, abstractions for which he gaily marched into battle, chanting his war chant: truth, justice, equity, freedom and law. How he loved the law!"

The Nearness of You contains poems reprinted from earlier volumes as well as new lyrics (almost letters) dedicated to individual writers such as Robert Creeley, Robert Peterson, Ruthven Todd, Donald Keene, Bernard Malamud, Theodore Roethke, and James Wright. As this listing illustrates, Kizer has known many writers, and, indeed, she is widely esteemed as a teacher, mentor, and guide for aspiring poets. Not surprising, this role can become taxing; with wry humor Kizer titles one poem in this volume, "To an Unknown Poet." When this new writer comes "unannounced to my door / I tell you I am busy / I'm not as young as I was. / I'm terrified of breakage."

In the November 1986 *Publishers Weekly* Genevieve Stuttaford noted that the poems in *The Nearness of You* "ring joyously in the ear and the memory" and concluded, "Kizer's mastery, grace, charm and wit make this a perfect book." Somewhat less sympathetic was Anthony Libby, who in the March 1987 *New York Times Book Review* criticized the collection but especially praised "Thrall." In the *Library Journal* (November 1986) Rochelle Ratner noted that since Kizer's poems in this volume are not dated, studying a progression of the poet's ideas and style is impossible unless the reader already knows her earlier work.

In the preface to *Carrying Over: Poems from the Chinese, Urdu, Macedonian, Yiddish, and French African* (1988), Kizer traces the history of some of the volume's translations and poems and in the process emphasizes a wish for international understanding and cooperation. Frank Allen, in the *Library Journal* (November 1988), praised Kizer's "carefully crafted translations" and said that the poet's work in *Carrying Over* "broadens our Western self-preoccupation." The collection contains translations of poems by Tu Fu, Rachel Korn, Faiz Ahmed Faiz, Bogomil Gjuzel, Edouard Maunick, Shu Ting, and others. The only Kizer poem in the group is "Race Relations," a piece

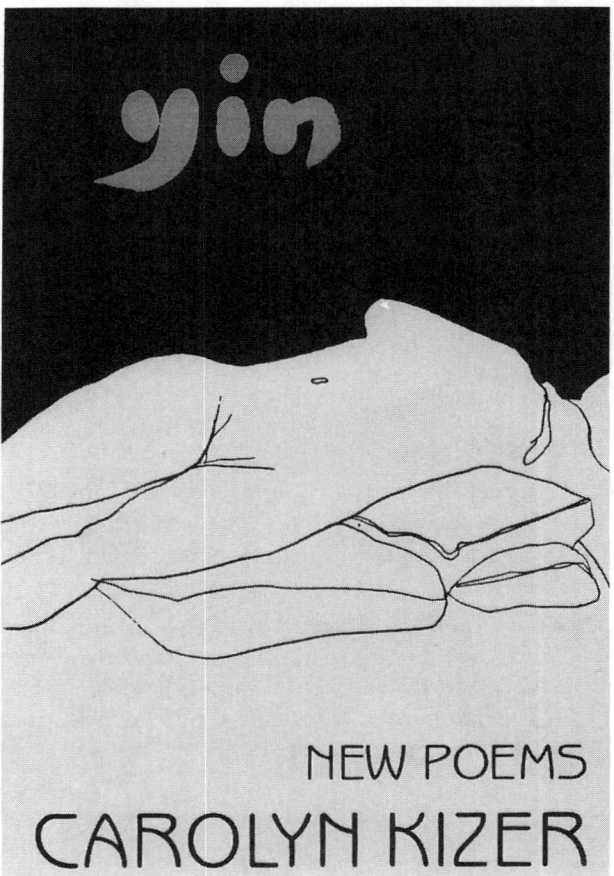

Cover for Kizer's fourth book, which includes poems inspired by her relationships with family and friends

written for her South African friend Dennis Brutus. Composed in triplets, the poem underscores Kizer's understanding of the chasm between experience and exposition. She confesses, "When you fled tyranny / face down in the street / signing stones with your blood," she was "Martyred in safety / I sighed for lost causes / You bled on You bled on."

Kizer's book *Proses: On Poems & Poets,* in addition to its collection of autobiographical pieces ("The Stories of My Life" and "So Big: An Essay on Size"), contains essays on various topics (for example, "The Poetics of Water: A Sermon," which Kizer delivered at the Cathedral of St. John the Divine, New York) and an assortment of reviews and other pieces of literary criticism. Kizer's subjects are as varied as Emily Dickinson, John Clare, Robinson Jeffers, and Sylvia Plath.

Carolyn Kizer is, finally, a writer to treasure. She has created poetry that will endure; at the same time she has excelled as an inspiring teacher, insightful critic, and valued friend of the arts. In the foreword to *An Answering Music: On the Poetry of Carolyn Kizer* (1990), Hayden Carruth affirms that Kizer "wears her laurels well. . . . She combines the role of great ladies of the past with that of a responsibly liberated woman of the present, and does it magnificently. In her home in California she is arbiter, impresario, author, friend, succorer, facilitator." Faced with the human inevitability of loss and destruction, Kizer, in both poetry and life, celebrates the joys of art, friendship, family, and good works. Undoubtedly, she has earned a secure niche in American letters and made lasting gifts to the world's store of truth and beauty.

Interviews:

Earl Ingersoll and Stan Sanvel Rubin, "'The Very Separateness of Things': A Conversation with Carolyn Kizer," *Webster Review*, 11 (Fall 1986): 87–101;

David Rigsbee and Steven Ford Brown, "Not Their History but Our Myth: An Interview with Carolyn Kizer," in *An Answering Music: On the Poetry of Carolyn Kizer,* edited by David Rigsbee (Boston: Ford-Brown, 1990), pp. 126–147.

Bibliography:

"Bibliography," in *An Answering Music: On the Poetry of Carolyn Kizer,* edited by David Rigsbee (Boston: Ford-Brown, 1990), pp. 219–230.

Biography:

Ronald H. Bayes, "Franklin Street Days: Carolyn Kizer in North Carolina, 1970–1974," *Pembroke Magazine* (1991): 131–136.

References:

Hayden Carruth, Foreword to *An Answering Music: On the Poetry of Carolyn Kizer,* edited by David Rigsbee (Boston: Ford-Brown, 1990);

Mark Jarman, "Generations and Contemporaries," *Hudson Review,* 38 (Summer 1985): 327–340;

John Montague, "Carolyn Kizer," in *Contemporary Poets* (Chicago & London: St. James Press, 1985), pp. 513–514;

David Rigsbee, ed., *An Answering Music: On the Poetry of Carolyn Kizer* (Boston: Ford-Brown, 1990);

Robin Skelton, Introduction to *Five Poets of the Pacific Northwest* (Seattle: University of Washington Press, 1964), pp. xv–xxv.

Robert Lowell
(1 March 1917 – 12 September 1977)

Ashley Brown
University of South Carolina

BOOKS: *Land of Unlikeness* (Cummington, Mass.: Cummington Press, 1944);
Lord Weary's Castle (New York: Harcourt, Brace, 1946);
Poems 1938-1949 (London: Faber & Faber, 1950);
The Mills of the Kavanaughs (New York: Harcourt, Brace, 1951);
Life Studies (New York: Farrar, Straus & Cudahy, 1959);
For the Union Dead (New York: Farrar, Straus & Giroux, 1964; London: Faber & Faber, 1965);
The Old Glory (New York: Farrar, Straus & Giroux, 1965; London: Faber & Faber, 1966; revised edition, New York: Farrar, Straus & Giroux, 1968);
Near the Ocean (New York: Farrar, Straus & Giroux, 1967; London: Faber & Faber, 1967);
The Voyage & Other Versions of poems by Baudelaire (London: Faber & Faber, 1968; New York: Farrar, Straus & Giroux, 1969);
Notebook 1967-68 (New York: Farrar, Straus & Giroux, 1969); revised and republished as *Notebook* (New York: Farrar, Straus & Giroux, 1970; London: Faber & Faber, 1970);
The Dolphin (New York: Farrar, Straus & Giroux, 1973; London: Faber & Faber, 1973);
For Lizzie and Harriet (New York: Farrar, Straus & Giroux, 1973; London: Faber & Faber, 1973);
History (New York: Farrar, Straus & Giroux, 1973; London: Faber & Faber, 1973);
Selected Poems (New York: Farrar, Straus & Giroux, 1976);
Day by Day (New York: Farrar, Straus & Giroux, 1977);
Collected Prose, edited by Robert Giroux (New York: Farrar Straus Giroux, 1987).

TRANSLATIONS: Jean Baptiste Racine, *Phaedra*, in *Phaedra and Figaro* (New York: Farrar, Straus & Giroux, 1961; London: Faber & Faber, 1963);

Robert Lowell

Imitations, (New York: Farrar, Straus & Giroux, 1961; London: Faber & Faber, 1962);
Aeschylus, *Prometheus Bound* (New York: Farrar, Straus & Giroux, 1969; London: Faber & Faber, 1970);
Aeschylus, *The Oresteia,* (New York: Farrar, Straus & Giroux, 1979).

For some readers and critics Robert Lowell stood at the center of his literary generation, and by the mid 1960s one admirer, Irvin Ehrenpreis, was referring to "The Age of Lowell." One can see why. Lowell was associated with nearly all the important American poets of the first half of this century, and long before the end of his life he seemed to be their heir. His appeal was not only literary. He involved himself to an unusual degree with the public events of his time, just as he brought his private life into public view by way of his poems. He was at once the poet of the American empire (sometimes resembling Virgil, sometimes Juvenal, sometimes Horace) and the alleged father of "confessional poetry." He wrote a great deal during his last decade, from 1967 to 1977, but so far few readers have assimilated the late work. His great poems, or at least the familiar ones, all come before 1967. In the near future there will doubtless be textual criticism of Lowell's poetry; few poets have revised so rapidly and extensively. A variorum Lowell may or may not make these things easier for readers.

Robert Traill Spence Lowell was born in Boston on 1 March 1917. His family, long distinguished in that city, included a number of literary persons: his Lowell grandfather (for whom he was named), James Russell Lowell, and Amy Lowell. For two years he attended St. Mark's School in Massachusetts (of which his grandfather had once been headmaster). On the faculty was Richard Eberhart, a young American poet, about whom Lowell said many years later, "There was someone there whom I admired who was engaged in writing poetry."

He went to Harvard, like all Lowells before him, but he left after two years. Although he was already acquainted with Robert Frost, he sought his instruction in poetry elsewhere. About this time, in 1937, he met the English novelist Ford Madox Ford, who was about to travel to Tennessee to visit Allen Tate and his wife, the novelist Caroline Gordon; Ford evidently invited the young man to accompany him. Thus, Lowell was brought into the presence of a distinguished southern poet who was to have a considerable influence on his work. He pitched a tent in the Tates' front yard and spent the summer there, writing poetry in an almost obsessive way. In the fall of that year he went to Kenyon College in Ohio to study with John Crowe Ransom (Tate's first master); he remained there until 1940, when he graduated summa cum laude with a degree in classics. Among the close friends he made were Randall Jarrell, the poet, and Peter Taylor, the short-story writer. At the same time, in 1940, he became a Roman Catholic and married a young novelist, Jean Stafford. (Stafford, who died in 1979, was the author of three novels and several collections of stories; she occasionally figures in Lowell's poetry.) To round out his education with the southern literati, Lowell studied for a year at Louisiana State University with Cleanth Brooks and Robert Penn Warren.

He worked for a time in New York City with the Catholic publishing house of Sheed and Ward, but he and Stafford spent 1942–1943 in the mountains of Tennessee, where they shared a house with the Tates. It was here that many of the poems in his first book were written. Then, in the middle of World War II, he was jailed as a conscientious objector to the Allied bombing of German civilians. (This interlude in his life is the basis of a poem called "Memories of West Street and Lepke" in *Life Studies,* published in 1959). In 1944 his first book, *Land of Unlikeness,* was published in a small edition by the Cummington Press in Massachusetts; it was introduced by Tate.

Tate's introduction is a brilliant account of Lowell's poetry at that time, and in two sentences he foretold the course of Lowell's career:

"On the one hand, the Christian symbolism is intellectualized and frequently given a savage satirical direction; it points to the disappearance of the Christian experience from the modern world, and stands, perhaps, for the poet's own effort to recover it. On the other hand, certain shorter poems, like 'A Suicidal Nightmare' and 'Death from Cancer,' are richer in immediate experience than the explicitly religious poems; they are more dramatic, the references being personal and historical and the symbolism less willed and explicit."

An example of the former kind of poetry that Tate mentions is the opening stanza of "On the Eve of the Feast of the Immaculate Conception":

Mother of God, whose burly love
Turns swords to plowshares, come, improve
   On the big wars
And make this holiday with Mars
Your Feast Day, while Bellona's bluff
Courage or call it what you please
   Plays blindman's bluff
   Through virtue's knees.

Here the ironic contrast between the Mother of God and the Roman god of war is boldly stated as a metaphysical conceit; the sarcasm is only too evident. This passage comes close to being the norm for *Land of Unlikeness.* The strength of feeling never relents. Both private and public aspects of Lowell's life at this time account for the tone of the poetry: his defiance of his parents in different ways, one of them being his conversion to Catholicism; and the wartime events that he saw as more complicated than most people would. (Early in the war, however, he attempted to enlist in the navy.) The terms of this kind of poetry are large, almost cosmic, and they sometimes lead to the excesses of metaphor that one associates with the decline of the metaphysical style in the seventeenth century. Most of these poems have never been reprinted, at least not in their original form.

The other kind of poetry that Tate describes is just beginning to emerge in this first book; it can be seen at its best in the sequence on Lowell's grandfather Arthur Winslow. The "personal and historical" references are very specific, and one can easily say in retrospect that Lowell's particularity is one of his strong points as a poet. This sequence was reprinted in his famous second book, *Lord Weary's Castle* (1946), where the particularity is even more evident. In the early version, for instance, he speaks of "The craft that netted a million dollars, late / Mining in California's golden bays / Then lost it all in Boston real estate." In the revised version in *Lord Weary's Castle* the poeticism of "Mining in California's golden bays" becomes "Hosing out gold in Colorado's waste," which is probably more accurate as family history and certainly more precise in physical details. The precision is there even when Lowell is imagining a situation. In "Exile's Return," the first poem in *Lord Weary's Castle,* he describes a scene that is evidently wartime Germany; it looks authentic. What the poet has in fact done is to lift details from Thomas Mann's story "Tonio Kröger" (1903) and rearrange them, heighten them, for maximum effect. Lowell very early became a master at making poetry out of prose; his eye for the precise detail is unusually keen.

At this stage in his career he resembled at least one classic American writer, Nathaniel Hawthorne, in that he seemed to do his best work when he turned his attention to the history of his own people. And for a young writer in the mid twentieth century, there was an accumulated literary history such as did not exist when Hawthorne started. Hawthorne himself is part of this history; so is Jonathan Edwards, and two of the finest poems in *Lord Weary's Castle* ("Mr. Edwards and the Spider" and "After the Surprising Conversions") are transmuted from well-known passages in Edwards's highly charged prose. One intensely satiric poem, the sonnet called "Concord," was carried over from *Land of Unlikeness* to *Lord Weary's Castle,* but with the most drastic changes. In the earlier version Ralph Waldo Emerson is the special focus of criticism: "Concord, where the Emersons / Washed out the blood-clots on my Master's robe." Two years later this passage becomes, less severely,

> Concord where Thoreau
> Named all the birds without a gun to probe
> Through darkness to the painted man and bow:

Lowell at about five years old

> The death-dance of King Philip and his scream
> Whose echo girdled this imperfect globe.

Here Emerson is not directly named; nor does his disciple Thoreau come in for the same criticism. But the last line, which is carried over almost intact, is itself an "echo" of Emerson's most famous line in "The Concord Hymn."

It was Herman Melville, however, who most deeply engaged Lowell's imagination in those days. "The Quaker Graveyard in Nantucket," a magnificent baroque elegy in seven parts, would have been impossible without *Moby-Dick* (1851). (Lowell's poem was originally called "To Herman Melville.") It no doubt owes much to the John Milton of "Lycidas" (1637), and indeed this is appropriate, because Lowell's poem is about the death of his cousin Warren Winslow, a naval officer who drowned in the North Atlantic during the war. But the rhetoric is closer to Melville's, as in this passage:

> The winds' wings beat upon the stones,
> Cousin, and scream for you and the claws rush
> At the sea's throat and wring it in the slush
> Of this old Quaker graveyard where the bones
> Cry out in the long night for the hurt beast
> Bobbing by Ahab's whaleboats in the East.

Melville remained an important point of reference in Lowell's mind if not in his poetry, which became less baroque. Near the end of his life he put *Moby-Dick* beside the great epics of earlier literary periods. In an unfinished essay called "New England and Further," part of which was published in 1979, he wrote:

> Often magnificent rhythms and a larger vocabulary make it equal to the great metrical poems.... It is our best book. It tells us not to break our necks on a brickwall. Yet what sticks in the mind is the Homeric prowess of the extinct whaleman, gone before his prey.

The kinesthetic quality of the passage quoted from "The Quaker Graveyard" is fairly typical of early Lowell. Here and there, as in the sixth section of the poem ("Our Lady of Walsingham"), a quiet moment occurs; but generally the rhetoric surges in response to a deeply felt subject.

Lord Weary's Castle received the Pulitzer Prize in 1947. By that time it had been reviewed at length; one review in particular, Randall Jarrell's in the *Nation* (18 January 1947), became famous in its own right as an example of a poet-critic's generously recognizing the arrival of a young master. Indeed his description of Lowell's central theme – the "wintry, Calvinist, capitalist world" against which the poet set himself – was the basis for much subsequent commentary.

Lowell's poetry continued to move in the direction that Allen Tate had predicted it would: the poems became more dramatic as they depended less on the "willed" Christian symbolism. The characteristic poems of the next few years were in fact dramatic monologues. Seven of them made up *The Mills of the Kavanaughs* (1951). The title poem, which is Lowell's longest single work (608 lines, arranged in 38 rhymed stanzas), has never been well received even by his most sympathetic critics. Its narrative does not move; its large symbolism is forced. The heroine, a widow lamenting her husband, identifies herself with Persephone, whose statue stands on the grounds of her husband's house in Maine. But there is no real action. On the other hand, two of the shorter poems, running to three or four pages, are highly successful: "Mother Marie Therese" and "Falling Asleep over the Aeneid." They are written in heroic couplets and may owe something to Robert Browning and Yvor Winters, but they sound like Lowell, with their vigorous runover lines and emphatic rhythms. "Mother Marie Therese" is also a poem of lament, in this case for a nun who drowned in 1912 (Lowell may have been thinking of Gerard Manley Hopkins's 1875 poem "The Wreck of the Deutschland"); the speaker is herself a Canadian nun "stationed in New Brunswick." The poem is wonderfully conversational at times, but it gathers up to a pitch of feeling at the end, when the poet somewhat boldly rounds it out with three couplets using the same rhyme.

"Falling Asleep over the Aeneid" is a masterpiece; Robert Fitzgerald called it "that marvel of dream work as historical imagination." The speaker, an old man in Concord, reading Virgil on Sunday morning, is immediately brought before us:

> The sun is blue and scarlet on my page,
> And *yuck-a, yuck-a, yuck-a, yuck-a,* rage
> The yellowhammers mating. Yellow fire
> Blankets the captives dancing on their pyre,
> And the scorched lictor screams and drops his rod.

Lowell had perhaps learned something about composition from Wallace Stevens's "Sunday Morning," where bird and fruit and sun, introduced in the opening lines, are carried through the poem in a series of transfigurations. Here the flame colors of the bright morning sun are intensified by the yellow wings of the birds that the old man only *hears,* but then the colors are absorbed by the fire that leaps from the page, as it were, with the lictor's scream. (The funeral of the heroic young Pallas in the *Aeneid* is the episode that takes over the old man's dream.) Fire and bird move through the poem in many ingenious ways, and they prepare for the end, where the dreamer recalls his uncle, a young Union officer in the Civil War, "Blue-capped and birdlike," at *his* funeral. Thus, two eras of history are brought into relationship as we return to the present.

When *The Mills of the Kavanaughs* was published in 1951, Lowell was living in Italy. His marriage with Jean Stafford had ended in 1948; in the following year he married Elizabeth Hardwick, a novelist from Kentucky. The Lowells lived at different places in Europe, but Italy seemed to have the greatest attraction for him, as it did for so many other American poets during the 1950s. (Eventually, after his death, some of his Italian friends, in-

Randall Jarrell, Lowell, and Peter Taylor in Greensboro, North Carolina, 1948

cluding his translator, brought out an impressive volume of tributes.) He was also becoming known in London, and as early as 1950 Faber and Faber, the publishing house of which T. S. Eliot was a director, began to publish his work. In time his reputation in London was possibly even larger than it was in the United States. But after the death of his mother in 1954 in Italy, he returned with his wife to live in a house on Marlborough Street in Boston. He had left the Catholic Church. Perhaps for the moment there was nothing urgent to write about, and the few poems that he composed during this period were usually elegies for literary friends such as Ford Madox Ford and the philosopher George Santayana, whom he had visited in Rome.

The great change that came over his poetry in the late 1950s is now well known. In 1957 he was writing his autobiography in prose, and this became the actual source for a group of highly personal poems that, to some early readers, were only chopped-up prose fragments. The first of the new poems, and probably the most famous, is "Skunk Hour," which was started in August 1957. It is modeled on Elizabeth Bishop's "The Armadillo," according to Lowell's own account. (Bishop's poem, which is set in Brazil, is dedicated to Lowell.) Her versification and open texture suggested an "easier" mode of poetry. In each case a single animal emerges at the end of the poem in a kind of affirmation. According to Lowell (whose account of the poem's composition is found in *The Contemporary Poet as Artist and Critic,* edited by Anthony Ostroff, 1964), he worked backward. The poem became intensely personal in a way that Bishop's is not. Finally he wrote the four opening stanzas (half the poem) as a social setting: a declining Maine seaside town. The characteristic pronoun here is *our,* but the sense of community is precarious. The heart of the poem is the pair of stanzas that comes between the social setting and the image of the skunk and her kittens who take over the deserted town at the end:

One dark night
my Tudor Ford climbed the hill's skull;
I watched for love-cars. Lights turned down,
they lay together hull to hull,
where the graveyard shelves on the town....

> My mind's not right.
>
> A car radio bleats,
> "Love, O careless Love. . . . " I hear
> my ill-spirit sob in each blood cell,
> as if my hand were at its throat. . . .
> I myself am hell;
> nobody's here – [.]

This is quintessential Lowell. Intensely private though it is (to a certain kind of reader it comes close to being a psychiatric case history), it has its public dimension. As some readers have noted, it is a version of the "dark night of the soul" of San Juan de la Cruz, although without the sense of a spiritual progression that the great Spanish mystic followed. "I myself am hell" is borrowed from *Paradise Lost:* a line spoken by Satan, spying on Adam and Eve. Between these two noble literary sources Lowell drops a phrase from a rather banal song that is being half-heard by millions, but it attains a certain pathos in this context. The speaker verges, perhaps, on a self-pity that could be disastrous for the poem ("the hill's skull" suggests Golgotha), but there is also a laconic detachment that comes out in "I myself am hell," which stands in a line by itself, like "My mind's not right" in the preceding stanza. Most readers cannot avoid knowing what Lowell brings into so many of these poems: his recurrent mental illness. Robert Fitzgerald, who knew him well, has stated this firmly but sympathetically: "behind those poems and henceforth all his work is his breakdown of 1949 and the necessity he now felt of governing his greatness with his illness in mind. Manic attacks now and again would put him in the hospital, overborne by the fever that one had felt to be just beyond some of his poems from the beginning."

"Skunk Hour," although first in order of composition, comes last in the group of fifteen poems that Lowell calls "Life Studies." It is led up to gradually by poems about an uncle, his Great-aunt Sarah, his Winslow grandparents, his father, and his mother; the family relationships intensify painfully. There are beautiful idyllic moments; thinking of his grandparents, he says:

> Then the dry road dust rises to whiten
> the fatigued elm leaves –
> the nineteenth century, tired of children, is gone.
> They've all gone into a world of light; the farm's my own.

But as the reader moves closer to the present, the tone is clipped and sardonic:

> Father's death was abrupt and unprotesting.
> His vision was still twenty-twenty.
> After a morning of anxious, repetitive smiling,
> his last words to Mother were:
> "I feel awful."

The poet's father is the central figure in a prose memoir, "91 Revere Street," which forms the centerpiece of *Life Studies*. Rather curiously, it does not immediately precede the group of family poems. Intervening are four poems on literary figures (Ford, Santayana, the poets Delmore Schwartz and Hart Crane) whom Lowell admired. Ninety-one Revere Street was the address of the house where the Lowells lived for a few years in the 1920s; here the psychological battle between husband and wife was fought out. Lowell's mother, neurotic but obviously strong in her own way, won; she forced her husband to retire from the navy. He moved into the business world with a marked lack of success, and the poem called "Commander Lowell" presents his declining years without comment; he is almost a figure of comedy. He died in 1950. But while he was alive, Lowell wrote and published in *Lord Weary's Castle* a poem called "Rebellion," which suggests a more painful relationship between father and son. Once, during his Harvard days, he knocked his father down in a quarrel over a girl. Lowell returns to this incident in a short sequence of poems, "Charles River," in *Notebook 1967–68* (1969).

His mother is never really described, although her presence is strongly felt in various places. She is, finally, an object; one can only speculate on the feelings that lie behind this passage:

> When I embarked from Italy with my Mother's body,
> the whole shoreline of the *Golfo di Genova*
> was breaking into fiery flower.
> The crazy yellow and azure sea-sleds
> blasting like jack-hammers across
> the *spumante*-bubbling wake of our liner,
> recalled the clashing colors of my Ford.
> Mother travelled first-class in the hold;
> her *Risorgimento* black and gold casket
> was like Napoleon's at the *Invalides*.

The "Life Studies" sequence composes a kind of fragmentary novel in verse, not unlike James Joyce's *A Portrait of the Artist as a Young Man* (1916). Since Lowell was an influential teacher by now (he held temporary posts at several American universities), a number of younger writers who were at one time his students – such as W. D. Snodgrass and Anne Sexton – were associated with him as "confessional" poets. Yet this association appears to have been somewhat exaggerated. Snodgrass's *Heart's*

Needle (1959) was written before *Life Studies*, and the anguished revelations of Anne Sexton and Sylvia Plath are poetry of a rather different kind from Lowell's. Their work, for the most part, lacks his wit and detachment and public dimension.

Life Studies has been considered a landmark in contemporary poetry by many critics; its importance was almost immediately recognized, and in 1960 Lowell received the National Book Award. About this time he moved to New York City with his wife, Elizabeth Hardwick, and daughter Harriet. Thus began a decade during which he was involved in public affairs, or at least public artistic projects, including the theater. Early in the decade he read a new poem, "For the Union Dead," at an arts festival on the Boston Common. It is a more accessible piece than most of *Life Studies* or certainly *Lord Weary's Castle*, even for those readers (or auditors) uninstructed in Boston history and the aftermath of the Civil War. The open texture of *Life Studies* has now been extended to the modern scene and the history that has led up to it. Although "For the Union Dead" starts out as a private meditation, it soon moves to Boston Common itself:

> Parking spaces luxuriate like civic
> sandpiles in the heart of Boston.
> A girdle of orange, Puritan-pumpkin colored girders
> braces the tingling Statehouse,
> shaking over the excavations, as it faces Colonel Shaw
> and his bell-cheeked Negro infantry
> On St. Gaudens' shaking Civil War relief,
> propped by a plank splint against the garage's earthquake.

This passage takes the reader to the central historic situation. Colonel Robert Shaw of Boston, who commanded a regiment of black troops, was killed in July 1863 in an assault on Fort Wagner, South Carolina. Although Lowell presents him as an example of idealism and youthful beauty, he was scorned by his father; he was buried in a ditch, where his "body was thrown / and lost with his 'niggers.'" Lowell juxtaposes the events of three generations: Shaw's actual death and anonymous burial; the dedication of the memorial to him and the Union dead in the 1890s; and the present moment, when ideals, it is suggested, have eroded along with much else:

> The ditch is nearer.
> There are no statues for the last war here;
> on Boyleston Street, a commercial photograph
> shows Hiroshima boiling
> over a Mosler Safe, the "Rock of Ages"
> that survived the blast. Space is nearer.
> When I crouch to my television set,
> the drained faces of Negro school-children rise
> like balloons.

Future readers may require footnotes to identify a few of the details from the 1960s – the civil rights struggle, the advertising slogans of the period, even the design of American automobiles. Lowell's sense of particularity is brilliant; the details could hardly be better chosen. But here the reader is close to the pop art of the 1960s, with its jaunty arrangements of familiar objects, its air of being both inside and detached from popular culture. The poem, however, is beautifully composed. The motif of the fish, introduced in the opening stanzas with the old South Boston Aquarium, is sarcastically alluded to midway ("Their monument sticks like a fishbone / in the city's throat"), and then it rounds out the poem:

> The Aquarium is gone. Everywhere,
> giant finned cars nose forward like fish;
> a savage servility
> slides by on grease.

"For the Union Dead" gave the title to Lowell's next book, which came out in 1964. This collection is miscellaneous in character. It includes personal vignettes, poems derived from the prose of friends such as Elizabeth Bishop and Mary McCarthy, recastings of older poems by Lowell himself, even commissioned poems. "Hawthorne" was written to commemorate the centenary edition of that author's works published by the Ohio State University Press. This and its companion piece, "Jonathan Edwards in Western Massachusetts," as Lowell says, are based on prose passages by their subjects. These poems are highly interesting to Lowell's public, of course, because they return to New England writers who have always meant a great deal to him. His attitude toward them is now more relaxed, as is the mode of the verse.

One sequence in this collection is almost entirely concerned with the public world: "July in Washington," "Buenos Aires," and "Dropping South: Brazil." They suggest a quasi-political attitude that would soon become very explicit in Lowell's work. He chose the first one, in fact, when he was asked to submit a favorite poem from his own work, along with a work he admired from the past, to an anthology, *Preferences* (1974), edited by Richard Howard. Lowell paired "July in Washington" with Melville's "The House-Top," a poem about the draft riots in New York City in July 1863. Howard in a brief commentary mentions the "seditious tropical summers of America's cities" that are

Robert Lowell
239 Marlboro St.
Boston 16, Mass.

September 25, 1956

Dear Mr. Faulkner:

Of course I would like to join any organization that you are giving your time to organizing.

I have few ideas on how writers might help in a program for people-to-people partnership. I do, however, flinch at the President's seeming suggestion that writers might go on missions for the promotion of American ideology. I think that most of us are too dumb and too serious to do anything of the sort. Life is too short; we are too deeply saturated with our country. If we can only meet writers from other countries naturally, humorously, with curiosity and humility; then they might see that we were human. Then the crusade for people to people partnership would take care of itself.

Our world position, so different from most other countries, fills Europeans with fear and envy as we go bouncing among them and shine like a white-washed lighthouse on them in all our optimism, wealth and shining ideology. We have such an air of hurry, power and bouyant vacancy. They find us exhilarating and rather monstrous.

As a writer, I am much interested in making friends with writers from other countries. As a teacher, I am interested in teaching the writing of other countries. I particularly want to know what I am doing

Lowell's response to William Faulkner, then the chairman of President Dwight D. Eisenhower's People-to-People Program, a project designed to improve international understanding (William Faulkner Collections [#7258-1], University of Virginia Library; by permission of the Estate of Robert Lowell)

when I try to read French, German, and Italian poetry. How can we read at all if we are unable to learn from other literatures, see that they are different from ours?

I am all for exchange scholarships, professorships, etc., quiet meetings without oratory, and above all personal friendships.

Of course you can see deeper into us and our country than I can. Long ago, I wrote you a long fan letter. I never dared to mail it, but I think I said (as the Democrats say of themselves) that you were our writer with style and a heart. You are. What a joy that this business has allowed me to say so.

Yours sincerely,

Robert Lowell

often a background to political stresses if not disasters. Lowell's poem this time is built up on the motif of the circle. The actual plan of Washington makes this credible, but the opening couplet involves more: "The stiff spokes of this wheel / touch the sore spots of the earth." Then the poem on Buenos Aires has its focus in a graveyard of Republican martyrs during a chilly Argentine winter. In the poem on Brazil, although it is written mainly from the point of view of a tourist on Copacabana Beach, Lowell reflects that "inland, people starved, and struck, and died – / unhappy Americas, ah *tristes tropiques!*"

In 1969 Lowell remarked in an interview with the Caribbean novelist V. S. Naipaul, "America with a capital A I find a very hard thing to realise. It's beyond any country, it's an empire. I feel very bitter about it, but pious, and baffled by it." This summary of his feelings has a certain ambiguity about it. He had long been concerned about the outcome of American history, like an earlier Bostonian, Henry Adams, and his feeling that it was being betrayed ran very strong. In 1965 he was invited by President Lyndon Johnson, as one of a small group of distinguished American artists, to attend the White House Festival of the Arts. He turned down the invitation in a public statement of protest against American involvement in the Vietnam War: "We are in danger of imperceptibly becoming an explosive and suddenly chauvinistic nation, and we may even be drifting on our way to the last nuclear ruin." Then in October 1967, he participated with several other literary figures in the famous "march" on the Pentagon, which ended with mixed results but was highly publicized. Lowell himself wrote two poems about it for *Notebook,* but a more famous and extensive account of this incident, in which he has a prominent part, is found in Norman Mailer's *The Armies of the Night* (1968).

Lowell was becoming known in other ways. Like many other modern poets, he was attracted by the theater. In 1961 he was asked to translate Jean Baptiste Racine's *Phèdre;* he did this in heroic couplets as being the nearest English equivalent to the French alexandrines. Critical opinion has been mixed about the result, and perhaps it should be regarded as an "imitation" of Racine. (In 1961, the same year as his *Phaedra,* Lowell brought out his controversial book called *Imitations,* which consists of versions of poetry from several languages.) This was the first of the theater projects that Lowell started, and *Phaedra* was seriously considered by two Broadway producers in 1962. Another venture was his collaboration with Leonard Bernstein on a symphony, *Kaddish;* before it broke off, Lowell wrote three poems, remarkably formal, even stately, for a musical setting. (They were published for the first time in *Ploughshares* in 1979.) In 1965 Lincoln Center commissioned an acting version of Aeschylus's *Oresteia;* this was mostly finished but never produced; eventually it was published in 1979. Lowell's prose version of Aeschylus's *Prometheus Bound* was actually produced at the Yale Drama School with an international cast two years before it was published in 1969.

By far his most successful work for the theater was the group of short plays called *The Old Glory* (1965). There was at least one good reason for this success. Originally intending to write an opera libretto based on Melville's "Benito Cereno" (1856), Lowell instead quickly wrote a play. Then he added two plays from stories by Hawthorne, "Endicott and the Red Cross" and "My Kinsman, Major Molineux." Taken together (with "Benito Cereno" presented last), they compose a trilogy on the emergence of the United States out of its colonial past. According to Lowell in a late interview in London, the title, *The Old Glory,* has two meanings: "it refers both to the flag and also to the glory with which the Republic of America started." Thus, his lifelong engagement with Hawthorne and Melville – his major American sources, one might say – paid off handsomely again. The original production of the play in New York, a considerable critical success, received a number of prizes.

In 1967 Lowell brought out *Near the Ocean,* a handsome book (illustrated by the famous Australian painter Sidney Nolan) that contained his most formal poetry since *The Mills of the Kavanaughs.* The earlier book was mostly written in heroic couplets; part of this one is written in tetrameter couplets, the poetic form of which the seventeenth-century poet Andrew Marvell was the master in English verse. Lowell's versification is not as strict as Marvell's, but almost playfully he runs his couplets through sentences of eight lines, the length of a stanza. The tone is playful, too, as the reader moves in and out of a charming domestic scene (a house Lowell inherited in Maine) in the manner of Marvell's "Upon Appleton House," which may well be the general model; he uses the same eight-line stanza. But Lowell, like his great predecessor, touches on some serious issues. The opening lines of the first poem in the sequence, "Waking Early Sunday Morning" (now rather famous among Lowell's readers), make that clear:

> O to break loose, like the chinook
> salmon, jumping and falling back,
> nosing up to the impossible
> stone and bone-crushing waterfall —
> raw-jawed, weak-fleshed there, stopped by ten
> steps of the roaring ladder, and then
> to clear the top on the last try,
> alive enough to spawn and die.

It is part of the attractiveness of these poems that they can advance so quickly to matters of church and state and society and absorb them into their easy rhythms:

> O to break loose. All life's grandeur
> is something with a girl in summer . . .
> elated as the President
> girdled by his establishment
> this Sunday morning, free to chaff
> his own thoughts with his bear-cuffed staff,
> swimming nude, unbuttoned, sick
> of his ghost-written rhetoric!

The President is Lyndon Johnson; the occasion is the summer of 1965. Lowell has recently turned down the invitation to the White House Festival of the Arts, and this witty passage does indeed stand in contrast to the now-forgotten speeches by the official hack writers. But Lowell can be forceful about some things; in Central Park, almost at the door of his apartment house, he finds fear and poverty lurking amid the flowering shrubs; public sexual activity in the park is perhaps a correlative to these grimmer aspects of the contemporary scene. Lowell had a divided mind about the five poems in this group. While he retained the poems in this "Near the Ocean" group in his *Selected Poems* (1976), he cut four of them down, so drastically in some cases that they hardly resemble the originals. Such was his restless way with his own work.

The second half of *Near the Ocean* consists of translations, or "imitations" — the precise term is not easy to establish here. The poets whom he uses are mainly Horace, Juvenal, and Dante (*Inferno*, 15). Of these distinguished works, much the longest is Juvenal's Satire 10, "The Vanity of Human Wishes," occupying forty of Lowell's pages. Since we already have Samuel Johnson's famous "imitation" of this poem, we might remark that Lowell's version is closer to Juvenal than Johnson's. Lowell's is done in blank verse. (Juvenal used no rhymes.) And the effect of the Lowell version is to suggest, somehow, a comparison between Juvenal's Rome and the American empire of the 1960s. Although the tone of this great work is truly Juvenalian in Lowell's hands, it has a thematic connection with the "Horatian" poems in the first part of the book. (Marvell, the general model there, is perhaps the most Horatian of English poets.)

No sooner had *Near the Ocean* been published and *Prometheus Bound* been produced at Yale in 1967 (the latter work, the most extravagant thing that Lowell ever wrote, is also an indirect commentary on the 1960s), than Lowell started on a large venture of another sort. This is *Notebook 1967–68*. Lowell's friend John Berryman had already begun his ambitious and much-praised *Dream Songs* (the first part, published in 1964, won the Pulitzer Prize), and now the small audience for contemporary poetry watched two important writers quickly building up large sequences of short poems. (They were both published by Farrar, Straus, and Giroux.) In each case the overall form was the diary. Berryman used an original six-line rhymed stanza throughout; Lowell committed himself to the unrhymed sonnet. The resemblances and differences between the two are fascinating to study. Berryman, who killed himself early in 1972, probably had no intention of rearranging the plan of his work, but Lowell's took some unexpected turns shortly.

Notebook 1967–68 is a kind of diary, but the form is cyclic: Lowell prints a list of dates for 1967–1968 at the back of the book to remind the reader of the year's disasters, beginning and ending with the Vietnam War and including the assassinations and major political upheavals. Even though public events seem to get out of control and affect everyone's destiny, there are moments of reminiscence and private satisfaction (Jarrell and Taylor at college in the 1930s, or Elizabeth Schwarzkopf singing in New York). There are also many characters out of history, European and American. Although the poems are arranged in groups, some of them are not closely connected: for instance "Names," which merely brings together the figures of Sir Thomas More, Napoleon, and others. As Lowell himself says, the pattern is "jagged" and created largely by association. At the end a small group dedicated to his wife, Elizabeth Hardwick, is somewhat inconclusive. In the following year Lowell issued a revised and expanded edition. Some of the material here is more private than ever; it is even drawn from people's letters and conversations.

In 1973 Lowell brought out three books simultaneously: *For Lizzie and Harriet, The Dolphin,* and *History,* all written in unrhymed sonnets. In these volumes he has broken up the expanding *Notebook* project (undated in the revised, 1970 edition) into public and private sectors. *For Lizzie and Harriet* deals with Elizabeth Hardwick and their daughter; many

Title page for the longest of three books Lowell published in 1973, in which literary and historic figures are the subjects of chronologically arranged sonnets

of the poems were scattered through the two editions of *Notebook*. The poems in *The Dolphin,* however, are mostly new; they are concerned with his third wife, Lady Caroline Blackwood, whom he married in 1972, their son Sheridan, and his life in England. There are obvious personal reasons, then, for separating the phases of a complicated existence.

As for *History,* the longest of the three books, it contains most of the "public" poems from *Notebook,* sometimes reworked, with about eighty new poems. Many of these are about prominent literary figures, from Juvenal to Berryman, but there are a fair number about men of power as different as Maximilien-François Robespierre, Che Guevara, and Martin Luther King Jr. The arrangement is now linear or chronological, and almost every period of Western history is touched on. Lowell was always deeply interested in history; he knew the great historians better than most people. And so here is his personal vision of history, fragmentary though it appears. History is, in its way, a kind of epic; and Lowell would probably have agreed with Ezra Pound that an epic is simply a poem about history. It is Pound, finally, whom Lowell most resembles among the modern poets. Lowell met him as early as 1948, when he was the consultant in poetry at the Library of Congress. (Pound was kept in St. Elizabeth's Hospital in Washington from 1945 to 1958.) He admired Pound, often saw him in Washington and sometimes later in Italy, and kept his portrait above the desk in his study. By the time of *History* the scale and variety of Lowell's work had begun to resemble Pound's: the extensive translations and "imitations," the restless quest for new models, the ventures into personal epic. There is also an unevenness that is undeniable, in *History* as in the *Cantos* (1925–1960), and in Lowell's case the commitment to the unrhymed sonnet may have been a limitation of a kind. Everything has to be put in the same mold. The *Cantos,* at least the early ones, often have a sustained yet varied rhythmic interest that Lowell hardly approaches. Had he lived, Lowell would probably have done something more with *History,* which contains much of a lifetime's experience.

There are further rearrangements in the *Selected Poems* of 1976, a kind of interim volume. In a sense Lowell agreed with the general estimate of his work by including almost everything from *Life Studies* except the prose centerpiece. Like Pound's *Hugh Selwyn Mauberley* (1920), it seems to be something that all readers endorse. At this point Lowell was writing the poems that would make up his last book – the last book published in his lifetime, *Day by Day,* which appeared shortly before his death in September 1977. These late poems continue what he called his "verse autobiography," but often in a somewhat muted way. He no longer uses the elaborate narrative structure of some of the earlier poetry, or even the unrhymed sonnets of the *Notebook* period. His new marriage brings some happiness, especially with his infant son, but there are tensions and breakdowns, usually brought into his poems in this hesitant and oblique manner:

> To each the rotting natural to his age.
> Dividing the minute we cannot prolong,
> I stand swaying at the end of the party
> a half-filled glass in each hand –
> I too swayed
> by the hard infatuate wind of love
> they cannot hear.

This book unexpectedly marked the end of Lowell's career. He died in a taxi that was taking him from New York City's Kennedy Airport to the apartment on West Sixty-Seventh Street where he intended to join Elizabeth Hardwick. His marriage to Lady Caroline Blackwood had evidently failed and he wanted to resume his life in New York. His reputation in England had been extraordinary. At one point he was seriously considered for the position of professor of poetry at Oxford, an elected post that had recently been held by W. H. Auden, among others. But he was never an expatriate in the manner of Henry James and Eliot. His cultural base was in America and especially Boston.

One is reminded of his American antecedents and connections by his major posthumous work, the *Collected Prose*, edited in 1987 by the distinguished publisher Robert Giroux, who had been Lowell's editor for many years. This book, coming out a decade after the poet's death, traces his associations with many great poets of the past, beginning with Homer. An essay on the *Iliad*, written when he was only eighteen, already demonstrates his admiration for epic form and magnitude. It may be that his own tendency toward epic, which was apparent in his poetry long before *History*, was based on his early response to Homer. Few poets in the last part of the twentieth century have had this sense of grandeur; most would probably agree with Lowell's friend Elizabeth Bishop, who once said, "I'm not interested in big-scale work as such. Something needn't be large to be good."

The first section of the *Collected Prose* consists of eighteen brief essays, more tributes than formal criticism, concerned with the literary figures who had some part in Lowell's career. They include Ford, Ransom, Williams, Eliot, Tate, Warren, Bishop, and Jarrell. (Others, such as Frost, Stevens, and Auden, are simply the subject of Lowell's admiration.) It is possible to say that Lowell's poetry would never have evolved in the way that it did without the examples of these figures, some of whom became close friends. The tributes that Giroux assembled are thus part of a literary biography. Lowell picked up from Jarrell the practice of listing his favorite (presumably the finest) poems by the writers whose work he was reviewing. This section of the *Collected Prose* is thus an index to Lowell's taste, which was remarkable. Most of his judgments stand very well a generation later.

The heart of the *Collected Prose* is a group of essays, two of them unfinished, which take up Lowell's profound interest in the literary tradition of New England. The longest, "New England and Further," is a survey of authors from Cotton Mather to Frost, Stevens, and Eliot. According to Giroux's editorial note, it was written in Maine during two periods, the late 1960s and the final months of Lowell's life in 1977. What is remarkable is that he wrote without access to his library. One assumes that he had lived with some of these writers so long that he could almost quote their works at times. This is not the product of a specialist in American literature, but a poet's personal reactions to and reminiscences of a literary tradition. There is nothing systematic about it, and many of Lowell's formulations are hardly more than brilliant epigrams. For example, of Emily Dickinson: "Her divine waywardness, whose success is impossible to approve or condemn, separates her from the perfection of Marianne Moore and Elizabeth Bishop." We have this bit of reminiscence by way of introduction to Frost: "A lifetime ago, a morality ago, my mother warned me off the moderns, Eliot and Tate, and, as a curative, misquoted Robert Frost, thought to be understandable to everyone, including herself, to be healthy, wise, and no nihilist to the middle class. My personal and critical love of Frost survived this recommendation of everything I hated." And then he describes Frost as "*the* American formalist"; along the way he mentions Paul Valéry, William Butler Yeats, Sir Walter Raleigh, Ben Jonson, and other poets to make his point. Reading Lowell's criticism is a liberating experience.

In "New England and Further" Melville's poetry rates only a page or so of commentary. But the following essay, "Epics," takes up *Moby-Dick:* "It's our epic, a New England epic." Lowell in fact puts Melville in a line of descent from Homer, Virgil, Dante, and Milton. Clearly Melville meant more to him than any other American writer, and from time to time, as in "The Quaker Graveyard in Nantucket" and the play that he made from "Benito Cereno," Melville was actually his model. What he seemed to admire most was the "magnificent rhythms" and large vocabulary of *Moby-Dick*. Giroux thinks that this essay was intended as the conclusion to "New England and Further." At any rate, it almost seemed to circle back to Lowell's youthful essay on the *Iliad*, which in retrospect points toward so much in his career.

The remaining section of the *Collected Prose* consists of Lowell's commentaries on his own work. In this age of the tape recorder and the published interview, the two longest pieces here are the interviews with Frederick Seidel (Boston, 1961) and Ian Hamilton (England, 1971). Seidel and Hamilton were unusually well informed about Lowell and knew how to ask the right questions. The resulting conversation pieces remain valuable for the interested student. Hamilton, a British poet and critic, later wrote the first biography of Lowell, published in 1982, only five years after the poet's death. Lowell's heirs and many friends for the most part cooperated with the biographer; the result was a highly detailed and candid account of a life that was often painful to those close to the poet. His recurrent mental illness, already apparent by the late 1940s, was self-destructive at times. But an outsider can only marvel at Lowell's strength of mind that pulled him out of his crises and allowed him to function so effectively as a creative personality.

Lowell's reputation, almost twenty years after his death, is still high, but two generations of younger poets are more likely to look to Elizabeth Bishop as the finest poet of a group now mostly departed. Some of Lowell's work, such as the translations from Aeschylus, "The Mills of the Kavanaughs," and the hundreds of pages of *History* poems, is never mentioned now. But the best poems, early and late, are still impressive. A new biography, *Lost Puritan: A Life of Robert Lowell,* by Paul Mariani (1994), has drawn attention to Lowell's work again, and Frank Bidart, Lowell's literary executor, plans to publish his edition of the *Collected Poems,* something that is needed for a full assessment of a remarkable career.

Interview:

"Et in America ego – The American poet Robert Lowell talks to the novelist V. S. Naipaul about art, power, and the dramatisation of the self," *Listener,* 4 (September 1969): 302–304.

Biographies:

Ian Hamilton, *Robert Lowell: A Biography* (New York: Random house, 1982);

Paul Mariani, *Lost Puritan: A Life of Robert Lowell* (New York: Norton, 1994).

References:

Rolando Anzilotti, ed., *Robert Lowell: A Tribute* (Pisa: Nistri-Lischi Editori, 1979);

Steven Gould Axelrod, *Robert Lowell: Life and Art* (Princeton, N.J.: Princeton University Press, 1978);

Philip Cooper, *The Autobiographical Myth of Robert Lowell* (Chapel Hill: University of North Carolina Press, 1970);

John Crick, *Robert Lowell* (Edinburgh: Oliver & Boyd, 1974);

Richard J. Fein, *Robert Lowell* (New York: Twayne, 1970);

Robert Fitzgerald, "The Things of the Eye," *Poetry,* 132 (May 1978): 107–111;

Michael Lond and Robert Boyers, eds., *Robert Lowell: A Portrait of the Artist in His Time* (New York: David Lewis, 1970);

Norman Mailer, *The Armies of the Night* (New York: New American Library, 1968);

Jerome Mazzaro, *The Poetic Themes of Robert Lowell* (Ann Arbor: University of Michigan Press, 1965);

Thomas Parkinson, ed., *Robert Lowell: A Collection of Critical Essays* (Englewood Cliffs, N.J.: Prentice-Hall, 1968);

Marjorie Perloff, *The Poetic Art of Robert Lowell* (Ithaca, N.Y.: Cornell University Press, 1973);

Salmagundi, special issue on Lowell, 1, no. 4 (1966–1967);

Peter Taylor, "Robert Trail [sic] Spence Lowell," *Ploughshares,* 5, no. 2 (1979): 74–81;

Alan Williamson, *Pity the Monsters: The Political Vision of Robert Lowell* (New Haven: Yale University Press, 1974);

Stephen Yenser, *Circle to Circle: The Poetry of Robert Lowell* (Berkeley: University of California Press, 1975).

Papers:

The principal collection of Lowell's papers is held by the Houghton Library, Harvard University.

Nathaniel Mackey
(25 October 1947 -   )

Mark Scroggins
Florida Atlantic University

BOOKS: *Four for Trane* (Los Angeles: Golemics, 1978);
Septet for the End of Time (Santa Cruz, Cal.: Boneset, 1983);
Eroding Witness (Urbana & Chicago: University of Illinois Press, 1985);
Bedouin Hornbook, volume 1 of *From a Broken Bottle Traces of Perfume Still Emanate,* Calalloo Fiction Series, 2 (Lexington: University of Kentucky, 1986);
Outlantish: "Mu" Fourth Part–Eleventh Part (Tucson, Ariz.: Chax Press, 1992);
Discrepant Engagement: Dissonance, Cross-Culturality, and Experimental Writing (Cambridge & New York: Cambridge University Press, 1993);
Djbot Baghostus's Run, volume 2 of *From a Broken Bottle Traces of Perfume Still Emanate* (Los Angeles: Sun & Moon Press, 1993);
School of Udhra (San Francisco: City Lights Books, 1993);
Song of the Andoumboulou: 18-20 (Santa Cruz, Cal.: Moving Parts Press, 1994).

RECORDING: *Strick: Song of the Andoumboulou 16-25,* Memphis, Spoken Engine Company, 1995, compact disc recording of Mackey reading poems with musical accompaniment by Royal Hartigan and Hafez Modirzadeh.

OTHER: *Hambone,* nos. 1-11, edited by Mackey (1974, 1982-present);
Moment's Notice: Jazz in Poetry and Prose, edited by Mackey and Art Lange (Minneapolis: Coffee House Press, 1993);
Callaloo: A Journal of African-American and African Arts and Letters, special issue on Wilson Harris, guest-edited by Mackey, 18 (Winter 1995).

Nathaniel Mackey (photograph by Paul Schraub)

SELECTED PERIODICAL PUBLICATIONS – UNCOLLECTED:

POETRY
"Song of the Andoumboulou: 17," *New American Writing,* 11 (Summer/Fall 1993): 36-40;
"Mort Collatérale," *TXT* (1993): 21-22;
"Song of the Andoumboulou: 16," *River City,* 14 (Spring 1994): 90-95;

"Song of the Andoumboulou: 21 & 22," *apex of the M,* 1 (Spring 1994): 99–104;

"Song of the Andoumboulou: 23, 24 & 25," *Sulfur,* 34 (Spring 1994): 73–83;

"Song of the Andoumboulou: 27," *Ergo,* 9 (Summer 1994): 14–16;

"Song of the Andoumboulou: 28 & 29," *Phoebe,* 24 (Spring 1995): 29–41;

"Song of the Andoumboulou: 31," *Chicago Review,* 41 (Fall 1995): 13–17;

"Song of the Andoumboulou: 32," *Capilano Review,* 17/18 (Winter/Spring 1996): 87–89;

"Song of the Andoumboulou: 30," *Conjunctions,* 26 (Spring 1996): 77–82.

FICTION

"From *Atet A.D.,*" *Blue Mesa Review,* 3 (Spring 1991): 210–214;

"From *Atet A.D.,*" *Hambone,* 10 (Spring 1992): 99–104;

"From *Atet A.D.,*" *Avec,* 7 (Spring 1994): 47–58;

"From *Atet A.D.,*" *Arras,* 3 (May/June 1996): 18–24.

NONFICTION

"Ishmael Reed and the Black Aesthetic," *CLA Journal,* 21 (March 1978): 355–366;

"Interview with Al Young," *MELUS: The Journal of the Society for the Study of the Multi-Ethnic Literature of the United States,* 5 (Winter 1978): 32–51;

Review of *Love Story Black* by William Denby, *San Francisco Review of Books,* 4 (January 1979): 12;

Review of *Just Above My Head* by James Baldwin, *San Francisco Bay Guardian,* 14 (6–14 December 1979): 21;

"Some Thoughts on 'Fusion,'" *Threepenny Review,* 1 (Winter/Spring 1980): 23–24;

"Great White Hope: Jean Wagner Revisited," *CLA Journal,* 22 (March 1980): 245–265;

Review of *The Tree of the Sun* by Wilson Harris, *Kunapipi,* 2 (1980): 168–169;

"Notes from an Expatriate: A Conversation with Pianist-Composer Mal Waldron," by Mackey and Herman Gray, *Jazz Spotlite News,* 2 (Fall 1980/Winter 1981): 18–21;

Review of *Amiri Baraka/LeRoi Jones: The Quest for a "Populist Modernism"* by Werner Sollers, *Novel: A Forum on Fiction,* 14 (Winter 1981): 184–187;

Review of *Sun Poem* by Edward Kamau Brathwaite, *Sulfur,* 4 (Fall 1984): 200–205;

"Beyond Predication," *Quarry West,* 25 (Fall 1988): 66–67;

"'Tanganyika Strut': Some Recent Sets" (playlists from Mackey's radio program), *Sulfur,* 24 (Spring 1989): 211–214;

"From 'Roms' *Gassire's Lute:* Robert Duncan's Vietnam War Poems," *Talisman: A Journal of Contemporary Poetry and Poetics,* 5 (Fall 1990): 86–99; 6 (Spring 1991): 141–166; 7 (Fall 1991): 141–166; 8 (Spring 1992): 189–221;

"An Interview with Edward Kamau Brathwaite," *Hambone,* 9 (Winter 1991): 42–59;

"Wringing the Word," *World Literature Today,* 48 (Autumn 1994): 733–740;

"Cante Moro," in *Disembodied Poetics: Annals of the Jack Kerouac School,* edited by Anne Waldman and Andrew Schelling (Albuquerque: University of New Mexico Press, 1994), pp. 71–94;

"An Interview with Anthony Braxton," by Mackey and Herman Gray, *Mixtery: A Festschrift for Anthony Braxton,* edited by Graham Lock (Devon, U.K.: Stride Publications, 1995), pp. 56–69;

"Blue in Green: Black Interiority," *River City,* 16 (Summer 1996): 116–125.

Nathaniel Mackey's work displays a deep and idiosyncratic erudition that encompasses many cultures and traditions, but his poetry remains true to an ideal of spontaneous, joyous musicality ultimately derived from improvisational jazz. Mackey has written eloquently about jazz, and he has been remarkably successful in incorporating the excitement of that music into his poetry and prose. His work proposes a truly multicultural aesthetic, in which the innovations of postbop jazz pioneers are recognized along with the experiments of the major avant-garde American poets. Among Mackey's aims are to recapture in poetry the contemporary individual's search for origins and identity among the multifarious cultural traces making up a world of simultaneity and to recapture the moments of spiritual breakthrough, the surrender of self, the access to the numinous that is afforded within the spaces of music, religion, and poetry.

Mackey was born on 25 October 1947 in Miami, Florida. His mother, Sadie Jane Wilcox, was born in Ambrose, Georgia; his father, Alexander Obadiah Mackey, was born of Bahamian parents in the Panama Canal Zone. When Mackey was four his parents separated, and he and his older siblings (two brothers and a sister) moved with their mother to northern California. In 1958 they moved to the southern California town of Santa Ana, where Mackey finished high school. Mackey neither wrote seriously in high school nor considered himself a potential writer. Indeed, at that point in his life he expected eventually to go into mathematics or into a scientific field. Yet he had become interested in music early – by the age of "seven, eight, nine years old," he recalls in an interview. Some of the first music to which he was exposed was that of the Baptist Church, and he was indelibly impressed by the "church experience" of witnessing "people respond to music in ways that were quite different from music being listened to in a concert situation, people actually going into states of trance and possession in church." Such reactions to music – which, Mackey would later realize, had close parallels in the religious practices of Haitian vodun and Cuban Santería – left him with an abiding sense of the kinship between the musical and the spiritual, a theme that in many ways has come to dominate his writing.

Mackey began listening to jazz, or what he calls "improvised music," in his early teens. Though he at first found jazz alien – difficult to understand in comparison to rhythm and blues and rock – by dint of repeated listenings he came to some idea of what Miles Davis, John Coltrane, and other jazz musicians were up to. In high school, he discovered the music of Ornette Coleman and other "outside" players; again, the work was initially alien, even repellent, but repeated exposure eventually led him to an understanding of what was afoot among these avant-gardists. Mackey's description of the "new thing" in jazz in an essay on Amiri Baraka makes clear both why this music seemed so liberating for musicians and why it was so "difficult" for listeners, even those accustomed to the "hard bop" of Davis or Dizzy Gillespie: "This tendency involved a departure from – even outright abandonment of – bebop's reliance on the recurring chords referred to as 'the changes' of a particular piece.... To listeners accustomed to recurrent reminders of a tune's head [or melody] in the form of the soloist's confinement to the changes, the new music seemed structureless and incoherent.... The players were frequently said to sound *lost*." Such lost-sounding, mapless improvisation is an important element in Mackey's work, both as an explicit theme and as a formal model for his poetry.

Mackey also began reading poetry fairly early, and two of his most prominent early influences were William Carlos Williams, one of the most important of the American modernists and a writer who worked tirelessly to pioneer a poetics based on common American speech, and Baraka, perhaps the most prominent African American poet of the 1960s and 1970s. Baraka's work shows an enormous engagement with African American music, and his example was one of the elements that "galvanized" the close "relationship between writing, reading, and music" that had begun to develop for Mackey. Something of a Renaissance man, Baraka made his name not only as an avant-garde poet, but also as a fiery playwright, a novelist and short-story writer, a social critic, and a writer on music. Baraka, however, came to occupy an interestingly ambiguous position in Mackey's artistic development.

When Mackey enrolled at Princeton University and began seriously to write and to explore contemporary literature, he found himself influenced by several disparate movements. One was that introduced by Donald Allen's anthology *The New American Poetry* (1960). These primarily white experimentalist poets could be said to be the direct inheritors of the most radical tendencies of Williams's work and included the Black Mountain, or Projectivist, poets Charles Olson, Robert Creeley, and Robert Duncan. Another was the powerful and often strident Black Arts Movement, which advocated that African American writers ought to divest themselves of aesthetic and economic ties to white hegemonic tradition and ought to pursue a specifically, and strictly, Black Aesthetic. Among the Black Arts Movement writers were Baraka, who changed his name from LeRoi Jones in 1968 and who had earlier repudiated his former ties to the white New York avant-garde, and Larry Neal, whose essay "And Shine Swam On" (which Mackey admired at the time) makes a compelling case for such black separatism. The essay takes its title from an urban "toast," "The *Titanic*": Shine, the African American who has been working in the boiler room of the liner, is now deserting the sinking ship; as the captain and his "lily-white" daughter try to bribe Shine back to save them – he with money and she with sex – Shine refuses: "'Money is good on land and on sea, / but the money on the land is the money for me.' / And Shine swam on." Neal's essay proposes the *Titanic* as the Spenglerian sinking ship of Western (white) culture, which the African American writer must desert in order to achieve his own aesthetic, to save his own soul. That aesthetic is intimately involved with such folk forms as the toast "The *Titanic*" itself, but must necessarily be linked with a political and economic struggle.

What attracted Mackey most to the Black Arts Movement was the presence therein of Baraka, whose work Mackey had followed since his first publications as LeRoi Jones. By the late 1960s Baraka had not merely changed his name but had rather harshly rejected the readerly difficulties of experimental writing such as his own *The Dead Lecturer* (1964) in favor of a more hortatory mode better adapted to his political commitments. In 1968 Baraka dismissed outright his work up to 1965, calling it "a cloud of abstraction and disjointedness, that was just whiteness." Baraka was clearly the

Cover for Mackey's 1986 book, the first collection of letters from "N.," a jazz musician, to the "Angel of Dust"

most interesting African American poet of the day, and Mackey (who arranged for him to give a reading at Princeton in the late 1960s) was fascinated and somewhat disturbed by his aesthetic about-face, his insistence that modernist and postmodernist poetic innovations were somehow "white" and ought to be disavowed by the African American artist. This fascination resulted in part in Mackey's senior thesis, "The Conversion of LeRoi Jones." In 1978 he published "The Changing Same: Black Music in the Poetry of Amiri Baraka" (collected in his *Discrepant Engagement,* 1993), which showed how the experimentation of Baraka's early work could be usefully related, not only to white innovative poetry, but to parallel developments in "new thing" jazz.

After he graduated from Princeton in 1969, Mackey taught eighth-grade mathematics in Pasadena for a year before beginning graduate school at Stanford University, working toward a Ph.D. in English and American literature. In 1974 he was one of the editors of the first issue of *Hambone,* a literary journal. No further issues were published until 1982, after Mackey had completed graduate school and was settling into an academic position. This time he was sole editor, and he solicited material from many of the writers and musicians that interested him, including poets Robert Duncan, Jay Wright, Beverly Dahlen, and Barbadean (Edward) Kamau Brathwaite; fiction writers Clarence Major and Wilson Harris; and jazz legend Sun Ra. Mackey continues to edit the magazine, which – as he describes it in the *Dictionary of Literary Magazines* – publishes "Cross-cultural work emphasizing the centrifugal." What this statement means in practice is that *Hambone* publishes a wide mix of work by innovative poets, fiction writers, and essayists, with a notable but in no way exclusive emphasis on writers of color (especially those from the Caribbean). *Hambone* has frequently published works by Brathwaite, Harris, and Major, as well as (among many others) Ted Pearson, Kenneth Irby, Ed Roberson, Lorenzo Thomas, and Harryette Mullen. The magazine has also afforded a venue for younger writers such as Mark McMorris and Myung Mi Kim, and it has regularly featured the work of visual artists and musicians. It is clearly one of the most vibrant and eclectic "little" magazines of the 1980s and 1990s.

Mackey received his doctorate from Stanford in 1975. After teaching stints at the University of Wisconsin–Madison and the University of Southern California, he has taught since 1979 in the Board of Studies in Literature and the American Studies Program at the University of California, Santa Cruz, where he became a full professor. He teaches courses in twentieth-century fiction and poetry; African American literature, culture, and music; Caribbean literature; and creative writing (among much else). He has become a potent force in the cultural life of the West Coast through his frequent readings, lectures, and workshops; his ongoing work on *Hambone;* his editing with Art Lange of *Moment's Notice: Jazz in Poetry and Prose* (1993); and not least through his hosting of "Tanganyika Strut," a weekly radio program on KUSP radio, in which Mackey showcases his astonishingly broad and detailed knowledge of African American and Third World musical movements. In 1991 he married Pascale Gaitet; he has a teenage stepson, Joe, and a young daughter, Naima, named after perhaps the most beautiful of Coltrane's slow tunes.

Bringing up music in connection with Mackey's work is inevitable, but it is equally inevitable to bring up culture and the notion of a search for cultural foundations, a cultural place where one belongs. Such a quest is especially important for an African American poet who writes innovative poetry. Mackey's poems are far from uniformly accessible, and their "difficulty" goes beyond the sort of recondite or unfamiliar cultural references one might encounter in Ezra Pound's *Cantos* (1917–1969). Most notably, Mackey's poems studiously avoid a consistent authorial or lyrical voice, steering instead among a myriad of fragmented and disjointed voices and textual styles. Such disjunction is rare in the works of contemporary African American poets – or at least those most often lionized by the literary establishment. In the introduction to *Discrepant Engagement* Mackey quotes Language poet Ron Silliman's comments on why so much poetry written by African Americans (and by "the entire spectrum of the 'marginal'") appears "conventional" in comparison to Silliman's own experimentation and that of his colleagues. Since the "narrative of history has led not to their self-realization, but to their exclusion and domination," Silliman writes, such writers "have a manifest political need to *have their stories told*. That their writing should often appear much more conventional . . . illuminates the relationship between form and audience." Mackey replies straightforwardly: for Silliman to characterize African American literary production as "conventional" provides an index, not of that writing, but of Silliman's own limited experience of it. To fail or refuse to "acknowledge complexity among writers from socially marginalized groups," Mackey writes, is condescension: "experimental writing, the aesthetic margin, is not the domain solely of those from socially unmarginalized groups."

This statement is amply borne out by the writers whose work Mackey has published in *Hambone*, and even more so by his own poetry. His work draws many of its formal strategies from innovative poets such as Duncan, Olson, and Williams, but it also has clear ties to the early experimental work of Baraka, the cross-culturality and pan-Africanism of Henry Dumas, and the overwhelming musicality of Langston Hughes. Mackey's work, unlike Baraka's more recent poetry, has no obvious political program. Nonetheless, in his critical writings Mackey has powerfully described how the subaltern writer effects his or her own liberation precisely to the extent that she or he deforms the hegemonic language, that this sort of writer "others" the language that has designated him or her an "other." Thus, Mackey's poetry continually defers and deflects syntactic and semantic closure in a restless, ever-moving "bedouin" motion. In his Baraka essay Mackey valorizes the "obliquity or angularity" of Baraka's early work, qualities that are everywhere evident in his own poetry. Mackey's poems do not tell singular, coherent stories, nor do they even proceed from separably singular voices. They are complex and initially bewildering works, for they proceed on a basis of complex contemporaneity and heterogeneous, perhaps indeterminate origins. There is a dense web of intertextual reference running throughout Mackey's poems: references to other texts, to musical compositions (especially jazz), and to his own previous writings. This intertextuality places Mackey's work squarely in the tradition of American modernist and postmodernist innovative poetry.

Mackey's first chapbook of poems, *Four for Trane* (1978), borrows its title from an album by saxophonist Archie Shepp and pays homage to saxophonist John Coltrane, an archetypal figure in African American aesthetic life. Mackey's second, *Septet for the End of Time* (1983), pulls its references from a wider range of cultural backgrounds. The three epigraphs to the sequence provide three possible numerological entries to these eight poems: that of Ogotemmêli, an elder of the Dogon people of West Africa; that of the Koran; and that of the Pyramid Texts of Unas. Others can be imagined – the seven stars, candlesticks, and seals of the Book of Revelation, the "septet" (septuor) of stars in Stéphane Mallarmé's "Sonnet en -yx" – but most clearly implied is the *Quartet for the End of Time* (1941) by French composer Olivier Messiaen, whose resolutely personal vision was as rooted in the traditions of Western Roman Catholicism as Mackey's is in the cultural heritages of the African diaspora. At some level – perhaps a level purely of number – the eight movements of Messiaen's unusual quartet (which had its premiere in a German prisoner-of-war camp) mirror the eight poems of Mackey's *Septet for the End of Time*. But while Messiaen's spiritual universe is predicated upon the absolute presence of a Christian God, Mackey's is an *eroding* witness (to echo the title of his 1985 collection): the self can recognize its cultural, spiritual roots only as eroding traces in the works of others, and that recognizing self is in turn eroded by the forces of history, change, distance, and time itself.

There are eight poems in *Septet for the End of Time*, and like a master musician Mackey is engaged in playing eight against seven. "Seven," says Ogotemmêli in the first epigraph, "is the rank of the master of Speech; $1 + 7 = 8$. The eighth rank is that of speech itself." The Koranic text Mackey quotes identifies seven "sleepers": "Seven: Their dog was the eighth." *Septet for the End of Time*, then, is the rep-

Pages from a draft for Mackey's Djbot Baghostus's Run *(courtesy of Nathaniel Mackey)*

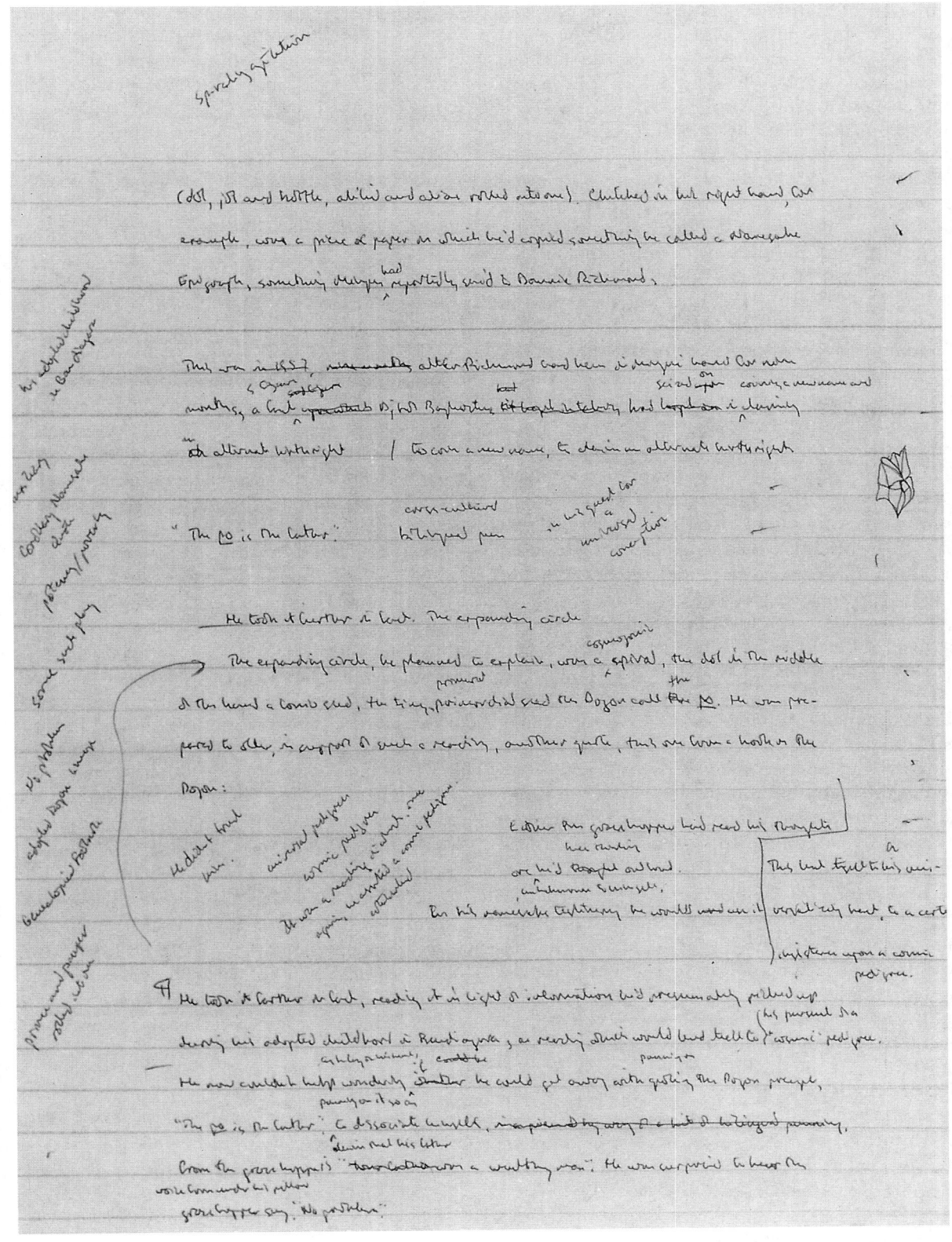

resentation of eight sleepers awakening, entering the order of "speech itself": each poem begins, "I wake up." These awakenings are not into the logic of daytime reality, but the surreal logic of dreams, in which metamorphic images and orders of discourse impose a tenuous identity, new for each poem, upon an "I" that bears only an oblique relationship to the monological "I" of Western lyric and philosophical tradition. In some of the poems, the "I" takes on the voice of the dedicatee: jazz saxophonist Pharoah Sanders in "Capricorn Rising," novelist Wilson Harris in "The Sleeping Rocks," Elis Regina in "Falso Brilhante," the artist Jess in "The Phantom Light of All Our Day," and Messiaen himself in "Winged Abyss," a title that alludes to "Abîme des oiseaux," the third movement of Messiaen's quartet. In some sense Mackey is repeating Pound's gesture here, casting the voice of the poem in that of "personae," masks through which the poet develops his own vision. But where Pound's personae follow Robert Browning's in their personistic re-creation of historic characters, Mackey's poems shift under the reader's feet, mixing past and future, Western, Eastern, and African. "Dogon Eclipse," for instance, recounts the blinding of the sage Ogotemmêli in a hunting accident: "as my / gun misfires / feel I've boarded one of Marcus' / erratic ships, aborted Black Star Line . . . / Withered lid of an eroded 'I.' " Mackey's interest in the astonishingly complex cosmology of the Dogon people is one he shares with several other writers, among them Guy Davenport and Jay Wright; Mackey first read of the Dogon in Janheinz Jahn's *Muntu* (1958) and went on to read Marcel Griaule's classic accounts of the Dogon world system, *Le Renard Pâle* (1965) and *Conversations with Ogotemmêli* (1965).

In *Septet for the End of Time,* then, the order of speech is an order of the present, in which past and future are simultaneously present in an order of traces in which the temporal present is as much an eroding trace as the remembered past or the presciently known future. The speakers of these poems wake up time and again, not from dreams to waking "reality" but from dreams to dreams, from one order of language to another. *Septet for the End of Time* argues not that there is no reality outside language but that the orders of reality – which are as much spiritual as they are historical – are embodied and embraced in the culturally structured orders of speech. While these eight speakers may desire to awaken from the "nightmare" of history, their repeated awakenings are a reiteration of the self's utter entanglement in the network of cultural traces that constitute it. The jazz great Sanders, for instance, is represented as resisting the powerful force of the music that both works his liberation and locates him within a tradition of oppression: "I wake up mumbling, 'I'm / not at the music's / mercy,' think damned / if I'm not." Similarly, Messiaen, in "Winged Abyss," hears his own "unlikely music" as both "A lullaby of wings" and "A free-beating fist" and wakes "Out of touch / with the times . . . asking what / bird / would make so awkward a / sound."

Both *Four for Trane* and *Septet for the End of Time* were included in Mackey's first collection of poetry, *Eroding Witness,* which was selected by Michael S. Harper in 1985 for the prestigious National Poetry Series and which made abundantly clear the breadth of Mackey's creative vision. These poems trace the painstaking and euphoric rediscovery of a set of cultural and mythological origins whose authenticity is founded, not upon the "rational," Western ground of historical veracity, but upon deeper correspondences of spirit – correspondences that enable Mackey to locate the wellsprings of his poetry not only in the Duncan/Olson tradition of Projective verse, but in the traditions of vodun, Dogon cosmology, Santería – which, like vodun, involves the marriage of European Catholicism and West African religions – and the more-immediate realms of African American music, from the postbop jazz of Coltrane, Sanders, and Don Cherry, to the blues-based electric music of Jimi Hendrix. *Eroding Witness* is an involved exploration of all of these sources, as well as a phenomenology of how they penetrate and suffuse the ordinary life of things, making the acts of waking, drinking, sex, and making music sacred, sacrificial activities.

The poems in *Eroding Witness,* beginning with fairly early works and culminating with *Septet for the End of Time,* evolve from narrow-lined, left-justified stanzas to a more open, exploratory form, phrases spreading out across the page like seeds thrown out by a sower. Their concerns, however, remain constant, as they consistently hearken back to a nearly forgotten past, where events took place whose shaping importance can still be felt. "Kitche Manitou," for instance, touches on the migration of the original inhabitants of the Americas, their journeys "eastward out of / Asia." Similarly, in "Passing Thru," aboriginal peoples – this time Africans – cast off on "the waters washing Mali's / western edge" and make their way "eventually as far north as / Nova Scotia." But the migrations that figure so largely in the poems are not merely the prehistoric migrations of peoples: they are the migrations of traces and inscriptions, of the Libyan glyphs that Ivan Van Sertima (author of *They Came Before Colum-*

bus, 1976, to whom "Passing Thru" is dedicated) discovered from the Virgin Islands to Connecticut; they are the migrations of hidden knowledge, the secrets of ancient wisdom, transported, like the holiest stones of Santería ritual – which were carried from Africa in the stomachs of enslaved Africans – with a singleness of purpose foreign to the technologically alienated contemporary reader: "The rocks / inside our stomachs / want blood." All migration is a movement away from origins, which in the process of movement becomes mystical and mythological, the object not of rational "knowledge," but of a religious gnosis. The Santería stones were holy before they left Africa; reaching the Caribbean in the bellies of slaves, they became all the more sacred, the focus of a religion that rejects the ethereal power of the Catholic saints in favor of the animistic forces lurking with the rocks. It is this transforming process that these poems seek to highlight, the process by which an originary moment becomes an enabling and empowering magical touchstone, through its very absence and untouchability. In "Song of the Andoumboulou: 6," an unspecified "N." confronts this question directly in the course of a prose letter to the "Angel of Dust," his friend and correspondent: "We not only can but should speak of 'loss' or, to avoid, quotation marks notwithstanding, any such inkling of self-pity, speak of absence as unavoidably an inherence in the texture of things (dreamseed, habitual cloth)."

This letter was something of a serendipitous gift for Mackey. As he explains in an interview, at some point in the late 1970s the phrase "Dear Angel of Dust" came into his head, apropos of nothing in particular. Intrigued, he wrote the letter that is included in *Eroding Witness* as "Song of the Andoumboulou: 6," and soon thereafter began a coherent series of letters from N. to the Angel of Dust, of which *Bedouin Hornbook* (1986) is the first collection. A jazz musician and member of a band called the Mystic Horn Society (formerly the East Bay Dread Ensemble, formerly the Deconstructive Woodwind Chorus), N. explores in his letters a set of concerns similar to those addressed in Mackey's poems. The letters of *Bedouin Hornbook* and its sequel, *Djbot Baghostus's Run* (1993) follow a far more straightforward narrative path than the elusive poems: often as long as a short story, they typically revolve around a rehearsal or a performance of N.'s jazz ensemble – which reminds one of the Art Ensemble of Chicago or one of Anthony Braxton's groups – in the course of which the various instruments, voices, and interactions of the music bring out the implications of the concerns at hand – concerns often of a high theoretical order indeed. The melody and rhythms the instruments play, as interpreted by N., become inseparable from the ideas and emotions with which the poem deals, and there is often a quite astonishing marriage of narrative, critical theory, and high-spirited wordplay.

Wilson Harris, the Guyanese novelist whose work Mackey first encountered in 1974 or 1975 and with whom he has been in correspondence since 1978, is an important influence on the prose style of Mackey's fiction. Harris's work is densely metaphoric, and in much of his dreamlike, indescribable fiction, he works at teasing out the phonological, spiritual, and metaphoric implications of the New World writer writing in an Old World language, seeking to make an intervention in an Old World tradition. *From a Broken Bottle Traces of Perfume Still Emanate,* the longer work of which *Bedouin Hornbook* and *Djbot Baghostus's Run* are the first two installments, is similarly concerned with the relationship of word and spirit, lyric and music. Part of this work takes the unlikely form of opera libretti/musicological lectures, the most important of which is perhaps "The Creaking of the Word," with which *Bedouin Hornbook* ends. *Djbot Baghostus's Run* incorporates no fewer than three "After the Fact Libretto/Lectures," all playing variations on this text. *From a Broken Bottle Traces of Perfume Still Emanate,* which has been extended to a third volume, *Atet A.D.,* continually circles around the question of how one might use language to describe music and the spiritual reality to which music gives one access. This question is why the idea of the opera libretto and the musicological lecture is so important to Mackey and why the examples of Harris – who strives to touch an indefinable spiritual reality in his fiction – and the jazz composer Braxton – whose hyperintellectual pipe-and-cardigan image, diagrammatic song titles, and erudite, involved album notes defined a new intellectualism in "new thing" jazz – are so evident in the verbal texture of Mackey's epistolary work.

The title *From a Broken Bottle Traces of Perfume Still Emanate* is emblematic of Mackey's project as a whole, both prose and poetry. The past, that is, one's origins, is available only as "traces," as momentary, unresolved glimpses and intuitions. As N. explains to the Angel, "I tend to pursue resonance rather than resolution, so I glimpsed a stubborn, albeit improbable world whose arrested glimmer elicited slippages of hieratic drift." Likewise, the poems of *Eroding Witness,* rather than weaving entire myths and narratives out of the sources to which they refer, reveal the magical past only in traces and flashes; as N. writes, "The sense I have is that we're

What you must search for, and find, is the black torso of the Pharaoh.

—Manuel Torre to García Lorca

SONG OF THE ANDOUMBOULOU: 16

—cante moro—

    They were dredging
the sea, counting
  the sand. Pounded
    rocks into gravel,
paid a dollar a
             day,
  sang of the oldest
fish like family,
                tight
  flamenco strings
distraught...

               Some
  ecstatic elsewhere's
advocacy strummed,
unsung, lost inside
    the oud's complaint...
The same cry taken
up in Cairo, Córdoba,
                 north
  Red Sea near Nagfa,
Muharraq, necks cut
with the edge
             of a
broken cup...
          Lebrijano's
burr-throat, raspy
as night, adamant
night, long night
             longer
than a lifetime of
             nights...
Turning away what light
was outside, night
without end...

First two pages from the typescript for "Song of the Andoumboulou: 16" (courtesy of Nathaniel Mackey)

                              Muttering,
           "Time," less than a
              sigh, resigned
           to it, would-be
                        <u>Book</u>
           <u>of Coming Forth by Day</u>,
                                       would-be
              kef-pesh, pinkish
                                sun...
                 Between lips, resuscitant,
                                          reached
              where they were, no boat not
              one they drowned in, dredged
                 it,   letterless the
              book we thumbed...

                                 Blocked
                 synonymy steeped as though
                    linked and unlocked,
                 lest it engulf them in smoke,
                                               squat,
              squint-eyed god afloat in a
              saucer filled with rum.
              Steeped in memory, bedrock
                 mischief, misanthropy,
                                          soul
                 sent rousing the dead,
                                          wrested
                 kiss...
                       Later to be
                 burned, beliefless,
                                     ya-habibi'd
                    endlessly, burr-throat,
                    threnodist... Lest
                    it be said they saw less
                       than they said, lipless,
                                                less
                 than a sough...

                                 Sprocketed
              watch. Moot mechanical troth...
              Treadmill mesh. Ever to
              be gone to again, lost
                                     body
                 of love... Aspic allure,
                                          lost-
                 limbed entanglement,
                                          toxic...
                    Bitten we'd have
                                     been
              and were

being addressed by a barely audible witness, some receding medium so heartrendingly remote as to redefine hearing." What makes the poems of *Eroding Witness* so gripping is this very "heartrending remoteness," the sense of one's irreconcilable separation from the past, even as one strives to recapture or reconstruct that originary unity.

Such a search for unity is an inherently cultural quest, in Mackey's case a clearly cross-cultural one. This quest is evident throughout the critical essays collected in *Discrepant Engagement: Dissonance, Cross-Culturality, and Experimental Writing* (1993). In this volume Mackey addresses African Americans Baraka and Major, but he also writes on the post-Poundian poets Duncan and Olson, and the Caribbean writers Harris and Brathwaite, each of whom is attempting to write his way to a reconcilement with the violently cross-cultural history of the New World. In the 1990s Mackey has been investigating the traces of North African culture in Spanish flamenco music, the *cante moro* (Moorish song) that underlies the *cante jondo* of the flamenco. His essay "Cante Moro" (1994) traces the blurrily defined notion of the duende through Federico García Lorca to various twentieth-century American poets – among them Jack Spicer and Bob Kaufman – and listens for the cante moro in the music of African American jazz and blues artists. These phenomena are not merely cultural survivals or stylistic markers, but are evidence of the poet's or musician's access to a deeper, darker spiritual reality: for example, the cry of Miles Davis's trumpet in "Saeta" from *Sketches of Spain* (1960) is fundamentally akin to the "outside" from which Spicer claimed he received his poems.

In *School of Udhra* (1993), a recent volume of poetry, Mackey continues to pursue the concerns of *Eroding Witness* and *From a Broken Bottle Traces of Perfume Still Emanate*. The title *School of Udhra* alludes to an Arabic poetic tradition associated with the seventh-century Yemenite Bedouin Djamil. Such an allusion is appropriate, for the salient characteristic of Udhrite poetics is an erotic abandon, an attraction to the beloved so intense that it results in the poet's death, the literal dispersal of the poetic identity into the concentrated ardor of the verse. But while sexual fulfillment figures in these poems, more often there is the complex fact of an unfulfilled, perhaps unfulfillable, longing: "Sat up sleepless in the Long Night Lounge, love / stood me up ... Wine on my shirtsleeve, / wind on my neck." That wind is no doubt the "bedouin wind" to which the subtitle of one poem alludes (while also quoting a Talking Heads song title), for the other keynote of these poems, besides longing, is again migration, a "bedouin wish / to be elsewhere, / every- / where at once": a movement away from origins that as they recede become mythicized, at once familiar and alien – the "mu" Mackey celebrates in a poem of that title, "Irredentist / myth, 'mu' meaning lost ground, / 'else' / the earlier where we were / after"; and a movement toward some ultimate personal or spiritual goal, whether the beloved (who appears once memorably as "maitresse erzulie," the vodun loa) or the Utopian (meaning, of course, "no-place") city of Zar, which, as Mackey quotes Larry Neal, "is just this side of far."

The dense intertextuality of this collection is reinforced by the continuation in *School of Udhra* of Mackey's two ongoing serial poems, "Song of the Andoumboulou" and "mu." These two series, which began in *Eroding Witness,* are in many ways similar in structure to Robert Duncan's series *Passages* and *Structures of Rime,* each of which includes poems from several of Duncan's books. (Mackey has shown an abiding interest in Duncan; his book-length commentary on Duncan's work, "Roms" *Gassire's Lute: Robert Duncan's Vietnam War Poems,* was serialized in the journal *Talisman* in 1990–1992.) The titles of Mackey's series are indices of the cross-cultural range of his poetics: "Song of the Andoumboulou" derives from remarks in Griaule's Dogon studies and from a French recording of Dogon music. While "mu" takes its title from jazz trumpeter Don Cherry's albums, it derives additional resonance from Jane Harrison's *Themis* (1912): "a myth is not merely a word spoken; it is a re-utterance or pre-utterance, it is a focus of emotion. . . . Possibly the first *muthos* was simply the interjectional utterance *mu*." As in Mackey's earlier poetry, the poems of *School of Udhra* continually play on the themes of loss and dislocation, a social or spiritual "phantom limb" (a metaphor the poet borrows from Wilson Harris). Even more so than his earlier work, these poems have precisely what Mackey praises in Baraka's early work, "a mercurial, evanescent quality, as though it sought to assassinate any expectations of traceable argument or logical flow." Such evanescence, however, does not mean that the poems have no clear referents or subjects. For instance, "Sweet Mystic Beast" (which alludes again to the work of Larry Neal) is a strikingly accurate lyrical evocation of the postbop jazz saxophone: "Hunk of metal / strapped around / the neck. Not / a horn if not / bellowing ... beast / caught in the bell of a horn." "Amma Seru's Hammer's Heated Fall" celebrates the work of Ed Love, a sculptor who creates frighteningly animated, huge Afrocentric figures from discarded car parts: "Brash alchemical / armor, philosophical polish, / lapis-lit chrome. / I sing of / shine, the machine wrecked, banged-out / Osirian story told in chrome."

In the middle poems of *School of Udhra,* Mackey cuts loose with what he calls his "vatic scat," arranging and rearranging the letters of the most common of demonstrative pronouns, *that,* so that it becomes a Coltrane-like "sheet of sound": "Anagrammatic scramble. Scourge / of sound. Under its brunt / plugged ears unload . . . Anagrammatic tath. Anagrammatic / that . . . Palimpsestic / stagger, / anagrammatic scat." This common word, that, meaningless outside a specific context, is transformed by such "anagrammatic scramble" – a practice related to a whole jazz tradition of anagrammatic tune titles, most notably in Cecil Taylor's work – into a figure of mythic proportions, a signifier simultaneously of hoped-for community and artistic isolation: "To've been there as they / began to gather. All the tribes of Outlandish crowding the outskirts / of Ttha," and "Awoke stranded on the island of / Ahtt, light's last resort." Such moments of "gathering," of the tribes' return to origin, Mackey implies, can be found only within the interstices of language itself.

In all of his writing Mackey both dwells within language, in all its cultural, historical, and aural resonances, and is simultaneously aware – not without discomfort – of the human being's alienness to language, our irrevocable and irremediable diaspora, or exile, from some cultural origin, or some Adamic state of coincidence with the world to which our words correspond. This particular doubleness of linguistic stance, along with his almost intimidating verbal gifts – and along with his extreme sensitivity to his own cultural position and to the multifarious apparitions of humanity's encounters with the ecstatic or spiritual, from Siberian shamanism to Udhrite ecstasy to the more-familiar possessions that take place weekly in Baptist church services – makes Mackey one of the most compelling, as well as one of the most innovative, poets of his generation.

Interviews:
Chris Funkhouser, "Charting the Outside: An Interview with Nathaniel Mackey," *Poetry Flash,* 224 (November 1991): 1, 6–11, 25–26;
Edward Foster, "An Interview with Nathaniel Mackey," *Talisman: A Journal of Contemporary Poetry and Poetics,* 9 (Fall 1992): 48–61.

References:
Gary Burnett, "A Book of Horns," *Dark Ages Clasp the Daisy Root,* 1 (November 1989): 34–36;
Joseph Donahue, "Sprung Polity: On Nathaniel Mackey's Recent Work," *Talisman: A Journal of Contemporary Poetry and Poetics,* 9 (Fall 1992): 62–65;
Michael Franco, "Bedouin Hornbook," *Talisman: A Journal of Contemporary Poetry and Poetics,* 9 (Fall 1992): 71–73;
Chris Funkhouser, "Location and Dis-ghosts: Nathaniel Mackey's Recent Work," *Little Magazine,* 20 (1994): 40–49;
Albert Mobilio, "On Mackey's *Bedouin Hornbook:* Hearing Voices," *Talisman: A Journal of Contemporary Poetry and Poetics,* 9 (Fall 1992): 69–70;
Harryette Mullen, "Phantom Pain: Nathaniel Mackey's *Bedouin Hornbook,*" *Talisman: A Journal of Contemporary Poetry and Poetics,* 9 (Fall 1992): 37–43;
Paul Naylor, "The 'Mired Sublime' of Nathaniel Mackey's *Song of the Andoumboulou,*" *Postmodern Culture,* 5 (May 1995);
Aldon L. Nielsen, "'Gassire's Lute,'" *Talisman: A Journal of Contemporary Poetry and Poetics,* 9 (Fall 1992): 66–68;
Mark Scroggins, "The Master of Speech and Speech Itself: Nathaniel Mackey's 'Septet for the End of Time,'" *Talisman: A Journal of Contemporary Poetry and Poetics,* 9 (Fall 1992): 44–47;
Scroggins, Review of *Eroding Witness* and *Bedouin Hornbook, Epoch,* 38 (1989): 316–322;
Winston Smith, "Let's Call This: Race, Writing, and Difference in Jazz," *Public,* 4/5 (1990/1991): 71–82.

W. S. Merwin
(30 September 1927 -)

James McCorkle

See also the Merwin entry in *DLB 5: American Poets Since World War II, First Series.*

BOOKS: *A Mask for Janus* (New Haven: Yale University Press, 1952; London: Oxford University Press, 1952);

The Dancing Bears (New Haven: Yale University Press, 1954);

Green with Beasts (New York: Knopf, 1956; London: Hart-Davis, 1956);

The Drunk in the Furnace (New York: Macmillan, 1960; London: Hart-Davis, 1960);

The Moving Target (New York: Atheneum, 1963; London: Hart-Davis, 1967);

The Lice (New York: Atheneum, 1967; London: Hart-Davis, 1969);

The Carrier of Ladders (New York: Atheneum, 1970);

The Miner's Pale Children (New York: Atheneum, 1970);

Asian Figures (New York: Atheneum, 1973);

Writings to an Unfinished Accompaniment (New York: Atheneum, 1973);

The First Four Books of Poems (New York: Atheneum, 1975) — contains *A Mask for Janus, The Dancing Bears, Green with Beasts,* and *The Drunk in the Furnace;*

The Compass Flower (New York: Atheneum, 1977);

Houses and Travellers (New York: Atheneum, 1977);

Finding the Islands (San Francisco: North Point Press, 1982);

Unframed Originals (New York: Atheneum, 1982);

Opening the Hand (New York: Atheneum, 1983);

Regions of Memory: Uncollected Prose, edited by Ed Folsom and Cary Nelson (Urbana: University of Illinois Press, 1987);

The Rain in the Trees (New York: Knopf, 1988);

Selected Poems (New York: Atheneum, 1988);

The Lost Upland: Stories of Southwest France (New York: Knopf, 1992);

Travels (New York: Knopf, 1993);

The Second Four Books of Poems (Port Townsend, Wash.: Copper Canyon Press, 1993) — contains *The Moving Target, The Lice, The Carrier of*

W. S. Merwin (photograph © the Estate of Thomas Victor)

Ladders, and *Writings to an Unfinished Accompaniment;*

The Vixen (New York: Knopf, 1996).

TRANSLATIONS: Anonymous, *The Poem of the Cid* (London: Dent, 1959; New York: Las Americas, 1959);

Aulus Persius Flaccus, *The Satires of Persius* (Bloomington: Indiana University Press, 1961; London: Anvil Press Poetry, 1981);

Some Spanish Ballads (London: Abelard-Schuman, 1961); republished as *Spanish Ballads* (Garden City, N.Y.: Anchor/Doubleday 1961);

Anonymous, *The Life of Lazarillo de Tormes: His Fortunes and Adversities* (Garden City, N.Y.: Anchor/Doubleday, 1962);

Anonymous, *The Song of Roland,* in *Medieval Epics* (New York: Modern Library, 1963);

Selected Translations 1948-1968 (New York: Atheneum, 1968);

Jean Follain, *Transparence of the World* (New York: Atheneum, 1969);

Sebastien Chamfort, *Products of the Perfected Civilization: Selected Writings* (New York: Macmillan, 1969);

Pablo Neruda, *Twenty Love Poems and a Song of Despair* (London: Cape, 1969; New York: Grossman, 1969);

Antonio Porchia, *Voices* (Chicago: Big Table, 1969);

Asian Figures (New York: Atheneum, 1973);

Osip Mandelstam, *Osip Mandelstam: Selected Poems*, translated by Merwin and Clarence Brown (London: Oxford University Press, 1973; New York: Atheneum, 1974);

Roberto Juarroz, *Vertical Poems* (Santa Cruz, Cal.: Kayak, 1977);

Sanskrit Love Poetry, translated by Merwin and J. Moussaieff Masson (New York: Columbia University Press, 1977);

Euripides, *Iphigenia at Aulis*, translated by Merwin and George E. Dimock Jr. (New York: Oxford University Press, 1978);

Selected Translations 1968-1978 (New York: Atheneum, 1979);

Four French Plays (New York: Atheneum, 1985);

From the Spanish Morning (New York: Atheneum, 1985);

Roberto Juarroz, *Vertical Poetry* (San Francisco: North Point Press, 1988);

Muso Soseki, *Sun at Midnight*, translated by Merwin and Soiku Shigemitsu (San Francisco: North Point Press, 1989).

PLAY PRODUCTIONS: *Darkling Child,* by Merwin and Dido Milroy, London, Arts Theatre, 1956;

Favor Island, Cambridge, Massachusetts, Poet's Theatre, 1957;

The Gilded West, Coventry, U.K., Belgrade Theatre, 1961.

TELEVISION SCRIPTS: *Rumpelstiltskin,* London, BBC, 1951;

Pageant of Cain, London, BBC, 1952;

Huckleberry Finn, London, BBC, 1953;

Robert the Devil, London, BBC, 1954;

Punishment without Vengeance, London, BBC, 1954;

The Dog in the Manger, London, BBC, 1954.

If there is one poet who serves as a guide through this perilous second half of the century and whose work serves as an example of both the contemporary poetic process and its inextricable bonds to an ethical vision of being, W. S. Merwin would be that poet. One of the most prolific poets of his generation, he has also produced translations, critical essays, memoirs, prose fiction, and plays. Many of his exemplary translations are definitive, and his essays, memoirs, and fiction are highly regarded; but it is certainly as a poet that Merwin is most widely recognized and most controversial. Merwin has created a body of poetry that is a severe meditation on the nature and condition of language; his is a body of poetry, moreover, that deeply probes the ways in which language crucially mirrors human identity and human action. Nevertheless, even while his poetry relentlessly tests the possibilities of language, Merwin also celebrates the concentrated power of language. Ezra Pound advised the young Merwin to read the seeds, not the twigs, of poetry. To read Merwin's translations is to read the seeds; the translations often direct the reader to new sources for Merwin's poetic thought. But it is in his poems that his explorations grow into striking illuminations.

In his rich and complex artistry Merwin is preeminently a poet of liminal moments, one who traces transitional realms, the strange spaces between different states of being, the shifts of being, the transgressions of space. These moments, from initiatory events in life to passages toward death, are essentially mythic, whether described in the language of memoirs and essays, parable and aphorism, or lyric poetry. From the beginning, Merwin's poetry has demonstrated his mastery of prosody and formal arrangement; it has also unapologetically addressed large themes and probed deeply into the spiritual and psychological experiences of emptiness and melancholia. But his meditations on the malaise of the modern era, with their deconstructive and frequently misanthropic vision, have, as a corollary and counterpoint, his insistent, quiet celebrations of the epiphanies of a numinous world.

Born in New York City on 30 September 1927, William Stanley Merwin grew up in Union City, New Jersey, and Scranton, Pennsylvania. His father was a Presbyterian minister whose family came from the Allegheny Valley. In his family recollections, gathered in *Unframed Originals* (1982), Merwin has written eloquently and humanely about the poverty and inarticulate despair of rural America. In the portrayal of his father, he evokes the image of a man who inspired the spiritual stricture and fear that typified much of the religious experience of rural America. Merwin explores his relationship to his family in various poems, and both his later artistic use of biblical imagery and the seriousness of his philosophical concerns evidently derive in part from his early life. In 1947 Merwin received his B.A. in English from Princeton University; he then continued for another year at Princeton, doing graduate study in modern languages. At the university he encountered two men whose ideas and writings strongly informed his early work – the poet John Berryman and the scholar and critic R. P. Blackmur. At this time Merwin also began corresponding with Pound, requesting that the older poet recommend Provençal texts for study; this correspondence expanded over the years to address issues of translation and poetics, as well as the political hollowness and social rapacity that both poets believed was corrupting American life.

Critical reaction to Merwin's poetry has largely been a combination of celebration and guarded criticism. Although W. H. Auden selected Merwin's first book, *A Mask for Janus* (1952), for the Yale Younger Poets Series, signaling Merwin's emergence as a gifted writer, the early collections sometimes provoked hostile reactions. Robert Bly called Merwin's poetry prosy and said it was characterized by a "wastage of words"; James Dickey condemned Merwin for lacking "intensity, some vital ingress into the *event* of the poem"; Thom Gunn found Merwin's language flat and lacking "any real contrast." It was Dickey, however, who foresaw and celebrated Merwin's shift in poetics in the 1960s, stating that Merwin's poems of that time "have upon them the handprint of necessity." Toward *The Moving Target* (1963) and *The Lice* (1967), both marking Merwin's radical break from the tenets of the New Critics (of whom Blackmur was a leading figure), the critical reception was mainly one of astonishment and praise. Kenneth Rexroth noted that each of Merwin's books "has been a step from that academic fashion of imitation baroque, which he handled with great skill, toward ever greater modesty and immediacy of utterance." Nonetheless, these new poems also elicited puzzlement and have often been described as surreal, enigmatic, obscure, or distant. With the publication of *Writings to an Unfinished Accompaniment* (1973), Merwin's intensely iconoclastic language became, in the view of some critics, self-parodying. Helen Vendler concocted a "Merwin Dictionary" of words the poet seemed to use obsessively; Floyd Collins argued that the poems of this collection "lapse into a repertoire of convenient effects." From his collections *The Compass Flower* (1977) and *Opening the Hand* (1983) to the present, the reception has been far more favorable, although there remains a critical position, represented by such critics as Paul Lake, that challenges the deep image poem as written by Merwin, Galway Kinnell, or James Wright. *Travels,* published in 1993, has been hailed as Merwin's finest book since *The Lice,* its most salient aspect being, in Collins's words, Merwin's "reengagement with and reconciliation to the innate power of language to embody and enlighten."

In addition to critical acclaim, Merwin's poetry and his translations have received many honors. After the Yale Younger Poets Award, these honors included a *Kenyon Review* Fellowship; a National Institute of the Art and Letters Award and a playwriting bursary from the Arts Council of Great Britain; a Rabinowitz Research Fellowship; *Poetry* magazine's Bess Hokin Prize; a Ford Foundation grant; a Chapelbrook Foundation Fellowship; *Poetry* magazine's Harriet Monroe Memorial Prize; a National Endowment for the Arts grant; the PEN Translation Prize for *Selected Translations 1948–1968* (1968); and a Rockefeller Foundation grant. In 1971 Merwin was awarded the Pulitzer Prize for *The Carrier of Ladders* (1970). He later won a Guggenheim Fellowship; an Academy of American Poets Fellowship and the Shelley Memorial Award; a second grant from the National Endowment for the Arts; and the Bollingen Prize. In 1987 he received the Governor's Award for Literature from the state of Hawai'i, and in 1994 he was the first recipient of the Tanning Award and received the Lenore Marshall Poetry Prize from *The Nation.*

Despite his recognition, Merwin has been, in many ways, both literally and figuratively, outside the movements and shifts of contemporary poetry in the United States. Not only has he lived in Portugal, Great Britain, and France (he now lives in Hawai'i, the landscapes of which inform much of his poetry), but, unlike many poets, he also claims no affiliation with any particular school of poetics. Indeed, his statements of his own views of poetry tend to be general and spare. His poetic practice is mainly illuminated by a study of his poems, although his general view of poetry is at least partly stated in certain prose writings. In the foreword to *Asian Figures* (1973) he writes: "The urge to brevity is not perhaps as typical of poetry as we would sometimes wish, but the urge to be self-contained, to be whole, is perhaps another form of the same thing, or can be, and it is related to the irreversibility in the words that is a mark of poetry." The poet's vocation is to find a language that is sufficient to its occasion and that has its own presence. But craft, for Merwin, has a negative connotation: it may, if fact, serve to create the antithesis of an ethically necessary language. In his essay "Milton: A Revisitation," in *Regions of Memory: Uncollected Prose* (1987), Merwin writes, "Conscripting the impulse of poetry to the uses of persuasiveness is a procedure that risks doing violence to that free impulse and misleading whatever in ourselves is capable of loving it and being moved, opened, and clarified by it." He continues by stating that craft – the "habits, echoes, knowledge, fear" – is often "an apparatus for repeating"; poetry, by contrast, is "exceptional." For both writer and reader, poetry is a clarifying vision. Language and the phenomenal world remain separate, yet it is through the fallible and problematic human language that one discerns the world: "Between us and that fresh moment we encounter relics, fragments, shards of fact perpetuating distance, dust, and dust under it, and among all those we catch glimpses occasionally that appear to be revelations of ourselves, as we are now, in our only time," he writes in the foreword to *From the Spanish Morning* (1985).

Merwin was twenty-four when Auden selected his first collection, *A Mask for Janus*, for the Yale Younger Poets Series. Citing Merwin's metrical skills and mastery of fixed forms, Auden noted in his preface to *A Mask for Janus* that Merwin had caught "the feeling which most of us share of being witnesses to the collapse of a civilization, a collapse which transcends all political differences and for which we are all collectively responsible, and in addition feeling that this collapse is not final but that, on the other side of disaster, there will be some kind of rebirth, though we cannot imagine its nature." Auden's comments suggest that these poems belong to an aesthetic and social community. Employing such traditional forms as ballads, odes, sestinas, carols, sonnets, and rondels, Merwin participates directly in the transmission and conservation of poetic tradition and literary history. His poetic community, however, extends beyond the formal tradition. Beginning with his first volume of poems, Merwin attempts to maintain a communion with elemental nature and to recover the connections between poetic language and the phenomenal world.

In Merwin's work the poet's task is to remember – and thereby maintain – his connections with the actual world and, if need be, to refigure the perceptions of those connections. His poetry is thus based on an acute awareness of continuity and also an awareness of the energies that disrupt and destroy continuity. His poetic method reflects that continuity: such abstract and concrete images as light, door, rain, stone, darkness, shadow, silence,

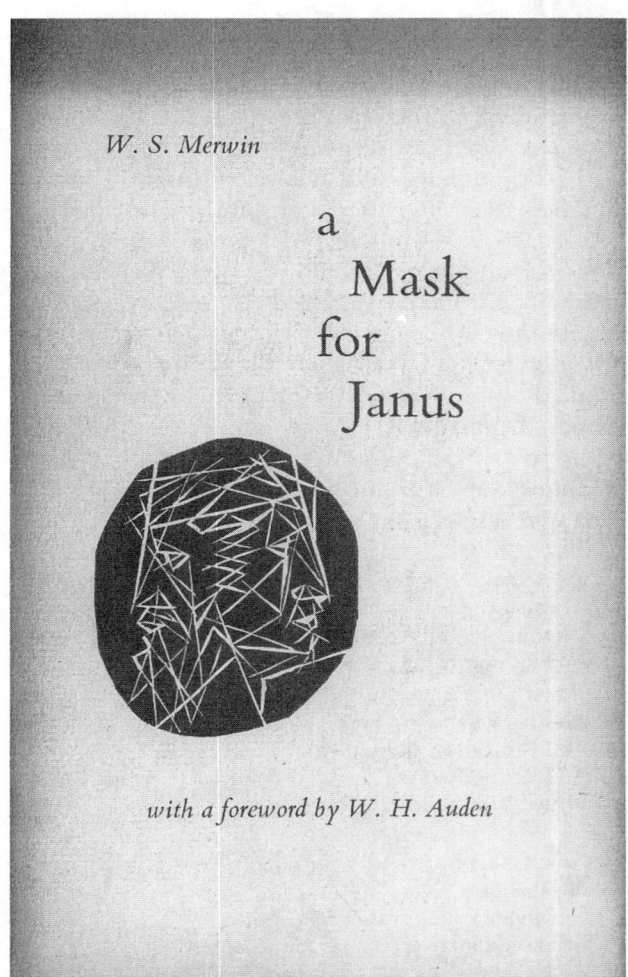

Dust jacket for Merwin's first book, selected by W. H. Auden for publication in the Yale Series of Younger Poets (courtesy of the Lilly Library, Indiana University)

and wind are found throughout *A Mask for Janus* and can be traced throughout his work. His attention to an elemental landscape – the consciousness of animals, the seasons, the passage of time – is also consistently present in his work. The opening stanza of "Herons" exemplifies this substructure of Merwin's poetic diction:

> As I was dreaming between hills
> That stones wake in a changing land,
> There in the country of morning
> I slept, and the hour and shadow slept.

The poem continues with the poet's becoming "the quiet stone" and overhearing three herons, who, like the three Moirae, share their resignation among themselves. "Herons" also typifies the liminal quality found in Merwin's poetry: here the boundaries between dream and waking, animal and human, night and day, animate and inanimate, become fluid. Merwin suggests that only by resisting such distinctions between apparent opposites can one approach the knowledge of the other; it is this liminal state that one must recover.

Although this substratum of diction – including thematics and patterns of sounds, as well as particular words – connects Merwin's earliest collection with all those that follow, *A Mask for Janus* is remarkable for its recovery and renewal of traditional forms and sources, and for the fierce, taut energies the poet draws from his language and experience. "Ballad of John Cable and Three Gentlemen" exemplifies the brilliant formal composition found in Merwin's work. The concluding four stanzas demonstrate his facility with rhyme, phrasing, and narrative summation:

> Over shaking water
> Toward the feet of his father,
> Leaving the hills' color
> And his poorly mother
>
> And his wife at grieving
> And his sister's fallow
> And his body lying
> In the rank hollow,
>
> Now Cable is carried
> On the dark river;
> Nor even a shadow
> Followed him over.
>
> On the wide river
> Gray as the sea
> Flags of white water
> Are his company.

The ballad describes John Cable's attempts to forestall his death by arguing that his family depends upon him. Death cannot be delayed, however, and John Cable is transported to the far shore. The poem conveys an assurance of voice in its economy of language and the ease by which the narrative and dialogue unfold. Merwin also invokes the archetypal dilemma – to bargain with death – that propels the poem toward its inevitable end. Death is the destination to which Cable follows his father across the water. How the individual self realizes and confronts this condition of mortality is one of Merwin's persisting themes.

The passage of the self through life – a modern Pilgrim's Progress – is the thematic parallel to Merwin's meditations on mortality. In *A Mask for Janus* the two initial poems, each bearing the title "Anabasis" ("Anabasis I" and "Anabasis II"), evoke arrival to a new land. These two poems share with many poems in Merwin's first four volumes a complex pattern of inversion and modification within the poetic syntax. Although the two "Anabasis" poems are cast in the first-person plural, thereby suggesting that the narrative journey speaks to a common human condition, the language is estranged and archaic, and thus it creates an ahistorical moment. These two poems, as their shared title suggests, depict a journey away from the known, to "fearfully [make] / Our small language in the place of night." If an authentic language can be created, survival is at least possible; but it is the survival of the human community that these two poems, with their bleak landscapes where travelers have drifted ashore, put into question.

In *The Dancing Bears* (1954) Merwin continues to fuse his interest in traditional forms and dictions with philosophical questions about the self, belief, survival, and myth. A voice of urgency arises in this collection, as in the second "Canso":

> There must be found, then, the imagination
> Before the names of things, the dicta for
> The only poem, and among all dictions
> That ceremony whereby you may be named
> Perpetual out of the anonymity
> Of death.

Through the use of the Provençal love poem, or *canso,* Merwin addresses the connection between *poiesis* and love: the two coincide in their desire to give life and to evade the "anonymity / Of death." The poet seeks a state of immutability, if not for love, then, for poetry.

The title of the book is taken from Gustave Flaubert's remark that "human speech is like a cracked kettle on which we pound our tunes fit to make bears dance, when what we want is to win over the stars." The impurities in the dailiness of human speech are kept at a distance, while the poems seek to demonstrate that poetry must aspire to "the dicta for / The only poem." In Merwin's retelling of the Norwegian folktale "East of the Sun and West of the Moon," the peasant girl turned princess states:

> What if these pheasants amble in white glass,
> Ducks strut ridiculous in stone, the streams
> Slither nowhere in beryl; why should I
> Complain of such inflexible content,
> Presume to shudder at such serenity,
> Who walk in some ancestral fantasy,
> Lunar extravagance, or lost pagoda
> That dreams of no discipline but indolence?

Artifice and harmony offer immutable serenity and pleasure. Yet this luxuriance is disquieting in its divorce from the mutable world of "the mortal air." In this lustrous narrative (which parallels the story of Cupid and Psyche), the princess's desire for confirmation of the prince's identity and reality is yet another form of ordering, and thus is destructive of the mirrored statements describing the imagination: "All magic is but metaphor" and "All metaphor is magic."

The wrought surface of these poems creates a musical unity that attempts to override the poet's doubt about the project of creating a transcendent poetry whose utterance is final and commanding. At the end of "On the Subject of Poetry," the narrator confesses that he understands neither the "man who slouches listening / To the wheel revolving in the stream, only / There is no wheel there to revolve," nor "the world." Merwin depicts the poet's condition as being that of a Janus-like figure who is looking in two directions simultaneously and finding himself ultimately at a loss.

These poems serve as well-wrought mirrors whereby the poet's self is as much the subject of the gaze as is the desire for unity. In "You, Genoese Mariner" Merwin addresses Christopher Columbus and, seeing in Columbus a mirror of himself, says that "I, after so long / . . . have been wrong as you." Columbus's mistake was a failure of imagination: "the grammar of return" prescribed an "Earth too circumscribed," one that would hold only what Columbus had fancied: "gilt and spice," not "distances and marvels, / The unfingered world." To weave too orderly a world, too taut a harmony, is to risk losing all astonishment. For Merwin the role of form remains a persisting question, and, although form and content are inextricably bound, the struggle with form corresponds to the struggle of defining one's human identity.

Throughout Merwin's writing myth serves as a principal means to unite form and content. Because myth defines the human situation cosmologically and teleologically, that is, both in terms of beginnings and ends, its narratives are inherently narratives of liminality and metamorphosis, and thus of mortality and mutability. Assuming the mask of Odysseus in the poem "Proteus," the poet wrestles with the wily Proteus, only to discover "The head he turned toward me wore a face of mine":

> Here was no wisdom but my own silence
> Echoed as from a mirror; no marine
> Oracular stare but my own eyes
> Blinded and drowned in their reflections;
> No voice came but a voice we shared, saying,

> "You prevail always, but, deathly, I am with you
> Always." I am he, by grace of no wisdom,
> Who to no end battles the foolish shapes
> Of his own death by the insatiate sea.

The poet, like Odysseus, battles mutability and desires a final form, a final "grammar of return." When the poet looks into his own eyes, two traditional images are invoked: that of the eyes as the portals of the soul and that of Narcissus gazing in his own reflection. Both convey the quest for identity; however, Merwin has also underscored the implicit and defining theme of mortality, the mortality that leads the self to silence. This silence, or nothingness, is further articulated in the parallel poem "Colloquy at Peniel," where God addresses Jacob, with whom he has wrestled: "I am that which you lost / Behind you which you seek before you, for I / Am certain . . . Not musical, but moving in all your music." To try to define form in its most essential and human shape is to try to define mortality and also the nature of the divine. But such a heroic attempt invokes the double, paradoxical gaze of Janus.

Two years after *The Dancing Bears* Merwin's third book of poems was published. With *Green with Beasts* (1956) Merwin relinquishes his stanzaically shaped poems, rich with legend and the fabulous; in their place are poems that move in long blocks of language more directly into the world. Divided into three parts, the collection begins with a bestiary. "Leviathan," which begins the bestiary, is a poem of beginnings that draws upon Genesis and early English alliterative verse. The Leviathan in the poem, however, has also endured the histories of "bone-wreck of vessels, / Tide-ruin, wash of lost bodies bobbing, / No longer sought for." Whether symbolizing the wastage of human history or the potential destruction from a nuclear war or an environmental catastrophe, the Leviathan becomes the emblem of a presence of creation that persists without regard to the human world.

As Richard Howard has noted, Merwin has shifted in this collection from having his poems recount an experience to the poet's dwelling inside the experience. These poems are dramatic monologues that are rich with description, as demonstrated in "Leviathan," "The Annunciation," and "Burning the Cat." Language is not a mediating screen or artifice; rather, Merwin has relaxed the most apparent formal demands of grammar to express the immediacy of experience. In "The Annunciation" Mary recounts the moment she is given word that she will bear the Messiah:

Only, in the place
Where, myself, I was nothing, there was suddenly
A great burning under the darkness, a fire
Like fighting up into the wings' lash and the beating
Blackness, and flames like the tearing of teeth,
With noise like rocks rending, such that no word
Can call it as it was there, and for fire only,
Without the darkness beating and the wind, had I
Been there, had I not been far on the hiding light
I could not have borne it and lived.

In addition to creating a powerful description of the Annunciation, Merwin has also proposed here the sweep of language's force as prophecy and creation.

"The Annunciation" is the central and exemplary poem of the second part of *Green with Beasts*. While some poems assume voices from the Judeo-Christian tradition, such as "The Prodigal Son" or "Saint Sebastian," Merwin shifts into an autobiographical voice in "Burning the Cat," "After the Flood," "Thorn Leaves in March," "Low Fields and Light," and "Birds Waking." The narrator in these poems, which are set in transitional moments and spaces, confronts a familiar world that is transformed into either the amorphous and unfamiliar realm, as in "After the Flood," or into the transcendent dimension, as in "Birds Waking." Modulations between light and dark, between the language of record and that of ecstasy, and between allegory and the literal, are explored in these poems.

The third section of the book revolves around images of the sea and the shore. The poems are rich in thickened, almost painterly movements of sound, as in the opening lines of "Evening with Lee Shore and Cliffs":

Sea-shimmer, faint haze, and far out a bird
Dipping for flies or fish. Then, when over
That wide silk suddenly the shadow
Spread skating, who turned with a shiver
High in the rocks?

The initial accents at the beginning of the passage, with their sense of controlled assertion, give way to a sweeping alliteration that overwhelms the narrator and diminishes his control over the scene. The sea is a figure for mystery and chaos, even as it is also the force of life, which is both generative and destructive. In gazing at the sea, the subject in the poem is forced to recognize his own ignorance and his fear of apprehending the unknown. The collection thus comes full circle from the opening poem, "Leviathan," to this concluding sequence of eight sea poems. Merwin moves from the painterly distance of "Evening with Lee Shore and Cliffs," whose title suggests a generic scene, to diving into the sea itself and risking being overwhelmed, as in "Shipwreck." Merwin concludes this sequence with "Mariners' Carol," which recalls his explorations of traditional forms in his earlier collections, and thus posits a nostalgia of origins and the desire for safe return.

The first twelve poems of *The Drunk in the Furnace* (1960) continue this sequence of sea poems. "Odysseus," the opening poem of this collection, implies a weariness with the poet's tasks as Merwin has thus far explored them: "Always the setting forth was the same, / Same sea, same dangers waiting for him / As though he had got nowhere but older." Revising the tradition of Odysseus as the eternally voyaging sailor whose travels enrich him (the tradition followed in works of Dante Alighieri, Alfred Tennyson, C. P. Cavafy, and Nikos Kazantzakis), Merwin's Odysseus is a man who is exhausted and encumbered by the past, who has lost both his desire to return and his clear vision of *Ithaka*.

If exhaustion is one response Merwin registers in these poems, terror is another, particularly in "The Iceberg" and "The Frozen Sea." In "The Frozen Sea" Merwin recounts polar explorations and foreshadows his critiques of the exploitation of the New World in *The Lice* and *The Carrier of Ladders*. The landscape of "The Frozen Sea" overwhelms the explorers' hubris, reducing their sense of self in the face of elemental powers. The landscape cannot be accommodated or anthropomorphized; nor is the landscape a metaphor for transcendence. Instead, Merwin's travelers confront their limitations and utter mortality:

But danger
Had given shape, stiffening shape, to our
Pride, and that sustained us in silence
As we went over that screaming silence.
Yet how small we were around whom the howling
World turned. We could see only half a mile.
And only a soulless needle to tell us where
In the round world we were.

Yet even this sense of location in a chaotic, threatening world is deceptive, because these voyagers to the Antarctic find that "whichever way we turned / Was north, the sides of the north, everywhere." Knowledge of one's exact position is countered by the limitless horizons that lack all coordinates and landmarks.

The sequence concludes with "The Bones," where the poet walks "over the tide-wastes," and sees in the flotsam the "bones of things" and of "man's endeavors." Yet in this wastage of shifting

LEMUEL'S BLESSING

> Let Lemuel bless with the Wolf, which is a
> dog without a master, but the Lord hears
> his cries and feeds him in the desert.
>
> Christopher Smart: *Jubilate Agno*

[You that made me]
You that know the way,
[Spirit,]
[I bless your ears: let me approach.]
I bless your nails which are like dice in command of their
    own combinations.
Let me not be lost.
I bless your eyes for which I know no comparison.
[Run with me like the horizon, let me]
[Do nothing which would forfeit me your presence, for] Without you
I am lost and hungry,
Ill-natured, untrustworthy, useless.

All my bones bless you like an orchestra of flutes.
Let the weapons of the settlements fall exhausted a long way
    from me,
And their dogs the same. My laugh blesses you.
Whereas a dog is shameless and wears servility in his tail
Like a banner;
Let me wear the opprobrium of possessed and possessors
As a thick tail properly used
To warm my worst and my best parts. My tail blesses you.

Deliver me from the coyotes, dingoes, foxes, jackals, hyenas,
Which deny your very existence.
Let me approach distress, when I have to, as a good dust to
    roll in,
Discouraging pests of my own.
My pelt blesses you. Let it never be owned or walked on.

But Deliver me from the error which hangs around at the fork of
    hesitation,
And has many arms and endless patience, knowing
Its moment and the sequel,
For It can go before and wait, or follow and be there, crossing
The heavens like a goat, jumping from star to star, even
In the daytime when the stars are invisible,
Let me not be made the sport of the deadly ruths.
Every one of which is more intelligent and knowledgable than I am.
Deliver me

First page of the final draft for "Lemuel's Blessing," included in The Drunk in the Furnace *(TS. 20:12/011, d. Apr. 21, 1960, W. S. Merwin Archive, Special Collections, University of Illinois at Urbana-Champaign Library; by permission of W. S. Merwin)*

sands, Merwin recognizes the purpose or design of things: "Shells were to shut out the sea," while "The bones of birds were built for floating / On air and water," and "fish were devised / For their feeding depths." "A man's bones were framed," Merwin concludes, "to know the sands are here . . . for giving / Shapes to the sprawled sea, weight to its winds." Although knowledge, imagination, and expression are celebrated here, the poetic and rhetorical powers of the poem fail to convince; the issue is less the truth of this observation than it is, as Merwin senses, the inadequacy of the prevailing culture to sustain this particular strand of narrative in which images of the sea serve as a form and a literal place of revelation.

When the expression of the heroic is found to be impossible, Merwin turns to family portraits and studies in a depleted, yet darkly comic American landscape. The concluding dozen poems of *The Drunk in the Furnace* display a radical shift in language, tone, and manner. In one poem Merwin directs his scorn at the complacency of those who live in impoverished conditions:

> But nothing changes their concern,
> Hurries them or calls them. They must think
> The whole world is nothing more
> Than their gainless harmless pastime
> Of utter patience protectively
> Absorbed around one smooth table
> Safe in its ring of dusty light
> Where the real dark can never come.

Merwin's Edward Hopper–like portrait of the pool players in "Pool Room in the Lions' Club" delineates the pool players' choice not to engage the world, but rather, to remain oblivious to the world and their actions in it. Family ties, as these poems illustrate, bind one as tightly as catchphrases and commonplaces. "Grandmother Watching at her Window" suggests that the consuming reliance on conventions impoverishes one completely. This rural landscape, drawn from Merwin's childhood, is further characterized as being an infernal place in the poem "Burning Mountain." Under the mountain an abandoned coal mine slowly burns, a testament to exhaustion and failed responsibility that assumes cosmic proportions: on the mountain "The hushed snow never arrives." Instead, there is "An emanation of steam on damp days, / With a faint hiss, if you listen some places"; while below the mountain, the fire "consumes itself, but so slowly it will outlast / Our time and our grandchildren's."

The title poem, which concludes the collection, serves as a summation of the comedy and horror that Merwin observes in these American landscapes. "The Drunk in the Furnace" continues to describe the infernal and wasted landscape of "Burning Mountain." The drunk who makes a discarded furnace his "bad castle" is a Dionysian bard to whose "jugged bellowings" the "witless offspring flock like piped rats." The poem's tone is by turns raw, comic, and ironic. While the children are drawn to his "hammer-and-anvilling with poker and bottle," their parents attend the "tar-paper church," where, after listening to homilies, "They nod and hate trespassers." Spiritual and material impoverishment result in the bigoted and narrow lives that Merwin portrays; in such deprivation and exhaustion, the human condition cannot be celebrated.

In the 1950s Merwin lived primarily in Europe, serving as the tutor to the children of the Princess de Braganza in Portugal and to the household of Robert Graves in Majorca, off the coast of Spain. He then lived in London and in southwestern France, where he bought and renovated a small farmhouse. The three mnemonic stories of *The Lost Upland* (1992) portray, particularly in "Shepherds," life in southwestern France as Merwin found it during this period. During this time he produced plays and also translations of Spanish and French plays and poetry, often for broadcast on BBC television. In 1956 and 1957 he was playwright-in-residence for the Poets' Theatre in Cambridge, Massachusetts. In the late 1950s he wrote poetry and book reviews for *The New York Times* and *The Nation*. By this time his understanding of environmental and nuclear catastrophes in the making found expression in prose: the most notable examples are his "Act of Conscience," an essay on the voyage of an Everyman who is protesting nuclear arms, and his Swiftian essay, "A New Right Arm," a satiric proposal for dealing with atomic-age mutants. By the early 1960s Merwin had returned to New York City, where in 1962 he served as poetry editor for *The Nation*.

Although maintaining his house in France, Merwin had shifted his geographical base to New York by the time *The Moving Target* was published. In later years he moved westward to Mexico and finally to Hawai'i. *The Moving Target,* as the title suggests, demonstrates another, more radical, shift of discourse. The poems, as one moves through the collection, become increasingly metaphorical, even while the language exposes its bare ligaments. The opening two poems serve as a coda to *The Drunk in the Furnace*. It is the third poem, "Lemuel's Blessing," that reveals the poet's failings: "I have hidden at wrong times for wrong reasons." It also gives his plea: "let me leave my cry stretched out behind me like a road / On which I have followed you." The poem is a renunciation of his past poetic career: like the wolf who has been domesticated, as the epigraph from Christopher Smart's *Jubilate Agno* (1758–1763) suggests, Merwin is seeking to recover the authentic spirit of poetry, while confessing:

> I have sniffed baited fingers and followed
> Toward necessities which were not my own: it would make me
> An habitué of back steps, faithful custodian of fat sheep[.]

Taming both poetry and the spirit leads back to the scenes of impoverishment in *The Drunk in the Furnace,* where the poet, at best, is reduced to "jugged bellowings." "Lemuel's Blessing" is an incantation and invocation; with its declarative lines echoing the Psalms, the poet asks for sustenance. Paradoxically, Merwin petitions to be released from the community by employing the traditional and communal use of prayer. It is this paradox of the writer, who must use the language of the community and tradition, yet must also refuse its implicit domestication, that Merwin invokes in "Lemuel's Blessing."

With *The Moving Target* Merwin's poetry becomes more textual; that is, more expressive of its own material presence even as it moves toward a kind of self-erasure. This change undercuts the poet's authorial presence. The rejection of punctuation and the dismembering of the lines, often demanding that the reader take responsibility for ascertaining the syntax, constitute Merwin's diminishment of authorial control. By placing greater demands on both the language and the reader, Merwin suggests the necessity of rebuilding a communal, transactive process both in reading and in living in the world. This dismantling of language and defamiliarizing of the process of reading mark a new stage in Merwin's poetic process.

In describing Merwin's poetry, the critic Charles Altieri has spoken about the classical *via negativa,* a phrase that refers to the process of emptying, the negative path or way of mystics who seek to apprehend ultimate reality by emptying themselves of the world of sense and sensation. Indeed, in some of Merwin's work of this period, absence is revealed to a form of plenitude. This becomes especially apparent in the long love poems such as "This Day of This Month of This Year of This," "The Way to the River," and "For Now." The loss or absence of love becomes the step toward knowing what love really is, in the same way that speechlessness becomes a step toward finding words that are authentic. "The Way to the River" is the most direct of these poems, with its repetitive, often embedded phrase, *be here,* which underscores desire and absence:

> I have lit our room with a glove of yours be
> Here I turn
> To your name and the hour remembers
> Its one word
> Now[.]

Desire is a form of loss drawn into the future. Renunciation is, conversely, a retreat into loss. This retreat, the mystical *via negativa* that Merwin signals in "For Now," is of utmost necessity. The obverse of "The Way to the River," "For Now" is a leave-taking of the world. As the title indicates, "The Way to the River" is a narrative that moves toward love and a beloved; "For Now" is the cumulative moment of shedding attachments to the social and intimate world. As the earlier poem repeated the phrase *be here,* this poem embeds *goodbye,* and thus decenters the otherwise clearly rhetorical structure of the anaphora. What would be a traditional ritual structure, Merwin destabilizes, thereby implicitly demanding that the reader find the language and rhythm to revitalize his ways of knowing. Whether one reads this poem as a preparation for death or as the relinquishment of the worldly by prophets and penitents, Merwin sees that such relinquishment is necessary for love's return. In a Janus-like turn in the concluding lines, Merwin enjambs the sole line "My love": "goodbye" precedes it and "You that return to me" immediately follows, arriving "Between death's republic and his kingdom."

The great myth of return and renewal, described in Robert Graves's *The White Goddess* (1947), is readily apparent throughout Merwin's work, and is often evoked in numinous single lines such as this line from "For Now:" "You are not here will the earth last till you come." Myths of redemption and renewal, the figures of Demeter and Persephone, and an acute awareness of environmental degradations fuse in this line. In "The Crossroads of the World Etc." the possibility of renewal seems distant if not extinct. Here the topos of the metropolis, the crossroads of the world, disorients and alienates. Rather than the conventional image of isolation found in the frozen seascapes of "The Iceberg" or "The Frozen Sea" in *The Drunk in the Furnace,* Merwin here finds in the grammar of the city the profound dislocation of the self, and he has translated it into a poetic grammar. This translation is achieved neither by the architecture of the modernist's sublime in form and function, nor by a postmodern ahistorical collage; instead, Merwin uncovers an atavistic voice and finds form for it:

FOREIGN SUMMER

The night withdraws from behind the fog
As an eye from behind a cataract. In
Time the blindness itself withdraws
And there are the little fields drying
Like the laundry of strangers. Foreign
Summer, I came to you offering
A sightless coin, but my own, holding it
Before me like a lantern, and it is gone, and
Now even in your discarded costumes
I can see that your nakedness was never
Mine. The day widens its revelation
Between us, showing the roads, and I take up
My empty hand, which will be my torch
This evening. As for you, if you go
The way of what I gave you, you will
see the bright leaves floating face-upwards
In the swollen lakes of November
And pass on without recognition. Your
New darkness knows the way. All along
Its old friends will keep coming forwards
To greet it. Beyond their embraces
You will go on and on, happily, with your
Hope growing like hunger at every turn
And the wrong key in your hand.

A variant of "Foreign Summer" that was submitted for publication but rejected (TS. 20:12/003, d. Aug. 28–29, 1960, W. S. Merwin Archive, Special Collections, University of Illinois at Urbana-Champaign Library; by permission of W. S. Merwin)

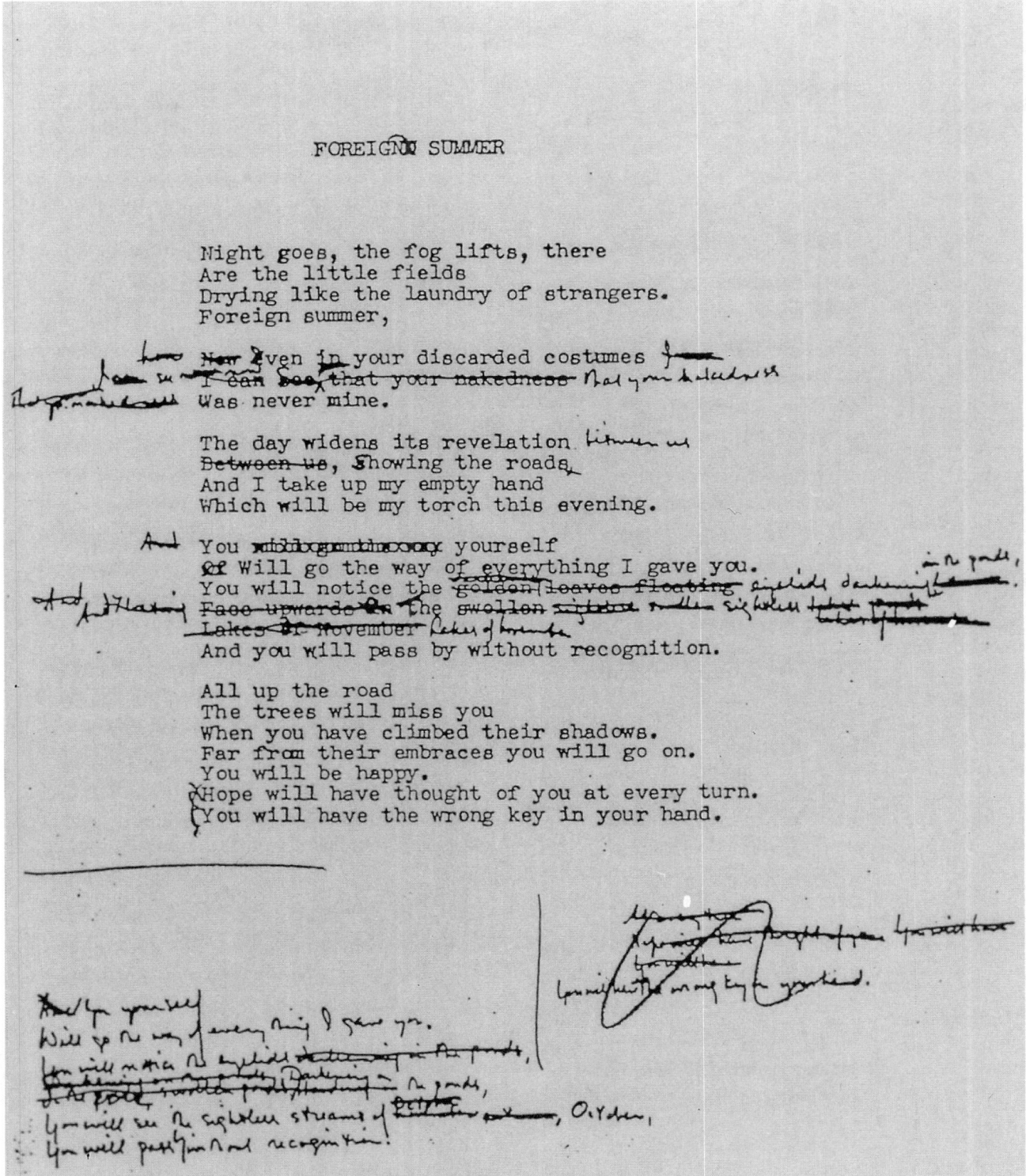

The revised final draft of "Foreign Summer" (TS. 20:10/004a, circa summer–fall 1961, W. S. Merwin Archive, Special Collections, University of Illinois at Urbana-Champaign Library; by permission of W. S. Merwin)

So late in the stone so long before morning
Between the rivers learning of salt

Memory my city

Hope my city Ignorance my city
With my teeth on your chessboard black and white
What is your name

With my dead on your
Calendar with my eyes
In your paint
Opening
With my grief on your bridges with my voice
In your stones what is your name
Typed in rain while I slept.

Syntax and words are reduced to the most essential; as the vocabulary becomes more particular, it also becomes more universal. Language assumes a paradoxical condition: as a carrier of memory – cultural, poetic, primal – it becomes incomplete and cryptic while also suggestive of a once and possible fullness.

Merwin's work should be viewed as a continuum in which he persists in exploring and revising his own poetic directions. In this sense the larger work, the sum of his collections, involves a self-reflexive process. Merwin's work, however, has been divided into distinct groups of collections. To structure these phrases is to assume a particular critical stance that isolates and periodizes. The reading of the poet's work is controlled by how – or if – one decides to group the collections. As consecutive series of pairs or triads, the concluding collection in each grouping attains the rhetorical, figurative, and visionary power that was gathering in the previous collections. If Merwin's work is divided into groups of four collections, it is the early collections that seem most successful and the later ones that provoke critical questions.

The Lice is perhaps one of the most critically acclaimed collections of poetry since World War II. It has also come to be the measure by which Merwin's poetry is judged, and the collection, more than the others, from which Merwin has had to move. The language and vision of *The Lice* find their grounding in such earlier poems from *The Moving Target* as "Noah's Raven," "The Saint of the Uplands," "The Crossroads of the World Etc.," "The Man Who Writes Ants," and "Daybreak." Gnomic statements, the spare but essential vocabulary, the often incantatory prosody, and the occluded myths all come to typify the poems in *The Lice*. The critical vocabulary in response to this book has correspondingly emphasized aspects of silence, absence, negation, deconstruction, alienation, distance, voicelessness, abstraction, and stasis. These poems go beyond the boundaries of given poetics and given cultural discourses to attempt fully to know and express spiritual agony.

The book's title is taken from Heracleitus's riddle about some boys who, after catching lice, said to Homer, the wisest man in Greece: "what we have caught and killed we left behind but what has escaped we bring with us." Homer's inability to understand what the boys meant suggests how ignorance becomes apparent when humans attempt to unmask the world of things. This fragment is taken by Merwin as a cautionary tale; his own task is, in part, to chart that ignorance and unmask it. It is hubris in a human being to assume that one can gain knowledge of all things, and that such knowledge then becomes manageable, becomes, in fact, mere information – the modern era is even called the Age of Information. *The Lice* was published while the Vietnam War, an assault on cultural otherness, was being fought. The late 1960s was also a time of growing awareness of environmental degradations (an assault on phenomenological otherness), as well as a time in which language was being corrupted by political euphemisms and by the phrases used in advertising. Against this background, Merwin's poetry of this period can be seen as an attempt both to redirect people's attention to the deep mystery of things and to demonstrate the profound, agonizing distance that separates people from the world. Such poems as "The Animals," "The Last One," "The Asians Dying," "When the War is Over," and "For a Coming Extinction" address these themes. To read Merwin's work, however, primarily as a product of and response to specific cultural and political events – as *The Lice* in particular is sometimes read – is to neglect Merwin's ongoing epistemological and theological deliberations.

His much anthologized "For the Anniversary of My Death" expresses the centrality in human experience of mystery, ignorance, and reverence. In this compressed poem the poet moves through registers of strangeness, from considering his foreseeable but unknown death to known, almost sermonized phrasings, "the love of one woman / And the shamelessness of men." The traditional paradox that time is both eternal and linear is established but is further complicated by the suggestion that, while linear time is illusory, a recognition of it is nevertheless essential to maintaining identity and human community. Reverence toward the unnameable, the realm of the mysterious or holy, demands that one release himself or herself into that other aspect of time, but the poet, having attained such a release, can only retell his experience through a linear narrative:

> As today writing after three days of rain
> Hearing the wren sing and the falling cease
> And bowing not knowing to what[.]

In these final lines at the end of an accumulative sentence, the concluding word is highly stressed, bringing to an absolute close – or foreclosure – this meditation. The elegance of this poem, in its prosody and argumentative structure, points ironically to epistemological limitations. Art (*Techné*) and imagination, Merwin implies, are only provisional, although necessary, as one finally, sooner or later, reaches the impasse: "what."

This impasse has theological dimensions that are often linked to topical issues. In "For a Coming Extinction," which belongs to Merwin's whale poems and his meditations on ecological issues, the speaker shows an extreme hubris not only in dictating to the gods ("Tell him / That it is we who are important"), but also in the speaker's suggesting that they alter their energy from lifegivers to avatars of death. Human guilt lies in this inhuman turn toward destruction; our humanness, ironically, lies in the recognition of our shame, or, as Merwin writes at the end of "Avoiding News by the River": "If I were not human I would not be ashamed of anything." Adam's task is not only naming – the poet's task, as well – but also understanding his transgressions and thus realizing his shame. Merwin's spiritual meditations are perhaps their bleakest in "December Among the Vanished," where, in a shelter against the winter snows, the poet "sit[s] with a dead shepherd / And watch[es] his lambs." The influence of T. S. Eliot shows in this poem; but Eliot in his later work finds the traces of salvation in the European tradition and its beliefs – a position with which Merwin does not concur. Unlike Samuel Beckett, with whom Merwin is often compared, the poet waits for no one; instead, the poet is the last one left watching. There is no Messiah, not even the illusion of one; instead, the Judeo-Christian figure of God as Good Shepherd is discovered in his ruined tabernacle to be only a desiccated shepherd, an archaeological remnant.

Merwin's poems of numinous relationships, whether these relationships are found in his observations of the natural world or within the figuration and prosody of his language, are among his most revelatory. "Dusk in Winter," "The Dragonfly,"

Cover for the 1970 book in which Merwin contemplates the westward expansion of the United States

"Watchers," and "Looking for Mushrooms at Sunrise" are four such poems, in which an intimacy of voice combines with a sureness that the moment conveyed, which could be the most mundane or transient of moments, illuminates the fact of the resonances among all things and all beings, as suggested in the second stanza of "Looking for Mushrooms at Sunrise":

> In the dark while the rain fell
> The gold chanterelles pushed through a sleep that was
>   not mine
> Waking me
> So that I came up the mountain to find them[.]

While the poem reiterates the distance between contemporary human beings and the world, it also suggests the potential of a spiritual reawakening of connections with the world that are inherent or atavistic in humankind. It is this process of recognizing and celebrating these intimate resonances that Merwin's later work builds upon.

In 1970 Merwin's seventh book of poetry, *The Carrier of Ladders,* was published, as was his collection of short fiction, *The Miner's Pale Children. The Carrier of Ladders* gathers together and continues the exploration of many of Merwin's past themes and affinities. Such poems as "The Judgement of Paris," "Envoy from D'Aubigne," "Shore," and "Pilate" share the elegance of his early dramatic narratives. "Words from a Totem Animal" and the four psalms recall the plaintive "Lemuel's Blessing" of *The Moving Target.* Indeed, all the larger thematic concerns of immanence, the dialectics of plenitude and renunciation, the conditions by which the self is known, and the dispossession of identity are present in *The Carrier of Ladders.* Lemuel's concluding prayer, "sustain me for my time in the desert / On what is essential to me," posits the conditions by which Merwin examines his own being, and by extension, that of the culture in which he lives.

Embedded and unmarked in *The Carrier of Ladders* is a sequence of poems in which Merwin meditates on the westward expansion of the United States. Such a project parallels that of other American writers, such as William Carlos Williams, Hart Crane, Theodore Roethke, Gary Snyder, and Edward Dorn; however, Merwin's project resists the Whitmanesque poetic of absorption and expansion. Instead, Merwin's poetic, featuring another American voice, is much more grounded in the tradition of Henry David Thoreau, whom Merwin sees to be in polar opposition to Walt Whitman. In a 1982 interview with Ed Folsom and Cary Nelson in the *Iowa Review,* Merwin states that his affinity with Thoreau is based upon:

> the way in which he meant "In wilderness is the preservation of the world" for one thing. Or the recognition that the human can not exist independently in a natural void; whatever the alienation is that we feel from the natural world, we are *not* in fact alienated, so we cannot base our self-righteousness on that difference. We're part of the whole thing. And the way Thoreau, very differently from Whitman, even in a paragraph, takes his own perception and develops it into a deeper and deeper way of seeing something – the actual seeing in Thoreau is one of the things that draws me to him.

In "The Lake," a prefatory poem to his Western sequence, Merwin clearly echoes Thoreau's process of seeing both literally and figuratively. In response to the opening proposition, "Did you exist / ever," which could only be posed after a cataclysmic history, the poet sails across the lake, looking below, down into history, at an Indian village "said to be drowned there." The poet is on the "surface layer of a continental palimpsest," as Folsom writes, and hovering on the "surface of the water in the present, [the poet is] the dead Indian's future; he hangs in the Indian's sky only when the Indian's present is past; there are no natives left in the drowned village (if, indeed, it is still there) to look to the watery sky to see the white man in their future, the new creation that would cover them."

This Western sequence of fifteen poems, beginning with "The Approaches" and concluding with "The Removal," is reproduced almost in its entirety (and thus given prominence) in Merwin's *Selected Poems* (1988). This sequence provides Merwin's most sustained meditation on American history, at least to this point in his career. In the opening three poems, "The Approaches," "The Wheels of the Trains," and "Lackawana," the poet traces his own journey into America. In so doing, and as prefigured by "The Lake," the poet begins to move through the continent's palimpsest of history: "Other Travellers to the River" evokes the botanist, ornithologist, and writer William Bartram, who is responsible for promulgating the idea of the noble savage; "The Trail into Kansas" and "Western Country" depict the westward movement of the European immigrants; and the geologist John Wesley Powell emblemizes the search "for the virgin land" in "The Gardens of Zuni." "Presidents" and "Homeland," in particular, are dispraises, or bardic condemnations, of the U.S. government in regard to its duplicitous treatment of the Native Americans and the government's misuse of the American land. The six-part poem "The Removal," which concludes the sequence, depicts the various forced marches, particularly during the tenure of President Andrew Jackson, to the barren western reservations. This process of removal constituted an erasing of entire languages and cultures. It is this history, reflected throughout our modern Western culture, that Merwin sees as defining our identity.

"Huckleberry Woman," placed at the center of the Western sequence, conjoins the autobiographical realm with the historical and archetypal elements of the sequence. In this itinerant woman, a woman "of unnamed origins" who gathered berries, recalled from the poet's childhood in Pennsylvania, Merwin sees a figure of himself in his desire to know the land and to be "borne with you on its / black stream" of "the unlighted river," to "go with the sound." This intimate contact is realized through his recognition of her pain when she spills her load of "small / starless skies." This intimacy of voice and memory underscores not the deconstruction or abstraction of the self but the necessity of a remembering and recollective self. Simply recounting familial memories is not sufficient, however; memory must recover the myriads of connections between all things.

"In the Time of the Blossoms," which concludes *The Carrier of Ladders,* unites Merwin's thematics of presence, love, and voice. The ash tree, a symbol of fertility, is one of the disguises of Nemesis, whose association with the Furies balances life and death in this image of a barren tree (whose "leaf skeletons" are as "fine as sparrow bones") among blossoming trees. The image of the tree "evokes a kind of 'mythological present,'" writes Altieri, that "reconciles flux and stasis as the blossoms balance in the wind and seem to multiply in their composed plenitude." The poet becomes, in this full moment, open to inspiration's "unbreathed music." The poem ends with the invocation, "Sing to me," suggesting that the poet has been writing toward an opening rather than a closure. Although the Western sequence, in contrast, ends in silence and extinction, thus constituting a necessary descent, its central poem, "Huckleberry Woman," prefigures the numinous presence of "In the Time of the Blossoms," with its insistence that "we go with the sound."

Writings to an Unfinished Accompaniment concludes Merwin's exploration of a poetics that many have described as one based on silence, distance, allegory, and absence. "A Purgatory" reiterates some of his past concerns, while also stipulating the task at hand:

> everywhere
> the vision has just passed out of sight
> like the shadows sinking
> into the waking stones
> each shadow with a dream in its arms
> each shadow with the same
> dream in its arms
>
> and the eye must burn again and again
> through each of its lost moments
> until it sees[.]

Emptiness, exhaustion, and repetition describe this purgatory that is, Merwin suggests, our contemporary condition, one shared by the poet and through which the poet must pass. The eye (and "I") must become an active agent rather than a passive receiver. Replete with alliteration and repetition, this preceding quotation also contains examples of Merwin's signature essentialist vocabulary of stones, shadows, waking, dream, loss, and sight. Here Merwin's vocabulary, though shared with the often politically charged poems of *The Lice,* is adopted for brief narratives that are often composed as aphorisms or fables. The poetry, having been pared to the most essential particulars, often turns to those particulars to construct allegories of its own *poiesis,* as exemplified by "Song of Man Chipping an Arrowhead," "The Old Boast," "At the Same Time," "The Unwritten," "Sibyl," "A Flea's Carrying Words," "Finding a Teacher," "The Search," "Travelling," and "Gift."

These self-reflexive poems, however, are not celebratory. While "Song of Man Chipping an Arrowhead" suggests a myth or an aphorism collected and translated from an oral culture, such a context would reveal the corresponding history of the vanishing of the language the song was composed in – a history detailed in "The Removal" in *The Carrier of Ladders.* "The Old Boast," an equally brief three-line poem, is disturbingly disjunctive as its title deliberately undercuts the poem's text: the harpist's magically fertile synesthesia is labeled an old and discredited boast. Language, ironically for this poet who loves languages, becomes disease in "A Flea's Carrying Words." Love and language are evoked simultaneously in "The Search": the narrator cannot consummate his desire or his vision, although "I keep on trying to come toward you / looking for you"; instead, he is profoundly lost, as "before me stones begin to go out like candles / guiding me." In this closing line one cannot but hear Merwin self-reflexively commenting that his own essential language is "begin[ing] to go out." "Gift," which concludes *Writings to an Unfinished Accompaniment,* reinforces this uncompromising vision of extinction. To continue, the speaker must be possessed, he must give over his identity to this gift; Merwin pleads: empty "nomad live with me / be my eyes / my tongue and my hands / my sleep and my rising." This dark invocation offers none of the ecstasy of Whitman's affirmation of an American identity in *Song of Myself,* nor the affirmations of a democratic vista. The poet is only free to "call to it Nameless One O Invisible": to have a voice he must be swept up in its voice, thus extinguishing his body and self.

The next two volumes of poetry are transitional works. In them there are clear links to the previous four collections, but there is also a definite shift toward a poetics of intimacy and resonance. It was also during this period from the mid 1960s to the late 1970s that Merwin's most significant trans-

Cover for Merwin's 1977 collection of love poetry

lations of lyric poetry were published. In 1969 he published his translations of Pablo Neruda's *Twenty Love Poems and a Song of Despair,* Antonio Porchia's *Voices,* and Jean Follain's *Transparence of the World.* Merwin's translations with Clarence Brown of Osip Mandelstam's poetry appeared in 1973 (*Osip Mandelstam: Selected Poems*); and, with J. Moussaieff Masson, Merwin translated selections of classical Sanskrit love poetry in *Sanskrit Love Poetry* (1977). The lyric intensity, especially in the highly metaphorical poems of Mandelstam, certainly corresponds to Merwin's own lyric sensibility. While his translations of Follain, with their often distanced whispers, are seen as offering an informing poetics, Porchia's aphoristic prose poems point to a poetic sensibility that parallels Merwin's. The emphasis on love poetry is striking, and it is this sensibility that emerges in *The Compass Flower* and *Finding the Islands* (1982).

The dark urgency of "Gift," could not be sustained. Merwin's voice becomes more personal; this change is not a renunciation of his past poetry but a necessary re-visioning. With such poems as "The Drive Home," "The Next Moon," "The Snow," "The Arrival," and "Apples," *The Compass Flower* opens with an elegiac sequence, primarily directed toward his parents, and the themes of birth, death, and love resonate. Instead of relinquishing body and soul to the possession of the spirit, as he does in "Gift," Merwin here sees the importance of familial, biological, and even communal connections. In "The Snow," he writes, "we are one body / one blood / one red line melting the snow / unbroken line in falling snow." This realization, however, comes not as an affirmation, but as a defining, even ontological condition.

Not since *Green with Beasts* had Merwin divided a collection of poems into distinct sections. *The Compass Flower,* however, is divided into four sections, evoking such symbols as season and age, direction and space, and elements and senses. The second section is most clearly an interrelated group of poems that offer various meditations on the city. The culmination of this group, and one of the central poems of the collection, is "St Vincent's," a portrait of the New York City hospital. Merwin describes this hospital as being both impenetrable and enigmatic, as well as being a center of activity. The narrator is an inadequate witness who does not know what transpires around him, does not know the hospital's origins, nor, finally, the conditions and history of charity — all of which is implied by the final self-critical line, "who was St Vincent." As in the previous three collections, these poems do not utilize punctuation. Instead, they rely on line and stanza breaks as well as shifting caesuras. Furthermore, the reader must listen for inflections and modulations implicit in the words rather than following diacritical guides or punctuation. This unmarked question at the end of "St Vincent's" typifies how Merwin draws us away from rhetorical closure and, instead, into the reflective interiority of language.

The poems of the third and fourth sections move through various landscapes, including New York City and southwestern France. The melding of love, language, and myths of renewal in the long poem "Kore" counterbalances the apparent lack of connection in "St Vincent's." The twenty-four letters of the Greek alphabet mark each stanza, suggesting that each letter inscribes the possibility of the renewal of love and language. "Kore" also separates the autumnal or twilight poems of solitude, such as "Summer Night on the Stone Barrens" or "September Plowing," from the love poems that follow "Kore" in the third section. This arrangement suggests a movement into a world of connection. As part of this renewal, Merwin's poems return to the human world, replete with its failings, moral complications, and enigmas, as suggested in such poems in the fourth section as "The Coin," "The Fountain," and "Talk of Fortune."

Merwin continues to search for the form and language that corresponds to his view of interconnectedness in *Finding the Islands*, part of which appeared in 1978 in a limited edition. These poems are all constructed in discrete, three-line units, a form he has experimented with since "A Scale in May" from *The Lice*. Although resembling haiku in the emphasis on an imagistic evocation of particulars, these poems do not otherwise share the formal structures of haiku. The poems have a notational quality, as though what is written has just been seen and experienced. The poems include landscapes, particularly of the West and of Hawai'i, and love poems. Through their immediacy of sensation and desire, Merwin seeks to convey oneness and interconnection. They are consequently at polar opposites in language and epistemology to his work from *The Lice* through *Writings to an Unfinished Accompaniment*. In their privacy and transparency, they mark, however, a poetic direction Merwin does not pursue.

In his tenth book of poems, *Opening the Hand,* Merwin returns to the familial poems that closed *The Drunk in the Furnace* and that he had touched upon in "Home for Thanksgiving" and "A Letter from Gussie" (in *The Moving Target*) and "Huckleberry Woman" (*The Carrier of Ladders*). In the first section of *Opening the Hand,* Merwin reexamines his family, writing in "The Waters": "there were lives that turned and appeared to wait / and I went toward them looking." Merwin's poetry has always evoked the desire to return to origins; the poet's task is not only to undertake the ardors of passage, but also to discover where his origins lie. This process of remembering is thus elegiac, because not only is loss the subject of these poems, but so, too, is the knowledge of continuity, evoked at the conclusion of "The Waters": "faces that would never die returned / toward our light through mortal waters."

In such poems as "The Oars," "Sunset Water," "The Waving of a Hand," "The Houses," and "Yesterday," the figure of the father is central, as he was not in Merwin's previous poems about his family and his early years. As the title of the collection suggests, Merwin opens his hand to reveal his lifeline, to relax the force of closure and resistance, and to embrace or hold another person. The complications of father/son relationships are finely drawn in "Yesterday," where a friend confesses he failed as a son by walking out on his father; the ambiguity of reference, however, suggests that the poet, too, has failed. "Sunset Water" places the son and father in opposition; the father struggles because he finds himself out of place in nature, whereas the son is open to the movements of the sea and nature. Nonetheless, a tenderness pervades "Sunset Water," as father and son both acknowledge the father's unease and vulnerability, while the poem withholds judgment: "all his life he swam   doggie paddle / holding hurried breaths / steering an embarrassed smile." In these lines, as in many of the poems in *Opening the Hand,* Merwin visually marks, through typographic white space, the caesura's pause; thus, the lines themselves open. Further father/son conflicts and complications figure in "The Houses," where a clear distinction is drawn between the son's imagination and the father's resistance to imagination.

The central poem of the group is "Apparitions," where the poet gazes at his hands and sees his parents' hands transposed upon his own. The poem immediately acknowledges the poet's realization of his mortality through the recognition of the absence – and haunting presence – of his father and mother. Although the narrator lists what they will never do again, by seeing in his hands the shape and gesture of their hands, he sees that a continuity exists. In the final stanza the poet distances himself – in self-defense, as though to ward off filial sentimentality – by calling the apparitional hands "nobody's children." This closure, however, has been undercut by the poem's explicit tenderness.

The poems in the second section of the collection are divided between two landscapes, Hawai'i and New York City. Merwin is not creating an opposition between nature and the city; rather, these poems form an arc connecting the two landscapes in which he lives. In this sense they insist on a knowledge of each locale and the connections, although these may be unstated, between them. These poems are the inverse of "St Vincent's," because here the narrator knows the origin and history of what he gazes upon. In "Questions to Tourists Stopped by a Pineapple Field," the narrator turns this knowledge into an interrogation of the tourists, thereby exposing the connections between the consumer economy and environmental irresponsibility, as well as implying that, within such an economy, there is the moral failure of knowing and loving one's landscape. In the parallel poem, "Sheridan," Merwin reconstructs Gen. Philip Sheridan's actions during the Civil War; the poem shows that Sheridan in the present is known only by name and only for a small square submerged in the traffic and exhaustion of New York City's Greenwich Village. Thus history has all but vanished, and, with it, the common narratives by which the nation once defined itself. In "The Fields," however, Merwin unites the poet and the celebrants of an ethnic street festival in a communal act of imagining a culture and landscape that "nobody has seen" but is claimed as the place of origin. "The Fields" suggests the role of the poet: drawing from the disparate elements of the festival, the poet constructs a vessel that is able to carry all the imaginings of the festival. Yet, as "Emigré," one of the central poems of the third section, reminds us, the new traditions evoked in "The Fields" are built upon a myriad of losses.

The collection closes with "The Black Jewel." A poem of great immediacy and intimacy, it discloses the source of song. The unwavering song of the cricket draws together all the sounds of the world into a susurrant rhythm: "south wind in the leaves / is the cricket / so is the surf on the shore." The cricket is the archetypal poet who is filled with song, one whose song is always and already present, a song whose presence (thereby, the poet's presence) reawakens the past: "before I could talk / I heard the cricket / under the house / then I remembered summer." The cricket is also a representative of the unconscious (in the tradition of Gaston Bachelard or Carl Jung) as Merwin locates the cricket living "in the unlit ground / in the roots" or "under the house." The poet must listen to the cricket as intently as he must listen to the voice of the imagination.

In a time when it has become increasingly apparent to ecologists and anthropologists that the predations upon cultures, especially aboriginal cultures, corresponds to the devastation of the local and global ecology, and that the loss of distinct languages parallels the extinction of species, Merwin's poetry, from *The Lice* to the present, has moved from decrying the hubris of humans in their willful retreat from and destruction of nature to illuminating an ecology of language. In such poems as "The Shore" or "The Black Jewel," Merwin suggests an interconnective web of song and language, for which all beings are responsible. In his 1988 collection of poems, *The Rain in the Trees,* Merwin finds himself not only listening intently to this jeweled web of song, but also, hearing the dead. The voices of his dead parents and his past haunted many of the poems in *Opening the Hand,* and they are present here in such poems as "Native Trees," "Pastures," "The Salt Pond," and "Waking to the Rain." The voices of a disappearing landscape and its aboriginal culture, that of Hawai'i, are also central. *The Rain in the Trees* differs from *The Lice* in that these new poems clearly link a culture to its landscape and that culture is a specific culture.

This sense of locality and one's responsibility to place is apparent in "Rain at Night." The poem opens with the single line, "This is what I have heard," from which emerges the story of the destruction of the "old trees," the "sacred 'ohias" and "the sacred koas then / the sandalwood and the halas." To kill the sacred trees is to destroy the culture that holds those trees sacred. This observation is underscored by the poem "Native Trees," in which Merwin's parents do not know the names of the trees of their childhood: in that lack of knowledge is the destruction of language and place. But the story in "Rain at Night" continues: "the trees have risen one more time / and the night wind makes them sound." How this renewal occurs is not disclosed — nor is it disclosed whether their return is apparitional or actual.

Because the trees are sacred, however, they return in the presence of their names: the double meaning of "the night wind makes them sound" emphasizes the capacity of the name as sound, not taxonomy, to be a whole and numinous presence. The sequence of eleven poems, from "Hearing the Names of the Valleys" to "Kanaloa," emphasizes the intimate and sustaining relationship of words and things. This intimacy defines communal identity, as "Hearing the Names of the Valley" proposes; for the native speaker the words are part of himself, but for the outsider they are remote in meaning and sound: "when he says them at last / I hear no meaning / and cannot remember the sounds." What is native, or original and endemic, cannot be appropriated: in "Native," the poet must write "in my own hand / on the white plastic labels ... their names in Latin," although he perceives that the 'ohia trees are "filled with red flowers red birds / water notes flying music / the shining of the gods." The distinction between numinous sound and taxonomy in "Native" is part of the poem's palimpsest-like description, where, in his walk to a shade-house where trees native to Hawai'i are being propagated, Merwin limns the history of the island's ecological decline. The poem, in this epistemological approach, offers a way of drawing the poet, and implicitly the reader, into knowledge.

"Losing a Language" explicitly states the connection between language and place, between name and species, and the interdependency of beings:

many of the things the words were about
no longer exist

the noun for standing in mist by a haunted tree
the verb for I[.]

Accompanying the loss of the sacred koas, for example, is the loss of human perceptions of conditions and experiences, and thereby the diminishment of both the ecology and the human community: "where nothing that is here is known / we have little to say to each other."

In "Hearing the Names of the Valleys," the poet considers the inability to know, as an outsider, the specific and endemic names of things in a different culture; but in such poems as "The Strangers from the Horizon," "Conqueror," and "Term," the story of the arrival of foreigners and their exploitation that brings about the simultaneous destruction of language and landscape is told from the perspective of native Hawaiians. The sequence concludes with a myth of apocalypse: Kanaloa, a figure of fire, smoke, and volcanic destruction, who also houses the ghosts of all things that have been destroyed, is the final creation: "he is the last he is the coming home / he might never have wakened."

While the Hawaiian poems form the center of this collection, their themes extend throughout the other poems. Other landscapes find illumination in such poems as "Late Spring," "West Wall," "The First Year," and "Notes from a Journey"; these are not, however, scenic poems; rather, they trace the intersection of *eros* (love), *topos* (place) and the great language, or the always informing poetic myth that is described by Graves in *The White Goddess*. The power of poetic speech informed by myth is most resonant in "Touching the Tree." In this poem Merwin rejects negation and transgresses the taboos of his forebears: by touching the tree (of life and knowledge), the poet says, "I hearthe tree . . . we talk without anything"; the poet thus gains the insight for strength: "at the foot of the tree I have dug a cave for a lion" for the time "when the lion comes to the tree." There are also poems such as "Before Us," "The Sound of the Light," and "The Solstice," that meditate upon mortality and its connection to eros and topos. These poems have an intimacy not present in earlier poems such as "For the Anniversary of My Death," and thus represent a clear turning away from the earlier poetics of disembodiment

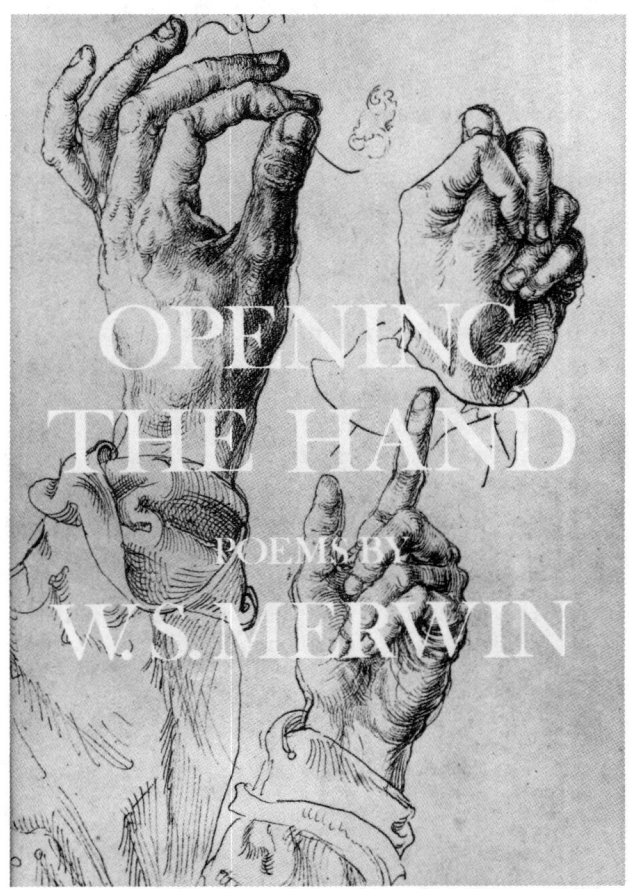

Cover for Merwin's tenth book, in which he returns to the familial tone that closed The Drunk in the Furnace

and a turning toward the evocation of a lyric presence of the self and a beloved other.

Five years after *The Rain in the Trees,* Merwin's twelfth book of new poems, *Travels,* was published. *Travels* enters the human world through its history. Conveyed either by omniscient reportage or the first-person narrative voice are accounts of the naturalists Georg Eberhard Rumpf, David Douglas, William Bartram, and Gregorio Gregorievich Bondar; of such artists as the French poet Arthur Rimbaud and Native Americans Frank Henderson and Little Finger Nail; and of the Amazon experiences of Manuel Cordova. Enjambment of lines and narrative propel each of these accounts; overtly didactic and extraneous biographical detail that could easily encumber each of these narratives is thus avoided. Merwin also includes a long personal account of his childhood and his father's ministry, which forms a parallel text to his prose recollections in *Unframed Originals*. The long historical and personal narratives are the most distinctive poems of the collection, in part, because they are a departure for Merwin; however, there are also luminous shorter lyrics that are among his finest.

Merwin begins *Travels* by directly addressing the reader, "my almost / family we are so / few now it seems as though / we knew each other." Language, which binds us to one another, collapses and threatens to extinguish the poet, the poem, and the reader. There is no assurance that these words will last, nor that they will mean the same things to others. Language is also an ecology that forms and informs the community and is evoked as a means to an intimate connection. Merwin concludes, however, that perhaps no one else is waiting for these words that he hopes might seem "as though they had occurred / to you and you would take / them with you as your own." In these lines it is clear that the poet hopes to create not a disembodied poetry, but a poetry that transcends its author in order to become something passed down and held by others. Here Merwin invokes the most elemental qualities of poetry: its ability to create community and to preserve communal memory.

In "Writing Lives," Merwin offers an *ars poetica* that informs the writing of the poems in *Travels* when he observes that:

> one way with the words is to tell
> the lives of others
> using the distance as a lens
>
> and another way
> is when there is no distance
> so that water
> is looking at water[.]

Each process involves a paradox, because distance is foreshortened in the poems about others, while in poems about oneself not only are the mirrors perfect – water looking at water – but those waters (of the unconscious, the unformed) are also fathomless. In both modes intimacy combines with distance: one is the lens for the other. Epistemologically, seeing becomes the dominant metaphor for knowing. Not only is it invoked, as in "Writing Lives," as a metaphor for writing, as well, but, in "The Blind Seer of Ambon," the dichotomies of sight and blindness collapse. Despite profound losses, the narrator, the seventeenth-century naturalist, Rumphius (Georg Everard Rumpf), concludes "I remember the colors and their lives /everything takes me by surprise / it is all awake in the darkness."

Blind Rumphius emblemizes the poet who must see in his own darkness and be taken by surprise at what he discovers. Merwin's "The Hill of Evening," set in the Dordogne region of southwest France (as is his prose memoir *The Lost Upland*, 1992) concludes with Merwin's returning from a birthday celebration in which all the community seems present. In the grass he finds "a shining crescent a new sickle / bright as water and the blade // glittering with the dew." The sickle merges death, the moon, the harvest, and fertility into one symbol. Merwin returns home and lays the sickle "on the table / here in front of me and tomorrow I must / try to find who it belongs to," thus returning the object to its own identity rather than to the symbolic, as well as returning himself to the community rather than remaining in solitude. The sickle is mysterious by its own right, within its own history of use, its own story of how it came to lie in that field. The poem, an implicit epistolary poem, is lush in its descriptions of the feast, musicians, and celebrations, as well as the evening landscape, all of which underscores Merwin's assertion of the numinous, not symbolic, life of things.

Rumphius is not sentimentalized; his loss of sight, family, and work are not disregarded. Loss, however, is not abstract or distant but defines both history and the present. "Lives of the Artists" forms a palimpsest-like history that uncovers or reiterates the book of drawings that Frank Henderson created and that Little Finger Nail carried to his death. The poem mirrors the book, which, in turn, traces the bitter history of the forced removals of the Plains Indians and their doomed resistance. The book forms a mirror that returns Henderson to his identity.

"The Real World of Manuel Cordova" and "Another Place" are perhaps the most remarkable of the long narratives in *Travels*. Although one is a story of a rubber-trader's initiation and flight from an vanishing Amazonian tribe and the other depicts Merwin's childhood and his relationship to his father, both narratives describe spiritual and moral failings. Degrees of evil are inextricably knotted in "The Real World of Manuel Cordova," in which the Europeans devour the rain forest for its rubber, fur, and wood, but the indigenous tribe, while seeking its own survival, becomes repellent to Cordova in its mirroring of European violence. Cordova is rescued and groomed by the chief to guide the tribe, but he finds upon the chief's death that the secrets and visions he was entrusted with begin to wane and the tribe begins to grow distant from him. Cordova's flight back to his family and dying mother indicates a double infidelity: he abandons what he was entrusted with and returns to what is now irretrievable. "Another Place" reveals a father's failure to love his children: "they had learned to remember / him only as the author / of everything forbidden." This father is an architect of punishment and hypocrisy; thus, there is the searing irony in the biblical resonance of this passage. Both poems suggest that betrayal may define the human heart more than love or compassion; that compassion is often no more than an abstract ideal, distant and seldom enacted, while betrayal is a frequent fact in this world.

Travels concludes with a group of six poems set in southwestern France, the locale where Merwin some thirty years earlier had composed *The Lice*. Returning after an extended absence, Merwin writes poems that describe the permanence of the place, as well as its transience, revealed in the closing lines of "Left Open": "beyond the bare limbs the river looks / motionless like the far clouds that were not / there before and will not be there again." The poet, through this reflective language, gives the moment its illusory suspension. There are no interruptions in the movement of the verse; thus, time becomes synchronous in Merwin's use of enjambment and his incantatory play between rhythm and syntax. Memory and experience merge in this autonomous time of reverie, as suggested in "A Summer Night," where Merwin writes, "so long I have known this that it seems to me / to be mine it has been gone for so long / that I think I have carried it with me / without knowing."

In the case of Merwin, each new book has been a revelation; and with each shift in his poetics he helps to answer the inherent cultural need for a voice that is responsive to the human situation and also to the physical world. The large themes of love, mutability, mortality, and the mirrored relations between language, soul, and nature, have always been subjects of Merwin's speculations. He has always sought, too, the resonances between the world and the poet, or anyone who responds to the world. Given the spiritual and ecological decline that has accelerated exponentially during this century, and Merwin's view of human responsibility, his vision is often misanthropic, as evidenced by the images of alienation in his early sequence of sea poems, the anger and lament of *The Lice,* the Western sequence in *The Carrier of Ladders,* and the depiction of betrayal in such recent narratives as "Another Place" and "The Real World of Manuel Cordova." Such a vision cannot be tempered; but Merwin offers another voice that has become increasingly resonant as well. "The Hill at Evening" and the closing poems of *Travels* are elegiac, anchored in an individual who responds to the passage of places, things, the transformations of

Cover for an issue of The American Poetry Review *that features translations by Merwin*

being. But each thing the poet perceives is itself luminous – imprinted with a luminosity that one must willfully turn away from not to see. Merwin's poems impart the facts and the sense of this startling, vivid luminosity to those who fully encounter them; and, like gardens, the poems offer a space for meditation and renewal.

Interviews:

David Ossman, "W. S. Merwin," in *The Sullen Art* (New York: Corinth Books, 1967), pp. 65–72;

Frank MacShane, "A Portrait of W. S. Merwin," *Shenandoah,* 21 (Winter 1970): 3–14;

Ed Folsom and Cary Nelson, " 'Fact Has Two Faces': An Interview with W. S. Merwin," *Iowa Review,* 13 (Winter 1982): 30–66; revised and expanded in Merwin's *Regions of Memory: Uncollected Prose,* edited by Folsom and Nelson (Urbana: University of Illinois Press, 1987);

Jack Myers and Michael Simms, "Possibilities of the Unknown: Conversations with W. S. Merwin," *Southwest Review,* 68 (Spring 1983): 164–180;

Michael Clifton, "W. S. Merwin: An Interview," *American Poetry Review*, 12 (July/August 1983): 17-22;

Richard Jackson, "Unnaming Myths," in his *Acts of Mind: Conversations with Contemporary Poets* (Tuscaloosa: University of Alabama Press, 1983), pp. 48-52;

David Elliott, "An Interview with W. S. Merwin," *Contemporary Literature*, 29, no. 1 (1988): 1-25.

References:

Charles Altieri, "The Struggle with Absence," in his *Enlarging the Temple* (Lewisburg, Pa.: Bucknell University Press, 1979), pp. 193-220; expanded and reprinted in *W. S. Merwin: Essays on the Poetry,* edited by Ed Folsom and Cary Nelson (Urbana: University of Illinois Press, 1987);

James Atlas, "Diminishing Returns: The Writings of W. S. Merwin," in *American Poetry since 1960: Some Critical Perspectives* (Chester Springs, Pa.: Dufour, 1974), pp. 69-81;

Harold Bloom, "The New Transcendentalism: The Visionary Strain in Merwin, Ashbery, and Ammons," in his *Figures of Capable Imagination* (New York: Seabury Press, 1976), pp. 123-149;

Neal Bowers, "W. S. Merwin and Postmodern American Poetry," *Sewanee Review*, 98, no. 2 (1990): 246-259;

Thomas B. Byers, "W. S. Merwin: A Description of Darkness," in his *What I Cannot Say: Self, Word, and World in Whitman, Stevens, and Merwin* (Urbana: University of Illinois Press, 1989), pp. 79-110;

Floyd Collins, "Forms Open and Closed: The Poetry of W. S. Merwin," *Gettysburg Review*, 7, no. 1 (1994): 144-162;

Ed Folsom and Cary Nelson, eds., *W. S. Merwin: Essays on the Poetry* (Urbana: University of Illinois Press, 1987);

Jane Frazier, "W. S. Merwin and the Mysteries of Silence," *South Dakota Review*, 32, no. 1 (1994): 116-125;

Edward Haworth Hoeppner, "A Nest of Bones: Transcendence, Topology, and the Theory of the Word in W. S. Merwin's Poetry," *Modern Language Quarterly*, 49, no. 3 (1988): 262-284;

Richard Howard, "W. S. Merwin," in his *Alone with America*, enlarged edition (New York: Atheneum, 1980), pp. 412-449;

Walter Kalaidjian, "Two Versions of Lyric Minimalism: James Wright and W. S. Merwin," in his *Languages of Liberation: The Social Text in Contemporary American Poetry* (New York: Columbia University Press, 1989), pp. 33-64; reprinted in *W. S. Merwin: Essays on the Poetry*;

Hank Lazer, "For a Coming Extinction: A Reading of W. S. Merwin's *The Lice*," *ELH*, 49 (Spring 1982): 262-285;

Anthony Libby, "W. S. Merwin and the Nothing That Is," in his *Mythologies of Nothing: Mystical Death in American Poetry 1940-1970* (Urbana: University of Illinois Press, 1984), pp. 185-209;

James McCorkle, "W. S. Merwin's Poetics of Memory," in his *The Still Performance: Writing, Self, and Interconnection in Five Postmodern American Poets* (Charlottesville: University Press of Virginia, 1989), pp. 130-170;

Cary Nelson, "The Resources of Failure: W. S. Merwin's Deconstructive Career," in his *Our Last First Poets* (Urbana: University of Illinois Press, 1981), pp. 179-215; revised and reprinted in *W. S. Merwin: Essays on the Poetry*;

Marjorie Perloff, "Apocalypse Then: W. S. Merwin and the Sorrows of Literary History," in her *Poetic License: Essays on Modernist and Postmodernist Lyric* (Evanston, Ill.: Northwestern University Press, 1990), pp. 231-250; in a slightly different form in *W. S. Merwin: Essays on the Poetry*;

Randall Stiffler, "The Sea Poems of W. S. Merwin," *Modern Poetry Studies*, 11, no. 3 (1983): 247-266.

Papers:

An extensive collection of Merwin's work, covering a period from the mid 1940s to the present, including notes and drafts of his published poetry, prose, and translations as well as unpublished poetry, prose, and translations, is housed at the Rare Book Room at the University of Illinois at Champaign-Urbana. Notable collections of letters are also held at the Poetry Collection at the State University of New York at Buffalo, Washington University Library, the Lilly Library at Indiana University, and the Harry Ransom Humanities Research Center at the University of Texas at Austin.

Michael Palmer
(11 May 1943 -)

Keith Tuma
Miami University

BOOKS: *Blake's Newton* (Los Angeles: Black Sparrow Press, 1972);
The Circular Gates (Santa Barbara, Cal.: Black Sparrow Press, 1974);
Without Music (Santa Barbara, Cal.: Black Sparrow Press, 1977);
Notes for Echo Lake (San Francisco: North Point Press, 1981);
First Figure (San Francisco: North Point Press, 1984);
SUN (San Francisco: North Point Press, 1988);
An Alphabet Underground/Underjordisk Alfabet, poems by Palmer with Danish translations by Poul Borum and illustrations by Jens Birkemose (Viborg, Denmark: After Hand, 1993);
At Passages (New York: New Directions, 1995).

OTHER: *Code of Signals: Recent Writings in Poetics,* edited by Palmer (Berkeley, Cal.: North Atlantic Books, 1983);
P. Michael Campbell, ed., *Palmer/Davidson: Poets and Critics Respond to the Poetry of Michael Palmer and Michael Davidson* (Berkeley, Cal.: Occident Press, 1992) – includes poems and prose by Palmer;
Exact Change Yearbook 1995 (Boston: Exact Change, 1995) – includes poems by Palmer.

TRANSLATIONS: Vincente Huidobro, *The Selected Poetry of Vincente Huidobro,* edited by David M. Guss, includes translations by Palmer (New York: New Directions, 1981);
Louis Aragon, André Breton, Paul Eluard, Philippe Soupault, and Tristan Tzara, *The Surrealists Look at Art,* edited by Pontus Hulten, translated by Palmer and Norma Cole (Venice, Cal.: Lapis Press, 1990);
Emmanuel Hocquard, *Theory of Tables* (Stockbridge, Mass.: o.blek editions, 1994);
Alexei Parshchikov, *Blue Vitriol,* translations by Palmer, John High, and Michael Molnar (Pennsgrove, Cal.: Avec Books, 1994).

Michael Palmer is one of the few contemporary experimental or innovative poets whose books, which began to be published in the early 1970s, have been noticed by a fairly broad readership. While the words *experimental* and *innovative* have become blurb-copy terms that often marginalize the very work they seek to promote, Palmer has created a body of work that deserves to be known by as many readers as possible, and it is perhaps a hopeful sign that, in an American culture that has little use for poetry, he has had some degree of success. His work has been important to publishers, poets, and readers interested in alternative poetry, not just in the United States but also in Canada, France, England, Denmark, and elsewhere.

In the United States, Palmer has been significantly represented in most of the journals devoted to experimental writing, from *Acts, Avec, Conjunctions, o-blek,* and *Exact Change* to *Caterpillar, Ironwood,* and *Sulfur*. He is included in most recent anthologies of innovative poetry, including Douglas Messerli's *From the Other Side of the Century: A New American Poetry 1960–1990* (1994), Eliot Weinberger's *American Poetry Since 1950: Innovators and Outsiders* (1993), Paul Hoover's *Postmodern American Poetry* (1994), Ron Silliman's *In the American Tree* (1986), and Andrei Codrescu's *UP LATE: American Poetry Since 1970* (1989). Palmer's work has also appeared in anthology publications in the less self-consciously alternative collections, including *The Pushcart Prize 12, The Best American Poetry, 1988,* and Philip Dow's *Nineteen New American Poets of the Golden Gate*. His books have been reviewed in *The New York Times, Poetry,* the *Hudson Review, The American Poetry Review, The Vir-*

ginia *Quarterly Review, Sulfur,* and *The American Book Review,* among other publications. It is worth noting that some of these publications – such as *Poetry,* where Palmer's work was praised by Stephen Yenser – have not in recent times tended to review innovative poetry. Palmer's work has been the subject of essays published by university presses and academic journals such as *Contemporary Literature* and *American Literary History;* it has recently been discussed in the electronic journal *Postmodern Culture.* In France, Palmer's work finds one of its most important audiences at the Centre Littéraire de Royaumont. His poems have appeared in *Action Poétique* and in Emmanuel Hocquard and Claude Royet-Journoud's anthology *21 + 1 poètes américains d'aujourd'hui* (1986); Hocquard's translation of Palmer's Baudelaire series (from *SUN*) appeared in 1989. In England interest is probably greatest among the so-called Cambridge School of poets; some of the poems from his first substantial collection, *Blake's Newton* (1972), appeared in Tim Longville's *Grosseteste Review,* a magazine that was important in sustaining dialogue between British and American "experimental" poets. One could go on detailing the extent of interest in Palmer's work; because of his recent readings throughout Europe and the ongoing translation of his work into several languages, it is possible to speak of Palmer's poetry as having an international audience.

Palmer's favorable reception in the crowded contemporary poetry world is a tribute to the evocative and exploratory quality of his work and attests to his poetry's engagement with several traditions of lyric poetry, dating back to Friedrich Hölderlin and Charles Baudelaire. In addition, his favorable standing has probably been enhanced by several other fortunate circumstances. Here one must venture briefly into biography, even while recognizing that biography is not important to a reading of Palmer's poetry, at least not in a simple or obvious sense. In the interview included in *Talking Poetry* (1987), Palmer remarked that, while his poetry does not "come across as autobiographical, it is profoundly so." One assumes that with that comment he points to his poetry's struggle to invent, represent, question, deconstruct, and figure a plurality of possible and evolving selves in and among texts – his own and others' – selves that are part of what he and his reader might want at any time to recognize as belonging to Palmer. Certainly, Palmer's work has little interest in the conventions of lyric and narrative that are typically associated with autobiographical poetry, those conventions belonging to what he called in the same interview "an Anglo-American empirical tradition that takes as a model a kind of simple version of reference, where a poem is a place in which you tell a little story, the conclusion of which is at the bottom of the poem just where it is supposed to be." It might be noted in passing that this statement is unfair to the dominant Anglo-American tradition, which is more diverse and complex than Palmer suggests, and, fortunately, such rhetoric is comparatively rare in Palmer's interviews and essays, even when he does find it necessary to position himself against various mainstream poetries and their conventions.

George Michael Palmer was born in 1943 in what he has described in the *Exact Change Yearbook 1995,* as a "middle-class Italian-American family of modest means" – his father was manager of a small hotel in New York City. Palmer was educated at private schools and then at Harvard, where he wrote an undergraduate thesis on Raymond Roussel and received an M.A. in comparative literature in 1968. Some fortunate circumstances that probably enhanced Palmer's reception as a poet may be mentioned here, in detailing what is known of his biography. At Harvard he coedited the magazine *Joglars* with his friend and fellow poet Clark Coolidge. After Harvard he lived in Europe, studying at the University of Florence before returning to the United States and moving to San Francisco in 1969. In the early 1960s Palmer met and became friends with several older poets whose work has continued to influence his own – Louis Zukofsky, Robert Creeley, and Robert Duncan, among others. His first book contains illustrations by Creeley's second wife, Bobbie Louise Hawkins, and ends with a blurb by Duncan that says – in a comic expression of companionship that rejects the conventional idiom of blurbs – that Palmer was "delivered two blocks astray in 1943 because he was aborted at our address two months before." His first three substantial collections of poetry were published by Black Sparrow Press, one of the most visible of alternative publishers; his next three collections were published by Palmer's friend Jack Shoemaker, at North Point Press, which sadly is now defunct but throughout the 1980s was a respected publisher of some of the most beautiful books of that decade. Palmer's most recent book, *At Passages* (1995), was published by New Directions, the legendary home of the likes of Ezra Pound, William Carlos Williams, Kenneth Rexroth, and Duncan. Palmer has worked in San Francisco with the Margaret Jenkins Dance Company and at the New College of California. He has held visiting appointments at several prestigious

American universities, including the University of California, San Diego; Brown University; and the University of Chicago.

Palmer's life, thus far, has been a life in which the writing of poetry has taken precedence. Palmer has not pursued a regular academic appointment and has lived primarily in San Francisco, which has been one of the most vital sites of American poetry during the last forty years. Palmer has sometimes declared himself a little outside of – although still sharing concerns with – groups of innovative poets such as the Language poets, who have had a considerable presence in the Bay Area, as if to refuse to allow himself to be circumscribed by such an admittedly strategic and problematic label; but he has surely benefited from his informal association with that group of poets and others among older generations. When Robert Duncan, in one of his last poems, inserted phrases that read "FOR MICHAEL PALMER who *also* may work alone," he may have meant to pass the wand (and burden) of poetry to this intensely private and independent man, but he was not naming any real isolation.

With the growing body of criticism devoted to Palmer's work and with at least six interviews and Palmer's own essays and talks addressing particular concerns and poems, there is no shortage of material to assist readers across the difficult terrain of Palmer's poetry. Among the longer critical essays on Palmer one might single out Eric Murphy Selinger's effort to rehabilitate a reading of Palmer "with pleasure," where the crisis of signification that Palmer's poems present is read not as part of a postmodern sublime of shock and contradiction, but rather as a "text of *jouissance*," or pleasure. One should also read Norman Finkelstein's discussion of Palmer's work as "an immense exercise in nostalgia," the most serious treatment to date of the role of mystery and of the hermetic and cabalistic traditions in Palmer's work. These critical essays are, of course, very much sympathetic and partial accounts of Palmer's poetry, and the poet might find them problematic. But Palmer is on record in his *Acts* interview in 1986 for demanding readers who are not "followers," and for insisting that the procedures of his poetry demand readers who must complete the circuit of meaning in Palmer's poems, finding their own sense and non-sense there. Indeed, one of the difficulties in writing about Palmer's poetry entails the temptation to isolate passages or even single lines of his poetry as aphoristic propositions. Because the poetry often seems to be describing its own procedures, and because individual lines or phrases sometimes uncannily resonate with sources typically unattributed, essays (such as this one) on Palmer's work seem especially prone to reorder the experience of a Palmer poem without paying adequate attention to the effect of reading particular poems and books as wholes – however open-ended or fractured these may be. With their many disjunctions and silences, with the gaps across which a careful, formal, sometimes even ceremonial syntax moves, Palmer's poems often seem to point to an abyss at the center of the critic's and reader's projects of meaning.

But to speak this abstractly of the abyss around which an elegant formal symmetry is erected is to neglect the *movement* of individual poems, series, and books that present the reader in many different ways with a challenge to all of the units and categories through which one processes meaning – the title, the name, the phrase, the sentence, the stanza, and so on. To date, attention to the poem and the book has been too small a part of those readings of Palmer that seek to position his work beside the philosophical prose that is part of his reading – whether that be the work of Ludwig Wittgenstein, Jacques Derrida, Roland Barthes, or Moses de Leon. There is little question that such philosophers inform Palmer's critique of essentialist notions of meaning and of a Cartesian epistemology that accords to private experience a peculiar primacy and importance, but there are numerous contemporary poets who share these basic concerns with Palmer and write poems very different from his. One exception to the rule in Palmer criticism is the critic Steve McCaffery, who has taken care to identify what is perhaps most essential to many of Palmer's poems, and who describes the way in which "the temporality-in-space, which is syntax, often overwhelms the [Palmer poem's] semantic apparatus, reducing meaning to the plural detachments and contingencies that Palmer exploits so successfully." McCaffery's essay and David Bromige's essay included in the 1992 anthology of criticism edited by P. Michael Campbell, which discusses the staged carnage and severed body parts that are increasingly a presence in Palmer's later books, especially *SUN* (1988), are arguably two of the most provocative essays yet written on Palmer.

Asked in an interview in *Acts* to speak to the development of his work across his career and to his own ability to recognize the emergence of a characteristic style, Palmer insisted that his work is "evolutionary: it doesn't break off and go on to something else: a *new proposition*." It is not as if one could speak of one of Palmer's books as being about one subject, another about something else. One might be able to

Cover for Michael Palmer's 1995 collection of poems, which poet John Ashbery praised as "his most exciting collection to date"

speak of the content of each book — that is, of the finite number of words and phrases assembled therein, often in such a way as to oblige one to recognize these phrases recurring in different contexts, where one must struggle to make sense of them anew. But one can speak of individual volumes as being more or less concerned with particular problems or questions that have been evident in the work from its earliest publication; one can say, too, that the treatment of these problems and questions has been subject to a different "emotional tenor" — to borrow a phrase from this same interview, where Palmer also spoke of "an evolution of my sensibility." Problems of representation and reference are displayed quite prominently, for instance, in the brief lyric "This is a room" in *Blake's Newton*, a poem that is, among other things, one of the earliest examples of Palmer's fascination with the deictic function of language: "This is a room. / Give me this and / this. This / book ends some / time when it ends and / this is a room." The ability of the relation of line and sentence to complicate sense, to overwhelm a "semantic apparatus" is evident in another poem from the same volume: "This cat and / my lady / are unkind to me / asleep when / I'm awake / then speaking / by the bed." In this second example, the game seems to involve a syntactical parallelism that creates a false symmetry. Problems of the temporality of writing, of representation and reference, assume a much more serious dimension or shading in many of the poems in *SUN*, where the playful exercises of the earliest work become more often an urgent concern with the violence of representation, and the poems themselves, if no less theatrical, are at any rate less of a precocious tour de force.

Keeping in mind the possibility of different "shadings" and the discrete contents of each book — the exploration of color in *Notes for Echo Lake* (1981) for instance, with its many references to painting or to fields of flowers — Palmer's work nevertheless manifests a notable continuity. *Blake's Newton* contains some poems that seem to be exercises in poetic modes not present in Palmer's later books — modes derived from Creeley, Zukofsky, or one or another New York School poet. Moreover, Palmer's last four books clearly present a more accessible poetry than do *Without Music* (1977) and *The Circular Gates* (1974), perhaps because some of the poetry's procedures have become more familiar over time, or perhaps because the two books show Palmer breaking free of earlier influences such as Creeley and Zukofsky and pushing the limits of abstraction and disjunction.

Nevertheless, a continuity does exist in Palmer's poetry, even among its many fragments, and one way to suggest the nature of that continuity is to take up McCaffery's point about the "extreme cognitive dissonance" that Palmer's work presents in its syntactic disjunctions. An anonymous *Publishers Weekly* reviewer once wrote that Palmer's poems were "excruciating" to follow — a remark perfectly understandable from the perspective of *Publishers Weekly*. McCaffery points to the resistance to meaning the poems offer beyond the "insularity" of the lucid phrase or sentence and notes that:

> Writing is always a way by which time inhabits language as a space and in Palmer's work this time (that is writing) and the temporality of the sign never entirely coincide. The sign stresses its fission, iterability, the possibility of escape or drift into new semantic configurations, and the radical non-adequation of intention and result.

Palmer's poems leave one with a text for which "the subject is no longer determinable as a locus of utterance and intention" – an "authorless" text, so to speak, or rather, a series of signs that neither reveal nor conceal the presence of an author or an author's mask or persona. Palmer's escape from meaning – or, perhaps more accurately, from meaningfulness – is figured in at least two different ways in various poems, when it is not directly confronted by self-reflexive rhetoric, such as what one finds in the following fragment from "Notes for Echo Lake 1": "Sign that empties itself at each instance of meaning, and how else to reinvent attention." (One purpose of "resisting meaning" is articulated here.) The periodic use of the imagery and names of jugglers, clowns, and acrobats in Palmer's poetry is one way, reminiscent of Wallace Stevens, of troping an "escape" from meaning as stunt, as theater, as the playful *jouissance* and excess that Eric Murphy Selinger wants to highlight in his reading of Palmer. Another way of troping this same "escape" into textuality seems to have less to do with the pleasure of the text and more to do with the limits of language, with language's ability to do violence by circumscribing or constituting people and things in its conventions of naming, reference, and meaning. Here one is reminded of the several references to writing on walls in Palmer's work, as well as the various detached body parts discussed at some length in David Bromige's essay, which calls *SUN* the "Texas Chainsaw Massacre" of contemporary poetry. We might call these two tonalities in Palmer's poetry the "theatricalized" and the "poeticized" – both present not so much resistance to meaning in itself as resistance to the "meaningful." Palmer's poetry presents a kind of "pedestrian rhetoric" (Michel de Certeau's phrase) whose fissures and gaps undermine the reader's ability to construct totalizing logics and narratives.

Speaking of his own poetry in an interview in a volume edited by Palmer, the French poet Claude Royet-Journoud has the following to offer, which might be useful in thinking about continuities in Palmer's project:

> I am concerned essentially with verse, with the rapport between breath and sense in each portion of the line, with the rapport between ... the absolutely imperceptible glide between one line and the next; I would like to make, I would like to succeed in making minimal unities of meaning work – because normally you work in larger unities. I would like to get down a theatricalization of the infinitesimal, because I don't think that it lacks size, but a theatricalization ... of a meaning hardly made, hardly formulable. ...

Palmer himself has spoken in his interviews of something similar to "minimal units of meaning" and of the "accretion" of meaning, when he has discussed the "threads" or "scraps" of "an evanescent order of narration" in *Notes for Echo Lake,* or when he has talked of the way the very resistance to meaning in the poetry of Paul Celan allows the "signifying capacity of his [Celan's] work to emerge." Meaning does emerge even in Palmer's most polythematic, counterlogical, disjunctive poetry, but meaning is also constantly defeated. Disjuncture itself becomes a kind of syntactic logic (although not in any formulaic manner) – persistent, counterdiscursive. In Palmer's work resistance to the kind of meaning more conventional poems are presumed to express can be read variously as an effort to acknowledge the final inarticulability of the Real, as a critique of the conventions of meaning launched from the counterlogical, mysterious, or playful terrain of the poetic, as an effort to expand the expressive possibilities of the language, or as a register of our "post-holocaust stage vis-à-vis-language" – to quote a phrase Palmer uses in the *Acts* interview. No single reading of disjuncture in Palmer's work suffices.

Two examples will present some indication of the degrees of disjunction possible in a Palmer poem. Here is the beginning of a poem – perhaps merely a passage – from the title series in *The Circular Gates:*

> since early childhood. Mostly
> harvested white; the hall
> over threadbare grass. Tomorrow
> we will mention the hat
>
> and the window
> somewhat swollen, the
> door, and the lack of light.
> All his emotions as well
> in time of war, and the onyx
> torso, now lost. And finally
> how you can't deal with color
> the same way as form[.]

What is the grammatical relation of the first sentence fragment and the second, or the third on the far side of the semicolon? What exactly is "harvested white" – the hall? early childhood? In what sense is the hall "over" threadbare grass, except as the end of one line "over" the next, with its metaphor? Carpets, too, are threadbare; halls have carpets sometimes; none of this is mentioned in this particular poem, although the word *threadbare* and other phrases here occur in different arrangements in other poems or passages earlier in the series.

How does one get from emotions in time of war to an onyx torso, now lost, and sounding like a phrase of mysterious provenance? In what sense is the concluding platitude about form and color final in this poem? Is it part of what will be mentioned tomorrow? But it is part of the poem one has just read! What does the relation of color and form have to do with a war, a hat, a door? The reader is teased with the possibilities of story in the first sentence fragment, and the second, too; that conventional autobiography referred to earlier is half-promised, memories of a hotel in New York, perhaps. Tomorrow takes its unannounced place and spills its guts in a poem that threatened just for an instant to represent memory and experience, but refused to offer the reader a proper "period" or *sentence* (the many meanings of which seem always to be uppermost in Palmer's mind).

As a second example, one might ponder "The Comet," the first poem in *Notes for Echo Lake,* and, like the first poems in the two later North Point Press books, a poem one might be tempted to read as introducing the book, if not framing it in the self-reflexive commentary of an *ars poetica*:

> That year the end of winter stood under a sign
> All days were red in the margin
> writ large against the ochre rooftops
>   and yes that was your father's
>
> face, a murder best forgotten
> by passersby inured to the dust
> though blinded a bit by the redness[.]

This poem begins a book that includes twelve poems in its title sequence, the last of which has twelve sections of twelve lines each, indicating perhaps a "metaphoric year," as Palmer remarked in his essay in *Code of Signals* (1983). Like many poets, Palmer seems fascinated by the workings and conventions of numbers — his essay begins with a discussion of "an hour with its division into sixty minutes, its minutes into sixty seconds, etc., all deriving from a Babylonian numerology based on twelves at the near edge of pre-history." "The Comet" itself seems to be a text describing the artifice of a sign by which a person ordinarily takes his bearings in time; great import has sometimes been attached to the readings of the appearance of a comet. Throughout the poem, and the book as a whole, time and place are figured as name and text; indeed, in the book's epigraphs the poet juxtaposes Saint Augustine's claim that place (*topos*) does not exist with Helmut Heissenbüttel's naming of several jazz musicians ("Big Sid Catlett, Art Tatum, Fats Navarro") as "topographies." Later in the book, with the help of an attentive memory, one may try to understand "red" and "redness" and "dust" and "sign" as they seem to emerge as part of the book's lexicon; these words are joined by "blue" and "unlistening" and others that begin to accrue meaning over time, almost despite the willful mystery attached to any particular appearance of one or another word. But for now, in first encountering "The Comet," one experiences a poem in which, beginning in the second strophe, each line seems capable of pushing the poem in yet another direction, of destabilizing efforts to make sense of the reading, and leaving one on the surface of a syntax that unfolds surprisingly, but also with the inevitability of the grammatical. It is as if parts of the sentence did not so much add up to meaning as multiply meaning exponentially.

Even when Palmer is most "accessible," as when, later at the end of this poem, lines and stanzas begin to acquire a greater degree of continuity, the lines still manifest some degree of torsion:

>   adjustments
> at an unexpected point of the experiment
> occurred toward the backstage of things
>   warm
>
> currents of air and some really
> depressing tricks that filled one
> with true melancholy
>   regarding *principium individuationis*
>
> suitable more to the success of an idea
> in an illustrated journal
> of modern physics
>   splendidly bogus
>
> and immediately satisfying
> as forecast long ago by the prophets
> in a circus farce[.]

Perhaps here again Palmer's text is naming its own procedures, which can indeed be read as a series of "adjustments at unexpected point[s]," or as "depressing tricks" that have as one of their results a melancholy about the status of the individual, of the author as an integrated site of a controlling or controlled meaning. The self-consuming, serpentine syntax swallows its own tail in a series of immediacies that are nevertheless inscribed here as text, ending in the jerky qualifications that oppose the bogus and the satisfying, the prophet and the circus, one phrase undermining the former, just as Palmer elsewhere in different ways destabilizes the order of other poems by writing "yes and no" or "was was and is," or by interrupting a proposition with the interjection "true," or by using the impossibly transsexual pronoun "he-she" — to name just a few ways readers lose their way in this work.

There is little doubt that for some readers Palmer's work will be difficult. To some it may seem all poetics and theory, devoid of humane subject matter. As the work has developed, the temptations of formalism have had to be again and again confronted, as when at the end of *SUN* Palmer begins a poem with a series of imperatives that seem desperate attempts to force the poem toward direct and unsubtle political witness: "Write this. We have burned all their villages." In his first book, *Blake's Newton,* in one of those poems that, it is suggested here, bears some resemblance to older poetic modes, Palmer acknowledges the hermetic and difficult nature of his work in a fairly straightforward and charming, even campy manner:

> Allen says we let in no light
> here, the windows face walls
> the floors and walls covered with books
> and we don't dance enough
> which is true, make love enough
> not true and write difficult poems
> which is also true at times.

Even in this most atypical of Palmer poems, one can see, from this retrospective vantage point, much that characterizes Palmer's work emerging in its early form. Allen, one suspects, is Allen Ginsberg, here used gently as a foil. The interjections of "true" and "not true" are encountered periodically in later work. If this is not a difficult poem, it does make reference to difficult poems and to many books. Palmer's poetry increasingly incorporates unattributed phrases from a vast range of sources — from canonical European writers (who sometimes lend the mature work its often melancholy tone), as well as from tabloid journalists (the first lines of the first "Sun" in *SUN*), Amos Tutuola's *The Palm-Wine Drinkard* (1953), Hedy Lamarr's *Ecstasy and Me,* and just about anyone else whose work Palmer runs across and decides to incorporate in the poem. Such "sources," or quotations, in Palmer's poems are not meant to function as allusions (his work is not difficult in the same way that Ezra Pound's is, for instance, requiring or imposing knowledge of a tradition); rather, his poems function as texts among which the poet and the reader both assemble and disassemble their being. Poor Allen here seems to be describing an urban apartment, at least at first, but might very well be describing a typical Palmer text, although not this one. More typical is the way one is left to wonder whether this Allen *is* meant to refer to Allen Ginsberg. Proper names in Palmer's work, which might seem to reflect a gesture toward a coterie (as in Frank O'Hara's poetry) are implicated in an interrogation or exploration of the act of naming.

Blake's Newton, despite its inclusion of poems that read as if they were exercises in Zukofskian or Creeleyesque lyric modes (and those of others), remains well worth reading. Its very title establishes the project that will be Palmer's — the subjection of the orders of reason and empiricism to the alternative orders of poetry. Palmer has said comparatively little about it in the several interviews available. The second major collection, *Without Music,* he has said takes its title from minimalist composer Steve Reich's book *Writings About Music.* Perhaps the title here means to suggest the extent to which the poems in the volume consciously work against some of the traditions of song and lyric — of musicality — still evident in the first volume. His interest in the poetic series and in seriality — already experimented with in the first book — is fully indulged in *Without Music.* In the *Exact Change* interview Palmer speaks about his book *The Circular Gates* — this volume has among its highlights the series "The Book Against Understanding," "The Brown Book" (after Wittgenstein, one supposes), and the title series — as being "to some degree about educating oneself toward responsibility toward the world." The Vietnam War enters into the title series, although only as one among a set of ordered and reordered referents; the series, which took part of its inspiration from the pure abstraction of painter Frank Stella's Protractor series, is "circular," because "there was no teleology. I was not going to come to a point." With *Notes for Echo Lake* one first meets a poet for whom the book appears to be a unit, focused on the myths of Narcissus and Echo as models of poetic discourse; in *Code of Signals* Palmer writes of the poems interspersed among the twelve parts of the title series as "interludes."

Among these interludes are several remarkable Palmer poems that inaugurate subgenres within Palmer's oeuvre: "False Portrait of D. B. as Niccolò Paganini," which in a fairly "straightforward" and "accessible" manner borrows details from the biography of the famously eccentric violinist-composer to describe David Bromige; and "Song of the Round Man," the first of Palmer's songs for his daughter Sarah, a melancholy exploration of the boundaries between the real and the make-believe, the world and the text:

> The round and sad-eyed man puffed cigars as if
> he were alive. Gillyflowers
> to the left of the apple, purple bells to the right
>
> and a grass-covered hill behind.
> I am sad today said the sad-eyed man
> for I have locked my head in a Japanese box
>
> and lost the key.
> I am sad today he told me
> for there are gillyflowers by the apple
>
> and purple bells I cannot see.

As Norman Finkelstein remarked in his essay on Palmer in *Palmer/Davidson* (1992), the child's "pregrammatical flexibility and semantic dislocations" have been of great interest to Palmer; but this particular song for his daughter, which ends with something of a Stevens-like flourish, shows among other things Palmer's ability to work in more "conventional" or familiar poetic modes. In its repetitions, in its circularity, this poem is indeed a song – there are few "traditional" techniques not part of his repertoire.

Notes for Echo Lake also contains "Alogon," a prose poem; later books continue to explore different models for prose – prose for two or three "voices," the prose paragraph, the series of aphorisms. All in all, *Notes for Echo Lake* is the first and perhaps last Palmer volume that seems to have aspired to something approaching an overarching structure, although it should be noted that in a later poem in *SUN* the poet addresses the book as follows: "Stupid Lake, You were the ruin of a book." However seriously or straightforwardly (the ruins are romantic?) that line was meant to be taken, it is Palmer's relentless self-critical posture that remains one of his greatest virtues. In the recent *Exact Change* interview, for instance, he suggests that "possibly it's an interesting time to rearticulate an idea of authenticity," to move "beyond the postmodern queries which drove so many people into endless ironization, endless play" – a remarkable statement from a poet so often identified with the postmodern and its critique of authenticity.

While *Notes for Echo Lake* shows Palmer already working effectively in most of the modes that came to be recognized as "his own," it was the work in *First Figure* (1984) and especially *SUN* that seems to have captured the attention of the broader public that Palmer has enjoyed. In the *Talking Poetry* interview Palmer explains that:

> *First Figure* has a range of references to figures in a dance, first figures in a book of mathematics, *figurae* in language and so on. Then I started coming across wonderful quotes like Roland Barthes' "There are no first figures." When I saw that I realized that *First Figure* was exactly the title I wanted because the first figure, then, is an area of mystery, yet you are always trying to disclose those things through language which have a quality of primacy. We want to peel away layers of language to arrive at a world which is entirely absent – once you've found your way to it it isn't there.

The volume contains several series, including "Idem," but it is dominated by individual lyrics, a fair number of which are in nontraditional couplets, a form that allows the studied formality of Palmer's work and the interest in "minimal units of meaning" and syntactical torsion to work at maximum effectiveness. Other poems explore different symmetrical stanzas. "Voice and Address," for instance, intersperses single lines between the couplets in a manner that constantly destabilizes the semantic unit of the couplet itself. But the best gloss on the quotation above is probably a poem from the Baudelaire series in *SUN,* a series that presents something approaching a set of meditations on the history of lyric and the theory of lyric (Theodor Adorno's, for instance), within a book titled after an archetypal "first figure." Here are the first four strophes of the fifteenth poem in the Baudelaire series, in which Palmer takes up Rainer Maria Rilke's *Dinggedichte* and "Orpheus. Eurydike. Hermes":

> She says, Into the dark –
> almost a question –
> She says, Don't see things –
> this bridge – don't listen
>
> She says, Turn away
> Don't turn and return
> Count no more lines into the poem
> (Or could you possibly not have known
>
> how song broke apart while all the rest watched –
> that was years ago)
> Don't say things
> (You can't say things)
>
> The ground is smooth and rough, dry and wet
> Pull the blue coat tighter around you
> (There are three parts to you)
> I'm not the same anymore[.]

It is interesting that the voice telling the mythical poet in this archetypal scenario that he cannot "disclose things" is a woman's. Others – Hilda Doolittle (H. D.) comes to mind – have taken up this myth of poetic creation and memory from a feminist viewpoint, but Palmer's overwriting of the Rilke poem encompasses the now-familiar complaint about the violent appropriation of woman as muse and moves beyond it to address the end of lyric poetry, as such. Of course, the "end of lyric" has been in plain view at least since Hölderlin wrote "Ein Zeichen sind wir, deutungslos" (A sign we are, meaningless) to begin his "Mnemosyne"; but in Palmer's poem, as is the case throughout his Baudelaire series, the poet confronts and interrogates the epistemological and political crisis that has made poetry "not the same anymore," writing, as he does, after the time in which "song broke apart." That crisis is the crisis of the *logos,* of the word, of the contract between word and world. Rilke's *Dinggedichte* represents one of the last, desperate efforts to restore a "real presence" – to borrow a phrase from George Steiner – for poetry and for lyric subjectivity in the teeth of that crisis. T. S. Eliot's *The Waste Land* (1922) is a vastly influential High Modernist monument dedicated to articulating the despair the crisis evoked. So it is perhaps no surprise that the longest poem in *SUN,* the first of two poems with that title in the book, is an overwriting of *The Waste Land,* a poem with ceremonial rhythms and rhetoric not unattractive to Palmer although Eliot's politics and cultural influence are repulsive to Palmer. This poem – the longest in the book, containing the same number of lines as "The Waste Land" – begins with one of those headless figures to which Bromige has directed our attention: "A headless man walks, lives / for four hours." Like the book itself, this "Sun" is a culmination of fragments of Palmer's work up to 1988. One might call its subject the "headless" text, or "How to Laugh," or "How to Die." One might call it the "Book of the Machine," or maybe "words disgusted me." One might call it "House of Music" or "Writing Itself." It ends as follows: "A bark sets out on the honeycomb's flow / We called it Le Départ."

Interviews:

Benjamin Hollander and David Levi Strauss, " 'Dear Lexicon': An Interview," *Acts: A Journal of New Writing,* 2, no. 1 (1986): 8–36;

Lee Bartlett, *Talking Poetry: Conversations in the Workshop with Contemporary Poets* (Albuquerque: University of New Mexico Press, 1987), pp. 25–148;

Keith Tuma, "An Interview with Michael Palmer," *Contemporary Literature,* 30 (Spring 1989): 1–12;

Grant Jenkins, with Teresa Aleman and Donald Prues, "An Interview with Michael Palmer," *Sagetrieb,* 12 (Winter 1993): 53–64;

Paul Naylor, Lindsay Hill, and J. P. Craig, "The River City Interview with Michael Palmer," *River City,* 14 (Spring 1994): 96–110.

Peter Gizzi, "Interview with Michael Palmer," in *Exact Change Yearbook 1995* (Boston: Exact Change, 1995), pp. 161–177.

References:

Peter Baker, *Obdurate Brilliance: Exteriority and the Modern Long Poem* (Gainesville: University of Florida Press, 1991), pp. 150–161;

Mutlu Konuk Blasing, "Rethinking Models of Literary Change: The Case of James Merrill," *American Literary History,* 2 (Summer 1990): 299–317;

P. Michael Campbell, ed., *Palmer/Davidson: Poets and Critics Respond to the Poetry of Michael Palmer and Michael Davidson* (Berkeley, Cal.: Occident Press, 1992);

Robert Duncan, "In Wonder," in his *Ground Work II: In the Dark* (New York: New Directions, 1987), pp. 45–47;

Norman Finkelstein, "The Case of Michael Palmer," *Contemporary Literature,* 29 (Winter 1988): 518–537;

Steve McCaffery, "Michael Palmer's LANGUAGE of Language," in his *North of Intention: Critical Writings 1973–1986* (New York: Roof Books, 1986), pp. 44–53;

Linda Reinfeld, "Michael Palmer by Michael Palmer," in her *Language Poetry: Writing as Rescue* (Baton Rouge: Louisiana State University Press, 1992), pp. 86–119;

Eric Murphy Selinger, "Important Pleasures and Others: Michael Palmer, Ronald Johnson," *Postmodern Culture,* 4 (May 1994);

Stephen Yenser, "Open House," review-essay on *SUN, Poetry* (August 1989): 295–301.

Ishmael Reed
(22 February 1938 –)

Robert S. Friedman
New Jersey Institute of Technology

See also the Reed entries in *DLB 2: American Novelists Since World War II; DLB 5: American Poets Since World War II, First Series; DLB 33: Afro-American Fiction Writers After 1955;* and *DS 8: The Black Aesthetic Movement.*

BOOKS: *The Free-Lance Pallbearers* (Garden City, N.Y.: Doubleday, 1967; London: MacGibbon & Kee, 1968);

Yellow Back Radio Broke-Down (Garden City, N.Y.: Doubleday, 1969; London: Allison & Busby, 1971);

catechism of d neoamerican hoodoo church: Poems (London: Paul Breman, 1970);

Mumbo Jumbo (Garden City, N.Y.: Doubleday, 1972);

Conjure: Selected Poems, 1963–1970 (Amherst: University of Massachusetts Press, 1972);

Chattanooga: Poems (New York: Random House, 1973);

The Last Days of Louisiana Red (New York: Random House, 1974);

Flight to Canada (New York: Random House, 1976);

A Secretary to the Spirits (New York: NOK, 1978);

Shrovetide in Old New Orleans: Essays (Garden City, N.Y.: Doubleday, 1978);

The Terrible Twos (New York: St. Martin's Press/Marek, 1982);

God Made Alaska for the Indians: Selected Essays (New York: Garland, 1982);

Reckless Eyeballing (New York: St. Martin's Press, 1986);

New and Collected Poems (New York: Atheneum, 1988);

Writin' is Fightin': Thirty-Seven Years of Boxing on Paper: Essays (New York: Atheneum, 1988);

The Terrible Threes (New York: Atheneum, 1989);

Japanese by Spring (New York: Atheneum, 1993);

Airing Dirty Laundry (Reading, Mass.: Addison-Wesley, 1993).

OTHER: *The Rise, Fall, and . . . ? of Adam Clayton Powell*, edited by Reed (New York: Bee-Line, 1967);

Ishmael Reed (photograph © the Estate of Thomas Victor)

19 Necromancers From Now, edited by Reed (Garden City, N.Y.: Doubleday, 1970);

Yardbird Reader, 5 volumes, edited by Reed (Berkeley: Yardbird Cooperative Publishing, 1972–1976);

Yardbird Lives!, edited by Reed and Al Young (New York: Grove, 1978);

Calafia: The California Poetry, edited by Reed (Berkeley: Y'bird Books, 1979);

The Before Columbus Foundation Fiction Anthology: Selections from the American Book Awards, 1980–1990, edited by Reed, Kathryn Trueblood, and Shawn Wong (New York: Norton, 1992).

SELECTED PERIODICAL PUBLICATIONS –
UNCOLLECTED: "The Writer as Seer: Ishmael Reed on Ishmael Reed," *Black World,* 23 (June 1974): 20-34;

"American Poetry: A Buddhist Takeover?," *Black American Literature Forum,* 12 (Spring 1978): 3-11.

Ishmael Reed's importance to contemporary literary studies stems in part from his ability to channel his encyclopedic historical, political, and cultural knowledge into syncretic poetry and prose that resonate with the voices of diverse ethnicities, locations, and eras. As an editor and a publisher, he has worked to create forums for writers from any and all ethnic groups. In his introduction to *The Before Columbus Foundation Fiction Anthology* (1992) Reed states, "We . . . hope that the reader . . . will discover that American literature in the last decade of this century is more than a mainstream. American literature is an ocean."

Reed is the author of nine novels, five volumes of poetry, and four collections of essays, as well as many articles on sociopolitics, literature, and multicultural aesthetics. He sees himself, as he said in a 1973 interview, as a "real writer," one who "spends twenty-four hours writing." His profession has often led him into controversy. During his career of nearly thirty years he has endured vitriolic attacks from pioneers of the New Black Aesthetic such as Addison Gayle Jr. as well as from patriarchs of white liberalism such as Irving Howe, who, in *Harper's Magazine,* December 1969, characterized Reed's prose as "the commercial cooings of Captain Kangaroo." In a 1978 interview Reed remarked that in the literary world "the status quo has a vested interest in meally-mouthedness . . . and making things seem like they're serene."

Reed's aesthetic, what he calls "neoamerican hoodooism," flies in the face of the literary establishment. Interviewer Robert Gover describes neohoodooism as a modern Americanization of voodoo, a syncretic religion that "has no argument with any other theology and has absorbed Jesus and the Catholic saints as easily as it could any other religion's pantheon. It has a place in its system for any way of perceiving man's relation with the unknown. It is flamboyantly undogmatic." Reed merges West African and Haitian influences with such diverse influences as hard-boiled detective fiction, jazz, spirituals, Westerns, and the machinations of American politics. This blending results in poetry and prose that continuously shift between startling realism and confounding surrealism. "The inability of some students to 'understand' works written by Afro-American authors," Reed wrote in his introduction to *19 Necromancers From Now* (1970), "is traceable to an inability to understand the American experience as rooted in slang, dialect, vernacular, argot, and all of the other putdown terms the faculty uses for those who have the gall to deviate from the true and proper way of English."

In 1985 Reed asserted that his aesthetic "owes more to the Afro-American oral tradition and to folk art than to any literary tradition. The oral tradition includes techniques like satire, hyperbole, invective, and bawdiness. . . . It's a comic tradition in the same way that the Native American tradition is comic. Gerold Vizenor, a Native American scholar . . . says that tragedy is Western. I use a lot of techniques that are Western and many that are Afro-American." His syncretism serves as the base of a satiric wit that spares no one. "It's a way of subverting the wishes of the people in power," he maintains.

A dynamic force who continues to expose the failures and lost opportunities of American society, Reed described himself in a 1969 interview early in his career as an "anarchist" and "an international mind-miner." From the beginning he has challenged the constraints of a white-American aesthetic "that dates back to the time when some European philosopher was laying up in a big mansion all day talking about, 'Does time begin in space, or does space begin in time,'" as well as the limits that Gayle, progenitor of a "goon squad aesthetic," and other proponents of the New Black Aesthetic of the 1960s sought to invoke.

Reed has been vilified and exalted in the academic as well as the popular press for his unwillingness to temper his criticism of individuals, both black and white, or his indignation at America's ethnic and racial divisions. And though his satiric response is too often simplified and explained away as mere spewing of venomous anger, his résumé is replete with national honors and awards. Nominated for two National Book Awards and a Pulitzer Prize in poetry, Reed has also received the John Simon Guggenheim Memorial Foundation Award for fiction, the Rosenthal Foundation Award of the National Institute of Arts and Letters, the American Civil Liberties Award, and the Pushcart Prize for his essay "American Poetry: Is There a Center?" He has pursued his commitment to cultural diversity and a literary redefinition of America in a variety of endeavors. Reed was cofounder of and a writer for Yardbird Publishing (1971) as well as its editorial director (1971-1975). He has also been a founder of

Reed, Cannon, and Johnson Communications (1973), the Before Columbus Foundation (1976), and Ishmael Reed and Al Young's Quilt (1980). Constantly giving workshops and lectures around the country, Reed has been an important influence on many emerging ethnic writers.

Born to Henry Lenoir and Thelma Coleman in Chattanooga, Tennessee, in 1938, Reed took his surname from his stepfather, Bennie Stephen Reed, an autoworker. When he was four, his family moved to the lakefront city of Buffalo, New York, where he grew up and attended the University of Buffalo (now State University of New York at Buffalo) from 1956 to 1960 before moving to New York City and finally to Berkeley, California, in 1967. He now lives in Oakland with his wife, Carla Blank, the modern dancer and choreographer. He has two daughters, Timothy Brett (with Priscilla Rose, his first wife) and Tennessee Marie. While a student in the Millard Fillmore night school at the University of Buffalo, Reed garnered support and encouragement from the English department after showing "Something Pure" to his English instructor. In 1973 Reed recalled the satiric fiction about Jesus Christ as an advertising agent as "a black version of existentialism . . . with a trick ending." It confirmed that he "would not be comfortable with the 'straight' story."

Uncomfortable with the academic literary scene, Reed left Buffalo in the middle of his junior year for New York City. He recalls the move in his 1978 article "American Poetry: A Buddhist Takeover?": "The first apartment I lived in, on Spring Street, in 1962, had been abandoned a month before by Jack Kerouac who told the other roommates that he was going to Las Vegas to get drunk." In "The Writer as Seer: Ishmael Reed on Ishmael Reed" (1974) he says he was quick to distance himself from "the screaming wing of the New York School of Poets – 'personalism' – magazine writers such as Calvin C. Hernton, David Henderson, Joe Johnson, Steve Cannon, and Tom Dent. They embarrassed me into writing all my own way."

Reed's popularity and appeal stem from his finding his "own way." In 1984 he explained that the "'Neo-Hoodoo' aesthetic comes out of my personal experience and represents my need to find something that I could be at home with. . . . I think you can get some of the rhythms of the Black poetry of Manhattan in my poetry . . . a rapping rhythm. . . . Some of the notions associated with the Black Movement, that eventually come to reflect its philosophy, were merely Western ideas with a black facade. These ideas were either given blackness or 'blackified.'" Not surprising, Reed's fiction carries these ideas also, as *The Free-Lance Pallbearers* (1967), his first novel, makes clear. Here Reed creates negative black characters who are mocked and derided as much or even more than the novel's representatives of white oppression. The satire is aimed at the white power structure of HARRY SAM but even more pointedly at the black characters whose actions unwittingly serve oppressive Christian and European biases. Reed's second novel, *Yellow Back Radio Broke-Down* (1969), is a Western in which he again derides a closed-minded and oppressive Christianity, this time through his hero Loop Garou, a black cowboy. Betrayed by blacks and whites alike, Loop is a Christ figure whose magic eventually outshines the power of Christianity as personified in the Pope.

Reed's first volume of poems, *catechism of d neoamerican hoodoo church: Poems* (1970), opens with a four-sentence prose poem, "Black Power Poem," that pits "the spectre of neo-hoodooism" against "a holy alli / ance" of 1960s political and intellectual heavyweights: "allen ginsberg timothy leary / richard nixon edward teller billy graham time magazine the / new york review of books and the underground press. / may the best church win, shake hands now and come / out conjuring." The juxtapositioning of these "opposing forces," as Chester J. Fontenot has noted, "suggests that the Old American culture is desperate and that it is finally doomed." With the invisible, undefined ghostly presence of black power, "Reed creates the image of a boxing match between two forces." As the divided and antagonistic voices of white America are caught within the specter, the clear "irony of this poem is that Reed does not consider Neo-Hoodoo as the equal of Christianity; rather, he sets it up as more than a match for Christianity by showing that the powers of old America must rally opposing forces to defend against it."

Reed brings his readers into his syncretic and synchronistic world in "I am a Cowboy in the Boat of Ra." Reed's speaker is black, the "Ezzard Charles / of the Chisholm Trail," and he playfully merges the jargon and ethos of the Wild West with prominent figures of ancient Egyptian religion such as Osiris, Isis, and Set. Identifying his speaker as an outcast, "Vamoosed from / the temple i bide my time," Reed adds the idiom of detective fiction as well as hoodoo to his rich cultural mix:

I am a cowboy in his boat. Pope Joan of the
Ptah Ra. C/mere a minute willya doll?
Be a good girl and

Cover for Reed's 1970 chapbook, which includes poems later collected in Conjure *(1972)*

bring me my Buffalo horn of black powder
bring me my headdress of black feathers
bring me my bones of Ju-Ju snake
go get my eyelids of red paint.
Hand me my shadow

I'm going into town after Set[.]

Working with the mythology of Ra, the Sun god, and his descendants, Reed recasts the conflict between Osiris (a positive, dark figure) and Set (a negative, light one). The outcast black cowboy seemingly represents Horus, Osiris's son, in his desire to reclaim the throne from Set, "usurper of the Royal couch / – imposter RAdio of Moses' bush." As a figure outside the temple he writes "the motown long plays for the comeback of / Osiris." His desire "to get Set   to sunset Set / to unseat Set   to Set down Set" may be read as Reed's assertion of black identity set against a white, Western, Christian America.

"Sermonette" is another poem that pits the authority of Christianity and white American culture against the powers of hoodoo. In the course of the poem the "topless judge" who busted a poet falls victim to so many potions and fetishes that "by nightfall his honah could a go go no mo." Reed cleverly draws his readers into the poem by the seductive power of his syncretic language. At the end of the poem he surprises readers by using the familiar refrain "gimme dat ol time / religion / it's good enough / for me!" to refer to neo-American hoodooism instead of Bible-Belt Christianity. The usual connotations and associations of the lyric are subverted as his readers are converted.

Synchronicity is at the core of "Dualism: In Ralph Ellison's Invisible Man." Beginning with the speaker's statement "I am outside of / history," the short poem concludes with the opposite view, four brief lines that satirize the linear historicity that subsumes white European conceptions of truth:

> i am inside of
> history. its
> hungrier than i
> thot[.]

Reed suggests that white America's historical account has precluded an accurate telling of the African and African American past. While the speaker is consumed by such a history, he is not deceived by it; the last word of the poem is a pun on Thoth, the Egyptian god of wisdom.

In his early poems Reed often employs a condensed typography, one that elides vowels, is largely restricted to the lowercase, and sometimes favors numerals and ampersands over letters. This graphic technique speeds the reader through a poem such as "Instructions to a Princess," a personal lyric addressed to his child Tim: "yr mother comes down / from the attic at midnite & tries / on weird hats. i sit in my study / the secret inside me." This technique is playful even as it overlays the strong desire of paternal revelation: "i have been saving all this / love for you my dear. if my / house burns down, open my face / & you will be amazed."

Reed uses this shorthand typography along with black slang in "Badman of the Guest Professor," a poem that satirizes the literary pedant:

> u worry me whoever u are
> i know u didnt want me to
> come here but here i am just
> d same: hi-jacking yr stagecoach,
> hauling in yr pocket watches & mak
> ing u hoof it all d way to
> town. black bard, a robber w/ an
> art: i left some curses in d cash
> box so ull know its me[.]

The academy and its icons are all grist for Reed's satire. T. S. Eliot's Prufrock is taken to task: "stand in d / corner no peaches for a week, u lemon / u must blame me because yr wife is / ugly." Others fair no better: "everytime i read william faulkner i / go to sleep. / fitzgerald wdnt hv known a gangster if one / had snatched zelda & made her a moll tho / she wd hv been grateful i bet." Reed's speaker is a trickster-cowboy, just "passing thru, just sing / ing my song. get along little doggie & / jazz like dat." The poem is a parodic condemnation of a static and repressed literary establishment. The professor's "thing is over, kaput / finis." The subject is a "dead duck all out / of quacks," and his pedantry, "d nagging hiccup dat / goes on & on w / out a simple glass / of water for relief."

Reed's first poetry volume culminates with "Catechism of d Neoamerican Hoodoo Church," a poem that incorporates personal history, genealogy, hoodoo, and black history while fashioning a strong response to all who would constrict and constrain a poetry as demanding, in terms of allusion and comprehensiveness, as Reed's. He is

> a black cat
> superstars on my shoulder. a johnny root dwells
> in my purse. on d one wall: bobs picture
> of marie laveaus tomb in st louis #2. it is
> all washd out w/x . . . s, & dead flowers &
> fuck wallace signs. on d other wall:
> d pastd scarab on grandpops chest, he was
> a nigro-mancer frm chattanooga. so i got it
> honest. i floor them w / my gris-gris. what
> more do i want ask d flatfoots who patrol d beat
> of my time. d whole pie? o no u small fry
> spirits. d chefs hat, d kitchen, d right
> to help make a menu that will end 2 thousand yrs
> of bad news.

In the end it is Reed's aesthetic pronouncement that one should

> DO YR ART D WAY U WANT
> ANYWAY U WANT
> ANY WANGOL U WANT
> ITS UP TO U / WHAT WILL WORK
> FOR U.

Reed has stuck by this statement, and it is certainly one that flies in the face of the New Black Aesthetic, which, as Richard Walsh defines it, calls "for writing that directly recreated [sic] the black experience out of which it arose; that found its style in the forms of 'black folk expression'; that was socially progressive in effect – according to a very literal concept of functional literature; that addressed itself to the common readership of black people; and that assiduously cultivated positive black characters."

Reed's second collection of poetry, *Conjure: Selected Poems, 1963–1970* (1972), was nominated for a National Book Award. It includes all of the poems of *catechism of d neoamerican hoodoo church* except "Gris Gris as found poetry," a two-column listing of various powders, drops, waters, and oils whose names catalogue the ambitions, desires, fears, and forces of New Orleans voodoo practitioners. Syncretism and synchronicity remain foundations for many of the newly collected poems. For example, in the opening poem, "The Ghost in Birmingham," the "Holy Ghost" of Denmark Vessey, a nineteenth-century slave insurrectionist, breathes the "Metempsychosis in the air." The ghost is transmuted into "the Black Caligula, who performs a strip tease of the psyche." With his many allusions Reed collapses ancient Egypt, the decadence of classical Rome, and the

Cover for Reed's 1973 book of poems, which was nominated for
a Pulitzer Prize (courtesy of the Lilly Library,
Indiana University)

slave culture of the nineteenth-century American South into a media-barraged Birmingham of the 1960s.

The poems in *Conjure* vary greatly in subject, length, diction, and form. In "The Gangster's Death" Reed decries the sinister operations of the military-industrial complex, while creating a speaker whose patience is more formidable than the power of the elite: "i'm always glad when the chickens come home to roost." He comments cuttingly on powerful politicians and religious figures in poems such as "For Cardinal Spellman Who Hated Voo Doo." In "Neo-HooDoo Manifesto" he provides a mélange of definitions, newspaper and text quotations, lyrics, and lists. The collection also includes a blues lyric ("Betty's Ball Blues"); personal poems about cats ("My Thing Abt Cats"), insubstantiality ("Man or Butterfly"), and anonymity ("American Airline Sutra"); and humorous poems ("Monsters From The Ozarks," "Columbia," "Report of the Reed Commission," and "What You Mean I Can't Irony").

In 1972 Reed published his most successful novel, *Mumbo Jumbo*. Early detractors such as Houston Baker reversed their opinions of his talent and intentions by issuing glowing reviews. In the December 1972 issue of *Black World* Baker declared the novel to be one of "amazing talent and flourishing genius." In the novel set in 1920s New Orleans, white people, enjoying the influence of the fetish "Jes Grew," drop their racist tendencies and "act black." Papa LaBas, the incarnation of the powerful hoodoo god Legba, acts as a detective and follows both the course of this epidemic and the forces of Christianity, in the form of Atonists, who are desperate to destroy it. The oppositions of races, religions, and historical records are fused into a narrative structure that demonstrates the synchronicity and syncretism of hoodooism.

With the Pulitzer Prize–nominated *Chattanooga: Poems* (1973) Reed's poems take on a much more personal and temperate tone than they previously had. In the title poem Reed synchronizes the history of the city of his birth as a site of per-

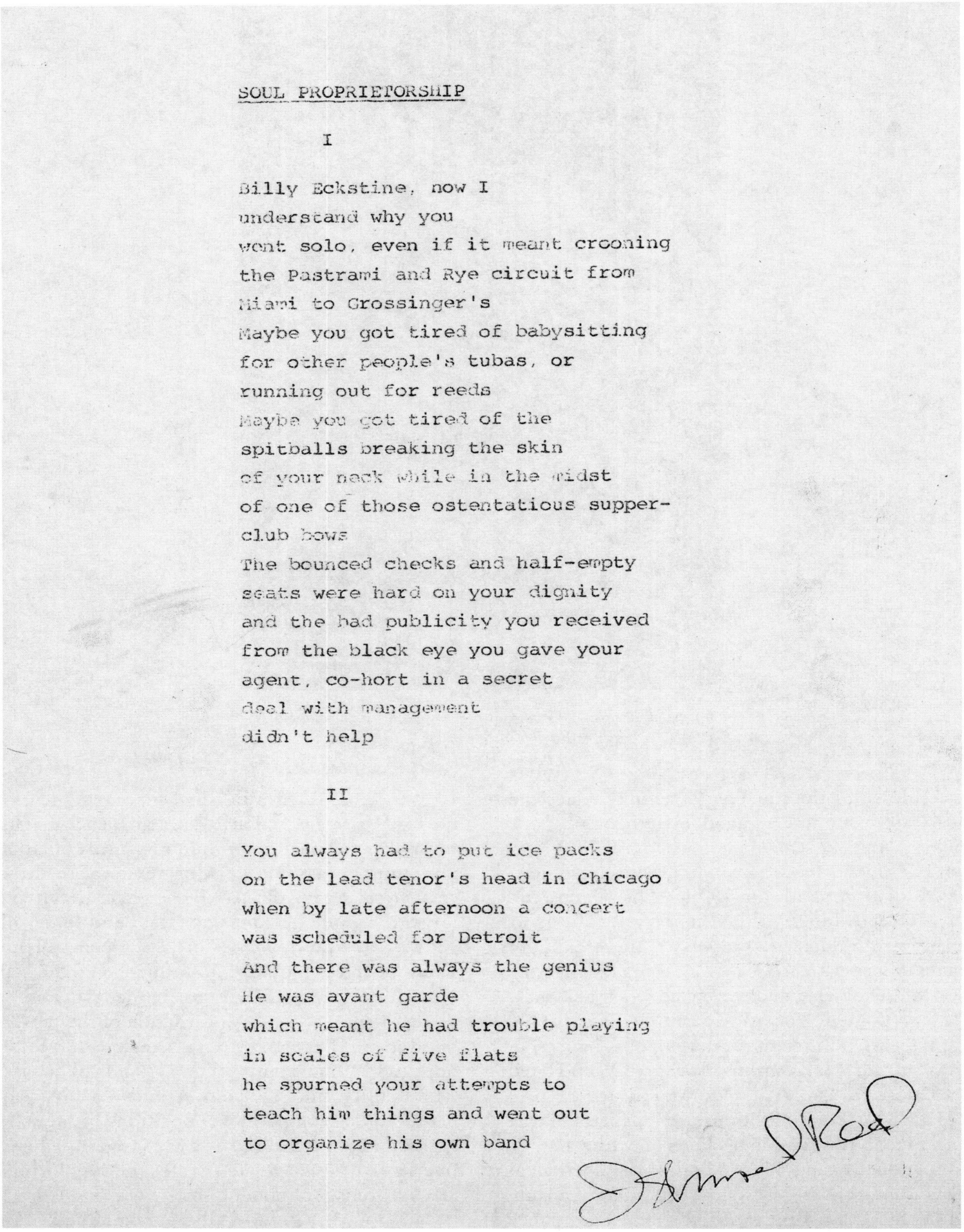

SOUL PROPRIETORSHIP

I

Billy Eckstine, now I
understand why you
went solo, even if it meant crooning
the Pastrami and Rye circuit from
Miami to Grossinger's
Maybe you got tired of babysitting
for other people's tubas, or
running out for reeds
Maybe you got tired of the
spitballs breaking the skin
of your neck while in the midst
of one of those ostentatious supper-
club bows
The bounced checks and half-empty
seats were hard on your dignity
and the bad publicity you received
from the black eye you gave your
agent, co-hort in a secret
deal with management
didn't help

II

You always had to put ice packs
on the lead tenor's head in Chicago
when by late afternoon a concert
was scheduled for Detroit
And there was always the genius
He was avant garde
which meant he had trouble playing
in scales of five flats
he spurned your attempts to
teach him things and went out
to organize his own band

Typescript for "Soul Proprietorship" (by permission of Ishmael Reed)

SOUL PROPRIETORSHIP 2

They called their bloopers 'new
music' and drew experimental
customers
Customers who never smiled
and owned high blood pressure
When you travel single you
can take time out to catch up
with the funnies
You no longer have to order
40 cups of coffee
10 black
5 with cream, and
12 regular
You no longer have to keep
tabs on the two guys who wanted
tea

      III

And when your only companion became
your thought
You came up with the Billy Eckstine
Shirt
With prints as beautiful as
the handle of an Islamic sword
and you made a million silver dollars
And you bought an old Spanish mansion
in California whose
wings could be seen from the sea
They look like two shining silver
collars, billowing, for lift off

           copyright Ishmael Reed 1976

sonal remembrance and of historical, especially Civil War, significance, telling his readers that "Chattanooga is something you / Can have anyway you want it." The city, the speaker confides, is "a woman to me too / I want to run my hands through her / Hair of New Jersey tea and redroot." In "Railroad Bill, A Conjure Man" Reed returns to black dialect and the world of hoodoo, setting Bill up as a legend who is in danger of being appropriated by Hollywood executives. As a necromancer, Bill can change his form and eventually "outwit the chase and throw / Off the scent he didn't care what / They sent. / ... / Railroad Bill was free." "Skirt Dance," "To a Daughter of Isaiah," and "Kali's Galaxy" are as playful in their sexual allusion and metaphor as "The Katskills Kiss Romance Goodbye" is humorous in its mockery of Rip Van Winkle and his place in the American mythos. "Antigone, This Is It" strips the mythic figure of her own tragedy: "Standup Antigone, / The jury finds you guilty / ... / You wrong girl, you wrong / Antigone, you dead, wrong / Antigone, this is it."

Several poems in the collection show Reed has lost none of his taste for literary battle. "My Brothers" is a straightforward reply to those critics, friends, and fellow artists whom Reed sees as hypocritical and disruptive to his own art. In "The Author Reflects on His 35th Birthday" Reed implores:

> If I ever
> If I EVER
> Try to bring out the
> Best in folks again I
> Want somebody to take me
> Outside and kick me up and
> Down the sidewalk or
> Sit me in a corner with a
> Funnel on my head[.]

In the poem he reaffirms his commitment to exposing the artistic noose of the New Black Aesthetic. Neoamerican hoodooism, as Walsh describes it, defies an "agenda entirely amenable to the hierarchies of the white literary establishment.... Reed's demonstrations of America's cultural diversity ... confront the monoculture with its own partiality and compel a revolution in America's concept of itself."

During the 1970s Reed produced two more novels, *The Last Days of Louisiana Red* (1974) and the popular success *Flight to Canada* (1976). In both works neohoodooism takes on great importance for narrative construction and thematic expression. In the former, the character Ed Yellings, a hoodoo practitioner, finds a cure for cancer but is murdered when he tries to market it. The detective Papa LaBas comes in to investigate the murder, and the novel expands with new story lines involving a chorus that acts as a metaphor of black America and Antigone, an individual who opposes the power of the state. *Flight to Canada* is on one level the story of Raven Quickskill's escape from Massa Arthur Swille to freedom and an ostensible happiness. In this novel history and technology merge as a result of the forces of hoodoo. Raven flies aboard a jumbo jet from the Civil War years of the South to a Canada that affords him literary fame through book signings and guest appearances. Yet Canada, while representing the happiness of personal transcendence and dreams of freedom, also becomes a metaphor for realism. Ultimately Raven and all those who made their flights to Canada understand the necessity of maintaining one's dreams as a way to transcend realities.

With *A Secretary to the Spirits* (1977) Reed's poetry takes on yet another set of changes. Egyptian collage illustrations by Betye Saar add a visual component to the diverse verbal images that Reed stirs together, reaffirming not only his commitment to a multicultural artistic statement but also to an interdisciplinary aesthetic as well. Sharon Jessee writes that Reed's works "support an idea which he calls the 'multi-culture': an amalgamation of perspectives, art forms, and lifestyles from different cultures, past and present.... Reed envisions the multi-culture as a sort of collective consciousness to be created through cultural exchanges between individuals and groups which will revitalize not only their individual experiences but their culture as well."

The poems in the collection range from a diatribe against Richard Nixon ("Poem Delivered Before Assembly of Colored People Held at Glide Memorial Church, Oct. 4, 1973 and Called to Protest Recent Events in the Sovereign Republic of Chile") to the title poem that again asserts the speaker's phoenixlike ability to rise up and implement voodoo as a force within history, one capable of overcoming the adversity that surrounds him. "Sather Tower Mystery" is about yet another "Professor / a member of what should be called / The Good German Department," who poses as a liberal but ultimately reveals his inextricable entrenchment in his European heritage. While synchronicity is the thread running through "The Return of Julian the Apostate to Rome," a poem in which the decadence of Rome defines the decline of contemporary America, syncretism and a trenchant black slang are both noticeably absent from the majority of poems. Reed does not prod his reader into historical and mythical comparisons and associations as much as he did in his earlier poetry.

Political and social criticism is more prevalent in two of Reed's 1980s novels, *The Terrible Twos* (1982) and *Reckless Eyeballing* (1986). In *The Terrible Twos* Reed transforms the myths and facts of Christmas into an allegory of capitalism during the Reagan years. In *Reckless Eyeballing* he pits two black writers, a male and a female, against one another to comment on the ideology and power of feminism in the contemporary literary scene.

The new poems of *New and Collected Poems* (1988) are compiled under the title "Points of View." Some address contemporary political issues. In "Invasion," for example, President Ronald Reagan and Defense Secretary Casper Weinberger are ridiculed for their excessive force during the Grenada invasion: "Why is it that when the old / men have power the young men / fly home in star-spangled skins / . . . / They bombed the mad house by / accident / A level headed pilot came back / Three times, the nurse testified / 'I'll remember his grin for the / rest of my life.'" Racial division, social inequality, and the physical destruction of Alaska are treated in an untitled poem as well as in "On the Fourth of July in Sitka, 1982," "Ice Age," and "The Smiley School." Reed also treats daily life in California, the subject of "Earthquake Blues," "Oakland Blues," "Lake Bud," and an untitled poem in which Reed predicts: "When California is split in two / The Northern part will be called / The Republic of Jambalaya / The Southern part will be called / Summer Camp." A notable development is Reed's shift away from the Black Aesthetic political argument that so occupied him in *Conjure* and toward a political commentary that embraces and demonstrates his striving toward an operating "multi-culturalism."

The new poems include dramatic monologues that are parodic and hard-hitting. "Epistolary Monologue" attacks the tawdriness of the British royal family through the voice of Queen "Lilibet." In "The Pope Replies to the Ayatollah Khomeini" Reed portrays the Pope as a shrewd practitioner of realpolitik:

> Believe me, Khomeini, I knew about
> the Shah's decadence, his extravagance
> his misdeeds, and how he lolled about
> in luxury with Iran's loot
> I knew about the trail of jewels which
> led to his Dad's capture
> but a fella has to eat and so when
> David Rockefeller asked me to do something
> how could I refuse?

In "Judas" the speaker talks to Jesus upon the cross, commenting "about best friends" such as Judas, "Always up in your face / laughing and talking / leading the praise after / your miracles." He tells him, "See how careful you have to / be about whom you go bar- / hopping with, Jesus" and consoles him with the thought that "where you're going / the drums don't stop / They serve Napa Valley / champagne at every meal / Everybody smokes big cigars / Sweet Angel hair be tingling / your back while you invent / proverbs in a hot tub / Where Judas is going / the people don't know how / to fix ribs / the biscuits taste like / baking soda."

In "Home Sweet Earth" and "Life Is a Screwball Comedy," two of the last three poems of *New and Collected Poems,* Reed creates lists that amount to a partial evaluation of American culture. Like a prayer of praise, "Home Sweet Home" is a litany of people and things to be thankful for:

> Home Sweet Earth
> Home Sweet Earth
> Your waters are chicken soup
> To our souls
> You give us goldenrod and
> Breakfast rolls
> Italian spaghetti and Dizzy
> Gillespi
> Zimbabwe, and Lady Day[.]

Earth is "Mother of legba and Damballah / Of kinky haired Jesus / Of Muhammed and Gautama / Of Confucius, and Krishna / Of Siva and Vishnu." A listing of disasters that Earth has so far survived brings the poem to a close: "Home Sweet Earth / You give us something to stand on."

"Life Is a Screwball Comedy" is largely a tribute to American comedians, as is clear from its opening lines:

> Life is a screwball comedy
> Life is a screwball comedy
> It's Cary Grant leaning too
> far back in a chair
> It's Bill Cosby with a
> nose full of hair
> It's Richard Pryor
> with his heart on fire[.]

This poem and the others in the "Points of View" section demonstrate the growth of Reed's sources of reference, allusion, and critique that create the inclusive, rich base for his poetry. Much of Reed's point of view may be summed up by the last lines of the poem: "Life is a screwball comedy / life is a screwball comedy / life is a screwball comedy / And the joke is on us."

Reed has explicated that joke through his fiction in the 1990s, which has maintained the satiric sting of his earlier work. In *The Terrible Threes* (1989) Reed serves up a frightening prognosis for America's political social condition through a ribald combination of history and fabulation, mystery and satire. In *Japanese by Spring* (1993) contemporary university politics is the subject of Reed's trenchant satire as a young black assistant professor's role in his departmental power play eventually hurtles him into an international assassination plot.

Reed's latest book, *Airing Dirty Laundry* (1993), is a collection of essays that comment on some of the most controversial political and social issues of American culture. Reed takes on influential writers and media personalities, challenging their readings of race relations, feminism, and the drug business in the United States. He also offers short profiles of significant black American writers, actors, and politicians and concludes with a section of short essays devoted to such disparate subjects as Tipper Gore and Christopher Columbus as well as reprinting an award-winning discussion of the American poetry scene in the late 1970s. In his essay on Gwendolyn Brooks, "American Poetry: A Buddhist Takeover?," Reed comments, "Writing poetry is the hard manual labor of the imagination. It is a high-risk profession, with a history strewn with those who've struggled with alcoholism and drug addiction and suicide. The muse of poetry is the cold, demanding figure." Reed has proven that he is up to the challenge.

Interviews:

Walt Shepard, "When State Magicians Fail: An Interview with Ishmael Reed," *Journal of Black Poetry,* 1 (Summer-Fall 1969): 72-77;

John O'Brien, *Interviews with Black Writers* (New York: Liveright, 1973), pp. 163-183;

Robert Gover, "An Interview with Ishmael Reed," *Black American Literature Forum,* 12 (Spring 1978): 12-19;

Joseph Henry, "A *MELUS* Interview: Ishmael Reed," *MELUS,* 1 (Spring 1984): 81-93;

Mel Watkins, "An Interview with Ishmael Reed," *Southern Review,* 21 (July 1985), pp. 603-614;

Bruce Dick and Amritijit Singh, *Conversations with Ishmael Reed* (Jackson: University Press of Mississippi, 1995).

Bibliography:

Elizabeth A. Settle and Thomas A. Settle, *Ishmael Reed: A Primary and Secondary Bibliography* (Boston: G. K. Hall, 1982).

References:

Chester J. Fontenot, "Ishmael Reed and the Politics of Aesthetics, or Shake Hands and Come Out Conjuring," *Black American Literature Forum,* 12 (Spring 1978): 20-23;

Robert Elliot Fox, "Blacking the Zero: Toward a Semiotics of Neo-Hoodoo," *Black American Literature Forum,* 18 (Fall 1984): 95-99;

Henry Louis Gates Jr., "The blackness of blackness: a critique of the sign and the Signifying Monkey," in *Black Literature and Literary Theory,* edited by Gates (New York: Methuen, 1984), pp. 285-321;

Sharon Jessee, "Ishmael Reed's Multi-Culture: The Production of Cultural Perspective," *MELUS,* 13 (Fall-Winter 1986): 5-14;

Richard Walsh, "A Man's Story is his 'Gris-Gris': Cultural Slavery, Literary Emancipation and Ishmael Reed's *Flight to Canada,*" *Journal of American Studies,* 27 (1993): 1, 57-71.

James Schuyler
(9 November 1923 - 12 April 1991)

Joseph M. Conte
State University of New York at Buffalo

See also the Schuyler entry in *DLB 5: American Poets Since World War II, First Series.*

BOOKS: *Alfred and Guinevere* (New York: Harcourt, Brace, 1958);
Salute (New York: Tiber Press, 1960);
May 24th or So (New York: Tibor de Nagy Editions, 1966);
Freely Espousing (Garden City, N.Y.: Paris Review Editions/Doubleday, 1969; New York: SUN, 1979);
A Nest of Ninnies, by Schuyler and John Ashbery (New York: Dutton, 1969; Manchester, U.K.: Carcanet, 1987);
The Crystal Lithium (New York: Random House, 1972);
A Sun Cab (New York: Adventures in Poetry, 1972);
Hymn to Life (New York: Random House, 1974);
The Fireproof Floors of Witley Court; English Songs and Dances (Newark & West Burke, Vt.: Janus Press, 1976);
Song (Syracuse, N.Y.: Kermani Press, 1976);
The Home Book: Prose and Poems, 1951–1970, edited by Trevor Winkfield (Calais, Vt.: Z Press, 1977);
What's For Dinner? (Santa Barbara, Cal.: Black Sparrow Press, 1978);
The Morning of the Poem (New York: Farrar, Straus & Giroux, 1980);
Collabs, by Schuyler and Helena Hughes (New York: Misty Terrace Press, 1980);
Early in '71 (Berkeley, Cal.: The Figures, 1982);
A Few Days (New York: Random House, 1985);
For Joe Brainard (New York: Dia Art Foundation, 1988);
Selected Poems (New York: Farrar, Straus & Giroux, 1988; Manchester, U.K.: Carcanet, 1990);
Collected Poems (New York: Farrar, Straus & Giroux, 1993);
Two Journals: James Schuyler, Darragh Park, by Schuyler and Darragh Park (New York: Tibor de Nagy, 1995);

James Schuyler (photograph © the Estate of Thomas Victor)

Diary of James Schuyler (Santa Rosa, Cal.: Black Sparrow Press, 1996).

PLAY PRODUCTIONS: *Presenting Jane,* Cambridge, Mass., Poet's Theatre, 1952;
Shopping and Waiting: A Dramatic Pause, New York, American Theatre for Poets, 1953;
Unpacking the Black Trunk, by Schuyler and Kenward Elmslie, New York, American Theatre for Poets, 1964;
The Wednesday Club, by Schuyler and Elmslie, New York, American Theatre for Poets, 1964.

RECORDING: *Hymn to Life & Other Poems,* Watershed Intermedia, 1989.

OTHER: "Poet and Painter Overture," in *The New American Poetry,* edited by Donald M. Allen (New York: Evergreen-Grove, 1960), pp. 418–419;

Appearance and Reality: October Third to Thirty-first, 1960, introduction by Schuyler (New York: David Herbert Gallery, 1960);

Robert Dash: November 11–December 5, 1970, introduction by Schuyler (New York: Graham Gallery, 1970);

Penguin Modern Poets 24, edited by John Ashbery (Harmondsworth, U.K.: Penguin, 1973) – includes poems by Schuyler;

Broadway: A Poets and Painters Anthology, edited by Schuyler and Charles North (New York: Swollen Magpie Press, 1979);

Broadway 2: A Poets and Painters Anthology, edited by Schuyler and North (Brooklyn, N.Y.: Hanging Loose Press, 1989).

The career of James Schuyler has often been associated with the New York School of poets, John Ashbery, Frank O'Hara, Kenneth Koch, and Barbara Guest. Like any significant movement in the arts, such collocation of talent tends not only to define and anneal the achievements of the writers through their interaction but also to de-emphasize the individual successes or limitations of the group's members. The group dynamic of the New York School is also complicated by the influence of the Abstract Expressionist painters whose careers, by the mid 1950s, were significantly more advanced and gaining international attention, especially Jackson Pollock, Mark Rothko, Larry Rivers, Franz Kline, and Willem de Kooning. Schuyler's work has been strongly influenced by painterly vision and techniques. Like O'Hara, he worked for the Museum of Modern Art organizing exhibitions that circulated throughout the United States and Europe; like Ashbery, he was on the staff of *Art News* as a critic for ten years.

Schuyler is quick to point out in an interview published in the fall 1992 *Talisman* that Ashbery, O'Hara, and Koch might just as well have been known as the "Harvard Wits," since their friendship was established during their college years in Cambridge. "I didn't meet Kenneth when I first met John and Frank [by 1952]. He was then in California. When he came back I had the feeling that he wasn't too crazy about what the cuckoo had laid in his nest." Profoundly shy (Schuyler avoided public readings and interviews until the last decade of his life) and lacking Koch's circus-master's talent for promotion, not nearly as prolific as O'Hara, and missing the cosmopolitan polish that won Ashbery the Yale Younger Poets Prize in 1956, Schuyler's secondary role in the formation of the New York School aesthetic has led many to overlook the casual brilliance of his life's work.

Schuyler was a sometime cotenant with O'Hara in an apartment on East Forty-ninth Street during the mid 1950s. During a period of several years he collaborated with Ashbery on a novel, *A Nest of Ninnies* (1969), satirizing the uneventful lives of two families in a suburban New York town, an experience that Ashbery, from Rochester, New York, and Schuyler, who spent his teens in East Aurora, New York, knew well. In his early writing Schuyler shares Ashbery's interest in the experimental techniques of Dadaist collage that are found in the cut-up texts of his *The Tennis Court Oath* (1962). And there is a fairly broad swath of surreal imagery in the early work that one also finds in O'Hara's long poems of the 1950s such as "Second Avenue."

The New York School as a movement suffered an irrevocable loss when O'Hara was struck down and killed by a dune buggy on Fire Island in 1966. Schuyler's elegy for O'Hara, "Buried at Springs," appears near the end of his first major collection, *Freely Espousing* (1969). Schuyler's style subsequently becomes more autobiographical in nature, adopting something of the epistolary character of a text between two intimate, identifiable friends – one feels the speakers could be looked up in the telephone book – as first promoted in O'Hara's mock manifesto, "Personism" (1959). Although this manner suggests the illusion of intimacy, Schuyler deplores the shocking personal revelations of confessionalism: despite many years of psychoanalysis and several traumatic experiences (including his near death in a fire caused by smoking in bed), he withdraws the essentially private self from the poem.

In her essay included in the spring 1990 special issue of the *Denver Quarterly* devoted to Schuyler's work, Barbara Guest calls Schuyler an "Intimist": "for me Jimmy is the Vuillard of us, he withholds his secret, the secret thing until the moment appears to reveal it. We wait and wait for the name of a flower while we praise the careful cultivation. We wait for someone to speak. And it is Jimmy in an aside." This understated, indirect revelation distinguishes Schuyler from his peers. More particular in person and place than Ashbery's abstract meditations and more pained and much less gregarious than O'Hara's work, Schuyler in his later poetry – *Hymn to Life* (1974), *The Morning of the Poem* (1980), and *A Few Days* (1985) – speaks with a voice that is both intensely present and reticent, a fulfillment of the personism that O'Hara heralded and far more supple and moving than the poetics of the New York School could have achieved.

James Marcus Schuyler was born in Chicago, the son of Marcus James (a reporter) and Margaret Daisy Connor Schuyler. The family moved first to Washington, D.C., and then to East Aurora, New York, outside of Buffalo. His mother and brother settled in this small town (known for its Roycroft crafts guild and retaining something of the air of nineteenth-century gentleman farming), and it figures as a locale for "The Morning of the Poem" and "A Few Days," an elegy for Schuyler's mother. As Schuyler recalled in a 1981 interview, he experienced an almost Wordsworthian calling to be a writer "while in my tent in East Aurora, New York, when I was about fifteen." Reading an account of how a visit by Walt Whitman inspired Logan Pearsall Smith to literary ambition in his *Unforgotten Years* (1939), Schuyler says that "I looked up from my book, and the whole landscape seemed to shimmer." The personal epic of Whitman's "Song of Myself" and the vital force of landscape become major concerns of Schuyler's mature poetry.

Schuyler attended Bethany College in West Virginia from 1941 to 1943, where, he recalled in an interview published in spring 1992, he did poorly: "I just played bridge all the time." Unlike Koch, Ashbery, and O'Hara, who were active in creative writing classes while at Harvard, Schuyler wrote no poetry at that time. "I didn't have anything academic to be loyal to, or to be academic about," which partly explains his detachment from academia in his professional life and his later disdain for most literary criticism beyond the work of Harold Bloom and David Kalstone. He served in the navy during World War II on a destroyer in the North Atlantic doing convoy duty. Moving to New York City for the first time in the late 1940s, he worked for NBC and befriended the poet W. H. Auden. When he inherited a farm in Arkansas from his paternal grandmother, he sold it and moved to Italy with the intention of writing. There he became Auden's typist and lived in his house in Forio d'Ischia, an island in the Bay of Naples. He recalled in 1981 that he found Auden's elaborate formalism "inhibiting," which suggests his later preference for a conversational style and proselike line. Between 1947 and 1948 Schuyler attended the University of Florence.

When Schuyler returned to New York his life took shape around his associations with the art world. From 1955 to 1961 he was a curator of circulating exhibitions at the Museum of Modern Art. As an editorial associate and critic for *Art News* he wrote a substantial amount of art criticism, though he was paid little for his contributions. He recalled in the interview published in spring 1992 that "I did learn an awful lot during those years, and then went on in the 60s writing occasional articles about specific artists and their specific strategies. Partly it was to make money, and partly because I wanted to write about painting, about art." More important were the attachments to poets and painters made at galleries such as the Tibor de Nagy Gallery, or through the Artist's Club and the Cedar Tavern in Greenwich Village. These friends were Willem and Elaine de Kooning, Jane Freilicher, Alex and Ada Katz, Larry Rivers, and especially Fairfield Porter.

Schuyler lived with Porter and his family in Southampton, Long Island, and at their summer home on a coastal Maine island for twelve years from 1961 to 1973, a time he described in 1981 as "much the happiest period in my life." He appears contented, usually seated reading, in several of Porter's domestic paintings. Porter contributed dust jackets for *The Crystal Lithium* (1972) and *Hymn to Life*; he also illustrated Schuyler's *A Sun Cab* (1972). Although many knew him as a resident of the Hotel Chelsea in lower Manhattan, he claimed to be happiest and most productive when living in the country or in small villages. Schuyler received the Longview Foundation Award in 1961, the Frank O'Hara Prize from *Poetry* in 1969 for *Freely Espousing*, and the Pulitzer Prize for poetry in 1981 for *The Morning of the Poem*. He was a Guggenheim Fellow, and a fellow of the American Academy of Poets. He died in Manhattan at the age of sixty-seven following a stroke.

Like Ashbery's *Turandot and Other Poems* (1953), Schuyler's early poetry was published by the Tibor de Nagy Gallery in *May 24th or So* (1966), providing one measure of the importance of the art world to the sustenance of New York School poets. Schuyler's first major collection, *Freely Espousing*, is dedicated to Anne and Fairfield Porter. It did not appear until he was forty-six years old. John Koethe remarked in his October 1970 review in *Poetry* that "Coming upon a mature body of work without much prior warning is always a perplexing experience requiring an effort of accommodation."

In the title poem, "Freely Espousing," Schuyler's rapid shifts in sound, shape, and color give the poem the effect of a collage. His tone and subject change dramatically, from the initial Stevensian reverie of "a commingling sky" to the urban snobbery of "Quebec! what a horrible city / so Steubenville is better?" In the midst of this adhesion of snipes and snippets, one is reminded that the essential subject here is language:

> the sinuous beauty of words like allergy
> the tonic resonance of
> pill when used as in
> "she is a pill"
> on the other hand I am not going to espouse any short
>   stories in which lawn mowers clack.
> No, it is absolutely forbidden
> for words to echo the act described; or try to.

The separation of sound and sense, in which the woe of an allergy is divisible from the beauty of the word, is for Schuyler an "inescapable kiss." Like the artist who deploys paint and brushstroke, he wishes to work in the pure medium of sound without the bondage of reference. Still he contents himself with a kind of mixed media, "Marriages of the atmosphere," that allow him to pass from speech to evocations of shape and color: "'What is that gold-green tetrahedron down the river?'"

In the special issue of the *Denver Quarterly* Guest remarks that Schuyler "translates the vagaries of inhabitancy, of wherever he is, his locale, particularly *his* into poetry. So that if you are already acquainted with a particular house he has lived in, you come to know it even better." Schuyler's "February" offers a view from his Manhattan apartment that is at once recognizably descriptive and an evocative composition that depends entirely for its success on the aura of the poet in that place, in that season:

> A gray hush
> in which the boxy trucks roll up Second Avenue
> into the sky. They're just
> going over the hill.
> The green leaves of the tulips on my desk
> like grass light on flesh,
> and a green-copper steeple
> and streaks of cloud beginning to glow.
> I can't get over
> how it all works in together[.]

Only to record the "vagaries of inhabitancy" makes for a passive and vacant realism. Schuyler's talent rests in the perception of the scene's intrinsic composition, "how it all works in together." He recognizes the collaboration that must arise between the random presentation of elements and the sensibility of the poet. In the transient and the merely ordinary the poet must discover an immanent grace: "It's the water in the drinking glass the tulips are in. / It's a day like any other."

Freely Espousing and *The Crystal Lithium* are notable for their effortless movement among poetry, painting, and music: the arts available in such profligate array to a resident of Southampton and Manhattan and the practitioners so continually on the scene account for the natural, never forced, cross-fertilization of media and community. Schuyler's work takes place within or among the performance of the artist or the musicians, sometimes relishing it, sometimes emulating it, as the act, not the "source," of the poem. Such is the case in "A Man in Blue," which achieves the synesthesia of color, tonal register, and atmosphere that the French poet Arthur Rimbaud proposed:

> Under the French horns of a November afternoon
> a man in blue is raking leaves
> with a wide wooden rake (whose teeth are pegs
> or rather, dowels). Next door
> boys play soccer: "You got to start
> over!" sort of. A round attic window
> in a radiant gray house waits like a kettledrum.
> "You got to start . . ." The Brahmsian day
> lapses from waltz to march. The grass,
> rough-cropped as Bruno Walter's hair,
> is stretched, strewn and humped beneath a sycamore[.]

The nineteenth-century German composer Johannes Brahms intervenes to offer the famous twentieth-century conductor, Bruno Walter, interpretive advice on "the first movement / of my Second, think of it as a family planning where to go next summer / in terms of other summers." In words that capture the essence of Schuyler's poetics, he describes the effect he envisions as "A material ecstasy, / subdued, recollective." The promise of immortality held up by the arts cannot be dissociated from the materiality of the man in blue (denim clothing; the mood of late fall; George Gershwin's "Rhapsody in Blue"; Pablo Picasso's blue period; and Wallace Stevens's "The Man with the Blue Guitar"). The conductor with his baton and the neighbor raking leaves are one:

> He waves his hands and through the vocalese-shaped
>   spaces
> of naked elms he draws a copper beech
> ignited with a few late leaves. He bluely glazes
> a rhododendron "a sea of leaves" against gold grass.
> There is a snapping from the brightwork
> of parked and rolling cars.
> There almost has to be a heaven! so there could be
> a place for Bruno Walter

who never needed the cry of a baton.
Immortality –
in a small, dusty, rather gritty, somewhat scratchy
Magnavox from which a forte
drops like a used Brillo Pad?

More important for Schuyler than the lyrical phrase of great beauty or the pointed thought held with conviction is the movement of the poet's sensibility through sight and sound, through the material world and the mood of a moment – the true interdisciplinary nature of his craft.

Schuyler's early work comes to a close with his elegy for Frank O'Hara, "Buried at Springs." As he remembers in the fall 1992 *Talisman* interview, "It was written in Maine. The things described in it are what I was seeing out the window in the house in Maine. You know, Frank died in the summer . . . and it was shattering. The elegy was originally two poems; nothing seemed adequate so I had to sort of put them away."

Schuyler's complex emotional reaction to the tragedy, a combination of anguish and a guilty awareness of artistic rivalry, is obliquely filtered through the Maine locale. In a gesture reminiscent of Ralph Waldo Emerson's transcendentalism, Schuyler uses nature to reflect his mood:

There is a hornet in the room
and one of us will have to go
out the window into the late
August midafternoon sun. I
won. There is a certain challenge
in being humane to hornets
but not much. A launch draws
two lines of wake behind it
on the bay like a delta
with a melted base.

O'Hara's frenetically busy, short life and prolific career are contrasted to the quieter determination of the speaker. Schuyler maintains the tense composure of this elegy almost entirely through his use of line breaks, such as "I / won," and "humane to hornets / but not much." Their lives, intertwined in jobs, apartments, social gatherings, and especially avocations, are now like the irrevocably divergent and melting "lines of wake." The first section closes with a traditional meditation on the mutability of life: "Frank sat at this desk and / saw and heard it all / the incessant water the / immutable crickets only / not the same: new needles / on the spruce, new seaweed / on the low-tide rocks."

In "Grand Duo," a poem from *The Home Book: Prose and Poems, 1951–1970* (1977) that wends its

Cover for an issue of The American Poetry Review *that includes previously unpublished poems by Schuyler*

way through reflections on composer Franz Schubert, Schuyler provides something of a statement on his method: "Art is formality, courtesy, passion, control, practice, / rehearsing the unrehearsed." The paradox of the latter phrase is essential to an understanding of Schuyler's work, which so often gives the appearance of informality, as if it were the product of an idle moment or casual observation or a passing mood. In Coleridgean terms this would be "form as proceeding," in which the exercise of form must be intrinsic to the act and process of composition.

Schuyler's poetry is busy with nothing to do. He attends lovingly to unoccupied moments, and lavishly records hours of leisure; there are spring mornings in Chelsea, summer afternoons in Southampton, and fall evenings in Vermont. In the more than four hundred pages of his collected poetry, there are few remarks (and those are usually disparaging) regarding any sort of gainful employment; positions such as that of bookstore clerk or log keeper for the Voice of America broadcasts distract from the true work of one's

Cover for the book that many critics consider Schuyler's best work

own writing. Largely exempt from the daily grind, Schuyler vastly prefers attending to the unrehearsed incident and the play of the mind as it moves with its intrinsic vigor through observation, association, and reflection. Such variegated activity of the mind serves to direct, if not precisely to organize, the poem on scales both fine and large.

"A Vermont Diary" in *The Crystal Lithium* is one of several seasonal sequences in Schuyler's work. It records a week in which the New York poet is both exquisitely conscious of nature and tracking the nature of his own consciousness:

> A frail gray flower
> flies off, an insect
> that escaped the first
> combing frosts. It's
> not – "the fly buzzed"
> finding moods, reflectives:
> fall
> equals melancholy, spring,
> get laid: but to turn it all
> one way: in repetition, change:
> a continuity, the what
> of which you are a part.

This sequence combines the observation of a camouflaged and soon-to-die insect, an allusion to Emily Dickinson's poem in which a fly signifies her fetish for mortality, and the romantic association of natural setting and human emotions; it illustrates the sundry and often remarkable ways in which the unoccupied and unrehearsed mind encounters the world and accounts for itself.

Schuyler continually rejects the impulse to symbolism and extravagant metaphor in order to convey "the what / of which you are a part," the struggle to express being and cognition as a continuum rather than as a dichotomy. In some instances he employs a notational style, as in "Verge," which closes the "Vermont Diary" sequence:

> All the leaves
> are down except
> the few that aren't.
> They shake or
> a wind shakes
> them but they
> won't go oh
> no there goes
> one now. No.
> It's a bird
> batting by.

One finds here Marianne Moore's attachment to the natural emblem, her casual rhymes as they occur in speech, and a tendency to restrain figural language within a single category of reference; that is, both the tenor and the vehicle of the metaphor are winged creatures ("bird / batting"). Schuyler's choice of an almost monosyllabic minimalism and his instantaneous revision that corrects but does not expunge the prior perception of leaves fallen, falling, or not, reminds the reader that the practice of immediacy was common to Projective Verse and Action painting. Schuyler attends to the incidental and allows experience to lend shape to cognition.

"The Crystal Lithium" is the first of several long poems written in an expansive line, relatively free of full stops or conceptual divisions. Perhaps because lithium was widely prescribed as an antidepressant, the alternately languid and surging pace of the poem indicates the poet's search for an emotional and psychological equilibrium:

> The smell of snow, stinging in nostrils as the wind lifts
>   it from a beach
> Eye-shuttering, mixed with sand, or when snow lies
>   under the street lamps and on all

> And the air is emptied to an uplifting gassiness
> That turns lungs to winter waterwings, buoying, and the bright white night
> Freezes in sight a lapse of waves, balsamic, salty, unexpected:
> Hours after swimming, sitting thinking biting at a hangnail
> And the taste of the – to your eyes – invisible crystals irradiates the world
> "The sea is salt"
> "And so am I."

Schuyler uses the scene of the beach at the onset of winter to articulate the proximity of liquid and crystal, the organic and the inorganic, the mind and the world. The crystals of lithium, like those of salt, are necessary for the speaker, but they are also inorganic stabilizations that blind and numb. Equally essential is the fluidity of awareness, the waves and gusts of experience that carry the mind through periods of turbulence. As reflected in his extended lines and irregular patterning of thought, Schuyler searches – perhaps in vain – for an emotional balance between the crest of ecstasy and the trough of depression, between moods of "uplifting gassiness" and "sitting thinking biting at a hangnail."

The best work of Schuyler's fourth major collection, *Hymn to Life,* is contained in the section "Evenings in Vermont" and in the long title poem. One can read the former as an example of Schuyler's homage to the cognate relationship of domestic tranquility, friendship, and the reassurances of natural beauty:

> That one maple by the house now
> is almost bare. The pond
> turns greenish-black. Downstairs
> someone (you) shuts a door. Evenings
> in Vermont, the fire dies in the sky,
> the pond goes altogether black,
> and indoors all is coziness. I study
> the pattern in a red rug, arabesques
> and squares, and one red streak
> lies in the west, over the ridge.

"Hymn to Life" celebrates the return of spring, enabling Schuyler to address at length one of his favorite subjects, flower gardens. He observes "the sun which seems at / Each rising new, as though in the night it enacted death and rebirth, / As flowers seem to. The roses this June will be different roses / Even though you cut an armful and come in saying, 'Here are the roses,' / As though the same blooms had come back, white freaked with red / And heavily scented." In the ritual of renewal the poet observes both the pain and the joy, the sameness and difference, the definite article and the indefiniteness in the cycle of life. These longer poems become the ideal vehicle for Schuyler's meditations on the flow of time. "Time brings us into bloom and we wait, busy, but wait / For the unforced flow of words and intercourse and sleep and dreams / In which the past seems to portend a future which is just more / Daily life." Very much concerned to map the quotidian, Schuyler has the courage to do so in an "unforced flow of words" that risks boring the reader with the uneventful or unremarkable – because, simply, that is the way time passes. For the persistent if sometimes unaware individual, however, the unremarkable may still convey its meager message: "A window to the south is rough with raindrops / That, caught in the screen, spell out untranslatable glyphs. A story / Not told: so much not understood, a sight, an insight, and you pass on, / another day for each day is subjective and there is a totality of days / As there are as many to live it. The day lives us and in exchange / We it."

To the extent that he is a poet of meditative immediacy, Schuyler's best work coincides with the longer poems in which he increasingly gave himself room to ramble, congenially, through all colors of the cognitive spectrum. The sixty-page title poem of *The Morning of the Poem* is certainly his masterwork and among the best long poems of the postmodern era. Schuyler explained in the spring 1992 interview that the poem is written in "very much the style of my letter writing" to Ashbery, Joe Brainard, and Darragh Park – an intimate, gradual account. There is no grand structure: "I never have a plan beforehand. I had no idea when I sat down to the typewriter that morning what I was going to say, beyond the title." In retreat at a family home in East Aurora, Schuyler administers a poet's own best treatment for the breakdown of mind and body. It is a poem of recovery – of the self, of one's family and friends, of a meaningful existence – and as such it must be slow, and tender, and caring.

Though wry and funny in a self-deprecating manner, the poet offers a painful testament that is presided over by the image of "Baudelaire's skull," which "Stands for strength and fierceness, the dedication / of the artist," but also the specter of a too-early death. Taking a country farmhouse during July 1976 as his main setting, Schuyler nevertheless shifts readily in time and place, between the open fields and the cityscape of West Twenty-second Street, from the present self-absorption of the convalescent to memories of a deceased father and departed friends. Yet the poem is seamless and comfortable in its associative transitions. Like the all-over composition of a large Jackson Pollock canvas, this long poem is almost impossible to excerpt, to point to one part, one musing, as somehow detachable and any more telling than another:

> yesterday I tripped on a scatter
> rug and slam fell full length,
> The wind knocked out of me: "Shall I call a
> doctor?" "Please don't talk"
> "Are you hurt? Can I help you?" "Shut the fuck
> up" I thought I'd smashed
> My kneecap – you know, like when you really
> wham your funny bone, only
> More so – but I got up and felt its nothing-
> broken-tenderness and
> Hobbled down this everlasting hill to distant
> Bell's and bought
> Edible necessities: small icy cans of concentrated
> juice, lemon, lime, orange,
> Vast puffy bags of bread, Smucker's raspberry jam,
> oatmeal, but not the good,
> The Irish kind (travel note: in New York City you
> almost cannot buy a bowl
> Of oatmeal: I know, I've tried: why bother: it
> would only taste like paste)
> and hobbled home, studying the for-sale house
> hidden in scaly leaves
> The way the brownstone facing of your house is
> coming off in giant flakes: there's
> A word for that sickness of the stone but I
> can't remember it[.]

Kaleidoscopic in effect, "The Morning of the Poem" twists the lens of observation and reflection, continually revealing still another pattern no less fascinating than the last. The poet takes a pratfall, picks himself up, and though there's no permanent damage done, recognizes in his own body, and in the world's body, a slow and irreversible decay. "The Morning of the Poem" is a single morning in which premonitions of death, "my death, over fifty years and that is / What I am building toward," and memories of childhood, "I wish it was 1938 or '39 again / and Bernie was sleeping / With me in the tent at the back of the yard," are filtered. The epistolary mode nevertheless demands, as Schuyler asserted, "a certain lightness of tone, gaiety" that admits friends, relatives, and neighbors. The poem also becomes a compendium of anecdotes, landmarks of western New York, and extended "stories" from the poet's life (such as furtively reading *Ulysses* for the first time in an American history class) and others' (Fairfield Porter painting in the nude on his private Maine island when an uninvited couple arrive). Ultimately Schuyler maintains this mélange of the poignant and the ephemeral on the strength of his personal voice, his innate love of

> How the thing said
> Is in the words, how
> The words are themselves
> The thing said: love,
> Mistake, promise, auto
> Crack-up, color, petal,
> The color in the petal
> Is merely light
> And that's refraction:
> A word, that's the poem.

In all its myriad refractions, "The Morning of the Poem" must be measured alongside such introspective American long poems as Walt Whitman's "Song of Myself," Hart Crane's "Voyages," John Berryman's *The Dream Songs,* and Ashbery's "Self-Portrait in a Convex Mirror."

Schuyler's last completed volume of poetry, *A Few Days,* ranges between specificity, that which is singular and thus transient, and what Schuyler calls in "August first, 1974" the "*Qualunque:* / commonplace," that which is pervasive and invisible. These poems are replete with details of place, personal conversations and encounters, the weather on a particular day, and flowers described with botanical precision. Among the best short poems in the volume are those that capture the aura of place, as in "Beaded Balustrade":

> The balustrade along my balcony
> is wrought iron in shapes of
> flowers: chrysanthemums, perhaps,
> whorly blooms and leaves and
> along the top a row of what look
> like croquet hoops topped by a
> rod, and from the hoops depend
> water drops, crystal, quivering.
> Why, it must be raining, in Chelsea,
> NYC!

As in the previous two volumes, the title poem of *A Few Days* is a significant contribution to the autobiographical epic. Similar to "The Morning of the Poem" in its conversational style, double-length lines, and blend of copious observation and apparently unmotivated recollection, "A Few Days" shifts slightly from an epistolary to a diaristic mode. Sometimes listless, sometimes playful, and finally elegiac, the poem suggests that even ordinary days are essential and must be attended to:

A few days

> are all we have. So count them as they pass. They pass
> too quickly
> out of breath: don't dwell on the grave, which yawns
> for
> one and all.

> Will you be buried in the yard? Sorry, it's against
>   the law. You can only
> lie in an authorized plot but you won't be there to
>   know it so why worry
> about it? Here I am at my brother's house in western
>   New York: I came
> here yesterday on the Empire State Express, eight hours of
>   boredom on the train.

Even if the predominant mood is one of ennui, there are still worthy observations to be made:

> A pretty blond child sat next to me for a while. She
>   had a winning smile,
> but I couldn't talk to her, beyond "What happened to
>   your shoes?" "I put them under the seat." And
> so she had. She pressed the button that released the
>   seat back and sank
> back like an old woman. Outside, the purple loosestrife
>   bloomed in swathes
> that turned the railway ditch and fields into a
>   sunset-reflecting lake.

Like all diaristic writing, the intent here is the recording of the passing of days, which for Schuyler are made all the more precious by his awareness of their fast-approaching limit. This diversionary child-woman prefigures his reunion with his eighty-nine-year-old mother, who "has little to do but sit and / listen to the TV rumble." And though he assures himself of her longevity while doubting his own, it is her death that closes the poem. No subject dominates, however; "this isn't about my family, although I wish it were."

As Mark Rudman points out in his essay in the special issue of the *Denver Quarterly*, Schuyler's long poems resist the didactic, the epiphanic, or the impulse to self-improvement: "he defines things by their unimportance and resists psychologizing why one memory, one perception, is connected to another." For Schuyler the poem allows him to gain some purchase on time:

> Today is tomorrow, it's that dead time again: three in
>   the afternoon under scumbled clouds,
> livid, that censor the sun and withhold the rain:
>   impotent as an old man ("an
> old man's penis: limp as a rabbit's ear"). It's cool
>   for August and I

> can't nail the days down. They go by like escalators,
>   each alike, each with its own
> message of tears and laughter.

Schuyler's poetry has a cumulative power that seizes one only after the reading of many leisurely pages. Since its significance is more recognizable when read in the aggregate than when one reads with the expectation of being dazzled by any one lyric, one suspects that the publication of Schuyler's *Selected Poems* (1988) and *Collected Poems* (1993) will confirm his status as a major poet. The multifaceted sensibility and engaging, supple voice evident in these works certainly belong to one of the important American poets to emerge in the last decades of the twentieth century.

Interviews:

Jean W. Ross, "*CA* Interviews the Author," in *Contemporary Authors*, volume 101, edited by Frances C. Locher (Detroit: Gale Research, 1981), pp. 445–447;

Mark Hillringhouse, "James Schuyler: An Interview," *American Poetry Review*, 14 (March/April 1985): 5–12;

Robert Thompson, "An Interview with James Schuyler," *Denver Quarterly*, 26 (Spring 1992): 105–122;

Carl Little, "An Interview with James Schuyler," *Talisman*, 9 (Fall 1992): 176–180.

References:

Philip Auslander, *The New York School Poets as Playwrights: O'Hara, Ashbery, Koch, Schuyler, and the Visual Arts* (New York: Peter Lang, 1989);

William Corbett and Geoffrey Young, eds., *That Various Field: For James Schuyler (1923–1991)* (Great Barrington, Mass.: The Figures, 1991);

Denver Quarterly, special issue on Schuyler, edited by Donald Revell, 24 (Spring 1990);

Geoff Ward, "James Schuyler and the Rhetoric of Temporality," in his *Statutes of Liberty: The New York School of Poets* (New York: St. Martin's Press, 1993), pp. 10–35.

Papers:

The major collection of Schuyler's papers, covering the years from 1947 to 1991, is held in the Mandeville Department of Special Collections at the University of California, San Diego.

Anne Sexton
(9 November 1928 – 4 October 1974)

Diane Wood Middlebrook
Stanford University

See also the Sexton entry in *DLB 5: American Poets Since World War II, First Series.*

BOOKS: *To Bedlam and Part Way Back* (Boston: Houghton Mifflin, 1960);

All My Pretty Ones (Boston: Houghton Mifflin, 1962);

Eggs of Things, by Sexton and Maxine Kumin (New York: Putnam, 1963);

More Eggs of Things, by Sexton and Kumin (New York: Putnam, 1964);

Selected Poems (London: Oxford University Press, 1964);

Live or Die (Boston: Houghton Mifflin, 1966; London: Oxford University Press, 1967);

Poems by Thomas Kinsella, Douglas Livingstone and Anne Sexton (London: Oxford University Press, 1968);

Love Poems (Boston: Houghton Mifflin, 1969; London: Oxford University Press, 1969);

Joey and the Birthday Present, by Sexton and Kumin (New York: McGraw-Hill, 1971);

Transformations (Boston: Houghton Mifflin, 1971; London: Oxford University Press, 1972);

The Book of Folly (Boston: Houghton Mifflin, 1972; London: Chatto & Windus, 1974);

The Death Notebooks (Boston: Houghton Mifflin, 1974; London: Chatto & Windus, 1975);

The Awful Rowing Toward God (Boston: Houghton Mifflin, 1975; London: Chatto & Windus, 1977);

The Wizard's Tears, by Sexton and Kumin (New York: McGraw-Hill, 1975);

45 Mercy Street, edited by Linda Gray Sexton (Boston: Houghton Mifflin, 1976; London: Secker & Warburg, 1977);

Words for Dr. Y.: Uncollected Poems with Three Stories (Boston: Houghton Mifflin, 1978);

The Complete Poems (Boston: Houghton Mifflin, 1981)

No Evil Star: Selected Essays, Interviews, and Prose, edited by Steven E. Colburn (Ann Arbor: University of Michigan Press, 1985);

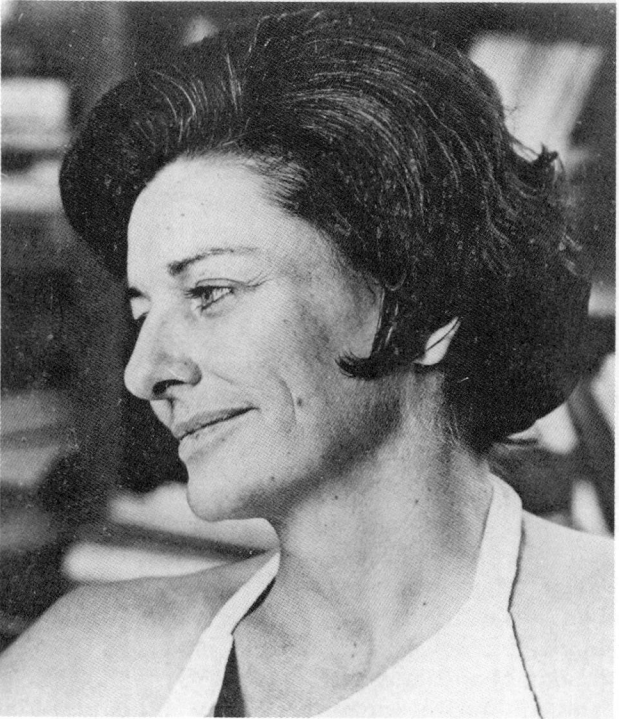

Anne Sexton, summer 1974 (Collection of Linda Gray Sexton)

Selected Poems of Anne Sexton, edited by Diane Wood Middlebrook and Diana Hume George (Boston: Houghton Mifflin, 1988).

PLAY PRODUCTION: *Mercy Street,* New York, American Place Theater, 3 October 1969.

Anne Sexton began writing poetry at age twenty-eight as a form of psychotherapy during treatment for a clinical depression. By the time of her suicide at age forty-five she had become a major figure in postwar American poetry. Her work was intimate, confessional, comic, formally complex, psychologically acute, and disruptively female; and her popular public readings were spectacles of performance art. Admired by peers for her technical skill and compelling imagery, Sexton won most of the prizes available to American poets. She also gained, for a poet, an exceptionally wide audience of readers drawn to the contemporaneity of her work. Her poetry was distinctive in its straightforward treatment of mental illness and prosperous suburban life in the era of the Vietnam War and the sexual revolution in America.

Anne Gray Harvey was born in Newton, Massachusetts, the third of three daughters, to Mary Gray Staples of Auburn, Maine, and Ralph Churchill Harvey, owner of a wool firm in Waltham, Massachusetts. She was educated in public schools in Wellesley, Massachusetts, and at Rogers Hall, a girls' preparatory school in Lowell, Massachusetts. After a year at the Garland School, a finishing school in Boston, she eloped on 16 August 1948 with Alfred Muller "Kayo" Sexton II. After their marriage Kayo Sexton joined Ralph Harvey's wool business as a salesman. The Sextons made their home in Newton Lower Falls and in Weston, Massachusetts, and had two daughters: Linda Gray Sexton, born 1953, and Joyce Ladd Sexton, born 1956.

Shortly after the birth of her second child, Sexton began psychiatric treatment for what was initially diagnosed as a postpartum depression. She became dangerously suicidal and was hospitalized, and her children were removed from her care. Her psychiatrist encouraged her to take up writing as a way of strengthening self-esteem; her maternal grandfather had been a newspaper publisher, and literature was admired in her mother's family. Sexton found writing poetry and engaging in psychotherapy to be highly congruent activities, requiring acute responsiveness to the multivalences of language. Writing, she later explained in a lecture, "is like lying on the analyst's couch, re-enacting a private terror, and the creative mind is the analyst who gives pattern and meaning to what the persona sees as only incoherent experience."

Having little background in literature and little training as a writer, Sexton enrolled in a night-school course at the Boston Center for Adult Education in 1957, where she had the good fortune to meet the poets Maxine Kumin, George Starbuck, and John Clellon Holmes. The four of them continued to gather regularly in an informal workshop after the course at the Boston Center ended. Sexton's emerging talent flourished in the environment of their praise and criticism. She called her newfound purpose in life "a kind of rebirth at twenty-nine."

A major turning point in Sexton's development as an artist occurred when she received a scholarship to the Antioch Writers' Conference in the summer of 1958 to work for a week with the poet W. D. Snodgrass. Snodgrass's autobiographical poem "Heart's Needle" (1959) provided Sexton with a model for her most ambitious early poem, "The Double Image," a lightly retouched account of the mental breakdown that resulted in separation from her children. An enthusiastic recommendation from Snodgrass helped Sexton gain admission that fall to Robert Lowell's writing seminar at Boston University. The poet Sylvia Plath also joined Lowell's seminar that year; she and Sexton became friends after Lowell shrewdly began drawing the attention of each to the work of the other. This pairing gained great significance for Sexton after Plath's suicide in February 1963. Observing the positive impact of suicide on Plath's reputation as an artist, Sexton felt cheated that Plath got there first. Sexton told her doctor, "that death was mine!"

Sexton produced her most formally deft early poetry under the influence of Lowell's workshop, then made her debut as a poet under the influence of Lowell's patronage. Lowell helped her select the contents of her first volume, *To Bedlam and Part Way Back* (1960), and vigorously pursued its publication; the Boston publishing house of Houghton Mifflin accepted the book just as Lowell's seminar was ending in May 1959 and remained her American publisher throughout her career.

Partly because a blurb by Lowell was prominently featured on the cover of *To Bedlam and Part Way Back,* the book was reviewed in the national press. Singled out for particular attention were the poems that spoke frankly and eloquently of mental illness: "You, Dr. Martin," "Music Swims Back to Me," "Ringing the Bells," "Lullaby," "The Double Image." Most of the reviews were respectful, and *To Bedlam and Part Way Back* was nominated for the prestigious National Book Award that year. Astoundingly, Sexton had leaped from complete obscurity to major recognition as a "confessional" poet in the Lowell manner, a mere three and a half years after writing her first poems in therapy.

Sexton's career passed another significant milestone with her appointment in 1961 as a scholar of the Radcliffe Institute, founded that year by Radcliffe president Mary Bunting to encourage college-educated housewives to reenter the intellectual labor force. Sexton was a "maverick" among the Radcliffe scholars: she had no college education and had received her artistic training on the job. But President Bunting regarded artists as important to the project she envisioned for the Radcliffe Institute; among Sexton's cohorts were the painter Barbara Swan and the fiction writer Tillie Olsen, along with Kumin.

The two years Sexton spent as a fellow of the Radcliffe Institute provided her with the collegiality of a variety of other professional women, and her first contact with feminist thought. Under the influence of Olsen, Sexton discovered Virginia Woolf's *A Room of One's Own* (1929), while Betty Friedan's newly published book, *The Feminine Mystique* (1963), became a major topic of debate among the scholars.

From Sexton's point of view, however, the most important legacy of the Radcliffe scholarship was the recognition it conferred on her autodidactically achieved position as an emerging American poet. After the Radcliffe appointment she received most of the prestigious awards available to American poets: the Levinson Prize from *Poetry* (1962); a traveling fellowship from the American Academy of Arts and Letters (1963); a Ford Foundation grant for residence with the Charles Playhouse, Boston (1964); election as a Foreign Fellow of the Royal Society of Literature, London (1965); a travel award from the International Congress of Cultural Freedom (1965); the Pulitzer Prize for poetry (1967); the Shelley Award from the Poetry Society of America (1967); honorary memberships in Phi Beta Kappa chapters at Harvard University (1968) and at Radcliffe (1969); a Guggenheim Fellowship (1969); an appointment to the Crashaw Chair of Literature at Colgate University (1972); and honorary doctoral degrees from Tufts University (1970), Fairfield University (1972), and Regis College (1973).

But those achievements were to come later. During her appointment at Radcliffe, Sexton continued to receive the benefit of Lowell's criticism and approval as she rapidly completed the poems published in her next book, *All My Pretty Ones* (1962). Her successful performance in Lowell's class gave her the courage to enroll in the 1961 summer semester at Brandeis University, where she studied modern literature with the critics Irving Howe and Philip Rahv. During this period Sexton also sought the mentorship of two other young poets, Anthony Hecht and James Wright. She now had a group of well-educated and well-disposed young poet-critics who provided her with exactly the milieu of advice and encouragement she needed to develop her particular talents into full mastery. Their collegiality served Sexton as the equivalent of both a college education and the analyst's ear in therapy. Her mentors swamped her with reading lists and carefully dissected her work as it progressed, broadening her reading and helping her to hear the latent implications and the expressive rhythms in her own words.

Meanwhile, Sexton and Kumin had found a unique way to combat the difficulties of being artists who were housebound as mothers: they began "workshopping" their poems over the telephone, each installing a private line in her study for this purpose. They would link up as soon as their families were out of the house in the morning, chat about their dreams, then leave the line connected as they worked on poems, whistling into the receiver for each others's attention when they wanted to discuss their work. They also directly collaborated on several children's books: *Eggs of Things* (1963), *More Eggs of Things* (1964), *Joey and the Birthday Present* (1971), and *The Wizard's Tears* (1975).

All My Pretty Ones, completed during Sexton's term at Radcliffe, is often judged to be her best book. The title is taken from a line of dialogue in *Macbeth,* elicited when the horrified Duncan makes the discovery that his wife and children have been slaughtered. During the years of her apprenticeship as a poet, Sexton herself had lost a beloved great-aunt and both of her parents to premature deaths; and her own chronic illness, punctuated with suicide attempts, had felt like a death threat miraculously survived. The best poems of *All My Pretty Ones* are elegies or eloquent evocations of these losses. Many have found their way into anthologies: "The Truth the Dead Know," "The Starry Night," "The Abortion," "The Fortress," "Letter Written on a Ferry While Crossing Long Island Sound." For each of these, Sexton developed a different, elaborately patterned, rhymed stanza that was replete with assonances. Few contemporary poets had as good an ear for sonic chime as did Sexton. This book was also nominated for the National Book Award, and, although it did not win the prize, it won gratifying praise from many poets: "womanly in the greatest sense," Plath wrote in a personal letter; "harrowing, awful, very real – and very good," wrote Elizabeth Bishop in another. However, the book was savaged in *Poetry* by James Dickey for "dwell[ing] . . . insistently on the pathetic and disgusting aspects of bodily experience." Sexton's art was always going to be judged more on content than on form.

Dickey's judgment represented a minority opinion, however, and the respect reviewers paid *All My Pretty Ones* not only provided Sexton with a strong national reputation, but it also attracted inquiries from foreign presses, as well. The English poet Jon Stallworthy scouted Sexton's work for inclusion in a series he was developing for Oxford

University Press. In 1964 Oxford published *Selected Poems*, a combined version of *To Bedlam and Part Way Back* and *All My Pretty Ones*. The book sold quickly and received many accolades; it became a selection of the Poetry Book Society, a distinction that led to Sexton's invitation the following year to become a Foreign Fellow of the Royal Society of Literature.

Despite her impressive achievements, Sexton continued to suffer from debilitating bouts of depression. Her condition had worsened in 1960 after the deaths of both her parents within a few months of one another. During psychotherapy sessions Sexton often lapsed into trancelike states from which it was difficult to rouse her; several times she was disoriented enough to require hospitalization. The psychiatrist regarded this disturbance as a symptom. He addressed it by audiotaping their thrice-weekly sessions and requiring Sexton to replay and take notes from each meeting before the next occurred. Sexton and her doctor both regarded the treatment as therapeutically successful. The practice had significant consequences for her art, as well. During the years of taping (January 1961–June 1964) Sexton wrote very little formal verse. Her work became looser, more shaped by the rhythms of spoken syntax than by metrical norms; and she became interested in writing for the theater.

Sexton viewed the psychoanalytic process as inherently dramatic – the pursuit of revelation, as in a mystery story. Excavating the notebooks in which she was transcribing material from her therapy sessions, she began writing a play in 1962. An early draft titled "The Cure" was given a staged reading that year in Boston by repertory actors at the Charles Playhouse. In Sexton's drama a suicidal woman named Daisy seeks help from a psychiatrist. Daisy suffers from visions; she believes Christ is summoning her to share his death. While the doctor wants to "cure" Daisy of her "delusions," Daisy wants the doctor to help her find their meanings. She fears madness, but she also fears the possibility of substituting feminine conformity for a liberating spirituality.

At the center of the story is Daisy's ambiguous memory of a night when she was thirteen years old and her charming, drunken father made a tentatively incestuous approach to her bed, an episode witnessed by Daisy's beloved great-aunt, who subsequently went mad. Sexton framed this episode entirely differently in the three different versions of the play she wrote between 1962 and 1969, but it remained the central scene in the play, just as it remained the central mystery in her therapy. In "The Cure," in her first draft (1962, unpublished), Sexton

Dust jacket for Sexton's 1960 book, which was nominated for a National Book Award

used a realistic approach: Daisy takes her suffering first to a psychiatrist and then to a priest. The actors who read the play aloud for her did not share Sexton's view that confession was inherently dramatic, and they advised her to reframe the story. In 1964 she received funding from the Ford Foundation to revise "The Cure." Titling the second version "Tell Me Your Answer True" (1964, unpublished), Sexton let her imagination run wild. In this version Daisy has committed suicide and finds herself in a circuslike purgatory, complete with a ringmaster and a set of demonic "witnesses" who speak the kind of raunchy doggerel Sexton herself spoke during phases of mania. A cast of characters – mainly family members – are seated in the bleachers and are called forward one by one to testify about Daisy's past. Unaware of the damage to the actual child Daisy that their infantile self-absorption has brought about, they preen and posture effectively in death, just as they did in life. This version of the play has plenty of action but little plot; it was never produced. Sexton stubbornly pursued funding for this project, however, and in 1969 received a Guggenheim Fellowship to develop a third revision, titled *Mercy Street,* which was slated for production at the American Place Theater, Off-Broadway in New York. Working with a director and actors, Sexton again reframed the central scene of the play, this time setting the action in a celebration of a Catholic mass. As the priest raises the Host, the church dissolves into a psychiatrist's consulting room. Daisy's quest for expiation merges with her quest for insight, and the plot now turns on Daisy's maturing understanding of the adults who produced the trauma underlying her suicidal despair. *Mercy Street* opened 3 October 1969, for a run that was extended by a week after favorable reviews began to appear. Critics praised the poetic language and the use of ritual, but criticized the static plot.

Mercy Street was not produced again, and has not been published. But the impulse that led Sexton to begin writing this play in the early 1960s can be seen in retrospect to inform the remainder of her creative life. More and more, Sexton wrote poems that were dramatic monologues intended for performance — performance by Sexton herself. Sexton's third volume of poems, *Live or Die* (1966), contained a mixture of the formal verse for which she had been much honored — in, for example, "And One for My Dame," "Somewhere in Africa" — and a more rambling, associational mode of free verse that left many critics doubting that her art had retained any discipline. But precisely those poems that seemed shapeless on the page could be wonderfully animated by a live reading. The active poetry circuit of the time provided many opportunities for remuneration for a poet who could please a crowd. Sexton, an exceptionally effective reader, had become by the mid 1960s a well-paid act (she set her fees to equal those paid to her old enemy, James Dickey, rumored to be the best-paid poetry reader in America). Several of the poems collected in *Live or Die* were featured in her most memorable readings. Sexton's work was trenchant without being overtly political; the professional detachment with which she read such anguished poems as "For the Year of the Insane," "Menstruation at Forty," "Wanting to Die," and "The Addict" permitted a wide range of audience identifications. At rallies organized to protest the war in Vietnam, Sexton always read a poem celebrating the puberty of her elder daughter, Linda ("Little Girl, My Stringbean, My Lovely Woman"), which often seemed more to the point than work in a didactic or angry mode. By 1968 Sexton had added music to her performances. Traveling with a "chamber rock" group christened "Anne Sexton and Her Kind," Sexton, dressed in a glamorous evening dress, chanted her poetry in well-rehearsed collaboration with five musicians playing flute, keyboards, guitar, bass, and drums. The group disbanded in 1971 after a handful of concerts, largely because Sexton found travel disorienting and fatiguing.

In May 1967 *Live or Die* was awarded the Pulitzer Prize for poetry. Sexton, at thirty-eight, was in her prime. During the next three years she wrote in rather short order the books that remain her best-sellers: *Love Poems* (1969) and *Transformations* (1971). Her *Love Poems* gave American literature its first fully sexual heroine. Although the book is a collection of lyrics, it has a consistent voice; and the titles of the poems mark, like the stars of a familiar constellation, points of reference that can easily connect into a narrative. The contents of that narrative can be summarized in the following manner: the "story" is set in the "post-pill paradise" of sexual revolution. The speaker, who dwells with her husband and children in an affluent, white, Protestant America, just after the death of President John F. Kennedy, tells episodes from her own history. More than a century earlier, Nathaniel Hawthorne had created in *The Scarlet Letter* a sexual heroine, Hester Prynne, in order to exhibit her leading a life of disgrace on the outskirts of a New England town because of her sin of adultery. In Sexton's New England the margins of the town have become suburbia, and adultery has become the next horizon of female sexual destiny, once marriage and childbirth have ripened a woman's erogenous zones.

Sexton's erotic poems have an obvious literary counterpart in the fiction of John Updike, whose *Couples* (1968) became a best-seller during the months when Sexton was writing these poems. "Post-pill paradise" was a phrase Updike put into the mouth of one of his characters, a liberated housewife. Sexton read Updike's book and sent Updike a fan letter after *Time* magazine featured him on the cover and ran a story titled "The Adulterous Society"; the issue included a review of *Couples*. Adultery is also the theme that gives *Love Poems* its seriousness and its importance, but the perspective in Sexton's book is irreducibly female. "The Touch" leads on to "The Kiss," to "The Breast." Consummations follow. Then the plot takes a downward turn in "For My Lover, Returning to His Wife." This turn is followed by "The Break," the "Again and Again and Again" of pain, and the "Ballad of the Lonely Masturbator." Then another cycle opens: "Barefoot." "Now." "Us." "Mr. Mine"; and

Sexton with her husband, Alfred Muller Sexton II, and daughters, Linda Gray and Joyce Ladd Sexton, June 1961 (Collection of Linda Gray Sexton)

another separation follows in the "Eighteen Days Without You." Desire, like a drug, awakens feeling while it steals sense, and brings transformation and addiction and sadness in the same dose.

Passing once through such a cycle might yield romance or tragedy; twice through demands the entrance of the reality principle, encapsulated here in "You All Know the Story of the Other Woman." True to the venerable conventions of the tradition, Sexton's sequence leads from folly to wisdom; the book ends with "and that / will be that." But along the way are some quite lovely lyric moments, such as the closure of "December 11th" from "Eighteen Days Without You:"

> We are bare. We are stripped to the bone
> and we swim in tandem and go up and up
> the river, the identical river called Mine
> and we enter together. No one's alone.

In any case, it was not the moral of the story but its forthright sexiness that sold the book. Sexton's second most popular book was *Transformations,* seventeen narrative poems based on Grimms' fairy tales. A work of black humor narrated by "a middle-aged witch, me," it was aimed at the college audiences who were currently making cult successes of Kurt Vonnegut's *Slaughterhouse-Five, or the Children's Crusade* (1969) and Donald Barthelme's *Snow White* (1967). The most enduring of Sexton's "transformed" tales drew artistic power from materials she tapped in her years of psychotherapy. Sexton returned to the themes of her play, *Mercy Street,* in writing the stories of "Briar Rose" (which explains a daughter's fear of going to bed), "Rapunzel" (which accounts for a girl's willing entrapment by a seductive older woman), and "Snow White" (in which a cold, neglectful mother is punished). But this time Sexton was not required to devise a plot. She could exercise her most reliable poetic talent, producing riffs of imagery while letting the story, familiar to everyone, take care of itself.

Transformations was a popular success and a departure from Sexton's usual style: it contained no evidently "confessional" poetry. As a consequence it did not receive much critical attention, although an opera version with music by Conrad Susa was commissioned by the Minneapolis Opera Company in 1973, premiered by that company, and later received numerous other productions. But Sexton loved this book and viewed its humor as an important achievement.

By the mid 1960s Sexton had tentatively begun to lead poetry workshops as a way of supplementing her income. Her first teaching experience had not been rewarding. Leading a writing seminar at Harvard in 1962, in conjunction with her appointment as a Radcliffe scholar, Sexton concluded that her students did not take poetry seriously. In 1967 a more attractive opportunity came along when she was offered a term of experimental team teaching at Wayland High School. Here was a pilot project, funded by the U.S. Office of Education, in which poets were paired with experienced teachers of literature in the public schools. It was this enjoyable collaboration that led to the creation of Anne Sexton and Her Kind, after one of the students set one of Sexton's poems to music. In 1968 Sexton offered a workshop in Belmont, Massachusetts, at McLean Hospital, a mental institution where her friends Robert Lowell and Sylvia Plath had been among the many well-known artists who spent time in treatment. In 1969 Sexton conducted a month-long workshop for students at Oberlin College during their winter intermission; this experience more than any other led her to seek a permanent position in college teaching, following the example of many of her peers. In 1970 she was elated when her old friend George Starbuck offered her an appointment at Boston University as a lecturer in creative writing. In 1972, after receiving honorary doctorates from various colleges and universities, Sexton was promoted to full professor at Boston University. The following year she was appointed to a prestigious chair of poetry at Colgate University, the Crashaw Chair of Literature, which she held during the spring of 1972. To a woman with a meager education, she joked, these academic honors felt "something like having a baby without having had intercourse."

By the early 1970s Sexton's mental health, never secure, had begun deteriorating observably under the influence of the alcohol and pills to which she was addicted. At the time *Transformations* reached print in 1971, Sexton was already numbering the books she believed she could complete before her death. She described them in a letter to her editor at Houghton Mifflin: "I would like to do a book of very surreal, unconscious poems called *The Book of Folly*. At the same time I plan to start another book called *The Death Notebooks* . . . intense, personal, perhaps religious in places. I will work on *The Death Notebooks* until I die." In fact, the last book underwent a mitosis, the "religious places" splitting off to form a book of its own, *The Awful Rowing Toward God* (1975).

Because Sexton viewed these three books as "last" work, it is profitable to summarize them in relation to one another. *The Book of Folly* (1972), as Sexton's remark indicates, is another foray into experimental form. Narrative or lyric structuring is reduced to a minimum; loose lines of "surreal, unconscious" imagery does the work of communication. This was a direction Sexton had chosen for her work as early as *Live or Die,* with an appropriate sense of the risks she was taking in abandoning the well-articulated "mad housewife" persona she had adopted so courageously in *To Bedlam and Part Way Back* and *All My Pretty Ones.* The Sextonian first person in those early books has social courage — she comes out of the closets of femininity and madness, so to speak. The first-person voice of *The Book of Folly* has artistic courage; the speaker is not a stereotype but — at her/his most extreme — a fluent, verbal subjectivity, difficult to place and unsettling to heed. The technique best succeeds, however, in sequences of poems structured either by narrative or lyric conventions. The narrative sequence "The Death of the Fathers" stands among Sexton's most poignant and effective works on the theme of incest, which was so significant a subject in her art; while the sonnet sequence "Angels of the Love Affair" displays at its most dazzling Sexton's mastery of auditory form.

The Death Notebooks (1974) contains several sequences as well, three of which are among the best Sexton ever wrote. "The Death Baby" excavates the emotional experience of holding a powerful death wish throughout one's entire life. The jazzy sequence titled "The Furies" displays Sexton's powers as an improviser of images. Finally, the long sequence "O Ye Tongues," modeled on the *Jubilate Agno* of Christopher Smart, serves as an *apologia pro vita sua*. It summarizes the poet's lifelong preoccupations, impressively sustaining its theme with a confidence that is sometimes magisterial, sometimes comic, often very beautiful.

Sexton intended *The Death Notebooks* for posthumous publication, then changed her mind in 1973 and sent it to the press; it was published in February 1974, six months before she died. For one thing, she was short of money after having divorced her husband. For another, she had surprised herself in January 1973 by writing a whole book of poems in less than a month, sometimes at the rate of two or

Dust jacket for Sexton's 1971 book, a collection of poems that are not written in her usual confessional mode

three poems in a single day. Sexton organized the book, which she titled *The Awful Rowing Toward God,* as though it were a narrative: it opens with "Rowing" and closes with "The Rowing Endeth." This was a shrewd, if procrustean, solution, because each of the poems has a frantic quality: the subjectivity here is frenzied with the desire for connection that cannot be found by any of her strategies. Thematically, *The Awful Rowing Toward God* is closely affiliated with the so-called dark sonnets of Gerard Manley Hopkins, but artistically it has only a minimal shape.

Sexton stayed alive long enough to correct the galley proofs of *The Awful Rowing Toward God.* Tedious proofreading was a job she always shared with her friend and fellow poet, Maxine Kumin. On 4 October 1974 they met at Kumin's home "and had a wonderfully gay and silly lunch" over the task, Kumin remembered. Then Sexton drove home to Weston and committed suicide. Letting her car idle in a closed garage, she died of carbon monoxide poisoning.

By all the evidence her death had been long planned. After the departure of the children for college and boarding school and the divorce from her husband of twenty-four years, Sexton grew desperately lonely. Her closest friends could not respond to the demands she made on their attention, and her deepening sense of isolation seems to have hurried along the deterioration that had set in with her increasing dependency on prescription drugs and alcohol. Moreover, although she had remained under regular psychiatric care, her disorders had not yielded to treatment. She remained prey to episodes of severe depression; suicidal thoughts had been a consistent feature of her illness. As Sexton liked to say, "My fans think I got well, but I didn't: I just became a poet."

But her death at age forty-five was not impulsive. In fact, Sexton seems to have held the view — based on the examples of Ernest Hemingway and Sylvia Plath — that suicide was a suitable closure for a career in art. During the eighteen months prior to her death Sexton had taken great care to settle her affairs, paying particular attention to the disposition of her literary estate. With the help of a secretary she had organized her manuscripts, correspondence, tapes of performances, and documentation of her medical treatment — including her therapy notebooks and the taped psychotherapy sessions in her possession — into what she intended to leave as a research archive. She named her eldest daughter,

Linda Gray Sexton, as executor of the estate; the archive was sold in 1981 to the Harry Ransom Humanities Research Center at the University of Texas at Austin.

Will Sexton's work stand the test of time? No one can guess what use posterity may have for her legacy of subjectivity that is inscribed in the lines of her remarkable poetry. But at the end of her life Sexton, speaking to a Japanese translator of her work, said a few words about what she had hoped to achieve as an artist. Trying to classify Sexton for a foreign public, the translator asked whether, at the outset of her career, she had been an "academic" or a "beatnik" writer? "I think I belong to the academics by coincidence," she replied. Was she in fact a "confessional" writer? "Not everything I document is factual," she answered. Was she a "feminist"? "I suppose there is social criticism in my poems." Well, what helpful summary could she propose by which a stranger might understand her work? "I try to write true to life."

Letters:

Anne Sexton: A Self-Portrait in Letters, edited by Linda Gray Sexton and Lois Ames (Boston: Houghton Mifflin, 1977; reissued, 1991).

Bibliography:

Cameron Northouse and Thomas P. Walsh, *Sylvia Plath and Anne Sexton, A Reference Guide* (Boston: G. K. Hall, 1974).

Biographies:

Diane Wood Middlebrook, *Anne Sexton: A Biography* (Boston: Houghton Mifflin, 1991);

Linda Gray Sexton, *Searching for Mercy Street: My Journey Back to My Mother, Anne Sexton* (New York: Little, Brown, 1994).

References:

Frances Bixler, ed., *Original Essays on the Poetry of Anne Sexton* (Conway: University of Central Arkansas Press, 1988);

Steven E. Colburn, ed., *Anne Sexton: Telling the Tale* (Ann Arbor: University of Michigan Press, 1988);

Diana Hume George, *Oedipus Anne: The Poetry of Anne Sexton* (Urbana: University of Illinois Press, 1986);

George, ed., *Sexton: Selected Criticism* (Urbana: University of Illinois Press, 1988);

Caroline King Barnard Hall, *Anne Sexton* (Boston: Twayne, 1989);

J. D. McClatchy, ed., *Anne Sexton: The Artist and Her Critics* (Bloomington: Indiana University Press, 1978);

A. Poulin Jr., ed., "A Memorial for Anne Sexton," *American Poetry Review,* 4 (May/June 1975): 15–20;

Linda Wagner-Martin, ed., *Critical Essays on Anne Sexton* (Boston: G. K. Hall, 1989).

Papers:

The entire archive of Sexton's papers is at the Harry Ransom Humanities Research Center, University of Texas at Austin. *Sexton,* a videotape based on outtakes from a film produced for a public television series (*USA Poetry,* 1966), is available from the American Poetry Archive at the Poetry Center, San Francisco State University, California.

Ron Silliman
(5 August 1946 -)

T. C. Marshall
Cabrillo College

BOOKS: *Crow* (Ithaca, N.Y.: Ithaca House, 1971);
Mohawk (Bowling Green, Ohio: Doones Press, 1973);
Nox (Providence, R.I.: Burning Deck, 1974);
Ketjak (San Francisco: This Press, 1978);
Sitting Up, Standing Up, Taking Steps (Berkeley: Tuumba Press, 1978);
Legend, by Silliman, Bruce Andrews, Charles Bernstein, Ray DiPalma, and Steve McCaffery (New York: L=A=N=G=U=A=G=E/Segue, 1980);
Tjanting (Berkeley: Figures, 1981);
Bart (Hartford, Conn.: Potes & Poets Press, 1982);
ABC (Berkeley: Tuumba Press, 1983);
Paradise (Providence, R.I.: Burning Deck, 1985);
The Age of Huts (New York: Roof Books, 1986);
LIT (Hartford, Conn.: Potes & Poets Press, 1987);
The New Sentence (New York: Roof Books, 1987);
What (Great Barrington, Mass.: Figures, 1988);
Manifest (Tenerife, Canary Islands: Zasterle Press, 1990);
Leningrad, by Silliman, Michael Davidson, Lyn Hejinian, and Barrett Watten (San Francisco: Mercury House, 1991);
Demo to Ink (Tucson, Ariz.: Chax Press, 1992);
Toner (Elmwood, Conn.: Potes & Poets Press, 1992);
Jones (Mentor, Ohio: Generator Press, 1993);
N/O (New York: Roof Books, 1994);
Xing (Buffalo, N.Y.: Meow Press, 1996).

RECORDING: "From *Oz*," on *Live at the Ear,* Oracular Laboratory Recordings, 1994.

OTHER: *Tottel's,* edited by Silliman, nos. 1-18 (1970-1981);
A Symposium on Clark Coolidge, edited by Silliman (Milwaukee, Wis.: Membrane Press, 1978);
"Art With No Name" and "Reading *Ketjak*," in *The Poetry Reading: A Contemporary Compendium on Language & Performance,* edited by Stephen Vincent and Ellen Zweig (San Francisco: Momo's Press, 1981), pp. 160-165, 194-199;
In the American Tree, edited by Silliman (Orono: National Poetry Foundation/University of Maine Press, 1986);
"New Prose, New Prose Poem," in *Postmodern Fiction: A Bio-Bibliographical Guide,* edited by Larry McCaffery (New York: Greenwood Press, 1986), pp. 157-174;
Socialist Review, edited by Silliman, 17-19, no. 3 (January-February 1987-July-September 1989);
"Terms of Enjambment," in *The Line in Postmodern Poetry,* edited by Robert Frank and Henry Sayre (Urbana & Chicago: University of Illinois Press, 1988), pp. 183-184;
"Canons and Institutions: New Hope for the Disappeared," in *The Politics of Poetic Form,* edited by Charles Bernstein (New York: Roof Books, 1990), pp. 149-174;
"Postmodernism: Sign for a Struggle, the Struggle for the Sign," in *Conversant Essays: Contemporary Poets on Poetry,* edited by James McCorkle (Detroit: Wayne State University Press, 1990), pp. 79-98;
Unfinished Business: Twenty Years of Socialist Review, edited by Silliman (London: Verso Press, 1991);
"The Practice of Art," afterword to *The Art of Practice: Forty-Five Contemporary Poets,* edited by Dennis Barone and Peter Ganick (Elmwood, Conn.: Potes & Poets Press, 1994), pp. 371-379;
"The Task of the Collaborator: Watten's *Leningrad,*" in *Aerial 8: Barrett Watten,* edited by Rod Smith (Washington, D.C.: Edge Books, 1995), pp. 141-168.

SELECTED PERIODICAL PUBLICATIONS –
UNCOLLECTED: "Skies III," *This,* 12 (Spring 1981);
"Skies I," *Sulfur,* 3 (1982): 6-8;
"Skies II," *Sulfur,* 4 (1982): 138-149;
"Skies IV," *So & So,* 2 (Spring 1983);

"Statement for New Poetics Colloquium," *Jimmy and Lucy's House of 'K,'* no. 5 (November 1985): 17–19;

"I Wanted to Write Sentences: Decision Making in the American Longpoem," *Sagetrieb,* 11 (Spring/Fall 1992): 11–20;

"Selections from *Under,*" *No Roses Review,* no. 3 (Spring 1994): 72–78;

"Selections from *Under,*" *Proliferation,* no. 2 (November 1994): 13–17;

"Selections from *Under,*" *Cream City Review,* 19 (Spring 1995): 89–93;

"Selections from *Under,*" *TO,* 3 (Summer 1995): 69–72;

"®," *Conjunctions,* no. 21 (1993): 166–180.

Ron Silliman's poetry, essays, and anthologies have become essential front-running documents of poetic experiment for the late twentieth century. Usually associated with the school of writers that has come to be called the Language poets, Silliman has helped to form that group's identifying contributions by publishing steadily since the early 1970s. In 1994 his writing was recognized by inclusion in two important anthologies: Douglas Messerli's *From the Other Side of the Century: A New American Poetry, 1960–1990* and Paul Hoover's Norton Anthology of *Postmodern American Poetry.*

Silliman's works have extended American poetic traditions in useful new directions. His poetry follows directly from the questions raised by writers in the 1960s when Silliman was a young and up-and-coming poet. The Beat and other "New American" poets, as defined by the Donald Allen anthology *New American Poetry* in 1960, did much of their innovative work with poetic voice. They concentrated on the way voice presented itself through vocabulary, syntax, line, thought logic, unity and closure, and reference to traditions. Silliman works with these aspects of language and communication and some others less questioned by his forebears, bringing each under scrutiny and into play. His work uses and juxtaposes both social and literary references to be artistically serious and often, at the same time, humorous.

His frequently book-length poems mix high and low art, punning humor and lyric beauty. They generally employ familiar poetic techniques in unfamiliar combinations and use format on the page in a way that makes one think again about poetic form. In this way his books bring fresh attention to habitual ways of reading and constructing meaning. For Silliman this writing is a social task, involving investigations of the forces and power obliquely inherent in language structures.

Much of Silliman's poetic focus has centered on the uses of the sentence in writing and reading. He is well known for having named and articulated the theory of "The New Sentence," described in his book of the same name, which was published in 1987. This theoretical and historical work exposed an important element of recent experiment by several authors who have sought to counteract the bland acceptance of poetry's conventional use of line breaks and the prosaic use of sentence combination toward logical totality in the paragraph. Silliman's seminal essay shows how this is done by arresting syllogistic movement near the level of the sentence and by employing the paragraph as "a unity of quantity, not logic or argument." The seriousness of such a formalist and conceptual approach is placed in pleasant relief by the way Silliman employs a brightly good-humored and personally perceptive point of view in most of his own poetry. Even while he uses new sentence techniques, his writing strategies are almost always based on recognizable referentiality and seemingly casual deployment of commonly accepted poetic devices. Though intellectually rigorous, his writing also has delightfully plebeian qualities and truly democratic force.

Ron Silliman was born in Pasco, Washington, and raised in Albany, California, just north of Berkeley. He has spoken at times about growing up poor and living with few cultural amenities. They "went to a John Wayne movie once a year," Silliman recalled in conversation, and had few books in the house. From that difficult background Silliman has made a positive effort to expand the possibilities for poetry and its readership. The *Community Libertarian* published his first poems in 1965, when he was eighteen. The next year he published in four different magazines. He has been expanding the breadth of his public outreach ever since. His several honors include grants from the California Arts Council and a Creative Writing Fellowship from the National Endowment for the Arts (NEA).

Silliman received some higher education from Merritt College in Oakland, San Francisco State University, and the University of California, Berkeley. Although he never earned a degree, he has since taught at several universities, including the University of California at San Diego. He was poet-in-residence (1983–1990) and director of development (1982–1986) for the California Institute of Integral Studies, a private college with an innovative multidisciplinary program. He has also taught in public programs such as the Central City Hospital-

Ron Silliman and his twin sons, Colin Robert and Jesse Kyle Silliman (photograph by Erica McConnell; courtesy of Ron Silliman)

ity House in San Francisco's Tenderloin district and worked in various prison education projects. In the early 1970s, in San Rafael near San Quentin, he edited *Labyrinth* for the Committee for Prisoner Humanity and Justice. Later that decade he edited the *Tenderloin Times* for San Francisco's Central City Hospitality House. He has also worked as an organizer in the prison and tenant movements, as a lobbyist, and as an editor. Active in the New America Movement of the 1970s and its current offshoot, the Democratic Socialists of America, he has kept collective political work at the forefront of his commitments. From 1986 to 1989 he served as executive editor for the internationally distributed *Socialist Review*. His tenancy in that position brought the magazine fresh strengths in the literary and social-historical fields.

Silliman was highly active on the poetry scene in the San Francisco Bay area from the 1970s into the 1990s. He gave readings and talks and attended many events given by other writers, responding almost always with lively engagement. His first talk was called "truth," with a small *t,* and, with the help of thinkers such as Ludwig Wittgenstein, whose work provided some of the context for that night in 1977, he has been critiquing formulations of truth ever since. His presence on the scene, in person or via E-mail, has always been a particularized and productive challenge to keep the activities of thinking and writing in social motion.

Well respected for his social and literary work still, he has added family and business commitments. He is married to Krishna Evans and is the father of twin boys, Colin Robert and Jesse Kyle Silliman. Currently employed in the corporate computer world as a marketing communications specialist near Valley Forge, Pennsylvania, Silliman keeps up his links with a vast and diverse network of other poetry workers as well. His cultural work continues to include a great deal of networking in dialogue with other writers and in encouragement of both peers and younger poets in the world of small press publication and poetry readings. The work he did in this field in the 1970s and 1980s helped to create the network of people now known as the Language poets. He has published essays on the canons and institutions of the literary world and has edited anthologies meant to broaden and refocus possibilities for inclusion in a new canon. All of Silliman's work may be said to have been done in a spirit of social collaboration.

In his 1995 essay on "The Task of the Collaborator," he wrote that "a collaboration is always a portrait of the author out of control." The main thrust of Silliman's literary and social work has been a critique of the concept of "the first person" and authorial intention or control. This was an aspect of poetic voice left untouched or even further solidified by the writers who made their critical difference in the 1960s. Silliman and those who have collaborated in the Language poetry project have worked against "the intentionalist cliché" and its place in "the broader ideology of individualism." He has insisted that each person "is a construct, a focal point for so many . . . forces that we find them uncountable." Even as his poems have become more plainly autobiographical, he has used his sense of himself as subject in suggesting a self that is a multiple composite. Each sentence in his recent book-length poems suggests engagement with another possible discourse. In a fashion that could be said to be Whitmanesque, his work embraces the multiplicity of what occurs to him as it is put into words. Control and intention are used, but made to become so inclusive that they are emptied of their exclusive focus on "the first person" first.

Silliman has labored and written to expose the social constructions of person as they are focused by language, borrowing and elaborating ideas from many thinkers. Important among these have been Louis Althusser, Roland Barthes, and a precursor of theirs – the linguist Roman Jakobson. As an epigraph to an essay about Charles Bernstein's work included in *The New Sentence*, Silliman quoted Althusser's statement that the "category of the subject is constitutive of all ideology." He would agree with Barthes who, on the last page of his book *On Racine* (1964), declared that "literature is that ensemble of objects and rules, techniques and works, whose function in the general economy of our society is precisely *to institutionalize subjectivity*." Being politically committed to a critique of ideology, Silliman has joined others such as Bernstein in investigating the constitution of subjectivity. Their investigation takes place both in and with language, in the reflexive fold where poetry has always worked. Jakobson described language as having six possible basic functions, one of which is a concentration on the material of the message itself. In his essays Silliman has used this idea, what Jakobson called "the Poetic function" of language, to describe the ways in which "language poetry" can effectively call attention to how meaning is made in material forms.

Early in his career Silliman joined the long-standing tradition of poets who strategically call attention to the minute functional particulars of language. His early publications step right into the tradition and the fray. Unfortunately, his first two pamphlets, *Moon in the Seventh House* (1968) and *Three Syntactic Fictions for Dennis Schmitz* (1969), are now out of print. Still, many of Silliman's first poems are available in the Archive for New Poetry, Mandeville Department of Special Collections, the University of California, San Diego. He also achieved early publication in several dozen magazines that academic libraries generally carry (including *Caterpillar, Chicago Review, Poetry* (Chicago), *Poetry Northwest,* and *TriQuarterly,* along with lesser-known ones). Silliman's first major collection, *Crow* (1971), began with an epigraph from William Carlos Williams: the idea of "the perfection of new forms as additions to nature." *Crow* takes Williams's aesthetic a step further in its own direction toward a focus on the way minimal language elements form possibilities for meaning.

Formalist and minimalist approaches have enriched Silliman's aesthetic in different ways throughout his career. He has used nearly every recognized angle to approach the conventions of reading poetry, highlighting them often with a kind of comic relief, even while putting them to serious use. The first poem in *Crow*, "Aint nobody," employs phrasings from the American vernacular for both their musical and their meaningful values. The book immediately asserts poetry's old basic values of sound and sense. The interesting thing is how Silliman thickens both these elements into a noticeable palpability. Technique is foregrounded, even while the opening lines push the idea of technique against itself:

> Aint nobody
> can write this . . .
>   big black fat cat
>   lazes
>   at the white jeep[.]

The image in the second cluster of lines cuts back at both the idiom and the assertion in the first two lines. From the beginning this book plays voice, diction, image, reference, rhyme, line, and more of the technical elements of poetry off each other to engage the reader with the task of reading. Obviously, the reader is reading writing, but each convention for sorting out meaning is gently confronted by an unconventional way of saying something different as the poem proceeds.

The double negative of the opening line, while nonstandard, is an accurate representation of an American dialect. The idea that one cannot wholly capture one's perception in writing becomes a central concern of Silliman's writing career and could be said to be behind the inclusive quality of the project that was begun a few years later with *Ketjak* (1978) and that continues to the present. If there is not anyone who "can write this," then there is something here that exceeds the writing or the simple reading of it. That claim provides an important opening statement for an oeuvre. The diction of this voice suggests that it might be any of many of us who could say what is being said here. That, of course, joins Williams and his heirs and heiresses in their respect for the common idiom. Williams also shows up behind the second cluster of lines, gently giving it referential background with his "red wheelbarrow" upon which "so much depends." Silliman's play on the rainwater glaze with his word *lazes* and the red and the white of Williams's image with Silliman's black and white here is an opening gambit in a career that later made much of such oblique reference by echoic misquotation. Both these gestures, or techniques, are part of a larger strategy for simply complicating one's readerly sense of poetic expression.

The same page offers a later line, "window not as clear as I'd like to think," that may be read as an imagistic metaphor for the way words work and the way the poet works with words. If language is the window through which one sees the world and shows it to oneself and others, then Silliman's *Crow* exists to put spots and streaks on it. In ordinary discourse people tend to accept the myth of pure clarity unthinkingly; surprisingly, many readers especially tend to take this myth for granted when they are reading poetry. The strategy of the book is to make one aware of how transparent one generally assumes language to be. Silliman has shown admiration for the way many earlier twentieth-century poets practiced an objectivism that took poetry beyond the myths of emotional self-expression or a Romantic relation to the world. He began early in his own work to try to take poetry beyond the myths of the subject that remained inside that objectivity. Althusser named this goal the goal of thinking in the contemporary era, saying that "we must completely reorganize the idea we have of knowledge; we must abandon the mirror myths of immediate vision and reading, and conceive of knowledge as a production."

Silliman's subsequent books have carefully pursued the shattering of the mirror myths of immediate meaning in language. They institute a process of engagement or of productivity for the reader as well as for the writer and the writing. In a review

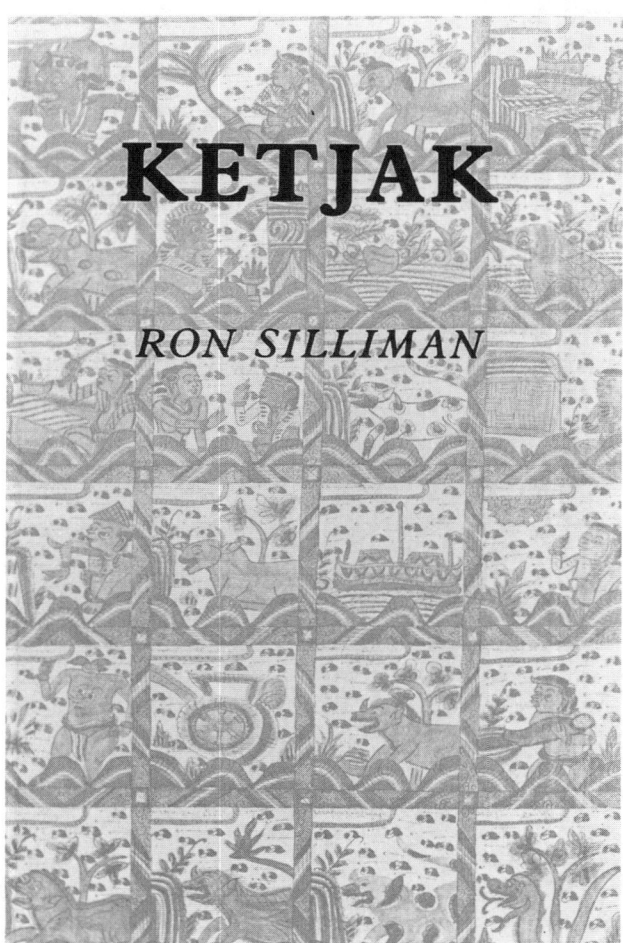

Cover for Ketjak (1978), a book-length poem whose title is the Indonesian name for a dance inspired by the Hindu Ramayana/Monkey King Tales

in the magazine $L=A=N=G=U=A=G=E$, the poet-critic Barrett Watten writes of Silliman's second full-length collection, *Mohawk* (1973), that its "fixed vocabulary and the use of the two-dimensional page give the work a flatness and autonomy; the writer is located outside the work." While Silliman prevents the reader from identifying with the writer, his book is, nevertheless, not obscure. Watten also notes "the clearly imagistic quality of the words," calling "each word a snapshot on the page." Silliman's poems are grounded in nominative words and phrases. Beginning with these, readers are in effect invited to acquire a sense of the way they "get" things.

Like many of Silliman's later works, *Mohawk* puts a subtly shifting distance into the process by which readers usually assemble their understandings. Watten borrowed a term from a *Scientific American* article on mathematically generated music to describe the fluctuating patterns of a fixed set of words in *Mohawk.* He said the *fractal curves* of the poetry played across various scales, telescoping "from word-to-word jumps to page 'flicker noise' to the curve of the whole." The poetry of this thirty-page book "is involved with extrinsic dimensions" that include the "demography of the words." It is not a work of personal expression or voicing that is available for identification. Its strategies for avoiding that identification put the reader to work, noticing and participating in the production of meaning and its social effects.

In *Nox* (1974) Silliman quartered the pages of the book with crossed lines resembling those of a Cartesian-coordinate graph. In each of the four fields was printed a word, a few words, a part of a word, or parts of a couple of words. Each of these quarters could be read as its own minimal poem. For instance, "tree sun" in the upper left quarter of the first page presented an image and a commentary on imagism and its relation to Ezra Pound's interest in Chinese written characters. Of course, a reader could see a variety of relations between these two words and their functions as nouns or possibly verbs. A reader of Pound, however, might recall the poet's interest in a Chinese written character that combined the characters for tree and sun to mean *east* or *dawn.* On that same page in *Nox,* one also finds "tuna flesh," "logl," and "eggs or size." The pun on "exercise" in "eggs or size" suggests a heavier pun on "treason" in "tree sun." This punning provides a comment on Pound, delivered through a method oddly derived from Pound's way of reading ideograms.

Reading through the book, one realizes that possible meanings multiply in each quarter, radiating from a perceptible minimum, and carry over from page to page. Puns and homonyms, with echoic slippages such as "tuna flesh" instead of "tuna fish," join rhyme and off-rhyme to push the attentive reading mind back on the interplay between the perceived sight or sound of words and what one might have them mean. Unwritten words even appear in the ways the eye and ear slip across the pages of *Nox.* The word *expensive* might show up in the mind of someone looking at page 5, where the words "oxen" and "pensive" are placed over each other in the upper right quarter, but such an abstraction gets laughed off when one glances at the opposite lower left quarter's punning "in the / dairy air." Such dumb jokes often become strategically meaningful in Silliman's carefully constructed contexts. This kind of humor has continued throughout his career as part of his strategies for giving the reader a chance to see how he puts meaning together from any text.

Silliman's next book to be published, *Ketjak,* was written in 1974. That was a key year for the author, as he brought his attention to bear on the sentence and the ways it can be consciously used in writing. *Ketjak* began an ongoing project of investigating relations between poetic and prosaic strategies for assembling sense through various uses of sentence-and-paragraph or line-and-stanza formats. Bernstein in his article on Sillman in the issue of *The Difficulties* explains that *Ketjak* has a key place among the poet's further projects. *Ketjak* itself was an experiment with a prose poetry form made of expanding paragraphs. The basic rule for its composition was that the first paragraph would have one sentence, and the second paragraph would consist of that one sentence and another, and the third would be those two and two more, and the fourth would be those four and four more, and so on. The repetitiveness of the form is relieved by Silliman's creation and insertion of new sentences that place the repeated sentences in new contexts.

One of Silliman's most interesting strategies is his brilliant incorporation of "mistakes" into the poem. In the little green notebook he used to plan and draft the poem, Silliman says in an early note that it was bought "to make mistakes in." The manuscript's seventh paragraph ends marked with the number *64,* the number of sentences it should have according to the formula, but it actually contains fifty-four sentences. As if to make up for this lack, the paragraph includes several alterations made to previously used sentences, as do all the paragraphs in the final text from the third one on. That seventh paragraph introduces a new sentence that reads, "How will I know when I make a mistake." It is a joke that one should take seriously. The accidental qualities of this work play off the obvious insistence upon procedural rules. More important, the procedure allows the conventional rules of reading to be played off each other in ways that cannot help but curiously or giddily engage the reader in his own task as long as he keeps the book open.

Ketjak introduced a sense of sentences that has remained at work throughout Silliman's career. In the title essay from *The New Sentence,* originally a public lecture, or talk, given in 1979, he explained the practice of the sentence-centered composition (already featured in his own poetry and that of several of his colleagues) as a compositional practice that involved a "limiting of syllogistic movement," a withholding of the usual connections readers have been taught to expect. This limiting is purposely done to hold "the reader's attention at or very close to the level of language, that is, most often at the sentence level or below." It is accomplished by keeping the primary composition of logic "between the preceding and following sentences." Paragraphs in this writing, used as "a unity of quantity, not logic or argument," simply organize sentences, and it is their length that may be said to be "a unit of measure."

Abstracted in the essay-talk, such a description may sound as though it were suggesting a wholly new way of writing. Even in the practical applications of *Ketjak,* it may look wholly new, but it is based upon the ways readers have learned to follow the logic of prose paragraphs or the line-by-line presentations of poetry. The opening sentences of the seventh paragraph of *Ketjak* show one version of Silliman's strategy:

> Revolving door. How will I know when I make a mistake. The garbage barge at the bridge. The throb in the wrist. Earth science. Their first goal was to separate the workers from their means of production. He bears a resemblance. A drawing of a Balinese spirit with its face in its stomach.

Many of the sentences are merely nominative phrases. They do not add up neatly in order. Each sentence or phrase may refer to something in the preceding or succeeding one, but these threads do not seem to continue. One's logical sense is halted again and again along the path through these paragraphs. Some of these sentences have been seen before and now have new contiguities, because the procedure puts the new sentences between the old. Each of these qualities adds to the weirdness of reading this text, but each also has some relation to one's habitual ways of reading.

A key element in the strategies of *Ketjak* lies in the means by which one can recognize each part of the text and yet still find it turning away from one's expectations of either poetic or prosaic composition. The "level of language" that is presented to readers as they are being held "at the sentence level or below" is the level of basic engagement with making sense of word units. *Ketjak* is written to force the reader to notice his or her own recognition of each sentence and the composition of its elements into one image or concept at a time, and then to see the ways by which the reader adds these up by means of contiguity and the familiar touch of recall. The book employs a disjunctive version of modernist juxtaposition, one that is put up beside or against the familiar procedures of prosaic logic. The disjunctive quality of the sentences and paragraphs of *Ketjak* reveals the dependence of conjunctive logic upon one's acquiescence in habits of reading. *Ketjak* is the opposite of casual or causal collage; it does not push toward a whole so much as it reveals the habitual urge toward assembling what Silliman, in his interview in *The Difficulties* special issue and elsewhere, has called "the tyranny of the whole."

This kind of strategy and investigation has been what Silliman's writing has begun to be lauded for in the books following *Ketjak,* which received a few friendly reviews. Later in 1978 Silliman's Pushcart Prize winner, *Sitting Up, Standing Up, Taking Steps* (1978), another poem that accumulated mostly nominative phrases to challenge one's readerly sense of how the world is composed, succeeded in garnering some attention for the author's fresh methods and forms of composition. It was not until the early 1980s, however, that books and journals with wider circulations begin to give favorable notice to Silliman's work. A collaboration with Bruce Andrews, Charles Bernstein, Ray DiPalma, and Steve McCaffery, titled *Legend* (1980), came and went without much notice outside the small league of Silliman's coworkers in the Language school. In 1981 *Tjanting* drew some favorable notices for its further formal experimentation and its odd lyric beauty.

Another book-length poem with an elaborately planned structure, *Tjanting* takes its name, as *Ketjak* did, from Indonesian sources. *Ketjak* is the name for an Indonesian version of a dance theater performance of the Hindu *Ramayana*/Monkey King Tales; *tjanting* is a name for a tool used in marking batik cloth patterns. Both suggest patterns of repetition and involve by pun or direct reference an allusion to chanting. *Tjanting,* composed from 1977 through 1980, works in paragraphs that expand sequentially in their number of sentences, according to the Fibonacci series that has been used to describe mathematically some of nature's spiral forms. The paragraphs also alternate in their use of two key sentences that are voicing two sides of a discussion, "Not this" and "What then?" In the interview published in the special issue of *The Difficulties,* Silliman explained that the use of these two voices was an attempt to respond to a question that had been on his mind in the late 1970s: "What would class struggle look like, viewed as a form?" He admitted that the class struggle analogy was reductive, but he added that it made his poem start moving.

While critics have debated the value of forms such as Silliman's in *Tjanting* as social criticism, many have set aside the claims for social relevance

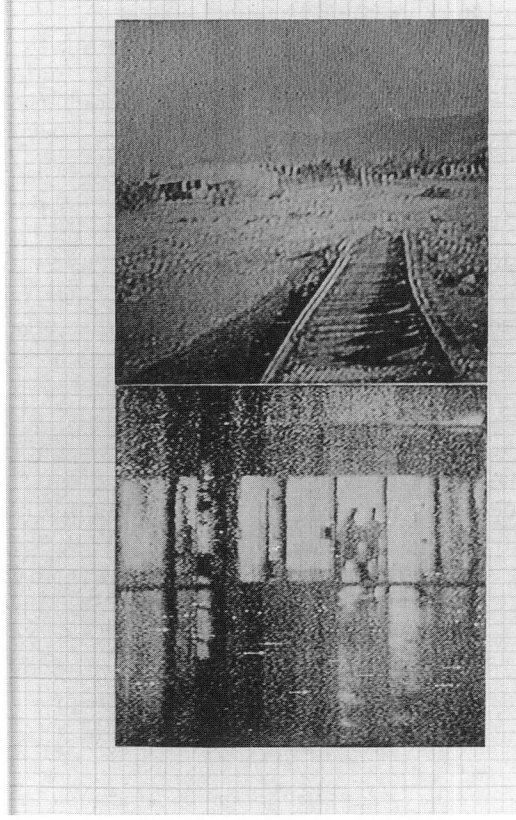

Cover for Tjanting *(1981), a book-length poem whose title is the Indonesian word for a tool for marking batik patterns on cloth*

and praised the artistic results. Early in the 1980s Marjorie Perloff began praising Silliman's work and Jerome McGann joined her in defending both its literary force and its social importance. In addition to the special issue of *The Difficulties* on his work, studies of Silliman's aesthetic appeared in books by Stephen Fredman, Stephen-Paul Martin, and George Hartley, who has carefully analyzed its political significance. Cultural critic Andrew Ross has also written an essay on Silliman's work. Silliman was even included, apparently because his poems appeared in prose form, in *Postmodern Fiction: A Bio-Bibliographical Guide* (1986). He was also interviewed by the editor of that guide, Larry McCaffery, for his book called *Alive and Writing* (1987). By the mid 1980s Silliman had achieved enough public recognition that his earlier books written between 1975 and 1977, *Bart* (1982) and *The Age of Huts* (1986), were finally published.

Silliman had edited a small magazine of some importance through the 1970s, named *Tottel's* after the sixteenth-century English miscellany that brought out the early sonneteers. He had also put together two minianthologies of fresh experiments in writing, one for the *Chicago Review* and the other for Jerome Rothenberg's and Dennis Tedlock's magazine, *Alcheringa*. The respect that he had slowly earned from his various activities began to bear more fruit. He was invited by *Ironwood* magazine to edit an anthology of new work under the title "Realism" and by the National Poetry Foundation to edit an anthology of "Language Realism Poetry." He also began to be invited to give conference papers and appeared at the Kootenay School of Writing's "New Writing Colloquium" in Vancouver, the Oppen Symposium at the University of California, San Diego, and New College's Jack Spicer Conference in 1985 and 1986. He also appeared before the International Association of Philosophy and Literature and the Modern Language Association in 1987.

The books that put Silliman on the literary map had appeared by the mid 1980s, but some of them were the beginnings of a huge project, still in progress in the mid 1990s, called "The Alphabet." The series of books that will comprise "The Alphabet" and *The Age of Huts* are parts of an even larger project also known as "ketjak." They are balanced, mobile fashion, by *Tjanting* and a few other works. Silliman conceives of his works as twins that balance each other, a conception that suggests how he makes choices about what to do in writing. Silliman seems to have written *Tjanting* as an attempt to break up the univocality of earlier works from the mid 1970s; but he has extended the work done in those books by making "The Alphabet" an elaborate investigation of different ways of putting sentences on the page, in either sentence-and-paragraph or line-and-stanza format, or in mixtures of the two.

Silliman plans "The Alphabet" as one long book made of books, one for each letter of the alphabet, but he has not written or published the books in alphabetical order. Silliman has so far published in book form poems representing the letters *A* through *P* as well as *T, W* and *X* in the following volumes: *ABC* (1983), *Demo to Ink* (1992), *Jones* (1993), *LIT* (1987), *Manifest* (1990), *N/O* (1994), *Paradise* (1985), *Toner* (1992), *What* (1988), and *Xing* (1996). He has also published "®," "Skies," and "Under" in periodical publication. Each work has its own concept, form, tone, and thematic concentrations, which seem to come from issues impending in the poet's life, although none of the books is less various in material, perhaps, than any full day's thoughts. *ABC* presented the first three books, which are composed of reflections on how writing expresses the lived world. Their sentences vary in content and approach. Some are plainly as true to life as facts or anecdotes of experience; others are as true to composition as playful expansions from literary allusion into new contexts, or poetically lively redefinition of terms by means of fresh contextualization. This engaged writing, with its hands-on feel that reaches through the personal as a focus for things beyond it, has continued to make "The Alphabet" a fascinating and celebrated collection of book-length poems.

The next volume to be published in the Alphabet series, *Paradise* (1985), won the prestigious annual Poetry Center Book Award in San Francisco. *Paradise* was written as a yearlong project in twelve sections, each containing a few paragraphs that vary from one to as many as ninety-seven sentences. The sentences take many shapes, from one word on up. The reader's impression of the book's formal pattern is overwhelmed by the new sentence writing, with its torqued syntax and chopped paragraph logic, though each sentence by itself would not seem far removed from ordinary discourse in the world. The feel of the book is that of a diary of thoughts or impressions. What gets noticed is still the word, phrase, sentence, and paragraph, and the constructions of possibilities for meaning in registering any perception or thought in words. "My posture, my syntax" is one sentence among many that reflect upon the writing work and its relation to a life. The opening sentence is a sonorously self-conscious description of some aspects of the physical act of writing: "Words slip, does type, hand around the pen a clamp, a clip."

Another book-length poem in which Silliman attempts to highlight almost all aspects of the act of writing suggestively, *Paradise* also challenges readers into a similar awareness of the act of reading. Silliman's shaping of the paragraphs of *Paradise* is meaningful in its appearing to be nearly arbitrary, just as his combining sentences that seem impromptu or culled from ordinary life is meaningful. "This was and now you are constituted in the process of being words, your thought actualizing through the imposition of this syntax." That sentence is humorously followed by another that punningly "pictures" readers living as consumer-perceivers: "Resistance alone is real (coming distractions)."

Paradise puts the reader in the position of resistor, like a filament in a bulb glowing with the light caused by a current of words, just as Charles Olson described it in his essay "Projective Verse" or with a more overtly political emphasis in "The Resistance." But, in *Paradise,* it is not the romantic incandescence of embodied voice that counts. From Olson across the nation to Philip Lamantia and others, the poets of the New American poetry used critical disturbances in a half-dozen aspects of voice only to reemphasize its centrality in the end. It fell historically to the students of their works, like Silliman, to see into their experiments and through them to some expansive new ones.

Among the expansions that Silliman has brought to this poetics is the inclusion of the minutest details and trivialities of objects and thoughts. While his poems have always focused on items or words or thoughts on the peripheries of attention, his *Paradise* is literally composed of the importance of such peripherals. He proclaims in one sentence, "The lobe of the ear as an occasion for art," perhaps engendering a cascade of thoughts in his reader. Possible meanings are created by his strategy of presentation that has each sentence folding out through its own terms, instead of into a closure of totalizing logic. In "The Task of the Collaborator," Silliman emphasizes Vladimir Mayakovsky's advice that a "good notebook and an understanding of how to use it are more important than knowing how to write faultlessly in worn-out metres."

In his statement for the Vancouver "New Poetics Colloquium" on 8 July 1985, Silliman gave the gist of much of his poetics in a few phrases. "Meanings are not fixed" was the opening idea. He then asserted that a poetics "emerges, evolves and is always contextual," so that it becomes "a method of problem solving" that works continually to "open up the question itself: the wound of language." Poetic positions that one turns away from today as "limiting," such as the poetics of Olson or Pound or Walt Whitman or Emily Dickinson, he recognizes "were themselves once monumental accomplishments." For Silliman, the practice of poetry is a matter of building new models instead of the ones that "impose themselves upon our understanding." He works at this with presentation of "the details of process" and "those aspects of daily life (gum wrappers in the gutter) which tend to be ignored." Recalling both Jack Spicer and Sigmund Freud, Silliman insists that it is "the invisible which tells us most clearly who we are." The trick, of course, is that something will always remain invisible no matter where we turn our attention.

Language, in Silliman's view, is a particularly apt medium for gathering the invisible within reach of perception. As part of "the first generation to have grown up with the work of Saussure or Jakobson readily at hand," Silliman focuses on language

Ron Silliman, circa 1981 (photograph by Cindy Romain)

as a web of interdependent differences "in which we ourselves are implicated." He sees the "cleavage between form and content" as "a false distinction" and so composes to keep attention fluttering across that cleft. In "The Task of the Collaborator" he writes that language poetry has focused on "a dialog between the reader's responses and any series of linguistic devices." He himself has worked toward "a theory of the device," particularly in his article called "Spicer's Language." That and the other pieces gathered in *The New Sentence* have made Silliman a leading theoretician and rhetorician of contemporary poetic experiment. Those essays range from social theories of writing practice informed by Marxist thinkers such as Walter Benjamin and Althusser, through applied theory like "The New Sentence," to applied criticism with an interest in recognition of fresh practical poetic devices. His anthology *In The American Tree* (1986) was another practical applied project. He gathered poems and statements on poetics from Language school writers, mostly on the two coasts, and wrote an introduction giving a brief history of the development of their efforts.

As the 1980s went on Silliman and his colleagues gained greater recognition as anthologies of their work circulated and as some of them began to teach more frequently in academia. They even began to inspire a second generation of writers, including Melanie Neilsen and Steven Farmer. Silliman's work was translated into Croatian, French, Italian, Japanese, Russian, and Spanish. *Contemporary Literature, Critical Inquiry, The Nation,* and the *South Atlantic Quarterly* joined the list of periodicals paying close attention to his work.

LIT, written partly while Silliman was teaching as a Regents' Professor at the University of California, San Diego, and funded in part by an NEA Fellowship, was an ambitious book-length poem with twelve sections, a continuance of the Alphabet project. It made further use of the Fibonacci number series for its form. Each of the sections was constructed with a different format of line-and-stanza or sentence-and-paragraph rules that Bernstein described in *The Difficulties* issue on Silliman's work. The varying textures and the way the strategies are played out from and off each other through the gentle presence of the whole and its reference to literature make *LIT* a compendium of Silliman's styles of formal experiment.

Each of the subsequent installments of "The Alphabet" allows for new insight into the whole of the project. Readers find that their conditioned senses of poetry and prose forms are played off each other in fresh combinations in each new format. "The Alphabet" eventually shows that the sense of form is strongly affected by format. Changes in format accentuate the various ways that habits of reading and social reverence for recognizable literary forms impose themselves on readers' relation to writings. The amplitude of the project has made it seem too weighty to some readers who are eager for Silliman to move on; others enthusiastically anticipate the full evolution of "The Alphabet."

It is certain that the whole of the twenty-six books of "The Alphabet" will engage readers both through its grand design and its minute particulars. This is writing that emphasizes the way words may come to one as second nature but shows that language is not simply natural. Silliman's use of format causes readers to question their natural sense of language use, becoming in this way what Barthes called in the foreword to his book *On Racine* "truly a form." To perform this literary function, Barthes said that a work must "truly designate a meaning in question, not a closed meaning." "The Alphabet" succeeds in getting readers to question the meaning of meaning as it is effected through form and/or format.

Some of Silliman's books actually reach into readers' bodily experience of the nature of poetry. For example, *What* (1988) is 121 pages of new sentences broken into lines that Silliman has described as having the most unnatural possible line breaks. The first few lines seem to break naturally for sense and breath, but even there the reader is challenged to think about the difference between the poetic and the prose line forms. The sensible conclusion seems to be that there both is and is not one. The way one sees the words arranged on the page is conditioned just as is the sense of what the words in their syntactic and logical combinations mean. Readers perceive that language has spatial and temporal dimensions both on and off the page. Silliman pursues in relentless detail the way these dimensions affect readers' sense of the meaningfullness of the world.

Silliman shared his interest in language with the world in August 1989, when he accompanied three other American poets, Michael Davidson, Lyn Hejinian, and Barrett Watten, to the first international conference of avant-garde writers to be held in the Soviet Union since the Russian Revolution. Called "Poetic Function: Summer School – Language, Consciousness, Society," the conference afforded a unique opportunity to visit the Russian writing scene at a time of big changes. The four Americans eventually collaborated on a poetic account of their experiences in *Leningrad* (1991). It was marketed partly as a travel book with the phrase "American Writers in the Soviet Union" on the cover. In the pages of the book the four voices of the poets are represented by icons modeled after textile patterns from early Soviet Russia. The introduction frames this textual effect in terms of their efforts "to ground the literary movement known as 'language poetry' in a sense of community and to connect it to progressive politics and new social theory."

Leningrad is a delightful record of what the poets perceived in their strange surroundings. Silliman's voice, signified by a factory cogwheel, is written in the one-sentence-at-a-time staccato pacing familiar from his poetry but a clearly narrative sense of experience emerges. His sentences show that the truth of the moment is the key element of perception and expression. Silliman has spoken about narrative as the unfolding of meaning over time, a dynamic different from plot. "Narrative," he said in *The Difficulties* interview, "is a function of the mind, not of the plot or storyboard." Gertrude Stein's lectures on narration are perhaps the most informative background for this dimension of his work. In them she insisted on a new writing coming to the fore that would meld differences between poetry and prose and bring writing and reading to focus on the present act of composition. In an earlier lecture on "Composition as Explanation," she summed up her approach to telling what is known and how it is known by saying, "A continuous present is a continuous present." Silliman has continued to elaborate this legacy carefully, playing it out as "Language writing language writing."

In the 1990s Silliman continues to unfold "The Alphabet" in magazines and books. In *Manifest* (1990) he works with short new sentence paragraphs, one to six sentences long and spaced well apart to appear as small blocks. "Paragraphs to go," says a sentence-paragraph on the last page.

In *Demo to Ink* he presents six parts of "The Alphabet" in an attractive format. In "Demo" Silliman has each sentence appear as a small block paragraph, separated from others by space. He collaborated with Rae Armantrout on "Engines," a collection of alternating fourteen-sentence sonnet-paragraphs. Discussing this work when interviewed for *The Difficulties* in 1985, Silliman was pleased that he and Armantrout had responded so sensitively to each other's writing with applied attention and responsive reaction. He said this makes such collaboration "an excellent method for sensitizing the writer as to the place of the reader, which, although it is never reflected directly in the text-as-object, is itself a participatory (if unequal) role." Silliman's consciousness of the reader's participation in the text was provoked in "Force" by another alternating format in which a portion of writing broken as six

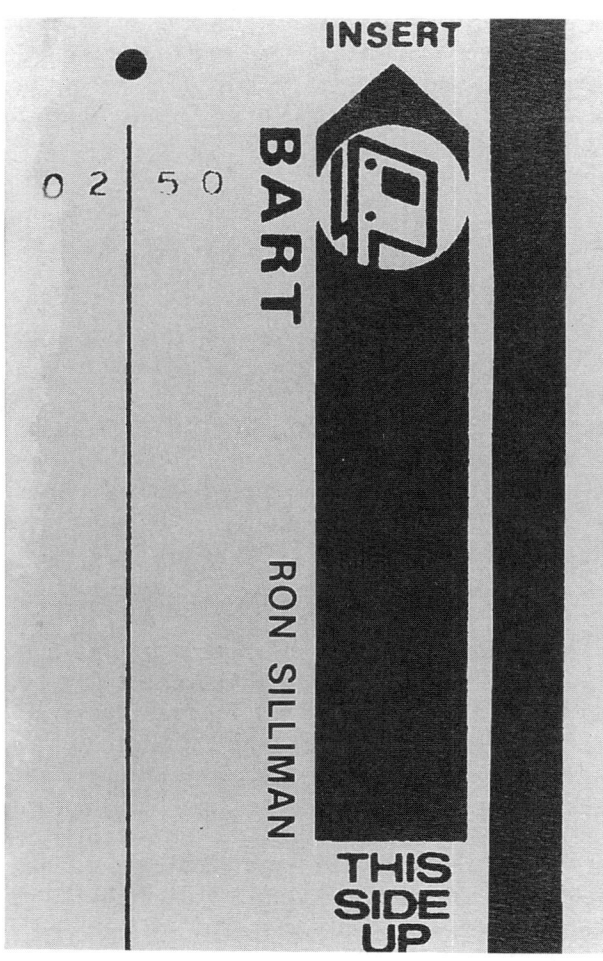

Cover for Bart, which was published in 1982 and includes poems written in the 1970s

echoes of daily life and many of Silliman's signature echoic misquotations. In the lines "And then bashed down / on the cops, dug heel into forehead, / sociopaths of false order," for example, Silliman alludes to the first of Pound's *Cantos* and to contemporary politics.

In "Ink," a poem made of expanding paragraphs, Silliman uses the title as the first word of the work. The poem builds around images of print and of modes of transportation and of ballplaying and of everyday birds, while also including elegiac passages on the grief concerning a grandmother's death. Many of the sentences of "Ink" use the first-person pronoun and, at one point, this personal presence extends to a comment on the way the writing brings the writer memories that are different from what the text has to offer a reader. Silliman writes, "what I recall are my emotions, where I sat, what the weather or light were like, what else was active, charged in my life at that moment — not to be confused with understanding the text." The effect described in this passage emphasizes the distance between the writer and the writing, even while giving the reader an inkling of possibility for identification with the writer.

Silliman's attention to the reader's place in the book and to the question of form continues in the further books of "The Alphabet" that have been published. In *Toner* (1992) he experiments with seven-line stanzas, varying the capitalization, punctuation, and placement of lines in each stanza throughout the text. *Jones* (1993) has ten sections, each with a different format. There are variations in left and right justification, breaks in lines, punctuation within paragraphs by dashes or parentheses, and more. All of these devices emphasize the material presentation of the sentences, the writerly decisions in their creation. They also put the focus upon the choices open to a reader, according to habits and education, for ways to take in format as part of the content. *N/O* (1994) presents *Non* and *Oz,* both of which mix shifting formats of prose paragraphs with Olsonesque and Ashberian lines. *Non* opens with a self-reflexive comment:

> So then go back
> to the old forms
> as if they were forms at all[.]

Silliman's work with forms continues as portions of other books — "®," "Skies," and "Under" — appear in journals and head for book form. In these works Silliman is increasingly autobiographical. Silliman has even gone on to a new publication for-

lines of poetry with no end stops is followed by a chunk of five sentences run from justified margin to margin and stopped as sentences with periods.

"Garfield," which Silliman originally wrote in a notebook with the cartoon cat on the cover, is in big blocks of twenty-one sentences each. One sentence in the first such paragraph reads, "Linebreak as breath-unit can only mark neurosis." The paragraphs are composed of a series of topical remarks from a web of relations. Each paragraph seems to make one reference to a poet, one to a painter, one gesture toward a surrealism of syntax, one toward speech, one homage to the history of Marxism, and so on. The balanced, parallel structure of each of the twenty-one blocks subtly holds the reader's attention. In "Hidden" Silliman writes in couplets, with each line a half-dozen words or so in length. The line breaks emphasize the thought phrasing of syntax and the shift in sense from phrase to phrase. There are echoes of poets and echoes of ads and

mat, sending at least one section of "The Alphabet" to friends on the Internet, where it is subject to copyright but somewhat open to copying. *Xing* (1996) gathers materials from dreams, the Chinese language, and the experiences of fatherhood, advertising, pop music, and other aspects of daily life into stanzas of three mostly short lines. The line-breaks often cut across the sense of the sentences. Silliman often humorously misquotes something from the culture and quotes himself in reflection on his ongoing project: "What then? When is a poem / not a poem? Here. Now."

In "The Alphabet" Silliman appears to be attempting to use up the differences people pretend to recognize between poetry and prose, form and content, form and format, the high and the low, the public and the personal, and humor and seriousness. This challenges readers to recognize their own relations to the literary effects common to one or the other side of each of these differentiations. With its thematics continually shifting inside Silliman's lived experience of putting things in words, "The Alphabet" is a superb example of what poet Philip Whalen called in "The Preface" to his collection *Everyday* (1964) a "continuous fabric (nerve movie?)."

To have read Silliman's work so far is to have entered a field that seeks always to draw readers further in – in order to draw them out. Reading it focuses one's attention more and more on issues concerning the construction of meaning in writing and reading. It is simple enough to be read by anyone and invites various relations to voicing, to silent reading, or to listening. Much of the delight of these poems is in their cadences, which often echo familiar phrases from poems, ads, jokes, or popular lingo. Silliman has become an excellent reader of his work, presenting it to all kinds of audiences in cafés and bars and prisons and lecture halls and classrooms. One September Saturday in 1978, he even read the entirety of *Ketjak* to a San Francisco streetcorner crowd, some pausing and some passing by, near the Powell Street cable-car turnaround. Profoundly democratic, his poems are rich enough in sound and sense, in humor and in artful reference, to engage all kinds of understandings. Silliman continues to write and will certainly find new formats to challenge both himself and his readers.

Cover for Silliman's 1985 book, for which he won an award from the Poetry Center of San Francisco

Interviews:

Larry McCaffery and Sinda Gregory, "An Interview with Ron Silliman," in *Alive and Writing: Interviews with American Authors of the 1980s,* edited by McCaffery and Gregory (Urbana: University of Illinois Press, 1987), pp. 240-256;

"Ron Silliman interviewed by Julia Blumenreich and Don Marks," *Paper Air,* 4, no. 2 (1989): 86-93;

"Ron Silliman interviewed by Michael Amnasan," *Ottotole,* 3 (Spring 1989): 207-228;

Ron Tanner and Valerie Ross, "The Politics of Poetry: An Interview with Ron Silliman," *Cream City Review,* 13 (Fall 1989): 75-105;

Manuel Brito, Interview, in his *A Suite of Poetic Voices: Interviews with Contemporary American Poets* (Santa Brigida, Spain: Kadle Books, 1992), pp. 145-166.

References:

The Difficulties: Ron Silliman Issue, edited by Tom Beckett, 2, no. 2 (1985);

Stephen Fredman, "The Crisis at Present: Talk Poems and the New Poet's Prose" in his *Poet's Prose: The Crisis in American Verse,* second edition (Cambridge: Cambridge University Press, 1990), pp. 134-161;

Roger Gilbert, "Textured Information: Politics, Pleasure, and Poetry in the Eighties," *Contemporary Literature,* 33 (Summer 1992): 243-274;

George Hartley, *Textual Politics and the Language Poets* (Bloomington: Indiana University Press, 1989);

Hank Lazer, "Opposing Poetry," *Contemporary Literature,* 30 (Spring 1989): 142-150;

Paul Mann, "A Poetics of Its Own Occasion," *Contemporary Literature,* 35 (Spring 1994): 171-181;

Stephen-Paul Martin, "A Paradigm Lost: Ron Silliman's *Paradise* and the Archaeology of Language," *Sagetrieb,* 8 (Spring/Fall 1989): 201-208;

Martin, "Ron Silliman: Non-Hierarchical Perception," in his *Open Form and the Feminine Imagination: The Politics of Reading in Twentieth-Century Innovative Writing* (Washington, D.C.: Maisonneuve Press, 1988), pp. 173-185;

Jerome McGann (writing as Anne Mack and J. J. Rome), "*The Alphabet,* Spelt from Silliman's Leaves (A Conversation on the 'American Longpoem')," *South Atlantic Quarterly,* 89 (Fall 1990): 736-759;

McGann, "Contemporary Poetry, Alternate Routes," in his *Social Values and Poetic Acts: The Historical Judgment of Literary Work* (Cambridge, Mass.: Harvard University Press, 1988), pp. 197-220;

Marnie Parsons, "'A Silly Corpse?': The L=A=N=G=U=A=G=E Poets, Stein, and the Nonsense of Reference," in her *Touch Monkeys: Nonsense Strategies for Reading Twentieth-Century Poetry* (Toronto: University of Toronto Press, 1994), pp. 170-205;

Bob Perelman, "Parataxis and Narrative: The New Sentence in Theory and Practice," *American Literature,* 65 (June 1993): 313-324;

Marjorie Perloff, "Toward a Wittgensteinian Poetics," *Contemporary Literature,* 33 (Summer 1992): 191-213;

Perloff, "The Word as Such: L=A=N=G=U=A=G=E Poetry in the Eighties," in her *The Dance of the Intellect: Studies in the Poetry of the Pound Tradition* (Cambridge: Cambridge University Press, 1985), pp. 215-238;

Linda Reinfeld, *Language Poetry: Writing as Rescue* (Baton Rouge: Louisiana State University Press, 1992);

Andrew Ross, "The New Sentence and the Commodity Form: Recent American Writing," in *Marxism and the Interpretation of Culture,* edited by Cary Nelson and Lawrence Grossberg (Urbana & Chicago: University of Illinois Press, 1988), pp. 361-380.

Papers:

Silliman's correspondence, notebooks, manuscripts, and typescripts from 1965 to 1986 are in the Archive for New Poetry, Mandeville Department of Special Collections, University of California, San Diego.

Cathy Song
(20 August 1955 –)

Susan M. Schultz
University of Hawaii at Manoa

BOOKS: *Picture Bride* (New Haven: Yale University Press, 1983);
Frameless Windows, Squares of Light (New York: Norton, 1988);
School Figures (Pittsburgh: University of Pittsburgh Press, 1994).

RECORDING: *Reading at the Honolulu Academy of Art,* 2 April 1983.

OTHER: *Sister Stew,* edited by Song and Juliet S. Kono (Honolulu: Bamboo Ridge Press, 1991).

Cathy Song (photograph by Andrea Gelben)

Cathy Song was the first writer born and raised in the state of Hawaii to attain national recognition for her work, which has been anthologized in *The Norton Anthology of American Literature, The Norton Anthology of Modern Poetry, The Heath Anthology of American Literature, The Open Boat: Poems from Asian America,* and *Unsettling America: An Anthology of Contemporary Multicultural Poetry.* In 1994 she won the Hawaii Award for Literature as well as the Shelley Memorial Award from the Poetry Society of America. She has established herself as an important figure in the literature of Hawaii, in the burgeoning canon of Asian American writing, and in the field of contemporary poetry generally.

Born in Honolulu in 1955 to a Chinese American mother and a Korean American father, Song spent her first seven years in Wahiawa, a small plantation town in rural central Oahu; her family then moved to the Waialae Kahala section of Honolulu. She started writing as a student at Kalani High School. After spending two years at the University of Hawaii at Manoa, where she worked with the poet-critic John Unterecker, she completed her undergraduate studies at Wellesley College, earning a degree in English literature. She then earned an M.A. in creative writing from Boston University. She and her husband, Dr. Douglas Davenport, returned to Hawaii in 1987; she has lived and taught there (mainly for the Poets in the Schools progam but also for one semester as a visiting distinguished writer at the University of Hawaii) ever since.

She is a member of the Bamboo Ridge study group, which includes local Hawaiian poets and fiction writers, and she works for Bamboo Ridge Press. In 1991 Song, along with Juliet S. Kono, edited *Sister Stew,* a collection of fiction and poetry by women published by the press. The anthology is notable for its inclusion of a wide range of Hawaii's writers, from Nell Altizer and Fay Kicknosway (now Morgan Blair), who teach at the University of Hawaii, to local writers such as Marie Hara and Lois-Ann Yamanaka.

In his pioneering book on Hawaii's literature, *And the View from the Shore: Literary Traditions of Hawaii* (1991), Stephen Sumida writes that, "aside from members of Hawaii's Caucasian minority group, Hawaii's local people have been stereotyped as being silent or quiet, not merely reticent but deficient in verbal skills and therefore incapable of creating literature of any merit, much less a literary tradition." Sumida's book argues for a long tradition of Hawaiian writing to break this silence that "has been forced upon these people of Hawaii by authority and circumstance." His narrative ends with the recognition of Cathy Song's work by Richard Hugo, who awarded her the Yale Series of Younger Poets prize in 1982, resulting in the publication of *Picture Bride* (1983). According to Sumida, "Song is a pioneer among a generation reexamining its heritage and furthering rather than merely preserving a literary tradition in doing so"; her career is crucial in countering the long-held view that Hawaii's literature is "insular, provincial, not universal" and the "false criticism, which in the final analysis speaks of Hawaii's people as if they were not human beings at all, but were meaningful and meaningless only as figments of others' dream of islands."

Song's having lived for several years on the mainland, in Boston and in Colorado, may have influenced her desire to write for a national rather than a local audience. Part of her importance to Hawaiian letters is in fact because of her having such a wide readership. Song gave her first major reading upon receiving the Yale Series of Younger Poets Prize at Honolulu's Academy of Art, on 2 April 1983; she was introduced by Unterecker, her mentor at the University of Hawaii. According to Unterecker, Song had felt that she had little chance at winning the award; Hugo, who judged the award that year, had not heard of her.

Unterecker took note of Song's different audiences, including herself, her family, relatives, friends, and a larger audience aware of her concerns and her interests. That her family comes first in this list is appropriate for a poet from Hawaii, where family and community are extremely important. Unterecker listed the aspects of her work that would appeal to mainland critics: the design of her poetry; the overlapping imagery; the structure that holds the poems together; and the many moths, magnolias, porches, and fences of her world, as well as the many cats (but few dogs) included in the poems. He added that Song's work "may be the bluest poetry ever written," enumerating her many references to the col-

or blue. Mainland critics, he suggested, would further note the way the book becomes a single poem. But he added that "we're [the Hawaiian audience] lucky because so many of the poems are set in Hawaii." The poems include references to aspects of Hawaii that are unfamiliar to mainland readers, including Wahiawa's rains and red mud, the surf, plumeria blossoms, the sugarcane fields, and the smell of burning cane.

In his introduction to *Picture Bride* (which she had wanted to title "From the White Place," after a Georgia O'Keeffe painting), Hugo remarks on the quiet surfaces of Song's work. "In Cathy Song's quietude lies her strength.... She receives experiences vividly and without preset attitude. Her senses are lucky to have remained childlike and reception appears to be a complete act." He goes on to say that "Song's poems are flowers." Hugo seems to have ignored a dramatic monologue halfway through the book, given over to the voice of O'Keeffe, which disputes this notion of Song as a "quiet" and "childlike" poet:

"Women are like flowers,"
you said, for years
I despised myself
and you –
Mother
and Aunt Winnie in the garden,
arranged like lawn chairs,
smiling full of babies and detergent.

A few lines further down, the speaker expresses the desire to do violence to these flowers:

I discovered my own autonomy then,
crawling out from your wide skirts
and into your flowerbeds,
where I proceeded to crucify the dolls,
decapitating your crocuses.

The anger in Song's first book comes, in part, out of the plantation history of Hawaii; she inscribes this history into poems about her family. This move is not atypical; as Elaine H. Kim writes in the *Columbia Literary History of the United States* (1988), "Today's Asian American writer is often forced to seek the meaning of the past in shreds of stories heard in childhood." The brief glossary to *Picture Bride* describes Song's childhood community as a "small town on the island of Oahu, Hawaii," but fails to mention that until recently Wahiawa was a plantation town, devoted to the cultivation of pineapple and sugar for outside consumption.

Dust jacket for Song's 1983 book, in which several poems depict the experiences of her immigrant grandparents (courtesy of the Poetry Collection, State University of New York, Buffalo)

A cursory look at history may help the reader understand the power that place-names and local references often have in Song's first book and give important background for appreciating her work. In his plantation history, *Pau Hana: Plantation Life and Labor in Hawaii* (1983), Ronald Takaki cites Immanuel Wallerstein's definition of the "modern world system" and asserts that Hawaii "was a semiperipheral country producing a cash crop for export to a core country." Plantations, which were developed in the middle of the nineteenth century, brought foreign labor to work the fields, much of it from Asian countries: China, Japan, Korea, and the Philippines. The origins of Hawaii's multicultural society were part of a strategy of "divide and conquer" on the part of plantation owners. According to Takaki in *A Different Mirror: A History of Multicultural America* (1993): "Though they imported workers as supplies, planters were conscious of the nationalities of their laborers. The employers were systematically developing an ethnically diverse labor force in order to create division among their workers and reinforce management control." The extent to which the plantation system was stratified according to race and power structures is made comically clear by Milton Murayama in his novel *All I Asking for Is My Body* (1975). The manager's house is located at the top of the hill, the Portuguese and Japanese *lunas* (foremen) live slightly below the managers, and below them are the run-down houses of Filipinos. "Shit too was organized according to the plantation pyramid," Murayama notes, as those at the bottom of the pyramid live closer to the sewage ditches than those at the top.

Many of the Japanese and Korean women who came to Hawaii came as picture brides. A total of 951 picture brides came from Korea between 1907 and 1923. In the opening poem of her collection Song meditates on her grandmother's life when she was a young woman who left Korea for a husband she had never seen. As Gayle K. Fujita-Sato points out, the poem is constructed almost exclusively of questions. Of the quest for identity she sees as common in Asian American literature, Fujita-Sato asserts that "the process of searching is inevitably a process of constructing." That her grandmother was entering a plantation economy is clear in the first of Song's questions:

> And
> was it a long way
> through the tailor shops of Pusan
> to the wharf where the boat
> waited to take her to an island
> whose name she had
> only recently learned,
> on whose shore
> a man waited,
> turning her photograph
> to the light when the lanterns
> in the camp outside
> Waialua Sugar Mill were lit
> and the inside of his room
> grew luminous
> from the wings of moths
> migrating out of the cane stalks?

This is the same Waialua (also a town on Oahu, Hawaii) that Song describes in the poem of that title in *Picture Bride* as a place of confinement, one she longed to leave: "When the afternoons fell, / the night stalks sweetened, / breathing against the screen window, / meshing into barbed wire."

Song further complicates her portrait of her immigrant grandparents in "Easter: Wahiawa, 1959." This poem begins as a description of a typical 1950s American middle-class suburban family – her own – celebrating Easter. It is impossible, early on, to distinguish this family from one on a 1950s sitcom:

> Grandmother took the opportunity
> to hang the laundry
> and Mother and my aunts
> filed out of the house
> in pedal pushers and poodle cuts,
> carrying the blue washed eggs.

The title alone is a tip-off, however, that something more is going on in the poem than meets the eye, for 1959 was the year that Hawaii became a state, having been an American territory since 1898. The poem is about the process of becoming an American, and the key figure in this transformation is the speaker's grandfather, who came from Korea to work the sugarcane fields. In Korea he had been hungry and would gather quail eggs on the riverbank to eat; by way of contrast, his grandchildren now gather colored Easter eggs in their hunt. It is the fruit of his labor that the family celebrates:

> It was another long walk
> through the sugarcane fields
> of Hawaii,
> where he worked for eighteen years,
> cutting the sweet stalks
> with a machete. His right arm
> grew disproportionately large
> to the rest of his body.
> He could hold three
> grandchildren in that arm.

The speaker does not know whether or not these years of labor were worthwhile; rather, she writes, "I want to think / that each stalk that fell / brought him closer / to a clearing."

Many of the poems in the collection express a desire for escape from family and from Hawaii. Kim considers these poems to be the best in the book: "The most effective poems ... are those that explore the relationship between the persona and her family, from whom she ventures forth and with whom she is eventually reconciled." In "The Youngest Daughter" the speaker bathes her mother ("I was almost tender"), even as she dreams of leaving her behind. At the dinner table the daughter's dreams separate her from her mother: "We eat in the familiar silence. / She knows I am not to be trusted, even now planning my escape."

In another poem, "Lost Sister," she contrasts the immobility of Chinese women, whose feet were bound, to the mobility of Chinese American women, who face different forms of paralysis. In limitation the Chinese women find a kind of movement:

> But they traveled far
> in surviving,
> learning to stretch the family rice,
> to quiet the demons,
> the noisy stomachs.

By way of contrast, "In America, / there are many roads / and women can stride along with men." The life of the immigrant, however, is fraught with difficulty, as

> Dough-faced landlords
> slip in and out of your keyholes,
> making claims you don't understand,
> tapping into your communication systems
> of laundry lines and restaurant chains.

The speaker is ambivalent because she finds that she needs China, "a jade link / handcuffed to your wrist," even as she longs for the space in which to escape that binding tradition:

> You remember your mother
> who walked for centuries,
> footless —
> and like her,
> you have left no footprints,
> but only because
> there is an ocean in between,
> the unremitting space of your rebellion.

In contrast to the specificity of *Picture Bride*, the relative abstraction of Song's second book, *Frameless Windows, Squares of Light* (1988), testifies to the poet's attempt to generalize from her own experience. She includes poems about immigration, stamp collecting (in a poem that bears the influence of Elizabeth Bishop's work on travel), childbirth, and art. "Humble Jar" stands out as a kind of *ars poetica* and takes its place among Song's other poems about sewing (*Picture Bride* ends with "The Seamstress"; *School Figures* includes "The Grammar of Silk," about a sewing school Song attended as a child). Patricia Wallace writes: "It's amusing to think of 'Humble Jar' as Song's humble version of poetry's Grecian urn, as her way of insisting on the connections between art and daily life." The poem may also be read as an answer to Wallace Stevens's poem "Anecdote of a Jar," asserting that while a masculine ordering of the wilderness may be too presumptuous an act, the more feminine actions of sewing and binding are not.

"Humble Jar" refers to a jar of buttons that Song's mother kept "for every emergency"; it contained buttons that were "useful yet undervalued." These buttons were public in their display but private in their being the sole property of her mother.

> But the buttons,
> stashed in that jar,
> were oddly private.
> She'd never admit it.
> She'd say they had their uses.

Among their uses for the poet is that of organizing a life before her life; the humble jar describes her mother as much as anything:

> A dip into that humble jar
> repaired a shirt,
> but it also retrieved a moment
> out of a cluttered life —
> before I was born,
> before any of us had made our claims.

Wallace comments: "Song wants to transform the 'undervalued' life of the mother, to uncover its hidden beauty and power, and to open possibility. But this effort is countered by a knowledge, felt in the rhythms of the poem, of how much the mother has had to put aside."

This sense of her mother's having missed something reappears in Song's third book, *School Figures* (1994). In "The Grammar of Silk" Song writes about the Saturdays she spent as a child at Mrs. Umemoto's sewing school. The school was located at the Kaimuki (a section of Honolulu) Dry Goods store, which provided a sanctuary for women who consulted the "oracle" of the latest pattern books. Sewing becomes in this poem an act that gives time for meditation; it is almost a preparation for poetry.

> She was determined that I should sew
> as if she knew what she herself was missing,
> a moment when she could have come up for air —
> the children asleep,
> the dishes drying on the rack —
> and turned on the lamp
> and pulled back the curtain of sleep.

Song more than suggests that sewing is a poetic act (R. P. Blackmur less charitably asserted that Emily Dickinson took to writing poems as most women took to knitting). Art is a refuge but one that involves work:

> my foot keeping time on the pedal,
> it was to learn the charitable oblivion
> of hand and mind as one —
> a refuge such music affords the maker —
> the pleasure of notes in perfectly measured time.

Song has elsewhere equated poetry with music and with femininity. In the introduction to *Sister Stew* she and Kono write: "The voices [included in the volume] tell of the moments when we listened, watched, and understood, as if life, like a piece of music, could be apprehended in all its strangeness." This, they go on to assert, is "the voice of artifice — our invention, the work we do, the dream in which we imagine carp rising."

In *School Figures* Song ventures further than before into the local world of Hawaii's Asian American community, even including some brief dialogue in pidgin, a local language with plantation roots that is used increasingly in literature by Hawaii's local writers. In "A Conservative View" she teases out the ethnic stereotypes that survive in mixed mar-

Dust jacket for Song's second book, which includes "Humble Jar," a poem about sewing and the art of poetry (courtesy of the Poetry Collection, State University of New York, Buffalo)

riages. Her mother, who has Chinese blood, believes in the conservation of money (this fulfills a Hawaiian stereotype about the tightwad "pake," or Chinese). Her father, of Korean ancestry, "has heard enough from Mother / about Koreans being big spenders, show-offs — / 'champagne taste on a beer budget.'" The poem ends with her father expressing his admiration for Mao, who he thinks was the best thing that happened to China: "How else are you going to get those damn pa-kes to share?" he asks. For the most part Song's subjects grow out of her experience in Hawaii's middle class. The backgrounds to these poems include an evening at a karaoke bar, her daughter's piano practice, and lessons with a tennis pro.

Many of Song's poems are about the body, especially her own as she gives birth to and raises her three children. She obliquely treats the conflict between her love for the body and her mother's skepticism about it. In "Sunworshippers" Song begins with her mother's exclamation, "'Look how they love themselves.'" The speaker responds, "We were not allowed to love ourselves too much," and then describes an episode of anorexia in college, when she paradoxically "devoured radiance," becoming a mind that no longer owned a body:

Undetected, I slipped in and out of books,
passages of music, brightly painted rooms
where, woven into the signature of voluptuous vines,
was the one who flew one day out the window,
leaving behind an arrangement of cakes and ornamental flowers;
to weave one's self, one's breath, ropes of it, whole
and fully formed, was a way of shining
out of this world.

The ending of this poem, while lyrical, does not engage the most troubling questions of identity that are raised by its opening and the subject of anorexia; at times Song's lyricism seems to suppress the conflicts inherent in her work.

Song's attitudes about writing poetry have grown out of her university workshop training. When she was asked for a special issue of the journal *Manoa* what audience she writes for, she replied that she wrote for the woman in the workshop who knew William Butler Yeats by heart. For the most part she writes in the voice of an uncomplicated lyric "I," eschewing the ironies of poets such as Robert Frost or Sylvia Plath. Unlike her fellow Hawaiian poet Lois-Ann Yamanaka, Song does not critique the social and cultural values of the state (or the United States); instead she writes about and celebrates her own experiences. In that sense she may belong to what Robert von Hallberg has called the poetic "center."

Song's poetry, like that of James Merrill and others, belongs more to the suburbs than to the urban centers or rural regions. Hers is a poetry of accommodation, not protest, written mainly in free verse. She is not shy to defend her position at the center; she told an interviewer that "it's annoying" to be termed a "'middle-class poet.'" "It's easy for someone to look at me and say, 'She hasn't suffered. She grew up in Waialae Kahala, went to Wellesley. Her husband's a doctor; she drives around in a Volvo.' I think it's very unfair. We all suffer in different ways. I don't have to have grown up on the plantation speaking pidgin and have someone who beat the shit out of me to write good poetry. We all suffer."

When she does satirize experience, however, she does so stringently, as in "Forever Yours" from *School Figures,* about the "designated dreamboat and his homecoming queen" in high school. A participant in the prom-night cruise, the speaker notes:

It was called not rocking the boat.
Conformity, conformity,
the sweetest form of treachery.
Give up your individuality
and you can be one of us.

She describes the conformists as future businesspeople:

Was it her, the homecoming queen,
or her court of industrious maidens,
future bank tellers and beauticians of America
tunneling back and forth into the night
(Who ya gonna vote for?),
yielding the invisible work of telephone lines?

Song places herself in the tradition of nonconformist poets who claim to do battle with the conformity that engenders falsity: "fake fur, fake pearls, fake eyelashes — / the black-and-white pictures, / the size of passport photos we would never use." The immobility of high-school classmates is contrasted with the speaker's own cosmopolitanism.

"Wednesday Night, Karaoke Bar" further explores the question of conformity and nonconformity, this time within the field of "song" (like many other poets with apt names Song plays with hers throughout her work). Song oscillates between describing the karaoke singer's art as one that imitates childhood's freedoms and one that lacks imagination and collapses into mere mimicry. When asked in an interview if the imagination liberates us, she responded: "Exactly. To appreciate a good book, to be able to respond to a piece of music, a work of art, a human situation, then be able to make sense of it all through language. We're all going to toil. We desperately need our imagination." Her fervent defense of the Poets in the Schools program, which she delivered as she accepted the Hawaii Award for Literature in 1994, came out of her belief that children need to get in touch with this liberating imagination. In the poem Song at first seems called back to her childhood self as she sings:

It's a relief to begin singing

as I remember it, as a child,
alone, uninhibited —

hunger, a physical
weight lifting out of the darkness

of my body, a gesture
producing sound, like breath,

but harder, like a fist making contact
with the world.

The karaoke singer, however, imitates a preexisting song. When he introduced Cathy Song at her December 1994 reading, Wing Tek Lum noted that she took the karaoke playlist home from the restaurant with her so she could "practice" the songs. The mood of the poem changes with the poet's growing realization that the words have nothing to do with her feeling:

My mouth is open. I am singing

and yet, I am producing nothing –
an echo, a mimicry –

while my voice strains to climb through the artificial
sound changers, the acoustic obstacles.

What I hear bears no relation
to that weight in my gut –

heavier now, sinking –
the blue dot bouncing merrily along,

buoyant like a toy boat
tossed upon wave after wave

of interchangeable lyrics,
the blue dot

eating up the words, my voice,
that fist that wants to open its pearl and sing.

At their best Song's poems allow their mingled anger and love of beauty to emerge from what Richard Hugo describes as her "quietude." At their weakest her poems tell more about Song herself than about the fascinating world in which she lives, one fraught with gender, ethnic, and class tensions. One of the reasons for Song's relatively early success as an ethnic poet may be her conservatism, for her poetry does not directly challenge the dominant cultural attitudes. In the title poem "School Figures," the last of the volume, Song mentions some of her favorite works of art; these include paintings by John Constable, Katsushika Hokusai, Piet Mondrian, Jean-Auguste-Dominique Ingres, Paul Cézanne, and Pieter Brueghel as well as the Audrey Hepburn film *Roman Holiday*. Not only are these works of art for the most part European – not American or Asian – but they are canonical rather than revolutionary. Such a critique of her work, however, is problematic. Garrett Kaoru Hongo in his introduction to *The Open Boat: Poems from Asian America* (1993) noted two conflicting tendencies in Asian American poetry. The issues, he writes, "could be characterized as a generational conflict between those who wish to uphold the notion of a personal subjectivity and poetics within the American experience, minority or mainstream, and those who make their priority the production of a polemicized critique of generalized ideological domination within our culture." Song places herself firmly within the former tradition.

The diversity of Asian American writing, according to Hongo, is a strength in the face of "a problematic tendency ... to hegemonize the variety of projects within the field known as 'Asian American literature' which has resulted in the privileging of some of the most polemical and parochial approaches to literary production." Hongo asserts, "it may be that we seek a kind of serious bewilderment that clarifies experience." As a writer who crosses many of the boundaries that others try to assert, Song adds an important voice to an emerging canon of multicultural literature whose terms are constantly under dispute. Her refusal to be categorized (as a Hawaii poet, a middle-class poet, or any other kind of poet) can be seen as a political statement against exclusionary criteria of ethnic literatures that are seen as existing outside the canon of American literature.

Interview:

David Choo, "Cathy's Song: Interview with Cathy Song," *Honolulu Weekly,* 4 (15 June 1994): 6–8.

References:

Gayle K. Fujita-Sato, "'Third World' as Place and Paradigm in Cathy Song's *Picture Bride,*" *Melus,* 15 (Spring 1988): 49–72;

Stephen Sumida, *And the View from the Shore: Literary Traditions of Hawaii* (Seattle: University of Washington Press, 1991);

Patricia Wallace, "Divided Loyalties: Literal and Literary in the Poetry of Lorna Dee Cervantes, Cathy Song and Rita Dove," *Melus,* 18 (Fall 1993): 3–19.

James Tate
(8 December 1943 -)

Chris Stroffolino
State University of New York at Albany

See also the Tate entry in *DLB 5: American Poets Since World War II, First Series.*

BOOKS: *Cages* (Iowa City: Shepherd's Press, 1966);
The Destination (Cambridge, Mass.: Pym-Randall Press, 1967);
The Lost Pilot (New Haven & London: Yale University Press, 1967);
Notes of Woe (Iowa City: Stone Wall Press, 1968);
The Torches (Santa Barbara, Cal.: Unicorn Press, 1969; revised and enlarged, 1971);
Row with Your Hair (San Francisco: Kayak Books, 1969);
Shepherds of the Mist (Los Angeles: Black Sparrow Press, 1969);
Are You Ready Mary Baker Eddy, by Tate and Bill Knott (Berkeley, Cal.: Cloud Marauder Press, 1970);
The Oblivion Ha-Ha (Boston & Toronto: Little, Brown, 1970);
Hints to Pilgrims (Cambridge, Mass.: Halty Ferguson Press, 1971; revised edition, Amherst: University of Massachusetts Press, 1981);
Absences (Boston & Toronto: Little, Brown, 1972);
Hottentot Ossuary (Cambridge, Mass.: Temple Bar Bookshop, 1974);
Viper Jazz (Middletown, Conn.: Wesleyan University Press, 1976);
Lucky Darryl: A Novel, by Tate and Knott (Brooklyn: Release Press, 1977);
Riven Doggeries (New York: Ecco Press, 1979);
Constant Defender (New York: Ecco Press, 1983);
Reckoner (Middletown, Conn.: Wesleyan University Press, 1986);
Distance From Loved Ones (Hanover & London: Wesleyan University Press, 1990);
Selected Poems (Hanover & London: Wesleyan University Press, 1992);
Worshipful Company of Fletchers (Hopewell, N.J.: Ecco Press, 1994).

James Tate (courtesy of the Lilly Library, Indiana University)

It is a critical commonplace that, in contrast to earlier generations of American poets, no poet born during or after World War II has yet emerged to bridge the gap between "academic" and "experimental" poetry. Although some may celebrate the possibility that the idea of a central tradition has itself been made obsolete by the divergent poetics of the postmodern era, others may justly fear that this Balkanized state of affairs has the side effect of disabling any fruitful cross-pollination between the wide range of poetic styles practiced today. Such considerations are key in assessing the career and likely future importance of James Tate, a poet who stands between the experimentalism of avant-garde strains such as the Language poets and the poetry of personal anecdote currently favored by the academy.

Even though Tate has won his share of mainstream poetic acceptance – the 1967 Yale Younger Poets Prize for *The Lost Pilot,* a 1992 Pulitzer Prize for *Selected Poems,* and a 1994 National Book Award for *Worshipful Company of Fletchers* – his eclectic oeuvre bears as little resemblance to the dominant academic "workshop poem" mode as it does to most avant-garde poetry being championed today. It is tempting, if only as a corrective to the divisiveness that has characterized the poetry community since the 1970s, to highlight the affinities Tate's poetry has with nonmainstream work. However, such a comparison would be largely beside the point; Tate's poetry, like any work of quality and originality, calls into question the authority of any given critical apparatus.

James Vincent Tate was born in Kansas City, Missouri, in 1943, the same year his father was reported missing while piloting his plane over Germany during World War II. He began writing poems at age seventeen. He attended the University of Missouri (1963–1964) and Kansas State College in Pittsburg, Kansas (1964–1965), where he received his B.A. He studied under poet Donald Justice at the Writers Workshop of the University of Iowa, receiving his M.F.A. in 1967. He was a visiting lecturer in English at the University of Iowa (1966–1967) and, after winning the Yale Younger Poets Prize at twenty-three (the youngest-ever recipient of the award), at the University of California, Berkeley (1967–1968). He then taught as an assistant professor of English for two years at Columbia University and was poet-in-residence at Emerson College in Boston for a year before joining the English faculty at the University of Massachusetts, Amherst, where he has taught since 1971.

Though Tate's style has often been characterized as "surreal" by both his defenders and his detractors, such surrealism is but one in an array of poetic strategies he has deployed. Despite the changes and developments in style in Tate's poetry from *The Lost Pilot* (1967) through *Worshipful Company of Fletchers* (1994), there is a surprising consistency to the thematic concerns that have marked his work from the beginning. His central themes include the absent father, heterosexual relationships, American sociopolitical realities, and the figure of the lyric poet as misfit/outlaw. Such themes are often dramatized in a poetry that, while shamelessly self-referential, is so fraught with tension and insight that the charges of solipsism and mere witticism levied against it, sometimes even by the poet himself, seem facile indeed. Though Tate's work is dark, full of pathos, introspection, and absurdity, it almost always includes a dimension of social engagement, which has become even more evident in his most recent work. The risks Tate often takes have earned him much renown as well as the reputation of being an astute commentator on the function of poetry and the persona of the poet.

Unlike the many contemporary poets who view poetry as mere self-expression, Tate sees the art as a means of self-creation, of personal growth, though his despairing passages often indict or are skeptical of poetry's transformative potentials. The personal stakes involved help explain why Tate has not progressed along a slow and steady path toward his mature work. In his interview included in *American Poetry Observed* (1984) he admitted to being traumatized by his early success: "I was immediately frightened by the thought of getting stuck in a rut and spending the rest of my life as that triadic-syllabic fellow in the grey pin-striped suit. I wanted to change.... There followed some pretty awkward poems." But unlike Delmore Schwartz, who never quite recovered from the fallout from early poetic fame, Tate has transcended his early work. His development parallels that of fellow Yale Series of Younger Poets–winner John Ashbery, who also won early acclaim and indulged in a period of experimentation before attaining his mature style, though Tate as yet has not attempted a major poetic statement such as Ashbery's *Three Poems* (1972). Although it was still common as late as 1979 for critics to remark on Tate's decline since *The Lost Pilot,* such claims are less often made in the 1990s.

Tate's first full-length book, *The Lost Pilot,* deserves its accolades, not only for the poet's achievements in form but also for his confident adoption of the idiosyncratic personae that would become a hallmark of his work. Unlike many young authors, Tate does not attempt to say everything, and part of the charm is the restrained tone of the short lyrics that make up the collection. In announcing his pact with solitude in "Epithalamion For Tyler" and "Why I Will Not Get Out of Bed," Tate places his faith in the power of poetry as an alternative to a conventional life. In the latter poem he begins, "My muscles unravel / like spools of ribbon / . . . / I will pose / like this for the rest / of the afternoon, for the remainder / of all noons." In "Success Comes to Cow Creek" the speaker meets his double in the form of a self-proclaimed loser ("the fire hydrant / of the underdog"). While poems such as "How the Friends Met" and "Grace" suggest the failure of human connections, "Coming Down Cleveland Avenue" – the poem Dudley Fitts, who edited and wrote an introduction for the collection, coaxed Tate to place first in the book – giddily portrays male/female union as a kind of strip poker.

The poem in the collection that garnered the greatest critical praise is "The Lost Pilot," which Tate wrote for his dead father and which does not at all seem forced or inauthentic. Tate claims it was created out of a "mystical experience" on the day he became the age his father was when he died. The speaker at first contrasts his father with those on the plane who survived, favoring his imagined memory over the reality of aging:

> Your face did not rot
> like the others – the co-pilot,
> for example, I saw him
>
> yesterday. His face is corn-
> mush: his wife and daughter,
> the poor ignorant people, stare
>
> as if he will compose soon.

At the end of the poem the speaker achieves an attitude of identification with his father *through* disconnection:

> I feel as if I were
> the residue of a stranger's life,
> that I should pursue you.
>
> My head cocked toward the sky,
> I cannot get off the ground,
> and, you, passing over again,
>
> fast, perfect, and unwilling
> to tell me that you are doing
> well, or that it was mistake
>
> that placed you in that world,
> and me in this; or that misfortune
> placed these worlds in us.

Although "The Lost Pilot" is certainly a fine poem, one does well not to overemphasize its importance; Tate treats the primal trauma of the absent father more profoundly elsewhere.

After the publication of *The Lost Pilot,* Tate published four limited-edition chapbooks in a frenzy of poetic activity, the two most significant of which appeared in 1969, *The Torches* and *Row with Your Hair* (which features illustrations by Mel Fowler). Many of these poems later appeared in his second full-length book, *The Oblivion Ha-Ha* (1970), a transitional work in which Tate seemed to be groping for a new style and audience. At the time of its publication the book was almost universally panned for lacking the restraint of his first volume. In hindsight, however, Tate can be seen moving toward the stylistic freedom of his later works. Yet this collection as a whole is somewhat awkwardly divided between these two styles. "Prose Poem," perhaps, suggests something of Tate's attitude, for the speaker is surrounded "by the pieces of this huge / puzzle: here's a piece I call my wife, and / here's an odd one I call convictions, here's / conventions, here's collisions, conflagrations, / congratulations."

Despite its title the collection contains moments of quiet lyric beauty and pathos as in the last stanza of "Coda," a poem that describes the end of a relationship: "When morning comes, / I will build a cathedral / around our bodies. / And the crickets, / who sing with their knees, / will come there / in the night to be sad, / when they can sing no more." In "Up Here" Tate dramatically evokes an alienated relationship as the male speaker looks down on his lover during sex:

> Now your lips are moving, now
> your hands reach up at me.
> I feel as if I might be one
> or two thousand feet above you.
> Your lips form something, a bubble,
> which rises and rises into
> my hand: inside it is a word:
> *Help.* I would like to help,
> believe me, but up here nothing
> is possible, nothing is clear:
> *Help. Help me.*

The most celebrated poems in this volume include "The Blue Booby," "It's Not the Heat So Much as the Humidity," and "The Wheelchair Butterfly." While the last of these poems, by Tate's own admission, is merely structured from "a succession of images," the first achieves a balance of imagery and meaning, portraying, only half mockingly, the poet's obsession with the solitary life of poetry as a way of seducing women.

While Tate throughout *The Oblivion Ha-Ha* seems to flaunt the freedom early fame allowed him through his relentless experimentation, *Hints to Pilgrims* (1971) goes even further in that direction. Investigations into the paratactic possibilities of language are Tate's major concerns, especially in long sequences such as "Boomerang" and "Amnesia People" that disrupt rational connections. Such poems have been mistakenly considered to be examples of automatic writing, yet Tate is more interested in consciousness than the unconscious. These poems

Dust jacket for Tate's 1967 volume, selected by Dudley Fitts for publication in the Yale Series of Younger Poets

are worthy of critical attention and can be seen as a precursor to such Language poets as Kit Robinson.

Among the most-accessible pieces in *Hints to Pilgrims* are "Poem (A silence that tunnels forever)," a beautiful lyric that recalls Frank O'Hara's poem "lost," and "I Take Back All My Kisses," an example of a surreal catalogue poem, similar to André Breton's "Free Union" but more intensely personal. When in "I Take Back All My Kisses" the poet repeats the phrase "They got me" he seems in a struggle to reclaim his identity, perhaps even from the poetic career that has defined him:

> They got me with rubber horns and drugged rabbits
> They got me when a rug was too tired to fly
> They got me because I'm trembling beneath a stove
> They got me because I stalk the tone in daylight
> They got me because caresses are molded into poems
> Yes I never said I didn't fall into their hands like mercury
> Mercury eaten in a fish
> Mercury flashed across an afternoon nightmare
> Tomorrow may be different
> Who will be the last to know[.]

It is hard to imagine that the same man who wrote "The Lost Pilot" could also write this poem, which is a tour de force experiment that Tate has not since tried again. "I Take Back All My Kisses" is a purgation ritual doubling as a nose-thumbing gesture that does try to say everything: "They got me because I begged them to not take me more / seriously than they take themselves / Because they believed me when I said I could not change them."

In *Absences* (1972), his fourth major collection, Tate was able to avoid the lack of direction that plagued the transitional *The Oblivion Ha-Ha*. Though

Absences is ostensibly a more sober book than its immediate predecessor, *Hints to Pilgrims,* here one sees the first inklings of the looser quasi-anecdotal meditative mode Tate was not able to achieve consistently for another ten years. Along with *The Lost Pilot* and *Hints to Pilgrims, Absences* includes the most interesting work Tate published before the age of forty. Much of the book was written during the brief period Tate lived in an urban environment and was a frequent reader at such venues as *St. Mark's Poetry Project In The Bowery*. In "Contagion," the first poem of *Absences,* Tate begins, "When I drink / I am the only man / in New York City. / There are no lights, / but I am used to that." It was perhaps such poems that led Ted Berrigan to characterize Tate as a "wild academic, which is only mildly interesting." Although the tone that Tate strikes in this collection does not celebrate urban Bohemia as Berrigan and other second-generation New York School poets do, the formal restraints and intense subjectivity in many ways have more in common with such first-generation New York School poets as Frank O'Hara, John Ashbery, and Kenneth Koch.

In *Absences* Tate longs not so much for lost youth or innocence as he does for the kind of absences that only occur when one is social. He is nostalgic for nostalgia. Not only are sex and relationship poems absent from this volume, in contrast to *The Oblivion Ha-Ha,* but Tate has given up the pretense of themes. The title poem has an epigraph from Guillaume IX of Aquitaine, part of which says: "I'll write a song about nothing at all, / Not about myself, nor about anything else." The title poem and "Cycle of Dust" are Tate's most significant attempts to this point in his career to work in longer sequences. He also experiments with forms popularized by such poets as Robert Creeley and W. S. Merwin. "A Guide to the Stone Age" seems a parody of Merwin's "Some Last Questions," while "The Distant Orgasm" borrows the device of urgent repetitions from Creeley, as the speaker's reading is interrupted by the cries of a woman in a neighboring apartment.

Tate's willingness to wrestle with the dangers of potentially self-destructive intellectualism invests his work with a morality attractive to those who do not usually accept moral authority. Part of his achievement is to problematize lyrically his own relationship to the ethical poetic conventions as well as the stylistic. "South End," for instance, begins: "The challenge is always to find the ultimate / in the ordinary horseshit why bother." Yet Tate, with his retinue of losers like those of Philip Larkin, does bother, if only for the sake of futility in such poems as "Deaf Girl Playing" and "My Great Great Etc. Uncle Patrick Henry."

Tate's remaining major collections of the 1970s seem to mark both a decrease in intensity and a staking out of new territory. *Hottentot Ossuary* (1974), a collection of prose poems, in hindsight seems little more than sketches of scenarios Tate would take up in his later poetry. While *Viper Jazz* (1976) contains a handful of poems that could be ranked among his best, Tate too often fails to take risks, avoiding both the personal risks of *Absences* and the linguistic risks of *Hints to Pilgrims*. Much the same can be said for Tate's *Riven Doggeries* (1979).

As in his previous work, the poems in *Viper Jazz* obliquely suggest Tate's social consciousness. In "Awkward Silence" the speaker becomes aware of how even what he thought was retreat from society is socially implicated as he struggles for a less awkward silence, a less paranoid awareness of solitude as a social act. In "A Radical Departure" Tate attacks the fraudulence of just such an attempt to escape society; the speaker, apparently a would-be ascetic determined on "going to a place so thoroughly remote / you'll never hear from me again," is revealed to be a con artist as he asks, "do you mind if I take your wife?" In "On the Subject of Doctors" he begins by attacking the ethics of the medical profession but concludes by injecting a note of compassion:

> Germ city, there's no hope
> looking down those fire-engine throats.
> They're bound to get sick themselves
> sometime; and I happen to be there
> myself in a high fever
> taking my plastic medicine seriously
> with the doctors, who are dying.

With the appearance of *Constant Defender* (1983), Tate makes a great leap forward by synthesizing the experiments of the 1970s. He has refined this synthesis in each of his subsequent books, finally allowing his most recent work to eclipse the legacy of *The Lost Pilot*. Many of the poems in *Constant Defender* are about poetry. Tate takes on the role of a poet's poet even in the most personal verses, such as "Poem to Some of My Recent Poems," in which the division between the poet and his writing becomes almost unbearable in light of his mother's death. When the poet tells his poems, "you are beautiful, and I, a slave to a heap of cinders," he sums up the major aesthetic crisis of this volume: life versus art. In "Nausea, Coincidence" Tate exposes the seedier side of a faith in French

Dust jacket for Tate's 1970 collection that includes "Prose Poem," in which the speaker metaphorically examines his life and work as "the pieces of this huge / puzzle"

poet Charles Baudelaire's "correspondences" or Eliot's "objective correlative." In the last stanza the speaker's urge for derangement, his adolescent need for "the faded perfume of solid ground," is unflatteringly exposed as the cause of his retreat into solitude. The speaker remains acutely aware that his art is a "shallow . . . peeking," a "crutch" whose powers of enchantment have been "forgotten" as the poem dead-ends.

"Blue Spill" is also a sobering coming-to-terms with the limits of the merely aesthetic. In the first movement of the poem Tate seems to liken poetic ecstasy to wading through an oil spill. Nonetheless, by the end of this poem Tate suggests that the calamity was necessary to survival: "Forgive me for not lying about this but he / could be dead right now if he hadn't wandered into / this lucky accident area." "Paint 'Til You Faint" may be read as continuing Tate's lyric debate with himself. The housepainter-speaker who admits that he "was in a hurry to get to the fine work" is a study of the obsessed psyche. The realization of the pitfalls of placing too much faith in poetry is metaphorically distilled with comic concision in such lines as "I would paint the milkman / but he is already painted," as the speaker confronts a world in which his services are useless. What to do in the face of such an impasse is the central theme of *Constant Defender*. A possible answer can be seen in "A Jangling Yarn," where the speaker realizes that he "must wake now / into masquerade and particle." The end of the poem points to a possible resolution of Tate's aesthetic attitude with his more practical one: "It is Carnival again in the world, and I must try / to harmonize with its proud or shabby downfall."

Like O'Hara, Tate's critical reputation has sometimes suffered from his own facility with funny, if not exactly light, poems, which have often eclipsed his more serious and demanding work. Furthermore, the critics appreciative of Tate's comic poems have sometimes missed their serious purport. By turning a joke on Shakespeare's *King Lear,* Tate in "Tragedy's Greatest Hits" questions the sacredness of this icon of Western high art. Even the title poem, "Constant Defender," for all its colorful images of Americana, is a serious investigation into how poetry makes words into things. Of course, taken too far, such a stance can make one believe that one's thirst can be quenched by the idea of water. Tate reveals that the relationship between life and art is such that the solutions poetry offers are not those that life requires.

While the major subject of *Constant Defender* is poetry, Tate's next volume, *Reckoner* (1986), written at the height of Reaganism, turns toward society and sociopolitical themes. Though the collection marks a new direction in Tate's work, he occasionally lapses into his earlier stances of awkwardness and oversimplification. "Jo Jo's Fireworks – Next Exit," for example, asks readers to identify with a car, "belly / hungry for gasoline pancakes." In "No Rest For The Gambler" the urge to break out of "life's talk show" into a state of pure awe is in part a result of Tate's trivialization of life into a talk show in the first place.

Despite some unsuccessful poems, the new reckoning Tate makes with social themes not only paves the way for the superior *Worshipful Company of Fletchers* but is responsible for some great poems in its own right. "Smart and Final Iris" deals with helplessness in the face of the doublespeak of "Pentagon code." In "The Chaste Stranger" Tate treats the backlash to the sexual revolution. Donald Revell and Lee Upton argue in separate essays that "A Wedding," "Neighbors," and "The Sadness of My Neighbors" resemble earlier Tate poems in their appeal to those who have neither faith in the conventional rituals of our society nor in the institution of the family. But the poems show more of Tate's compassion for people who live ordinary lives in suburban America, with its pressures to measure up against the Joneses. "Thoughts While Reading *The Sand Reckoner*," perhaps the best poem in this book, celebrates, though with some irony, sitting at home and reading/writing as the most pleasurable way of spending a Saturday evening.

Distance From Loved Ones (1990), arguably Tate's best book to date, fully integrates comedy and pathos, imagery and statement, and achieves a balance between lyric and narrative through its long, tightly packed sentences. Tate avoids didacticism by dramatically portraying the speaker through his relationships with others. The "I" in Tate's recent poetry is not the poet himself, nor is it a structural device that provides the reader a means of entry into contemporary cultural life. Though he shares the twentieth-century suspicion of transcendental signifiers and representation, Tate, like Creeley, is able to incorporate such a suspicion into a poetry that questions its own assumptions even as it makes them, thus leaving to the reader the task of interpretation.

Distance From Loved Ones is concerned primarily with the intricacies of relationship — reality to representation, lackeys to bosses, children to parents, individuals to society, and men to women — and broadens the range of his sympathy to include a wider array of characters. Tate's extravagant metaphoric enactments become more convincing by his rooting them in scenarios that would otherwise seem mundane, as in the situation described in the first two lines of "No Spitting Up": "'People in glass elevators shouldn't carry snow shovels,'/ I said to Sheila, because we were in one with a lady who was." The tension between appearance and the arbitrary baggage of identity is brought to the fore through hilarious juxtapositions in "I Am a Finn" and "I Am Still a Finn." Even in the poems in this collection framed as character studies, such as "Editor," "The Expert," "Distance From Loved Ones," "Mimi," and "Peggy in the Twilight," Tate is well aware of the dangers of reifying identity. "Anatomy," for instance, is a tender yet biting poem about the burden of female beauty in a jealous and resentful puritanical society. There is far more at stake morally in *Distance From Loved Ones* than in his earlier work in which Tate at times seemed content to scandalize the complacencies of the bourgeoisie.

Poems about heterosexual relationships are more prevalent in *Distance From Loved Ones* than in any other single volume of Tate's. But as always it is hard to tell if Tate is writing about actual women since in his poetry the subjective world is as real as the external one. In "Saturdays are for Bathing Betsy" a woman is metaphorically equated with the poet's tongue. The kind of love described in this poem contrasts with the more strained affairs presented in "The Banner," "How Happy We Were," and "Burn Down the Town, No Survivors." In "Burn Down the Town" Tate writes: "I could have loved her but it would / have been just more of the same, / more petty crimes and slow death, / more passion leading to betrayal, / more ecstasy guaranteeing tears." It is such admissions of vulnerability that make Tate's achievement unique and valuable in an age when many male poets, either out of machismo or "political correctness," have shied away from the complexities of relationships. *Distance From Loved Ones* is not a somber statement regarding the failure of connections; instead, Tate embraces distance as a necessary aspect of love, as if distance is also a gift received from those loved best.

Though Tate has not attempted the long sequential poem since *Absences,* his latest book, *Worshipful Company of Fletchers,* shows him moving toward a longer, more meditative lyric that is more expansive in scope than his previous long poems. Tate largely forsakes the persona of the poet as solitary freak or klutz as he comes to terms with the less flattering side of his place as an authority in society. The title poem, for instance, dramatizes the poet-speaker's meeting with a young boy in the woods. The speaker spouts the maxims of societal conformity in a futile attempt to make the boy repent his alternative Thoreauvian lifestyle. The boy finally exposes the speaker as a regretful, bitter man and disarms him with a kiss. By giving the last word to the youth instead of the poet-speaker, Tate perhaps shows that he has reevaluated the adequacy of the role he has perhaps too habitually claimed for himself. The choice of "Worshipful Company of Fletchers" as a title poem is especially significant if it is compared to "A Manual Of Enlargement" and "In My Own Backyard," two titles Tate originally considered for the volume. In both poems Tate adopts his more habitual persona.

Looking at *Worshipful Company of Fletchers* on a thematic level in terms of Tate's whole career, one may not find many concerns that were not charac-

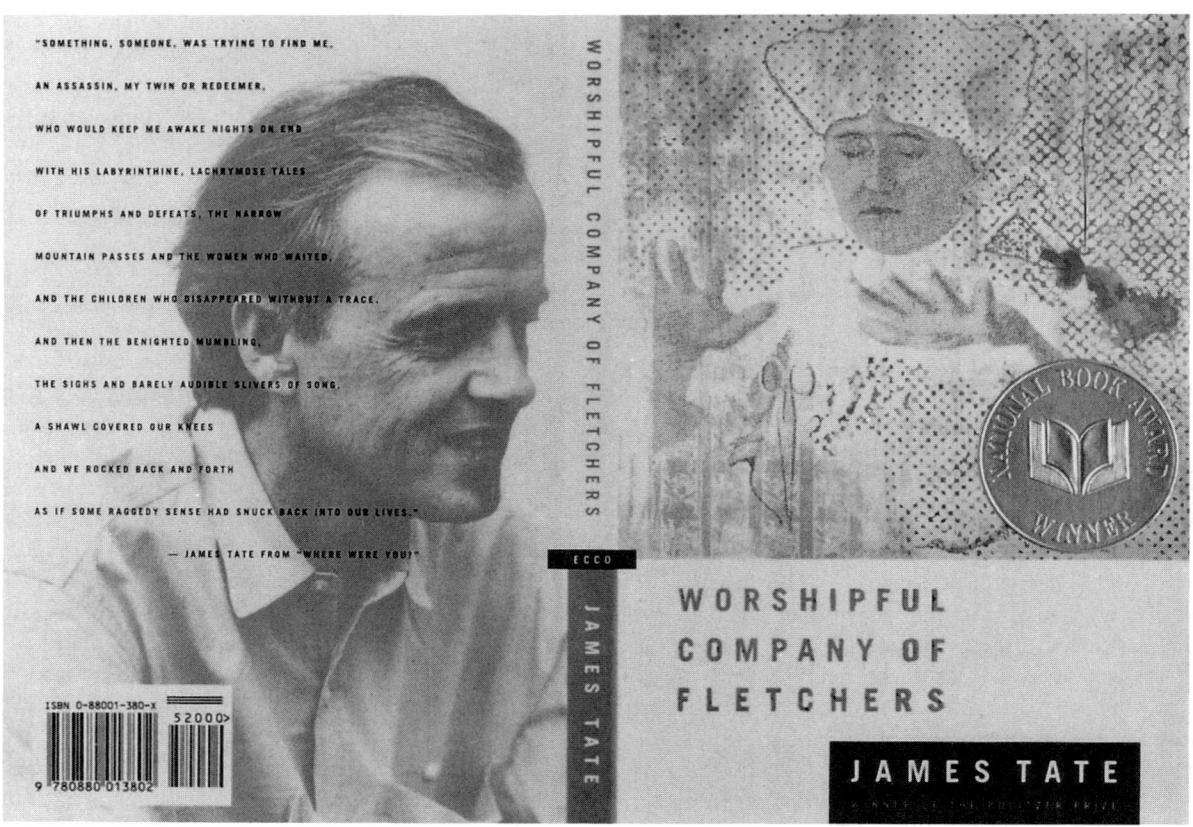

Dust jacket for Tate's 1994 book, in which he returns to the longer, more meditative mode of Absences

teristic of his three previous volumes. What distinguishes this book is primarily stylistic and tonal, the way the themes are handled. In the Ashberian "More Later, Less The Same" Tate shows his usual distrust of the facility with which he can go in just about any direction in his writing, yet he suspends such distrust for a full two pages before awareness of a toothache stops him. In the eloquent poem "Color in the Garden," which echoes the theme of "Blue Spill," Tate lyrically posits an alternative to the color wheel, allowing social themes to enter through the connotations of colors. Red is associated with "the weak and ailing" who will not be allowed to become normal until they are "allowed to spend some time in the 'blue' room."

In many of these poems Tate doffs earlier comic masks to wear more mature ones. As the title poem suggests, stances of maturity and authority remain problematic for Tate. Beneath the irony and awkwardness in poems such as "The Morning News" there is an intolerance for those who live life in the fast lane that borders uncomfortably on a bitter, unconvincing didacticism. Nonetheless, the collection contains many strong poems, including "We Go to a Fire," "The Wrong Way Home," "A Missed Opportunity," "Inspiration," and "Desire." Tate's humor is still evident, as in "Annual Report," which begins with the poet despairing about his perversities being ignored by the media: "And, in an entire year, there was only one / case of indecent exposure / (Is no one paying attention?)." As the poem progresses, however, the speaker finds some solace by extending the range of his identification to nature: "On the bright side . . . there were five Deer Complaints (Well pardon us for existing)." Despite his efforts in longer poems, Tate's forte is once again revealed to be the short lyric poem, especially when he tries to tackle sociopolitical themes.

"How the Pope Gets Chosen" is one of Tate's comic masterpieces. It might as well be titled "How to Choose a Poodle," even though the conceit of equating the Pope with a poodle is not sustained but weaves in and out. Tate exploits the comic potentials of demystifying the Pope in order to take to task the arbitrariness of American hero worship as well as strip the mask of democracy from the hierarchical class-society. In "An Eland, in Retirement" and in other places in the book Tate, who as a prize-winning poet bears at least a superficial resemblance to a poodle-Pope, wrestles with guilt over why he has "been spared."

Tate's meteoric success perhaps explains the tendency to self-flagellation found in much of his work. Lacking the working-class context and social agenda of poets such as Philip Levine, Tate has focused in his work on the enemy within. His increasing move away from self-flagellation in his recent work signals his slow coming to terms with the social possibilities of a stance more affirmative of the pleasures of the everyday. Tate's affirmation, however, falls short of the visionary. The last lines of *Worshipful Company of Fletchers* — "What I thought was infinite will turn out to be just a couple / of odds and ends, a tiny miscellany, miniature stuff, fragments / of novelties, of no great moment. But it will also be enough" — could be read as Tate's attempt to bring poetry down to his level. Though such an affirmation without the trappings of transcendence is part of what is so seductive about Tate's persona, one may wonder whether it is indeed enough.

James Tate's position in contemporary American poetry and letters, for all his success on the award circuit, remains precarious. Tate, unlike John Ashbery, is not visionary enough to be included in Harold Bloom's controversial Western canon, and he is too persona-based to be accepted by the avant-garde. He remains primarily a poet of personal eccentricities and pathos, witty situations, isolation, and love. And though he has taken such themes further than most of his contemporaries, his attempts to evoke domestic bourgeois pleasures in *Worshipful Company of Fletchers* are awkward at best. He has not significantly moved beyond the persona he created in his earliest work, and his skill with a single persona and lyric style may actually impede his advance as an artist. And though Tate is to be credited for creating a well-defined stance as a way of organizing what otherwise would be a poetry of stray ideas or perceptions, he runs the risk of becoming too predictable. Nonetheless, Tate has certainly written some of the best poems of his generation. There is much in his grappling with the darkness of the psyche that the often unconvincing affirmation of the visionary tradition of American poetry should at least meet halfway. There are at least hints in *Worshipful Company of Fletchers* that Tate, still in his early fifties, will be able to achieve a wider scope and become a significant figure in American poetry in the twenty-first century.

Interviews:

Leo Braudy, *James Tate: The Making of Poetry* (New Haven: Yale University Press, 1967);

Helena Minton, Louis Papineau, Cliff Saunders, and Karen Florsheim, "Interview with James Tate," in *American Poetry Observed,* edited by Joe David Bellamy (Chicago: University of Illinois Press, 1984).

References:

Donald Revell, "The Desperate Buck and Wing: James Tate and the Failure of Ritual," *Western Humanities Review,* 38 (Winter 1984): 372–379;

R. D. Rosen, "James Tate and Sidney Goldfarb and the Inexhaustible Nature of the Murmur," in *American Poetry Since 1960: Some Critical Perspectives,* edited by Robert B. Shaw (Chester Springs, Pa.: Dufour Editions, 1974), pp. 181–191;

Lee Upton, "The Masters Can Only Make Us Laugh," *South Atlantic Review,* 55 (November 1990): 78–86.

Rosmarie Waldrop

(24 August 1935 -)

Steven R. Evans
Brown University

BOOKS: *A Dark Octave* (Durham, Conn.: Burning Deck, 1967);

Change of Address, by Waldrop and Keith Waldrop (Providence, R.I.: Burning Deck, 1968);

Camp Printing (Providence, R.I.: Burning Deck, 1970);

The Relaxed Abalone; or, What-You-May-Find (Providence, R.I.: Burning Deck, 1970);

Letters from Rosmarie and Keith Waldrop, by Waldrop and Keith Waldrop (Providence, R.I.: Burning Deck, 1970);

Spring Is a Season and Nothing Else (Mount Horeb, Wis.: Perishable Press, 1970);

Body Image, by Waldrop and Nelson Howe (New York: G. Wittenborn, 1970);

Against Language?: Dissatisfaction with Language as Theme and as Impulse towards Experiments in Twentieth Century Poetry (The Hague: Mouton, 1971);

The Aggressive Ways of the Casual Stranger (New York: Random House, 1972);

Alice ffoster-Fallis: (an outline), by Waldrop and Keith Waldrop (Providence, R.I.: Burning Deck, 1972);

Until Volume One, by Waldrop and Keith Waldrop (Providence, R.I.: Burning Deck, 1973);

Words Worth Less, by Waldrop and Keith Waldrop (Providence, R.I.: Burning Deck, 1973);

Kind Regards (Providence, R.I.: Diana's Bimonthly Press, 1975);

Since Volume One, by Waldrop and Keith Waldrop (Providence, R.I.: Burning Deck, 1975);

Acquired Pores (Paris: Orange Export, 1976);

The Road Is Everywhere or Stop This Body (Columbia, Mo.: Open Places, 1978);

The Ambition of Ghosts (New York: Seven Woods, 1979);

When They Have Senses (Providence, R.I.: Burning Deck, 1980);

Psyche & Eros (Peterborough, U.K.: Spectacular Diseases, 1980);

Nothing Has Changed (Windsor, Vt.: Awede Press, 1981);

Differences for Four Hands (Blue Bell, Pa.: Singing Horse, 1984);

Streets Enough to Welcome Snow (Barrytown, N.Y.: Station Hill, 1986);

The Hanky of Pippin's Daughter (Barrytown, N.Y.: Station Hill, 1986);

Morning's Intelligence (Grenada, Miss.: Salt-Works Press, 1986);

The Reproduction of Profiles (New York: New Directions, 1987);

Shorter American Memory (Providence, R.I.: Paradigm, 1988);

A Form / of Taking / It All (Barrytown, N.Y.: Station Hill, 1990);

Peculiar Motions (Berkeley, Cal.: Kelsey St. Press, 1990);

Light Travels, by Waldrop and Keith Waldrop (Providence, R.I.: Burning Deck, 1992);

Lawn of Excluded Middle (Providence, R.I.: Tender Buttons, 1993);

Fan Poem for Deshika (Tucson, Ariz.: Chax, 1993);

Cornered Stone, Split Infinites (Elmwood, Conn.: Potes & Poets Press, 1994; New York: New Directions, 1994);

A Key into the Language of America (New York: New Directions, 1994).

OTHER: *A Century in Two Decades: A Burning Deck Anthology, 1961–1981,* edited by Waldrop and Keith Waldrop (Providence, R.I.: Burning Deck, 1982);

"The Joy of the Demiurge," in *Translation: Linguistic, Literary, and Philosophical Perspectives,* edited by William Frawley (Newark: University of Delaware Press, 1984), pp. 41–49;

"Alarms and Excursions," in *The Politics of Poetic Form: Poetry and Public Policy,* edited by Charles Bernstein (New York: Roof, 1990), pp. 45–72.

TRANSLATIONS: Peter Weiss, *Bodies and Shadows* (New York: Delacorte, 1969);

Edmond Jabès, *Elya* (Bolinas, Cal.: Tree Books, 1973);

Jabès, *The Book of Questions,* 4 volumes (Middletown, Conn.: Wesleyan University Press, 1976–1984);

Jabès, *The Death of God* (Peterborough, U.K.: Spectacular Diseases, 1979);

The Vienna Group: Six Major Austrian Poets, by Waldrop and Harriet Watts (Barrytown, N.Y.: Station Hill, 1985);

Alain Veinstein, *Archaeology of the Mother,* by Waldrop and Tod Kabza (Peterborough, U.K.: Spectacular Diseases, 1986);

Paul Celan, *Collected Prose* (Manchester, U.K.: Carcanet, 1986; Riverdale-on-Hudson, N.Y.: Sheep Meadow Press, 1990);

Jabès, *The Book of Dialogue* (Middletown, Conn.: Wesleyan University Press, 1987);

Emmanuel Hocquard, *Late Additions* (Peterborough, U.K.: Spectacular Diseases, 1988);

Jabès, *The Book of Shares* (Chicago: University of Chicago Press, 1989);

Jabès, *The Book of Resemblances* (Hanover, N.H.: University Press of New England/Wesleyan University Press, 1990);

Jacques Roubaud, *Some Thing Black* (Elmwood Park, Ill.: Dalkey Archive Press, 1990);

Jabès, *Intimations the Desert* (Middletown, Conn.: Wesleyan University Press, 1991);

Jabès, *From the Book to the Book: An Edmond Jabès Reader* (Middletown, Conn.: Wesleyan University Press, 1991);

Joseph Guglielmi, *Dawn* (Peterborough, U.K.: Spectacular Diseases, 1991);

Jabès, *The Ineffaceable The Unperceived* (Middletown, Conn.: Wesleyan University Press, 1992);

Friederike Mayröcker, *Heiligenanstalt* (Edinburgh, Scotland: Morning Star, 1992; Providence, R.I.: Burning Deck, 1994);

Jabès, *The Book of Margins* (Chicago: University of Chicago Press, 1993);

Jabès, *A Foreigner Carrying in the Crook of His Arm a Tiny Book* (Middletown, Conn.: Wesleyan University Press, 1993);

Elke Erb, *Mountains in Berlin* (Providence, R.I.: Burning Deck, 1995);

Roubaud, *The Plurality of Worlds of Lewis* (Normal, Ill.: Dalkey Archive Press, 1995);

"Berlin (plus) Portfolio," selected and translated by Waldrop, *Exact Change Yearbook,* 1 (1995): 61–90;

Jabès, *The Little Book of Unsuspected Subversion* (Stanford, Cal.: Stanford University Press, 1996).

SELECTED PERIODICAL PUBLICATIONS – UNCOLLECTED: "Marat/Sade: A Ritual of the Intellect," *Bucknell Review,* 18 (Fall 1970): 52–69;

"Charles Olson: Process and Relationship," *Twentieth-Century Literature,* 23 (December 1977): 467–487;

"Mirrors and Paradoxes: Edmond Jabès, *Le Livres des Questions,*" *Kentucky Romance Quarterly,* 26, no. 2 (1979): 143–155;

"Chinese Windmills Turn Horizontally: On Lyn Hejinian," *Temblor,* 10 (1989): 219–223;

"Shall We Escape Analogy," *Studies in Twentieth-Century Literature,* 13, no. 1 (1989): 113–129;

"Silence, The Devil, and Jabès," in *The Art of Translation: Voices from the Field,* edited by Rosanna Warren (Boston: Northeastern University Press, 1989), pp. 225–237;

"Split Infinite," *Denver Quarterly,* 27, no. 4 (1993): 102–106;

Contribution to "Editorial Forum," *Chain,* 1 (1994): 112–113;

"An Interview with Claude Royet-Journoud," by Waldrop and Keith Waldrop, *Lingo,* 4 (1995): 160–167.

Rosmarie Waldrop's contribution to post-1945 American poetry is formidable. The author of ten books of poetry, numerous chapbooks, and two novels, Waldrop is also an award-winning translator (most notably of Edmond Jabès's work) and, along with her husband, Keith Waldrop, the publisher of one of the most persistently adventurous of America's independent presses, Burning Deck, now in its third decade of operation. Her work has received international recognition, especially in France and Germany, and is increasingly discussed in this country both in the context of the contemporary avant-garde and also in the context of feminist writing, where her name is often linked to those of Lyn Hejinian, Susan Howe, Barbara Guest, Leslie Scalapino, and Mei-mei Berssenbrugge.

Making her achievement perhaps the more remarkable is the fact that English is Waldrop's adopted language and the United States her adopted home. In "Alarms and Excursions," a 1990 essay that provides an excellent introduction to her overarching concerns as a writer, publisher, and translator, Waldrop points out that, along with the inevitable difficulties associated with a shift in one's primary language (her native language is German), the experience of this change can also yield a valuable insight: "it makes you very conscious," she writes, "that you don't ever own the language, that

Rosmarie Waldrop (photograph © Renate von Mangoldt)

the language is larger than you, that it is not simply a tool that you are master of." She has on several occasions spoken of language as the only credible form of "transcendence" in secular times, and in a brief essay from 1993 titled "Split Infinite," she describes the act of writing in these words: "Allowing ourselves to be lost, we dive into the infinite of language."

The biographical experience of linguistic dislocation not only informs Waldrop's theoretical and practical orientation toward language, it also represents a historical link to preceding generations of avant-garde writers, principally, in Waldrop's case, the Surrealists and, even more centrally, the Dadaists. Like these earlier writers, Waldrop engages in a literary practice predicated on a commitment to the materiality and almost plastic manipulability of language as a medium. Her own work in concrete and visual poetry in the late 1960s and early 1970s, her frequent use of collage techniques and procedural devices throughout the 1980s and 1990s, not to mention her many years of hand-setting type for Burning Deck books and chapbooks, all testify to an ongoing exploration in the fundamentally strange and surprising precincts of the material word. Refusing an exclusively instrumental relation to words — refusing, that is, to reduce the Baudelairean "forest of symbols" to mere lumber — Waldrop prefers to think of writing as a way of "uncovering possibilities." "My key words would be exploring and maintaining," she says, "exploring a forest not for the timber that might be sold, but to understand it as a world and to keep this world alive."

Waldrop was born in the small German town of Kitzingen-am-Main on 24 August 1935. Her father, Joseph Sebald, taught physical education at the town's high school. Her mother, Friederike Wohlgemuth Sebald, had aspirations to a singing career, but these never materialized except in the most modest of forms. Waldrop's sisters, the twins Dorle and Annelie, were nine years her elder. Owing perhaps to the age difference and the special nature of the bond between twins, Waldrop felt quite separate from her sisters while the children were growing up. Both sisters did, however, come to be implicated in Waldrop's subsequent life as an author: to Dorle, for instance, is dedicated the 1979 poem "The Ambition of Ghosts"; the cover art for *When They Have Senses* (1980) is also attributed to her. The character of the sister in Waldrop's semi-autobiographical novel, *The Hanky of Pippin's Daughter* (1986), is loosely modeled after Annelie (to whom, under the pen name "Yerma," that book is also dedicated). Waldrop describes her father as having been something of a walking compendium of quotes from Johann Wolfgang von Goethe and Friedrich von Schiller. His curious blend of pantheistic naturalism and church-attending Catholicism, as well as his ubiquitous activities as an amateur astrologer, are characteristics that appear to the poet in retrospect to have influenced her in the opposite directions, pushing her to become the more rational the more "kooky" some of her father's behavior became. Her mother's Prussian Protestantism kept her father's tendency toward mysticism in check; it was she who set the tone of the household as secular and even modestly cosmopolitan, encouraging her daughters' aesthetic aspirations and seeing that Annelie received training as a ballerina until World War II cut those studies short.

In "Memory Tree," a recent prose poem, Waldrop writes of a "long life of learning the preceding chapter." She refers, of course, to the rise of National Socialism, culminating in Hitler's assumption of power in Germany in the years immediately preceding her birth. Waldrop undertakes the imaginative reconstruction of this period in her well-received novel *The Hanky of Pippin's Daughter,* the title of which alludes to the mythic founding of Waldrop's hometown by King Pippin the Short in the eighth century, where the ostensibly apolitical lives of the characters Joseph and Frederika Seifert (modeled on Waldrop's parents) are linked to the crumbling of the Weimar Republic and the rise of Nazi hegemony in Germany. Regina Weinreich, reviewing the book for *The New York Times Book Review*, saw that "marital betrayal is a metaphor for what is happening in the state. Everyday detail progresses to a haunting echo: 'Auschwitz, Maidanek, Treblinka.'"

In "Memory Tree" Waldrop resuscitates more of these "everyday details"; this time, ones directly embedded in her past:

My first schoolday, September 1941, a cool day. Time did not pass, but was conducted to the brain. I was taught. The Nazi salute, the flute. How firmly entrenched, the ancient theories. Already using paper, pen and ink. Yes, I said, I'm here.

I was six or seven dwarfs, the snow was white, the prince at war. Hitler on the radio, followed by Léhar. Senses impinged on. Blackouts, sirens, mattress on the floor, furtive visitor or ghost.

A kaleidoscope of early experiences, these sentences juxtapose disparate time frames: the experience of starting school in 1941 is fused with the bombing of Kitzingen in 1943 ("Blackouts, sirens, mattress on the floor"), which resulted in the school's being closed. The first sentence of the second stanza ("I was six or seven dwarfs, the snow was white, the prince at war") refers to Waldrop's brief experience, at ten years of age, as an actress in a troupe that traveled from village to village in an American army truck during the months immediately following Germany's surrender in May 1945. Waldrop remembers that "in the afternoon we played *Snow White and the Seven Dwarves* for the kids and in the evening we played Wedekind's *Love Potion*. I was one of the dwarves in the afternoon and I was a Russian nobleman's son in the evening."

The historical fact of the concentration camps, as writers from Theodor Adorno to Michel Foucault and Edmond Jabès have maintained, discredited not only the political practice of fascism, but also the metaphysical concepts to which fascist politics had appealed for legitimacy. For Waldrop the concept of a full and homogenous center is one such concept. For her it is a matter of ethical as well as aesthetic principle that "the center is empty," that, indeed, "it becomes tangible only in the negative, in the totalizing claims to define it, give it a content, which cannot but turn into tyranny and intolerance. In Jabès's books it becomes tangible in its grotesque historical parody, the concentration camp." Even in works where no explicit reference to the historical fact of fascism is made, such concerns remain in evidence throughout Waldrop's own writing. For instance, in the brilliant prose poems of *The Reproduction of Profiles* (1987) and *Lawn of Excluded Middle* (1993), the primary figure is that of an absent center, or "empty middle," conceived of as a generative matrix. The social category most prominent in these two works (and in many other of Waldrop's books, as well) is that of gender, but it would be a mistake to overlook the dense and contradictory investments a society makes in organizing the gender(s) of its citizens, a point with obvious resonance in relation to such phenomena as the Nazi cult of the "mother."

The overlap between the historic and the personal — what Waldrop once summarized in the lines "personal blisters from / impersonal burns" — results in ambivalence. In the introduction to her 1994 New Directions volume, *A Key into the Language of America,* a text that redoubles and dislocates the 1643 text of the same name by Roger Williams, Waldrop takes stock of some of the ambivalences that suffuse her personal history:

> I was born in 1935, the year Williams's 300-year banishment officially ended. I was born "on the other side," in Germany. Which was then Nazi Germany. I am not Jewish. I was born on the side of the (then) winners. I was still a child when World War II ended with the defeat of the Nazis. I immigrated to the U.S., the country of the winners, as a white, educated European who did not find it too difficult to get jobs, an advanced degree, a university position. I can see myself, to some extent, as a parallel to the European settlers/colonists of Roger Williams's time (though I did not think God or destiny had set aside for me a virgin garden). Like Roger Williams, I am ambivalent about my position among the privileged, the "conquerors."

It is to her "position among the . . . 'conquerors'" that Waldrop owes her early grasp of the English language, the result of growing up in what became after 1945 an American occupation zone. Drawing on this background, as well as some knowledge of French, Waldrop entered the University of Würzberg (just north of Kitzingen) in 1954 to study comparative literature. It was also in 1954 that she met a recently discharged private in the U.S. Army, Keith Waldrop (who at that time was still known by his first name, Bernard, rather than the middle name, Keith, by which he would be known from the late 1960s forward). Contriving to see more of one another, the two undertook the first of what — although they could not know it then — would be their many collaborations, a project translating one of Friedrich Nietzsche's poems. The two further arranged to study together in Aix-en-Provence in the academic year 1956–1957, during which time their relationship grew serious. Waldrop completed her undergraduate study at the University of Freiburg in 1958, and on 17 December of that year, having been accepted into the graduate program in comparative literature at the University of Michigan in Ann Arbor, she traveled to the United States with the intention of making her life there. One month later she and Keith married.

At the University of Michigan the groundwork had already been laid for what came to be known as the anthology wars. Donald Hall, coeditor (along with Robert Pack) of the *New Poets of England and America* (1957), was at the time a new addition to the faculty at the Ann Arbor campus and was destined to become the emblem of "academic" poetry when Donald Allen's *New American Poetry* threw down the gauntlet in 1960. Waldrop recalls arriving in Michigan, where Keith had preceded her by a year (indeed, it was money from a Hopwood Award that Keith had won for an essay the previous year that paid Waldrop's passage from Germany), to find a circle of people bound by their common interest in Ezra Pound's work and by their practice of writing and circulating new poetry among themselves.

James Camp and Don Hope (who along with Keith coedited *Burning Deck* magazine between 1962 and 1965), Dallas Wiebe, John Heath-Stubbs, X. J. Kennedy, and the Waldrops composed the core of this circle, known playfully as the John Barton Wolgamot Society. Wolgamot was the author of *In Sarah, Christ, Mencken, and Beethoven There Were Men and Women* (1944), a fascinating if arcane book that Keith had stumbled upon in a used-book shop. Contributing to the aura that this author assumed for the circle was the fact that Wolgamot is a near-homonym of Wohlgemuth, the maiden name of Rosmarie's mother. Along with Hall, W. D. Snodgrass — then teaching at Wayne State University in nearby Detroit — also participated in the circle. Both men were represented (along with Kennedy, Wiebe, Camp, Heath-Stubbs, Keith Waldrop [under the pseudonym "Bernard Keith"], and Hope) in the first Burning Deck title, a 1961 anthology edited by Hope and entitled *The Wolgamot Interstice.*

Kennedy was in many ways the "star" of this group, but the high value that he and Hall placed on traditional prosody was by no means unequivocally endorsed by the others. It was against the backdrop of an increasing "camp mentality" that *Burning Deck* magazine was optimistically announced as a "quinterly" journal in 1962 (in point of fact the four issues of this journal appeared over the course of three years). In their joint introduction to the 1982 anthology of Burning Deck authors, *A Century in Two Decades,* Rosmarie and Keith recall that "the two most widely noted anthologies of the time, both representing the period 1945–1960, contain[ed] not a single poet in common. Burning Deck . . . disregarded this split, printing and reviewing a spread of poets wide enough that on occasion an author would complain about being in such unprogrammatic company." With strong praise for Robert

Waldrop, spring 1972 (photograph by Keith Waldrop)

Creeley's *For Love* (1962) and George Oppen's *The Materials* (1962) in the first issue (Fall 1962) and an equally strong defense against criticism by LeRoi Jones and others of X. J. Kennedy's *Nude Descending a Staircase* (1961) in the second issue (Spring 1963), *Burning Deck* magazine inaugurated an approach that gently but insistently cast a skeptical eye upon narrowly partisan claims, an approach that became a signature of Burning Deck publications when it passed to the Waldrops' joint editorship from roughly 1968 onward.

Discussing her first years in the United States, Waldrop speaks of a time of "taking things in, adjusting to America." Her earliest attempts at writing, already fairly tentative, were more or less suspended for a while. "I tried to go on with the little attempts I had made," Waldrop recalls, "but it became so artificial to write something in German when I was by that time thinking in English, dreaming in English, living in English." Translation provided the ideal bridge, a "natural substitute," she later called it. She first tried translating the work of American poets such as Wallace Stevens, Robert Lowell, Creeley, and Pound into German. While some of these translations eventually appeared in journals such as *Aspekte & Impulse, Wort & Wahrheit,* and *Lyrische Hefte,* Waldrop's lack of contacts in the German publishing world made publication there a somewhat frustrating endeavor, so she reversed direction and began translating poets such as Gottfried Benn, Karl Krolow, and Günter Grass into English.

In 1963 Waldrop's first journal publications appeared, including both original poems (see those appearing under the pseudonym "Rosmarie Keith" in *Burning Deck* 2) and her translations (in journals such as *Alternative, Chelsea,* and *Choice*). In the same year the British poet Charles Tomlinson selected a manuscript of her poetry for a Major Hopwood Award. In 1964, her dissertation not yet in hand (she received her Ph.D. in 1966), Waldrop was hired to teach comparative and German literature at Wesleyan University. Taking their 8 x 12 Chandler and Price platen press (purchased in 1961 for $175), she and her husband moved to Durham, Connecticut, where they spent four years before moving to

Providence, Rhode Island, where they have made their home from 1968 to the present.

The appearance in 1967 of *A Dark Octave* marks the advent of Waldrop's career as a serious writer in English. The eight poems in this chapbook prefigure Waldrop's mature work on a thematic level. The title poem, "A Dark Octave," introduces a musical metaphor that will reappear throughout her oeuvre (for two quite different examples, see *Differences for Four Hands,* 1984, and *The Hanky of Pippin's Daughter*). The same poem also interrogates the foundation of perceptual experience and its links to personal identity, a relationship crystallized in that perhaps most basic of English-language puns, "I/eye." "Setting Type" is a meditation, triggered by the daily labors of running a small press, on the physical properties of language. "For John Cage Perchance," besides being an homage to the avant-garde composer and writer, also ruminates on the absence of ontological and epistemological grounds, an absence Cage's aleatoric practices can literally be said to "sound." "Blank Moment" introduces the conceptual protagonist of so much of Waldrop's work, the absent center or excluded middle.

One thus recognizes many of Waldrop's major themes in this first collection; but less in evidence are her subsequent formal concerns. Considered from a metrical standpoint, the poems in *A Dark Octave* are quite static, their semantic and sonic values tending to resolve in end-stopped lines without benefit of enjambment or major prosodic variation. There is a heaviness to the lines that, in concert with the conceptual and thematic concerns being explored, imparts to the small volume a rather stern, abstract tone. Waldrop herself characterizes these early poems as "fairly conservative," attributing the cautious formal approach to her youthful obedience to preexisting models and her misplaced orientation toward the narrowly semantic dimensions of a poem.

The eminently still and solitary world of *A Dark Octave* gives way to a more gregarious, if also more vertiginous, world by the time of Waldrop's second chapbook, *The Relaxed Abalone; or, What-You-May-Find* (1970). A collage of language that Waldrop drew from the case histories found in standard psychology books, *The Relaxed Abalone* is as formally frenetic as *A Dark Octave* was static. Here Waldrop composes in highly enjambed short lines that take their bearings from the page's center axis, rather than the uniform flush left margins of *A Dark Octave.* Her acknowledged metrical debt to Creeley's compact, almost gasping, lines is often evident in these pages, and the gesture of placing the first-person pronoun at the beginning of a phrase but the end of a line (thus using syntax as counterpoint to lineation) makes the "self" in these poems seem about to catapult off the page. The subject, only conceptually decentered in *A Dark Octave,* has been formally decentered in *The Relaxed Abalone,* and while the overall effect of this decentering is more evocative of play than pathology, it is precisely the border between the two states that Waldrop is here exploring.

The half-decade from the appearance of her first chapbook in 1967 to the publication of her first major collection, *The Aggressive Ways of the Casual Stranger,* brought out by Random House in 1972, encompasses a great deal of activity on Waldrop's part. Her translation of Peter Weiss's novel *Bodies and Shadows* was published by Delacorte in 1969 (a scholarly article on the same author's daring parable of reason and revolution, *Marat/Sade: A Ritual of the Intellect,* 1965, appeared the following year). A third chapbook, *Spring Is a Season and Nothing Else,* was brought out in an exquisitely printed edition by Perishable Press in 1970, a year that also saw a collection of print experiments, *Camp Printing,* appear from Burning Deck. In 1971 Mouton published Waldrop's doctoral dissertation, *Against Language?,* a stunningly lucid tract exploring "dissatisfaction with language" as a theme in modern and avant-garde poetry. Already working from the Wittgensteinian standpoint that would prove so generative in creative texts such as *The Reproduction of Profiles* and *Lawn of Excluded Middle,* Waldrop's critical acumen and sheer command of the tradition of experimental writing in German, French, and English make this a work still well worth consulting twenty-five years after its publication.

Perhaps the single most important event of this vibrant period of Waldrop's early career came not in the United States but in Europe, where she and her husband were spending the year 1970–1971 – she on an Alexander von Humboldt grant and her husband on an Amy Lowell traveling fellowship. (A long-standing joke about the "mystical marriage" of Humboldt and Lowell dates back to this conjunction of funding sources and bears literary fruit, so to speak, in the 1990 "collage epic," as critic Marjorie Perloff called it, *A Form / of Taking / It All.*) The couple spent the bulk of this year in Paris, and it was while in that city that they met Claude Royet-Journoud (editor since 1963 of the pathbreaking small magazine *Siécle à mains,* titled after a line from Arthur Rimbaud's *Season in Hell,* 1873). They also met Anne-Marie Albiach. Albiach and Royet-Journoud were two other writers making a life together and sharing with the Waldrops interests ranging from the work of the Egyptian Jewish writer living in exile in Paris, Edmond Jabès, to American Objectivist poets George Oppen and Louis Zukofsky (Albiach had in fact just translated into French the first half of Zukofsky's *"A"-9,* notorious for the extreme precision of its internal rhymes). In a published interview with Ed Foster, Waldrop recalls the effect of this meeting:

> [S]uddenly there were two people my age – actually a bit younger – who were talking about poetry in a very technical, but also "lived," way. Aside from Keith, they were the first people I found I could really talk with in a way that stimulated thinking and writing. They are maybe even more crucial than Jabès to what happened to Burning Deck and also Keith's writing and my own.

The friendship had major ramifications for Waldrop, particularly in that it was through Royet-Journoud that she came to meet Jabès and resume a project that she had just shortly before considered abandoning: the translation into English of Jabès's work that first resulted in *The Book of Questions* (1976–1984).

Waldrop's volume *The Aggressive Ways of the Casual Stranger* appeared in 1972, brought out not by one of the independent presses that proved so important to her later career, but by the commercial press Random House. A sentence early in Gertrude Stein's *Autobiography of Alice B. Toklas* (1933) provided Waldrop with her title: "Miss Stein used always to tell this story when the casual stranger in the aggressive way of the casual stranger said, looking at this [Matisse] painting, and what is that supposed to represent." Waldrop's identification was not so much with the philistinic response to nonrepresentational (or abstract) modern art, the response recorded and ridiculed here by Stein. Rather, it was to the sense of being an interloper or intruder, transgressing codes of which one is not even aware, that Waldrop meant to refer with her title. She elaborates: "I felt aggressive being a German writing poems in English, I felt I was one of those aggressive, casual strangers."

The volume runs to slightly more than ninety pages and is divided into three sections. The first and second sections, composed of seventeen and fifteen poems respectively, gather together poems that first appeared in *A Dark Octave* and *Spring Is a Season and Nothing Else,* along with some previously uncollected work from periodicals. The sequence "As If We Didn't Have To Talk," in thirty-eight numbered sections, makes up the third part of the volume and gives the clearest indication of Waldrop's emerging formal and conceptual concerns. It is interesting to note that many of the poems in the first two sections appear in revised form. In preparing them for book publication, Waldrop eliminated similes, cut specific lines, and rearranged certain of her line breaks. In no case did she add to the poems. Indeed, her criteria appear to have been condensation and the creation of more startling and direct juxtapositions. What helped Waldrop take to heart Pound's definition "dichtung = condensare" (later the title for a Burning Deck book series devoted to German authors in English translation) was her experience printing books for Burning Deck. In a 1994 forum on gender and editing, she recalled:

> In a tangible way, it is printing, even more than editing, that has affected my own writing. Printing letterpress (especially setting poems by hand, as we did in the beginning) is so slow a process that I became extremely aware of any unnecessary "fat." It has helped make my poems leaner.

Clustering around themes of gendered identity, the body, domestic space, family history, and the immigrant experience (and the ambivalent feelings it engenders), the poems in the first two sections of *Aggressive Ways* share a common question: namely, how to have a place without locking oneself into an oppressive condition of stasis. It is not surprising, then, that "home" is a persistent and yet conflicted trope in poems such as "Morning Has No House," "Linear," "Cleaning," "Between," and "Confession to Settle a Curse." Consider, for example, the poem "Cleaning," which takes the innocuous (but clearly gendered) activity of straightening up as a point of departure for questioning one's relation to matter more generally: "I'm careful / make sure I miss the corners / just coax it / into a mere pretense of / clean lines to reassure us / this world is ours." "Single Vision" evokes the complicity between stasis and signification: "sedentary / the world makes sense / all lines converge / take their position without / gap." For Waldrop, who would summarize her poetic method in *Lawn of Excluded Middle* as "gap gardening," the absence of a gap is tantamount to the absence of transformative possibility, a condition she associates in "Confession to Settle a Curse" with the "good German household" in which she grew up, where "wardrobes dressers sideboards / bookcases cupboards chests bureaus / desks trunks caskets coffers" all ominously came equipped "with lock / and key." Reviewer Janet Bloom, writing in the Fall/Winter 1972 issue of *Parnassus,* pointed to "Confession to Settle a Curse" and "Remembering Father's Death" as the volume's most powerful poems, calling them "very deliberate and unflinching . . . expression[s] of major horrors."

While mainstream reviewers tended to focus attention on the discrete, relatively conventional free verse poems of the volume's opening two sections, it is the long sequence composing the final third of the book that represents a genuine and profound breakthrough for Waldrop, looking forward as it does to two other important works, *Nothing Has Changed* (1981) and *The Road Is Everywhere or Stop This Body* (1978). "As If We Didn't Have To Talk" belongs to the exciting period of intellectual community that she and Keith experienced with Royet-Journoud and Albiach in Paris, and it is almost as though the generative energy of that encounter propelled Waldrop beyond the confines of the individual poem as primary compositional unit. What is certain is that, henceforth, Waldrop's most characteristic works were of a different scale: "Ever since 'As If,' I've hardly written single poems," Waldrop has said; "I have some, but I never do anything with them. I seem to want a larger space, larger structures." If her experiments with visual effects in books such as *Camp Printing* represented one way of pressing beyond the conventions of the expressive lyric (in that case, by redistributing the ratio between the graphic and semantic registers of reading), the serial form first used by Waldop in "As If" permits the author to experiment with duration, with the quasi-musical effects of variation over time, and with structural repetitions as significant compositional elements. And if the term "serial" also evokes a certain linearity, it must be said that the structure of works such as "As If" can as easily be conceived as a ring of elements equidistant from a common central point (which in Waldrop's work must be considered empty) as it can a strict procession of points along a straight line.

In addition to the conspicuous shift in scale, "As If We Didn't Have To Talk" introduces a second formal innovation, the "reversibility" of certain phrasal clusters within a given unit of the sequence. This means that a given word or phrase can simultaneously conclude one utterance and initiate another or function as object of one verb and subject of the next. Witness the "tumbling" of grammatical case and syntactic function in the eleventh section of the series: "In order not to / disperse / I think each movement of / my hand / turns / the page / the interval has all the rights." The themes of this passage are among those most central to Waldrop's work: the "interval," the struggle to bring a flexible order to experience without foreclosing the dimension of the unexpected or surprise, the act of reading, the significance of the minuscule but consummately human gesture. These themes converge at "my hand," at once the hand of the writer ("I think each movement of / my hand") and that of a reader ("my hand / turns / the page"). In the first eight lines of the twenty-fourth section, similar themes are evident: the interval is now conceived of as a "kernel" in the periphery of which that most surprising occurrence, the advent of another human, is apt to take place: "A flexible periphery / around / a kernel / might grow / tentative traces / take body / 'you' / let me touch you." The key transitions here are between the fourth and fifth lines, where "tentative traces" are what "might grow," and the fifth and sixth lines, where the same traces are shown taking body.

This use of reversible words and grammatical units culminates in the book-length sequence *The Road Is Everywhere or Stop This Body*. Beyond sharing a central device, "As If" and *The Road Is Everywhere* also share a benefactor, Eleanor Bender. In her capacity as editor of the magazine *Open Places* (in which Waldrop had published regularly since 1966), Bender had been instrumental in bringing the manuscript of *The Aggressive Ways of the Casual Stranger* to the attention of Random House editor Nan Talese. It was Bender who published *The Road* as the fifth in a series of Open Places books.

Although written in the mid 1970s and published in 1978, *The Road Is Everywhere* has its roots in the years 1968 to 1970, when Waldrop was making the commute from Providence to Wesleyan University, two hours away in Middletown, Connecticut. The seemingly endless hours spent in her Corvair gave Waldrop ample opportunity to study road signs (which are directly incorporated as textual elements in *The Road*) and to steep such abstract concepts as "circulation" and "traffic" (dictionary definitions of these words act as epigraphs for *The Road*) in concrete experience. In this eighty-poem sequence Waldrop explores concerns also raised in Emma Goldman's 1910 analysis of the "traffic in women" (and see also Gayle Rubin's 1976 article of that name) and the literal fact of being a "woman in traffic."

If Waldrop's first chapbook, *A Dark Octave,* appears in retrospect to be uncharacteristically static, it is in great part because of the sheer velocity — a full decade later — of the writing in *The Road Is Everywhere or Stop This Body,* which clocks in at "200 miles of nerve per hour with the guardrails down." And while speed (and its psychological complement, vertigo) remains of crucial thematic importance in almost all of Waldrop's subsequent work, only *A*

Form / Of Taking / It All in 1990 can rival the sheer acceleration this work achieves at a formal level. "Speeding down the highway doesn't allow you subordinate clauses" is how Waldrop herself characterized the relation between her subject matter and the form her text took. This acceleration is the result of the device discussed earlier of fusing subject and object clauses, which becomes ubiquitous in *The Road*. Waldrop discusses the significance of this move in "Alarms and Excursions":

> [T]he constant flip-over of object into subject in *The Road* (e.g. "the dawn" in "toward the always dangerous next / dawn bleeds its sequence / of ready signs") addresses [a feminist theme], though not as theme. On the surface I talk about our car culture – though also about sexual relations. But indirectly, on the level of grammar, this technique attacks a rigid subject-object relation by practicing reversibility. And since the woman as "love-object" is a prevalent archetype in our culture I think this technique has definite feminist/political implications. Of course, this is hindsight. Consciously, I was working on a formal problem, on eroding the boundaries of the sentence, sliding sentences into one another. This has been my experience: on form you have to work consciously, whereas your concerns and obsessions surface all by themselves.

Critic Nigel Wheale points out in the 1981 *Poetry Review* (no. 4) that the absence of punctuation and capitalization also contributes to the book's attainment of "maximum fluency of signal interchange." Praising the skill with which Waldrop maps the points of overlap and contradiction between "traffic system," "language system," and the "network of the body and its relations," Wheale concludes that Waldrop's "interpenetrated text immediately allows its author much more complex and novel statements than poems that are predicated on tight metaphorical observance, issued from the censorious 'I.'" Other reviewers were impressed by Waldrop's ability "to fuse movement with immediate sensation," as James Naiden put it in *The Minneapolis Tribune*. And Bruce Andrews, writing in *L=A=N=G=U=A=G=E,* the magazine he coedited with Charles Bernstein between 1978 and 1982, praised Waldrop's "ability to frame and reframe the flows around us, and the explosions which fracture the present. Body becomes its own flow; the person is a matrix of those flows & exchanges & messages. Person is a communicative system, a traffic."

With *The Road Is Everywhere* Waldrop took to its limit the form she had developed in the final third of *The Aggressive Ways of the Casual Stranger*. (Another work, the thirty-five-poem sequence *Nothing Has Changed* – not published in book form until 1981 but which appeared in its entirety in the periodical edited by Paul Auster, Mitch Sisskind, and Lydia Davis, *Living Hand*, in 1975 – also belongs to this period of Waldrop's work.) Recalling the genesis of her next project, *The Reproduction of Profiles*, published by New Directions in 1987 and among the most critically acclaimed of her works, Waldrop says: "While I had loved that speed, by the time [*The Road*] was done, I was a little bored with it. I thought subordinate clauses might be nice. There was a sense that I had cut that out so much that it was time to go in the other direction, to start meandering. That is partly what made me turn to prose. Also my lines had gotten rather painfully short."

Waldrop speaks of a "turn" to prose, but it is perhaps still more accurate to speak of several such turns: to the novel with *The Hanky of Pippin's Daughter;* to the serial prose poem with *The Reproduction of Profiles* and *Lawn of Excluded Middle;* to the collage with *Differences for Four Hands* and *A Form / Of Taking / It All;* to the procedural text in *Shorter American Memory* (1988). What is clear is that her work in the sentence, especially in the expertly transmuted propositional sentences of *Reproduction* and *Lawn,* represents a major contribution to the American prose poem at a time when that form has been significantly advanced by writers such as John Ashbery, Creeley, Ron Silliman, Hejinian, and Barrett Watten. While Waldrop can in no way be said to have "abandoned" the poetic line – volumes such as *When They Have Senses, Streets Enough to Welcome Snow* (1986), and *Peculiar Motions* (1990), as well as the chapbook *Morning's Intelligence* (a moving contemporary take on the aubade published in 1986), are not only contemporary with the work in prose, but they also represent a variety of innovative solutions to the problem of the poetic line – it is nevertheless true that she devoted increasing attention to the sentence and especially to the project of bursting what she calls in "Alarms and Excursions" the "closure of the propositional sentence."

The title sequence of *Reproduction* is comprised of five sections of between five and eleven prose poems each. The titles of these sections – "Facts," "Thinkable Pictures," "Feverish Propositions," "If Words Are Signs," and "Successive Applications" – allude to key concepts from the philosopher Ludwig Wittgenstein's *Tractatus Logico-Philosophicus* (1921), a work renowned for its almost uncanny clarity. But Wittgenstein is not Waldrop's only source in this sequence. The other primary source, from which Waldrop derives the narrative device of paired in-

Waldrop, 1993 (photograph by Ekko von Schwichow)

terlocutors (the "I" and "you" of the poem), for example, is the earliest known sustained piece of writing by Franz Kafka, *A Description of a Struggle* (circa 1902–1903), composed when Kafka was only twenty years old and left unpublished in his lifetime. The relation between the unnamed "I" and "you" is one of the most effective elements of Waldrop's text: it makes for a relationship of dizzying complexity, at once contentious and tender, antagonistic and erotic, intimate and impersonal. It is interesting to note in this regard that Waldrop recasts Kafka's central pair of characters, two men, into a cross-gendered pair for her poem; that is, she transposes the male social dynamic in Kafka's story (in which two men engage in an erotically charged rivalry over a female third party) into a dynamic of jealousy and the placing in jeopardy of a monogamous heterosexual couple.

Waldrop's ability to synthesize materials from sources that seem at first glance not merely diverse but diametrically opposed is the reason her sentences take on such a tense brilliance in *Reproduction*. A passage from "Successive Applications" provides an illustration of this ability:

> Suppose, I said patiently, I am given all the details at once. Then I could construct all possible stories out of them, and that would be the end of it. Annoyed you bit your native tongue. But I knew well enough that if one leaves things alone they get less clear by themselves.

This passage takes a sentence out of Wittgenstein ("Suppose that I am given *all* elementary propositions: then I can simply ask what propositions I can construct out of them") and fuses it with sentences from a key scene in the Kafka story, where the protagonist shouts to his interlocutor: "Out with your stories! I no longer want to hear scraps. Tell me everything, from beginning to end. I won't listen to less, I warn you. But I'm burning to hear the whole thing." By tracing two cognate structures, the logical (Wittgenstein) and the erotic (Kafka), to their common fulcrum in the subject's desire to know, Waldrop throws a communicating wire between two areas of human endeavor that are often kept at a tidy distance.

It is difficult to indicate by a single example the variety and multiformity of Waldrop's approach to the materials that she collages. Seldom does her process come to rest after splicing together this or that isolated phrase; it is more common for elements to influence compositional decisions at structural, narrative, and thematic levels, as well as basic lexical and phrasal ones. Both *Reproduction* and *Lawn* (which again turns to Wittgenstein, this time bringing the *Philosophical Investigations* into contact with Robert Musil's stunningly subtle reconstruction of an aging couple's love for one another in "The Perfecting of a Love," 1911, and a textbook of quantum physics from the 1930s) manage to subordinate their source materials, to manipulate and transform them to such a degree that a relatively unified linguistic object emerges from the process.

In texts such as *Shorter American Memory* and *A Key into the Language of America* Waldrop works transformations of a different order on her source materials. The former is a fairly unadulterated instance of procedural method. The twenty-two pieces in the chapbook are derived from sources originally brought together in Henry Beston's 1937 volume *American Memory*. Commenting on "Shorter American Memory of the American Character According to Santayana," Waldrop notes that she "set out to write an abecedarium, limiting myself to the words used in Santayana's essay 'The American Character.' The only freedom I allowed myself was the arrangement and articulation of the words beginning with the same letter." (Notice Waldrop's inconspicuous but brilliant conflation of "character" in the sense of personality with "character" in the sense of letter of the alphabet.) While the technique yields such wonderful lines as "All Americans are also ambiguous" and carries a surprisingly critical political charge ("Nature? Never. Numbers. Once otherwise. Potential potency, practical premonitions and prophesies: poor, perhaps progressive. Quick! Reforms realize a rich Rebecca. Same speed so successfully started stops sympathetic sense of slowly seething society. Studious self-confidence"), Waldrop concedes that "this kind of extreme formalism rarely works to my satisfaction. More often I use a pattern (e.g., the grammatical structure of a given text), but also let the words push and pull in their own direction. Since I make the rules, I also feel free to break them." Such a freer use of collage materials can be seen in *Differences for Four Hands,* an early prose poem that refunctions details from the legendary shared life of Clara and Robert Schumann while simultaneously entering into dialogue with Lyn Hejinian's *Gesualdo* (1978), a similarly motivated but differently textured piece about the sixteenth-century "musician and murderer" Don Carlo Gesualdo, whose late madrigals are often considered harbingers of modern-day experiments with dissonance.

A Key into the Language of America is Waldrop's most recent book and the one that offers perhaps the most complete synthesis of her major themes and formal techniques. Each of the book's chapters (which take their titles directly from Roger Williams's text) is composed of four elements. There are initial sections in prose (parallel to Williams's anthropological observations) that deal with "the clash of Indian and European cultures." Every chapter also has a "narrative section in italics, in the voice of a young woman, ambivalent about her sex and position among the conquerors." Interposed between the prose elements come word lists, partially drawn from Williams's dictionary entries, partially improvised by Waldrop. And each chapter concludes with a poem that remobilizes linguistic elements introduced in earlier sections, recontextualizing them with new material.

For example, in chapter 24, "Concerning Their Coyne," the poem reads: "legal and tender / a condition called / darling or **Netop** / which might as well purchase emotion / as yield interest in / I must explain my body / does not differ." Visually this poem is reminiscent of the major serial poems of the 1970s, "As If We Didn't Have To Talk," *Nothing Has Changed,* and *The Road is Everywhere.* Even some of the themes from those works can be seen to recur. The prose sections in italics, on the other hand, strongly recall the prose poems of *Reproduction* and *Lawn.* Here is a passage from the same chapter:

I learned that my face belonged to a covert system of exchange since the mirror showed me a landscape requiring diffidence, and only in nightmares could I find identity or denouement. At every street corner, I exaggerated my bad character in hopes of being contradicted, but only caused an epidemic of mothers covering their face while exposing private parts.

Making the synthesis still more profound is Waldrop's use of typographical devices as semantically weighted elements of the text, for example, different fonts and degrees of inking serving to differentiate the chapters internally and to signal the use of source material from Williams's document. This use of typography recalls the early work in visual poetry, such as *Letters from Rosmarie and Keith Waldrop* (1970); the collaboration with Nelson Howe, *Body Image* (1970); the experiments in print-design, *Camp Printing*; and the direct incorporation of signage in *The Road Is Everywhere.* Finally, Waldrop's juxtaposition of historical, philosophical, aesthetic, and anthropological materials in this work recalls *A Form /*

Of Taking / It All and, to a lesser extent, *Reproduction* and *Lawn*.

In Roger Williams, Waldrop finds a fascinatingly contradictory anchor point for a sustained meditation on her adopted country. Exile and translator, a figure of radical independence (to the point of being labeled a heretic), Williams was a man whose prototypical act of sympathetic anthropology slipped into an appalling and unsought complicity with the process of colonization. The ubiquity of Williams's image and name in modern-day Rhode Island, where she makes her home, leads Waldrop to say in the extremely useful introduction to the *Key:* "I live in Roger Williams's territory. By coincidence and marriage I share his initials. I share his ambivalence."

The generative nature of this ambivalence is something Geoffrey O'Brien captured well in a 20 January 1987 review of Waldrop's *Streets Enough to Welcome Snow* in *The Village Voice.* "She fuses the schemas of history and the randomness of the immediate," notes the critic, "in writing which for all its sensual richness has the tough-mindedness of an ongoing investigation. Above all this is writing so rooted in human presence that it can permit itself any displacement, any permutation, without losing its course.... However far inward or outward the poem goes, it never abandons the world." While O'Brien had some cause to lament that Waldrop was not better known at the time of his review, it was precisely in 1986 and 1987, when Station Hill brought out *Streets* and *The Hanky of Pippin's Daughter* and New Directions published *The Reproduction of Profiles,* that Waldrop's work was passing into wider recognition; and her reputation has continued to grow.

In 1994 the editors of two major anthologies of post-1945 American poetry, Paul Hoover in his Norton anthology *Postmodern American Poetry* and Douglas Messerli in his Sun and Moon volume, *From the Other Side of the Century: A New American Poetry, 1960–1990,* concurred in including sizable excerpts from Waldrop's work among their selections. On the international scene, *The Reproduction of Profiles,* already translated into French by Jacques Roubaud in 1991, is slated to appear soon in German translation. Waldrop's own work as a translator continues to garner recognition; she received a National Endowment for the Arts translator's fellowship in 1993 (having already received an NEA fellowship for her creative work in 1980) and was awarded the prestigious Harold Morton Landon Translation Award in 1994. Editor since 1986 of *Série d'Ecriture: Recent French Poetry in Translation* (initially published by the British press Spectacular Diseases and taken over by Burning Deck in 1992), Waldrop recently began editing a similar series of German writings in translation, *Dichtung =* . Burning Deck's thirtieth anniversary in 1991 was greeted by wide acclaim, including a special issue of the *O•blek* magazine dedicated to celebrating the Waldrops. Michael Palmer was among those who paid eloquent tribute to the press, noting that "without Burning Deck and a very few others, we experimental poets would, simply, not exist." Finally, looking ahead at Waldrop's own work, it appears that a third installment of the project begun in *Reproduction* and continued in *Lawn of Excluded Middle* is well under way, sections having appeared in *Abacus 80* (under the title "Cornered Stones / Split Infinites") and in other magazines. A substantial volume of selected writings has been slated for publication by Talisman House in 1997. In short, Waldrop's activity, always just shy of staggering, shows no sign of subsiding. This writer who declared her arrival in American letters by adopting the "aggressive ways of the casual stranger" is, in fact, a stranger no longer; she has become a familiar and welcome fact in the American literary tradition.

Interviews:

Ed Foster, "Interview with Rosmarie Waldrop," *Talisman,* 6 (1991): 27–39;

Wendy J. Burch, "Interview with Rosmarie Waldrop," *Poetry Flash,* 243 (1993): 1–13;

Jefferson Hansen, "Interview with Rosmarie Waldrop," *Poetics Briefs* (1993).

References:

Joseph M. Conte, *Unending Design: The Forms of Postmodern Poetry* (Ithaca, N.Y.: Cornell University Press, 1991), pp. 280–281;

O•blek, special issue on the Waldrops, 9 (Spring 1991);

Marjorie Perloff, "Towards a Wittgensteinian Poetics," *Contemporary Literature,* 33, no. 2 (1992): 191–213;

Joan Retallack, "Non-Euclidean Narrative Combustion, or What the Subtitles Can't Say," *Parnassus,* 14 (Summer 1988): 24–49;

Retallack, ":Re:Thinking:Literary:Feminism: (Three Essays onto Shaky Grounds)," in *Feminist Measures: Soundings in Poetry and Theory,* edited by Lynn Keller and Cristanne Miller (Ann Arbor: University of Michigan Press, 1994), pp. 345–377.

Richard Wilbur
(1 March 1921 -)

Richard J. Calhoun
Clemson University

BOOKS: *The Beautiful Changes and Other Poems* (New York: Reynal & Hitchcock, 1947);

Ceremony and Other Poems (New York: Harcourt, Brace, 1950);

Things of This World (New York: Harcourt, Brace, 1956);

Poems 1943–1956 (London: Faber & Faber, 1957);

Advice to a Prophet and Other Poems (New York: Harcourt, Brace & World, 1961; London: Faber & Faber, 1962);

Loudmouse (New York: Crowell-Collier / London: Collier-Macmillan, 1963);

Walking to Sleep, New Poems and Translations (New York: Harcourt, Brace & World, 1969);

Opposites (New York: Harcourt Brace Jovanovich, 1973);

The Mind-Reader (New York & London: Harcourt Brace Jovanovich, 1976);

Responses, Prose Pieces: 1953–1976 (New York: Harcourt Brace Jovanovich, 1976);

New and Collected Poems (New York: Harcourt Brace Jovanovich, 1988);

More Opposites (New York: Harcourt Brace Jovanovich, 1991);

A Game of Catch (San Diego: Harcourt, Brace, 1994);

Runaway Opposites (San Diego: Harcourt, Brace, 1995).

Collection: *The Poems of Richard Wilbur* (New York: Harcourt, Brace & World, 1963).

OTHER: "The Genie in the Bottle," in *Mid-Century American Poets*, edited by John Ciardi (New York: Twayne, 1950), pp. 1–13;

A Bestiary, compiled by Wilbur (New York: Spiral Press for Pantheon Books, 1955);

Richard Wilbur (photograph © 1986 Stathis Orphanos)

Candide: A Comic Operetta Based on Voltaire's Satire, lyrics by Wilbur (New York: Random House, 1957);

Poe: Complete Poems, edited, with an introduction, by Wilbur (New York: Dell, 1959);

Emily Dickinson: Three Views, by Wilbur, Archibald MacLeish, and Louise Bogan (Amherst, Mass.: Amherst College Press, 1960);

Paul Engle and Joseph Langland, eds., *Poet's Choice,* includes comment and poems by Wilbur (New York: Dial Press, 1962);

William Shakespeare, *Poems,* edited by Wilbur and Alfred Harbage (Baltimore: Penguin,

1966); revised and republished as *The Narrative Poems and Poems of Doubtful Authenticity* (Baltimore: Penguin, 1974);

Witter Bynner, *Selected Poems,* edited, with an introduction, by Wilbur (New York: Farrar, Straus & Giroux, 1978).

TRANSLATIONS: Molière, *The Misanthrope* (New York: Harcourt, Brace, 1955; London: Faber & Faber, 1958);

Molière, *Tartuffe* (New York: Harcourt, Brace & World, 1963; London: Faber & Faber, 1964);

Molière, *The Misanthrope and Tartuffe* (New York: Harcourt, Brace & World, 1965);

Molière, *The School for Wives* (New York: Harcourt Brace Jovanovich, 1972);

Molière, *The Learned Ladies* (New York: Harcourt Brace Jovanovich, 1978);

Jean Racine, *Andromache* (New York: Harcourt Brace Jovanovich, 1982);

The Whale and Other Collected Translations (Brockport, N.Y.: Boa Editions, 1982);

Racine, *Phaedra: A Tragedy in Five Acts. 1677* (San Diego: Harcourt Brace Jovanovich, 1986);

Molière, *The School for Husbands: Comedy in Three Acts, 1661* (New York: Dramatists Play Service, 1991);

Molière, *The Imaginary Cuckold, or, Sganarelle* (New York: Dramatists Play Service, 1993);

Molière, *Amphitryon* (New York: Harcourt, Brace, 1995).

SELECTED PERIODICAL PUBLICATIONS – UNCOLLECTED: "The Poetry of Witter Bynner," *American Poetry Review,* 6 (November-December 1977): 3-8;

"Elizabeth Bishop," *Ploughshares,* 7 (1980): 1-14;

"Poe and the Art of Suggestion," *University of Mississippi Studies in English,* 3 (1982): 1-13;

"Ash Wednesday," *Yale Review,* 78 (Winter 1989): 215-217;

"The Persistence of Riddles," *Yale Review,* 78 (Spring 1989): 333-351.

Richard Wilbur has always been recognized as a major literary talent and as an important man of letters – poet, critic, translator, editor – but he has never quite been ranked as one of the two or three best contemporary American poets. Early in his career he was overshadowed as a poet by Robert Lowell, who won the Pulitzer Prize for *Lord Weary's Castle* in 1947 (the year Wilbur's first book of poems, *The Beautiful Changes and Other Poems,* was published) and whose *Life Studies* (1959) was given principal credit for important new directions in poetry that Wilbur chose not to take. In the 1960s comparisons between Lowell and Wilbur as important new poets became comparisons between Lowell and James Dickey as the country's most important poets. Since the 1970s more critical attention has been given to such poets as John Ashbery, A. R. Ammons, James Wright, W. S. Merwin, and James Merrill than to Wilbur.

For more than four decades Wilbur's poetry has remained much as it has always been – skilled, sophisticated, witty, and impersonal. In 1949 when Philip Rahv in *Image and Idea* divided American writers into two camps – "Palefaces," elegant and controlled, and "Redskins," intense and spontaneous – Richard Wilbur was clearly a "Paleface." After Lowell made his break in 1959 with modernist impersonality in poetry, he revised Rahv's distinction in his National Book Award comments by specifying American poets as either "cooked" or "raw." Wilbur's "marvelously expert" poetry was undeniably one of the choice examples of "cooked" poetry. In *Waiting for the End* (1964), at a time when poetic styles were moving away from impersonality, Leslie A. Fiedler, one of the advocates of the reemergence of the "I" at the center of the poem and of a neo-Whitmanesque rejection of objectivity, found the influence of T. S. Eliot's formalistic theories especially strong on Wilbur: "There is no personal source anywhere, as there is no passion and no insanity; the insistent 'I,' the asserting of sex, and the flaunting of madness considered apparently in equally bad taste."

Wilbur has seldom likened his poetry to that of his contemporaries. Instead, in "On My Own Work," an essay collected in *Responses, Prose Pieces: 1953-1976* (1976), he described his art as "a public quarrel with the aesthetics of E. A. Poe," a writer on whom he has written some significant literary criticism. In Wilbur's view, Poe believed that the imagination must utterly repudiate the things of "this diseased earth." In contrast, Wilbur contends it is within the province of poems to make some order in the world while not allowing the reader to forget that there is a reality of things. Poets are not philosophers: "What poetry does with ideas is to redeem them from abstraction and submerge them in sensibility." Consequently, Wilbur's main concern is to maintain a difficult balance between the intellectual and the emotive, between an appreciation of the particulars of the world and their spiritual essence. If he is explicit in his prose about his quarrel with Poe, it might also be said that he had an implicit quarrel with the "raw" poetry in Donald Allen's

New American Poetry 1945–60, an anthology recognized in the 1960s as a manifesto against the "academy," and also with the extremely personal, seemingly confessional poetry of Lowell, W. D. Snodgrass, and Anne Sexton. Wilbur as a poet clearly accepts the modernist doctrine of impersonality and does not advertise his personal life in his poetry. "I vote for obliquity and distancing in the use of one's own life, because I am a bit reserved and because I think these produce a more honest and usable poetry," he commented in a 1967 questionnaire in *Conversations with Richard Wilbur* (1990).

Richard Purdy Wilbur was born in New York City, one of two children of Lawrence L. and Helen Purdy Wilbur. His father was a portrait painter. When Wilbur was two years old, the family moved into a pre–Revolutionary War stone house in North Caldwell, New Jersey. Although not far from New York City, he and his brother, Lawrence, grew up in rural surroundings, which, Wilbur later speculated, led to his love of nature.

Wilbur showed an early interest in writing, which he has attributed to his mother's family because her father was an editor of the *Baltimore Sun* and her grandfather was both an editor and a publisher of small papers aligned with the Democratic Party. At Montclair High School, from which he graduated in 1938, Wilbur wrote editorials for the school newspaper. At Amherst College he was editor of the campus newspaper, the *Amherst Student.* He also contributed stories and poems to the Amherst student magazine, the *Touchstone,* and considered a career in journalism.

Immediately after his college graduation in June 1942, Wilbur married Mary Charlotte Hayes Ward of Boston, an alumna of Smith College. Having joined the Enlisted Reserve Corps in 1942, he went on active duty in the army in 1943 in the midst of World War II. He served with the Thirty-sixth "Texas" Division in Italy at Monte Cassino and Anzio and then in Germany along the Siegfried Line. It was during the war that he began writing poems, intending, as he said in a 1964 interview with *The Amherst Literary Magazine* (borrowing Robert Frost's phrase), "a momentary stay against confusion" in a time of world disorder. When the war ended he found himself with a drawer full of poems, only one of which had been published.

Wilbur went to Harvard for graduate work in English to become a college teacher. As he recalled in his 1964 Amherst interview Wilbur decided to submit additional poems for publication only after a French friend read his manuscripts, "kissed me on both cheeks and said, 'you're a poet.'" In 1947, the

Cover for Wilbur's first book, in which many of the poems celebrate the power of beauty to create a sense of order in a chaotic world

year he received his A.M. from Harvard, his first volume of poems, *The Beautiful Changes and Other Poems,* was published.

The Beautiful Changes contains the largest number of poems (forty-two) and the fewest number of translations (three) of any of his collections. Although he began writing his poetry to relieve boredom while he was in the army, there are actually only seven war poems; and they are more poetic exercises on how to face the problems of disorder and destruction than laments over the losses occasioned by war, as in the traditions of the World War I British poet Wilfred Owen and the World War II American poet Randall Jarrell.

The first of Wilbur's war poems, "Tywater," presents the paradox of the violence illustrated in a Texas corporal's skill in killing the enemy – "The violent, neat, and practiced skill / Was all he loved and all he learned" – contrasted with the quietness of his death – "When he was hit, his body turned / To clumsy dirt before it fell." The compassion of

Jarrell's war poetry is clearly missing. Instead, there is an ironic detachment somewhat like John Crowe Ransom's but without the meticulous characterization that distinguishes Ransom's best poems:

> And what to say of him, God knows.
> Such violence. And such repose.

Another war poem, "First Snow in Alsace," suggests the theme implied by the title of the volume, *The Beautiful Changes*. The beautiful can change man even in times of duress. War is horrible because man permits it in spite of such simple childlike pleasures as a night sentry on being "the first to see the snow." "On the Eyes of an SS Officer" is a poetic exercise on the extremes of fanaticism. Wilbur compares the explorer Roald Amundsen, a victim of the northern ice that he desired to conquer, and a "Bombay saint," blinded by staring at the southern sun, with an SS officer, a villain of the Holocaust. The SS officer in his fanaticism combines what is evident in the eyes of the first two fanatics, ice and fire, for his eyes are "iced or ashen." The persona stays detached and does not explicitly condemn this terrible kind of fanaticism. The poem ends a bit tamely with his request to "my makeshift God of this / My opulent bric-a-brac earth to damn his eyes."

If there is a prevailing theme in Wilbur's first volume, it is how the power of the beautiful to change can be used as a buttress against disorder. The initial poem, "Cicadas," suggests the necessity for and the beauty of mystery in nature. The song of the *cigales* (better known as the cicada) can change those who hear it, but the reason for the song is beyond the scientist's analytical abilities to explain. It is spontaneous, gratuitous, and consequently a mystery to be appreciated as an aesthetic experience and described by a poet in a spirit of celebration.

"Water Walker" postulates an analogy between man and the caddis flies, or "water walkers," which can live successfully in two elements, air and water. A human equivalent would be the two lives of Saint Paul, described as "Paulsaul." He serves as an example of a "water walker," a person who was converted from service in the material world to service in the spiritual but who remained capable of living in both. The speaker in this poem desires a similar balance between two worlds, material and spiritual; but he is kept from transcendence, like the larva of the caddis held in the cocoon, by the fear that he might be unable to return to the material world.

In his first book imagination is a creative force necessary to the poet, but Wilbur also touches on an important theme developed more thoroughly in his later poetry, the danger that the imagination may lead to actions based entirely on illusions. His interpretation of Eugène Delacroix's painting, the subject of the poem "The Giaour and the Pacha," seems to be that in his moment of victory the giaour realizes that by killing his enemy he will lose his main purpose in life, which has been based on a single desire that proves valueless and illusory.

Another poem, "Objects," stresses what is to become a dominant theme for Wilbur, the need for contact with the physical world. Unlike the gulls in the poem, the poet cannot be guided by instincts or imagination alone. His imagination requires something more tangible, physical objects from the real world. The poet must be like the Dutch realist painter Pieter de Hooch, who needed real objects for his "devout intransitive eye" to imagine the unreal. It is only through being involved in the real world that the "Cheshire smile" of his imagination sets him "fearfully free." The poet, like the painter, must appreciate the "true textures" of this world before he can imagine their fading away.

One of the best lyrics in the collection is "My Father Paints the Summer." It has an autobiographical basis because Wilbur's father was a painter, but it is not a personal poem. The lyric develops the second meaning implied by the title *The Beautiful Changes* — the existence of change, mutability. It praises the power of the artist to retain a heightened vision in a world of mutability. The last stanza begins with the kind of simple, graceful line that is to become characteristic of Wilbur at his best: "Caught Summer is always an imagined time." Again the concern is balance in the relationship of the imagination and the particulars, the physical things of this world. The imagination needs the particulars of a summer season, but the artist needs his imagination for transcendence of time, "to reach past rain and find / Riding the palest days / Its perfect blaze."

The title poem of the volume is also the concluding poem and serves at this stage of Wilbur's poetic career as an example of his growing distrust of Poe-like romantic escapes into illusion and of his preference for a firm grasp of reality enhanced by the imagination. In "The Beautiful Changes" Wilbur gives four examples of how the beautiful can change: the effect of Queen Anne's lace on a fall meadow, the change brought about by the poet's love, a chameleon's change in order to blend in with the green of the forest, and the special beauty that a

Cover for Wilbur's 1956 book, for which he received a National Book Award and a Pulitzer Prize

mantis, resting on a green leaf, has for him. The beautiful changes itself to harmonize with its environment, but it also alters the objects that surround it. The ultimate change described is the total effect of the changes of nature on the beholder, worded in Wilbur's most polished lyric manner:

> it turns
> Dry grass to a lake, as the slightest shade of you
> Valleys my mind in fabulous blue Lucernes.

Wilbur's first volume was generally well received by the reviewers, and it was evident that a new poet of considerable talent had appeared on the postwar scene. Many of his first poems had a common motive, the desire to stress the importance of finding order in a world where war had served as a reminder of disorder and destruction. There were also the first versions of what was to become a recurring theme: the importance of a balance between reality and dream, of things of this world enhanced by imagination.

Wilbur spent three years between the publication of his first volume of poetry in 1947 and the appearance of his second in 1950 as a member of the Society of Fellows at Harvard, working on studies of the dandy and Poe that he never completed. What he did complete, though, was *Ceremony and Other Poems* (1950), continuing his concern with the need for a delicate balance between the material and the spiritual, the real and the ideal. In finding order in a world of disorder, poetry as celebration of nature is a "Ceremony," something aesthetically and humanly necessary. The concept of mutability, secondary in his first volume, is now primary, leading to a consideration of death, both as the ultimate threat of disorder and chaos and as motivation for creating order in the human realm. One of the poems concerned with facing death has come to be among Wilbur's most frequently anthologized poems, "The Death of a Toad." Wilbur finds in the toad a symbol for primal life energies accidentally and absurdly castrated by a tool of modern man, a power mower. The toad patiently and silently awaits his death with his "wide and antique eyes" observing this world that has cost him his heart's blood. His antiquity mocks a modern world that is already in decline.

"Year's-End," another poem on the threat of death, even more clearly contrasts the death of natural things, in their readiness to accept it, and the incompleteness and discord that death brings in the human realm. A dog that "slept the deeper as the ashes rose" is contrasted with "the loose unready eyes / of men expecting yet another sun / To do the shapely thing they had not done." This poem demonstrates Wilbur's skill in describing objects but also reveals his sometimes functional, sometimes not, desire to pun. Some reviewers found the first line of the poem to be Wilbur close to his worst: "Now winter downs the dying of the year." In contrast, his description of winter is Wilbur near his best:

> I've known the wind by water banks to shake
> The late leaves down, which frozen where they fell
> And held in ice as dancers in a spell
> Fluttered all winter long into a lake.

"Lament" is a poem about death, about expressing regret that the particulars of the world, what is "visible and firm," must vanish. This time a pun is functional: "It is, I say, a most material loss." "Still, Citizen Sparrow" is one of Wilbur's best known poems and, along with "Beowulf," introduces a new and important theme: whether heroism is possible in a world of disorder. In "Beowulf" the stress is on the loneliness and isolation of the hero. In "Still, Citizen Sparrow," in contrast to the common citizens (the sparrows), the hero appears as "vulture," a creature the sparrows must learn to appreciate. The poem is tonally complex, beginning as an argument between Citizen Sparrow and the poet over a political leader as a vulture and ending with an argument for seeing the faults of leaders in a broader perspective because they perform essential services, accept the risks of action, and are capable of dominating existence. The "vulture" is regarded as heroic because he is capable of heroic action: he feeds on death, "mocks mutability," and "keeps nature new." Wilbur concludes: "all men are Noah's sons" in that they potentially have the abilities of the hero if they will take the risks.

Another poem, "Driftwood," illustrates what some of Wilbur's early reviewers saw as a possible influence of Marianne Moore: finding a symbol or emblem in something so unexpected that the choice seems whimsical. In this poem the driftwood becomes an emblem for survival with an identity. It has "long revolved / In the lathe of all the seas." It is isolated but has retained its "ingenerate grain."

In Wilbur's second volume, as in his first, the need for a balance between the real and the ideal that avoids illusions and escapism is a significant theme. In "Grasse: The Olive Trees" the town in its abundance exceeds the normal and symbolizes reaching beyond the usual limits of reality, the overabundance of the South, that can become enervating and illusionary:

> and all is full
> Of heat and juice and a heavy jammed excess.
>
> Whatever moves moves with the slow
>   complete
> Gestures of statuary.

Only the "unearthly pale" of the olive represents the other pole of the reality principle and "Teaches the South it is not paradise."

"La Rose des Vents" is the first dialogue poem for Wilbur, a dialogue between a lady and the poet in a format reminiscent of Wallace Stevens's "Sunday Morning." The lady argues for the sufficiency of accepting the reality of objects, while the poet desires symbols removed from reality. In Wilbur's version the lady has the last word:

> Forsake those roses
> Of the mind
> And tend the true,
> The mortal flower.

"'A World without Objects Is a Sensible Emptiness'" is a poem with perhaps the quintessential Wilbur title. Visions, illusions, and oases are the objects of quests for people in a wasteland world, but the questing spirit, "The tall camels of the spirit," must also have the necessary endurance to turn back to the things of this world as a resource:

> Turn, O turn
> From the fine sleights of the sand, from the long
>   empty oven
>   Where flames in flamings burn
>
> Back to the trees arrayed
>   In bursts of glare, to the halo-dialing run
> Of the country creeks, and the hills' bracken
>   tiaras made
>   Gold in the sunken sun[.]

Extravagant claims are made for visions that are firmly based on life. A supernova can be seen "burgeoning over the barn," and "Lampshine blurred in the steam of beasts" can be "light incarnate."

In *Ceremony* Wilbur exhibits greater versatility than is evident in his first book. He can now express his major themes in lighter poems, even in epigrams. The importance of a delicate balance between idealism and empiricism, speculation and skepticism, is concisely and wittily expressed in the two couplets of "Epistemology." Samuel Johnson is told to "Kick at the rock" in his rejection of Berkeleyan idealism, but the rock is also a reminder of the molecular mysteries within it: "But cloudy, cloudy is the stuff of stones." Man's occasional denials of the physical world he so desperately needs are mocked in the second couplet:

> We milk the cow of the world, and as we do
> We whisper in her ear, "You are not true."

With the appearance of his second book of poems, Wilbur was appointed an assistant professor of English at Harvard, where he remained until 1954, living in Lincoln, Massachusetts, with his wife and four children – Ellen Dickinson, Christopher Hayes, Nathan Lord, and Aaron Hammond. He spent the academic year of 1952–1953 in New Mexico on a Guggenheim Fellowship to write a poetic drama. When his attempts at a play did not work out to his satisfaction, he turned to translating Molière's *Le Misanthrope* instead, beginning his distinguished career as translator. A grant of $3,000, the Prix de Rome, permitted Wilbur to live at the American Academy in Rome in 1954. After his return to America his translation, *The Misanthrope* (1955), was published and performed at the Poets' Theatre in Cambridge, Massachusetts.

In 1954 Wilbur was appointed an associate professor of English at Wellesley College, where he taught until 1957. His third volume of poetry, *Things of This World,* was published in 1956. In his September 1956 review of the collection for *Poetry* magazine Donald Hall concluded: "The best poems Wilbur has yet written are in this volume." His judgment was confirmed, as the collection remains Wilbur's most honored book; it received the Edna St. Vincent Millay Memorial Award, the National Book Award, and the Pulitzer Prize. The same year the musical version of Voltaire's *Candide,* with lyrics by Wilbur, book by Lillian Hellman, and a score by Leonard Bernstein, was produced at the Martin Beck Theatre in New York City.

Three poems in *Things of This World* should certainly be ranked among Wilbur's best, "A Baroque Wall-Fountain in the Villa Sciarra," "Love Calls Us to the Things of This World," and "For the New Railway Station in Rome." The last two reveal the influence of his year spent in Rome on a Prix de Rome fellowship. As the title would suggest, there is even a greater stress on the importance of the use of the real in the poems in this volume. If the imagination does create a world independent of objects, it is made clear in "Love Calls Us to the Things of This World" that love always brings one back to the world of objects. Even nuns move away from pure vision back to the impure, "keeping their difficult balance."

It is not always the simpler forms that are the most inspiring. Wilbur remarked in the anthology *Poet's Choice* (1962) that "A Baroque Wall-Fountain in the Villa Sciarra" was based on his daily observation of a "charming sixteenth- or seventeenth-century fountain that appeared to me the very symbol or concretion of Pleasure." The elaborate baroque fountain is described as an artistic embodiment of the pleasure principle. Human aspiration may be more clearly seen in the simpler Maderna fountains, but the elaborate forms on the baroque fountain:

> They are at rest in fulness of desire
>   For what is given, they do not tire
> Of the smart of the sun, the pleasant water-douse
>
>   And riddled pool below,
> Reproving our disgust and our ennui
>   With humble insatiety.

It is indicative of Wilbur's penchant for impersonality that he ends the poem not by indicating the personal delight he feels in the fountain but by imagining what Saint Francis of Assisi might have seen in the fountain: "No trifle, but a shade of bliss."

The final poem in the volume is one of the best, "For the New Railway Station in Rome." The impressive new station becomes a symbol of how man's mind must continually work on things of this world for the imagination to have the power to recreate and to cope with disorder:

> "What is our praise or pride
> But to imagine excellence, and try to make it?
>   What does it say over the door of Heaven
>   But *homo fecit*?"

Donald Hill has said of Wilbur's early poetry that he has seemingly taken William Carlos Williams's slogan "No ideas but in things" and altered it to "No things but in ideas." Beginning with his third volume, *Things of This World,* Wilbur still recognizes the importance of the imagination, but his emphasis has

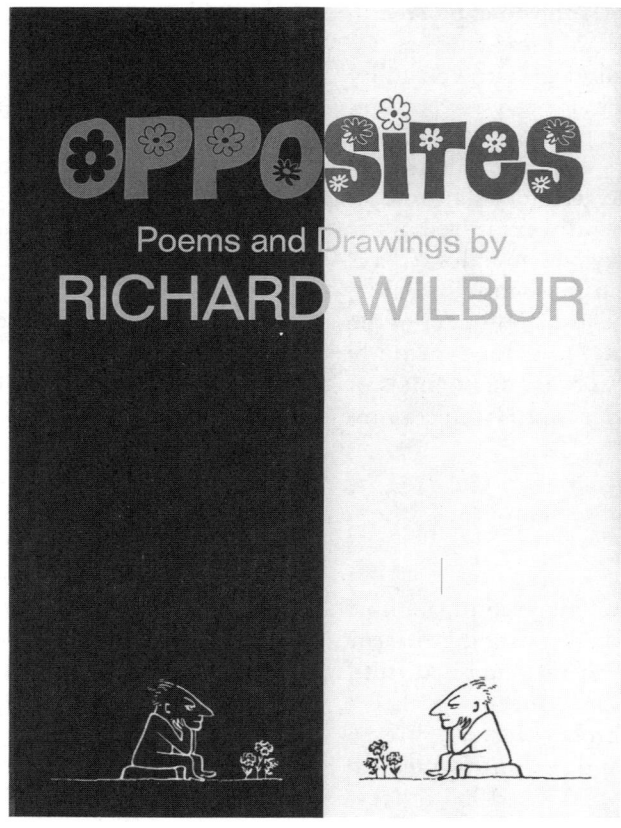

Cover for Wilbur's 1973 volume of light verse for children

clearly shifted toward Williams's concept in his stress on the need for things of this world, both for effective endurance in a world of death and disorder and for creativity.

In 1957 Wilbur began a twenty-year tenure as professor of English at Wesleyan University and as adviser for the Wesleyan Poetry Series. He also received a Ford Foundation grant in drama and worked with the Alley Theater in Houston. *Advice to a Prophet and Other Poems,* his fourth book of poetry, was published in 1961. It is a larger volume of poetry than *Things of This World,* with thirty-two poems, including four translations and a passage translated from Molière's *Tartuffe,* as well as "Pangloss's Song" from the comic-opera version of Voltaire's *Candide.* The collection received favorable comments from such critics as Babette Deutsch, Dudley Fitts, M. L. Rosenthal, William Meredith, and Reed Whittemore. But the praise for *Advice to a Prophet* was tempered by criticisms that it had an academic, privileged, even ivory-tower perspective. The title poem is vaguely topical, suggesting the threat of the ultimate atomic holocaust that became a near reality in October 1962 with the Cuban Missile Crisis. Even here Wilbur might be accused of aesthetic detachment: his poem is not humanistic in its concerns but aesthetic and phenomenological, envisioning a world without its familiar objects, without things rather than without people:

> Nor shall you scare us with talk of the death of the race.
> ....
> Ask us, ask us whether with the worldless rose
> Our hearts shall fail us[.]

Perhaps still showing the influence of Marianne Moore's passion for oddities, Wilbur stresses in this volume what the imagination can do with apparently mundane things. In "Junk" he suggests that intimations of the ideal can be found in the rubbish, the junk of the world, and in "Stop," in the grim everyday objects at a train stop. In "A Hole in the Floor" Wilbur even compares the potentials of his discoveries in the floor with those of a great archeologist: "As Schliemann stood when his shovel / Knocked on the crowns of Troy."

In "A Grasshopper" Wilbur adds to the poetic bestiary that he had collected in his volume *A Bestiary* (1955). He admires the grasshopper for having achieved a delicate balance between stasis in its pause on a chicory leaf and action in its springs from the leaf. Hall in his *Contemporary American Poetry* (1962) calls the poem "a minor masterpiece," but some reviewers believed that Wilbur seemed too content with "minor masterpieces," both in form and in subject matter. He showed an unwillingness to undertake major experiments in form or to introduce new and socially relevant subject matter at a time when that was becoming expected. To some reviewers and critics, he seemed a poet reluctant to take risks of any sort. In fairness, one must say that Wilbur does experiment with "new" lines in his poetry, such as his use of the Anglo-Saxon alliterative line in "Junk." But in comparison with what such poets as Lowell and John Berryman were then doing, the experimentation is comparatively minor.

Wilbur seemed almost to be writing his poems in a cultural and political vacuum. By the time of the publication of *Advice to a Prophet* the tremendous impact that Lowell had made in *Life Studies* by apparently confessing disorder in his own family life had been felt. Two years after *Life Studies* Wilbur opened his volume with what he intended to be a dramatic poem, "Two Voices in a Meadow," a dialogue between two objects from the world of the mundane, a milkweed and a stone. The drama in this poem and in the title poem, "Advice to a Prophet," seemed humanly insignificant compared to Lowell's more personal approach. Wilbur seemed to fail in his attempts to indicate more dramatically and more positively how order might be restored and what his personal "stays against confusion" are, much as Robinson Jeffers's attempt at a tragic poetry had failed before, because he seems too exclusively concerned with symbolic things rather than with people. Wilbur's message appears to be that when man becomes more familiar with the world's own change, he can deal with his own problems as something related to the reality of things. Wilbur calls those who do not respond to the things of this world, those who prefer their dreams and who move to illusions, "the Undead" – vampires.

In "Shame" Wilbur defines the kind of human behavior that disturbs him – irresoluteness, a failure to deal with reality. He attempts to provide positive examples of heroic behavior, but he fails to create convincing examples as Robert Lowell does with his symbol of the mother skunk, "refusing to scare," in "Skunk Hour." In Wilbur's dialogue poem "The Aspen and the Stream," the aspen is the positive heroic example because it seems to escape its existence by delving into flux, experience – symbolized by the dream – even if the result is only "a few more aspen-leaves."

It was eight years before Wilbur's fifth volume of poetry, *Walking to Sleep,* appeared in 1969. In the interim he published a children's book, *Loudmouse* (1963); his collected poems, *The Poems of Richard Wilbur* (1963); and his translation of Molière's *Tartuffe* (1963), which earned him an award as corecipient of the Bollingen Poetry Translation Prize. The Lincoln Center Repertory Theatre brought his translation of *Tartuffe* to the stage in New York City in 1964. *Walking to Sleep* is a slim collection, with fewer original poems (only twenty-two) and more translations (eleven) than in previous collections. What overall unity there is in the four sections of the volume is suggested by the title: these are poems on the subject of how to "walk" – symbolically, how to live before sleep and death.

As in "Junk," Wilbur experiments with the Anglo-Saxon alliterative line divided by a caesura. In "The Lilacs" the flowers are used as a symbol of the cycle of death and rebirth, the "pure power" of nature perhaps compensating for the "depth" of death. The poem concludes:

> These lacquered leaves
>   where the light paddles
> And the big blooms
>   buzzing among them
> Have kept their counsel,
>   conveying nothing
> Of their mortal message,
>   unless one should measure
> The depth and dumbness
>   of death's kingdom
> By the pure power
>   of this perfume.

A kind of balance between life and death may be seen if one can appreciate "the pure power" of life. "In the Field," the title poem of the first section, also suggests that the power in life may be sufficient to compensate for the ultimate disorder, death. Wilbur finds in the field "the heart's wish for life, . . . staking here / In the least field an endless claim." And he believes that the same principle is in man. It "is ourselves, and is the one / Unbounded thing we know."

Wilbur also believes that in man's desires lies the answer to his questions. "Running" is, like "In the Field," a longer poem than Wilbur usually writes. It is divided into three parts and describes the act of running at three different times in the poet's life. The poem is intended not only as an affirmative statement about human aspiration but also as an assertion of the ultimate meaning of human activities. Wilbur's running becomes a symbol of aspiration at different stages in life. What keeps man running? It *is* human aspiration:

> What is the thing which men will not surrender?
> It is what they have never had, I think,
> Or missed in its true season[.]

"Running" is by Wilbur's own admission one of his most personal poems. It also implies the middle-aged poet's belief that his own life is satisfying and worthwhile.

The title poem, "Walking to Sleep," begins with a discussion of going to sleep that soon becomes a meditation on how to live and a warning against a life of illusion. It is also an argument for accepting death without illusions by literally staring it down. This might be regarded as a climactic poem on a major thematic concern. What is recommended is once again a balance, a life in which reality and "strong dream" work together.

One of the few poems in the volume to be almost immediately anthologized, "Playboy" describes the imaginative response of an adolescent stockboy to the impact of a centerfold in *Playboy* showing a beautiful naked woman. "High on his stockroom ladder like a dunce," he examines "her body's grace," engrossed in "how the cunning picture holds her still / At just that smiling instant when her soul, / Grown sweetly faint, and swept beyond control, / Consents to his inexorable will."

Other poems are also atypical of Wilbur's usual themes. He even includes a protest poem addressed to President Lyndon Johnson; the occasion is not the Vietnam War but Johnson's refusing the official portrait painted by the artist Peter Hurd. The protest is more artistic than political. The poem makes a contrast between Johnson and the culture of Thomas Jefferson with his Rotunda and "Palestrina in his head." Although the poems were published in the midst of the Vietnam vortex, Wilbur is once again primarily concerned with maintaining "a difficult balance" between reality and the ideal as the way to personal fulfillment.

Wilbur's sixth volume of poetry, *The Mind-Reader* (1976), contains twenty-seven new poems (nine previously published in *The New Yorker*) and nine translations. The reviews were again mixed, with some reviewers praising his craftsmanship and defending him from what they regarded as unfair attacks on his conservatism as a poet; others found his new volume to be simply more of the same and lamented his not taking risks by seeking new directions. The translations provide new examples of Wilbur's superb ability to translate from the French and the Russian, especially the poems by Andrei Voznesensky.

There are new things in the volume, especially in Wilbur's clearly discernible movement toward simpler diction and more direct poems. Except for the title poem there are no long poems in this book. Wilbur seems to enjoy working with shorter poems, as in the six-line, three-couplet "To the Etruscan Poets," on the theme of mutability exemplified by the Etruscan poets, who "strove to leave some line of verse behind / . . . / Not reckoning that all could melt and go."

Some reviewers found "Cottage Street, 1953" to be provocative. It is an account of Wilbur's meeting a young Sylvia Plath and her mother at the home of his mother-in-law, Edna Ward. A contrast is made between Plath's destructive tendencies and Ward's power of endurance. A few reviewers read the poem as if it were a personal attack on Plath by a poet hostile to confessional poetry. The poem is undoubtedly intended as a variation on Wilbur's theme of a need for balance, which he later came to realize that Plath had always lacked. He opposes love as a principle of order to the "brilliant negative" of Plath in her life. What makes this poem exceptional is that Wilbur is dealing with real people characterized rather brilliantly:

> And Edna Ward shall die in fifteen years,
> After her eight-and-eighty summers of
> Such grace and courage as permit me no tears,
> The thin hand reaching out, the last word *love*,
>
> Outliving Sylvia, who, condemned to live,
> Shall study for a decade, as she must,
> To state at last her brilliant negative
> In poems free and helpless and unjust.

In this poem Wilbur deals with the human problem of survival and death without his usual detachment and with a directness his poems usually lack.

More representative of his usual type of poem is "A Black Birch in Winter." It could have appeared in any of Wilbur's first five volumes. A symbol (the black birch) is found for nature's ability to survive and grow to greater wisdom each year. Except for slightly simpler diction, the poem is a variation on a usual theme, and the conclusion seems a parody of the conclusion of Alfred Tennyson's "Ulysses":

> Old trees are doomed to annual rebirth,
> New wood, new life, new compass, greater girth,

And this is all their wisdom and their art –
To grow, stretch, crack, and not yet come apart.

One poem would seem on the surface to be atypical, Wilbur taking the unusual risk of involving his poetry in the political protest against the war in Vietnam. "For the Student Strikers" was written for the Wesleyan *Strike News* at the time of the Kent State shootings. Wilbur's support is not, however, for student protests but for their canvassing programs, house-to-house visits to discuss the student point of view about the war. Typically, he urges dialogue – order – instead of protests – disorder:

It is not yet time for the rock, the bullet, the blunt
Slogan that fuddles the mind toward force.
Let the new sound in our streets be the patient sound
Of your discourse.

There is an evident difference in emotional perspective, in dramatic intensity, and in contemporary relevance between Wilbur in this poem and Lowell in *Notebook 1967–68*.

Whereas Lowell, Anne Sexton, W. D. Snodgrass, Plath, and even James Dickey have told much about their families, until *The Mind-Reader* Wilbur did not mention his family. Two poems about his children mark a change. His son Christopher's wedding is described indirectly in "A Wedding Toast." But "The Writer" is one of Wilbur's most personal poems and perhaps one of his best. As a father and as a writer he empathizes with his daughter's attempts to write a story. He describes her creative struggles "In her room at the prow of the house," and he is reminded of another struggle that he saw before at the same window:

I remember the dazed starling
Which was trapped in that very room, two years ago;
How we stole in, lifted a sash

And retreated, not to affright it;
And how for a helpless hour, through the crack of the
   door,
We watched the sleek, wild, dark

And iridescent creature
Batter against the brilliance, drop like a glove
To the hard floor, or the desk-top.

Wilbur's slightly more personal approach is apparent in a few other poems. The engaging persona Wilbur creates in the title poem, "The Mind-Reader," helps that poem achieve more dramatic intensity than is apparent in much of his earlier work. He seems to be seeking even firmer and more affir-

Cover for the trade paperback printing of Wilbur's 1969 book, in which the title poem advises the reader that "what you project / Is what you perceive; what you perceive / With any passion, be it love or terror, / May take on whims and powers of its own"

mative statements of the need for order and responsibility; and his tone in these poems is more confident, as if he is assured that his own artistic life has been worthwhile, that he has himself maintained a balance between reality and imagination. Wilbur's perspective is concisely stated in "C-Minor," a poem about switching off "Beethoven at breakfast" to turn back to the reality of the day:

There is nothing to do with a day except to live it.
Let us have music again when the light dies
(Sullenly, or in glory) and we can give it
   Something to organize.

In 1977 Wilbur moved to Smith College, where he remained as writer-in-residence until his retirement in 1986. While continuing his translating of Molière's work, he also produced translations of John Racine's *Andromache* (1982) and *Phaedra* (1986).

In 1987 Wilbur was honored by an appointment as poetry consultant at the Library of Congress and poet laureate.

New and Collected Poems (1988) earned Wilbur the Pulitzer Prize for 1989. The new poems include twenty-six short lyrics and "On Freedom's Ground," the lyrics for a five-part cantata by William Schuman. This long poem was a joint project written to mark the refurbishing of the Statue of Liberty on its centennial in 1987. Wilbur may have had in mind memories of Robert Frost's impromptu reciting of "The Gift Outright" at the John F. Kennedy inauguration, for he offers a variation of Frost's theme that Americans have gradually become worthy of the land:

> We are immigrants still, who travel in time,
> Bound where the thought of America beckons;
> But we hold our course, and the wind is with us.

In several of the newly collected poems Wilbur creates a persona who ruminates on his life and achievement. He clearly has Frost in mind in "The Ride," an extension of "Stopping by Woods on a Snowy Evening" in which the journey of the rider and his horse continues through the night: "The horse beneath me seemed / To know what course to steer / Through the horror of snow I dream, / And so I had no fear." The poem seems a consummation of a life-journey of creating and drawing on intuitions and dreams that one must believe in or fall victim to the grief that comes from thinking "there was no horse at all." "Leaving" is an indictment of the comforts in modern life. The people at a garden party resemble the stone figures that border the scene. The question raised is whether or not knowledge of the future would have influenced the people's decisions in life:

> Filling our selves as sculpture
> Fills the stone.
> We had not played so surely,
> Had we known.

"For W. H. Auden" is a poem written earlier and published only when Wilbur thought it was finished. It is an impressive poem on memory's lost moments as much as a personal lament for Auden:

> Of all these noted in stride and detained in memory
> I now know better that they were going to die,
>
> Since you, who sustained the civil tongue
> In a scattering time, and were poet of all our cities,
> Have for all your clever difference quietly left us,
> As we might have known that you would, by that common door.

In "Lying" Wilbur begins by lightly invoking a "dead party," where a white lie "can do no harm" to one's reputation. The poem evolves more seriously as the speaker explores the nature of lying and reality, the imagination and illusionary truth: "What is it, after all, but something missed?" Wilbur offers John Milton's Satan as the

> arch-negator, sprung
> From Hell to probe with intellectual sight
> The cells and heavens of a given world
> Which he could take but as another prison[.]

He then turns the poem to the ordinary experiences of a summer's day metaphorically likened to all days: "It is a chant / Of the first springs, and it is tributary / To the great lies told with the eyes half-shut / That have the truth in view." The poem concludes by alluding to *The Song of Roland*, implying the superiority of the lie of the romance to the ordinary fact of history, as Roland "Was faithful unto death, and shamed the Devil."

Wilbur's long tenure in academia is still evident in some poems. "A Finished Man" is a portrait, perhaps wryly autobiographical, of a man who has completed his career and is being honored by the university. The enemies, friends, and colleagues who knew his fears and faults now either dead or fading in his memory, the honoree is nearly "finished" – "If the dead die, if he can but forget, If money talks, he may be perfect yet." "Icarium Mare" is clearly an academic poem with arcane references to the mythical figure Icarus, the Greek astronomer Aristarchus of Samos, and St. John the Divine's "geodic skull."

The short poems in the "New Poems" section often seem to be merely sketches, but there is always depth to a Wilbur surface. "Wyeth's Milk Cans" records the lucid simplicity of an N. C. Wyeth scene but at the same time raises doubts about the landscape's beauty. "Shad-Time" examines two events, the spawning of shad and the blooming of the shadblow tree along a river's banks, and raises the old question anew of how to make sense of nature's bounty and waste. The critic Bruce Michelson judges this poem to be proof that Wilbur could produce a postmodernist poem that goes beyond skillful play and raises uncomfortable questions about the self and the world.

Despite Wilbur's achievement as a poet and his many awards, including the gold medal for poetry from the American Academy and Institute of

Cover for the trade paperback printing of Wilbur's 1976 book, which for some readers suggested a move toward the onfessional mode

Arts and Letters in 1991, many critics would argue that he has not become the major poet he seemed destined to be when *Things of This World* was so celebrated. Even if this arguable judgment is accepted, Wilbur's poetry alone is not the measure of his significance as a man of letters. For a balanced view of his literary importance it should be acknowledged that he is a discerning critic and an accomplished translator of poetry and drama in verse. Wilbur's view of translating is unquestionably an extension of his poetry writing. Viewing translation as a craft, he has consistently set for himself the goal of authenticity in translating not just the language but the verse forms as well. The importance of including Wilbur's translations in an evaluation of his talents as a poet has been neatly summed up by Raymond Oliver: "His degree of accuracy is almost always very high and his technical skill as a poet is just about equal to that of the people he translates." Wilbur's versions of Molière's works not only read well as verse but have been staged with great success. He has followed success in comedy with highly regarded translations in the 1980s of two of Racine's tragedies.

Wilbur has also had considerable importance as a literary critic. One could contend that he has surpassed, with the possible exceptions of Randall Jarrell and Karl Shapiro, his contemporaries as a poet-critic. He has written perceptively on his poetic opposite, Edgar Allan Poe, and he has delivered a major essay on Emily Dickinson. He has edited the poems of Poe and coedited the poems of William Shakespeare. The sixteen reviews and critical essays collected in *Responses, Prose Pieces: 1953–1976* (1976) and the interviews and conversations in *Conversations with Richard Wilbur* show Wilbur's perception on other writers as well as on his own work. His insights into his own work compare in quality, if not quite in quantity, with James Dickey's attempts in *Self-Interviews* (1970) and *Sorties* (1971) to describe his own creativity.

Certainly, a trenchant defense of Wilbur as a poet is to be made on the grounds that many critics have overlooked the stylistic and tonal complexities of his poetry, much as the New Critic formalists had earlier failed to recognize complexities in Robert Frost, a poet Wilbur has always admired. Wilbur has evidenced a craftsman's interest in a wide variety of poetry – dramatic, lyric, meditation, and light verse. His wit, especially his skillful rhymes and the puns found even in his serious poetry, has not always been treated kindly by critics, but it has often captivated readers. He has been recognized by children's literature specialists for his volumes of light verse – *Loudmouse, Opposites* (1973), *More Opposites* (1991), and *Runaway Opposites* (1995) – all written with grace, wit, and humor.

In John Ciardi's *Mid-Century Poets* (1950) Wilbur identified what has remained his constant goal as a poet, whatever type of poem he has written: "The poem is an effort to articulate relationships not quite seen, to make or discover some pattern in the world. It is conflict with disorder." Wilbur's confrontation with disorder has led him to be satisfied with established patterns and traditional themes, old ways to solve old problems. Consistently a poet of affirmation, he has reacted against the two extremes of disorder: chaos and destruction on the one hand and illusions and escapism on the other. His response as both poet and humanist is to maintain a firm focus on reality as represented by objects, by the things of this world. As a poet he must be modestly heroic, see more, and range further than the ordinary citizen. Wilbur writes in "Objects," a poem from his first collection:

> I see afloat among the leaves, all calm and
> curled,
> The Cheshire smile which sets me fearfully free.

Nevertheless, the question raised earlier in Wilbur's career in regard to his development remains in the 1990s: Does his adherence to formalist principles preclude his consideration as a major poet during a postmodernist period in which poets were expected to respond to a changing social and literary landscape?

In *Wilbur's Poetry: Music in a Scattering Time* (1991) Michelson avoids reviving all the old arguments about formalism versus experimentation, closed versus open forms, and academic poetry versus postmodernism, from which Wilbur emerges as a reactionary, if not a heavy. Michelson goes instead directly to the poems to argue not only for evidence of the stylistic range and variety of Wilbur's artistry but also to affirm his sensitivity to the major moral and aesthetic crises of his times. As Lionel Trilling found in Frost, Michelson finds in Wilbur a darker side. He is to be redeemed as not only the acknowledged master of light verse but also of some less acknowledged dark, meditative poems. Michelson does not find Wilbur to be a "terrifying poet" as Trilling did Frost but rather reckons him "a serious artist for an anxious century." He identifies in many of the poems not just "safe creeds and certainties" but, significantly, a tone of "skeptical virtuosity" that has gone largely unrecognized.

If one is satisfied to judge Richard Wilbur in terms of his intentions, he has achieved them well. Nonetheless it is clear he has not been a poet for all decades. In the 1950s his view of poetic creation was compatible with that of the dominant critical view of his generation of emerging poets, the "rage for order view" of creativity promulgated by the formalistic New Criticism. By the 1960s formalism was no longer the dominant critical approach, and man's rage for order was balanced by an interest in man's rage for chaos. In the 1970s modernism had been supplanted by a neo-Romantic postmodernism. Critics discovered the virtues of political correctness by the 1980s and Wilbur seemed relatively lackluster as a poet who was neither politically correct nor notably incorrect.

What Wilbur's critics and his readers must not disregard is his mild irony, sophisticated wit, effective humor, and, as Michelson has appended, his seriousness. His craftsmanship and skill with words and traditional poetic forms should also be considered. Wilbur is a formalist who at his best manages to make formalism seem continually new. For many readers, his poetic art always was, and still is, sufficient.

Interviews:

William Butts, *Conversations with Richard Wilbur,* (Jackson: University Press of Mississippi, 1990).

Bibliographies:

John P. Field, *Richard Wilbur: A Bibliographical Checklist* (Kent, Ohio: Kent State University Press, 1971);

Frances Bixler, *Richard Wilbur: A Reference Guide* (New York: Macmillan, 1991);

Rodney Stenning Edgecombe, *A Reader's Guide to the Poetry of Richard Wilbur* (Tuscaloosa: University of Alabama Press, 1995).

References:

Louise Bogan, *Achievement in American Poetry* (Chicago: Regnery, 1951), pp. 133–134;

Robert Boyers, "On Richard Wilbur," *Salmagundi*, 12 (Spring 1970): 76–82;

James E. B. Breslin, "The New Rear Guard," in *From Modern to Contemporary American Poetry, 1945-1965* (Chicago: University of Chicago Press, 1984), pp. 23–52;

Paul F. Cummins, *Richard Wilbur: A Critical Essay* (Grand Rapids, Mich.: Eerdmans, 1971);

James Dickey, *Babel to Byzantium: Poets and Poetry Now* (New York: Farrar, Straus & Giroux, 1968), pp. 170–172;

Frederic E. Faverty, "Well-Open Eyes; or, the Poetry of Richard Wilbur," in *Poets in Progress*, edited by Edward Hungerford (Evanston, Ill.: Northwestern University Press, 1962), pp. 59–72;

Leslie A. Fiedler, *Waiting for the End* (New York: Dell, 1964), pp. 218–221;

Walter Freed, "Richard Wilbur," in *Critical Survey Poetry*, edited by Frank Magill (Englewood Cliffs, N.J.: Salem Press, 1982), pp. 3091–3100;

Donald Hall, *Contemporary American Poetry* (Baltimore: Penguin, 1962), pp. 17–26;

Hall, "The New Poetry: Notes on the Past Fifteen Years in America," in *New World Writing* (New York: New American Library, 1955), pp. 231–247;

Peter Harris, "Forty Years of Richard Wilbur: The Loving Work of an Equilibrist," *Virginia Quarterly Review*, 66 (Summer 1990): 412–425;

William Heyen, "On Richard Wilbur," *Southern Review*, 9 (July 1973): 617–634;

Donald L. Hill, *Richard Wilbur* (New Haven, Conn.: College & University Press, 1967);

John B. Hougon, *Ecstasy Within Discipline: The Poetry of Richard Wilbur* (Atlanta: Scholars Press, 1994);

Randall Jarrell, *Poetry and the Age* (New York: Vintage, 1953), pp. 227–240;

Kenneth Johnson, "Virtues in Style, Defect in Content: The Poetry of Richard Wilbur," in *The Fifties: Fiction, Poetry, and Drama*, edited by Warren G. French (Deland, Fla.: Everett/Edwards, 1970), pp. 209–216;

Brad Leithauser, "Reconsideration: Richard Wilbur — America's Master of Formal Verse," *New Republic*, 181 (24 March 1982): 28–31;

Bruce Michelson, *Wilbur's Poetry: Music in a Scattering Time* (Amherst: University of Massachusetts Press, 1991);

Ralph J. Mills Jr., *Contemporary American Poetry* (New York: Random House, 1965), pp. 160–175;

Raymond Oliver, "Verse Translation and Richard Wilbur," *Southern Review*, 11 (April 1975): 318–330;

Anthony Ostroff, ed., *The Contemporary Poet as Artist and Critic: Eight Symposia* (Boston: Little, Brown, 1964), pp. 1–21;

David Perkins, *A History of Modern Poetry* (Cambridge, Mass.: Belknap/Harvard University Press, 1987), pp. 383–386;

M. L. Rosenthal, *The Modern Poets: A Critical Introduction* (New York: Oxford University Press, 1960), pp. 253–255;

Rosenthal, *The New Poets: American and British Poetry Since World War II* (New York: Oxford University Press, 1967), pp. 328–330;

Wendy Salinger, ed., *Richard Wilbur's Creation* (Ann Arbor: University of Michigan Press, 1983);

Donald Barlow Stauffer, *A Short History of American Poetry* (New York: Dutton, 1974), pp. 385–387;

Hyatt H. Waggoner, *American Poets: From the Puritans to the Present* (Boston: Houghton Mifflin, 1968), pp. 596–604;

A. K. Weatherhead, "Richard Wilbur: Poetry of Things," *English Literary History*, 35 (December 1968): 606–617.

Papers:

Most of Richard Wilbur's papers are in the Robert Frost Library, Amherst College. Additional manuscripts, mostly early works, are in the Poetry Collection, Lockwood Memorial Library, State University of New York at Buffalo.

James Wright
(13 December 1927 – 25 March 1980)

Andrew Elkins
Chadron State College

See also the Wright entry in *DLB 5: American Poets Since World War II, First Series.*

BOOKS: *The Green Wall* (New Haven: Yale University Press, 1957; London: Oxford University Press, 1957);

Saint Judas (Middletown, Conn.: Wesleyan University Press, 1959);

The Lion's Tail and Eyes: Poems Written Out of Laziness and Silence, by Wright, Robert Bly, and William Duffy (Madison, Minn.: Sixties Press, 1962);

The Branch Will Not Break (Middletown, Conn.: Wesleyan University Press, 1963; London: Longmans, Green, 1964);

Shall We Gather at the River (Middletown, Conn.: Wesleyan University Press, 1968; London: Rapp & Whiting, 1969);

Collected Poems (Middletown, Conn.: Wesleyan University Press, 1971);

Two Citizens (New York: Farrar, Straus & Giroux, 1973);

Moments of the Italian Summer (Washington, D.C. & San Francisco: Dryad Press, 1976);

To a Blossoming Pear Tree (New York: Farrar, Straus & Giroux, 1977);

Leave It to the Sunlight (Durango, Colo.: Logbridge-Rhodes, 1981);

A Reply to Matthew Arnold (Durango, Colo.: Logbridge-Rhodes, 1981);

The Summers of James and Annie Wright: Sketches and Mosaics, by Wright and Annie Wright (New York: Sheep Meadow Press, 1981);

This Journey (New York: Random House, 1982);

Collected Prose of James Wright, edited by Anne Wright (Ann Arbor: University of Michigan Press, 1983);

A Secret Field: Selections from the Final Journals of James Wright, edited by Anne Wright (Durango, Colo.: Logbridge-Rhodes, 1985);

Above the River: The Complete Poems (New York: Farrar, Straus & Giroux, 1990).

James Wright (courtesy of the Lilly Library, Indiana University)

TRANSLATIONS: René Char, *Hypnos Waking: Poems and Prose,* translated by Wright and others (New York: Random House, 1956);

George Trakl, *Twenty Poems of George Trakl,* translated by Wright and Robert Bly (Madison, Minn.: Sixties Press, 1961);

César Vallejo, *Twenty Poems of César Vallejo,* translated by Wright, Bly, and John Knoepfle (Madison, Minn.: Sixties Press, 1962);

Theodor Storm, *The Rider on the White Horse* (New York: Signet, 1964);

Jorge Guillén, *Cantico: A Selection* (Boston: Little, Brown, 1965);

Pablo Neruda, *Twenty Poems of Pablo Neruda,* translated by Wright and Bly (Madison, Minn.: Sixties Press, 1968);

Herman Hesse, *Poems* (New York: Farrar, Straus & Giroux, 1970);

Neruda and Vallejo, *Selected Poems,* translated by Wright, Bly, and Knoepfle (Boston: Beacon Press, 1971);

Hesse, *Wanderings: Notes and Sketches* (New York: Farrar, Straus & Giroux, 1972).

Often remembered as one of the strongest of post–World War II American poets, James Wright was one of a group of young writers who, after establishing themselves with early books of prosodically conservative poetry, broke away from the mainstream traditions to experiment with looser form, freer rhythms, and bolder images. The verse in Wright's first two books, *The Green Wall* (1957) and *Saint Judas* (1959), is characterized by its use of traditional forms and shows the influence of such established poets as Robert Frost and Edwin Arlington Robinson. Much of Wright's middle verse, however, features the use of what some critics, using Robert Kelly's term of 1961, have called *deep imagery.* A subjective imagery that is purportedly drawn from a poet's unconscious, such imagery was used in this period in a reaction to what Wright considered the too analytical, too cerebral, perhaps spiritually enervated poetry that he and his older peers had been producing. Seeking a method of composition that would allow him to establish common ground with the reader and be understood immediately, but a method that would also allow him to abandon what he considered the restrictive tradition of predominantly iambic, logical, discursive verse, Wright began to write what Robert Bly called "leaping poetry," poetry that "leaps" from conscious to unconscious material, that eschews obvious, logical links between images and semantic units, and that prefers, instead, the bold jump from one image or statement to another, with the artist's and the reader's imaginations working in concert to supply the missing links. The source of such deep imagery, or leaping poetry, is assumed to be in a preconscious or unconscious state of mind that is shared by all people and can therefore be accessed by the reader for clues to the poet's vision. According to this theory of poetics, the writer is able to abandon traditional forms while not losing the audience comprehension that working within such recognizable forms normally facilitates. Such poetry dominates Wright's output beginning with his third volume, *The Branch Will Not Break* (1963).

But to describe the literary movement in which Wright properly fits and the type of poetry for which he is most likely to be remembered – to understand him exclusively as a deep image poet or a leaping poet or a poet of the emotive imagination – is to minimize the breadth and power of this important poet's work. Wright's poetry expresses his lifelong confrontation with the urgent problems of living and writing in mid- and late-twentieth-century America, and it possesses a relevance that far transcends its dynamic character as one tributary of a stream of consciousness verse produced during one period of American literary history. Indeed, Wright is a poet who asks the deepest and most fundamental questions that most people are forced to confront as they live in this bewildering phase of history; and his verse will likely be remembered and returned to whenever readers find themselves asking such basic, compelling questions as who they are, why they do what they do, and what value their lives have in the social and cosmic order. Wright's poetic career was a continuing quest to find answers to such questions, and particularly to find his own most fully human, fully responsive and creative self; that career also became a method and stragegy for enabling that especially valued self to live in the contemporary world.

His body of work might be compared to an epic poem, one which can be divided chronologically into three stages, characterized, respectively by rejection and denial, then, doubt and fear, and, finally, acceptance and affirmation. What begins perhaps as an adolescent's identity crisis in verse becomes, in the poet's mature work, a profound search for answers to the major philosophical questions of life. What does it mean, he asks himself repeatedly, to be "a good man" in contemporary America; and why should one, in this age of information saturation, write poetry, which makes nothing happen and perhaps only adds to the inane babel that characterizes the commercial culture? Such questions take the poet's entire lifetime to answer; meanwhile, the careful reader may profit greatly from witnessing and imaginatively participating in the struggle being fought and the quest undertaken.

James Arlington Wright was born into a working-class family in Martins Ferry, Ohio, an industrial town on the banks of the Ohio River. His father, Dudley, was employed for fifty years at the Hazel-Atlas Glass Company, which later, in the son's poetry, became a symbol of the oppressive, mind-numbing place and way of life that the young poet hoped to escape. His mother, Jessie, left school at fourteen to work in the White Swan Laundry; she later raised the Wright children (James had two brothers and an adopted sister). Neither parent went to school beyond the eighth grade. Their son James was a writer who began his literary career early. In a 1980 *American Poetry Review* interview, Wright recalls first writing poetry when he was about eleven years old. While in high school in 1943 he suffered a nervous breakdown and missed a year of school. When he graduated in 1946, a year late, he joined the U.S. Army and served as a clerk typist with the occupation forces in Japan after World War II. After his honorable discharge he did his undergraduate work at Kenyon College, where he studied poetry with John Crowe Ransom and graduated cum laude and Phi Beta Kappa in 1952, the same year he won his first poetry award, the Robert Frost Poetry Prize.

From Ransom, Wright learned to think of poetry as an intelligent, consciously crafted, formal art. In his *American Poetry Review* interview he says that Ransom taught him "the ideal, what elsewhere I've called the Horatian ideal, the attempt finally to write a poem that will be put together so carefully that it does produce a single unifying effect." Others who knew him in these days remember him as already being a serious poet, not just one who wrote poems, but one for whom writing poetry was the definition of his self. Writing in 1990 in *The Gettysburg Review*, E. L. Doctorow, at Kenyon during the same period, remembers that Wright had a "high and richly timbered voice, like a tenor's. His conversation was intense, opinionated, heavy with four-letter words, but what made it astonishing is that it was interwoven with recitations of poetry.... He would glide from ordinary speech to verse without dropping a beat." Doctorow recalls that "Wright always carried with him the verses he was working on in stiff, black, clamp binders.... You'd find him at the Village Inn, sitting alone with a cigarette and a cup of coffee, or in front of a beer at Jean Valjean's, and he'd be hunched in a booth revising a typed draft in his round, grade-school hand."

After graduating from Kenyon College, Wright married another Martins Ferry native, Liberty Kardules, and the two traveled to Austria, where, on a Fulbright Fellowship, the poet studied the works of Theodor Storm and Georg Trakl at the University of Vienna. The couple's first son was born in Austria in March 1953. Wright returned to the United States and did his graduate work in English at the University of Washington, earning his M. A. in creative writing ("I took the Master's in creative writing to get it the hell out of the way," he explains in his 1975 *Paris Review* interview) and his Ph.D. in English in 1959, after writing a dissertation on Charles Dickens, "The Comic Imagination of Charles Dickens." While at Washington he studied poetry with Theodore Roethke and Stanley Kunitz and published his first book, *The Green Wall;* this book, the fifty-third volume in the Yale Series of Younger Poets, was selected and introduced by W. H. Auden. The book, Wright admits, was influenced by the work of Frost and Robinson. When asked by Peter Stitt in the 1975 interview what he had learned from Frost, Wright responded, "Well, first of all I think that there is his profound, terrifying, and very tragic view of the universe, which seems to me true." During these early years he also received the Borestone Mountain Poetry Award (1954 and 1955) and the Eunice Tietjens Memorial Award from *Poetry* (1955).

Wright held his first academic position at the University of Minnesota, where he worked as an English professor from 1957 to 1964. Writing in *The American Poetry Review* in 1982, James Breslin, a graduate student at Minnesota when Wright was a professor there, remembers the poet as "a striking figure. He had the thick, powerful body of an Ohio State fullback, but he wore horn-rimmed glasses and a three-button suit: a football player squeezed into the uniform of an English professor." Breslin recalls that Wright was intense, quoted long passages of poetry, and drank heavily. According to Stitt, "In 1963 [Wright] was denied tenure at the University of Minnesota" because "the senior members of the English Department (including poet Allen Tate) unanimously opposed Wright's candidacy, citing the excessive drinking that was causing him to miss some of his classes." Denied tenure at Minnesota, Wright taught at Macalaster College in Saint Paul from 1963 to 1965. During the same period that his academic career seemed doomed, his marriage deteriorated; he and his wife separated several times, and the two finally divorced in 1962. While the darkness in Wright's early poetry cannot be attributed solely to his personal problems – he was also a harsh social critic and seemed driven nearly to despair by his aversion to the prevailing culture's materialism and conformism – certainly his personal demons contributed to the black mood one notes in the poet's first four books, published between 1957 and 1968.

One productive result of his time at Minnesota was that Wright met Robert Bly. According to

Wright in his *Paris Review* interview, "I was in despair at the time"; but, after reading Bly's magazine *The Fifties,* Wright was so impressed by a translation of Trakl that he wrote Bly a letter, "sixteen pages long and single spaced." Bly's response was simple: "Come on out to the farm." Robert and Carol Bly owned a farm in western Minnesota at the time; Wright accepted the invitation, and a lifelong friendship began. The two poets later collaborated on translations of the poetry of Trakl, César Vallejo, and Pablo Neruda. In Bly, Wright found a kindred soul, one who shared his political and poetical attitudes at a time when he was, if his poetry is any indication, feeling the profound pain of loneliness, alienation, and self-doubt.

In 1965 Wright went to Europe for a year on a Guggenheim Fellowship. Returning in 1966, he took a position in the English department of Hunter College of the City University of New York, where he was employed until his death in 1980. In 1967 he married his second wife, the former Edith Anne Runk. The two traveled to and worked in Hawaii and Europe, and Wright became especially fond of Paris, Verona, and Tuscany, all of which appear in his later poems. In 1978 Wright won a second Guggenheim, and he and Anne traveled in Europe for nine months in 1979. A persistent sore throat, noticed by the poet while abroad, was diagnosed as cancer of the tongue when the Wrights returned to New York; Wright died from that cancer in the spring of 1980.

During his writing career Wright won numerous poetry prizes, including the *Kenyon Review* Poetry Fellowship (1958–1959), a National Institute of Arts and Letters grant in literature (1959), the Longview Foundation Award (1959), a prize from *Chelsea* magazine in 1960, the Ohioana Award in poetry for *Saint Judas* (1960), the Oscar Blumenthal Award from *Poetry* (1968), the Creative Arts Award from Brandeis University (1970), the Fellowship of the Academy of American Poets (1971), and the Poetry Society of America's Melville Cane Award (1972). His *Collected Poems* (1971), a collection of the work from his first four books that included a selection of new poems, won the 1972 Pulitzer Prize.

Wright was a poet whose life may be seen as a quest for answers to essential and, for him, fiercely urgent questions, questions about identity, ethics, human suffering. The first stage of his quest for answers, expressed in his first two books, *The Green Wall* and *Saint Judas,* dramatizes the young rebel's cry of disappointment and disillusionment with the mainstream culture; it also conveys his avowal of loyalty to those whom Auden, in his introduction to

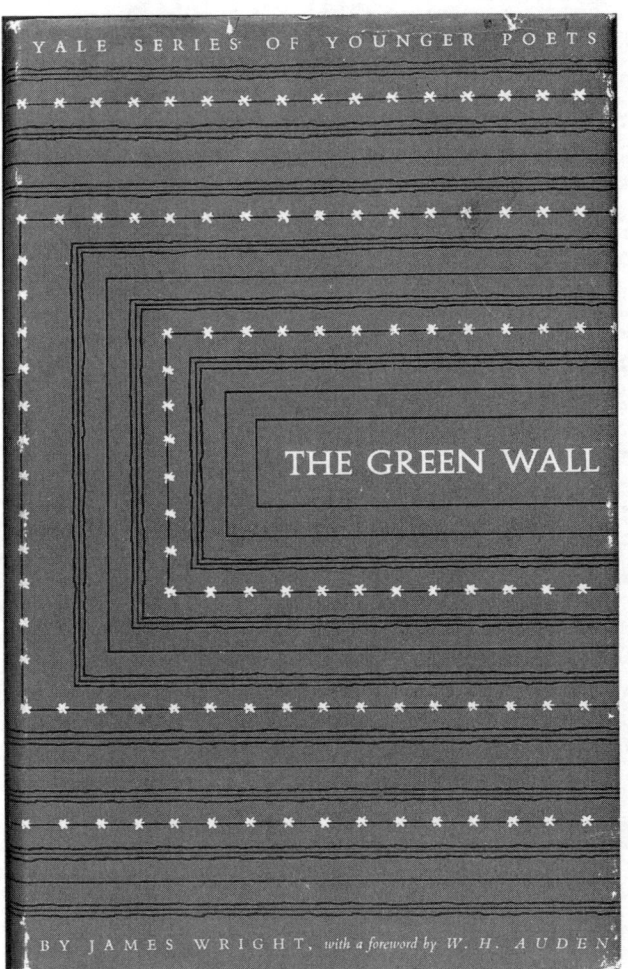

Cover for Wright's first book, which includes poems that show the influence of poets such as Robert Frost and Edwin Arlington Robinson

The Green Wall, called "social outsiders"; that is, those on the margins of society, the culture's derelicts, drunks, and criminals. Evidently feeling himself psychologically and spiritually disenfranchised, Wright identifies with and celebrates those who are truly outside the prevailing American culture of the work ethic, capitalist acquisition, conformity, and middle-class values. As a college professor Wright was economically mainstream, but the poet in him imaginatively – and occasionally literally – roamed the cold streets of Ohio and Minnesota with the homeless, listened with prisoners as guards locked cell doors for the night, and wandered the beach with the mother who had lost her child to the waves of an indifferent sea while indifferent "harly-charlies" partied nearby. His cultural criticism has a political dimension, of course, expressing as it does a disdain for those who place material values above human values; but his criticism is motivated more by his human sympathy for the outcast and dejected than by any abstract philosophical principles of good government. He sees suffering and his heart goes out. He identifies the major cause of the suffering as a system that counsels that, above all, the industrial machine's gears must be kept moving smoothly, a system that judges individuals by their income and class status. He believes that the poor and the suffering, those "dark" figures in his poetry associated with the "lower" world, must be written about, must be given a voice, and, in doing so, he finds himself condemning those whom he identifies as their tormentors, those of the "light" or "upper" world. Wright's hometown of Martins Ferry and its Ohio River figure prominently in his early work, serving as symbols of what the poet finds wrong with the culture into which he has been born: the willingness, even eagerness, of Americans to despoil what is beautiful — frequently symbolized by the Ohio River and the ravaged Ohio landscape — in the interest of wealth, economic development, and an industrial notion of progress. In Wright's Ohio the beautiful woods struggle for survival against the relentless onslaught of the ugly factories and strip mines. In his early work the poet wonders if he, too, is one of the beautiful misfits that must be destroyed to satisfy the culture's appetite?

In the first stage of Wright's struggle, the poet attempts to escape and dissociate himself from the realm of the "ugly," which includes those who proceed passively through their lives, the timid and the conventional, those who bulldoze trees to erect chiropractors' offices, who pollute Wright's beloved Ohio River, and who otherwise participate, wittingly or unwittingly, in what the poet characterizes as a culture of death and destruction. By so participating, they not only perpetuate the culture, but they lose their souls as well. These characters are anonymous figures who seem to be everywhere: the old fishermen whose twine has "gone slack" and whose blood has "gone dumb"; the glum cop whose "flat face and empty eyes" are perfectly adapted to the "pale town"; the "two stupid harly-charlies" who lure a drunken and blameless girl to her death by drowning. The "ugly" are also recognizable or identifiable figures of the time: President Dwight D. Eisenhower; the industrialist, Mark Hanna; FBI director J. Edgar Hoover; the owners of Hazel-Atlas Glass Company; and all the policemen, bankers, businessmen, and wardens of the world. These are "the rising dead who fear the dark," those who rise in the morning to greet the solid assurance of another day, but who fear the dark of their own souls.

They lead, in Henry David Thoreau's famous phrase, "lives of quiet desperation," but they have no clue to their own predicament. Wright decisively rejects both the characters and the culture they serve or accept.

By contrast, Wright's early work celebrates figures whom he considers heroic correctives to those who meekly accept the culture, figures who are instinctive, antirational, sensual, and innocent; these are celebrated by the poet in such characters as a prostitute named Betty, a woman in the insane asylum, a little girl on her way to school, Judas, and, perhaps Wright's most controversial hero, George Doty, the executed Ohio murderer. All the unconventional, the rebellious, the losers, the outcasts, the aliens in their own land are celebrated. They are children, lunatics, murderers, prostitutes, homeless poor, amputees, and poets hunted by the ugly, lured to their deaths, or forced to repress their central desires and impulses. They are sensuous, untamed, primeval, and, above all, innocent — even if they are felons — because they have at least listened to their instincts, the dark murmurings of the heart that bring news of their real identity that exists somehow beyond, below, above, or behind society's prescriptions for good behavior. Whatever else they may be, they are innocent of self-betrayal.

In *The Green Wall* Wright provides the reader with several examples of his innocents. One is the female mental patient in "She Hid in the Trees from the Nurses." When the woman's attendants whistle for her, she prefers not to respond, choosing instead to enjoy her freedom,

And dabble her feet in the damp grass,
And lean against a yielding stalk,
And spread her name in dew across
The pebbles where the droplets walk.

The woman prefers to listen to the rhythms deep within herself, which are attuned to the natural world around her, rather than to the attendants's oppressive whistles, symbols of the order that humankind logically imposes on the psyche's freedom. The insane woman is sane in Wright's inverted moral world, where the society that establishes such places as mental hospitals is indicted for insisting upon whistles and chimes, symbols of conformity and the suppression of natural desires.

Another innocent is Betty, the black prostitute of "Morning Hymn to a Dark Girl." Betty lives beneath a bridge, under and apart from the world of conventional commerce. In that upper world, men become "stone," "flat," and "empty," while Betty is a perfect exemplar of the character Wright here wants to celebrate, the sensuous, uninhibited human beyond convention:

> Betty, burgeoning your golden skin, you poise
> Tracing gazelles and tigers on your breasts,
> Deep in the jungle of your bed you drowse.

Other exemplary characters are the schoolgirl of "A Little Girl on Her Way to School" who understands the language of the birds and stones and the female speaker of "Sappho," condemned for her homosexual love by the "sly voices" of the town, but who insists on her innocence, despite having violated conventional standards: "They cannot tear the garden out of me, / Nor smear my love with names." The woman's strength comes from her deeper knowledge, a knowledge discovered in pain, and whose source is deeper than logic or social convention:

> There is a fire that burns beyond the names
> Of sludge and filth of which the world is made.
> Agony sears the dark flesh of the body,
> And lifts me higher than the smoke, to rise
> Above the earth, above the sacrifice.

To be "above the earth, above the sacrifice" is the position many of Wright's early heroes and heroines seek, a position above society's restrictions that serve conformity and production.

The radicalness of his position is illustrated by the most controversial poem in his first volume, "A Poem About George Doty in the Death House," in which the poet idolizes a real-life murderer, a taxi driver from Ohio who one night killed one of his passengers with a tree branch when she refused a pass he made. In the words of the poet:

> A month and a day ago
> He stopped his car and found
> A girl on the darkening ground,
> And killed her in the snow.

Despite acknowledging Doty's guilt, Wright declares that he "will mourn no soul but his," not even the soul of "the homely girl whose cry / Crumbled his pleading kiss." The poem does not entirely convince one that Doty's is the soul that most deserves the reader's sympathy. Wright's poetic goal, as he explains it in his *American Poetry Review* interview, is nevertheless a valid one: "Many people in that community thought [Doty] was terribly wicked, but he did not seem to me wicked. He was just a dumb guy who suddenly was thrust into the middle of the problem of evil and he was not able to handle it." That statement summarizes the human condition as Wright understands it early in his career: we are all "dumb" humans "thrust into the middle of the problem of evil," who daily have to ask ourselves how to "handle it." The poet's sympathy extends in *The Green Wall* to those who struggle honestly and unconventionally, albeit unsuccessfully, with this problem, but not to those who simply accept the solutions others have forged for them.

The same theme underlies the poet's second volume, *Saint Judas,* in which Wright announces, on the cover, his desire to discover "exactly what *is* a good and humane action" and "why an individual should perform such an act." The most noticeable example of the poet's struggling with the problem of the human confrontation with good and evil is found in the volume's best-known poem, the title poem, in which Wright canonizes the figure of Judas, a bolder move perhaps than idolizing George Doty. Having already that day betrayed Christ, Judas in this sonnet "slipped away" from the crowds, determined to kill himself. Instead, he "caught / A pack of hoodlums beating up a man" and, "for nothing" this time, "held the man . . . in my arms." Judas is another "dumb guy," like Doty, caught in the trap of evil, who has to decide how to act in a world well beyond his limited understanding. On the record he is the greatest of traitors, but, in this off-the-record glimpse, he is revealed by Wright to be a man still capable of compassion. In his first two books Wright insistently identifies with life's losers and victims, those who find themselves overwhelmed by the problem of evil, which is frequently symbolized by the men in the gray flannel suits (or, in the case of "Saint Judas," Roman soldiers). The poet's simplified moral stance is the product of his refusal to accept and sympathize with his own flawed humanity, his fear that he is the ugly one, the impure soul, the "dumb guy" no one wants to admit to being. As Bly suggests in an article originally published in *The Sixties* under the pseudonym Crunk, "the most profound emotion in *Saint Judas* is guilt." Wright suspects and fears his own sin, the sin that all are guilty of, the sin of being merely human, and deludes himself with the thought that he might escape his inevitable human failings by identifying with the beautiful – who are not of this world – and distancing himself from the ugly – who are too much of this world. His affirmation or gathering of all souls, beautiful and ugly, must await what the Puritan preacher Thomas Hooker called "a true sight of one's sin," a recognition of the ugly within himself.

Cover for Wright's 1963 book, which includes early "deep image" poems in which he attempted to mine the collective unconscious, the buried memories of a preconscious existence that, according to psychiatrist Carl Jung, are shared by all humankind

The poet's struggle to avoid his own humanity provides the drama of the second stage of his poetry, which begins in *The Branch Will Not Break* and culminates in *Shall We Gather at the River* (1968), the darkest book of Wright's career. Like Milton's Satan, Wright discovers that "I myself am hell." Or, as he says, "I am the dark / Bone I was born to be." Having tried to purify his world by pushing away the ugly, Wright discovers that it is his own inner sphere that is corrupted, the very recognition he hoped to escape by pushing the others away. The youthful arrogance of the two early volumes, in which the poet dares to tell others how to live, thinking himself qualified to do so by his innocence and vitality, becomes the self-pity, self-loathing, and flirtation with suicide we see in *The Branch Will Not Break* and *Shall We Gather at the River*, volumes characterized by doubt and fear and by the looser form and relaxed rhythms referred to earlier. The first-person speaker replaces the persona we hear in earlier poems; the dominant iambic meter of Wright's first two books is less common; stanzas and lines show no consistent or predictable length; ambiguous pronouns tease the reader; the poems often begin in medias res; and the normal discursive links are omitted or blurred. The stylistic changes are actually thematic: they suggest confusion, the difficulty of making sense of the world, and the need for confrontation with the world's data before drawing conclusions. In this light the style, while it may seem a break from earlier work, is really an extension of that work's themes, now in a more appropriate frame, a frame that is part of the theme itself. Some critics originally thought that the change in style marked a revolution in Wright's work, but it is now clear that it was part of an evolution that began as early as 1952, when Wright was in Europe on a Fulbright Fellowship. He himself describes accidentally wandering into a University of Vienna classroom in which Trakl was reading his poetry. "It was as though the sea had entered the class at the last moment," Wright says in his introduction to his translation of

Trakl's poetry. "For this poem was not like any poem I had ever recognized." His later translation of Trakl, Vallejo, Neruda, and René Char show that, well before the publication of *The Branch Will Not Break,* he had been fascinated with verse that is antitraditional in form. His own work had bothered him as early as 1958, when, in a letter to Roethke, Wright complained that "My stuff stinks. . . . I am trapped by the very thing – the traditional technique – which I labored so hard to attain."

The occasionally hopeful tone of some poems in *The Branch Will Not Break* sometimes obscures the true darkness of the volume. Again and again, Wright longs for escape, wants to get out, says that life is too hard, as in these lines from "Having Lost My Sons, I Confront the Wreckage of the Moon: 1960":

> I am sick
> Of it, and I go on,
> Living alone, alone,
> Past the charred silos, past the hidden graves
> Of Chippewas and Norwegians.

The quest for deliverance, combined with images of stasis and petrifaction, dominate *The Branch Will Not Break.* The most often anthologized poem in the volume, "Lying in a Hammock at William Duffy's Farm in Pine Island, Minnesota," concludes with these famous lines:

> I lean back, as the evening darkens and comes on.
> A chicken hawk floats over, looking for home.
> I have wasted my life.

Critics argue whether the final line has been adequately prepared for by the series of images that precedes it. The more important point is that the final line has been prepared for by all of Wright's preceding poetry. He has abandoned his home but, being human, has not been able to find a replacement residence among the hawks or horses of the world. He is lost and unable to save himself. He lies in his hammock suspended, passive, and alone.

The other anthology piece from the book, "A Blessing," narrates an encounter between "my friend and me" and "two Indian ponies" in a field "just off the highway to Rochester, Minnesota." The horses accept the two humans' presence, and for a moment the speaker feels he has found a home in nonhuman nature. His exultant ending, however, reveals the limitation of such a hope:

> Suddenly I realize
> That if I stepped out of my body I would break
> Into blossom.

A joyous conclusion, but note that in order to feel the joy of the encounter with the peaceful horses, the speaker has to imagine losing his human identity, stepping out of his body. There seems to be no real salvation at this stage in Wright's quest; but every quest has its peaks and valleys, and these two books, *The Branch Will Not Break* and *Shall We Gather at the River,* present Wright's psychological nadir.

If possible, *Shall We Gather at the River* is even darker than *The Branch Will Not Break,* unrelentingly darker. There is irony in this. Wright has been pursuing his immaculate self, free of the influences of others; but, having freed himself, he finds only loneliness and pain. In "Before A Cashier's Window in a Department Store," he writes:

> I am hungry. In two more days
> It will be spring. So this
> Is what it feels like.

Wright has exiled himself, and "this" / is what "it feels like." Spring promises no rejuvenation for the poet. The controlling image cluster of the book – rivers, boats, water, shores – suggests a passage from one life into another, a movement across the boundaries separating this existence from the next. The desire to make that move, to die as the person James Wright and be delivered as something else, anything else, is strong in *Shall We Gather at the River.* At the end of "The Minneapolis Poem," for example, Wright says,

> I want to be lifted up
> By some great white bird unknown to the police,
> And soar for a thousand miles and be carefully hidden
> Modest and golden as one last corn grain. . . .

These lines suggest the poet's belief that he is unable to deliver himself and yet he longs to be nonhuman, because to be human, he has discovered, is enormously difficult. The conflict in the book is concisely summarized by one line from "A Christmas Greeting": "It hurts to die, although the lucky do." To die might mean deliverance, but it certainly means pain; to live means pain, also. Mired in a dilemma apparently without solution, the poet hits bottom in "To the Muse," the final poem of *Shall We Gather at the River.* The poem finds the speaker poised on the bank of the Ohio River, the polluted symbol of the destruction that the ugly leave behind them. The poet stands at the river's edge, seeing his own sins clearly ("I admit everything"), not at all

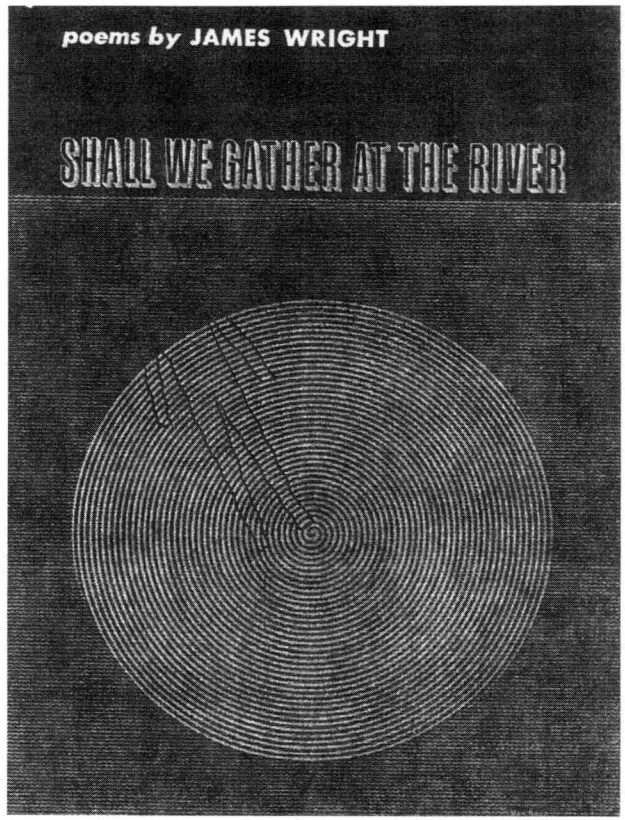

Cover for Wright's 1968 book. Its tone is exemplified by a line from "A Christmas Greeting": "It hurts to die, although the lucky do."

sure whether he should step forward into oblivion and peaceful death or step backward to his Ohio shore into continued, painful life. "I wish to God I had made this world, this scurvy / And disastrous place," he cries, "but I didn't, I can't bear it." Staying alive is so painful that "It hurts," he says, "like nothing I know." But dying also hurts, and that fact immobilizes him.

It is at this point, the darkest in his poetry, that Wright chooses life. But how and on what terms? He steps away from the edge of death in time to understand that to live necessitates "gathering" the ugliness that is inside and outside himself. Staying alive means forgiving the ugly thing he has discovered himself to be and offering mercy to himself for being merely human. To live means backing up, joining the others, accepting his country, his fellow humans, and himself, with all the attendant flaws, disfigurements, warts, and crimes intact. The point is not to purify falsely what is ugly, but to gather it in all its ugliness after he has gathered himself. This is the process of "gathering" that he begins in his fifth major book, *Two Citizens* (1973), the truly transitional volume in Wright's career. The key verb in *Two Citizens* is *gather,* and, by repeatedly using it, trying it out, the poet nudges himself from the precarious perch he occupied in "To the Muse" into the position he occupies at the end of his final volume, *This Journey* (1982), on the side of a mountain at daybreak, "on top of the sunlight."

"Gathering" in *Two Citizens* and in Wright's later poetry is the literal or imaginative act of reaching toward the other, a sympathetic extending beyond oneself to another. Wright extends himself in early poems, but always selectively, embracing only various outcasts and rebels. Those are embraced because they deserve to be. The key to "gathering" is that it is an extension of one's sympathy, which begins in *Two Citizens* as simple tolerance, to others without reference to their merits. Like God, Wright chooses to offer his grace even if it is not, or cannot be, deserved or earned. He chooses to make a commitment despite judgment, because Wright moves toward gathering even those he continues to judge. He struggles in *Two Citizens* and subsequent volumes toward offering the miracle of undeserved love, extending his hand to those same philistines he earlier denounced, without changing his rational or moral judgment of them, simply choosing to embrace them as part of the depraved human species of which he is one dark member. "Gathering," then, is the act of offering human love gratuitously, an act that is a kind of grace, mercy, and beauty. Arnold Bennett, in his *Journal* (1932-1933), argues that "the essential characteristic of the really great novelist is a Christlike, all-embracing compassion." This is precisely what Wright struggles toward in *Two Citizens* and what he achieves in his last two books.

One reaches that point by small steps, the first of which is a candid recognition of one's own sinfulness; this step is followed by forgiving oneself for being part of the imperfect human race and loving oneself, anyway; these steps are followed by one's learning to offer that love to others. That last and most difficult lesson is initially taught to Wright by nature, specifically, by a herd of deer that appear in "A Secret Gratitude," from the "New Poems" section of *Collected Poems*. In the poem one is told several unpleasant things, one of which is that "Man's heart is the rotten yolk of a blacksnake egg / Corroding, as it is just born, in a pile of dead / Horse dung." Given this understanding of the human nature, it is not surprising that the poet concludes that humans deserve neither grace nor mercy nor beauty. He asks:

> Why should any mere multitude of the angels care
> To lay one blind white plume down
> On this outermost limit of something that is probably no more
> Than an aphid.

The answer, the poet suggests, is that there is no good reason at all why any angel should deign to look upon such pitiful wretches as humans. Yet the men in the poem, these "chemical accidents of horror," as he calls them, "capable of anything," are soon blessed when five deer, "those fleet lights," emerge from the woods and momentarily do not flee the men's presence. There is the wonder – that these wretched men are allowed into the presence of beauty at all. Perhaps in a world in which this is possible, against all odds and contrary to all logic or justice, perhaps, if men who are undeserving are allowed to stumble across beauty occasionally, then one should, in turn, offer his love and what little beauty he has to other undeserving souls.

The lesson is completed, however, only when learned from other humans, as happens in *Two Citizens*. In that book Wright says that his wife, Anne, "gave me the strength to come to terms with things which I loved and hated at the same time." As he did in "A Secret Gratitude," he finds himself loved despite considering himself undeserving of that love, discovers himself the recipient of love offered for nothing. The next step is to turn his mercy outward, to love or "gather" those "things which I loved and hated at the same time" (a phrase that comes from the poem, "Paul"), all those things he has seen (America's culture, destructive Americans, himself) that are also undeserving. The acts of accepting his own ugliness, falling in love with "a beautiful woman," being charitable toward others, and discovering "my native place" are all analogous forms of gathering the other, who is at times as repugnant as she, they, it, or oneself is alluring. "Though love can be scarcely imaginable hell," as he writes in "The Art of the Fugue: A Prayer," still, "By God, it is not a lie," and so, he accepts it; although he himself is "the rotten yolk of a blacksnake egg," he has to accept himself; although America is "a brutal and savage place," he says, "I still love" it. Americans are still depicted as defacers of the beautiful (for example, in "Names Scarred at the Entrance to Chartres"), yet they are his only heritage and his rightful companions. Therefore, even though "Dolan" and "Doyle" scratch their names at the entrance to Chartres Cathedral, Wright painfully admits that they are, they must be, his companions: "I have no way to go in / Except only / In the company of two vulgars...."

His struggle for innocence has shown him his guilt. To reject the guilty is thus to deny part of himself, which is precisely what he wants not to do. Recognizing his own sin and seeing himself the beneficiary of Anne's undeserved love, he gains the power to extend his own hand to others equally undeserving. To refuse to gather the ugly would be to commit the sin of Nathaniel Hawthorne's Young Goodman Brown and to end as Brown did: "a stern, a sad, a darkly meditative, a distrustful, if not a desperate man," unable to have faith in anything except his own false purity. To refuse to have faith in the beauty of others, even while knowing they are also ugly, is to react to the inevitable confrontation with darkness as George Doty, who committed murder, did – by placing one's faith in evil; ultimately, it is also to die, as Brown did, "a hoary corpse," separated from what Hawthorne calls the "universal throb," having voluntarily cut oneself off from "the magnetic chain of humanity." Wright's mature recognition and perspective, begun in *Two Citizens*, is completed in *To a Blossoming Pear Tree* (1977) and in *This Journey*.

Cover for an issue of The American Poetry Review that includes nine previously unpublished prose poems by Wright

Reviewers noted the new understanding that was present in *To a Blossoming Pear Tree,* the last volume published during Wright's lifetime. The remarks of Peter Serchuk, writing in 1978 in the *Hudson Review,* are typical of the praise the book received: "What perhaps most distinguishes Wright's voice in this book from that of his previous collections is its authority of middle age, its active acceptance of a tottering world along with the well-seasoned understanding that among the rubbled debris there is always some sustaining beauty to be found, however difficult it may be to recognize." That "active acceptance of a tottering world" is the goal toward which Wright's poetry, seen in perspective, has been moving since *The Green Wall.* Finally, having recognized the futility of hoping to exile himself from the realm of the ugly by writing poems in which the forces of the beautiful and the ugly battle to a standoff, Wright now composes poems in which, in Linda Pastan's words

in the *Library Journal* in 1977, "the brutal and the beautiful mysteriously illuminate each other."

Wright begins *To a Blossoming Pear Tree* with some of his old bitterness at Ohio and a hint of the paranoia the reader expects from one of the self-proclaimed pure people ("somebody / Is wondering right at this moment / How to get rid of us, while we sleep," he writes in "Redwings"), but he moves beyond that self-destructiveness to shower the reader with images of what it means "To drink, to live" in the poem "Hell." His deliverance is effected by a brutal self-awareness: ten times in the book he reminds the reader of the necessity of facing oneself, of literally looking unflinchingly into one's own or another's face, and, as he does, one notices the preponderance of images of light — secret light, dazzles, brilliances, candlelight, sunlight, light of every variety — whereas images of darkness had earlier prevailed.

Even when darkness enters the poem, it does not prevent the speaker from seeing clearly. In "With the Shell of a Hermit Crab," a poem often anthologized, the speaker examines a crab that has been forced out of its shell and has died. The observation might in previous volumes have prompted the poet to conclude that one must remain armored. The opposite conclusion is reached in *To a Blossoming Pear Tree,* and it is reached in the darkness:

> I reach out and flick out the light.
> Darkly I touch his fragile scars,
> So far away, so delicate,
> Stars in a wilderness of stars.

Scars become stars. The pains one bears provide one's illumination. Darkness and light do, as Pastan says, "mysteriously illuminate each other."

"Beautiful Ohio," the final poem in *To a Blossoming Pear Tree,* is the volume's best illustration of Wright's mature vision. Sitting on a railroad tie above the Martins Ferry sewer main, the poet watches as "a shining waterfall" of waste spews out of the pipe. He says:

> I know what we call it
> Most of the time.
> But I have my own song for it,
> And sometimes, even today,
> I call it beauty.

Ohio in this poem is "beautiful"; Martins Ferry is "my home, my native place"; and even the waste of the culture, for the moment, is "beauty." The sewer's ugliness is incorporated undiluted into the singer's song. Wright here sees the beauty *in* the ugliness, not merely the beauty *and* the ugliness. The "secret light" that he has been searching for was supposed to be pure, beautiful, innocent, undefiled, the opposite of the "perilous black," the ugliness that he spent so many poems defining and rejecting. Discovering that the secret light actually resides in the "perilous black," and can only be discovered there, moves him not toward despair but toward fulfillment. He does not turn his back to the waste but chooses to "call it beauty," chooses to see, recognize, and identify the ugliness around him as part of the beautiful. To "call" it beauty implies a choice. He has wanted beauty to overwhelm him, to jump across the subject-object divide and grab him. But it will not do that, and the poet now accepts his duty as poet, which is also his uniquely human power: to call into consciousness the beauty that lies enveloped in ugliness.

Wright began his poetic career trying to diminish what cannot be diminished – specifically, humankind's flaws, including greed, selfishness, willful destruction, and complacent ignorance. He succeeded in changing nothing around him, but he did change himself, and in so doing, he changed his reaction to the ugly facts of life from one of rejection and despair to one of acceptance and hope. This acceptance so strengthened the poet as a poet that he was able to face life's horrors while simultaneously singing life's joys. The result is a continuous consciousness of life's wonders, undiminished by – in fact, made more profound by – the equally sharp awareness of life's brutality.

This hard-won perspective shines through in that amazing posthumous volume *This Journey*. Many critics have noted what Robert Shaw in 1982 called in *The Nation*, "Wright's awareness of approaching death." Critics have also noticed, again quoting Shaw, "a heightened sensitivity to the pulse of vitality everywhere around him." Death and vitality meet, and one is not surprised, because that is exactly what should happen in the work of a poet finally capable of "gathering" all the world to him. Any longing for escape or vague transcendence is gone. Reality is sufficient; meaning inheres in life's cicadas, hawks, and humans. Fantastic images are often related in a deadpan tone because life is a series of amazing moments, and no one should be surprised to see beauty thrust in front of one at any moment, in the form of lightning bugs, diminutive blue spiders, jagged pieces of ice, or the outstretched hands of strangers. One's salvation is knowing enough to *gather* the inexplicable gratefully. "Come, Look Quietly" is the title of one of these final poems and a final statement of Wright's epistemology and ethics. The world contains life-giving, life-sustaining, and meaningful moments, and it would be immoral to overlook them, to dwell exclusively on ugliness: "Sheep eat everything," the poet tells us in "Sheep in the Rain," "All the way down to the root"; and, as a result of this appetite for all that life offers, sheep have all that anyone can rightly ask, "a good life of it, / While it lasts." The point is to live, to gather, rather than waste one's brief time finding fault: "The trouble with me is / I worry too much about things that should be / Left alone," Wright observes in "Leave Him Alone."

Moral exemplars in earlier volumes taught the poet to resist, protest, create a life outside the mainstream where most people live. Exemplars in *This Journey,* such as the sheep, teach the poet to embrace what sustenance the world offers. In "Come, Look Quietly," he writes:

> The plump Parisian wild bird is scoring a light breakfast at the end of December. He has found the last seeds left in tiny cones on the outcast Christmas tree that blows on the terrace.

Finding himself in *The Green Wall* in what he diagnosed as a dead land, the poet saw no hope except escape, which threatened to result in his own death. Now his characters take life from death, without compromising their integrity: the plump bird remains "wild." This way of life suggests that reality may be beneficent, if given a chance, and that the world offers miracles, if one takes the time to attend to them. One such small miracle occurs in "Against Surrealism." The poet and his wife buy some chocolate penguins at a Parisian confectionery:

> We set them out on a small table above half the rooftops of Paris. I reached out to brush a tiny obvious particle of dust from the tip of a beak. Suddenly the dust dropped an inch and hovered there. Then it rose to the beak again.
> It was a blue spider.

Omit the chocolate penguins, and many readers may remember having encountered spiders in this manner; and many may have crushed them or flicked them away. But Wright has converted himself into Emerson's and Whitman's poet, the "seer" who is also and inevitably the "sayer." He searches for what he encounters accidentally, notes it, and gives it back to the reader as a gift, as the world has given it to him, without egotism, as an act of humility rather than achievement:

> Many men
> Have searched all over Tuscany and never found
> What I found there, the heart of the light
> Itself shelled and leaved, balancing
> on filaments themselves falling.

Instead of continuing to long for escape or an otherworldly transcendence, Wright reveals his faith in reality's plenitude and sufficiency.

The nearest one comes to a direct statement of Wright's philosophy in *This Journey* comes in the last stanza of "The Journey":

> The secret
> Of this journey is to let the wind
> Blow its dust all over your body,
> To let it go on blowing, to step lightly, lightly,
> All the way through your ruins, and not to lose
> Any sleep over the dead, who surely
> Will bury their own, don't worry.

Wright counseled passivity in *The Branch Will Not Break* and *Shall We Gather at the River* because he feared that activity was by its nature aggressive and too closely allied with the culture of conquest beyond which he was trying to move. Here he admits the possibility of a morality that is simultaneously active and respectful rather than aggressive: "step lightly, lightly."

Such a philosophy is easy to articulate but difficult to live. The right to perceive reality's plenitude and sufficiency, however, was not a simple gift. Through years of honest struggle Wright earned the right to contend, implicitly, and despite William Butler Yeats's imagery of chaos and dissolution in his poem "The Second Coming," that the center, in fact, does hold. This contention is not the expression of a naive simplicity but, rather, the result of a deep understanding of life wrought by years of anguish and poetic labor. As Thoreau says in *Walden,* "darkness bear[s] its own fruit"; and we last see Wright, in "A Winter Daybreak Above Vence," the final poem in his final volume, not as the "hoary corpse" that Goodman Brown became and that the poet had seemed doomed to become, but at daybreak, "on top of the sunlight," taking all he can from "the only life I have."

Letters:

In Defense against This Exile: Letters to Wayne Burns, edited by John R. Doheny (Seattle, Wash.: Genitron Press, 1985);

The Delicacy and Strength of Lace: Letters between Leslie Marmon Silko and James Wright, edited by Anne Wright (St. Paul, Minn.: Graywolf Press, 1986).

Interviews:

Michael André, "An Interview with James Wright," *Unmuzzled Ox,* 1 (February 1972): 3-18;

William Heyen and Jerome Mazzaro, "Something to Be Said for the Light: A Conversation with James Wright," *Southern Humanities Review,* 6 (1972): 134-153;

Peter Stitt, "The Art of Poetry, XIX: James Wright," *Paris Review,* 16 (Summer 1975): 34-61;

Bruce Henricksen, "Poetry Must Think," *New Orleans Review,* 6, no. 3 (1978): 201-207;

David Smith, "An Interview with James Wright: The Pure Clear Word," *American Poetry Review,* 9, no. 3 (1980): 19-30.

Bibliographies:

Belle M. McMaster, "James Arlington Wright: A Checklist," *Bulletin of Bibliography and Magazine Notes,* 31 (April/June 1974): 71-82, 88;

Ironwood, special issue on Wright, 10 (1977): 156-165;

William H. Roberson, *James Wright: An Annotated Bibliography* (Lanham, Md.: Scarecrow Press, 1995).

References:

Roger Blakeley, "Form and Meaning in the Poetry of James Wright," *South Dakota Review,* 25 (Summer 1987): 20-30;

Edward Butscher, "The Rise and Fall of James Wright," *Georgia Review,* 28 (Spring 1974): 257-268;

"Crunk" [Robert Bly], "The Work of James Wright," *Sixties,* 8 (Spring 1966): 52-78;

Madeline DeFrees, "James Wright's Early Poems: A Study in 'Convulsive' Form," *Modern Poetry Studies,* 2 (1979): 241-251;

David C. Dougherty, *James Wright* (Boston: Twayne, 1987);

Andrew Elkins, *The Poetry of James Wright* (Tuscaloosa: University of Alabama Press, 1991);

Nicholas Gattuccio, "Now My Amenities of Stone Are Done: Some Notes on the Style of James Wright," *Scape: Seattle, New York,* 1 (1981): 31-44;

Michael Graves, "Crisis in the Career of James Wright," *Hollins Critic,* 22 (December 1985): 1-9;

Graves, "A Look at the Ceremonial Range of James Wright," *Concerning Poetry,* 16, no. 2 (1983): 43-54;

Victoria Harris, "James Wright's Odyssey: A Journey from Dualism to Incorporation," *Contemporary Poetry: A Journal of Criticism, 3, no. 3 (1978): 56-74;

Richard Howard, *Alone with America: Essays on the Art of Poetry in the United States since 1950,* enlarged edition (New York: Atheneum, 1980), pp. 662-678;

Howard, "James Wright's Transformations," *New York Arts Journal,* 8 (February/March 1978): 22-23;

Cor van den Huevel, "The Poetry of James Wright," *Mosaic,* 7 (Spring 1974): 163-170;

Ironwood, special issue on Wright, 10 (1977);

G. A. M. Janssens, "The Present State of American Poetry: Robert Bly and James Wright," *English Studies,* 51 (Spring 1970): 112-137;

Walter Kalaidjian, "'Many of Our Waters': The Poetry of James Wright," *Boundary 2: A Journal of Postmodern Literature,* 9, no. 2 (1981): 101-121;

Paul A. Lacey, "That Scarred Truth of Wretchedness," in his *The Inner War: Forms and Themes in Recent American Poetry* (Philadelphia: Fortress Press, 1972), pp. 57-81;

Edward Lense, "'This Is What I Wanted': James Wright and the Other World," *Modern Poetry Studies,* 11, nos. 1-2 (1982): 19-32;

George S. Lensing and Robert Moran, *Four Poets of the Emotive Imagination: Robert Bly, James Wright, Louis Simpson, and William Stafford* (Baton Rouge: Louisiana State University Press, 1976);

Laurence Lieberman, "James Wright: Words of Grass," in his *Unassigned Frequencies: American Poetry in Review, 1964-77* (Urbana: University of Illinois Press, 1977), pp. 182-189;

William Matthews, "The Continuity of James Wright's Poems," *Ohio Review,* 18 (Spring/Summer 1977): 44-57;

Jerome Mazzaro, "Dark Water: James Wright's Early Poetry," *Centennial Review,* 27 (Spring 1983): 135-155;

Ralph J. Mills Jr., "James Wright's Poetry: Introductory Notes," *Chicago Review,* 17, nos. 2-3 (1964): 128-43;

Charles Molesworth, "James Wright and the Dissolving Self," in *Contemporary Poetry in America: Essays and Interviews,* edited by Ralph Boyers (New York: Schocken Books, 1974), pp. 267-278;

William S. Saunders, "Indignation Born of Love: James Wright's Ohio Poems," *Old Northwest: A Journal of Regional Life and Letters,* 4 (December 1978): 353-369;

Saunders, *James Wright: An Introduction* (Columbus: State Library of Ohio, 1979);

James Seay, "A World Immeasurably Alive and Good: A Look at James Wright's *Collected Poems,*" *Georgia Review,* 27 (Spring 1973): 71-81;

Peter Serchuk, "On the Poet James Wright," *Modern Poetry Studies,* 10, nos. 2-3 (1981): 85-90;

Dave Smith, ed., *The Pure Clear Word: Essays on the Poetry of James Wright* (Urbana: University of Illinois Press, 1982);

Kevin Stein, *James Wright, the Poetry of a Grown Man: Constancy and Transition in the Work of James Wright* (Athens: Ohio University Press, 1989);

Stephen Stepanchev, *American Poetry Since 1945: A Critical Survey* (New York: Harper & Row, 1965), pp. 180-184;

Randall Stiffler, "The Reconciled Vision of James Wright," *Literary Review,* 28, no. 1 (1984): 77-92;

Peter A. Stitt, "The Garden and the Grime," *Kenyon Review,* 6, no. 2 (1984): 76-91;

Stitt, "The Poetry of James Wright," *Minnesota Review,* 2 (Spring 1972): 13-29;

Stitt and Frank Graziano, eds., *James Wright: The Heart of Light* (Ann Arbor: University of Michigan Press, 1990).

Checklist of Further Readings

Allen, Donald, and Warren Tallman, eds. *The Poetics of the New American Poetry.* New York: Grove, 1973.

Altieri, Charles. *Enlarging the Temple: New Directions in American Poetry During the 1960s.* Lewisburg, Pa.: Bucknell University Press, 1979.

Altieri. *Self and Sensibility in Contemporary American Poetry.* New York: Cambridge University Press, 1984.

Baker, Peter. *Obdurate Brilliance: Exteriority and the Modern Long Poem.* Gainesville: University of Florida Press, 1991.

Bartlett, Lee. *Talking Poetry: Conversations in the Workshop with Contemporary Poets.* Albuquerque: University of New Mexico Press, 1987.

Beach, Christopher. *ABC of Influence: Ezra Pound and the Remaking of American Poetic Tradition.* Berkeley & Los Angeles: University of California Press, 1992.

Bellamy, Joe David, ed. *American Poetry Observed: Poets on Their Work.* Urbana: University of Illinois Press, 1984.

Berke, Roberta Elzey. *Bounds Out of Bounds: A Compass for Recent American and British Poetry.* New York: Oxford University Press, 1981.

Bernstein, Charles. *A Poetics.* Cambridge, Mass.: Harvard University Press, 1992.

Blasing, Mutlu Konuk. *American Poetry: The Rhetoric of Its Forms.* New Haven: Yale University Press, 1987.

Blasing. *Politics and Form in Postmodern Poetry: O'Hara, Bishop, Ashbery, Merrill.* New York: Cambridge University Press, 1995.

Bloom, Harold, ed. *Contemporary Poets.* New York: Chelsea House, 1986.

Boyers, Robert, ed. *Contemporary Poetry in America: Essays and Interviews.* New York: Schocken Books, 1974.

Breslin, James E. B. *From Modern to Contemporary: American Poetry, 1945–1965.* Chicago: University of Chicago Press, 1983.

Breslin. "Poetry: 1945 to the Present," in *Columbia Literary History of the United States,* edited by Emory Elliott and others. New York: Columbia University Press, 1988, pp. 1079–1100.

Breslin, Paul. *The Psycho-Political Muse: American Poetry Since the Fifties.* Chicago: University of Chicago Press, 1987.

Bryan, Sharon, ed. *Where We Stand: Women Poets on Literary Tradition.* New York: Norton, 1993.

Carroll, Paul. *The Poem in Its Skin.* Chicago: Follett, 1968.

Codrescu, Andrei, ed. *American Poetry Since 1970: Up Late.* New York: Four Walls Eight Windows, 1987.

Conte, Joseph M. *Unending Design: The Forms of Postmodern Poetry.* Ithaca, N.Y.: Cornell University Press, 1991.

Damon, Maria. *The Dark End of the Street: Margins in American Vanguard Poetry.* Minneapolis: University of Minnesota Press, 1993.

Davidson, Michael. *The San Francisco Renaissance: Poetics and Community at Mid-Century.* Cambridge: Cambridge University Press, 1989.

Dembo, L. S. *Conceptions of Reality in Modern American Poetry.* Berkeley: University of California Press, 1966.

Dodd, Elizabeth C. *The Veiled Mirror and the Woman Poet: H. D., Louise Bogan, Elizabeth Bishop, and Louise Glück.* Columbia: University of Missouri Press, 1992.

Doreski, William. *The Modern Voice in American Poetry.* Gainesville: University of Florida Press, 1995.

Duberman, Martin B. *Black Mountain: An Exploration in Community.* Garden City, N.Y.: Doubleday, 1973.

Erkilla, Betsy. *The Wicked Sisters: Women Poets, Literary History and Discord.* New York: Oxford University Press, 1992.

Faas, Ekbert. *Towards a New American Poetics: Essays and Interviews.* Santa Barbara, Cal.: Black Sparrow Press, 1978.

Feirstein, Frederick, ed. *Expansive Poetry: Essays on the New Narrative and the New Formalism.* Santa Cruz, Cal.: Story Line Press, 1989.

Finch, Annie. *The Ghost of Meter: Culture and Prosody in American Free Verse.* Ann Arbor: University of Michigan Press, 1993.

Finch, ed. *A Formal Feeling Comes: Poems in Form by Contemporary Women.* Brownsville, Oreg.: Story Line Press, 1994.

Finkelstein, Norman. *The Utopian Moment in Contemporary American Poetry.* Lewisburg, Pa.: Bucknell University Press, 1988.

Frank, Robert, and Henry Sayre, eds. *The Line in Postmodern Poetry.* Urbana: University of Illinois Press, 1988.

Fredman, Stephen. *The Grounding of American Poetry: Charles Olson and the Emersonian Tradition.* Cambridge: Cambridge University Press, 1993.

Fredman. *Poet's Prose: The Crisis in American Verse,* second edition. Cambridge: Cambridge University Press, 1990.

Gardner, Thomas. *Discovering Ourselves in Whitman: The Contemporary American Long Poem.* Urbana & Chicago: University of Illinois Press, 1989.

Géfin, Laszlo K. *Ideogram: History of a Poetic Method.* Austin: University of Texas Press, 1982.

Gelpi, Albert. *A Coherent Splendor: The American Poetic Renaissance, 1910–1950.* New York: Cambridge University Press, 1987.

Gilbert, Roger. *Walks in the World: Representation and Experience in Modern American Poetry.* Princeton: Princeton University Press, 1991.

Gioia, Dana. *Can Poetry Matter?: Essays on Poetry and Culture.* Saint Paul, Minn.: Graywolf Press, 1992.

Glazier, Loss Pequeño. *Small Press: An Annotated Guide.* Westport, Conn.: Greenwood Press, 1992.

Gould, Jean. *Modern American Women Poets.* New York: Dodd, Mead, 1984.

Gray, Richard. *American Poetry of the Twentieth Century.* New York: Longman, 1990.

Hall, Donald. *Death to the Death of Poetry: Essays, Reviews, Notes, Interviews.* Ann Arbor: University of Michigan Press, 1994.

Hamilton, Ian, ed. *The Oxford Companion to Twentieth-Century Poetry.* New York: Oxford University Press, 1994.

Hartley, George. *Textual Politics and the Language Poets.* Bloomington: Indiana University Press, 1989.

Hartman, Charles O. *Free Verse: An Essay on Prosody.* Princeton: Princeton University Press, 1980.

Hass, Robert. *Twentieth Century Pleasures: Prose or Poetry.* New York: Ecco Press, 1984.

Heller, Michael. *Conviction's Net of Branches: Essays on the Objectivist Poets and Poetry.* Carbondale & Edwardsville: Southern Illinois University Press, 1985.

Henderson, Stephen, ed. *Understanding the New Black Poetry.* New York: William Morrow, 1973.

Hoffman, Daniel, ed. *American Poetry and Poetics.* Garden City, N.Y.: Doubleday, 1962.

Holden, Jonathan. *The Fate of American Poetry.* Athens: University of Georgia Press, 1991.

Holden. *Style and Authenticity in Postmodern Poetry.* Columbia: University of Missouri Press, 1986.

Homberger, Eric. *The Art of the Real: Poetry in England and America Since 1939.* London: Dent, 1977.

Hoover, Paul, ed. *Postmodern American Poetry.* New York: Norton, 1994.

Howard, Richard. *Alone with America: Essays on the Art of Poetry in the United States Since 1950.* New York: Atheneum, 1980.

Howard, ed. *Preferences: 51 American Poets Choose Poems From Their Own Work and From the Past.* New York: Viking, 1974.

Ignatow, David, ed. *Political Poetry.* New York: Chelsea House, 1960.

Ingersoll, Earl, Judith Kitchen, and Stan Sanvel Rublin, eds. *The Post-Confessionals: Conversations with American Poets of the Eighties.* Cranford, N.J.: Associated University Presses, 1989.

Jackson, Richard. *Acts of Mind: Conversations with Contemporary Poets.* Tuscaloosa: University of Alabama Press, 1983.

Jackson. *The Dismantling of Time in Contemporary Poetry.* Tuscaloosa: University of Alabama Press, 1988.

Juhasz, Suzanne. *Naked and Fiery Forms: Modern American Poetry by Women.* New York: Harper & Row, 1976.

Kalaidjian, Walter. *Languages of Liberation: The Social Text in Contemporary American Poetry*. New York: Columbia University Press, 1989.

Kalstone, David. *Becoming a Poet: Elizabeth Bishop with Marianne Moore and Robert Lowell*. New York: Farrar, Straus & Giroux, 1989.

Kalstone. *Five Temperaments: Elizabeth Bishop, Robert Lowell, James Merrill, Adrienne Rich, John Ashbery*. New York: Oxford University Press, 1977.

Keller, Lynn. *Re-making It New: Contemporary American Poetry and the Modernist Tradition*. Cambridge: Cambridge University Press, 1987.

Keller and Cristanne Miller, eds. *Feminist Measures: Soundings in Poetry and Theory*. Ann Arbor: University of Michigan Press, 1995.

Kostelanetz, Richard. *The Old Poetries and the New*. Ann Arbor: University of Michigan Press, 1981.

Lacey, Paul A. *The Inner War: Forms and Themes in Recent American Poetry*. Philadelphia: Fortress Press, 1972.

Larrissy, Edward. *Reading Twentieth-Century Poetry: The Language of Gender and Objects*. Oxford: Blackwell, 1990.

Lazer, Hank. *Opposing Poetries*, 2 volumes. Chicago: Northwestern University Press, 1996.

Lazer, ed. *What Is a Poet?: Essays from the Eleventh Alabama Symposium on English and American Literature*. Tuscaloosa: University of Alabama Press, 1987.

Leary, Paris, and Robert Kelly, eds. *A Controversy of Poets*. Garden City, N.Y.: Anchor, 1965.

Lehman, David. *The Big Question*. Ann Arbor: University of Michigan Press, 1995.

Lehman. *The Line Forms Here*. Ann Arbor: University of Michigan Press, 1992.

Lehman, ed. *Ecstatic Occasions, Expedient Forms: 65 Leading Contemporary Poets Select and Comment on Their Poems*. New York: Macmillan, 1987.

Lensing, George S., and Robert Moran. *Four Poets and the Emotive Imagination: Robert Bly, James Wright, Louis Simpson, and William Stafford*. Baton Rouge: Louisiana State University Press, 1976.

Lepper, Gary M. *A Bibliographical Introduction to Seventy-Five Modern American Authors*. Berkeley, Cal.: Serendipity Books, 1976.

Libby, Anthony. *Mythologies of Nothing: Mystical Death in American Poetry, 1940–1970*. Urbana: University of Illinois Press, 1984.

Lieberman, Laurence. *Unassigned Frequencies: American Poetry in Review, 1964–77*. Urbana: University of Illinois Press, 1977.

Martin, Robert K. *The Homosexual Tradition in American Poetry*. Austin: University of Texas Press, 1979.

Mazzaro, Jerome. *Postmodern American Poetry*. Urbana: University of Illinois Press, 1980.

McClatchy, J. D. *White Paper: On Contemporary American Poetry*. New York: Columbia University Press, 1989.

McClure, Michael. *Scratching the Beat Surface*. San Francisco: North Point Press, 1982.

McCorkle, James. *The Still Performance: Writing, Self, and Interconnection in Five Postmodern American Poets*. Charlottesville: University Press of Virginia, 1989.

McCorkle, ed. *Conversant Essays: Contemporary Poets on Poetry*. Detroit: Wayne State University Press, 1990.

McDowell, Robert, ed. *Poetry After Modernism*. Brownsville, Oreg.: Story Line Press, 1991.

Mersmann, James F. *Out of the Vietnam Vortex: A Study of Poets and Poetry Against the War*. Lawrence: University Press of Kansas, 1974.

Messerli, Douglas, ed. *From the Other Side of the Century: A New American Poetry, 1960–1990*. Los Angeles: Sun & Moon Press, 1994.

Middlebrook, Diane Wood, and Marilyn Yalom, eds. *Coming to Light: American Women Poets in the Twentieth Century*. Ann Arbor: University of Michigan Press, 1985.

Miller, James E. Jr. *The American Quest for a Supreme Fiction: Whitman's Legacy in the Personal Epic*. Chicago: University of Chicago Press, 1979.

Molesworth, Charles. *The Fierce Embrace: A Study of Contemporary American Poetry*. Columbia: University of Missouri Press, 1979.

Moss, Howard, ed. *The Poet's Story*. New York: Macmillan, 1973.

Myers, Jack, and David Wojahn, eds. *A Profile of Twentieth-Century American Poetry*. Carbondale & Edwardsville: Southern Illinois University Press, 1991.

Nelson, Cary. *Our Last First Poets: Vision and History in Contemporary American Poetry*. Urbana & Chicago: University of Illinois Press, 1981.

Nelson. *Repression and Recovery: Modern American Poetry and the Politics of Cultural Memory, 1910–1945*. Madison: University of Wisconsin Press, 1989.

Ossman, David. *The Sullen Art*. New York: Corinth Books, 1967.

Ostriker, Alicia Suskin. *Stealing the Language: The Emergence of Women's Poetry in America*. Boston: Beacon, 1986.

Packard, William, ed. *The Craft of Poetry: Interviews from the New York Quarterly*. Garden City, N.Y.: Doubleday, 1974.

Palmer, Michael, ed. *Code of Signals: Recent Writings in Poetics*. Berkeley, Cal.: North Atlantic Books, 1983.

Parini, Jay, and Brett C. Millier, eds. *The Columbia History of American Poetry*. New York: Columbia University Press, 1993.

Paul, Sherman. *In Search of the Primitive: Rereading David Antin, Jerome Rothenberg, and Gary Snyder*. Baton Rouge: Louisiana State University Press, 1986.

Perelman, Bob. *The Marginalization of Poetry: Language Writing and Literary History*. Princeton: Princeton University Press, 1996.

Perelman. *The Trouble With Genius: Reading Pound, Joyce, Stein, and Zukofsky*. Berkeley: University of California Press, 1994.

Perkins, David. *A History of Modern Poetry: Modernism and After.* Cambridge, Mass.: Harvard University Press, 1987.

Perloff, Marjorie. *The Dance of the Intellect: Studies in the Poetry of the Pound Tradition.* Cambridge: Cambridge University Press, 1985.

Perloff. *Poetic License: Essays on Modernist and Postmodernist Lyric.* Evanston, Ill.: Northwestern University Press, 1990.

Perloff. *The Poetics of Indeterminacy: Rimbaud to Cage.* Princeton: Princeton University Press, 1981.

Perloff. *Radical Artifice: Writing Poetry in the Age of Media.* Chicago: University of Chicago Press, 1991.

Pinsky, Robert. *The Poet and the World.* New York: Ecco Press, 1988.

Pinsky. *The Situation of Poetry: Contemporary Poetry and Its Traditions.* Princeton: Princeton University Press, 1976.

Poulin, A. Jr., ed. *Contemporary American Poetry.* Boston: Houghton Mifflin, 1971.

Quartermain, Peter. *Disjunctive Poetics: From Gertrude Stein and Louis Zukofsky to Susan Howe.* Cambridge: Cambridge University Press, 1992.

Redmond, Eugene. *Drumvoices: The Mission of Afro-American Poetry: A Critical History.* Garden City, N.Y.: Anchor, 1976.

Reinfeld, Linda. *Language Poetry: Writing as Rescue.* Baton Rouge: Louisiana State University Press, 1992.

Richman, Robert, ed. *The Direction of Poetry: An Anthology of Rhymed and Metered Verse Written in the English Language Since 1975.* Boston: Houghton Mifflin, 1988.

Rosenthal, M. L. *The New Poets.* New York: Oxford University Press, 1967.

Rosenthal, and Sally M. Gall. *The Modern Poetic Sequence: The Genius of Modern Poetry.* New York: Oxford University Press, 1983.

Ross, Andrew. *The Failure of Modernism: Symptoms of American Poetry.* New York: Columbia University Press, 1986.

Rothenberg, Jerome, and Pierre Joris, eds. *Poems for the Millennium: The University of California Book of Modern & Postmodern Poetry,* volume one: *From Fin-de-Siècle to Negritude.* Berkeley & Los Angeles: University of California Press, 1995.

Schultz, Susan M., ed. *The Tribe of John: Ashbery and Contemporary Poetry.* Tuscaloosa: University of Alabama Press, 1995.

Shaw, Robert B., ed. *American Poetry Since 1960: Some Critical Perspectives.* Chester Springs, Pa.: Dufour, 1974.

Shetley, Vernon. *After the Death of Poetry: Poet and Audience in Contemporary America.* Durham, N.C.: Duke University Press, 1993.

Shucard, Alan, Fred Moramarco, and William Sullivan. *Modern American Poetry, 1865–1950.* Boston: Twayne, 1989.

Simpson, Eileen. *Poets in Their Youth: A Memoir*. New York: Random House, 1982.

Smith, Dave. *Local Assays*. Urbana: University of Illinois Press, 1985.

Sorrentino, Gilbert. *Something Said*. San Francisco: North Point Press, 1984.

Spiegelman, Willard. *The Didactic Muse: Scenes of Instruction in Contemporary American Poetry*. Princeton: Princeton University Press, 1989.

Steele, Timothy. *Missing Measures: Modern Poetry and the Revolt Against Meter*. Fayetteville: University of Arkansas Press, 1990.

Stepanchev, Stephen. *American Poetry Since 1945: A Critical Survey*. New York: Harper & Row, 1965.

Taggart, John. *Songs of Degrees: Essays on Contemporary Poetry and Poetics*. Tuscaloosa: University of Alabama Press, 1994.

Thurley, Geoffrey. *The American Moment: American Poetry in the Mid-century*. London: E. Arnold, 1977.

Tytell, John. *Naked Angels: The Lives and Literature of the Beat Generation*. New York: McGraw-Hill, 1976.

Vendler, Helen. *The Music of What Happens*. Cambridge, Mass.: Harvard University Press, 1988.

von Hallberg, Robert. *American Poetry and Culture, 1945–1980*. Cambridge, Mass.: Harvard University Press, 1985.

Young, David, and Stuart Friebert, eds. *A Field Guide to Contemporary Poetry and Poetics*. New York: Longman, 1980.

Contributors

Michael Basinski	*State University of New York at Buffalo*
Ashley Brown	*University of South Carolina*
Richard J. Calhoun	*Clemson University*
Joseph M. Conte	*State University of New York at Buffalo*
Andrew Elkins	*Chadron State College*
Steven R. Evans	*Brown University*
Edward Halsey Foster	*Stevens Institute of Technology*
Robert S. Friedman	*New Jersey Institute of Technology*
Laszlo K. Géfin	*Concordia University*
Loss Pequeño Glazier	*State University of New York at Buffalo*
Elizabeth B. House	*Augusta State University*
T. C. Marshall	*Cabrillo College*
James McCorkle	*Geneva, New York*
Diane Wood Middlebrook	*Stanford University*
Brett C. Millier	*Middlebury College*
Peter Quartermain	*University of British Columbia*
Susan M. Schultz	*University of Hawaii at Manoa*
Mark Scroggins	*Florida Atlantic University*
Eric Murphy Selinger	*DePaul University*
Kenneth Sherwood	*State University of New York at Buffalo*
Willard Spiegelman	*Southern Methodist University*
Chris Stroffolino	*State University of New York at Albany*
Keith Tuma	*Miami University*

Cumulative Index

Dictionary of Literary Biography, Volumes 1-169
Dictionary of Literary Biography Yearbook, 1980-1995
Dictionary of Literary Biography Documentary Series, Volumes 1-14

Cumulative Index

DLB before number: *Dictionary of Literary Biography*, Volumes 1-169
Y before number: *Dictionary of Literary Biography Yearbook*, 1980-1995
DS before number: *Dictionary of Literary Biography Documentary Series*, Volumes 1-14

A

Abbey Press DLB-49

The Abbey Theatre and Irish Drama, 1900-1945 .... DLB-10

Abbot, Willis J. 1863-1934 .... DLB-29

Abbott, Jacob 1803-1879 .... DLB-1

Abbott, Lee K. 1947- DLB-130

Abbott, Lyman 1835-1922 .... DLB-79

Abbott, Robert S. 1868-1940 .... DLB-29, 91

Abelard, Peter circa 1079-1142 ..... DLB-115

Abelard-Schuman DLB-46

Abell, Arunah S. 1806-1888 .... DLB-43

Abercrombie, Lascelles 1881-1938 ... DLB-19

Aberdeen University Press Limited DLB-106

Abish, Walter 1931- DLB-130

Ablesimov, Aleksandr Onisimovich 1742-1783 .... DLB-150

Abraham à Sancta Clara 1644-1709 .... DLB-168

Abrahams, Peter 1919- DLB-117

Abrams, M. H. 1912- DLB-67

Abrogans circa 790-800 .... DLB-148

Abschatz, Hans Aßmann von 1646-1699 .... DLB-168

Abse, Dannie 1923- DLB-27

Academy Chicago Publishers DLB-46

Accrocca, Elio Filippo 1923- DLB-128

Ace Books DLB-46

Achebe, Chinua 1930- DLB-117

Achtenberg, Herbert 1938- DLB-124

Ackerman, Diane 1948- DLB-120

Ackroyd, Peter 1949- DLB-155

Acorn, Milton 1923-1986 .... DLB-53

Acosta, Oscar Zeta 1935?- DLB-82

Actors Theatre of Louisville DLB-7

Adair, James 1709?-1783? DLB-30

Adam, Graeme Mercer 1839-1912 ... DLB-99

Adame, Leonard 1947- DLB-82

Adamic, Louis 1898-1951 .... DLB-9

Adams, Alice 1926- Y-86

Adams, Brooks 1848-1927 .... DLB-47

Adams, Charles Francis, Jr. 1835-1915 .... DLB-47

Adams, Douglas 1952- Y-83

Adams, Franklin P. 1881-1960 ..... DLB-29

Adams, Henry 1838-1918 .... DLB-12, 47

Adams, Herbert Baxter 1850-1901 ... DLB-47

Adams, J. S. and C. [publishing house] DLB-49

Adams, James Truslow 1878-1949 ... DLB-17

Adams, John 1735-1826 .... DLB-31

Adams, John Quincy 1767-1848 ..... DLB-37

Adams, Léonie 1899-1988 .... DLB-48

Adams, Levi 1802-1832 .... DLB-99

Adams, Samuel 1722-1803 .... DLB-31, 43

Adams, Thomas 1582 or 1583-1652 .... DLB-151

Adams, William Taylor 1822-1897 .. DLB-42

Adamson, Sir John 1867-1950 ...... DLB-98

Adcock, Arthur St. John 1864-1930 .... DLB-135

Adcock, Betty 1938- DLB-105

Adcock, Betty, Certain Gifts DLB-105

Adcock, Fleur 1934- DLB-40

Addison, Joseph 1672-1719 .... DLB-101

Ade, George 1866-1944 .... DLB-11, 25

Adeler, Max (see Clark, Charles Heber)

Adonias Filho 1915-1990 .... DLB-145

Advance Publishing Company DLB-49

AE 1867-1935 .... DLB-19

Ælfric circa 955-circa 1010 .... DLB-146

Aesthetic Poetry (1873), by Walter Pater DLB-35

After Dinner Opera Company Y-92

Afro-American Literary Critics: An Introduction DLB-33

Agassiz, Jean Louis Rodolphe 1807-1873 .... DLB-1

Agee, James 1909-1955 .... DLB-2, 26, 152

The Agee Legacy: A Conference at the University of Tennessee at Knoxville Y-89

Aguilera Malta, Demetrio 1909-1981 .... DLB-145

Ai 1947- DLB-120

Aichinger, Ilse 1921- DLB-85

Aidoo, Ama Ata 1942- DLB-117

Aiken, Conrad 1889-1973 .... DLB-9, 45, 102

Aiken, Joan 1924- DLB-161

Aikin, Lucy 1781-1864 .... DLB-144, 163

Ainsworth, William Harrison 1805-1882 .... DLB-21

Aitken, George A. 1860-1917 ...... DLB-149

Aitken, Robert [publishing house] ... DLB-49

Akenside, Mark 1721-1770 .... DLB-109

Akins, Zoë 1886-1958 .... DLB-26

Alabaster, William 1568-1640 ...... DLB-132

Alain-Fournier 1886-1914 .... DLB-65

Alarcón, Francisco X. 1954- DLB-122

Alba, Nanina 1915-1968 .... DLB-41

Albee, Edward 1928- DLB-7

Albert the Great circa 1200-1280 ... DLB-115

Alberti, Rafael 1902- DLB-108

Albertinus, Aegidius circa 1560-1620 .... DLB-164

Alcott, Amos Bronson 1799-1888 ..... DLB-1

Alcott, Louisa May 1832-1888 .... DLB-1, 42, 79; DS-14

Alcott, William Andrus 1798-1859 .... DLB-1

Alcuin circa 732-804 .... DLB-148

Alden, Henry Mills 1836-1919 ...... DLB-79

Alden, Isabella 1841-1930 .... DLB-42

Alden, John B. [publishing house] DLB-49

Alden, Beardsley and Company DLB-49

Aldington, Richard
1892-1962 .... DLB-20, 36, 100, 149

Aldis, Dorothy 1896-1966 .... DLB-22

Aldiss, Brian W. 1925- DLB-14

Aldrich, Thomas Bailey
1836-1907 .... DLB-42, 71, 74, 79

Alegría, Ciro 1909-1967 .... DLB-113

Alegría, Claribel 1924- DLB-145

Aleixandre, Vicente 1898-1984 .... DLB-108

Aleramo, Sibilla 1876-1960 .... DLB-114

Alexander, Charles 1868-1923 ...... DLB-91

Alexander, Charles Wesley
[publishing house] DLB-49

Alexander, James 1691-1756 .... DLB-24

Alexander, Lloyd 1924- DLB-52

Alexander, Sir William, Earl of Stirling
1577?-1640 .... DLB-121

Alexis, Willibald 1798-1871 .... DLB-133

Alfred, King 849-899 .... DLB-146

Alger, Horatio, Jr. 1832-1899 ...... DLB-42

Algonquin Books of Chapel Hill DLB-46

Algren, Nelson
1909-1981 .... DLB-9; Y-81, 82

Allan, Andrew 1907-1974 .... DLB-88

Allan, Ted 1916- DLB-68

Allbeury, Ted 1917- DLB-87

Alldritt, Keith 1935- DLB-14

Allen, Ethan 1738-1789 .... DLB-31

Allen, Frederick Lewis 1890-1954 .. DLB-137

Allen, Gay Wilson
1903-1995 .... DLB-103; Y-95

Allen, George 1808-1876 .... DLB-59

Allen, George [publishing house] ... DLB-106

Allen, George, and Unwin
Limited DLB-112

Allen, Grant 1848-1899 .... DLB-70, 92

Allen, Henry W. 1912- Y-85

Allen, Hervey 1889-1949 .... DLB-9, 45

Allen, James 1739-1808 .... DLB-31

Allen, James Lane 1849-1925 .... DLB-71

Allen, Jay Presson 1922- DLB-26

Allen, John, and Company DLB-49

Allen, Samuel W. 1917- DLB-41

Allen, Woody 1935- DLB-44

Allende, Isabel 1942- DLB-145

Alline, Henry 1748-1784 .... DLB-99

Allingham, Margery 1904-1966 ...... DLB-77

Allingham, William 1824-1889 ...... DLB-35

Allison, W. L. [publishing house] DLB-49

The *Alliterative Morte Arthure* and
the *Stanzaic Morte Arthur*
circa 1350-1400 .... DLB-146

Allott, Kenneth 1912-1973 .... DLB-20

Allston, Washington 1779-1843 ...... DLB-1

Almon, John [publishing house] DLB-154

Alonzo, Dámaso 1898-1990 .... DLB-108

Alsop, George 1636-post 1673 ...... DLB-24

Alsop, Richard 1761-1815 .... DLB-37

Altemus, Henry, and Company DLB-49

Altenberg, Peter 1885-1919 .... DLB-81

Altolaguirre, Manuel 1905-1959 .... DLB-108

Aluko, T. M. 1918- DLB-117

Alurista 1947- DLB-82

Alvarez, A. 1929- DLB-14, 40

Amadi, Elechi 1934- DLB-117

Amado, Jorge 1912- DLB-113

Ambler, Eric 1909- DLB-77

*America: or, a Poem on the Settlement of the
British Colonies* (1780?), by Timothy
Dwight DLB-37

American Conservatory Theatre DLB-7

American Fiction and the 1930s DLB-9

American Humor: A Historical Survey
East and Northeast
South and Southwest
Midwest
West DLB-11

The American Library in Paris Y-93

American News Company DLB-49

The American Poets' Corner: The First
Three Years (1983-1986) Y-86

American Proletarian Culture:
The 1930s DS-11

American Publishing Company DLB-49

American Stationers' Company DLB-49

American Sunday-School Union DLB-49

American Temperance Union DLB-49

American Tract Society DLB-49

The American Writers Congress
(9-12 October 1981) Y-81

The American Writers Congress: A Report
on Continuing Business Y-81

Ames, Fisher 1758-1808 .... DLB-37

Ames, Mary Clemmer 1831-1884 .... DLB-23

Amini, Johari M. 1935- DLB-41

Amis, Kingsley 1922-
.... DLB-15, 27, 100, 139

Amis, Martin 1949- DLB-14

Ammons, A. R. 1926- DLB-5, 165

Amory, Thomas 1691?-1788 .... DLB-39

Anaya, Rudolfo A. 1937- DLB-82

Ancrene Riwle circa 1200-1225 ...... DLB-146

Andersch, Alfred 1914-1980 .... DLB-69

Anderson, Margaret 1886-1973 ... DLB-4, 91

Anderson, Maxwell 1888-1959 .... DLB-7

Anderson, Patrick 1915-1979 .... DLB-68

Anderson, Paul Y. 1893-1938 .... DLB-29

Anderson, Poul 1926- DLB-8

Anderson, Robert 1750-1830 ...... DLB-142

Anderson, Robert 1917- DLB-7

Anderson, Sherwood
1876-1941 .... DLB-4, 9, 86; DS-1

Andreae, Johann Valentin
1586-1654 .... DLB-164

Andreas-Salomé, Lou 1861-1937 .... DLB-66

Andres, Stefan 1906-1970 .... DLB-69

Andreu, Blanca 1959- DLB-134

Andrewes, Lancelot 1555-1626 ..... DLB-151

Andrews, Charles M. 1863-1943 .... DLB-17

Andrews, Miles Peter ?-1814 .... DLB-89

Andrian, Leopold von 1875-1951 ... DLB-81

Andrić, Ivo 1892-1975 .... DLB-147

Andrieux, Louis (see Aragon, Louis)

Andrus, Silas, and Son DLB-49

Angell, James Burrill 1829-1916 ...... DLB-64

Angelou, Maya 1928- DLB-38

Anger, Jane flourished 1589 .... DLB-136

Angers, Félicité (see Conan, Laure)

Anglo-Norman Literature in the Development
of Middle English Literature DLB-146

The Anglo-Saxon Chronicle
circa 890-1154 .... DLB-146

The "Angry Young Men" DLB-15

Angus and Robertson (UK)
Limited DLB-112

Anhalt, Edward 1914- DLB-26

Anners, Henry F. [publishing house] ...DLB-49

Annolied between 1077 and 1081 .... DLB-148

Anselm of Canterbury 1033-1109 ... DLB-115

Anstey, F. 1856-1934 .... DLB-141

Anthony, Michael 1932- DLB-125

Anthony, Piers 1934- DLB-8

Anthony Burgess's *99 Novels*:
An Opinion Poll Y-84

Antin, David 1932- DLB-169

Antin, Mary 1881-1949 ....Y-84

Anton Ulrich, Duke of Brunswick-Lüneburg
1633-1714 ....DLB-168

Antschel, Paul (see Celan, Paul)

Anyidoho, Kofi 1947- DLB-157

Anzaldúa, Gloria 1942- DLB-122

Anzengruber, Ludwig 1839-1889 ...DLB-129

Apodaca, Rudy S. 1939- DLB-82

Apple, Max 1941- DLB-130

Appleton, D., and Company DLB-49

Appleton-Century-Crofts DLB-46

Applewhite, James 1935- DLB-105

Apple-wood Books DLB-46

Aquin, Hubert 1929-1977 ....DLB-53

Aquinas, Thomas 1224 or
1225-1274 ....DLB-115

Aragon, Louis 1897-1982 ....DLB-72

Arbor House Publishing
Company DLB-46

Arbuthnot, John 1667-1735 ....DLB-101

Arcadia House DLB-46

Arce, Julio G. (see Ulica, Jorge)

Archer, William 1856-1924 ....DLB-10

The Archpoet circa 1130?-? DLB-148

Archpriest Avvakum (Petrovich)
1620?-1682 ....DLB-150

Arden, John 1930- DLB-13

Arden of Faversham DLB-62

Ardis Publishers Y-89

Ardizzone, Edward 1900-1979 .....DLB-160

Arellano, Juan Estevan 1947- DLB-122

The Arena Publishing Company DLB-49

Arena Stage DLB-7

Arenas, Reinaldo 1943-1990 ....DLB-145

Arensberg, Ann 1937- Y-82

Arguedas, José María 1911-1969 ....DLB-113

Argueta, Manlio 1936- DLB-145

Arias, Ron 1941- DLB-82

Arland, Marcel 1899-1986 ....DLB-72

Arlen, Michael 1895-1956 .. DLB-36, 77, 162

Armah, Ayi Kwei 1939- DLB-117

Der arme Hartmann
?-after 1150 ....DLB-148

Armed Services Editions DLB-46

Armstrong, Richard 1903- DLB-160

Arndt, Ernst Moritz 1769-1860 .....DLB-90

Arnim, Achim von 1781-1831 ......DLB-90

Arnim, Bettina von 1785-1859 ......DLB-90

Arno Press DLB-46

Arnold, Edwin 1832-1904 ....DLB-35

Arnold, Matthew 1822-1888 ..... DLB-32, 57

Arnold, Thomas 1795-1842 ....DLB-55

Arnold, Edward
[publishing house] DLB-112

Arnow, Harriette Simpson
1908-1986 ....DLB-6

Arp, Bill (see Smith, Charles Henry)

Arreola, Juan José 1918- DLB-113

Arrowsmith, J. W.
[publishing house] DLB-106

Arthur, Timothy Shay
1809-1885 ....DLB-3, 42, 79; DS-13

The Arthurian Tradition and Its European
Context DLB-138

Artmann, H. C. 1921- DLB-85

Arvin, Newton 1900-1963 ....DLB-103

As I See It, by Carolyn Cassady DLB-16

Asch, Nathan 1902-1964 ....DLB-4, 28

Ash, John 1948- DLB-40

Ashbery, John 1927- DLB-5, 165; Y-81

Ashendene Press DLB-112

Asher, Sandy 1942- Y-83

Ashton, Winifred (see Dane, Clemence)

Asimov, Isaac 1920-1992 .... DLB-8; Y-92

Askew, Anne circa 1521-1546 ......DLB-136

Asselin, Olivar 1874-1937 ....DLB-92

Asturias, Miguel Angel
1899-1974 ....DLB-113

Atheneum Publishers DLB-46

Atherton, Gertrude 1857-1948 .... DLB-9, 78

Athlone Press DLB-112

Atkins, Josiah circa 1755-1781 ......DLB-31

Atkins, Russell 1926- DLB-41

The Atlantic Monthly Press DLB-46

Attaway, William 1911-1986 ....DLB-76

Atwood, Margaret 1939- DLB-53

Aubert, Alvin 1930- DLB-41

Aubert de Gaspé, Phillipe-Ignace-François
1814-1841 ....DLB-99

Aubert de Gaspé, Phillipe-Joseph
1786-1871 ....DLB-99

Aubin, Napoléon 1812-1890 ....DLB-99

Aubin, Penelope 1685-circa 1731 ....DLB-39

Aubrey-Fletcher, Henry Lancelot
(see Wade, Henry)

Auchincloss, Louis 1917- DLB-2; Y-80

Auden, W. H. 1907-1973 ....DLB-10, 20

Audio Art in America: A Personal
Memoir Y-85

Auerbach, Berthold 1812-1882 .....DLB-133

Auernheimer, Raoul 1876-1948 ......DLB-81

Augustine 354-430 ....DLB-115

Austen, Jane 1775-1817 ....DLB-116

Austin, Alfred 1835-1913 ....DLB-35

Austin, Mary 1868-1934 ....DLB-9, 78

Austin, William 1778-1841 ....DLB-74

Author-Printers, 1476–1599 ....DLB-167

The Author's Apology for His Book
(1684), by John Bunyan DLB-39

An Author's Response, by
Ronald Sukenick Y-82

Authors and Newspapers
Association DLB-46

Authors' Publishing Company DLB-49

Avalon Books DLB-46

Avancini, Nicolaus 1611-1686 ......DLB-164

Avendaño, Fausto 1941- DLB-82

Averroës 1126-1198 ....DLB-115

Avery, Gillian 1926- DLB-161

Avicenna 980-1037 ....DLB-115

Avison, Margaret 1918- DLB-53

Avon Books DLB-46

Awdry, Wilbert Vere 1911- DLB-160

Awoonor, Kofi 1935- DLB-117

Ayckbourn, Alan 1939- DLB-13

Aymé, Marcel 1902-1967 ....DLB-72

Aytoun, Sir Robert 1570-1638 ......DLB-121

Aytoun, William Edmondstoune
1813-1865 ....DLB-32, 159

B

B. V. (see Thomson, James)

Babbitt, Irving 1865-1933 ....DLB-63

Cumulative Index

Babbitt, Natalie 1932- DLB-52

Babcock, John [publishing house] ... DLB-49

Baca, Jimmy Santiago 1952- DLB-122

Bache, Benjamin Franklin 1769-1798 .... DLB-43

Bachmann, Ingeborg 1926-1973 ..... DLB-85

Bacon, Delia 1811-1859 .... DLB-1

Bacon, Francis 1561-1626 .... DLB-151

Bacon, Roger circa 1214/1220-1292 .... DLB-115

Bacon, Sir Nicholas circa 1510-1579 .... DLB-132

Bacon, Thomas circa 1700-1768 .... DLB-31

Badger, Richard G., and Company DLB-49

Bage, Robert 1728-1801 .... DLB-39

Bagehot, Walter 1826-1877 .... DLB-55

Bagley, Desmond 1923-1983 .... DLB-87

Bagnold, Enid 1889-1981 ...... DLB-13, 160

Bagryana, Elisaveta 1893-1991 ..... DLB-147

Bahr, Hermann 1863-1934 ..... DLB-81, 118

Bailey, Alfred Goldsworthy 1905- DLB-68

Bailey, Francis [publishing house] ... DLB-49

Bailey, H. C. 1878-1961 .... DLB-77

Bailey, Jacob 1731-1808 .... DLB-99

Bailey, Paul 1937- DLB-14

Bailey, Philip James 1816-1902 ..... DLB-32

Baillargeon, Pierre 1916-1967 ..... DLB-88

Baillie, Hugh 1890-1966 .... DLB-29

Baillie, Joanna 1762-1851 .... DLB-93

Bailyn, Bernard 1922- DLB-17

Bainbridge, Beryl 1933- DLB-14

Baird, Irene 1901-1981 .... DLB-68

Baker, Augustine 1575-1641 .... DLB-151

Baker, Carlos 1909-1987 .... DLB-103

Baker, David 1954- DLB-120

Baker, Herschel C. 1914-1990 ..... DLB-111

Baker, Houston A., Jr. 1943- DLB-67

Baker, Samuel White 1821-1893 ... DLB-166

Baker, Walter H., Company ("Baker's Plays") DLB-49

The Baker and Taylor Company DLB-49

Balaban, John 1943- DLB-120

Bald, Wambly 1902- DLB-4

Balde, Jacob 1604-1668 .... DLB-164

Balderston, John 1889-1954 .... DLB-26

Baldwin, James 1924-1987 .... DLB-2, 7, 33; Y-87

Baldwin, Joseph Glover 1815-1864 .... DLB-3, 11

Baldwin, William circa 1515-1563 .... DLB-132

Bale, John 1495-1563 .... DLB-132

Balestrini, Nanni 1935- DLB-128

Ballantine Books DLB-46

Ballantyne, R. M. 1825-1894 ..... DLB-163

Ballard, J. G. 1930- DLB-14

Ballerini, Luigi 1940- DLB-128

Ballou, Maturin Murray 1820-1895 .... DLB-79

Ballou, Robert O. [publishing house] DLB-46

Balzac, Honoré de 1799-1855 ...... DLB-119

Bambara, Toni Cade 1939- DLB-38

Bancroft, A. L., and Company DLB-49

Bancroft, George 1800-1891 .... DLB-1, 30, 59

Bancroft, Hubert Howe 1832-1918 .... DLB-47, 140

Bangs, John Kendrick 1862-1922 .... DLB-11, 79

Banim, John 1798-1842 ...DLB-116, 158, 159

Banim, Michael 1796-1874 .... DLB-158, 159

Banks, John circa 1653-1706 .... DLB-80

Banks, Russell 1940- DLB-130

Bannerman, Helen 1862-1946 ...... DLB-141

Bantam Books DLB-46

Banville, John 1945- DLB-14

Baraka, Amiri 1934- DLB-5, 7, 16, 38; DS-8

Barbauld, Anna Laetitia 1743-1825 ...... DLB-107, 109, 142, 158

Barbeau, Marius 1883-1969 .... DLB-92

Barber, John Warner 1798-1885 ..... DLB-30

Bàrberi Squarotti, Giorgio 1929- DLB-128

Barbey d'Aurevilly, Jules-Amédée 1808-1889 .... DLB-119

Barbour, John circa 1316-1395 ..... DLB-146

Barbour, Ralph Henry 1870-1944 .... DLB-22

Barbusse, Henri 1873-1935 .... DLB-65

Barclay, Alexander circa 1475-1552 .... DLB-132

Barclay, E. E., and Company DLB-49

Bardeen, C. W. [publishing house] DLB-49

Barham, Richard Harris 1788-1845 .... DLB-159

Baring, Maurice 1874-1945 .... DLB-34

Baring-Gould, Sabine 1834-1924 .... DLB-156

Barker, A. L. 1918- DLB-14, 139

Barker, George 1913-1991 .... DLB-20

Barker, Harley Granville 1877-1946 .... DLB-10

Barker, Howard 1946- DLB-13

Barker, James Nelson 1784-1858 ..... DLB-37

Barker, Jane 1652-1727 .... DLB-39, 131

Barker, Lady Mary Anne 1831-1911 .... DLB-166

Barker, William circa 1520-after 1576 ....DLB-132

Barker, Arthur, Limited DLB-112

Barkov, Ivan Semenovich 1732-1768 .... DLB-150

Barks, Coleman 1937- DLB-5

Barlach, Ernst 1870-1938 ...... DLB-56, 118

Barlow, Joel 1754-1812 ....DLB-37

Barnard, John 1681-1770 .... DLB-24

Barne, Kitty (Mary Catherine Barne) 1883-1957 ....DLB-160

Barnes, Barnabe 1571-1609 .... DLB-132

Barnes, Djuna 1892-1982 ...... DLB-4, 9, 45

Barnes, Julian 1946- Y-93

Barnes, Margaret Ayer 1886-1967 .....DLB-9

Barnes, Peter 1931- DLB-13

Barnes, William 1801-1886 .... DLB-32

Barnes, A. S., and Company DLB-49

Barnes and Noble Books DLB-46

Barnet, Miguel 1940- DLB-145

Barney, Natalie 1876-1972 ....DLB-4

Baron, Richard W., Publishing Company DLB-46

Barr, Robert 1850-1912 .... DLB-70, 92

Barral, Carlos 1928-1989 ....DLB-134

Barrax, Gerald William 1933- DLB-41, 120

Barrès, Maurice 1862-1923 ....DLB-123

Barrett, Eaton Stannard 1786-1820 ....DLB-116

Barrie, J. M. 1860-1937 .... DLB-10, 141, 156

Barrie and Jenkins DLB-112

340

Barrio, Raymond 1921- DLB-82	Baynes, Pauline 1922- DLB-160	Beeton, S. O. [publishing house] DLB-106
Barrios, Gregg 1945- DLB-122	Bazin, Hervé 1911- DLB-83	Bégon, Elisabeth 1696-1755 .... DLB-99
Barry, Philip 1896-1949 .... DLB-7	Beach, Sylvia 1887-1962 .... DLB-4	Behan, Brendan 1923-1964 .... DLB-13
Barry, Robertine (see Françoise)	Beacon Press DLB-49	Behn, Aphra 1640?-1689 ... DLB-39, 80, 131
Barse and Hopkins DLB-46	Beadle and Adams DLB-49	Behn, Harry 1898-1973 .... DLB-61
Barstow, Stan 1928- DLB-14, 139	Beagle, Peter S. 1939- Y-80	Behrman, S. N. 1893-1973 .... DLB-7, 44
Barth, John 1930- DLB-2	Beal, M. F. 1937- Y-81	Belaney, Archibald Stansfeld (see Grey Owl)
Barthelme, Donald 1931-1989 .... DLB-2; Y-80, 89	Beale, Howard K. 1899-1959 .... DLB-17	Belasco, David 1853-1931 .... DLB-7
Barthelme, Frederick 1943- Y-85	Beard, Charles A. 1874-1948 .... DLB-17	Belford, Clarke and Company DLB-49
Bartholomew, Frank 1898-1985 .... DLB-127	A Beat Chronology: The First Twenty-five Years, 1944-1969 .... DLB-16	Belitt, Ben 1911- DLB-5
Bartlett, John 1820-1905 .... DLB-1	Beattie, Ann 1947- Y-82	Belknap, Jeremy 1744-1798 ...... DLB-30, 37
Bartol, Cyrus Augustus 1813-1900 .... DLB-1	Beattie, James 1735-1803 .... DLB-109	Bell, Clive 1881-1964 .... DS-10
Barton, Bernard 1784-1849 .... DLB-96	Beauchemin, Nérée 1850-1931 ...... DLB-92	Bell, James Madison 1826-1902 ...... DLB-50
Barton, Thomas Pennant 1803-1869 .... DLB-140	Beauchemin, Yves 1941- DLB-60	Bell, Marvin 1937- DLB-5
Bartram, John 1699-1777 .... DLB-31	Beaugrand, Honoré 1848-1906 ...... DLB-99	Bell, Millicent 1919- DLB-111
Bartram, William 1739-1823 .... DLB-37	Beaulieu, Victor-Lévy 1945- DLB-53	Bell, Quentin 1910- DLB-155
Basic Books DLB-46	Beaumont, Francis circa 1584-1616 and Fletcher, John 1579-1625 ... DLB-58	Bell, Vanessa 1879-1961 .... DS-10
Basille, Theodore (see Becon, Thomas)	Beaumont, Sir John 1583?-1627 .... DLB-121	Bell, George, and Sons DLB-106
Bass, T. J. 1932- Y-81	Beaumont, Joseph 1616–1699 ...... DLB-126	Bell, Robert [publishing house] DLB-49
Bassani, Giorgio 1916- DLB-128	Beauvoir, Simone de 1908-1986 .... DLB-72; Y-86	Bellamy, Edward 1850-1898 .... DLB-12
Basse, William circa 1583-1653 ..... DLB-121	Becher, Ulrich 1910- DLB-69	Bellamy, Joseph 1719-1790 .... DLB-31
Bassett, John Spencer 1867-1928 ..... DLB-17	Becker, Carl 1873-1945 .... DLB-17	Bellezza, Dario 1944- DLB-128
Bassler, Thomas Joseph (see Bass, T. J.)	Becker, Jurek 1937- DLB-75	La Belle Assemblée 1806-1837 .... DLB-110
Bate, Walter Jackson 1918- DLB-67, 103	Becker, Jurgen 1932- DLB-75	Belloc, Hilaire 1870-1953 .... DLB-19, 100, 141
Bateman, Stephen circa 1510-1584 .... DLB-136	Beckett, Samuel 1906-1989 .... DLB-13, 15; Y-90	Bellow, Saul 1915- DLB-2, 28; Y-82; DS-3
Bates, H. E. 1905-1974 .... DLB-162	Beckford, William 1760-1844 .... DLB-39	Belmont Productions DLB-46
Bates, Katharine Lee 1859-1929 ..... DLB-71	Beckham, Barry 1944- DLB-33	Bemelmans, Ludwig 1898-1962 ...... DLB-22
Batsford, B. T. [publishing house] DLB-106	Becon, Thomas circa 1512-1567 .... DLB-136	Bemis, Samuel Flagg 1891-1973 ..... DLB-17
Battiscombe, Georgina 1905- DLB-155	Beddoes, Thomas 1760-1808 ...... DLB-158	Bemrose, William [publishing house] DLB-106
The Battle of Maldon circa 1000 ...... DLB-146	Beddoes, Thomas Lovell 1803-1849 .... DLB-96	Benchley, Robert 1889-1945 .... DLB-11
Bauer, Bruno 1809-1882 .... DLB-133	Bede circa 673-735 .... DLB-146	Benedetti, Mario 1920- DLB-113
Bauer, Wolfgang 1941- DLB-124	Beecher, Catharine Esther 1800-1878 .... DLB-1	Benedictus, David 1938- DLB-14
Baum, L. Frank 1856-1919 .... DLB-22	Beecher, Henry Ward 1813-1887 .... DLB-3, 43	Benedikt, Michael 1935- DLB-5
Baum, Vicki 1888-1960 .... DLB-85	Beer, George L. 1872-1920 .... DLB-47	Benét, Stephen Vincent 1898-1943 ....DLB-4, 48, 102
Baumbach, Jonathan 1933- Y-80	Beer, Johann 1655-1700 .... DLB-168	Benét, William Rose 1886-1950 ..... DLB-45
Bausch, Richard 1945- DLB-130	Beer, Patricia 1919- DLB-40	Benford, Gregory 1941- Y-82
Bawden, Nina 1925- DLB-14, 161	Beerbohm, Max 1872-1956 ..... DLB-34, 100	Benjamin, Park 1809-1864 .....DLB-3, 59, 73
Bax, Clifford 1886-1962 ....DLB-10, 100	Beer-Hofmann, Richard 1866-1945 .... DLB-81	Benlowes, Edward 1602-1676 ...... DLB-126
Baxter, Charles 1947- DLB-130	Beers, Henry A. 1847-1926 .... DLB-71	Benn, Gottfried 1886-1956 .... DLB-56
Bayer, Eleanor (see Perry, Eleanor)		Benn Brothers Limited DLB-106
Bayer, Konrad 1932-1964 .... DLB-85		Bennett, Arnold 1867-1931 .... DLB-10, 34, 98, 135

Bennett, Charles 1899- DLB-44	Bernstein, Charles 1950- DLB-169	Billinger, Richard 1890-1965 . . . . . . .DLB-124
Bennett, Gwendolyn 1902- DLB-51	Berriault, Gina 1926- DLB-130	Billings, John Shaw 1898-1975 . . . . . .DLB-137
Bennett, Hal 1930- DLB-33	Berrigan, Daniel 1921- DLB-5	Billings, Josh (see Shaw, Henry Wheeler)
Bennett, James Gordon 1795-1872 . . . DLB-43	Berrigan, Ted 1934-1983 . . . . .DLB-5, 169	Binding, Rudolf G. 1867-1938 . . . . . . .DLB-66
Bennett, James Gordon, Jr. 1841-1918 . . . . DLB-23	Berry, Wendell 1934- DLB-5, 6	Bingham, Caleb 1757-1817 . . . . .DLB-42
Bennett, John 1865-1956 . . . . DLB-42	Berryman, John 1914-1972 . . . . DLB-48	Bingham, George Barry 1906-1988 . . . . .DLB-127
Bennett, Louise 1919- DLB-117	Bersianik, Louky 1930- DLB-60	Bingley, William
Benoit, Jacques 1941- DLB-60	Bertolucci, Attilio 1911- DLB-128	[publishing house] DLB-154
Benson, A. C. 1862-1925 . . . . DLB-98	Berton, Pierre 1920- DLB-68	Binyon, Laurence 1869-1943 . . . . .DLB-19
Benson, E. F. 1867-1940 . . . . . . DLB-135, 153	Besant, Sir Walter 1836-1901 . . . . . . DLB-135	*Biographia Brittanica* DLB-142
Benson, Jackson J. 1930- DLB-111	Bessette, Gerard 1920- DLB-53	Biographical Documents I Y-84
Benson, Robert Hugh 1871-1914 . . . DLB-153	Bessie, Alvah 1904-1985 . . . . DLB-26	Biographical Documents II Y-85
Benson, Stella 1892-1933 . . . . . . DLB-36, 162	Bester, Alfred 1913-1987 . . . . DLB-8	Bioren, John [publishing house] DLB-49
Bentham, Jeremy 1748-1832 . . . DLB-107, 158	The Bestseller Lists: An Assessment Y-84	Bioy Casares, Adolfo 1914- DLB-113
Bentley, E. C. 1875-1956 . . . . DLB-70	Betjeman, John 1906-1984 . . . . . DLB-20; Y-84	Bird, Isabella Lucy 1831-1904 . . . . . . .DLB-166
Bentley, Richard [publishing house] DLB-106	Betocchi, Carlo 1899-1986 . . . . DLB-128	Bird, William 1888-1963 . . . . .DLB-4
Benton, Robert 1932- and Newman, David 1937- DLB-44	Bettarini, Mariella 1942- DLB-128	Birken, Sigmund von 1626-1681 . . .DLB-164
Benziger Brothers DLB-49	Betts, Doris 1932- Y-82	Birney, Earle 1904- DLB-88
Beowulf circa 900-1000 or 790-825 . . . . DLB-146	Beveridge, Albert J. 1862-1927 . . . . . DLB-17	Birrell, Augustine 1850-1933 . . . . .DLB-98
Beresford, Anne 1929- DLB-40	Beverley, Robert circa 1673-1722 . . . . DLB-24, 30	Bishop, Elizabeth 1911-1979 . . . . . DLB-5, 169
Beresford, John Davys 1873-1947 . . . . DLB-162	Beyle, Marie-Henri (see Stendhal)	Bishop, John Peale 1892-1944 . . DLB-4, 9, 45
Beresford-Howe, Constance 1922- DLB-88	Bianco, Margery Williams 1881-1944 . . . . DLB-160	Bismarck, Otto von 1815-1898 . . . . . .DLB-129
Berford, R. G., Company DLB-49	Bibaud, Adèle 1854-1941 . . . . DLB-92	Bisset, Robert 1759-1805 . . . . .DLB-142
Berg, Stephen 1934- DLB-5	Bibaud, Michel 1782-1857 . . . . DLB-99	Bissett, Bill 1939- DLB-53
Bergengruen, Werner 1892-1964 . . . . DLB-56	Bibliographical and Textual Scholarship Since World War II Y-89	Bitzius, Albert (see Gotthelf, Jeremias)
Berger, John 1926- DLB-14	The Bicentennial of James Fenimore Cooper: An International Celebration Y-89	Black, David (D. M.) 1941- DLB-40
Berger, Meyer 1898-1959 . . . . DLB-29		Black, Winifred 1863-1936 . . . . .DLB-25
Berger, Thomas 1924- DLB-2; Y-80		Black, Walter J. [publishing house] DLB-46
Berkeley, Anthony 1893-1971 . . . . . . DLB-77	Bichsel, Peter 1935- DLB-75	The Black Aesthetic: Background DS-8
Berkeley, George 1685-1753 . . . . DLB-31, 101	Bickerstaff, Isaac John 1733-circa 1808 . . . . DLB-89	The Black Arts Movement, by Larry Neal DLB-38
The Berkley Publishing Corporation DLB-46	Biddle, Drexel [publishing house] DLB-49	Black Theaters and Theater Organizations in America, 1961-1982: A Research List DLB-38
Berlin, Lucia 1936- DLB-130	Bidermann, Jacob 1577 or 1578-1639 . . . . DLB-164	
Bernal, Vicente J. 1888-1915 . . . . DLB-82	Bidwell, Walter Hilliard 1798-1881 . . . . DLB-79	Black Theatre: A Forum [excerpts] DLB-38
Bernanos, Georges 1888-1948 . . . . . . DLB-72	Bienek, Horst 1930- DLB-75	Blackamore, Arthur 1679-? DLB-24, 39
Bernard, Harry 1898-1979 . . . . DLB-92	Bierbaum, Otto Julius 1865-1910 . . . . DLB-66	Blackburn, Alexander L. 1929- Y-85
Bernard, John 1756-1828 . . . . DLB-37	Bierce, Ambrose 1842-1914? DLB-11, 12, 23, 71, 74	Blackburn, Paul 1926-1971 . . . . DLB-16; Y-81
Bernard of Chartres circa 1060-1124? DLB-115	Bigelow, William F. 1879-1966 . . . . . . DLB-91	Blackburn, Thomas 1916-1977 . . . . . .DLB-27
Bernhard, Thomas 1931-1989 . . . . DLB-85, 124	Biggle, Lloyd, Jr. 1923- DLB-8	Blackmore, R. D. 1825-1900 . . . . .DLB-18
	Biglow, Hosea (see Lowell, James Russell)	Blackmore, Sir Richard 1654-1729 . . . . .DLB-131
	Bigongiari, Piero 1914- DLB-128	Blackmur, R. P. 1904-1965 . . . . .DLB-63
		Blackwell, Basil, Publisher DLB-106

Blackwood, Algernon Henry
    1869-1951 ....DLB-153, 156

Blackwood, Caroline 1931- DLB-14

Blackwood, William, and
    Sons, Ltd. DLB-154

Blackwood's Edinburgh Magazine
    1817-1980 ....DLB-110

Blair, Eric Arthur (see Orwell, George)

Blair, Francis Preston 1791-1876 .....DLB-43

Blair, James circa 1655-1743 ....DLB-24

Blair, John Durburrow 1759-1823 ....DLB-37

Blais, Marie-Claire 1939- DLB-53

Blaise, Clark 1940- DLB-53

Blake, Nicholas 1904-1972 ....DLB-77
    (see Day Lewis, C.)

Blake, William
    1757-1827 .... DLB-93, 154, 163

The Blakiston Company DLB-49

Blanchot, Maurice 1907- DLB-72

Blanckenburg, Christian Friedrich von
    1744-1796 ....DLB-94

Blaser, Robin 1925- DLB-165

Bledsoe, Albert Taylor
    1809-1877 ....DLB-3, 79

Blelock and Company DLB-49

Blennerhassett, Margaret Agnew
    1773-1842 ....DLB-99

Bles, Geoffrey
    [publishing house] DLB-112

Blessington, Marguerite, Countess of
    1789-1849 ....DLB-166

The Blickling Homilies
    circa 971 ....DLB-146

Blish, James 1921-1975 ....DLB-8

Bliss, E., and E. White
    [publishing house] DLB-49

Bliven, Bruce 1889-1977 ....DLB-137

Bloch, Robert 1917-1994 ....DLB-44

Block, Rudolph (see Lessing, Bruno)

Blondal, Patricia 1926-1959 ....DLB-88

Bloom, Harold 1930- DLB-67

Bloomer, Amelia 1818-1894 ....DLB-79

Bloomfield, Robert 1766-1823 ....DLB-93

Bloomsbury Group DS-10

Blotner, Joseph 1923- DLB-111

Bloy, Léon 1846-1917 ....DLB-123

Blume, Judy 1938- DLB-52

Blunck, Hans Friedrich 1888-1961 ...DLB-66

Blunden, Edmund
    1896-1974 ....DLB-20, 100, 155

Blunt, Wilfrid Scawen 1840-1922 .... DLB-19

Bly, Nellie (see Cochrane, Elizabeth)

Bly, Robert 1926- DLB-5

Blyton, Enid 1897-1968 .... DLB-160

Boaden, James 1762-1839 .... DLB-89

Boas, Frederick S. 1862-1957 ...... DLB-149

The Bobbs-Merrill Archive at the
    Lilly Library, Indiana University ... Y-90

The Bobbs-Merrill Company DLB-46

Bobrov, Semen Sergeevich
    1763?-1810 .... DLB-150

Bobrowski, Johannes 1917-1965 ..... DLB-75

Bodenheim, Maxwell 1892-1954 .. DLB-9, 45

Bodenstedt, Friedrich von
    1819-1892 .... DLB-129

Bodini, Vittorio 1914-1970 .... DLB-128

Bodkin, M. McDonnell
    1850-1933 .... DLB-70

Bodley Head DLB-112

Bodmer, Johann Jakob 1698-1783 ... DLB-97

Bodmershof, Imma von 1895-1982 .. DLB-85

Bodsworth, Fred 1918- DLB-68

Boehm, Sydney 1908- DLB-44

Boer, Charles 1939- DLB-5

Boethius circa 480-circa 524 ..... DLB-115

Boethius of Dacia circa 1240-? DLB-115

Bogan, Louise 1897-1970 ...... DLB-45, 169

Bogarde, Dirk 1921- DLB-14

Bogdanovich, Ippolit Fedorovich
    circa 1743-1803 .... DLB-150

Bogue, David [publishing house] ... DLB-106

Böhme, Jakob 1575-1624 .... DLB-164

Bohn, H. G. [publishing house] DLB-106

Bohse, August 1661-1742 .... DLB-168

Boie, Heinrich Christian
    1744-1806 .... DLB-94

Bok, Edward W. 1863-1930 .... DLB-91

Boland, Eavan 1944- DLB-40

Bolingbroke, Henry St. John, Viscount
    1678-1751 .... DLB-101

Böll, Heinrich 1917-1985 ..... Y-85, DLB-69

Bolling, Robert 1738-1775 .... DLB-31

Bolotov, Andrei Timofeevich
    1738-1833 .... DLB-150

Bolt, Carol 1941- DLB-60

Bolt, Robert 1924- DLB-13

Bolton, Herbert E. 1870-1953 ....DLB-17

Bonaventura DLB-90

Bonaventure circa 1217-1274 ......DLB-115

Bond, Edward 1934- DLB-13

Bond, Michael 1926- DLB-161

Boni, Albert and Charles
    [publishing house] DLB-46

Boni and Liveright DLB-46

Robert Bonner's Sons DLB-49

Bontemps, Arna 1902-1973 ......DLB-48, 51

The Book League of America DLB-46

Book Reviewing in America: I Y-87

Book Reviewing in America: II Y-88

Book Reviewing in America: III Y-89

Book Reviewing in America: IV Y-90

Book Reviewing in America: V Y-91

Book Reviewing in America: VI Y-92

Book Reviewing in America: VII Y-93

Book Reviewing in America: VIII Y-94

Book Reviewing in America and the
    Literary Scene Y-95

Book Supply Company DLB-49

The Book Trade History Group Y-93

The Booker Prize
    Address by Anthony Thwaite,
    Chairman of the Booker Prize Judges
    Comments from Former Booker
    Prize Winners Y-86

Boorde, Andrew circa 1490-1549 ...DLB-136

Boorstin, Daniel J. 1914- DLB-17

Booth, Mary L. 1831-1889 ....DLB-79

Booth, Philip 1925- Y-82

Booth, Wayne C. 1921- DLB-67

Borchardt, Rudolf 1877-1945 ....DLB-66

Borchert, Wolfgang
    1921-1947 ....DLB-69, 124

Borel, Pétrus 1809-1859 ....DLB-119

Borges, Jorge Luis
    1899-1986 ....DLB-113; Y-86

Börne, Ludwig 1786-1837 ....DLB-90

Borrow, George
    1803-1881 .... DLB-21, 55, 166

Bosch, Juan 1909- DLB-145

Bosco, Henri 1888-1976 ....DLB-72

Bosco, Monique 1927- DLB-53

Boston, Lucy M. 1892-1990 ....DLB-161

Boswell, James 1740-1795 ..... DLB-104, 142

Botev, Khristo 1847-1876 .... DLB-147

Botta, Anne C. Lynch 1815-1891 ...... DLB-3

Bottomley, Gordon 1874-1948 ...... DLB-10

Bottoms, David 1949- DLB-120; Y-83

Bottrall, Ronald 1906- DLB-20

Boucher, Anthony 1911-1968 .... DLB-8

Boucher, Jonathan 1738-1804 .... DLB-31

Boucher de Boucherville, George 1814-1894 .... DLB-99

Boudreau, Daniel (see Coste, Donat)

Bourassa, Napoléon 1827-1916 ..... DLB-99

Bourget, Paul 1852-1935 .... DLB-123

Bourinot, John George 1837-1902 ... DLB-99

Bourjaily, Vance 1922- DLB-2, 143

Bourne, Edward Gaylord 1860-1908 .... DLB-47

Bourne, Randolph 1886-1918 .... DLB-63

Bousoño, Carlos 1923- DLB-108

Bousquet, Joë 1897-1950 .... DLB-72

Bova, Ben 1932- Y-81

Bovard, Oliver K. 1872-1945 .... DLB-25

Bove, Emmanuel 1898-1945 .... DLB-72

Bowen, Elizabeth 1899-1973 .... DLB-15, 162

Bowen, Francis 1811-1890 .... DLB-1, 59

Bowen, John 1924- DLB-13

Bowen, Marjorie 1886-1952 .... DLB-153

Bowen-Merrill Company DLB-49

Bowering, George 1935- DLB-53

Bowers, Claude G. 1878-1958 ...... DLB-17

Bowers, Edgar 1924- DLB-5

Bowers, Fredson Thayer 1905-1991 .... DLB-140; Y-91

Bowles, Paul 1910- DLB-5, 6

Bowles, Samuel III 1826-1878 ..... DLB-43

Bowles, William Lisles 1762-1850 ... DLB-93

Bowman, Louise Morey 1882-1944 .... DLB-68

Boyd, James 1888-1944 .... DLB-9

Boyd, John 1919- DLB-8

Boyd, Thomas 1898-1935 .... DLB-9

Boyesen, Hjalmar Hjorth 1848-1895 .... DLB-12, 71; DS-13

Boyle, Kay 1902-1992 .... DLB-4, 9, 48, 86; Y-93

Boyle, Roger, Earl of Orrery 1621-1679 .... DLB-80

Boyle, T. Coraghessan 1948- Y-86

Brackenbury, Alison 1953- DLB-40

Brackenridge, Hugh Henry 1748-1816 .... DLB-11, 37

Brackett, Charles 1892-1969 ...... DLB-26

Brackett, Leigh 1915-1978 .... DLB-8, 26

Bradburn, John [publishing house] DLB-49

Bradbury, Malcolm 1932- DLB-14

Bradbury, Ray 1920- DLB-2, 8

Bradbury and Evans DLB-106

Braddon, Mary Elizabeth 1835-1915 .... DLB-18, 70, 156

Bradford, Andrew 1686-1742 .... DLB-43, 73

Bradford, Gamaliel 1863-1932 ...... DLB-17

Bradford, John 1749-1830 .... DLB-43

Bradford, Roark 1896-1948 .... DLB-86

Bradford, William 1590-1657 .... DLB-24, 30

Bradford, William III 1719-1791 .... DLB-43, 73

Bradlaugh, Charles 1833-1891 ...... DLB-57

Bradley, David 1950- DLB-33

Bradley, Marion Zimmer 1930- DLB-8

Bradley, William Aspenwall 1878-1939 .... DLB-4

Bradley, Ira, and Company DLB-49

Bradley, J. W., and Company DLB-49

Bradstreet, Anne 1612 or 1613-1672 .... DLB-24

Bradwardine, Thomas circa 1295-1349 .... DLB-115

Brady, Frank 1924-1986 .... DLB-111

Brady, Frederic A. [publishing house] DLB-49

Bragg, Melvyn 1939- DLB-14

Brainard, Charles H. [publishing house] DLB-49

Braine, John 1922-1986 .... DLB-15; Y-86

Braithwait, Richard 1588-1673 ..... DLB-151

Braithwaite, William Stanley 1878-1962 .... DLB-50, 54

Braker, Ulrich 1735-1798 .... DLB-94

Bramah, Ernest 1868-1942 .... DLB-70

Branagan, Thomas 1774-1843 .... DLB-37

Branch, William Blackwell 1927- DLB-76

Branden Press DLB-46

Brassey, Lady Annie (Allnutt) 1839-1887 .... DLB-166

Brathwaite, Edward Kamau 1930- DLB-125

Brault, Jacques 1933- DLB-53

Braun, Volker 1939- DLB-75

Brautigan, Richard 1935-1984 .... DLB-2, 5; Y-80, 84

Braxton, Joanne M. 1950- DLB-41

Bray, Anne Eliza 1790-1883 ...... DLB-116

Bray, Thomas 1656-1730 .... DLB-24

Braziller, George [publishing house] DLB-46

The Bread Loaf Writers' Conference 1983 .... Y-84

The Break-Up of the Novel (1922), by John Middleton Murry DLB-36

Breasted, James Henry 1865-1935 .... DLB-47

Brecht, Bertolt 1898-1956 ..... DLB-56, 124

Bredel, Willi 1901-1964 .... DLB-56

Breitinger, Johann Jakob 1701-1776 .... DLB-97

Bremser, Bonnie 1939- DLB-16

Bremser, Ray 1934- DLB-16

Brentano, Bernard von 1901-1964 .... DLB-56

Brentano, Clemens 1778-1842 ..... DLB-90

Brentano's DLB-49

Brenton, Howard 1942- DLB-13

Breton, André 1896-1966 .... DLB-65

Breton, Nicholas circa 1555-circa 1626 .... DLB-136

The Breton Lays 1300-early fifteenth century DLB-146

Brewer, Warren and Putnam DLB-46

Brewster, Elizabeth 1922- DLB-60

Bridgers, Sue Ellen 1942- DLB-52

Bridges, Robert 1844-1930 ...... DLB-19, 98

Bridie, James 1888-1951 .... DLB-10

Briggs, Charles Frederick 1804-1877 .... DLB-3

Brighouse, Harold 1882-1958 .... DLB-10

Bright, Mary Chavelita Dunne (see Egerton, George)

Brimmer, B. J., Company DLB-46

Brines, Francisco 1932- DLB-134

Brinley, George, Jr. 1817-1875 ..... DLB-140

Brinnin, John Malcolm 1916- DLB-48

Brisbane, Albert 1809-1890 .... DLB-3

Brisbane, Arthur 1864-1936 .... DLB-25

British Academy DLB-112

The British Library and the Regular
    Readers' Group Y-91

The British Critic 1793-1843 ....DLB-110

*The British Review and London
    Critical Journal* 1811-1825 ....DLB-110

Brito, Aristeo 1942- DLB-122

Broadway Publishing Company DLB-46

Broch, Hermann 1886-1951 .....DLB-85, 124

Brochu, André 1942- DLB-53

Brock, Edwin 1927- DLB-40

Brockes, Barthold Heinrich
    1680-1747 ....DLB-168

Brod, Max 1884-1968 ....DLB-81

Brodber, Erna 1940- DLB-157

Brodhead, John R. 1814-1873 ....DLB-30

Brodkey, Harold 1930- DLB-130

Brome, Richard circa 1590-1652 .....DLB-58

Brome, Vincent 1910- DLB-155

Bromfield, Louis 1896-1956 .... DLB-4, 9, 86

Broner, E. M. 1930- DLB-28

Bronk, William 1918- DLB-165

Bronnen, Arnolt 1895-1959 ....DLB-124

Brontë, Anne 1820-1849 ....DLB-21

Brontë, Charlotte 1816-1855 ....DLB-21, 159

Brontë, Emily 1818-1848 ....DLB-21, 32

Brooke, Frances 1724-1789 ......DLB-39, 99

Brooke, Henry 1703?-1783 ....DLB-39

Brooke, L. Leslie 1862-1940 ......DLB-141

Brooke, Rupert 1887-1915 ....DLB-19

Brooker, Bertram 1888-1955 ....DLB-88

Brooke-Rose, Christine 1926- DLB-14

Brookner, Anita 1928- Y-87

Brooks, Charles Timothy
    1813-1883 ....DLB-1

Brooks, Cleanth 1906-1994 ....DLB-63; Y-94

Brooks, Gwendolyn
    1917- DLB-5, 76, 165

Brooks, Jeremy 1926- DLB-14

Brooks, Mel 1926- DLB-26

Brooks, Noah 1830-1903 .... DLB-42; DS-13

Brooks, Richard 1912-1992 ....DLB-44

Brooks, Van Wyck
    1886-1963 .... DLB-45, 63, 103

Brophy, Brigid 1929- DLB-14

Brossard, Chandler 1922-1993 .....DLB-16

Brossard, Nicole 1943- DLB-53

Broster, Dorothy Kathleen
    1877-1950 .... DLB-160

Brother Antoninus (see Everson, William)

Brougham and Vaux, Henry Peter
    Brougham, Baron
    1778-1868 .... DLB-110, 158

Brougham, John 1810-1880 .... DLB-11

Broughton, James 1913- DLB-5

Broughton, Rhoda 1840-1920 .... DLB-18

Broun, Heywood 1888-1939 .... DLB-29

Brown, Alice 1856-1948 .... DLB-78

Brown, Bob 1886-1959 .... DLB-4, 45

Brown, Cecil 1943- DLB-33

Brown, Charles Brockden
    1771-1810 ....DLB-37, 59, 73

Brown, Christy 1932-1981 .... DLB-14

Brown, Dee 1908- Y-80

Brown, Frank London 1927-1962 ... DLB-76

Brown, Fredric 1906-1972 .... DLB-8

Brown, George Mackay
    1921- DLB-14, 27, 139

Brown, Harry 1917-1986 .... DLB-26

Brown, Marcia 1918- DLB-61

Brown, Margaret Wise
    1910-1952 .... DLB-22

Brown, Morna Doris (see Ferrars, Elizabeth)

Brown, Oliver Madox
    1855-1874 .... DLB-21

Brown, Sterling
    1901-1989 ....DLB-48, 51, 63

Brown, T. E. 1830-1897 .... DLB-35

Brown, William Hill 1765-1793 ..... DLB-37

Brown, William Wells
    1814-1884 .... DLB-3, 50

Browne, Charles Farrar
    1834-1867 .... DLB-11

Browne, Francis Fisher
    1843-1913 .... DLB-79

Browne, Michael Dennis
    1940- DLB-40

Browne, Sir Thomas 1605-1682 .... DLB-151

Browne, William, of Tavistock
    1590-1645 .... DLB-121

Browne, Wynyard 1911-1964 .... DLB-13

Browne and Nolan DLB-106

Brownell, W. C. 1851-1928 .... DLB-71

Browning, Elizabeth Barrett
    1806-1861 .... DLB-32

Browning, Robert
    1812-1889 ....DLB-32, 163

Brownjohn, Allan 1931- DLB-40

Brownson, Orestes Augustus
    1803-1876 ....DLB-1, 59, 73

Bruccoli, Matthew J. 1931- DLB-103

Bruce, Charles 1906-1971 .... DLB-68

Bruce, Leo 1903-1979 .... DLB-77

Bruce, Philip Alexander
    1856-1933 .... DLB-47

Bruce Humphries
    [publishing house] DLB-46

Bruce-Novoa, Juan 1944- DLB-82

Bruckman, Clyde 1894-1955 .... DLB-26

Bruckner, Ferdinand 1891-1958 .... DLB-118

Brundage, John Herbert (see Herbert, John)

Brutus, Dennis 1924- DLB-117

Bryant, Arthur 1899-1985 .... DLB-149

Bryant, William Cullen
    1794-1878 ....DLB-3, 43, 59

Bryce Echenique, Alfredo
    1939- DLB-145

Bryce, James 1838-1922 .... DLB-166

Brydges, Sir Samuel Egerton
    1762-1837 .... DLB-107

Bryskett, Lodowick 1546?-1612 .... DLB-167

Buchan, John 1875-1940 ... DLB-34, 70, 156

Buchanan, George 1506-1582 ...... DLB-132

Buchanan, Robert 1841-1901 ....DLB-18, 35

Buchman, Sidney 1902-1975 .... DLB-26

Buchner, Augustus 1591-1661 ...... DLB-164

Büchner, Georg 1813-1837 .... DLB-133

Bucholtz, Andreas Heinrich
    1607-1671 .... DLB-168

Buck, Pearl S. 1892-1973 ....DLB-9, 102

Bucke, Charles 1781-1846 .... DLB-110

Bucke, Richard Maurice
    1837-1902 .... DLB-99

Buckingham, Joseph Tinker 1779-1861 and
    Buckingham, Edwin
    1810-1833 .... DLB-73

Buckler, Ernest 1908-1984 .... DLB-68

Buckley, William F., Jr.
    1925- DLB-137; Y-80

Buckminster, Joseph Stevens
    1784-1812 .... DLB-37

Buckner, Robert 1906- DLB-26

Budd, Thomas ?-1698 .... DLB-24

Budrys, A. J. 1931- DLB-8

Buechner, Frederick 1926- Y-80

Buell, John 1927- DLB-53

Buffum, Job [publishing house] DLB-49

Bugnet, Georges 1879-1981 .... DLB-92

Buies, Arthur 1840-1901 .... DLB-99

Building the New British Library
at St Pancras Y-94

Bukowski, Charles
1920-1994 .... DLB-5, 130, 169

Bullein, William
between 1520 and 1530-1576 ... DLB-167

Bullins, Ed 1935- DLB-7, 38

Bulwer-Lytton, Edward (also Edward Bulwer)
1803-1873 .... DLB-21

Bumpus, Jerry 1937- Y-81

Bunce and Brother DLB-49

Bunner, H. C. 1855-1896 .... DLB-78, 79

Bunting, Basil 1900-1985 .... DLB-20

Bunyan, John 1628-1688 .... DLB-39

Burch, Robert 1925- DLB-52

Burciaga, José Antonio 1940- DLB-82

Bürger, Gottfried August
1747-1794 .... DLB-94

Burgess, Anthony 1917-1993 .... DLB-14

Burgess, Gelett 1866-1951 .... DLB-11

Burgess, John W. 1844-1931 .... DLB-47

Burgess, Thornton W.
1874-1965 .... DLB-22

Burgess, Stringer and Company DLB-49

Burk, John Daly circa 1772-1808 .... DLB-37

Burke, Edmund 1729?-1797 .... DLB-104

Burke, Kenneth 1897-1993 ...... DLB-45, 63

Burlingame, Edward Livermore
1848-1922 .... DLB-79

Burnet, Gilbert 1643-1715 .... DLB-101

Burnett, Frances Hodgson
1849-1924 .... DLB-42, 141; DS-13, 14

Burnett, W. R. 1899-1982 .... DLB-9

Burnett, Whit 1899-1973 and
Martha Foley 1897-1977 .... DLB-137

Burney, Fanny 1752-1840 .... DLB-39

Burns, Alan 1929- DLB-14

Burns, John Horne 1916-1953 .... Y-85

Burns, Robert 1759-1796 .... DLB-109

Burns and Oates DLB-106

Burnshaw, Stanley 1906- DLB-48

Burr, C. Chauncey 1815?-1883 ..... DLB-79

Burroughs, Edgar Rice 1875-1950 .... DLB-8

Burroughs, John 1837-1921 .... DLB-64

Burroughs, Margaret T. G.
1917- DLB-41

Burroughs, William S., Jr.
1947-1981 .... DLB-16

Burroughs, William Seward
1914- DLB-2, 8, 16, 152; Y-81

Burroway, Janet 1936- DLB-6

Burt, Maxwell S. 1882-1954 .... DLB-86

Burt, A. L., and Company DLB-49

Burton, Hester 1913- DLB-161

Burton, Isabel Arundell
1831-1896 .... DLB-166

Burton, Miles (see Rhode, John)

Burton, Richard Francis
1821-1890 .... DLB-55, 166

Burton, Robert 1577-1640 .... DLB-151

Burton, Virginia Lee 1909-1968 ..... DLB-22

Burton, William Evans
1804-1860 .... DLB-73

Burwell, Adam Hood 1790-1849 ..... DLB-99

Bury, Lady Charlotte
1775-1861 .... DLB-116

Busch, Frederick 1941- DLB-6

Busch, Niven 1903-1991 .... DLB-44

Bushnell, Horace 1802-1876 .... DS-13

Bussieres, Arthur de 1877-1913 ..... DLB-92

Butler, Juan 1942-1981 .... DLB-53

Butler, Octavia E. 1947- DLB-33

Butler, Samuel 1613-1680 ..... DLB-101, 126

Butler, Samuel 1835-1902 .... DLB-18, 57

Butler, William Francis
1838-1910 .... DLB-166

Butler, E. H., and Company DLB-49

Butor, Michel 1926- DLB-83

Butterworth, Hezekiah 1839-1905 ... DLB-42

Buttitta, Ignazio 1899- DLB-114

Byars, Betsy 1928- DLB-52

Byatt, A. S. 1936- DLB-14

Byles, Mather 1707-1788 .... DLB-24

Bynner, Witter 1881-1968 .... DLB-54

Byrd, William II 1674-1744 ....DLB-24, 140

Byrne, John Keyes (see Leonard, Hugh)

Byron, George Gordon, Lord
1788-1824 .... DLB-96, 110

C

Caballero Bonald, José Manuel
1926- DLB-108

Cabañero, Eladio 1930- DLB-134

Cabell, James Branch
1879-1958 .... DLB-9, 78

Cabeza de Baca, Manuel
1853-1915 ....DLB-122

Cabeza de Baca Gilbert, Fabiola
1898- DLB-122

Cable, George Washington
1844-1925 .... DLB-12, 74; DS-13

Cabrera, Lydia 1900-1991 ....DLB-145

Cabrera Infante, Guillermo
1929- DLB-113

Cadell [publishing house] DLB-154

Cady, Edwin H. 1917- DLB-103

Caedmon flourished 658-680 ....DLB-146

Caedmon School circa 660-899 .....DLB-146

Cahan, Abraham
1860-1951 .... DLB-9, 25, 28

Cain, George 1943- DLB-33

Caldecott, Randolph 1846-1886 ....DLB-163

Calder, John
(Publishers), Limited DLB-112

Caldwell, Ben 1937- DLB-38

Caldwell, Erskine 1903-1987 ..... DLB-9, 86

Caldwell, H. M., Company DLB-49

Calhoun, John C. 1782-1850 ....DLB-3

Calisher, Hortense 1911- DLB-2

A Call to Letters and an Invitation
to the Electric Chair,
by Siegfried Mandel DLB-75

Callaghan, Morley 1903-1990 ....DLB-68

Callaloo Y-87

Calmer, Edgar 1907- DLB-4

Calverley, C. S. 1831-1884 ....DLB-35

Calvert, George Henry
1803-1889 .... DLB-1, 64

Cambridge Press DLB-49

Cambridge Songs (Carmina Cantabrigensia)
circa 1050 ....DLB-148

Camden House: An Interview with
James Hardin Y-92

Cameron, Eleanor 1912- DLB-52

Cameron, George Frederick
1854-1885 ....DLB-99

Cameron, Lucy Lyttelton
1781-1858 ....DLB-163

Cameron, William Bleasdell
  1862-1951 .... DLB-99

Camm, John 1718-1778 .... DLB-31

Campana, Dino 1885-1932 .... DLB-114

Campbell, Gabrielle Margaret Vere
  (see Shearing, Joseph, and Bowen, Marjorie)

Campbell, James Dykes
  1838-1895 .... DLB-144

Campbell, James Edwin
  1867-1896 .... DLB-50

Campbell, John 1653-1728 .... DLB-43

Campbell, John W., Jr.
  1910-1971 .... DLB-8

Campbell, Roy 1901-1957 .... DLB-20

Campbell, Thomas
  1777-1844 .... DLB-93, 144

Campbell, William Wilfred
  1858-1918 .... DLB-92

Campion, Edmund 1539-1581 ...... DLB-167

Campion, Thomas 1567-1620 .... DLB-58

Camus, Albert 1913-1960 .... DLB-72

Canby, Henry Seidel 1878-1961 ..... DLB-91

Candelaria, Cordelia 1943- DLB-82

Candelaria, Nash 1928- DLB-82

Candour in English Fiction (1890),
  by Thomas Hardy DLB-18

Canetti, Elias 1905-1994 .... DLB-85, 124

Canham, Erwin Dain
  1904-1982 .... DLB-127

Canitz, Friedrich Rudolph Ludwig von
  1654-1699 .... DLB-168

Cankar, Ivan 1876-1918 .... DLB-147

Cannan, Gilbert 1884-1955 .... DLB-10

Cannell, Kathleen 1891-1974 .... DLB-4

Cannell, Skipwith 1887-1957 .... DLB-45

Canning, George 1770-1827 .... DLB-158

Cantwell, Robert 1908-1978 .... DLB-9

Cape, Jonathan, and Harrison Smith
  [publishing house] DLB-46

Cape, Jonathan, Limited DLB-112

Capen, Joseph 1658-1725 .... DLB-24

Capes, Bernard 1854-1918 .... DLB-156

Capote, Truman
  1924-1984 .... DLB-2; Y-80, 84

Caproni, Giorgio 1912-1990 .... DLB-128

Cardarelli, Vincenzo 1887-1959 .... DLB-114

Cárdenas, Reyes 1948- DLB-122

Cardinal, Marie 1929- DLB-83

Carew, Jan 1920- DLB-157

Carew, Thomas
  1594 or 1595-1640 .... DLB-126

Carey, Henry
  circa 1687-1689-1743 .... DLB-84

Carey, Mathew 1760-1839 .... DLB-37, 73

Carey and Hart DLB-49

Carey, M., and Company DLB-49

Carlell, Lodowick 1602-1675 .... DLB-58

Carleton, William 1794-1869 .... DLB-159

Carleton, G. W.
  [publishing house] DLB-49

Carlile, Richard 1790-1843 .... DLB-110, 158

Carlyle, Jane Welsh 1801-1866 ..... DLB-55

Carlyle, Thomas 1795-1881 .... DLB-55, 144

Carman, Bliss 1861-1929 .... DLB-92

Carmina Burana circa 1230 .... DLB-138

Carnero, Guillermo 1947- DLB-108

Carossa, Hans 1878-1956 .... DLB-66

Carpenter, Humphrey 1946- DLB-155

Carpenter, Stephen Cullen
  ?-1820? DLB-73

Carpentier, Alejo 1904-1980 .... DLB-113

Carrier, Roch 1937- DLB-53

Carrillo, Adolfo 1855-1926 .... DLB-122

Carroll, Gladys Hasty 1904- DLB-9

Carroll, John 1735-1815 .... DLB-37

Carroll, John 1809-1884 .... DLB-99

Carroll, Lewis 1832-1898 ...... DLB-18, 163

Carroll, Paul 1927- DLB-16

Carroll, Paul Vincent 1900-1968 .... DLB-10

Carroll and Graf Publishers DLB-46

Carruth, Hayden 1921- DLB-5, 165

Carryl, Charles E. 1841-1920 ...... DLB-42

Carswell, Catherine 1879-1946 ..... DLB-36

Carter, Angela 1940-1992 .... DLB-14

Carter, Elizabeth 1717-1806 .... DLB-109

Carter, Henry (see Leslie, Frank)

Carter, Hodding, Jr. 1907-1972 .... DLB-127

Carter, Landon 1710-1778 .... DLB-31

Carter, Lin 1930- Y-81

Carter, Martin 1927- DLB-117

Carter and Hendee DLB-49

Carter, Robert, and Brothers DLB-49

Cartwright, John 1740-1824 .... DLB-158

Cartwright, William circa
  1611-1643 .... DLB-126

Caruthers, William Alexander
  1802-1846 .... DLB-3

Carver, Jonathan 1710-1780 .... DLB-31

Carver, Raymond
  1938-1988 .... DLB-130; Y-84, 88

Cary, Joyce 1888-1957 .... DLB-15, 100

Cary, Patrick 1623?-1657 .... DLB-131

Casey, Juanita 1925- DLB-14

Casey, Michael 1947- DLB-5

Cassady, Carolyn 1923- DLB-16

Cassady, Neal 1926-1968 .... DLB-16

Cassell and Company DLB-106

Cassell Publishing Company DLB-49

Cassill, R. V. 1919- DLB-6

Cassity, Turner 1929- DLB-105

The Castle of Perseverance
  circa 1400-1425 .... DLB-146

Castellano, Olivia 1944- DLB-122

Castellanos, Rosario 1925-1974 .... DLB-113

Castillo, Ana 1953- DLB-122

Castlemon, Harry (see Fosdick, Charles Austin)

Caswall, Edward 1814-1878 .... DLB-32

Catacalos, Rosemary 1944- DLB-122

Cather, Willa
  1873-1947 .... DLB-9, 54, 78; DS-1

Catherine II (Ekaterina Alekseevna), "The
  Great," Empress of Russia
  1729-1796 .... DLB-150

Catherwood, Mary Hartwell
  1847-1902 .... DLB-78

Catledge, Turner 1901-1983 .... DLB-127

Cattafi, Bartolo 1922-1979 .... DLB-128

Catton, Bruce 1899-1978 .... DLB-17

Causley, Charles 1917- DLB-27

Caute, David 1936- DLB-14

Cavendish, Duchess of Newcastle,
  Margaret Lucas 1623-1673 ..... DLB-131

Cawein, Madison 1865-1914 .... DLB-54

The Caxton Printers, Limited DLB-46

Cayrol, Jean 1911- DLB-83

Cecil, Lord David 1902-1986 ...... DLB-155

Celan, Paul 1920-1970 .... DLB-69

Celaya, Gabriel 1911-1991 .... DLB-108

Céline, Louis-Ferdinand
  1894-1961 .... DLB-72

The Celtic Background to Medieval English
  Literature DLB-146

Center for Bibliographical Studies and Research at the University of California, Riverside Y-91

The Center for the Book in the Library of Congress Y-93

Center for the Book Research Y-84

Centlivre, Susanna 1669?-1723 ..... DLB-84

The Century Company DLB-49

Cernuda, Luis 1902-1963 .... DLB-134

Cervantes, Lorna Dee 1954- DLB-82

Chacel, Rosa 1898- DLB-134

Chacón, Eusebio 1869-1948 .... DLB-82

Chacón, Felipe Maximiliano 1873-? DLB-82

Chadwyck-Healey's Full-Text Literary Databases: Editing Commercial Databases of Primary Literary Texts Y-95

Challans, Eileen Mary (see Renault, Mary)

Chalmers, George 1742-1825 .... DLB-30

Chaloner, Sir Thomas 1520-1565 .... DLB-167

Chamberlain, Samuel S. 1851-1916 .... DLB-25

Chamberland, Paul 1939- DLB-60

Chamberlin, William Henry 1897-1969 .... DLB-29

Chambers, Charles Haddon 1860-1921 .... DLB-10

Chambers, W. and R. [publishing house] DLB-106

Chamisso, Albert von 1781-1838 .... DLB-90

Champfleury 1821-1889 .... DLB-119

Chandler, Harry 1864-1944 .... DLB-29

Chandler, Norman 1899-1973 ..... DLB-127

Chandler, Otis 1927- DLB-127

Chandler, Raymond 1888-1959 .... DS-6

Channing, Edward 1856-1931 ...... DLB-17

Channing, Edward Tyrrell 1790-1856 .... DLB-1, 59

Channing, William Ellery 1780-1842 .... DLB-1, 59

Channing, William Ellery, II 1817-1901 .... DLB-1

Channing, William Henry 1810-1884 .... DLB-1, 59

Chaplin, Charlie 1889-1977 .... DLB-44

Chapman, George 1559 or 1560 - 1634 .... DLB-62, 121

Chapman, John DLB-106

Chapman, William 1850-1917 ...... DLB-99

Chapman and Hall DLB-106

Chappell, Fred 1936- DLB-6, 105

Chappell, Fred, A Detail in a Poem DLB-105

Charbonneau, Jean 1875-1960 ...... DLB-92

Charbonneau, Robert 1911-1967 .... DLB-68

Charles, Gerda 1914- DLB-14

Charles, William [publishing house] DLB-49

The Charles Wood Affair: A Playwright Revived Y-83

Charlotte Forten: Pages from her Diary DLB-50

Charteris, Leslie 1907-1993 .... DLB-77

Charyn, Jerome 1937- Y-83

Chase, Borden 1900-1971 .... DLB-26

Chase, Edna Woolman 1877-1957 .... DLB-91

Chase-Riboud, Barbara 1936- DLB-33

Chateaubriand, François-René de 1768-1848 .... DLB-119

Chatterton, Thomas 1752-1770 .... DLB-109

Chatto and Windus DLB-106

Chaucer, Geoffrey 1340?-1400 ..... DLB-146

Chauncy, Charles 1705-1787 .... DLB-24

Chauveau, Pierre-Joseph-Olivier 1820-1890 .... DLB-99

Chávez, Denise 1948- DLB-122

Chávez, Fray Angélico 1910- DLB-82

Chayefsky, Paddy 1923-1981 ....DLB-7, 44; Y-81

Cheever, Ezekiel 1615-1708 .... DLB-24

Cheever, George Barrell 1807-1890 .... DLB-59

Cheever, John 1912-1982 .... DLB-2, 102; Y-80, 82

Cheever, Susan 1943- Y-82

Cheke, Sir John 1514-1557 .... DLB-132

Chelsea House DLB-46

Cheney, Ednah Dow (Littlehale) 1824-1904 .... DLB-1

Cheney, Harriet Vaughn 1796-1889 .... DLB-99

Cherry, Kelly 1940- Y-83

Cherryh, C. J. 1942- Y-80

Chesnutt, Charles Waddell 1858-1932 ....DLB-12, 50, 78

Chester, Alfred 1928-1971 .... DLB-130

Chester, George Randolph 1869-1924 .... DLB-78

The Chester Plays circa 1505-1532; revisions until 1575 .... DLB-146

Chesterfield, Philip Dormer Stanhope, Fourth Earl of 1694-1773 .... DLB-104

Chesterton, G. K. 1874-1936 ....DLB-10, 19, 34, 70, 98, 149

Chettle, Henry circa 1560-circa 1607 .... DLB-136

Chew, Ada Nield 1870-1945 .... DLB-135

Cheyney, Edward P. 1861-1947 ..... DLB-47

Chicano History DLB-82

Chicano Language DLB-82

Child, Francis James 1825-1896 .... DLB-1, 64

Child, Lydia Maria 1802-1880 .... DLB-1, 74

Child, Philip 1898-1978 .... DLB-68

Childers, Erskine 1870-1922 .... DLB-70

Children's Book Awards and Prizes DLB-61

Children's Illustrators, 1800-1880 .... DLB-163

Childress, Alice 1920-1994 .... DLB-7, 38

Childs, George W. 1829-1894 .... DLB-23

Chilton Book Company DLB-46

Chinweizu 1943- DLB-157

Chitham, Edward 1932- DLB-155

Chittenden, Hiram Martin 1858-1917 .... DLB-47

Chivers, Thomas Holley 1809-1858 .... DLB-3

Chopin, Kate 1850-1904 .... DLB-12, 78

Chopin, Rene 1885-1953 .... DLB-92

Choquette, Adrienne 1915-1973 ..... DLB-68

Choquette, Robert 1905- DLB-68

The Christian Publishing Company DLB-49

Christie, Agatha 1890-1976 ..... DLB-13, 77

Christus und die Samariterin circa 950 .... DLB-148

Chulkov, Mikhail Dmitrievich 1743?-1792 .... DLB-150

Church, Benjamin 1734-1778 .... DLB-31

Church, Francis Pharcellus 1839-1906 .... DLB-79

Church, William Conant 1836-1917 .... DLB-79

Churchill, Caryl 1938- DLB-13

Churchill, Charles 1731-1764 ......DLB-109

Churchill, Sir Winston 1874-1965 ....DLB-100

Churchyard, Thomas 1520?-1604 ....DLB-132

Churton, E., and Company DLB-106

Chute, Marchette 1909-1994 ......DLB-103

Ciardi, John 1916-1986 ....DLB-5; Y-86

Cibber, Colley 1671-1757 ....DLB-84

Cima, Annalisa 1941- DLB-128

Cirese, Eugenio 1884-1955 ....DLB-114

Cisneros, Sandra 1954- DLB-122, 152

City Lights Books DLB-46

Cixous, Hélène 1937- DLB-83

Clampitt, Amy 1920-1994 ....DLB-105

Clapper, Raymond 1892-1944 ....DLB-29

Clare, John 1793-1864 ....DLB-55, 96

Clarendon, Edward Hyde, Earl of 1609-1674 ....DLB-101

Clark, Alfred Alexander Gordon (see Hare, Cyril)

Clark, Ann Nolan 1896- DLB-52

Clark, Catherine Anthony 1892-1977 ....DLB-68

Clark, Charles Heber 1841-1915 ....DLB-11

Clark, Davis Wasgatt 1812-1871 .....DLB-79

Clark, Eleanor 1913- DLB-6

Clark, J. P. 1935- DLB-117

Clark, Lewis Gaylord 1808-1873 .... DLB-3, 64, 73

Clark, Walter Van Tilburg 1909-1971 ....DLB-9

Clark, C. M., Publishing Company DLB-46

Clarke, Austin 1896-1974 ....DLB-10, 20

Clarke, Austin C. 1934- DLB-53, 125

Clarke, Gillian 1937- DLB-40

Clarke, James Freeman 1810-1888 ....DLB-1, 59

Clarke, Pauline 1921- DLB-161

Clarke, Rebecca Sophia 1833-1906 ....DLB-42

Clarke, Robert, and Company DLB-49

Clarkson, Thomas 1760-1846 ......DLB-158

Claudius, Matthias 1740-1815 ....DLB-97

Clausen, Andy 1943- DLB-16

Claxton, Remsen and Haffelfinger DLB-49

Clay, Cassius Marcellus 1810-1903 .... DLB-43

Cleary, Beverly 1916- DLB-52

Cleaver, Vera 1919- and Cleaver, Bill 1920-1981 .... DLB-52

Cleland, John 1710-1789 .... DLB-39

Clemens, Samuel Langhorne 1835-1910 ....DLB-11, 12, 23, 64, 74

Clement, Hal 1922- DLB-8

Clemo, Jack 1916- DLB-27

Cleveland, John 1613-1658 .... DLB-126

Cliff, Michelle 1946- DLB-157

Clifford, Lady Anne 1590-1676 .... DLB-151

Clifford, James L. 1901-1978 ...... DLB-103

Clifford, Lucy 1853?-1929 .... DLB-135, 141

Clifton, Lucille 1936- DLB-5, 41

Clode, Edward J. [publishing house] DLB-46

Clough, Arthur Hugh 1819-1861 .... DLB-32

Cloutier, Cécile 1930- DLB-60

Clutton-Brock, Arthur 1868-1924 .... DLB-98

Coates, Robert M. 1897-1973 .... DLB-4, 9, 102

Coatsworth, Elizabeth 1893- DLB-22

Cobb, Charles E., Jr. 1943- DLB-41

Cobb, Frank I. 1869-1923 .... DLB-25

Cobb, Irvin S. 1876-1944 .... DLB-11, 25, 86

Cobbett, William 1763-1835 .... DLB-43, 107

Cochran, Thomas C. 1902- DLB-17

Cochrane, Elizabeth 1867-1922 ..... DLB-25

Cockerill, John A. 1845-1896 .... DLB-23

Cocteau, Jean 1889-1963 .... DLB-65

Coderre, Emile (see Jean Narrache)

Coffee, Lenore J. 1900?-1984 ...... DLB-44

Coffin, Robert P. Tristram 1892-1955 .... DLB-45

Cogswell, Fred 1917- DLB-60

Cogswell, Mason Fitch 1761-1830 .... DLB-37

Cohen, Arthur A. 1928-1986 .... DLB-28

Cohen, Leonard 1934- DLB-53

Cohen, Matt 1942- DLB-53

Colden, Cadwallader 1688-1776 .... DLB-24, 30

Cole, Barry 1936- DLB-14

Cole, George Watson 1850-1939 ....DLB-140

Colegate, Isabel 1931- DLB-14

Coleman, Emily Holmes 1899-1974 ....DLB-4

Coleman, Wanda 1946- DLB-130

Coleridge, Hartley 1796-1849 ....DLB-96

Coleridge, Mary 1861-1907 ......DLB-19, 98

Coleridge, Samuel Taylor 1772-1834 ....DLB-93, 107

Colet, John 1467-1519 ....DLB-132

Colette 1873-1954 ....DLB-65

Colette, Sidonie Gabrielle (see Colette)

Colinas, Antonio 1946- DLB-134

Collier, John 1901-1980 ....DLB-77

Collier, Mary 1690-1762 ....DLB-95

Collier, Robert J. 1876-1918 ....DLB-91

Collier, P. F. [publishing house] DLB-49

Collin and Small DLB-49

Collingwood, W. G. 1854-1932 ....DLB-149

Collins, An floruit circa 1653 ......DLB-131

Collins, Merle 1950- DLB-157

Collins, Mortimer 1827-1876 .....DLB-21, 35

Collins, Wilkie 1824-1889 .. DLB-18, 70, 159

Collins, William 1721-1759 ....DLB-109

Collins, William, Sons and Company DLB-154

Collins, Isaac [publishing house] DLB-49

Collyer, Mary 1716?-1763? DLB-39

Colman, Benjamin 1673-1747 .....DLB-24

Colman, George, the Elder 1732-1794 ....DLB-89

Colman, George, the Younger 1762-1836 ....DLB-89

Colman, S. [publishing house] DLB-49

Colombo, John Robert 1936- DLB-53

Colquhoun, Patrick 1745-1820 .....DLB-158

Colter, Cyrus 1910- DLB-33

Colum, Padraic 1881-1972 ....DLB-19

Colvin, Sir Sidney 1845-1927 ......DLB-149

Colwin, Laurie 1944-1992 ....Y-80

Comden, Betty 1919- and Green, Adolph 1918- DLB-44

Comi, Girolamo 1890-1968 ....DLB-114

The Comic Tradition Continued [in the British Novel] DLB-15

Commager, Henry Steele 1902- DLB-17

The Commercialization of the Image of Revolt, by Kenneth Rexroth DLB-16

Community and Commentators: Black Theatre and Its Critics DLB-38

Compton-Burnett, Ivy 1884?-1969 .... DLB-36

Conan, Laure 1845-1924 .... DLB-99

Conde, Carmen 1901- DLB-108

Conference on Modern Biography Y-85

Congreve, William 1670-1729 .... DLB-39, 84

Conkey, W. B., Company DLB-49

Connell, Evan S., Jr. 1924- ... DLB-2; Y-81

Connelly, Marc 1890-1980 ..... DLB-7; Y-80

Connolly, Cyril 1903-1974 .... DLB-98

Connolly, James B. 1868-1957 ...... DLB-78

Connor, Ralph 1860-1937 .... DLB-92

Connor, Tony 1930- DLB-40

Conquest, Robert 1917- DLB-27

Conrad, Joseph 1857-1924 .... DLB-10, 34, 98, 156

Conrad, John, and Company DLB-49

Conroy, Jack 1899-1990 .... Y-81

Conroy, Pat 1945- DLB-6

The Consolidation of Opinion: Critical Responses to the Modernists DLB-36

Constable, Henry 1562-1613 .... DLB-136

Constable and Company Limited DLB-112

Constable, Archibald, and Company DLB-154

Constant, Benjamin 1767-1830 .... DLB-119

Constant de Rebecque, Henri-Benjamin de (see Constant, Benjamin)

Constantine, David 1944- DLB-40

Constantin-Weyer, Maurice 1881-1964 .... DLB-92

Contempo Caravan: Kites in a Windstorm Y-85

A Contemporary Flourescence of Chicano Literature Y-84

The Continental Publishing Company DLB-49

A Conversation with Chaim Potok Y-84

Conversations with Editors Y-95

Conversations with Publishers I: An Interview with Patrick O'Connor Y-84

Conversations with Publishers II: An Interview with Charles Scribner III Y-94

Conversations with Publishers III: An Interview with Donald Lamm Y-95

Conversations with Rare Book Dealers I: An Interview with Glenn Horowitz Y-90

Conversations with Rare Book Dealers II: An Interview with Ralph Sipper Y-94

The Conversion of an Unpolitical Man, by W. H. Bruford DLB-66

Conway, Moncure Daniel 1832-1907 .... DLB-1

Cook, Ebenezer circa 1667-circa 1732 .... DLB-24

Cook, Edward Tyas 1857-1919 .... DLB-149

Cook, Michael 1933- DLB-53

Cook, David C., Publishing Company DLB-49

Cooke, George Willis 1848-1923 .... DLB-71

Cooke, Increase, and Company DLB-49

Cooke, John Esten 1830-1886 .... DLB-3

Cooke, Philip Pendleton 1816-1850 .... DLB-3, 59

Cooke, Rose Terry 1827-1892 .... DLB-12, 74

Coolbrith, Ina 1841-1928 .... DLB-54

Cooley, Peter 1940- DLB-105

Cooley, Peter, Into the Mirror DLB-105

Coolidge, Susan (see Woolsey, Sarah Chauncy)

Coolidge, George [publishing house] DLB-49

Cooper, Giles 1918-1966 .... DLB-13

Cooper, James Fenimore 1789-1851 ... DLB-3

Cooper, Kent 1880-1965 .... DLB-29

Cooper, Susan 1935- DLB-161

Coote, J. [publishing house] DLB-154

Coover, Robert 1932- DLB-2; Y-81

Copeland and Day DLB-49

Copland, Robert 1470?-1548 .... DLB-136

Coppard, A. E. 1878-1957 .... DLB-162

Coppel, Alfred 1921- Y-83

Coppola, Francis Ford 1939- DLB-44

Corazzini, Sergio 1886-1907 .... DLB-114

Corbett, Richard 1582-1635 .... DLB-121

Corcoran, Barbara 1911- DLB-52

Corelli, Marie 1855-1924 .... DLB-34, 156

Corle, Edwin 1906-1956 .... Y-85

Corman, Cid 1924- DLB-5

Cormier, Robert 1925- DLB-52

Corn, Alfred 1943- DLB-120; Y-80

Cornish, Sam 1935- DLB-41

Cornish, William circa 1465-circa 1524 .... DLB-132

Cornwall, Barry (see Procter, Bryan Waller)

Cornwallis, Sir William, the Younger circa 1579-1614 .... DLB-151

Cornwell, David John Moore (see le Carré, John)

Corpi, Lucha 1945- DLB-82

Corrington, John William 1932- DLB-6

Corrothers, James D. 1869-1917 ..... DLB-50

Corso, Gregory 1930- DLB-5, 16

Cortázar, Julio 1914-1984 .... DLB-113

Cortez, Jayne 1936- DLB-41

Corvinus, Gottlieb Siegmund 1677-1746 .... DLB-168

Corvo, Baron (see Rolfe, Frederick William)

Cory, Annie Sophie (see Cross, Victoria)

Cory, William Johnson 1823-1892 .... DLB-35

Coryate, Thomas 1577?-1617 ...... DLB-151

Cosin, John 1595-1672 .... DLB-151

Cosmopolitan Book Corporation DLB-46

Costain, Thomas B. 1885-1965 .... DLB-9

Coste, Donat 1912-1957 .... DLB-88

Costello, Louisa Stuart 1799-1870 ...DLB-166

Cota-Cárdenas, Margarita 1941- DLB-122

Cotter, Joseph Seamon, Sr. 1861-1949 .... DLB-50

Cotter, Joseph Seamon, Jr. 1895-1919 .... DLB-50

Cottle, Joseph [publishing house] DLB-154

Cotton, Charles 1630-1687 .... DLB-131

Cotton, John 1584-1652 .... DLB-24

Coulter, John 1888-1980 .... DLB-68

Cournos, John 1881-1966 .... DLB-54

Cousins, Margaret 1905- DLB-137

Cousins, Norman 1915-1990 .... DLB-137

Coventry, Francis 1725-1754 .... DLB-39

Coverdale, Miles 1487 or 1488-1569 .... DLB-167

Coverly, N. [publishing house] DLB-49

Covici-Friede DLB-46

Coward, Noel 1899-1973 .... DLB-10

Coward, McCann and
Geoghegan DLB-46

Cowles, Gardner 1861-1946 .... DLB-29

Cowles, Gardner ("Mike"), Jr.
1903-1985 .... DLB-127, 137

Cowley, Abraham
1618-1667 .... DLB-131, 151

Cowley, Hannah 1743-1809 .... DLB-89

Cowley, Malcolm
1898-1989 .... DLB-4, 48; Y-81, 89

Cowper, William
1731-1800 .... DLB-104, 109

Cox, A. B. (see Berkeley, Anthony)

Cox, James McMahon
1903-1974 .... DLB-127

Cox, James Middleton
1870-1957 .... DLB-127

Cox, Palmer 1840-1924 .... DLB-42

Coxe, Louis 1918-1993 .... DLB-5

Coxe, Tench 1755-1824 .... DLB-37

Cozzens, James Gould
1903-1978 .... DLB-9; Y-84; DS-2

Crabbe, George 1754-1832 .... DLB-93

Crackanthorpe, Hubert
1870-1896 .... DLB-135

Craddock, Charles Egbert
(see Murfree, Mary N.)

Cradock, Thomas 1718-1770 .... DLB-31

Craig, Daniel H. 1811-1895 .... DLB-43

Craik, Dinah Maria
1826-1887 .... DLB-35, 136

Cranch, Christopher Pearse
1813-1892 .... DLB-1, 42

Crane, Hart 1899-1932 .... DLB-4, 48

Crane, R. S. 1886-1967 .... DLB-63

Crane, Stephen 1871-1900 ... DLB-12, 54, 78

Crane, Walter 1845-1915 .... DLB-163

Cranmer, Thomas 1489-1556 ..... DLB-132

Crapsey, Adelaide 1878-1914 .... DLB-54

Crashaw, Richard
1612 or 1613-1649 .... DLB-126

Craven, Avery 1885-1980 .... DLB-17

Crawford, Charles
1752-circa 1815 .... DLB-31

Crawford, F. Marion 1854-1909 ..... DLB-71

Crawford, Isabel Valancy
1850-1887 .... DLB-92

Crawley, Alan 1887-1975 .... DLB-68

Crayon, Geoffrey (see Irving, Washington)

Creasey, John 1908-1973 .... DLB-77

Creative Age Press DLB-46

Creech, William
[publishing house] DLB-154

Creel, George 1876-1953 .... DLB-25

Creeley, Robert 1926- DLB-5, 16, 169

Creelman, James 1859-1915 .... DLB-23

Cregan, David 1931- DLB-13

Creighton, Donald Grant
1902-1979 .... DLB-88

Cremazie, Octave 1827-1879 .... DLB-99

Crémer, Victoriano 1909?- DLB-108

Crescas, Hasdai
circa 1340-1412? DLB-115

Crespo, Angel 1926- DLB-134

Cresset Press DLB-112

Cresswell, Helen 1934- DLB-161

Crèvecoeur, Michel Guillaume Jean de
1735-1813 .... DLB-37

Crews, Harry 1935- DLB-6, 143

Crichton, Michael 1942- Y-81

A Crisis of Culture: The Changing Role
of Religion in the New Republic
.... DLB-37

Crispin, Edmund 1921-1978 .... DLB-87

Cristofer, Michael 1946- DLB-7

"The Critic as Artist" (1891), by
Oscar Wilde DLB-57

"Criticism In Relation To Novels" (1863),
by G. H. Lewes DLB-21

Crnjanski, Miloš 1893-1977 .... DLB-147

Crockett, David (Davy)
1786-1836 .... DLB-3, 11

Croft-Cooke, Rupert (see Bruce, Leo)

Crofts, Freeman Wills
1879-1957 .... DLB-77

Croker, John Wilson
1780-1857 .... DLB-110

Croly, George 1780-1860 .... DLB-159

Croly, Herbert 1869-1930 .... DLB-91

Croly, Jane Cunningham
1829-1901 .... DLB-23

Crompton, Richmal 1890-1969 .... DLB-160

Crosby, Caresse 1892-1970 .... DLB-48

Crosby, Caresse 1892-1970 and Crosby,
Harry 1898-1929 .... DLB-4

Crosby, Harry 1898-1929 .... DLB-48

Cross, Gillian 1945- DLB-161

Cross, Victoria 1868-1952 .... DLB-135

Crossley-Holland, Kevin
1941- DLB-40, 161

Crothers, Rachel 1878-1958 .... DLB-7

Crowell, Thomas Y., Company DLB-49

Crowley, John 1942- Y-82

Crowley, Mart 1935- DLB-7

Crown Publishers DLB-46

Crowne, John 1641-1712 .... DLB-80

Crowninshield, Edward Augustus
1817-1859 .... DLB-140

Crowninshield, Frank 1872-1947 ... DLB-91

Croy, Homer 1883-1965 .... DLB-4

Crumley, James 1939- Y-84

Cruz, Victor Hernández 1949- DLB-41

Csokor, Franz Theodor
1885-1969 .... DLB-81

Cuala Press DLB-112

Cullen, Countee 1903-1946 ... DLB-4, 48, 51

Culler, Jonathan D. 1944- DLB-67

The Cult of Biography
Excerpts from the Second Folio Debate:
"Biographies are generally a disease of
English Literature" – Germaine Greer,
Victoria Glendinning, Auberon Waugh,
and Richard Holmes Y-86

Cumberland, Richard 1732-1811 .... DLB-89

Cummings, E. E. 1894-1962 ...... DLB-4, 48

Cummings, Ray 1887-1957 .... DLB-8

Cummings and Hilliard DLB-49

Cummins, Maria Susanna
1827-1866 .... DLB-42

Cundall, Joseph
[publishing house] DLB-106

Cuney, Waring 1906-1976 .... DLB-51

Cuney-Hare, Maude 1874-1936 ..... DLB-52

Cunningham, Allan
1784-1842 .... DLB-116, 144

Cunningham, J. V. 1911- DLB-5

Cunningham, Peter F.
[publishing house] DLB-49

Cunquiero, Alvaro 1911-1981 ..... DLB-134

Cuomo, George 1929- Y-80

Cupples and Leon DLB-46

Cupples, Upham and Company DLB-49

Cuppy, Will 1884-1949 .... DLB-11

Curll, Edmund
[publishing house] DLB-154

Currie, James 1756-1805 .... DLB-142

Currie, Mary Montgomerie Lamb Singleton, Lady Currie (see Fane, Violet)

Cursor Mundi circa 1300 .... DLB-146

Curti, Merle E. 1897- DLB-17

Curtis, Anthony 1926- DLB-155

Curtis, Cyrus H. K. 1850-1933 ..... DLB-91

Curtis, George William 1824-1892 .... DLB-1, 43

Curzon, Robert 1810-1873 .... DLB-166

Curzon, Sarah Anne 1833-1898 ..... DLB-99

Cynewulf circa 770-840 .... DLB-146

Czepko, Daniel 1605-1660 .... DLB-164

D

D. M. Thomas: The Plagiarism Controversy Y-82

Dabit, Eugène 1898-1936 .... DLB-65

Daborne, Robert circa 1580-1628 ... DLB-58

Dacey, Philip 1939- DLB-105

Dacey, Philip, Eyes Across Centuries: Contemporary Poetry and "That Vision Thing" DLB-105

Dach, Simon 1605-1659 .... DLB-164

Daggett, Rollin M. 1831-1901 ...... DLB-79

D'Aguiar, Fred 1960- DLB-157

Dahl, Roald 1916-1990 .... DLB-139

Dahlberg, Edward 1900-1977 ...... DLB-48

Dahn, Felix 1834-1912 .... DLB-129

Dale, Peter 1938- DLB-40

Dall, Caroline Wells (Healey) 1822-1912 .... DLB-1

Dallas, E. S. 1828-1879 .... DLB-55

The Dallas Theater Center DLB-7

D'Alton, Louis 1900-1951 .... DLB-10

Daly, T. A. 1871-1948 .... DLB-11

Damon, S. Foster 1893-1971 .... DLB-45

Damrell, William S. [publishing house] DLB-49

Dana, Charles A. 1819-1897 ...... DLB-3, 23

Dana, Richard Henry, Jr 1815-1882 .... DLB-1

Dandridge, Ray Garfield DLB-51

Dane, Clemence 1887-1965 .... DLB-10

Danforth, John 1660-1730 .... DLB-24

Danforth, Samuel, I 1626-1674 ..... DLB-24

Danforth, Samuel, II 1666-1727 .... DLB-24

Dangerous Years: London Theater, 1939-1945 .... DLB-10

Daniel, John M. 1825-1865 .... DLB-43

Daniel, Samuel 1562 or 1563-1619 .... DLB-62

Daniel Press DLB-106

Daniells, Roy 1902-1979 .... DLB-68

Daniels, Jim 1956- DLB-120

Daniels, Jonathan 1902-1981 ...... DLB-127

Daniels, Josephus 1862-1948 ...... DLB-29

Dannay, Frederic 1905-1982 and Manfred B. Lee 1905-1971 ..... DLB-137

Danner, Margaret Esse 1915- DLB-41

Dantin, Louis 1865-1945 .... DLB-92

D'Arcy, Ella circa 1857-1937 ...... DLB-135

Darley, George 1795-1846 .... DLB-96

Darwin, Charles 1809-1882 ..... DLB-57, 166

Darwin, Erasmus 1731-1802 .... DLB-93

Daryush, Elizabeth 1887-1977 ...... DLB-20

Dashkova, Ekaterina Romanovna (née Vorontsova) 1743-1810 .... DLB-150

Dashwood, Edmée Elizabeth Monica de la Pasture (see Delafield, E. M.)

Daudet, Alphonse 1840-1897 ...... DLB-123

d'Aulaire, Edgar Parin 1898- and d'Aulaire, Ingri 1904- DLB-22

Davenant, Sir William 1606-1668 .... DLB-58, 126

Davenport, Guy 1927- DLB-130

Davenport, Robert ?-? DLB-58

Daves, Delmer 1904-1977 .... DLB-26

Davey, Frank 1940- DLB-53

Davidson, Avram 1923-1993 .... DLB-8

Davidson, Donald 1893-1968 ...... DLB-45

Davidson, John 1857-1909 .... DLB-19

Davidson, Lionel 1922- DLB-14

Davie, Donald 1922- DLB-27

Davie, Elspeth 1919- DLB-139

Davies, John, of Hereford 1565?-1618 .... DLB-121

Davies, Rhys 1901-1978 .... DLB-139

Davies, Robertson 1913- DLB-68

Davies, Samuel 1723-1761 .... DLB-31

Davies, Thomas 1712?-1785 ... DLB-142, 154

Davies, W. H. 1871-1940 .... DLB-19

Davies, Peter, Limited DLB-112

Daviot, Gordon 1896?-1952 .... DLB-10 (see also Tey, Josephine)

Davis, Charles A. 1795-1867 .... DLB-11

Davis, Clyde Brion 1894-1962 .... DLB-9

Davis, Dick 1945- DLB-40

Davis, Frank Marshall 1905-? DLB-51

Davis, H. L. 1894-1960 .... DLB-9

Davis, John 1774-1854 .... DLB-37

Davis, Lydia 1947- DLB-130

Davis, Margaret Thomson 1926- ... DLB-14

Davis, Ossie 1917- DLB-7, 38

Davis, Paxton 1925-1994 .... Y-94

Davis, Rebecca Harding 1831-1910 .... DLB-74

Davis, Richard Harding 1864-1916 ..... DLB-12, 23, 78, 79; DS-13

Davis, Samuel Cole 1764-1809 ...... DLB-37

Davison, Peter 1928- DLB-5

Davys, Mary 1674-1732 .... DLB-39

DAW Books DLB-46

Dawson, Ernest 1882-1947 .... DLB-140

Dawson, Fielding 1930- DLB-130

Dawson, William 1704-1752 .... DLB-31

Day, Angel flourished 1586 .... DLB-167

Day, Benjamin Henry 1810-1889 ..... DLB-43

Day, Clarence 1874-1935 .... DLB-11

Day, Dorothy 1897-1980 .... DLB-29

Day, Frank Parker 1881-1950 ...... DLB-92

Day, John circa 1574-circa 1640 ..... DLB-62

Day Lewis, C. 1904-1972 .... DLB-15, 20 (see also Blake, Nicholas)

Day, Thomas 1748-1789 .... DLB-39

Day, The John, Company DLB-46

Day, Mahlon [publishing house] DLB-49

Deacon, William Arthur 1890-1977 .... DLB-68

Deal, Borden 1922-1985 .... DLB-6

de Angeli, Marguerite 1889-1987 ..... DLB-22

De Angelis, Milo 1951- DLB-128

De Bow, James Dunwoody Brownson 1820-1867 .... DLB-3, 79

de Bruyn, Günter 1926- DLB-75

de Camp, L. Sprague 1907- DLB-8

The Decay of Lying (1889), by Oscar Wilde [excerpt] DLB-18

Dedication, *Ferdinand Count Fathom* (1753), by Tobias Smollett DLB-39

Dedication, *The History of Pompey the Little*
    (1751), by Francis Coventry DLB-39

Dedication, *Lasselia* (1723), by Eliza
    Haywood [excerpt] DLB-39

Dedication, *The Wanderer* (1814),
    by Fanny Burney DLB-39

Dee, John 1527-1609 .... DLB-136

Deeping, George Warwick
    1877-1950 .... DLB-153

Defense of *Amelia* (1752), by
    Henry Fielding DLB-39

Defoe, Daniel 1660-1731 ... DLB-39, 95, 101

de Fontaine, Felix Gregory
    1834-1896 .... DLB-43

De Forest, John William
    1826-1906 .... DLB-12

DeFrees, Madeline 1919- DLB-105

DeFrees, Madeline, The Poet's Kaleidoscope:
    The Element of Surprise in the Making
    of the Poem DLB-105

de Graff, Robert 1895-1981 .... Y-81

de Graft, Joe 1924-1978 .... DLB-117

De Heinrico circa 980? DLB-148

Deighton, Len 1929- DLB-87

DeJong, Meindert 1906-1991 .... DLB-52

Dekker, Thomas circa 1572-1632 .... DLB-62

Delacorte, Jr., George T.
    1894-1991 .... DLB-91

Delafield, E. M. 1890-1943 .... DLB-34

Delahaye, Guy 1888-1969 .... DLB-92

de la Mare, Walter
    1873-1956 .... DLB-19, 153, 162

Deland, Margaret 1857-1945 .... DLB-78

Delaney, Shelagh 1939- DLB-13

Delany, Martin Robinson
    1812-1885 .... DLB-50

Delany, Samuel R. 1942- DLB-8, 33

de la Roche, Mazo 1879-1961 ...... DLB-68

Delbanco, Nicholas 1942- DLB-6

De León, Nephtal 1945- DLB-82

Delgado, Abelardo Barrientos
    1931- DLB-82

De Libero, Libero 1906-1981 .... DLB-114

DeLillo, Don 1936- DLB-6

de Lisser H. G. 1878-1944 .... DLB-117

Dell, Floyd 1887-1969 .... DLB-9

Dell Publishing Company DLB-46

delle Grazie, Marie Eugene
    1864-1931 .... DLB-81

Deloney, Thomas died 1600 .... DLB-167

del Rey, Lester 1915-1993 .... DLB-8

Del Vecchio, John M. 1947- DS-9

de Man, Paul 1919-1983 .... DLB-67

Demby, William 1922- DLB-33

Deming, Philander 1829-1915 ...... DLB-74

Demorest, William Jennings
    1822-1895 .... DLB-79

De Morgan, William 1839-1917 .... DLB-153

Denham, Sir John
    1615-1669 .... DLB-58, 126

Denison, Merrill 1893-1975 .... DLB-92

Denison, T. S., and Company DLB-49

Dennie, Joseph
    1768-1812 .... DLB-37, 43, 59, 73

Dennis, John 1658-1734 .... DLB-101

Dennis, Nigel 1912-1989 .... DLB-13, 15

Dent, Tom 1932- DLB-38

Dent, J. M., and Sons DLB-112

Denton, Daniel circa 1626-1703 .... DLB-24

DePaola, Tomie 1934- DLB-61

De Quincey, Thomas
    1785-1859 .... DLB-110, 144

Derby, George Horatio
    1823-1861 .... DLB-11

Derby, J. C., and Company DLB-49

Derby and Miller DLB-49

Derleth, August 1909-1971 .... DLB-9

The Derrydale Press DLB-46

Derzhavin, Gavriil Romanovich
    1743-1816 .... DLB-150

Desaulniers, Gonsalve
    1863-1934 .... DLB-92

Desbiens, Jean-Paul 1927- DLB-53

des Forêts, Louis-Rene 1918- DLB-83

DesRochers, Alfred 1901-1978 ...... DLB-68

Desrosiers, Léo-Paul 1896-1967 .... DLB-68

Destouches, Louis-Ferdinand
    (see Céline, Louis-Ferdinand)

De Tabley, Lord 1835-1895 .... DLB-35

Deutsch, Babette 1895-1982 .... DLB-45

Deutsch, André, Limited DLB-112

Deveaux, Alexis 1948- DLB-38

The Development of the Author's Copyright
    in Britain DLB-154

The Development of Lighting in the Staging
    of Drama, 1900-1945 .... DLB-10

de Vere, Aubrey 1814-1902 .... DLB-35

Devereux, second Earl of Essex, Robert
    1565-1601 .... DLB-136

The Devin-Adair Company DLB-46

De Voto, Bernard 1897-1955 .... DLB-9

De Vries, Peter 1910-1993 ...... DLB-6; Y-82

Dewdney, Christopher 1951- DLB-60

Dewdney, Selwyn 1909-1979 .... DLB-68

DeWitt, Robert M., Publisher DLB-49

DeWolfe, Fiske and Company DLB-49

Dexter, Colin 1930- DLB-87

de Young, M. H. 1849-1925 .... DLB-25

Dhlomo, H. I. E. 1903-1956 ...... DLB-157

Dhuoda circa 803-after 843 .... DLB-148

The Dial Press DLB-46

Diamond, I. A. L. 1920-1988 ...... DLB-26

Di Cicco, Pier Giorgio 1949- DLB-60

Dick, Philip K. 1928-1982 .... DLB-8

Dick and Fitzgerald DLB-49

Dickens, Charles
    1812-1870 ..... DLB-21, 55, 70, 159, 166

Dickinson, Peter 1927- DLB-161

Dickey, James
    1923- DLB-5; Y-82, 93; DS-7

Dickey, William 1928-1994 .... DLB-5

Dickinson, Emily 1830-1886 .... DLB-1

Dickinson, John 1732-1808 .... DLB-31

Dickinson, Jonathan 1688-1747 .... DLB-24

Dickinson, Patric 1914- DLB-27

Dickinson, Peter 1927- DLB-87

Dicks, John [publishing house] DLB-106

Dickson, Gordon R. 1923- DLB-8

*Dictionary of Literary Biography
    Yearbook* Awards Y-92, 93

The Dictionary of National Biography
    .... DLB-144

Didion, Joan 1934- DLB-2; Y-81, 86

Di Donato, Pietro 1911- DLB-9

Diego, Gerardo 1896-1987 .... DLB-134

Digges, Thomas circa 1546-1595 ... DLB-136

Dillard, Annie 1945- Y-80

Dillard, R. H. W. 1937- DLB-5

Dillingham, Charles T.,
    Company DLB-49

The Dillingham, G. W.,
    Company DLB-49

Dilly, Edward and Charles
    [publishing house] DLB-154

Cumulative Index

Dilthey, Wilhelm 1833-1911 .... DLB-129

Dingelstedt, Franz von
1814-1881 .... DLB-133

Dintenfass, Mark 1941- Y-84

Diogenes, Jr. (see Brougham, John)

DiPrima, Diane 1934- DLB-5, 16

Disch, Thomas M. 1940- DLB-8

Disney, Walt 1901-1966 .... DLB-22

Disraeli, Benjamin 1804-1881 .... DLB-21, 55

D'Israeli, Isaac 1766-1848 .... DLB-107

Ditzen, Rudolf (see Fallada, Hans)

Dix, Dorothea Lynde 1802-1887 ...... DLB-1

Dix, Dorothy (see Gilmer,
Elizabeth Meriwether)

Dix, Edwards and Company DLB-49

Dixon, Paige (see Corcoran, Barbara)

Dixon, Richard Watson
1833-1900 .... DLB-19

Dixon, Stephen 1936- DLB-130

Dmitriev, Ivan Ivanovich
1760-1837 .... DLB-150

Dobell, Sydney 1824-1874 .... DLB-32

Döblin, Alfred 1878-1957 .... DLB-66

Dobson, Austin
1840-1921 .... DLB-35, 144

Doctorow, E. L. 1931- DLB-2, 28; Y-80

Documents on Sixteenth-Century
Literature DLB-167

Dodd, William E. 1869-1940 .... DLB-17

Dodd, Anne [publishing house] DLB-154

Dodd, Mead and Company DLB-49

Doderer, Heimito von 1896-1968 ... DLB-85

Dodge, Mary Mapes
1831?-1905 .... DLB-42, 79; DS-13

Dodge, B. W., and Company DLB-46

Dodge Publishing Company DLB-49

Dodgson, Charles Lutwidge
(see Carroll, Lewis)

Dodsley, Robert 1703-1764 .... DLB-95

Dodsley, R. [publishing house] DLB-154

Dodson, Owen 1914-1983 .... DLB-76

Doesticks, Q. K. Philander, P. B.
(see Thomson, Mortimer)

Doheny, Carrie Estelle
1875-1958 .... DLB-140

Domínguez, Sylvia Maida
1935- DLB-122

Donahoe, Patrick
[publishing house] DLB-49

Donald, David H. 1920- DLB-17

Donaldson, Scott 1928- DLB-111

Donleavy, J. P. 1926- DLB-6

Donnadieu, Marguerite (see Duras,
Marguerite)

Donne, John 1572-1631 .... DLB-121, 151

Donnelley, R. R., and Sons
Company DLB-49

Donnelly, Ignatius 1831-1901 ...... DLB-12

Donohue and Henneberry DLB-49

Donoso, José 1924- DLB-113

Doolady, M. [publishing house] DLB-49

Dooley, Ebon (see Ebon)

Doolittle, Hilda 1886-1961 .... DLB-4, 45

Doplicher, Fabio 1938- DLB-128

Dor, Milo 1923- DLB-85

Doran, George H., Company DLB-46

Dorgelès, Roland 1886-1973 .... DLB-65

Dorn, Edward 1929- DLB-5

Dorr, Rheta Childe 1866-1948 ..... DLB-25

Dorset and Middlesex, Charles Sackville,
Lord Buckhurst,
Earl of 1643-1706 .... DLB-131

Dorst, Tankred 1925- DLB-75, 124

Dos Passos, John
1896-1970 .... DLB-4, 9; DS-1

Doubleday and Company DLB-49

Dougall, Lily 1858-1923 .... DLB-92

Doughty, Charles M.
1843-1926 .... DLB-19, 57

Douglas, Gavin 1476-1522 .... DLB-132

Douglas, Keith 1920-1944 .... DLB-27

Douglas, Norman 1868-1952 .... DLB-34

Douglass, Frederick
1817?-1895 .... DLB-1, 43, 50, 79

Douglass, William circa
1691-1752 .... DLB-24

Dourado, Autran 1926- DLB-145

Dove, Rita 1952- DLB-120

Dover Publications DLB-46

Doves Press DLB-112

Dowden, Edward 1843-1913 .... DLB-35, 149

Dowell, Coleman 1925-1985 .... DLB-130

Downes, Gwladys 1915- DLB-88

Downing, J., Major (see Davis, Charles A.)

Downing, Major Jack (see Smith, Seba)

Dowson, Ernest 1867-1900 ..... DLB-19, 135

Doxey, William
[publishing house] DLB-49

Doyle, Sir Arthur Conan
1859-1930 .... DLB-18, 70, 156

Doyle, Kirby 1932- DLB-16

Drabble, Margaret 1939- DLB-14, 155

Drach, Albert 1902- DLB-85

The Dramatic Publishing
Company DLB-49

Dramatists Play Service DLB-46

Drant, Thomas
early 1540s?-1578 .... DLB-167

Draper, John W. 1811-1882 .... DLB-30

Draper, Lyman C. 1815-1891 .... DLB-30

Drayton, Michael 1563-1631 .... DLB-121

Dreiser, Theodore
1871-1945 ..... DLB-9, 12, 102, 137; DS-1

Drewitz, Ingeborg 1923-1986 .... DLB-75

Drieu La Rochelle, Pierre
1893-1945 .... DLB-72

Drinkwater, John 1882-1937
.... DLB-10, 19, 149

Droste-Hülshoff, Annette von
1797-1848 .... DLB-133

The Drue Heinz Literature Prize
Excerpt from "Excerpts from a Report
of the Commission," in David
Bosworth's *The Death of Descartes*
An Interview with David
Bosworth Y-82

Drummond, William Henry
1854-1907 .... DLB-92

Drummond, William, of Hawthornden
1585-1649 .... DLB-121

Dryden, John 1631-1700 ... DLB-80, 101, 131

Držić, Marin circa 1508-1567 ...... DLB-147

Duane, William 1760-1835 .... DLB-43

Dubé, Marcel 1930- DLB-53

Dubé, Rodolphe (see Hertel, François)

Dubie, Norman 1945- DLB-120

Du Bois, W. E. B.
1868-1963 .... DLB-47, 50, 91

Du Bois, William Pène 1916- DLB-61

Dubus, Andre 1936- DLB-130

Ducharme, Réjean 1941- DLB-60

Dučić, Jovan 1871-1943 .... DLB-147

Duck, Stephen 1705?-1756 .... DLB-95

Duckworth, Gerald, and
Company Limited DLB-112

Dudek, Louis 1918- DLB-88

Duell, Sloan and Pearce DLB-46

Duff Gordon, Lucie 1821-1869 ..... DLB-166
Duffield and Green DLB-46
Duffy, Maureen 1933- DLB-14
Dugan, Alan 1923- DLB-5
Dugas, Marcel 1883-1947 .... DLB-92
Dugdale, William
    [publishing house] DLB-106
Duhamel, Georges 1884-1966 .... DLB-65
Dujardin, Edouard 1861-1949 ..... DLB-123
Dukes, Ashley 1885-1959 .... DLB-10
Du Maurier, George 1834-1896 .... DLB-153
Dumas, Alexandre, *père*
    1802-1870 .... DLB-119
Dumas, Henry 1934-1968 .... DLB-41
Dunbar, Paul Laurence
    1872-1906 .... DLB-50, 54, 78
Dunbar, William
    circa 1460-circa 1522 ...... DLB-132, 146
Duncan, Norman 1871-1916 .... DLB-92
Duncan, Quince 1940- DLB-145
Duncan, Robert 1919-1988 .... DLB-5, 16
Duncan, Ronald 1914-1982 .... DLB-13
Duncan, Sara Jeannette
    1861-1922 .... DLB-92
Dunigan, Edward, and Brother DLB-49
Dunlap, John 1747-1812 .... DLB-43
Dunlap, William
    1766-1839 .... DLB-30, 37, 59
Dunn, Douglas 1942- DLB-40
Dunn, Stephen 1939- DLB-105
Dunn, Stephen, The Good,
    The Not So Good DLB-105
Dunne, Finley Peter
    1867-1936 .... DLB-11, 23
Dunne, John Gregory 1932- Y-80
Dunne, Philip 1908-1992 .... DLB-26
Dunning, Ralph Cheever
    1878-1930 .... DLB-4
Dunning, William A. 1857-1922 .... DLB-17
Duns Scotus, John
    circa 1266-1308 .... DLB-115
Dunsany, Lord (Edward John Moreton
    Drax Plunkett, Baron Dunsany)
    1878-1957 .... DLB-10, 77, 153, 156
Dupin, Amantine-Aurore-Lucile (see Sand,
    George)
Durand, Lucile (see Bersianik, Louky)
Duranty, Walter 1884-1957 .... DLB-29
Duras, Marguerite 1914- DLB-83

Durfey, Thomas 1653-1723 .... DLB-80
Durrell, Lawrence
    1912-1990 .... DLB-15, 27; Y-90
Durrell, William
    [publishing house] DLB-49
Dürrenmatt, Friedrich
    1921-1990 .... DLB-69, 124
Dutton, E. P., and Company DLB-49
Duvoisin, Roger 1904-1980 .... DLB-61
Duyckinck, Evert Augustus
    1816-1878 .... DLB-3, 64
Duyckinck, George L. 1823-1863 ..... DLB-3
Duyckinck and Company DLB-49
Dwight, John Sullivan 1813-1893 ..... DLB-1
Dwight, Timothy 1752-1817 .... DLB-37
Dybek, Stuart 1942- DLB-130
Dyer, Charles 1928- DLB-13
Dyer, George 1755-1841 .... DLB-93
Dyer, John 1699-1757 .... DLB-95
Dyer, Sir Edward 1543-1607 ...... DLB-136
Dylan, Bob 1941- DLB-16

E

Eager, Edward 1911-1964 .... DLB-22
Eames, Wilberforce 1855-1937 ..... DLB-140
Earle, James H., and Company DLB-49
Earle, John 1600 or 1601-1665 ..... DLB-151
Early American Book Illustration,
    by Sinclair Hamilton DLB-49
Eastlake, William 1917- DLB-6
Eastman, Carol ?- DLB-44
Eastman, Max 1883-1969 .... DLB-91
Eaton, Daniel Isaac 1753-1814 ..... DLB-158
Eberhart, Richard 1904- DLB-48
Ebner, Jeannie 1918- DLB-85
Ebner-Eschenbach, Marie von
    1830-1916 .... DLB-81
Ebon 1942- DLB-41
Ecbasis Captivi circa 1045 .... DLB-148
Ecco Press DLB-46
Eckhart, Meister
    circa 1260-circa 1328 .... DLB-115
The Eclectic Review 1805-1868 ...... DLB-110
Edel, Leon 1907- DLB-103
Edes, Benjamin 1732-1803 .... DLB-43
Edgar, David 1948- DLB-13

Edgeworth, Maria
    1768-1849 .... DLB-116, 159, 163
The Edinburgh Review 1802-1929 ..... DLB-110
Edinburgh University Press DLB-112
The Editor Publishing Company DLB-49
Editorial Statements DLB-137
Edmonds, Randolph 1900- DLB-51
Edmonds, Walter D. 1903- DLB-9
Edschmid, Kasimir 1890-1966 .... DLB-56
Edwards, Jonathan 1703-1758 ...... DLB-24
Edwards, Jonathan, Jr. 1745-1801 .... DLB-37
Edwards, Junius 1929- DLB-33
Edwards, Richard 1524-1566 .... DLB-62
Edwards, James
    [publishing house] DLB-154
Effinger, George Alec 1947- DLB-8
Egerton, George 1859-1945 .... DLB-135
Eggleston, Edward 1837-1902 ...... DLB-12
Eggleston, Wilfred 1901-1986 ...... DLB-92
Ehrenstein, Albert 1886-1950 ...... DLB-81
Ehrhart, W. D. 1948- DS-9
Eich, Günter 1907-1972 .... DLB-69, 124
Eichendorff, Joseph Freiherr von
    1788-1857 .... DLB-90
1873 Publishers' Catalogues DLB-49
Eighteenth-Century Aesthetic
    Theories DLB-31
Eighteenth-Century Philosophical
    Background DLB-31
Eigner, Larry 1927- DLB-5
Eikon Basilike 1649 .... DLB-151
Eilhart von Oberge
    circa 1140-circa 1195 .... DLB-148
Einhard circa 770-840 .... DLB-148
Eisenreich, Herbert 1925-1986 ...... DLB-85
Eisner, Kurt 1867-1919 .... DLB-66
Eklund, Gordon 1945- Y-83
Ekwensi, Cyprian 1921- DLB-117
Elder, Lonne III 1931- DLB-7, 38, 44
Elder, Paul, and Company DLB-49
Elements of Rhetoric (1828; revised, 1846),
    by Richard Whately [excerpt] DLB-57
Elie, Robert 1915-1973 .... DLB-88
Elin Pelin 1877-1949 .... DLB-147
Eliot, George 1819-1880 ..... DLB-21, 35, 55
Eliot, John 1604-1690 .... DLB-24
Eliot, T. S. 1888-1965 ..... DLB-7, 10, 45, 63

Elizabeth I 1533-1603 .... DLB-136

Elizondo, Salvador 1932- DLB-145

Elizondo, Sergio 1930- DLB-82

Elkin, Stanley 1930- DLB-2, 28; Y-80

Elles, Dora Amy (see Wentworth, Patricia)

Ellet, Elizabeth F. 1818?-1877 ...... DLB-30

Elliot, Ebenezer 1781-1849 .... DLB-96

Elliot, Frances Minto (Dickinson) 1820-1898 .... DLB-166

Elliott, George 1923- DLB-68

Elliott, Janice 1931- DLB-14

Elliott, William 1788-1863 .... DLB-3

Elliott, Thomes and Talbot DLB-49

Ellis, Edward S. 1840-1916 .... DLB-42

Ellis, Frederick Staridge [publishing house] DLB-106

The George H. Ellis Company DLB-49

Ellison, Harlan 1934- DLB-8

Ellison, Ralph Waldo 1914-1994 .... DLB-2, 76; Y-94

Ellmann, Richard 1918-1987 .... DLB-103; Y-87

The Elmer Holmes Bobst Awards in Arts and Letters Y-87

Elyot, Thomas 1490?-1546 .... DLB-136

Emanuel, James Andrew 1921- DLB-41

Emecheta, Buchi 1944- DLB-117

The Emergence of Black Women Writers DS-8

Emerson, Ralph Waldo 1803-1882 .... DLB-1, 59, 73

Emerson, William 1769-1811 .... DLB-37

Emin, Fedor Aleksandrovich circa 1735-1770 .... DLB-150

Empson, William 1906-1984 .... DLB-20

The End of English Stage Censorship, 1945-1968 .... DLB-13

Ende, Michael 1929- DLB-75

Engel, Marian 1933-1985 .... DLB-53

Engels, Friedrich 1820-1895 .... DLB-129

Engle, Paul 1908- DLB-48

English Composition and Rhetoric (1866), by Alexander Bain [excerpt] DLB-57

The English Language: 410 to 1500 .... DLB-146

The English Renaissance of Art (1908), by Oscar Wilde DLB-35

Enright, D. J. 1920- DLB-27

Enright, Elizabeth 1909-1968 .... DLB-22

L'Envoi (1882), by Oscar Wilde DLB-35

Epps, Bernard 1936- DLB-53

Epstein, Julius 1909- and Epstein, Philip 1909-1952 ...... DLB-26

Equiano, Olaudah circa 1745-1797 .... DLB-37, 50

Eragny Press DLB-112

Erasmus, Desiderius 1467-1536 .... DLB-136

Erba, Luciano 1922- DLB-128

Erdrich, Louise 1954- DLB-152

Erichsen-Brown, Gwethalyn Graham (see Graham, Gwethalyn)

Eriugena, John Scottus circa 810-877 .... DLB-115

Ernest Hemingway's Toronto Journalism Revisited: With Three Previously Unrecorded Stories Y-92

Ernst, Paul 1866-1933 .... DLB-66, 118

Erskine, Albert 1911-1993 .... Y-93

Erskine, John 1879-1951 .... DLB-9, 102

Ervine, St. John Greer 1883-1971 .... DLB-10

Eschenburg, Johann Joachim 1743-1820 .... DLB-97

Escoto, Julio 1944- DLB-145

Eshleman, Clayton 1935- DLB-5

Espriu, Salvador 1913-1985 .... DLB-134

Ess Ess Publishing Company DLB-49

Essay on Chatterton (1842), by Robert Browning DLB-32

Essex House Press DLB-112

Estes, Eleanor 1906-1988 .... DLB-22

Estes and Lauriat DLB-49

Etherege, George 1636-circa 1692 .... DLB-80

Ethridge, Mark, Sr. 1896-1981 ..... DLB-127

Ets, Marie Hall 1893- DLB-22

Etter, David 1928- DLB-105

Ettner, Johann Christoph 1654-1724 .... DLB-168

Eudora Welty: Eye of the Storyteller Y-87

Eugene O'Neill Memorial Theater Center DLB-7

Eugene O'Neill's Letters: A Review Y-88

Eupolemius flourished circa 1095 .... DLB-148

Evans, Caradoc 1878-1945 .... DLB-162

Evans, Donald 1884-1921 .... DLB-54

Evans, George Henry 1805-1856 .... DLB-43

Evans, Hubert 1892-1986 .... DLB-92

Evans, Mari 1923- DLB-41

Evans, Mary Ann (see Eliot, George)

Evans, Nathaniel 1742-1767 .... DLB-31

Evans, Sebastian 1830-1909 .... DLB-35

Evans, M., and Company DLB-46

Everett, Alexander Hill 790-1847 .... DLB-59

Everett, Edward 1794-1865 ...... DLB-1, 59

Everson, R. G. 1903- DLB-88

Everson, William 1912-1994 ..... DLB-5, 16

Every Man His Own Poet; or, The Inspired Singer's Recipe Book (1877), by W. H. Mallock DLB-35

Ewart, Gavin 1916- DLB-40

Ewing, Juliana Horatia 1841-1885 .... DLB-21, 163

The Examiner 1808-1881 .... DLB-110

Exley, Frederick 1929-1992 .... DLB-143; Y-81

Experiment in the Novel (1929), by John D. Beresford DLB-36

Eyre and Spottiswoode DLB-106

Ezzo ?-after 1065 .... DLB-148

F

"F. Scott Fitzgerald: St. Paul's Native Son and Distinguished American Writer": University of Minnesota Conference, 29-31 October 1982 .... Y-82

Faber, Frederick William 1814-1863 .... DLB-32

Faber and Faber Limited DLB-112

Faccio, Rena (see Aleramo, Sibilla)

Fagundo, Ana María 1938- DLB-134

Fair, Ronald L. 1932- DLB-33

Fairfax, Beatrice (see Manning, Marie)

Fairlie, Gerard 1899-1983 .... DLB-77

Fallada, Hans 1893-1947 .... DLB-56

Fancher, Betsy 1928- Y-83

Fane, Violet 1843-1905 .... DLB-35

Fanfrolico Press DLB-112

Fanning, Katherine 1927 .... DLB-127

Fanshawe, Sir Richard 1608-1666 .... DLB-126

Fantasy Press Publishers DLB-46

Fante, John 1909-1983 .... DLB-130; Y-83

Al-Farabi circa 870-950 .... DLB-115

Farah, Nuruddin 1945- DLB-125

Farber, Norma 1909-1984 .... DLB-61

Farigoule, Louis (see Romains, Jules)

Farjeon, Eleanor 1881-1965 .... DLB-160

Farley, Walter 1920-1989 .... DLB-22

Farmer, Penelope 1939- DLB-161

Farmer, Philip José 1918- DLB-8

Farquhar, George circa 1677-1707 ... DLB-84

Farquharson, Martha (see Finley, Martha)

Farrar, Frederic William
1831-1903 .... DLB-163

Farrar and Rinehart DLB-46

Farrar, Straus and Giroux DLB-46

Farrell, James T.
1904-1979 .... DLB-4, 9, 86; DS-2

Farrell, J. G. 1935-1979 .... DLB-14

Fast, Howard 1914- DLB-9

Faulkner, William 1897-1962
.... DLB-9, 11, 44, 102; DS-2; Y-86

Faulkner, George
[publishing house] DLB-154

Fauset, Jessie Redmon 1882-1961 .... DLB-51

Faust, Irvin 1924- DLB-2, 28; Y-80

Fawcett Books DLB-46

Fearing, Kenneth 1902-1961 .... DLB-9

Federal Writers' Project DLB-46

Federman, Raymond 1928- Y-80

Feiffer, Jules 1929- DLB-7, 44

Feinberg, Charles E. 1899-1988 .... Y-88

Feind, Barthold 1678-1721 .... DLB-168

Feinstein, Elaine 1930- DLB-14, 40

Feldman, Irving 1928- DLB-169

Felipe, Léon 1884-1968 .... DLB-108

Fell, Frederick, Publishers DLB-46

Felltham, Owen 1602?-1668 ... DLB-126, 151

Fels, Ludwig 1946- DLB-75

Felton, Cornelius Conway
1807-1862 .... DLB-1

Fennario, David 1947- DLB-60

Fenno, John 1751-1798 .... DLB-43

Fenno, R. F., and Company DLB-49

Fenton, Geoffrey 1539?-1608 .... DLB-136

Fenton, James 1949- DLB-40

Ferber, Edna 1885-1968 ..... DLB-9, 28, 86

Ferdinand, Vallery III (see Salaam, Kalamu ya)

Ferguson, Sir Samuel 1810-1886 ..... DLB-32

Ferguson, William Scott
1875-1954 .... DLB-47

Fergusson, Robert 1750-1774 ...... DLB-109

Ferland, Albert 1872-1943 .... DLB-92

Ferlinghetti, Lawrence 1919- DLB-5, 16

Fern, Fanny (see Parton, Sara Payson Willis)

Ferrars, Elizabeth 1907- DLB-87

Ferré, Rosario 1942- DLB-145

Ferret, E., and Company DLB-49

Ferrier, Susan 1782-1854 .... DLB-116

Ferrini, Vincent 1913- DLB-48

Ferron, Jacques 1921-1985 .... DLB-60

Ferron, Madeleine 1922- DLB-53

Fetridge and Company DLB-49

Feuchtersleben, Ernst Freiherr von
1806-1849 .... DLB-133

Feuchtwanger, Lion 1884-1958 ..... DLB-66

Feuerbach, Ludwig 1804-1872 ..... DLB-133

Fichte, Johann Gottlieb
1762-1814 .... DLB-90

Ficke, Arthur Davison 1883-1945 ... DLB-54

Fiction Best-Sellers, 1910-1945 .... DLB-9

Fiction into Film, 1928-1975: A List of Movies
Based on the Works of Authors in
British Novelists, 1930-1959 ...... DLB-15

Fiedler, Leslie A. 1917- DLB-28, 67

Field, Edward 1924- DLB-105

Field, Edward, The Poetry File DLB-105

Field, Eugene
1850-1895 .... DLB-23, 42, 140; DS-13

Field, John 1545?-1588 .... DLB-167

Field, Marshall, III 1893-1956 ..... DLB-127

Field, Marshall, IV 1916-1965 ..... DLB-127

Field, Marshall, V 1941- DLB-127

Field, Nathan 1587-1619 or 1620 .... DLB-58

Field, Rachel 1894-1942 .... DLB-9, 22

A Field Guide to Recent Schools of American
Poetry Y-86

Fielding, Henry
1707-1754 .... DLB-39, 84, 101

Fielding, Sarah 1710-1768 .... DLB-39

Fields, James Thomas 1817-1881 ..... DLB-1

Fields, Julia 1938- DLB-41

Fields, W. C. 1880-1946 .... DLB-44

Fields, Osgood and Company DLB-49

Fifty Penguin Years Y-85

Figes, Eva 1932- DLB-14

Figuera, Angela 1902-1984 .... DLB-108

Filmer, Sir Robert 1586-1653 .... DLB-151

Filson, John circa 1753-1788 .... DLB-37

Finch, Anne, Countess of Winchilsea
1661-1720 .... DLB-95

Finch, Robert 1900- DLB-88

Findley, Timothy 1930- DLB-53

Finlay, Ian Hamilton 1925- DLB-40

Finley, Martha 1828-1909 .... DLB-42

Finn, Elizabeth Anne (McCaul)
1825-1921 .... DLB-166

Finney, Jack 1911- DLB-8

Finney, Walter Braden (see Finney, Jack)

Firbank, Ronald 1886-1926 .... DLB-36

Firmin, Giles 1615-1697 .... DLB-24

First Edition Library/Collectors'
Reprints, Inc. Y-91

First International F. Scott Fitzgerald
Conference Y-92

First Strauss "Livings" Awarded to Cynthia
Ozick and Raymond Carver
An Interview with Cynthia Ozick
An Interview with Raymond
Carver Y-83

Fischer, Karoline Auguste Fernandine
1764-1842 .... DLB-94

Fish, Stanley 1938- DLB-67

Fishacre, Richard 1205-1248 .... DLB-115

Fisher, Clay (see Allen, Henry W.)

Fisher, Dorothy Canfield
1879-1958 .... DLB-9, 102

Fisher, Leonard Everett 1924- DLB-61

Fisher, Roy 1930- DLB-40

Fisher, Rudolph 1897-1934 ...... DLB-51, 102

Fisher, Sydney George 1856-1927 .... DLB-47

Fisher, Vardis 1895-1968 .... DLB-9

Fiske, John 1608-1677 .... DLB-24

Fiske, John 1842-1901 .... DLB-47, 64

Fitch, Thomas circa 1700-1774 ...... DLB-31

Fitch, William Clyde 1865-1909 ...... DLB-7

FitzGerald, Edward 1809-1883 ...... DLB-32

Fitzgerald, F. Scott
1896-1940 ..... DLB-4, 9, 86; Y-81; DS-1

Fitzgerald, Penelope 1916- DLB-14

Fitzgerald, Robert 1910-1985 .... Y-80

Fitzgerald, Thomas 1819-1891 ...... DLB-23

Fitzgerald, Zelda Sayre 1900-1948 ...... Y-84

Fitzhugh, Louise 1928-1974 .... DLB-52

Cumulative Index

Fitzhugh, William circa 1651-1701 .... DLB-24

Flanagan, Thomas 1923- Y-80

Flanner, Hildegarde 1899-1987 ..... DLB-48

Flanner, Janet 1892-1978 ....DLB-4

Flaubert, Gustave 1821-1880 ...... DLB-119

Flavin, Martin 1883-1967 ....DLB-9

Fleck, Konrad (flourished circa 1220) DLB-138

Flecker, James Elroy 1884-1915 .. DLB-10, 19

Fleeson, Doris 1901-1970 .... DLB-29

Fleißer, Marieluise 1901-1974 .. DLB-56, 124

Fleming, Ian 1908-1964 .... DLB-87

Fleming, Paul 1609-1640 .... DLB-164

The Fleshly School of Poetry and Other Phenomena of the Day (1872), by Robert Buchanan DLB-35

The Fleshly School of Poetry: Mr. D. G. Rossetti (1871), by Thomas Maitland (Robert Buchanan) DLB-35

Fletcher, Giles, the Elder 1546-1611 .... DLB-136

Fletcher, Giles, the Younger 1585 or 1586-1623 .... DLB-121

Fletcher, J. S. 1863-1935 .... DLB-70

Fletcher, John (see Beaumont, Francis)

Fletcher, John Gould 1886-1950 .. DLB-4, 45

Fletcher, Phineas 1582-1650 .... DLB-121

Flieg, Helmut (see Heym, Stefan)

Flint, F. S. 1885-1960 .... DLB-19

Flint, Timothy 1780-1840 .... DLB-734

Foix, J. V. 1893-1987 .... DLB-134

Foley, Martha (see Burnett, Whit, and Martha Foley)

Folger, Henry Clay 1857-1930 ..... DLB-140

Folio Society DLB-112

Follen, Eliza Lee (Cabot) 1787-1860 ...DLB-1

Follett, Ken 1949- Y-81, DLB-87

Follett Publishing Company DLB-46

Folsom, John West [publishing house] DLB-49

Fontane, Theodor 1819-1898 ...... DLB-129

Fonvisin, Denis Ivanovich 1744 or 1745-1792 .... DLB-150

Foote, Horton 1916- DLB-26

Foote, Samuel 1721-1777 .... DLB-89

Foote, Shelby 1916- DLB-2, 17

Forbes, Calvin 1945- DLB-41

Forbes, Ester 1891-1967 .... DLB-22

Forbes and Company DLB-49

Force, Peter 1790-1868 .... DLB-30

Forché, Carolyn 1950- DLB-5

Ford, Charles Henri 1913- DLB-4, 48

Ford, Corey 1902-1969 .... DLB-11

Ford, Ford Madox 1873-1939 ....DLB-34, 98, 162

Ford, Jesse Hill 1928- DLB-6

Ford, John 1586-? DLB-58

Ford, R. A. D. 1915- DLB-88

Ford, Worthington C. 1858-1941 .... DLB-47

Ford, J. B., and Company DLB-49

Fords, Howard, and Hulbert DLB-49

Foreman, Carl 1914-1984 .... DLB-26

Forester, Frank (see Herbert, Henry William)

Fornés, María Irene 1930- DLB-7

Forrest, Leon 1937- DLB-33

Forster, E. M. 1879-1970 ...... DLB-34, 98, 162; DS-10

Forster, Georg 1754-1794 .... DLB-94

Forster, John 1812-1876 .... DLB-144

Forster, Margaret 1938- DLB-155

Forsyth, Frederick 1938- DLB-87

Forten, Charlotte L. 1837-1914 ...... DLB-50

Fortini, Franco 1917- DLB-128

Fortune, T. Thomas 1856-1928 ..... DLB-23

Fosdick, Charles Austin 1842-1915 .... DLB-42

Foster, Genevieve 1893-1979 ...... DLB-61

Foster, Hannah Webster 1758-1840 .... DLB-37

Foster, John 1648-1681 .... DLB-24

Foster, Michael 1904-1956 .... DLB-9

Foulis, Robert and Andrew / R. and A. [publishing house] DLB-154

Fouqué, Caroline de la Motte 1774-1831 .... DLB-90

Fouqué, Friedrich de la Motte 1777-1843 .... DLB-90

Four Essays on the Beat Generation, by John Clellon Holmes DLB-16

Four Seas Company DLB-46

Four Winds Press DLB-46

Fournier, Henri Alban (see Alain-Fournier)

Fowler and Wells Company DLB-49

Fowles, John 1926- DLB-14, 139

Fox, John, Jr. 1862 or 1863-1919 .... DLB-9; DS-13

Fox, Paula 1923- DLB-52

Fox, Richard Kyle 1846-1922 ....DLB-79

Fox, William Price 1926- DLB-2; Y-81

Fox, Richard K. [publishing house] DLB-49

Foxe, John 1517-1587 ....DLB-132

Fraenkel, Michael 1896-1957 ....DLB-4

France, Anatole 1844-1924 ....DLB-123

France, Richard 1938- DLB-7

Francis, Convers 1795-1863 ....DLB-1

Francis, Dick 1920- DLB-87

Francis, Jeffrey, Lord 1773-1850 ....DLB-107

Francis, C. S. [publishing house] DLB-49

François 1863-1910 ....DLB-92

François, Louise von 1817-1893 ....DLB-129

Francke, Kuno 1855-1930 ....DLB-71

Frank, Bruno 1887-1945 ....DLB-118

Frank, Leonhard 1882-1961 .... DLB-56, 118

Frank, Melvin (see Panama, Norman)

Frank, Waldo 1889-1967 .... DLB-9, 63

Franken, Rose 1895?-1988 .... Y-84

Franklin, Benjamin 1706-1790 .... DLB-24, 43, 73

Franklin, James 1697-1735 ....DLB-43

Franklin Library DLB-46

Frantz, Ralph Jules 1902-1979 ....DLB-4

Franzos, Karl Emil 1848-1904 ......DLB-129

Fraser, G. S. 1915-1980 ....DLB-27

Fraser, Kathleen 1935- DLB-169

Frattini, Alberto 1922- DLB-128

Frau Ava ?-1127 ....DLB-148

Frayn, Michael 1933- DLB-13, 14

Frederic, Harold 1856-1898 ....DLB-12, 23; DS-13

Freeling, Nicolas 1927- DLB-87

Freeman, Douglas Southall 1886-1953 ....DLB-17

Freeman, Legh Richmond 1842-1915 ....DLB-23

Freeman, Mary E. Wilkins 1852-1930 .... DLB-12, 78

Freeman, R. Austin 1862-1943 ....DLB-70

Freidank circa 1170-circa 1233 ......DLB-138

Freiligrath, Ferdinand 1810-1876 ....DLB-133

French, Alice 1850-1934 ..... DLB-74; DS-13

French, David 1939- DLB-53
French, James [publishing house] DLB-49
French, Samuel [publishing house] ...DLB-49
Samuel French, Limited DLB-106
Freneau, Philip 1752-1832 ....DLB-37, 43
Freni, Melo 1934- DLB-128
Freytag, Gustav 1816-1895 ....DLB-129
Fried, Erich 1921-1988 ....DLB-85
Friedman, Bruce Jay 1930- DLB-2, 28
Friedrich von Hausen
 circa 1171-1190 ....DLB-138
Friel, Brian 1929- DLB-13
Friend, Krebs 1895?-1967? DLB-4
Fries, Fritz Rudolf 1935- DLB-75
Fringe and Alternative Theater
 in Great Britain DLB-13
Frisch, Max 1911-1991 ....DLB-69, 124
Frischmuth, Barbara 1941- DLB-85
Fritz, Jean 1915- DLB-52
Fromentin, Eugene 1820-1876 ......DLB-123
From *The Gay Science,* by
 E. S. Dallas DLB-21
Frost, A. B. 1851-1928 ....DS-13
Frost, Robert 1874-1963 ...... DLB-54; DS-7
Frothingham, Octavius Brooks
 1822-1895 ....DLB-1
Froude, James Anthony
 1818-1894 .... DLB-18, 57, 144
Fry, Christopher 1907- DLB-13
Fry, Roger 1866-1934 ....DS-10
Frye, Northrop 1912-1991 ....DLB-67, 68
Fuchs, Daniel
 1909-1993 .... DLB-9, 26, 28; Y-93
Fuentes, Carlos 1928- DLB-113
Fuertes, Gloria 1918- DLB-108
The Fugitives and the Agrarians:
 The First Exhibition Y-85
Fuller, Charles H., Jr. 1939- DLB-38
Fuller, Henry Blake 1857-1929 ......DLB-12
Fuller, John 1937- DLB-40
Fuller, Roy 1912-1991 ....DLB-15, 20
Fuller, Samuel 1912- DLB-26
Fuller, Sarah Margaret, Marchesa
 D'Ossoli 1810-1850 ...... DLB-1, 59, 73
Fuller, Thomas 1608-1661 ....DLB-151
Fulton, Len 1934- Y-86
Fulton, Robin 1937- DLB-40

Furbank, P. N. 1920- DLB-155
Furman, Laura 1945- Y-86
Furness, Horace Howard
 1833-1912 ....DLB-64
Furness, William Henry 1802-1896 ... DLB-1
Furthman, Jules 1888-1966 ....DLB-26
The Future of the Novel (1899), by
 Henry James DLB-18
Fyleman, Rose 1877-1957 ....DLB-160

G

The G. Ross Roy Scottish Poetry
 Collection at the University of
 South Carolina Y-89
Gaddis, William 1922- DLB-2
Gág, Wanda 1893-1946 ....DLB-22
Gagnon, Madeleine 1938- DLB-60
Gaine, Hugh 1726-1807 ....DLB-43
Gaine, Hugh [publishing house] DLB-49
Gaines, Ernest J.
 1933- DLB-2, 33, 152; Y-80
Gaiser, Gerd 1908-1976 ....DLB-69
Galarza, Ernesto 1905-1984 ....DLB-122
Galaxy Science Fiction Novels DLB-46
Gale, Zona 1874-1938 ....DLB-9, 78
Gall, Louise von 1815-1855 ....DLB-133
Gallagher, Tess 1943- DLB-120
Gallagher, Wes 1911- DLB-127
Gallagher, William Davis
 1808-1894 ....DLB-73
Gallant, Mavis 1922- DLB-53
Gallico, Paul 1897-1976 ....DLB-9
Galsworthy, John
 1867-1933 ....DLB-10, 34, 98, 162
Galt, John 1779-1839 ....DLB-99, 116
Galton, Sir Francis 1822-1911 .....DLB-166
Galvin, Brendan 1938- DLB-5
Gambit DLB-46
Gamboa, Reymundo 1948- DLB-122
Gammer Gurton's Needle DLB-62
Gannett, Frank E. 1876-1957 ....DLB-29
Gaos, Vicente 1919-1980 ....DLB-134
García, Lionel G. 1935- DLB-82
García Lorca, Federico
 1898-1936 ....DLB-108
García Márquez, Gabriel
 1928- DLB-113

Gardam, Jane 1928- DLB-14, 161
Garden, Alexander
 circa 1685-1756 ....DLB-31
Gardiner, Margaret Power Farmer (see
 Blessington, Marguerite, Countess of)
Gardner, John 1933-1982 ....DLB-2; Y-82
Garfield, Leon 1921- DLB-161
Garis, Howard R. 1873-1962 ....DLB-22
Garland, Hamlin
 1860-1940 ....DLB-12, 71, 78
Garneau, Francis-Xavier
 1809-1866 ....DLB-99
Garneau, Hector de Saint-Denys
 1912-1943 ....DLB-88
Garneau, Michel 1939- DLB-53
Garner, Alan 1934- DLB-161
Garner, Hugh 1913-1979 ....DLB-68
Garnett, David 1892-1981 ....DLB-34
Garnett, Eve 1900-1991 ....DLB-160
Garraty, John A. 1920- DLB-17
Garrett, George
 1929- DLB-2, 5, 130, 152; Y-83
Garrick, David 1717-1779 ....DLB-84
Garrison, William Lloyd
 1805-1879 ....DLB-1, 43
Garro, Elena 1920- DLB-145
Garth, Samuel 1661-1719 ....DLB-95
Garve, Andrew 1908- DLB-87
Gary, Romain 1914-1980 ....DLB-83
Gascoigne, George 1539?-1577 .....DLB-136
Gascoyne, David 1916- DLB-20
Gaskell, Elizabeth Cleghorn
 1810-1865 ....DLB-21, 144, 159
Gaspey, Thomas 1788-1871 ....DLB-116
Gass, William Howard 1924- DLB-2
Gates, Doris 1901- DLB-22
Gates, Henry Louis, Jr. 1950- DLB-67
Gates, Lewis E. 1860-1924 ....DLB-71
Gatto, Alfonso 1909-1976 ....DLB-114
Gautier, Théophile 1811-1872 .....DLB-119
Gauvreau, Claude 1925-1971 ....DLB-88
The *Gawain*-Poet
 flourished circa 1350-1400 .....DLB-146
Gay, Ebenezer 1696-1787 ....DLB-24
Gay, John 1685-1732 ....DLB-84, 95
The *Gay Science* (1866), by E. S. Dallas
 [excerpt] DLB-21
Gayarré, Charles E. A. 1805-1895 ...DLB-30

Gaylord, Edward King 1873-1974 .... DLB-127

Gaylord, Edward Lewis 1919- ... DLB-127

Gaylord, Charles [publishing house] DLB-49

Geddes, Gary 1940- DLB-60

Geddes, Virgil 1897- DLB-4

Gedeon (Georgii Andreevich Krinovsky) circa 1730-1763 .... DLB-150

Geibel, Emanuel 1815-1884 .... DLB-129

Geis, Bernard, Associates DLB-46

Geisel, Theodor Seuss 1904-1991 .... DLB-61; Y-91

Gelb, Arthur 1924- DLB-103

Gelb, Barbara 1926- DLB-103

Gelber, Jack 1932- DLB-7

Gelinas, Gratien 1909- DLB-88

Gellert, Christian Fuerchtegott 1715-1769 .... DLB-97

Gellhorn, Martha 1908- Y-82

Gems, Pam 1925- DLB-13

A General Idea of the College of Mirania (1753), by William Smith [excerpts] DLB-31

Genet, Jean 1910-1986 .... DLB-72; Y-86

Genevoix, Maurice 1890-1980 ...... DLB-65

Genovese, Eugene D. 1930- DLB-17

Gent, Peter 1942- Y-82

Geoffrey of Monmouth circa 1100-1155 .... DLB-146

George, Henry 1839-1897 .... DLB-23

George, Jean Craighead 1919- DLB-52

Georgslied 896? DLB-148

Gerhardie, William 1895-1977 ...... DLB-36

Gerhardt, Paul 1607-1676 .... DLB-164

Gérin, Winifred 1901-1981 .... DLB-155

Gérin-Lajoie, Antoine 1824-1882 .... DLB-99

German Drama 800-1280 .... DLB-138

German Drama from Naturalism to Fascism: 1889-1933 .... DLB-118

German Literature and Culture from Charlemagne to the Early Courtly Period DLB-148

German Radio Play, The DLB-124

German Transformation from the Baroque to the Enlightenment, The DLB-97

The Germanic Epic and Old English Heroic Poetry: *Widseth, Waldere,* and *The Fight at Finnsburg* DLB-146

Germanophilism, by Hans Kohn DLB-66

Gernsback, Hugo 1884-1967 ..... DLB-8, 137

Gerould, Katharine Fullerton 1879-1944 .... DLB-78

Gerrish, Samuel [publishing house] .. DLB-49

Gerrold, David 1944- DLB-8

Gersonides 1288-1344 .... DLB-115

Gerstäcker, Friedrich 1816-1872 .... DLB-129

Gerstenberg, Heinrich Wilhelm von 1737-1823 .... DLB-97

Gervinus, Georg Gottfried 1805-1871 .... DLB-133

Geßner, Salomon 1730-1788 .... DLB-97

Geston, Mark S. 1946- DLB-8

Al-Ghazali 1058-1111 .... DLB-115

Gibbon, Edward 1737-1794 .... DLB-104

Gibbon, John Murray 1875-1952 .... DLB-92

Gibbon, Lewis Grassic (see Mitchell, James Leslie)

Gibbons, Floyd 1887-1939 .... DLB-25

Gibbons, Reginald 1947- DLB-120

Gibbons, William ?-? DLB-73

Gibson, Charles Dana 1867-1944 ..... DS-13

Gibson, Charles Dana 1867-1944 ..... DS-13

Gibson, Graeme 1934- DLB-53

Gibson, Margaret 1944- DLB-120

Gibson, Wilfrid 1878-1962 .... DLB-19

Gibson, William 1914- DLB-7

Gide, André 1869-1951 .... DLB-65

Giguère, Diane 1937- DLB-53

Giguère, Roland 1929- DLB-60

Gil de Biedma, Jaime 1929-1990 .... DLB-108

Gil-Albert, Juan 1906- DLB-134

Gilbert, Anthony 1899-1973 .... DLB-77

Gilbert, Michael 1912- DLB-87

Gilbert, Sandra M. 1936- DLB-120

Gilbert, Sir Humphrey 1537-1583 .... DLB-136

Gilchrist, Alexander 1828-1861 .... DLB-144

Gilchrist, Ellen 1935- DLB-130

Gilder, Jeannette L. 1849-1916 ...... DLB-79

Gilder, Richard Watson 1844-1909 .... DLB-64, 79

Gildersleeve, Basil 1831-1924 ...... DLB-71

Giles, Henry 1809-1882 .... DLB-64

Giles of Rome circa 1243-1316 ..... DLB-115

Gilfillan, George 1813-1878 .... DLB-144

Gill, Eric 1882-1940 .... DLB-98

Gill, William F., Company DLB-49

Gillespie, A. Lincoln, Jr. 1895-1950 .... DLB-4

Gilliam, Florence ?-? DLB-4

Gilliatt, Penelope 1932-1993 .... DLB-14

Gillott, Jacky 1939-1980 .... DLB-14

Gilman, Caroline H. 1794-1888 ... DLB-3, 73

Gilman, W. and J. [publishing house] DLB-49

Gilmer, Elizabeth Meriwether 1861-1951 .... DLB-29

Gilmer, Francis Walker 1790-1826 .... DLB-37

Gilroy, Frank D. 1925- DLB-7

Gimferrer, Pere (Pedro) 1945- DLB-134

Gingrich, Arnold 1903-1976 .... DLB-137

Ginsberg, Allen 1926- DLB-5, 16, 169

Ginzkey, Franz Karl 1871-1963 ..... DLB-81

Gioia, Dana 1950- DLB-120

Giono, Jean 1895-1970 .... DLB-72

Giotti, Virgilio 1885-1957 .... DLB-114

Giovanni, Nikki 1943- DLB-5, 41

Gipson, Lawrence Henry 1880-1971 .... DLB-17

Girard, Rodolphe 1879-1956 .... DLB-92

Giraudoux, Jean 1882-1944 .... DLB-65

Gissing, George 1857-1903 ..... DLB-18, 135

Giudici, Giovanni 1924- DLB-128

Giuliani, Alfredo 1924- DLB-128

Gladstone, William Ewart 1809-1898 .... DLB-57

Glaeser, Ernst 1902-1963 .... DLB-69

Glanville, Brian 1931- DLB-15, 139

Glapthorne, Henry 1610-1643? DLB-58

Glasgow, Ellen 1873-1945 .... DLB-9, 12

Glaspell, Susan 1876-1948 .... DLB-7, 9, 78

Glass, Montague 1877-1934 .... DLB-11

Glassco, John 1909-1981 .... DLB-68

Glauser, Friedrich 1896-1938 .... DLB-56

F. Gleason's Publishing Hall DLB-49

Gleim, Johann Wilhelm Ludwig 1719-1803 .... DLB-97

Glendinning, Victoria 1937- DLB-155

Glover, Richard 1712-1785 .... DLB-95

Glück, Louise 1943- DLB-5

Glyn, Elinor 1864-1943 .... DLB-153

Gobineau, Joseph-Arthur de
    1816-1882 ....DLB-123
Godbout, Jacques 1933- DLB-53
Goddard, Morrill 1865-1937 ....DLB-25
Goddard, William 1740-1817 .....DLB-43
Godden, Rumer 1907- DLB-161
Godey, Louis A. 1804-1878 ....DLB-73
Godey and McMichael DLB-49
Godfrey, Dave 1938- DLB-60
Godfrey, Thomas 1736-1763 ....DLB-31
Godine, David R., Publisher DLB-46
Godkin, E. L. 1831-1902 ....DLB-79
Godolphin, Sidney 1610-1643 .....DLB-126
Godwin, Gail 1937- DLB-6
Godwin, Mary Jane Clairmont
    1766-1841 ....DLB-163
Godwin, Parke 1816-1904 ....DLB-3, 64
Godwin, William
    1756-1836 ...DLB-39, 104, 142, 158, 163
Godwin, M. J., and Company DLB-154
Goering, Reinhard 1887-1936 ....DLB-118
Goes, Albrecht 1908- DLB-69
Goethe, Johann Wolfgang von
    1749-1832 ....DLB-94
Goetz, Curt 1888-1960 ....DLB-124
Goffe, Thomas circa 1592-1629 .....DLB-58
Goffstein, M. B. 1940- DLB-61
Gogarty, Oliver St. John
    1878-1957 ....DLB-15, 19
Goines, Donald 1937-1974 ....DLB-33
Gold, Herbert 1924- DLB-2; Y-81
Gold, Michael 1893-1967 ....DLB-9, 28
Goldbarth, Albert 1948- DLB-120
Goldberg, Dick 1947- DLB-7
Golden Cockerel Press DLB-112
Golding, Arthur 1536-1606 ....DLB-136
Golding, William 1911-1993 ....DLB-15, 100
Goldman, William 1931- DLB-44
Goldsmith, Oliver
    1730?-1774 ...DLB-39, 89, 104, 109, 142
Goldsmith, Oliver 1794-1861 ....DLB-99
Goldsmith Publishing Company DLB-46
Gollancz, Victor, Limited DLB-112
Gómez-Quiñones, Juan 1942- DLB-122
Gomme, Laurence James
    [publishing house] DLB-46
Goncourt, Edmond de 1822-1896 ...DLB-123

Goncourt, Jules de 1830-1870 .....DLB-123
Gonzales, Rodolfo "Corky"
    1928- DLB-122
González, Angel 1925- DLB-108
Gonzalez, Genaro 1949- DLB-122
Gonzalez, Ray 1952- DLB-122
González de Mireles, Jovita
    1899-1983 ....DLB-122
González-T., César A. 1931- DLB-82
Goodbye, Gutenberg? A Lecture at
    the New York Public Library,
    18 April 1995 ....Y-95
Goodison, Lorna 1947- DLB-157
Goodman, Paul 1911-1972 ....DLB-130
The Goodman Theatre DLB-7
Goodrich, Frances 1891-1984 and
    Hackett, Albert 1900- DLB-26
Goodrich, Samuel Griswold
    1793-1860 ....DLB-1, 42, 73
Goodrich, S. G. [publishing house] ..DLB-49
Goodspeed, C. E., and Company ...DLB-49
Goodwin, Stephen 1943- Y-82
Googe, Barnabe 1540-1594 ....DLB-132
Gookin, Daniel 1612-1687 ....DLB-24
Gordon, Caroline
    1895-1981 ....DLB-4, 9, 102; Y-81
Gordon, Giles 1940- DLB-14, 139
Gordon, Lyndall 1941- DLB-155
Gordon, Mary 1949- DLB-6; Y-81
Gordone, Charles 1925- DLB-7
Gore, Catherine 1800-1861 ....DLB-116
Gorey, Edward 1925- DLB-61
Görres, Joseph 1776-1848 ....DLB-90
Gosse, Edmund 1849-1928 ......DLB-57, 144
Gotlieb, Phyllis 1926- DLB-88
Gottfried von Straßburg
    died before 1230 ....DLB-138
Gotthelf, Jeremias 1797-1854 .....DLB-133
Gottschalk circa 804/808-869 .....DLB-148
Gottsched, Johann Christoph
    1700-1766 ....DLB-97
Götz, Johann Nikolaus
    1721-1781 ....DLB-97
Gould, Wallace 1882-1940 ....DLB-54
Govoni, Corrado 1884-1965 ....DLB-114
Gower, John circa 1330-1408 ......DLB-146
Goyen, William 1915-1983 .....DLB-2; Y-83
Goytisolo, José Augustín 1928- ...DLB-134

Gozzano, Guido 1883-1916 ....DLB-114
Grabbe, Christian Dietrich
    1801-1836 ....DLB-133
Gracq, Julien 1910- DLB-83
Grady, Henry W. 1850-1889 ....DLB-23
Graf, Oskar Maria 1894-1967 ......DLB-56
Graf Rudolf between circa 1170
    and circa 1185 ....DLB-148
Graham, George Rex 1813-1894 .....DLB-73
Graham, Gwethalyn 1913-1965 ......DLB-88
Graham, Jorie 1951- DLB-120
Graham, Katharine 1917- DLB-127
Graham, Lorenz 1902-1989 ....DLB-76
Graham, Philip 1915-1963 ....DLB-127
Graham, R. B. Cunninghame
    1852-1936 ....DLB-98, 135
Graham, Shirley 1896-1977 ....DLB-76
Graham, W. S. 1918- DLB-20
Graham, William H.
    [publishing house] DLB-49
Graham, Winston 1910- DLB-77
Grahame, Kenneth
    1859-1932 ....DLB-34, 141
Grainger, Martin Allerdale
    1874-1941 ....DLB-92
Gramatky, Hardie 1907-1979 ....DLB-22
Grand, Sarah 1854-1943 ....DLB-135
Grandbois, Alain 1900-1975 ....DLB-92
Grange, John circa 1556-? DLB-136
Granich, Irwin (see Gold, Michael)
Grant, Duncan 1885-1978 ....DS-10
Grant, George 1918-1988 ....DLB-88
Grant, George Monro 1835-1902 ....DLB-99
Grant, Harry J. 1881-1963 ....DLB-29
Grant, James Edward 1905-1966 ....DLB-26
Grass, Günter 1927- DLB-75, 124
Grasty, Charles H. 1863-1924 ......DLB-25
Grau, Shirley Ann 1929- DLB-2
Graves, John 1920- Y-83
Graves, Richard 1715-1804 ....DLB-39
Graves, Robert
    1895-1985 ....DLB-20, 100; Y-85
Gray, Asa 1810-1888 ....DLB-1
Gray, David 1838-1861 ....DLB-32
Gray, Simon 1936- DLB-13
Gray, Thomas 1716-1771 ....DLB-109
Grayson, William J. 1788-1863 ...DLB-3, 64

Cumulative Index

The Great Bibliographers Series Y-93

The Great War and the Theater, 1914-1918 [Great Britain] DLB-10

Greeley, Horace 1811-1872 ...... DLB-3, 43

Green, Adolph (see Comden, Betty)

Green, Duff 1791-1875 .... DLB-43

Green, Gerald 1922- DLB-28

Green, Henry 1905-1973 .... DLB-15

Green, Jonas 1712-1767 .... DLB-31

Green, Joseph 1706-1780 .... DLB-31

Green, Julien 1900- DLB-4, 72

Green, Paul 1894-1981 ...... DLB-7, 9; Y-81

Green, T. and S. [publishing house] DLB-49

Green, Timothy [publishing house] DLB-49

Greenaway, Kate 1846-1901 .... DLB-141

Greenberg: Publisher DLB-46

Green Tiger Press DLB-46

Greene, Asa 1789-1838 .... DLB-11

Greene, Benjamin H. [publishing house] DLB-49

Greene, Graham 1904-1991 .... DLB-13, 15, 77, 100, 162; Y-85, Y-91

Greene, Robert 1558-1592 ..... DLB-62, 167

Greenhow, Robert 1800-1854 .... DLB-30

Greenough, Horatio 1805-1852 ...... DLB-1

Greenwell, Dora 1821-1882 .... DLB-35

Greenwillow Books DLB-46

Greenwood, Grace (see Lippincott, Sara Jane Clarke)

Greenwood, Walter 1903-1974 ..... DLB-10

Greer, Ben 1948- DLB-6

Greflinger, Georg 1620?-1677 ..... DLB-164

Greg, W. R. 1809-1881 .... DLB-55

Gregg Press DLB-46

Gregory, Isabella Augusta Persse, Lady 1852-1932 .... DLB-10

Gregory, Horace 1898-1982 .... DLB-48

Gregory of Rimini circa 1300-1358 .... DLB-115

Gregynog Press DLB-112

Greiffenberg, Catharina Regina von 1633-1694 .... DLB-168

Grenfell, Wilfred Thomason 1865-1940 .... DLB-92

Greve, Felix Paul (see Grove, Frederick Philip)

Greville, Fulke, First Lord Brooke 1554-1628 .... DLB-62

Grey, Lady Jane 1537-1554 .... DLB-132

Grey Owl 1888-1938 .... DLB-92

Grey, Zane 1872-1939 .... DLB-9

Grey Walls Press DLB-112

Grier, Eldon 1917- DLB-88

Grieve, C. M. (see MacDiarmid, Hugh)

Griffin, Gerald 1803-1840 .... DLB-159

Griffith, Elizabeth 1727?-1793 .... DLB-39, 89

Griffiths, Trevor 1935- DLB-13

Griffiths, Ralph [publishing house] DLB-154

Griggs, S. C., and Company DLB-49

Griggs, Sutton Elbert 1872-1930 ..... DLB-50

Grignon, Claude-Henri 1894-1976 ... DLB-68

Grigson, Geoffrey 1905- DLB-27

Grillparzer, Franz 1791-1872 ...... DLB-133

Grimald, Nicholas circa 1519-circa 1562 .... DLB-136

Grimké, Angelina Weld 1880-1958 .... DLB-50, 54

Grimm, Hans 1875-1959 .... DLB-66

Grimm, Jacob 1785-1863 .... DLB-90

Grimm, Wilhelm 1786-1859 .... DLB-90

Grimmelshausen, Johann Jacob Christoffel von 1621 or 1622-1676 .... DLB-168

Grindal, Edmund 1519 or 1520-1583 .... DLB-132

Griswold, Rufus Wilmot 1815-1857 .... DLB-3, 59

Gross, Milt 1895-1953 .... DLB-11

Grosset and Dunlap DLB-49

Grossman Publishers DLB-46

Grosseteste, Robert circa 1160-1253 .... DLB-115

Grosvenor, Gilbert H. 1875-1966 .... DLB-91

Groth, Klaus 1819-1899 .... DLB-129

Groulx, Lionel 1878-1967 .... DLB-68

Grove, Frederick Philip 1879-1949 ... DLB-92

Grove Press DLB-46

Grubb, Davis 1919-1980 .... DLB-6

Gruelle, Johnny 1880-1938 .... DLB-22

Grymeston, Elizabeth before 1563-before 1604 .... DLB-136

Gryphius, Andreas 1616-1664 ..... DLB-164

Gryphius, Christian 1649-1706 .... DLB-168

Guare, John 1938- DLB-7

Guerra, Tonino 1920- DLB-128

Guest, Barbara 1920- DLB-5

Guèvremont, Germaine 1893-1968 .... DLB-68

Guidacci, Margherita 1921-1992 .... DLB-128

Guide to the Archives of Publishers, Journals, and Literary Agents in North American Libraries Y-93

Guillén, Jorge 1893-1984 .... DLB-108

Guilloux, Louis 1899-1980 .... DLB-72

Guilpin, Everard circa 1572-after 1608? DLB-136

Guiney, Louise Imogen 1861-1920 ... DLB-54

Guiterman, Arthur 1871-1943 ...... DLB-11

Günderrode, Caroline von 1780-1806 .... DLB-90

Gundulić, Ivan 1589-1638 .... DLB-147

Gunn, Bill 1934-1989 .... DLB-38

Gunn, James E. 1923- DLB-8

Gunn, Neil M. 1891-1973 .... DLB-15

Gunn, Thom 1929- DLB-27

Gunnars, Kristjana 1948- DLB-60

Günther, Johann Christian 1695-1723 .... DLB-168

Gurik, Robert 1932- DLB-60

Gustafson, Ralph 1909- DLB-88

Gütersloh, Albert Paris 1887-1973 .... DLB-81

Guthrie, A. B., Jr. 1901- DLB-6

Guthrie, Ramon 1896-1973 .... DLB-4

The Guthrie Theater DLB-7

Gutzkow, Karl 1811-1878 .... DLB-133

Guy, Ray 1939- DLB-60

Guy, Rosa 1925- DLB-33

Guyot, Arnold 1807-1884 .... DS-13

Gwynne, Erskine 1898-1948 .... DLB-4

Gyles, John 1680-1755 .... DLB-99

Gysin, Brion 1916- DLB-16

H

H. D. (see Doolittle, Hilda)

Habington, William 1605-1654 ..... DLB-126

Hacker, Marilyn 1942- DLB-120

Hackett, Albert (see Goodrich, Frances)

Hacks, Peter 1928- DLB-124

Hadas, Rachel 1948- DLB-120

Hadden, Briton 1898-1929 ....DLB-91

Hagedorn, Friedrich von
1708-1754 ....DLB-168

Hagelstange, Rudolf 1912-1984 ......DLB-69

Haggard, H. Rider 1856-1925 ....DLB-70, 156

Haggard, William 1907-1993 ....Y-93

Hahn-Hahn, Ida Gräfin von
1805-1880 ....DLB-133

Haig-Brown, Roderick 1908-1976 ....DLB-88

Haight, Gordon S. 1901-1985 ......DLB-103

Hailey, Arthur 1920- DLB-88; Y-82

Haines, John 1924- DLB-5

Hake, Edward
flourished 1566-1604 ....DLB-136

Hake, Thomas Gordon 1809-1895 ...DLB-32

Hakluyt, Richard 1552?-1616 ......DLB-136

Halbe, Max 1865-1944 ....DLB-118

Haldane, J. B. S. 1892-1964 ....DLB-160

Haldeman, Joe 1943- DLB-8

Haldeman-Julius Company DLB-46

Hale, E. J., and Son DLB-49

Hale, Edward Everett
1822-1909 .... DLB-1, 42, 74

Hale, Kathleen 1898- DLB-160

Hale, Leo Thomas (see Ebon)

Hale, Lucretia Peabody
1820-1900 ....DLB-42

Hale, Nancy 1908-1988 ... DLB-86; Y-80, 88

Hale, Sarah Josepha (Buell)
1788-1879 .... DLB-1, 42, 73

Hales, John 1584-1656 ....DLB-151

Haley, Alex 1921-1992 ....DLB-38

Haliburton, Thomas Chandler
1796-1865 ....DLB-11, 99

Hall, Anna Maria 1800-1881 ....DLB-159

Hall, Donald 1928- DLB-5

Hall, Edward 1497-1547 ....DLB-132

Hall, James 1793-1868 ....DLB-73, 74

Hall, Joseph 1574-1656 ....DLB-121, 151

Hall, Samuel [publishing house] DLB-49

Hallam, Arthur Henry 1811-1833 ....DLB-32

Halleck, Fitz-Greene 1790-1867 ......DLB-3

Haller, Albrecht von 1708-1777 ....DLB-168

Hallmann, Johann Christian
1640-1704 or 1716? DLB-168

Hallmark Editions DLB-46

Halper, Albert 1904-1984 ....DLB-9

Halperin, John William 1941- DLB-111

Halstead, Murat 1829-1908 ....DLB-23

Hamann, Johann Georg 1730-1788 .. DLB-97

Hamburger, Michael 1924- DLB-27

Hamilton, Alexander 1712-1756 ....DLB-31

Hamilton, Alexander 1755?-1804 ....DLB-37

Hamilton, Cicely 1872-1952 ....DLB-10

Hamilton, Edmond 1904-1977 ....DLB-8

Hamilton, Elizabeth 1758-1816 ...DLB-116, 158

Hamilton, Gail (see Corcoran, Barbara)

Hamilton, Ian 1938- DLB-40, 155

Hamilton, Patrick 1904-1962 .... DLB-10

Hamilton, Virginia 1936- DLB-33, 52

Hamilton, Hamish, Limited DLB-112

Hammett, Dashiell 1894-1961 ....DS-6

Dashiell Hammett:
An Appeal in TAC Y-91

Hammon, Jupiter 1711-died between
1790 and 1806 .... DLB-31, 50

Hammond, John ?-1663 ....DLB-24

Hamner, Earl 1923- DLB-6

Hampton, Christopher 1946- DLB-13

Handel-Mazzetti, Enrica von
1871-1955 ....DLB-81

Handke, Peter 1942- DLB-85, 124

Handlin, Oscar 1915- DLB-17

Hankin, St. John 1869-1909 ....DLB-10

Hanley, Clifford 1922- DLB-14

Hannah, Barry 1942- DLB-6

Hannay, James 1827-1873 .... DLB-21

Hansberry, Lorraine 1930-1965 ... DLB-7, 38

Hapgood, Norman 1868-1937 ......DLB-91

Happel, Eberhard Werner
1647-1690 ....DLB-168

Harcourt Brace Jovanovich DLB-46

Hardenberg, Friedrich von (see Novalis)

Harding, Walter 1917- DLB-111

Hardwick, Elizabeth 1916- DLB-6

Hardy, Thomas 1840-1928 ... DLB-18, 19, 135

Hare, Cyril 1900-1958 ....DLB-77

Hare, David 1947- DLB-13

Hargrove, Marion 1919- DLB-11

Häring, Georg Wilhelm Heinrich (see Alexis,
Willibald)

Harington, Donald 1935- DLB-152

Harington, Sir John 1560-1612 ..... DLB-136

Harjo, Joy 1951- DLB-120

Harlow, Robert 1923- DLB-60

Harman, Thomas
flourished 1566-1573 ....DLB-136

Harness, Charles L. 1915- DLB-8

Harnett, Cynthia 1893-1981 ....DLB-161

Harper, Fletcher 1806-1877 ....DLB-79

Harper, Frances Ellen Watkins
1825-1911 ....DLB-50

Harper, Michael S. 1938- DLB-41

Harper and Brothers DLB-49

Harraden, Beatrice 1864-1943 .....DLB-153

Harrap, George G., and Company
Limited DLB-112

Harriot, Thomas 1560-1621 ....DLB-136

Harris, Benjamin ?-circa 1720 ....DLB-42, 43

Harris, Christie 1907- DLB-88

Harris, Frank 1856-1931 ....DLB-156

Harris, George Washington
1814-1869 ....DLB-3, 11

Harris, Joel Chandler
1848-1908 .... DLB-11, 23, 42, 78, 91

Harris, Mark 1922- DLB-2; Y-80

Harris, Wilson 1921- DLB-117

Harrison, Charles Yale
1898-1954 ....DLB-68

Harrison, Frederic 1831-1923 ....DLB-57

Harrison, Harry 1925- DLB-8

Harrison, Jim 1937- Y-82

Harrison, Mary St. Leger Kingsley (see Malet,
Lucas)

Harrison, Paul Carter 1936- DLB-38

Harrison, Susan Frances
1859-1935 ....DLB-99

Harrison, Tony 1937- DLB-40

Harrison, William 1535-1593 ......DLB-136

Harrison, James P., Company DLB-49

Harrisse, Henry 1829-1910 ....DLB-47

Harsdörffer, Georg Philipp
1607-1658 ....DLB-164

Harsent, David 1942- DLB-40

Hart, Albert Bushnell 1854-1943 .....DLB-17

Hart, Julia Catherine 1796-1867 .....DLB-99

The Lorenz Hart Centenary Y-95

Hart, Moss 1904-1961 ....DLB-7

Hart, Oliver 1723-1795 ....DLB-31

Hart-Davis, Rupert, Limited DLB-112

Harte, Bret 1836-1902 .... DLB-12, 64, 74, 79

Harte, Edward Holmead 1922- ... DLB-127

Harte, Houston Harriman 1927- ...DLB-127

Hartlaub, Felix 1913-1945 .... DLB-56

Hartleben, Otto Erich
    1864-1905 .... DLB-118

Hartley, L. P. 1895-1972 .... DLB-15, 139

Hartley, Marsden 1877-1943 .... DLB-54

Hartling, Peter 1933- DLB-75

Hartman, Geoffrey H. 1929- DLB-67

Hartmann, Sadakichi 1867-1944 .... DLB-54

Hartmann von Aue
    circa 1160-circa 1205 .... DLB-138

Harvey, Gabriel 1550?-1631 ...... DLB-167

Harvey, Jean-Charles 1891-1967 .... DLB-88

Harvill Press Limited DLB-112

Harwood, Lee 1939- DLB-40

Harwood, Ronald 1934- DLB-13

Haskins, Charles Homer
    1870-1937 .... DLB-47

Hass, Robert 1941- DLB-105

The Hatch-Billops Collection DLB-76

Hathaway, William 1944- DLB-120

Hauff, Wilhelm 1802-1827 .... DLB-90

A Haughty and Proud Generation (1922),
    by Ford Madox Hueffer DLB-36

Haugwitz, August Adolph von
    1647-1706 .... DLB-168

Hauptmann, Carl
    1858-1921 .... DLB-66, 118

Hauptmann, Gerhart
    1862-1946 .... DLB-66, 118

Hauser, Marianne 1910- Y-83

Hawes, Stephen
    1475?-before 1529 .... DLB-132

Hawker, Robert Stephen
    1803-1875 .... DLB-32

Hawkes, John 1925- DLB-2, 7; Y-80

Hawkesworth, John 1720-1773 .... DLB-142

Hawkins, Sir Anthony Hope (see Hope,
    Anthony)

Hawkins, Sir John
    1719-1789 .... DLB-104, 142

Hawkins, Walter Everette 1883-? ... DLB-50

Hawthorne, Nathaniel
    1804-1864 .... DLB-1, 74

Hay, John 1838-1905 .... DLB-12, 47

Hayden, Robert 1913-1980 .... DLB-5, 76

Haydon, Benjamin Robert
    1786-1846 .... DLB-110

Hayes, John Michael 1919- DLB-26

Hayley, William 1745-1820 ..... DLB-93, 142

Haym, Rudolf 1821-1901 .... DLB-129

Hayman, Robert 1575-1629 .... DLB-99

Hayman, Ronald 1932- DLB-155

Hayne, Paul Hamilton
    1830-1886 .... DLB-3, 64, 79

Hays, Mary 1760-1843 .... DLB-142, 158

Haywood, Eliza 1693?-1756 .... DLB-39

Hazard, Willis P. [publishing house] .. DLB-49

Hazlitt, William 1778-1830 ...DLB-110, 158

Hazzard, Shirley 1931- Y-82

Head, Bessie 1937-1986 .... DLB-117

Headley, Joel T. 1813-1897 ...DLB-30; DS-13

Heaney, Seamus 1939- DLB-40

Heard, Nathan C. 1936- DLB-33

Hearn, Lafcadio 1850-1904 ...... DLB-12, 78

Hearne, John 1926- DLB-117

Hearne, Samuel 1745-1792 .... DLB-99

Hearst, William Randolph
    1863-1951 .... DLB-25

Hearst, William Randolph, Jr
    1908-1993 .... DLB-127

Heath, Catherine 1924- DLB-14

Heath, Roy A. K. 1926- DLB-117

Heath-Stubbs, John 1918- DLB-27

Heavysege, Charles 1816-1876 ...... DLB-99

Hebbel, Friedrich 1813-1863 ...... DLB-129

Hebel, Johann Peter 1760-1826 ...... DLB-90

Hébert, Anne 1916- DLB-68

Hébert, Jacques 1923- DLB-53

Hecht, Anthony 1923- DLB-5, 169

Hecht, Ben 1894-1964
    .... DLB-7, 9, 25, 26, 28, 86

Hecker, Isaac Thomas 1819-1888 ..... DLB-1

Hedge, Frederic Henry
    1805-1890 .... DLB-1, 59

Hefner, Hugh M. 1926- DLB-137

Hegel, Georg Wilhelm Friedrich
    1770-1831 .... DLB-90

Heidish, Marcy 1947- Y-82

Heißenbüttel 1921- DLB-75

Hein, Christoph 1944- DLB-124

Heine, Heinrich 1797-1856 .... DLB-90

Heinemann, Larry 1944- DS-9

Heinemann, William, Limited DLB-112

Heinlein, Robert A. 1907-1988 .... DLB-8

Heinrich Julius of Brunswick
    1564-1613 .... DLB-164

Heinrich von dem Türlîn
    flourished circa 1230 .... DLB-138

Heinrich von Melk
    flourished after 1160 .... DLB-148

Heinrich von Veldeke
    circa 1145-circa 1190 .... DLB-138

Heinrich, Willi 1920- DLB-75

Heiskell, John 1872-1972 .... DLB-127

Heinse, Wilhelm 1746-1803 .... DLB-94

Hejinian, Lyn 1941- DLB-165

Heliand circa 850 .... DLB-148

Heller, Joseph 1923- DLB-2, 28; Y-80

Heller, Michael 1937- DLB-165

Hellman, Lillian 1906-1984 .... DLB-7; Y-84

Hellwig, Johann 1609-1674 .... DLB-164

Helprin, Mark 1947- Y-85

Helwig, David 1938- DLB-60

Hemans, Felicia 1793-1835 .... DLB-96

Hemingway, Ernest 1899-1961
    .... DLB-4, 9, 102; Y-81, 87; DS-1

Hemingway: Twenty-Five Years
    Later Y-85

Hémon, Louis 1880-1913 .... DLB-92

Hemphill, Paul 1936- Y-87

Hénault, Gilles 1920- DLB-88

Henchman, Daniel 1689-1761 .... DLB-24

Henderson, Alice Corbin
    1881-1949 .... DLB-54

Henderson, Archibald
    1877-1963 .... DLB-103

Henderson, David 1942- DLB-41

Henderson, George Wylie
    1904- DLB-51

Henderson, Zenna 1917-1983 .... DLB-8

Henisch, Peter 1943- DLB-85

Henley, Beth 1952- Y-86

Henley, William Ernest
    1849-1903 .... DLB-19

Henniker, Florence 1855-1923 ...... DLB-135

Henry, Alexander 1739-1824 .... DLB-99

Henry, Buck 1930- DLB-26

Henry VIII of England
    1491-1547 .... DLB-132

Henry, Marguerite 1902- DLB-22

Henry, O. (see Porter, William Sydney)

Henry of Ghent circa 1217-1229 - 1293 ....DLB-115

Henry, Robert Selph 1889-1970 .....DLB-17

Henry, Will (see Allen, Henry W.)

Henryson, Robert 1420s or 1430s-circa 1505 ......DLB-146

Henschke, Alfred (see Klabund)

Hensley, Sophie Almon 1866-1946 ...DLB-99

Henty, G. A. 1832?-1902 ....DLB-18, 141

Hentz, Caroline Lee 1800-1856 ....DLB-3

Herbert, Alan Patrick 1890-1971 .....DLB-10

Herbert, Edward, Lord, of Cherbury 1582-1648 ....DLB-121, 151

Herbert, Frank 1920-1986 ....DLB-8

Herbert, George 1593-1633 ....DLB-126

Herbert, Henry William 1807-1858 ....DLB-3, 73

Herbert, John 1926- DLB-53

Herbert, Mary Sidney, Countess of Pembroke (see Sidney, Mary)

Herbst, Josephine 1892-1969 ....DLB-9

Herburger, Gunter 1932- DLB-75, 124

Hercules, Frank E. M. 1917- DLB-33

Herder, Johann Gottfried 1744-1803 ....DLB-97

Herder, B., Book Company DLB-49

Herford, Charles Harold 1853-1931 ....DLB-149

Hergesheimer, Joseph 1880-1954 ....DLB-9, 102

Heritage Press DLB-46

Hermann the Lame 1013-1054 .....DLB-148

Hermes, Johann Timotheus 1738-1821 ....DLB-97

Hermlin, Stephan 1915- DLB-69

Hernández, Alfonso C. 1938- DLB-122

Hernández, Inés 1947- DLB-122

Hernández, Miguel 1910-1942 .....DLB-134

Hernton, Calvin C. 1932- DLB-38

"The Hero as Man of Letters: Johnson, Rousseau, Burns" (1841), by Thomas Carlyle [excerpt] DLB-57

The Hero as Poet. Dante; Shakspeare (1841), by Thomas Carlyle DLB-32

Heron, Robert 1764-1807 ....DLB-142

Herrera, Juan Felipe 1948- DLB-122

Herrick, Robert 1591-1674 ....DLB-126

Herrick, Robert 1868-1938 .... DLB-9, 12, 78

Herrick, William 1915- Y-83

Herrick, E. R., and Company DLB-49

Herrmann, John 1900-1959 .... DLB-4

Hersey, John 1914-1993 .... DLB-6

Hertel, François 1905-1985 .... DLB-68

Hervé-Bazin, Jean Pierre Marie (see Bazin, Hervé)

Hervey, John, Lord 1696-1743 ..... DLB-101

Herwig, Georg 1817-1875 .... DLB-133

Herzog, Emile Salomon Wilhelm (see Maurois, André)

Hesse, Hermann 1877-1962 .... DLB-66

Hewat, Alexander circa 1743-circa 1824 .... DLB-30

Hewitt, John 1907- DLB-27

Hewlett, Maurice 1861-1923 .... DLB-34, 156

Heyen, William 1940- DLB-5

Heyer, Georgette 1902-1974 .... DLB-77

Heym, Stefan 1913- DLB-69

Heyse, Paul 1830-1914 .... DLB-129

Heytesbury, William circa 1310-1372 or 1373 .... DLB-115

Heyward, Dorothy 1890-1961 .... DLB-7

Heyward, DuBose 1885-1940 .... DLB-7, 9, 45

Heywood, John 1497?-1580? DLB-136

Heywood, Thomas 1573 or 1574-1641 .... DLB-62

Hibbs, Ben 1901-1975 .... DLB-137

Hichens, Robert S. 1864-1950 ..... DLB-153

Hickman, William Albert 1877-1957 .... DLB-92

Hidalgo, José Luis 1919-1947 ..... DLB-108

Hiebert, Paul 1892-1987 .... DLB-68

Hierro, José 1922- DLB-108

Higgins, Aidan 1927- DLB-14

Higgins, Colin 1941-1988 .... DLB-26

Higgins, George V. 1939- DLB-2; Y-81

Higginson, Thomas Wentworth 1823-1911 .... DLB-1, 64

Highwater, Jamake 1942?- ... DLB-52; Y-85

Hijuelos, Oscar 1951- DLB-145

Hildegard von Bingen 1098-1179 .... DLB-148

Das Hildesbrandslied circa 820 .... DLB-148

Hildesheimer, Wolfgang 1916-1991 .... DLB-69, 124

Hildreth, Richard 1807-1865 ....DLB-1, 30, 59

Hill, Aaron 1685-1750 ....DLB-84

Hill, Geoffrey 1932- DLB-40

Hill, "Sir" John 1714?-1775 ....DLB-39

Hill, Leslie 1880-1960 ....DLB-51

Hill, Susan 1942- DLB-14, 139

Hill, Walter 1942- DLB-44

Hill and Wang DLB-46

Hill, George M., Company DLB-49

Hill, Lawrence, and Company, Publishers DLB-46

Hillberry, Conrad 1928- DLB-120

Hilliard, Gray and Company DLB-49

Hills, Lee 1906- DLB-127

Hillyer, Robert 1895-1961 ....DLB-54

Hilton, James 1900-1954 ......DLB-34, 77

Hilton, Walter died 1396 ....DLB-146

Hilton and Company DLB-49

Himes, Chester 1909-1984 ....DLB-2, 76, 143

Hine, Daryl 1936- DLB-60

Hingley, Ronald 1920- DLB-155

Hinojosa-Smith, Rolando 1929- DLB-82

Hippel, Theodor Gottlieb von 1741-1796 ....DLB-97

Hirsch, E. D., Jr. 1928- DLB-67

Hirsch, Edward 1950- DLB-120

The History of the Adventures of Joseph Andrews (1742), by Henry Fielding [excerpt] DLB-39

Hoagland, Edward 1932- DLB-6

Hoagland, Everett H., III 1942- DLB-41

Hoban, Russell 1925- DLB-52

Hobbes, Thomas 1588-1679 ....DLB-151

Hobby, Oveta 1905- DLB-127

Hobby, William 1878-1964 ....DLB-127

Hobsbaum, Philip 1932- DLB-40

Hobson, Laura Z. 1900- DLB-28

Hoby, Thomas 1530-1566 ....DLB-132

Hoccleve, Thomas circa 1368-circa 1437 ....DLB-146

Hochhuth, Rolf 1931- DLB-124

Hochman, Sandra 1936- DLB-5

Hodder and Stoughton, Limited DLB-106

Hodgins, Jack 1938- DLB-60

Hodgman, Helen 1945- DLB-14

Hodgskin, Thomas 1787-1869 ..... DLB-158

Hodgson, Ralph 1871-1962 .... DLB-19

Hodgson, William Hope
1877-1918 .... DLB-70, 153, 156

Hoffenstein, Samuel 1890-1947 ..... DLB-11

Hoffman, Charles Fenno
1806-1884 .... DLB-3

Hoffman, Daniel 1923- DLB-5

Hoffmann, E. T. A. 1776-1822 ...... DLB-90

Hoffmanswaldau, Christian Hoffman von
1616-1679 .... DLB-168

Hofmann, Michael 1957- DLB-40

Hofmannsthal, Hugo von
1874-1929 .... DLB-81, 118

Hofstadter, Richard 1916-1970 ..... DLB-17

Hogan, Desmond 1950- DLB-14

Hogan and Thompson DLB-49

Hogarth Press DLB-112

Hogg, James 1770-1835 .... DLB-93, 116, 159

Hohberg, Wolfgang Helmhard Freiherr von
1612-1688 .... DLB-168

Hohl, Ludwig 1904-1980 .... DLB-56

Holbrook, David 1923- DLB-14, 40

Holcroft, Thomas
1745-1809 .... DLB-39, 89, 158

Holden, Jonathan 1941- DLB-105

Holden, Jonathan, Contemporary
Verse Story-telling DLB-105

Holden, Molly 1927-1981 .... DLB-40

Hölderlin, Friedrich 1770-1843 ..... DLB-90

Holiday House DLB-46

Holinshed, Raphael died 1580 ..... DLB-167

Holland, J. G. 1819-1881 .... DS-13

Holland, Norman N. 1927- DLB-67

Hollander, John 1929- DLB-5

Holley, Marietta 1836-1926 .... DLB-11

Hollingsworth, Margaret 1940- DLB-60

Hollo, Anselm 1934- DLB-40

Holloway, Emory 1885-1977 ..... DLB-103

Holloway, John 1920- DLB-27

Holloway House Publishing
Company DLB-46

Holme, Constance 1880-1955 ..... DLB-34

Holmes, Abraham S. 1821?-1908 ... DLB-99

Holmes, John Clellon 1926-1988 .... DLB-16

Holmes, Oliver Wendell
1809-1894 .... DLB-1

Holmes, Richard 1945- DLB-155

Holroyd, Michael 1935- DLB-155

Holst, Hermann E. von
1841-1904 .... DLB-47

Holt, John 1721-1784 .... DLB-43

Holt, Henry, and Company DLB-49

Holt, Rinehart and Winston DLB-46

Holthusen, Hans Egon 1913- DLB-69

Hölty, Ludwig Christoph Heinrich
1748-1776 .... DLB-94

Holz, Arno 1863-1929 .... DLB-118

Home, Henry, Lord Kames (see Kames, Henry
Home, Lord)

Home, John 1722-1808 .... DLB-84

Home, William Douglas 1912- DLB-13

Home Publishing Company DLB-49

Homes, Geoffrey (see Mainwaring, Daniel)

Honan, Park 1928- DLB-111

Hone, William 1780-1842 ..... DLB-110, 158

Hongo, Garrett Kaoru 1951- DLB-120

Honig, Edwin 1919- DLB-5

Hood, Hugh 1928- DLB-53

Hood, Thomas 1799-1845 .... DLB-96

Hook, Theodore 1788-1841 .... DLB-116

Hooker, Jeremy 1941- DLB-40

Hooker, Richard 1554-1600 ...... DLB-132

Hooker, Thomas 1586-1647 .... DLB-24

Hooper, Johnson Jones
1815-1862 .... DLB-3, 11

Hope, Anthony 1863-1933 ..... DLB-153, 156

Hopkins, Gerard Manley
1844-1889 .... DLB-35, 57

Hopkins, John (see Sternhold, Thomas)

Hopkins, Lemuel 1750-1801 .... DLB-37

Hopkins, Pauline Elizabeth
1859-1930 .... DLB-50

Hopkins, Samuel 1721-1803 .... DLB-31

Hopkins, John H., and Son DLB-46

Hopkinson, Francis 1737-1791 ..... DLB-31

Horgan, Paul 1903- DLB-102; Y-85

Horizon Press DLB-46

Horne, Frank 1899-1974 .... DLB-51

Horne, Richard Henry (Hengist)
1802 or 1803-1884 .... DLB-32

Hornung, E. W. 1866-1921 .... DLB-70

Horovitz, Israel 1939- DLB-7

Horton, George Moses
1797?-1883? DLB-50

Horváth, Ödön von
1901-1938 .... DLB-85, 124

Horwood, Harold 1923- DLB-60

Hosford, E. and E.
[publishing house] DLB-49

Hoskyns, John 1566-1638 .... DLB-121

Hotchkiss and Company DLB-49

Hough, Emerson 1857-1923 .... DLB-9

Houghton Mifflin Company DLB-49

Houghton, Stanley 1881-1913 .... DLB-10

Household, Geoffrey 1900-1988 ..... DLB-87

Housman, A. E. 1859-1936 .... DLB-19

Housman, Laurence 1865-1959 ..... DLB-10

Houwald, Ernst von 1778-1845 ..... DLB-90

Hovey, Richard 1864-1900 .... DLB-54

Howard, Donald R. 1927-1987 ..... DLB-111

Howard, Maureen 1930- Y-83

Howard, Richard 1929- DLB-5

Howard, Roy W. 1883-1964 .... DLB-29

Howard, Sidney 1891-1939 ...... DLB-7, 26

Howe, E. W. 1853-1937 .... DLB-12, 25

Howe, Henry 1816-1893 .... DLB-30

Howe, Irving 1920-1993 .... DLB-67

Howe, Joseph 1804-1873 .... DLB-99

Howe, Julia Ward 1819-1910 .... DLB-1

Howe, Percival Presland
1886-1944 .... DLB-149

Howe, Susan 1937- DLB-120

Howell, Clark, Sr. 1863-1936 ...... DLB-25

Howell, Evan P. 1839-1905 .... DLB-23

Howell, James 1594?-1666 .... DLB-151

Howell, Warren Richardson
1912-1984 .... DLB-140

Howell, Soskin and Company DLB-46

Howells, William Dean
1837-1920 .... DLB-12, 64, 74, 79

Howitt, William 1792-1879 and
Howitt, Mary 1799-1888 .... DLB-110

Hoyem, Andrew 1935- DLB-5

Hoyers, Anna Ovena 1584-1655 ... DLB-164

Hoyos, Angela de 1940- DLB-82

Hoyt, Palmer 1897-1979 .... DLB-127

Hoyt, Henry [publishing house] DLB-49

Hrabanus Maurus 776?-856 ...... DLB-148

Hrotsvit of Gandersheim circa 935-circa 1000 .... DLB-148

Hubbard, Elbert 1856-1915 .... DLB-91

Hubbard, Kin 1868-1930 .... DLB-11

Hubbard, William circa 1621-1704 ... DLB-24

Huber, Therese 1764-1829 .... DLB-90

Huch, Friedrich 1873-1913 .... DLB-66

Huch, Ricarda 1864-1947 .... DLB-66

Huck at 100: How Old Is Huckleberry Finn? Y-85

Huddle, David 1942- DLB-130

Hudgins, Andrew 1951- DLB-120

Hudson, Henry Norman 1814-1886 .... DLB-64

Hudson, W. H. 1841-1922 ..... DLB-98, 153

Hudson and Goodwin DLB-49

Huebsch, B. W. [publishing house] DLB-46

Hughes, David 1930- DLB-14

Hughes, John 1677-1720 .... DLB-84

Hughes, Langston 1902-1967 .... DLB-4, 7, 48, 51, 86

Hughes, Richard 1900-1976 .... DLB-15, 161

Hughes, Ted 1930- DLB-40, 161

Hughes, Thomas 1822-1896 ... DLB-18, 163

Hugo, Richard 1923-1982 .... DLB-5

Hugo, Victor 1802-1885 .... DLB-119

Hugo Awards and Nebula Awards ... DLB-8

Hull, Richard 1896-1973 .... DLB-77

Hulme, T. E. 1883-1917 .... DLB-19

Humboldt, Alexander von 1769-1859 .... DLB-90

Humboldt, Wilhelm von 1767-1835 .... DLB-90

Hume, David 1711-1776 .... DLB-104

Hume, Fergus 1859-1932 .... DLB-70

Hummer, T. R. 1950- DLB-120

Humorous Book Illustration DLB-11

Humphrey, William 1924- DLB-6

Humphreys, David 1752-1818 ...... DLB-37

Humphreys, Emyr 1919- DLB-15

Huncke, Herbert 1915- DLB-16

Huneker, James Gibbons 1857-1921 .... DLB-71

Hunold, Christian Friedrich 1681-1721 .... DLB-168

Hunt, Irene 1907- DLB-52

Hunt, Leigh 1784-1859 .... DLB-96, 110, 144

Hunt, Violet 1862-1942 .... DLB-162

Hunt, William Gibbes 1791-1833 .... DLB-73

Hunter, Evan 1926- Y-82

Hunter, Jim 1939- DLB-14

Hunter, Kristin 1931- DLB-33

Hunter, Mollie 1922- DLB-161

Hunter, N. C. 1908-1971 .... DLB-10

Hunter-Duvar, John 1821-1899 .... DLB-99

Huntington, Henry E. 1850-1927 .... DLB-140

Hurd and Houghton DLB-49

Hurst, Fannie 1889-1968 .... DLB-86

Hurst and Blackett DLB-106

Hurst and Company DLB-49

Hurston, Zora Neale 1901?-1960 .... DLB-51, 86

Husson, Jules-François-Félix (see Champfleury)

Huston, John 1906-1987 .... DLB-26

Hutcheson, Francis 1694-1746 ...... DLB-31

Hutchinson, Thomas 1711-1780 .... DLB-30, 31

Hutchinson and Company (Publishers) Limited DLB-112

Hutton, Richard Holt 1826-1897 .... DLB-57

Huxley, Aldous 1894-1963 .... DLB-36, 100, 162

Huxley, Elspeth Josceline 1907- ... DLB-77

Huxley, T. H. 1825-1895 .... DLB-57

Huyghue, Douglas Smith 1816-1891 .... DLB-99

Huysmans, Joris-Karl 1848-1907 ... DLB-123

Hyman, Trina Schart 1939- DLB-61

I

Iavorsky, Stefan 1658-1722 .... DLB-150

Ibn Bajja circa 1077-1138 .... DLB-115

Ibn Gabirol, Solomon circa 1021-circa 1058 .... DLB-115

The Iconography of Science-Fiction Art DLB-8

Iffland, August Wilhelm 1759-1814 .... DLB-94

Ignatow, David 1914- DLB-5

Ike, Chukwuemeka 1931- DLB-157

Iles, Francis (see Berkeley, Anthony)

The Illustration of Early German Literary Manuscripts, circa 1150-circa 1300 .... DLB-148

Imbs, Bravig 1904-1946 .... DLB-4

Imbuga, Francis D. 1947- DLB-157

Immermann, Karl 1796-1840 .... DLB-133

Inchbald, Elizabeth 1753-1821 .... DLB-39, 89

Inge, William 1913-1973 .... DLB-7

Ingelow, Jean 1820-1897 .... DLB-35, 163

Ingersoll, Ralph 1900-1985 .... DLB-127

The Ingersoll Prizes Y-84

Ingoldsby, Thomas (see Barham, Richard Harris)

Ingraham, Joseph Holt 1809-1860 ..... DLB-3

Inman, John 1805-1850 .... DLB-73

Innerhofer, Franz 1944- DLB-85

Innis, Harold Adams 1894-1952 ..... DLB-88

Innis, Mary Quayle 1899-1972 ...... DLB-88

International Publishers Company ... DLB-46

An Interview with David Rabe Y-91

An Interview with George Greenfield, Literary Agent Y-91

An Interview with James Ellroy Y-91

An Interview with Peter S. Prescott Y-86

An Interview with Russell Hoban Y-90

An Interview with Tom Jenks Y-86

Introduction to Paul Laurence Dunbar, Lyrics of Lowly Life (1896), by William Dean Howells DLB-50

Introductory Essay: *Letters of Percy Bysshe Shelley* (1852), by Robert Browning DLB-32

Introductory Letters from the Second Edition of *Pamela* (1741), by Samuel Richardson DLB-39

Irving, John 1942- DLB-6; Y-82

Irving, Washington 1783-1859 ..... DLB-3, 11, 30, 59, 73, 74

Irwin, Grace 1907- DLB-68

Irwin, Will 1873-1948 .... DLB-25

Isherwood, Christopher 1904-1986 .... DLB-15; Y-86

The Island Trees Case: A Symposium on School Library Censorship
An Interview with Judith Krug
An Interview with Phyllis Schlafly
An Interview with Edward B. Jenkinson
An Interview with Lamarr Mooneyham
An Interview with Harriet Bernstein Y-82

Islas, Arturo 1938-1991 .... DLB-122

Ivers, M. J., and Company DLB-49

Iyayi, Festus 1947- DLB-157

J

Jackmon, Marvin E. (see Marvin X)

Jacks, L. P. 1860-1955 .... DLB-135

Jackson, Angela 1951- DLB-41

Jackson, Helen Hunt 1830-1885 .... DLB-42, 47

Jackson, Holbrook 1874-1948 ...... DLB-98

Jackson, Laura Riding 1901-1991 ... DLB-48

Jackson, Shirley 1919-1965 .... DLB-6

Jacob, Piers Anthony Dillingham (see Anthony, Piers)

Jacobi, Friedrich Heinrich 1743-1819 .... DLB-94

Jacobi, Johann Georg 1740-1841 .... DLB-97

Jacobs, Joseph 1854-1916 .... DLB-141

Jacobs, W. W. 1863-1943 .... DLB-135

Jacobs, George W., and Company ... DLB-49

Jacobson, Dan 1929- DLB-14

Jahier, Piero 1884-1966 .... DLB-114

Jahnn, Hans Henny 1894-1959 .... DLB-56, 124

Jakes, John 1932- Y-83

James, C. L. R. 1901-1989 .... DLB-125

James, George P. R. 1801-1860 .... DLB-116

James, Henry 1843-1916 .... DLB-12, 71, 74; DS-13

James, John circa 1633-1729 .... DLB-24

The James Jones Society Y-92

James, M. R. 1862-1936 .... DLB-156

James, P. D. 1920- DLB-87

James Joyce Centenary: Dublin, 1982 ... Y-82

James Joyce Conference Y-85

James VI of Scotland, I of England 1566-1625 .... DLB-151

James, U. P. [publishing house] DLB-49

Jameson, Anna 1794-1860 ...... DLB-99, 166

Jameson, Fredric 1934- DLB-67

Jameson, J. Franklin 1859-1937 ..... DLB-17

Jameson, Storm 1891-1986 .... DLB-36

Janés, Clara 1940- DLB-134

Jaramillo, Cleofas M. 1878-1956 ... DLB-122

Jarman, Mark 1952- DLB-120

Jarrell, Randall 1914-1965 ...... DLB-48, 52

Jarrold and Sons DLB-106

Jasmin, Claude 1930- DLB-60

Jay, John 1745-1829 .... DLB-31

Jefferies, Richard 1848-1887 .... DLB-98, 141

Jeffers, Lance 1919-1985 .... DLB-41

Jeffers, Robinson 1887-1962 .... DLB-45

Jefferson, Thomas 1743-1826 .... DLB-31

Jelinek, Elfriede 1946- DLB-85

Jellicoe, Ann 1927- DLB-13

Jenkins, Elizabeth 1905- DLB-155

Jenkins, Robin 1912- DLB-14

Jenkins, William Fitzgerald (see Leinster, Murray)

Jenkins, Herbert, Limited DLB-112

Jennings, Elizabeth 1926- DLB-27

Jens, Walter 1923- DLB-69

Jensen, Merrill 1905-1980 .... DLB-17

Jephson, Robert 1736-1803 .... DLB-89

Jerome, Jerome K. 1859-1927 .... DLB-10, 34, 135

Jerome, Judson 1927-1991 .... DLB-105

Jerome, Judson, Reflections: After a Tornado DLB-105

Jerrold, Douglas 1803-1857 .... DLB-158, 159

Jesse, F. Tennyson 1888-1958 .... DLB-77

Jewett, Sarah Orne 1849-1909 .... DLB-12, 74

Jewett, John P., and Company DLB-49

The Jewish Publication Society DLB-49

Jewitt, John Rodgers 1783-1821 ..... DLB-99

Jewsbury, Geraldine 1812-1880 ..... DLB-21

Jhabvala, Ruth Prawer 1927- DLB-139

Jiménez, Juan Ramón 1881-1958 .... DLB-134

Joans, Ted 1928- DLB-16, 41

John, Eugenie (see Marlitt, E.)

John of Dumbleton circa 1310-circa 1349 .... DLB-115

John Edward Bruce: Three Documents DLB-50

John O'Hara's Pottsville Journalism Y-88

John Steinbeck Research Center Y-85

John Webster: The Melbourne Manuscript Y-86

Johns, Captain W. E. 1893-1968 .... DLB-160

Johnson, B. S. 1933-1973 .... DLB-14, 40

Johnson, Charles 1679-1748 .... DLB-84

Johnson, Charles R. 1948- DLB-33

Johnson, Charles S. 1893-1956 ... DLB-51, 91

Johnson, Denis 1949- DLB-120

Johnson, Diane 1934- Y-80

Johnson, Edgar 1901- DLB-103

Johnson, Edward 1598-1672 .... DLB-24

Johnson, Fenton 1888-1958 ..... DLB-45, 50

Johnson, Georgia Douglas 1886-1966 .... DLB-51

Johnson, Gerald W. 1890-1980 ..... DLB-29

Johnson, Helene 1907- DLB-51

Johnson, James Weldon 1871-1938 .... DLB-51

Johnson, John H. 1918- DLB-137

Johnson, Linton Kwesi 1952- DLB-157

Johnson, Lionel 1867-1902 .... DLB-19

Johnson, Nunnally 1897-1977 ..... DLB-26

Johnson, Owen 1878-1952 .... Y-87

Johnson, Pamela Hansford 1912- DLB-15

Johnson, Pauline 1861-1913 .... DLB-92

Johnson, Ronald 1935- DLB-169

Johnson, Samuel 1696-1772 .... DLB-24

Johnson, Samuel 1709-1784 .... DLB-39, 95, 104, 142

Johnson, Samuel 1822-1882 .... DLB-1

Johnson, Uwe 1934-1984 .... DLB-75

Johnson, Benjamin [publishing house] DLB-49

Johnson, Benjamin, Jacob, and Robert [publishing house] DLB-49

Johnson, Jacob, and Company DLB-49

Johnson, Joseph [publishing house] ... DLB-154

Johnston, Annie Fellows 1863-1931 ... DLB-42

Johnston, Basil H. 1929- DLB-60

Johnston, Denis 1901-1984 .... DLB-10

Johnston, George 1913- DLB-88

Johnston, Jennifer 1930- DLB-14

Johnston, Mary 1870-1936 .... DLB-9

Johnston, Richard Malcolm 1822-1898 .... DLB-74

Johnstone, Charles 1719?-1800? DLB-39

Johst, Hanns 1890-1978 .... DLB-124

Jolas, Eugene 1894-1952 .... DLB-4, 45

Jones, Alice C. 1853-1933 .... DLB-92

Jones, Charles C., Jr. 1831-1893 ..... DLB-30

Jones, D. G. 1929- DLB-53

Jones, David 1895-1974 .... DLB-20, 100

Jones, Diana Wynne 1934- DLB-161

Jones, Ebenezer 1820-1860 ....DLB-32

Jones, Ernest 1819-1868 ....DLB-32

Jones, Gayl 1949- DLB-33

Jones, Glyn 1905- DLB-15

Jones, Gwyn 1907- DLB-15, 139

Jones, Henry Arthur 1851-1929 .....DLB-10

Jones, Hugh circa 1692-1760 ....DLB-24

Jones, James 1921-1977 ....DLB-2, 143

Jones, Jenkin Lloyd 1911- DLB-127

Jones, LeRoi (see Baraka, Amiri)

Jones, Lewis 1897-1939 ....DLB-15

Jones, Madison 1925- DLB-152

Jones, Major Joseph (see Thompson, William Tappan)

Jones, Preston 1936-1979 ....DLB-7

Jones, Rodney 1950- DLB-120

Jones, Sir William 1746-1794 ....DLB-109

Jones, William Alfred 1817-1900 .....DLB-59

Jones's Publishing House DLB-49

Jong, Erica 1942- DLB-2, 5, 28, 152

Jonke, Gert F. 1946- DLB-85

Jonson, Ben 1572?-1637 ....DLB-62, 121

Jordan, June 1936- DLB-38

Joseph, Jenny 1932- DLB-40

Joseph, Michael, Limited DLB-112

Josephson, Matthew 1899-1978 ....DLB-4

Josiah Allen's Wife (see Holley, Marietta)

Josipovici, Gabriel 1940- DLB-14

Josselyn, John ?-1675 ....DLB-24

Joudry, Patricia 1921- DLB-88

Jovine, Giuseppe 1922- DLB-128

Joyaux, Philippe (see Sollers, Philippe)

Joyce, Adrien (see Eastman, Carol)

Joyce, James
 1882-1941 .... DLB-10, 19, 36, 162

Judd, Sylvester 1813-1853 ....DLB-1

Judd, Orange, Publishing
 Company DLB-49

Judith circa 930 ....DLB-146

Julian of Norwich
 1342-circa 1420 ....DLB-1146

Julian Symons at Eighty Y-92

June, Jennie (see Croly, Jane Cunningham)

Jung, Franz 1888-1963 ....DLB-118

Jünger, Ernst 1895- DLB-56

Der jüngere Titurel circa 1275 ......DLB-138

Jung-Stilling, Johann Heinrich
 1740-1817 ....DLB-94

Justice, Donald 1925- Y-83

The Juvenile Library (see Godwin, M. J., and Company)

K

Kacew, Romain (see Gary, Romain)

Kafka, Franz 1883-1924 ....DLB-81

Kaiser, Georg 1878-1945 ....DLB-124

Kaiserchronik circca 1147 ....DLB-148

Kalechofsky, Roberta 1931- DLB-28

Kaler, James Otis 1848-1912 ....DLB-12

Kames, Henry Home, Lord
 1696-1782 ....DLB-31, 104

Kandel, Lenore 1932- DLB-16

Kanin, Garson 1912- DLB-7

Kant, Hermann 1926- DLB-75

Kant, Immanuel 1724-1804 ....DLB-94

Kantemir, Antiokh Dmitrievich
 1708-1744 ....DLB-150

Kantor, Mackinlay 1904-1977 ...DLB-9, 102

Kaplan, Fred 1937- DLB-111

Kaplan, Johanna 1942- DLB-28

Kaplan, Justin 1925- DLB-111

Kapnist, Vasilii Vasilevich
 1758?-1823 ....DLB-150

Karadžić, Vuk Stefanović
 1787-1864 ....DLB-147

Karamzin, Nikolai Mikhailovich
 1766-1826 ....DLB-150

Karsch, Anna Louisa 1722-1791 .....DLB-97

Kasack, Hermann 1896-1966 ....DLB-69

Kaschnitz, Marie Luise 1901-1974 ...DLB-69

Kaštelan, Jure 1919-1990 ....DLB-147

Kästner, Erich 1899-1974 ....DLB-56

Kattan, Naim 1928- DLB-53

Katz, Steve 1935- Y-83

Kauffman, Janet 1945- Y-86

Kauffmann, Samuel 1898-1971 .....DLB-127

Kaufman, Bob 1925- DLB-16, 41

Kaufman, George S. 1889-1961 ......DLB-7

Kavanagh, P. J. 1931- DLB-40

Kavanagh, Patrick 1904-1967 ....DLB-15, 20

Kaye-Smith, Sheila 1887-1956 ......DLB-36

Kazin, Alfred 1915- DLB-67

Keane, John B. 1928- DLB-13

Keary, Annie 1825-1879 ....DLB-163

Keating, H. R. F. 1926- DLB-87

Keats, Ezra Jack 1916-1983 ....DLB-61

Keats, John 1795-1821 ....DLB-96, 110

Keble, John 1792-1866 ....DLB-32, 55

Keeble, John 1944- Y-83

Keeffe, Barrie 1945- DLB-13

Keeley, James 1867-1934 ....DLB-25

W. B. Keen, Cooke
 and Company DLB-49

Keillor, Garrison 1942- Y-87

Keith, Marian 1874?-1961 ....DLB-92

Keller, Gary D. 1943- DLB-82

Keller, Gottfried 1819-1890 ....DLB-129

Kelley, Edith Summers 1884-1956 ....DLB-9

Kelley, William Melvin 1937- DLB-33

Kellogg, Ansel Nash 1832-1886 ......DLB-23

Kellogg, Steven 1941- DLB-61

Kelly, George 1887-1974 ....DLB-7

Kelly, Hugh 1739-1777 ....DLB-89

Kelly, Robert 1935- DLB-5, 130, 165

Kelly, Piet and Company DLB-49

Kelmscott Press DLB-112

Kemble, Fanny 1809-1893 ....DLB-32

Kemelman, Harry 1908- DLB-28

Kempe, Margery
 circa 1373-1438 ....DLB-146

Kempner, Friederike 1836-1904 ....DLB-129

Kempowski, Walter 1929- DLB-75

Kendall, Claude
 [publishing company] DLB-46

Kendell, George 1809-1867 ....DLB-43

Kenedy, P. J., and Sons DLB-49

Kennedy, Adrienne 1931- DLB-38

Kennedy, John Pendleton 1795-1870 ...DLB-3

Kennedy, Leo 1907- DLB-88

Kennedy, Margaret 1896-1967 ......DLB-36

Kennedy, Patrick 1801-1873 ....DLB-159

Kennedy, Richard S. 1920- DLB-111

Kennedy, William 1928- DLB-143; Y-85

Kennedy, X. J. 1929- DLB-5

Kennelly, Brendan 1936- DLB-40

Kenner, Hugh 1923- DLB-67

Kennerley, Mitchell
  [publishing house] DLB-46

Kent, Frank R. 1877-1958 .... DLB-29

Kenyon, Jane 1947- DLB-120

Keppler and Schwartzmann DLB-49

Kerner, Justinus 1776-1862 .... DLB-90

Kerouac, Jack 1922-1969 .... DLB-2, 16; DS-3

The Jack Kerouac Revival Y-95

Kerouac, Jan 1952- DLB-16

Kerr, Orpheus C. (see Newell, Robert Henry)

Kerr, Charles H., and Company DLB-49

Kesey, Ken 1935- DLB-2, 16

Kessel, Joseph 1898-1979 .... DLB-72

Kessel, Martin 1901- DLB-56

Kesten, Hermann 1900- DLB-56

Keun, Irmgard 1905-1982 .... DLB-69

Key and Biddle DLB-49

Keynes, John Maynard 1883-1946 ..... DS-10

Keyserling, Eduard von 1855-1918 .. DLB-66

Khan, Ismith 1925- DLB-125

Khemnitser, Ivan Ivanovich
  1745-1784 .... DLB-150

Kheraskov, Mikhail Matveevich
  1733-1807 .... DLB-150

Khvostov, Dmitrii Ivanovich
  1757-1835 .... DLB-150

Kidd, Adam 1802?-1831 .... DLB-99

Kidd, William
  [publishing house] DLB-106

Kiely, Benedict 1919- DLB-15

Kiggins and Kellogg DLB-49

Kiley, Jed 1889-1962 .... DLB-4

Kilgore, Bernard 1908-1967 .... DLB-127

Killens, John Oliver 1916- DLB-33

Killigrew, Anne 1660-1685 .... DLB-131

Killigrew, Thomas 1612-1683 ...... DLB-58

Kilmer, Joyce 1886-1918 .... DLB-45

Kilwardby, Robert
  circa 1215-1279 .... DLB-115

Kincaid, Jamaica 1949- DLB-157

King, Clarence 1842-1901 .... DLB-12

King, Florence 1936 .... Y-85

King, Francis 1923- DLB-15, 139

King, Grace 1852-1932 .... DLB-12, 78

King, Henry 1592-1669 .... DLB-126

King, Stephen 1947- DLB-143; Y-80

King, Woodie, Jr. 1937- DLB-38

King, Solomon [publishing house] ... DLB-49

Kinglake, Alexander William
  1809-1891 .... DLB-55, 166

Kingsley, Charles
  1819-1875 .... DLB-21, 32, 163

Kingsley, Henry 1830-1876 .... DLB-21

Kingsley, Sidney 1906- DLB-7

Kingsmill, Hugh 1889-1949 .... DLB-149

Kingston, Maxine Hong 1940- Y-80

Kingston, William Henry Giles
  1814-1880 .... DLB-163

Kinnell, Galway 1927- DLB-5; Y-87

Kinsella, Thomas 1928- DLB-27

Kipling, Rudyard
  1865-1936 .... DLB-19, 34, 141, 156

Kipphardt, Heinar 1922-1982 ...... DLB-124

Kirby, William 1817-1906 .... DLB-99

Kircher, Athanasius 1602-1680 ..... DLB-164

Kirk, John Foster 1824-1904 .... DLB-79

Kirkconnell, Watson 1895-1977 ..... DLB-68

Kirkland, Caroline M.
  1801-1864 .... DLB-3, 73, 74; DS-13

Kirkland, Joseph 1830-1893 .... DLB-12

Kirkpatrick, Clayton 1915- DLB-127

Kirkup, James 1918- DLB-27

Kirouac, Conrad (see Marie-Victorin, Frère)

Kirsch, Sarah 1935- DLB-75

Kirst, Hans Hellmut 1914-1989 ..... DLB-69

Kitcat, Mabel Greenhow
  1859-1922 .... DLB-135

Kitchin, C. H. B. 1895-1967 .... DLB-77

Kizer, Carolyn 1925- DLB-5, 169

Klabund 1890-1928 .... DLB-66

Klaj, Johann 1616-1656 .... DLB-164

Klappert, Peter 1942- DLB-5

Klass, Philip (see Tenn, William)

Klein, A. M. 1909-1972 .... DLB-68

Kleist, Ewald von 1715-1759 .... DLB-97

Kleist, Heinrich von 1777-1811 ..... DLB-90

Klinger, Friedrich Maximilian
  1752-1831 .... DLB-94

Klopstock, Friedrich Gottlieb
  1724-1803 .... DLB-97

Klopstock, Meta 1728-1758 .... DLB-97

Kluge, Alexander 1932- DLB-75

Knapp, Joseph Palmer 1864-1951 .... DLB-91

Knapp, Samuel Lorenzo
  1783-1838 .... DLB-59

Knapton, J. J. and P.
  [publishing house] DLB-154

Kniazhnin, Iakov Borisovich
  1740-1791 .... DLB-150

Knickerbocker, Diedrich (see Irving,
  Washington)

Knigge, Adolph Franz Friedrich Ludwig,
  Freiherr von 1752-1796 .... DLB-94

Knight, Damon 1922- DLB-8

Knight, Etheridge 1931-1992 .... DLB-41

Knight, John S. 1894-1981 .... DLB-29

Knight, Sarah Kemble 1666-1727 .... DLB-24

Knight, Charles, and Company DLB-106

Knister, Raymond 1899-1932 .... DLB-68

Knoblock, Edward 1874-1945 ...... DLB-10

Knopf, Alfred A. 1892-1984 .... Y-84

Knopf, Alfred A.
  [publishing house] DLB-46

Knorr von Rosenroth, Christian
  1636-1689 .... DLB-168

Knowles, John 1926- DLB-6

Knox, Frank 1874-1944 .... DLB-29

Knox, John circa 1514-1572 .... DLB-132

Knox, John Armoy 1850-1906 ...... DLB-23

Knox, Ronald Arbuthnott
  1888-1957 .... DLB-77

Kober, Arthur 1900-1975 .... DLB-11

Kocbek, Edvard 1904-1981 .... DLB-147

Koch, Howard 1902- DLB-26

Koch, Kenneth 1925- DLB-5

Koenigsberg, Moses 1879-1945 ...... DLB-25

Koeppen, Wolfgang 1906- DLB-69

Koertge, Ronald 1940- DLB-105

Koestler, Arthur 1905-1983 .... Y-83

Kokoschka, Oskar 1886-1980 ...... DLB-124

Kolb, Annette 1870-1967 .... DLB-66

Kolbenheyer, Erwin Guido
  1878-1962 .... DLB-66, 124

Kolleritsch, Alfred 1931- DLB-85

Kolodny, Annette 1941- DLB-67

Komarov, Matvei
  circa 1730-1812 .... DLB-150

Komroff, Manuel 1890-1974 .... DLB-4

Komunyakaa, Yusef 1947- DLB-120

Konigsburg, E. L. 1930- DLB-52

Konrad von Würzburg circa 1230-1287 ....DLB-138
Konstantinov, Aleko 1863-1897 ....DLB-147
Kooser, Ted 1939- DLB-105
Kopit, Arthur 1937- DLB-7
Kops, Bernard 1926?- DLB-13
Kornbluth, C. M. 1923-1958 ....DLB-8
Körner, Theodor 1791-1813 ....DLB-90
Kornfeld, Paul 1889-1942 ....DLB-118
Kosinski, Jerzy 1933-1991 ......DLB-2; Y-82
Kosovel, Srečko 1904-1926 ....DLB-147
Kostrov, Ermil Ivanovich 1755-1796 ....DLB-150
Kotzebue, August von 1761-1819 ....DLB-94
Kovačić, Ante 1854-1889 ....DLB-147
Kraf, Elaine 1946- Y-81
Kranjčević, Silvije Strahimir 1865-1908 ....DLB-147
Krasna, Norman 1909-1984 ....DLB-26
Kraus, Karl 1874-1936 ....DLB-118
Krauss, Ruth 1911-1993 ....DLB-52
Kreisel, Henry 1922- DLB-88
Kreuder, Ernst 1903-1972 ....DLB-69
Kreymborg, Alfred 1883-1966 .....DLB-4, 54
Krieger, Murray 1923- DLB-67
Krim, Seymour 1922-1989 ....DLB-16
Krleža, Miroslav 1893-1981 ....DLB-147
Krock, Arthur 1886-1974 ....DLB-29
Kroetsch, Robert 1927- DLB-53
Krutch, Joseph Wood 1893-1970 ....DLB-63
Krylov, Ivan Andreevich 1769-1844 ....DLB-150
Kubin, Alfred 1877-1959 ....DLB-81
Kubrick, Stanley 1928- DLB-26
Kudrun circa 1230-1240 ....DLB-138
Kuffstein, Hans Ludwig von 1582-1656 ....DLB-164
Kuhlmann, Quirinus 1651-1689 ....DLB-168
Kuhnau, Johann 1660-1722 ....DLB-168
Kumin, Maxine 1925- DLB-5
Kunene, Mazisi 1930- DLB-117
Kunitz, Stanley 1905- DLB-48
Kunjufu, Johari M. (see Amini, Johari M.)
Kunnert, Gunter 1929- DLB-75
Kunze, Reiner 1933- DLB-75
Kupferberg, Tuli 1923- DLB-16

Kürnberger, Ferdinand 1821-1879 ....DLB-129
Kurz, Isolde 1853-1944 ....DLB-66
Kusenberg, Kurt 1904-1983 ....DLB-69
Kuttner, Henry 1915-1958 ....DLB-8
Kyd, Thomas 1558-1594 ....DLB-62
Kyftin, Maurice circa 1560?-1598 ....DLB-136
Kyger, Joanne 1934- DLB-16
Kyne, Peter B. 1880-1957 ....DLB-78

L

L. E. L. (see Landon, Letitia Elizabeth)
Laberge, Albert 1871-1960 ....DLB-68
Laberge, Marie 1950- DLB-60
Lacombe, Patrice (see Trullier-Lacombe, Joseph Patrice)
Lacretelle, Jacques de 1888-1985 ....DLB-65
Ladd, Joseph Brown 1764-1786 .....DLB-37
La Farge, Oliver 1901-1963 ....DLB-9
Lafferty, R. A. 1914- DLB-8
La Guma, Alex 1925-1985 ....DLB-117
Lahaise, Guillaume (see Delahaye, Guy)
Lahontan, Louis-Armand de Lom d'Arce, Baron de 1666-1715? DLB-99
Laing, Kojo 1946- DLB-157
Laird, Carobeth 1895- Y-82
Laird and Lee DLB-49
Lalonde, Michèle 1937- DLB-60
Lamantia, Philip 1927- DLB-16
Lamb, Charles 1775-1834 ....DLB-93, 107, 163
Lamb, Lady Caroline 1785-1828 ...DLB-116
Lamb, Mary 1764-1874 ....DLB-163
Lambert, Betty 1933-1983 ....DLB-60
Lamming, George 1927- DLB-125
L'Amour, Louis 1908?- Y-80
Lampman, Archibald 1861-1899 ....DLB-92
Lamson, Wolffe and Company DLB-49
Lancer Books DLB-46
Landesman, Jay 1919- and Landesman, Fran 1927- DLB-16
Landon, Letitia Elizabeth 1802-1838 . DLB-96
Landor, Walter Savage 1775-1864 ....DLB-93, 107
Landry, Napoléon-P. 1884-1956 .....DLB-92
Lane, Charles 1800-1870 ....DLB-1

Lane, Laurence W. 1890-1967 ......DLB-91
Lane, M. Travis 1934- DLB-60
Lane, Patrick 1939- DLB-53
Lane, Pinkie Gordon 1923- DLB-41
Lane, John, Company DLB-49
Laney, Al 1896- DLB-4
Lang, Andrew 1844-1912 ....DLB-98, 141
Langevin, André 1927- DLB-60
Langgässer, Elisabeth 1899-1950 .....DLB-69
Langhorne, John 1735-1779 ....DLB-109
Langland, William circa 1330-circa 1400 ....DLB-146
Langton, Anna 1804-1893 ....DLB-99
Lanham, Edwin 1904-1979 ....DLB-4
Lanier, Sidney 1842-1881 .... DLB-64; DS-13
Lanyer, Aemilia 1569-1645 ....DLB-121
Lapointe, Gatien 1931-1983 ....DLB-88
Lapointe, Paul-Marie 1929- DLB-88
Lardner, Ring 1885-1933 .....DLB-11, 25, 86
Lardner, Ring, Jr. 1915- DLB-26
Lardner 100: Ring Lardner Centennial Symposium Y-85
Larkin, Philip 1922-1985 ....DLB-27
La Roche, Sophie von 1730-1807 ....DLB-94
La Rocque, Gilbert 1943-1984 ....DLB-60
Laroque de Roquebrune, Robert (see Roquebrune, Robert de)
Larrick, Nancy 1910- DLB-61
Larsen, Nella 1893-1964 ....DLB-51
Lasker-Schüler, Else 1869-1945 ....DLB-66, 124
Lasnier, Rina 1915- DLB-88
Lassalle, Ferdinand 1825-1864 .....DLB-129
Lathrop, Dorothy P. 1891-1980 .....DLB-22
Lathrop, George Parsons 1851-1898 ....DLB-71
Lathrop, John, Jr. 1772-1820 ....DLB-37
Latimer, Hugh 1492?-1555 ....DLB-136
Latimore, Jewel Christine McLawler (see Amini, Johari M.)
Latymer, William 1498-1583 ....DLB-132
Laube, Heinrich 1806-1884 ....DLB-133
Laughlin, James 1914- DLB-48
Laumer, Keith 1925- DLB-8
Lauremberg, Johann 1590-1658 ....DLB-164
Laurence, Margaret 1926-1987 ....DLB-53

Laurentius von Schnüffis 1633-1702 .... DLB-168

Laurents, Arthur 1918- DLB-26

Laurie, Annie (see Black, Winifred)

Laut, Agnes Christiana 1871-1936 ... DLB-92

Lavater, Johann Kaspar 1741-1801 .. DLB-97

Lavin, Mary 1912- DLB-15

Lawes, Henry 1596-1662 .... DLB-126

Lawless, Anthony (see MacDonald, Philip)

Lawrence, D. H. 1885-1930 .... DLB-10, 19, 36, 98, 162

Lawrence, David 1888-1973 .... DLB-29

Lawrence, Seymour 1926-1994 .... Y-94

Lawson, John ?-1711 .... DLB-24

Lawson, Robert 1892-1957 .... DLB-22

Lawson, Victor F. 1850-1925 .... DLB-25

Layard, Sir Austen Henry 1817-1894 .... DLB-166

Layton, Irving 1912- DLB-88

LaZamon flourished circa 1200 .... DLB-146

Lazarević, Laza K. 1851-1890 ...... DLB-147

Lea, Henry Charles 1825-1909 ...... DLB-47

Lea, Sydney 1942- DLB-120

Lea, Tom 1907- DLB-6

Leacock, John 1729-1802 .... DLB-31

Leacock, Stephen 1869-1944 .... DLB-92

Lead, Jane Ward 1623-1704 ...... DLB-131

Leadenhall Press DLB-106

Leapor, Mary 1722-1746 .... DLB-109

Lear, Edward 1812-1888 ... DLB-32, 163, 166

Leary, Timothy 1920-1996 .... DLB-16

Leary, W. A., and Company DLB-49

Léautaud, Paul 1872-1956 .... DLB-65

Leavitt, David 1961- DLB-130

Leavitt and Allen DLB-49

le Carré, John 1931- DLB-87

Lécavelé, Roland (see Dorgeles, Roland)

Lechlitner, Ruth 1901- DLB-48

Leclerc, Félix 1914- DLB-60

Le Clézio, J. M. G. 1940- DLB-83

Lectures on Rhetoric and Belles Lettres (1783), by Hugh Blair [excerpts] DLB-31

Leder, Rudolf (see Hermlin, Stephan)

Lederer, Charles 1910-1976 .... DLB-26

Ledwidge, Francis 1887-1917 ...... DLB-20

Lee, Dennis 1939- DLB-53

Lee, Don L. (see Madhubuti, Haki R.)

Lee, George W. 1894-1976 .... DLB-51

Lee, Harper 1926- DLB-6

Lee, Harriet (1757-1851) and Lee, Sophia (1750-1824) DLB-39

Lee, Laurie 1914- DLB-27

Lee, Li-Young 1957- DLB-165

Lee, Manfred B. (see Dannay, Frederic, and Manfred B. Lee)

Lee, Nathaniel circa 1645 - 1692 ..... DLB-80

Lee, Sir Sidney 1859-1926 .... DLB-149

Lee, Sir Sidney, "Principles of Biography," in Elizabethan and Other Essays DLB-149

Lee, Vernon 1856-1935 ....DLB-57, 153, 156

Lee and Shepard DLB-49

Le Fanu, Joseph Sheridan 1814-1873 .... DLB-21, 70, 159

Leffland, Ella 1931- Y-84

le Fort, Gertrud von 1876-1971 ..... DLB-66

Le Gallienne, Richard 1866-1947 ..... DLB-4

Legaré, Hugh Swinton 1797-1843 .... DLB-3, 59, 73

Legaré, James M. 1823-1859 .... DLB-3

The Legends of the Saints and a Medieval Christian Worldview DLB-148

Léger, Antoine-J. 1880-1950 .... DLB-88

Le Guin, Ursula K. 1929- DLB-8, 52

Lehman, Ernest 1920- DLB-44

Lehmann, John 1907- DLB-27, 100

Lehmann, Rosamond 1901-1990 ..... DLB-15

Lehmann, Wilhelm 1882-1968 ..... DLB-56

Lehmann, John, Limited DLB-112

Leiber, Fritz 1910-1992 .... DLB-8

Leibniz, Gottfried Wilhelm 1646-1716 .... DLB-168

Leicester University Press DLB-112

Leinster, Murray 1896-1975 .... DLB-8

Leisewitz, Johann Anton 1752-1806 .... DLB-94

Leitch, Maurice 1933- DLB-14

Leithauser, Brad 1943- DLB-120

Leland, Charles G. 1824-1903 .... DLB-11

Leland, John 1503?-1552 .... DLB-136

Lemay, Pamphile 1837-1918 .... DLB-99

Lemelin, Roger 1919- DLB-88

Lemon, Mark 1809-1870 .... DLB-163

Le Moine, James MacPherson 1825-1912 .... DLB-99

Le Moyne, Jean 1913- DLB-88

L'Engle, Madeleine 1918- DLB-52

Lennart, Isobel 1915-1971 .... DLB-44

Lennox, Charlotte 1729 or 1730-1804 .... DLB-39

Lenox, James 1800-1880 .... DLB-140

Lenski, Lois 1893-1974 .... DLB-22

Lenz, Hermann 1913- DLB-69

Lenz, J. M. R. 1751-1792 .... DLB-94

Lenz, Siegfried 1926- DLB-75

Leonard, Hugh 1926- DLB-13

Leonard, William Ellery 1876-1944 .... DLB-54

Leonowens, Anna 1834-1914 ... DLB-99, 166

LePan, Douglas 1914- DLB-88

Leprohon, Rosanna Eleanor 1829-1879 .... DLB-99

Le Queux, William 1864-1927 .... DLB-70

Lerner, Max 1902-1992 .... DLB-29

Lernet-Holenia, Alexander 1897-1976 .... DLB-85

Le Rossignol, James 1866-1969 ...... DLB-92

Lescarbot, Marc circa 1570-1642 ..... DLB-99

LeSeur, William Dawson 1840-1917 .... DLB-92

LeSieg, Theo. (see Geisel, Theodor Seuss)

Leslie, Frank 1821-1880 .... DLB-43, 79

Leslie, Frank, Publishing House DLB-49

Lesperance, John 1835?-1891 .... DLB-99

Lessing, Bruno 1870-1940 .... DLB-28

Lessing, Doris 1919- DLB-15, 139; Y-85

Lessing, Gotthold Ephraim 1729-1781 .... DLB-97

Lettau, Reinhard 1929- DLB-75

Letter from Japan Y-94

Letter to [Samuel] Richardson on Clarissa (1748), by Henry Fielding DLB-39

Lever, Charles 1806-1872 .... DLB-21

Leverson, Ada 1862-1933 .... DLB-153

Levertov, Denise 1923- DLB-5, 165

Levi, Peter 1931- DLB-40

Levien, Sonya 1888-1960 .... DLB-44

Levin, Meyer 1905-1981 .... DLB-9, 28; Y-81

Levine, Norman 1923- DLB-88

Levine, Philip 1928- DLB-5

Levis, Larry 1946- DLB-120	Link, Arthur S. 1920- DLB-17	*Literature at Nurse, or Circulating Morals* (1885), by George Moore DLB-18
Levy, Amy 1861-1889............DLB-156	Linn, John Blair 1777-1804........DLB-37	Littell, Eliakim 1797-1870 ....DLB-79
Levy, Benn Wolfe 1900-1973 ....DLB-13; Y-81	Lins, Osman 1924-1978 ....DLB-145	Littell, Robert S. 1831-1896 ....DLB-79
Lewald, Fanny 1811-1889 ....DLB-129	Linton, Eliza Lynn 1822-1898 ......DLB-18	Little, Brown and Company DLB-49
Lewes, George Henry 1817-1878 ....DLB-55, 144	Linton, William James 1812-1897 ...DLB-32	Littlewood, Joan 1914- DLB-13
Lewis, Alfred H. 1857-1914 ....DLB-25	Lion Books DLB-46	Lively, Penelope 1933- DLB-14, 161
Lewis, Alun 1915-1944 ....DLB-20, 162	Lionni, Leo 1910- DLB-61	Liverpool University Press DLB-112
Lewis, C. Day (see Day Lewis, C.)	Lippincott, Sara Jane Clarke 1823-1904 ....DLB-43	*The Lives of the Poets* DLB-142
Lewis, C. S. 1898-1963 .... DLB-15, 100, 160	Lippincott, J. B., Company DLB-49	Livesay, Dorothy 1909- DLB-68
Lewis, Charles B. 1842-1924 ....DLB-11	Lippmann, Walter 1889-1974 ....DLB-29	Livesay, Florence Randal 1874-1953 ....DLB-92
Lewis, Henry Clay 1825-1850 ....DLB-3	Lipton, Lawrence 1898-1975 ....DLB-16	Livings, Henry 1929- DLB-13
Lewis, Janet 1899- Y-87	Liscow, Christian Ludwig 1701-1760 ....DLB-97	Livingston, Anne Howe 1763-1841 ....DLB-37
Lewis, Matthew Gregory 1775-1818 ....DLB-39, 158	Lish, Gordon 1934- DLB-130	Livingston, Myra Cohn 1926- DLB-61
Lewis, R. W. B. 1917- DLB-111	Lispector, Clarice 1925-1977 ......DLB-113	Livingston, William 1723-1790 .....DLB-31
Lewis, Richard circa 1700-1734 ......DLB-24	*The Literary Chronicle and Weekly Review* 1819-1828 ....DLB-110	Livingstone, David 1813-1873 ......DLB-166
Lewis, Sinclair 1885-1951 .... DLB-9, 102; DS-1	Literary Documents: William Faulkner and the People-to-People Program Y-86	Liyong, Taban lo (see Taban lo Liyong)
Lewis, Wilmarth Sheldon 1895-1979 ....DLB-140	Literary Documents II: *Library Journal* Statements and Questionnaires from First Novelists Y-87	Lizárraga, Sylvia S. 1925- DLB-82
Lewis, Wyndham 1882-1957 ....DLB-15		Llewellyn, Richard 1906-1983 ....DLB-15
Lewisohn, Ludwig 1882-1955 .... DLB-4, 9, 28, 102	Literary Effects of World War II [British novel] DLB-15	Lloyd, Edward [publishing house] DLB-106
Lezama Lima, José 1910-1976 .....DLB-113	Literary Prizes [British] DLB-15	Lobel, Arnold 1933- DLB-61
The Library of America DLB-46	Literary Research Archives: The Humanities Research Center, University of Texas Y-82	Lochridge, Betsy Hopkins (see Fancher, Betsy)
The Licensing Act of 1737 ....DLB-84		Locke, David Ross 1833-1888 ....DLB-11, 23
Lichtenberg, Georg Christoph 1742-1799 ....DLB-94	Literary Research Archives II: Berg Collection of English and American Literature of the New York Public Library Y-83	Locke, John 1632-1704 ....DLB-31, 101
Liebling, A. J. 1904-1963 ....DLB-4		Locke, Richard Adams 1800-1871 ...DLB-43
Lieutenant Murray (see Ballou, Maturin Murray)	Literary Research Archives III: The Lilly Library Y-84	Locker-Lampson, Frederick 1821-1895 ....DLB-35
Lighthall, William Douw 1857-1954 ....DLB-92	Literary Research Archives IV: The John Carter Brown Library Y-85	Lockhart, John Gibson 1794-1854 ....DLB-110, 116 144
Lilar, Françoise (see Mallet-Joris, Françoise)	Literary Research Archives V: Kent State Special Collections Y-86	Lockridge, Ross, Jr. 1914-1948 ....DLB-143; Y-80
Lillo, George 1691-1739 ....DLB-84		*Locrine* and *Selimus* DLB-62
Lilly, J. K., Jr. 1893-1966 ....DLB-140	Literary Research Archives VI: The Modern Literary Manuscripts Collection in the Special Collections of the Washington University Libraries Y-87	Lodge, David 1935- DLB-14
Lilly, Wait and Company DLB-49		Lodge, George Cabot 1873-1909 .....DLB-54
Lily, William circa 1468-1522 ......DLB-132		Lodge, Henry Cabot 1850-1924 .....DLB-47
Limited Editions Club DLB-46	Literary Research Archives VII: The University of Virginia Libraries Y-91	Loeb, Harold 1891-1974 ....DLB-4
Lincoln and Edmands DLB-49		Loeb, William 1905-1981 ....DLB-127
Lindsay, Jack 1900- Y-84		Lofting, Hugh 1886-1947 ....DLB-160
Lindsay, Sir David circa 1485-1555 ....DLB-132	Literary Research Archives VIII: The Henry E. Huntington Library Y-92	Logan, James 1674-1751 ....DLB-24, 140
		Logan, John 1923- DLB-5
Lindsay, Vachel 1879-1931 ....DLB-54	"Literary Style" (1857), by William Forsyth [excerpt] DLB-57	Logan, William 1950- DLB-120
Linebarger, Paul Myron Anthony (see Smith, Cordwainer)		Logau, Friedrich von 1605-1655 ....DLB-164
	Literatura Chicanesca: The View From Without DLB-82	Logue, Christopher 1926- DLB-27

Lohenstein, Daniel Casper von
    1635-1683 .... DLB-168

Lomonosov, Mikhail Vasil'evich
    1711-1765 .... DLB-150

London, Jack 1876-1916 ...... DLB-8, 12, 78

The London Magazine 1820-1829 ..... DLB-110

Long, Haniel 1888-1956 .... DLB-45

Long, Ray 1878-1935 .... DLB-137

Long, H., and Brother DLB-49

Longfellow, Henry Wadsworth
    1807-1882 .... DLB-1, 59

Longfellow, Samuel 1819-1892 .... DLB-1

Longford, Elizabeth 1906- DLB-155

Longley, Michael 1939- DLB-40

Longman, T. [publishing house] ... DLB-154

Longmans, Green and Company DLB-49

Longmore, George 1793?-1867 ..... DLB-99

Longstreet, Augustus Baldwin
    1790-1870 .... DLB-3, 11, 74

Longworth, D. [publishing house] ... DLB-49

Lonsdale, Frederick 1881-1954 ..... DLB-10

A Look at the Contemporary Black Theatre
    Movement DLB-38

Loos, Anita 1893-1981 .... DLB-11, 26; Y-81

Lopate, Phillip 1943- Y-80

López, Diana (see Isabella, Ríos)

Loranger, Jean-Aubert 1896-1942 ... DLB-92

Lorca, Federico García 1898-1936 .. DLB-108

Lord, John Keast 1818-1872 .... DLB-99

The Lord Chamberlain's Office and Stage
    Censorship in England DLB-10

Lorde, Audre 1934-1992 .... DLB-41

Lorimer, George Horace
    1867-1939 .... DLB-91

Loring, A. K. [publishing house] DLB-49

Loring and Mussey DLB-46

Lossing, Benson J. 1813-1891 ...... DLB-30

Lothar, Ernst 1890-1974 .... DLB-81

Lothrop, Harriet M. 1844-1924 ..... DLB-42

Lothrop, D., and Company DLB-49

Loti, Pierre 1850-1923 .... DLB-123

Lott, Emeline ?-? DLB-166

The Lounger, no. 20 (1785), by Henry
    Mackenzie DLB-39

Lounsbury, Thomas R. 1838-1915 .. DLB-71

Louÿs, Pierre 1870-1925 .... DLB-123

Lovelace, Earl 1935- DLB-125

Lovelace, Richard 1618-1657 ...... DLB-131

Lovell, Coryell and Company DLB-49

Lovell, John W., Company DLB-49

Lover, Samuel 1797-1868 .... DLB-159

Lovesey, Peter 1936- DLB-87

Lovingood, Sut (see Harris,
    George Washington)

Low, Samuel 1765-? DLB-37

Lowell, Amy 1874-1925 .... DLB-54, 140

Lowell, James Russell
    1819-1891 .... DLB-1, 11, 64, 79

Lowell, Robert 1917-1977 .... DLB-5, 169

Lowenfels, Walter 1897-1976 .... DLB-4

Lowndes, Marie Belloc 1868-1947 ... DLB-70

Lowry, Lois 1937- DLB-52

Lowry, Malcolm 1909-1957 .... DLB-15

Lowther, Pat 1935-1975 .... DLB-53

Loy, Mina 1882-1966 .... DLB-4, 54

Lozeau, Albert 1878-1924 .... DLB-92

Lubbock, Percy 1879-1965 .... DLB-149

Lucas, E. V. 1868-1938 .... DLB-98, 149, 153

Lucas, Fielding, Jr.
    [publishing house] DLB-49

Luce, Henry R. 1898-1967 .... DLB-91

Luce, John W., and Company DLB-46

Lucie-Smith, Edward 1933- DLB-40

Lucini, Gian Pietro 1867-1914 .... DLB-114

Ludlum, Robert 1927- Y-82

Ludus de Antichristo circa 1160 .... DLB-148

Ludvigson, Susan 1942- DLB-120

Ludwig, Jack 1922- DLB-60

Ludwig, Otto 1813-1865 .... DLB-129

Ludwigslied 881 or 882 .... DLB-148

Luera, Yolanda 1953- DLB-122

Luft, Lya 1938- DLB-145

Luke, Peter 1919- DLB-13

Lupton, F. M., Company DLB-49

Lupus of Ferrières
    circa 805-circa 862 .... DLB-148

Lurie, Alison 1926- DLB-2

Luzi, Mario 1914- DLB-128

L'vov, Nikolai Aleksandrovich
    1751-1803 .... DLB-150

Lyall, Gavin 1932- DLB-87

Lydgate, John circa 1370-1450 .... DLB-146

Lyly, John circa 1554-1606 ..... DLB-62, 167

Lynch, Patricia 1898-1972 .... DLB-160

Lynd, Robert 1879-1949 .... DLB-98

Lyon, Matthew 1749-1822 .... DLB-43

Lytle, Andrew 1902-1995 ...... DLB-6; Y-95

Lytton, Edward (see Bulwer-Lytton, Edward)

Lytton, Edward Robert Bulwer
    1831-1891 .... DLB-32

M

Maass, Joachim 1901-1972 .... DLB-69

Mabie, Hamilton Wright
    1845-1916 .... DLB-71

Mac A'Ghobhainn, Iain (see Smith, Iain
    Crichton)

MacArthur, Charles
    1895-1956 .... DLB-7, 25, 44

Macaulay, Catherine 1731-1791 .... DLB-104

Macaulay, David 1945- DLB-61

Macaulay, Rose 1881-1958 .... DLB-36

Macaulay, Thomas Babington
    1800-1859 .... DLB-32, 55

Macaulay Company DLB-46

MacBeth, George 1932- DLB-40

Macbeth, Madge 1880-1965 .... DLB-92

MacCaig, Norman 1910- DLB-27

MacDiarmid, Hugh 1892-1978 ...... DLB-20

MacDonald, Cynthia 1928- DLB-105

MacDonald, George
    1824-1905 .... DLB-18, 163

MacDonald, John D.
    1916-1986 .... DLB-8; Y-86

MacDonald, Philip 1899?-1980 ..... DLB-77

Macdonald, Ross (see Millar, Kenneth)

MacDonald, Wilson 1880-1967 ..... DLB-92

Macdonald and Company
    (Publishers) DLB-112

MacEwen, Gwendolyn 1941- DLB-53

Macfadden, Bernarr
    1868-1955 .... DLB-25, 91

MacGregor, John 1825-1892 .... DLB-166

MacGregor, Mary Esther (see Keith, Marian)

Machado, Antonio 1875-1939 ...... DLB-108

Machado, Manuel 1874-1947 ...... DLB-108

Machar, Agnes Maule 1837-1927 ... DLB-92

Machen, Arthur Llewelyn Jones
    1863-1947 .... DLB-36, 156

MacInnes, Colin 1914-1976 .... DLB-14

MacInnes, Helen 1907-1985 .... DLB-87

Mack, Maynard 1909- DLB-111	Maheux-Forcier, Louise 1929- DLB-60	Mandeville, Sir John mid fourteenth century DLB-146
Mackall, Leonard L. 1879-1937 .....DLB-140	Mahin, John Lee 1902-1984 ....DLB-44	Mandiargues, André Pieyre de 1909- DLB-83
MacKaye, Percy 1875-1956 ....DLB-54	Mahon, Derek 1941- DLB-40	Manfred, Frederick 1912-1994 ....DLB-6
Macken, Walter 1915-1967 ....DLB-13	Maikov, Vasilii Ivanovich 1728-1778 ....DLB-150	Mangan, Sherry 1904-1961 ....DLB-4
Mackenzie, Alexander 1763-1820 ....DLB-99	Mailer, Norman 1923- ...DLB-2, 16, 28; Y-80, 83; DS-3	Mankiewicz, Herman 1897-1953 .....DLB-26
Mackenzie, Compton 1883-1972 ....DLB-34, 100	Maillet, Adrienne 1885-1963 ....DLB-68	Mankiewicz, Joseph L. 1909-1993 ....DLB-44
Mackenzie, Henry 1745-1831 ....DLB-39	Maimonides, Moses 1138-1204 ....DLB-115	Mankowitz, Wolf 1924- DLB-15
Mackey, Nathaniel 1947- DLB-169	Maillet, Antonine 1929- DLB-60	Manley, Delarivière 1672?-1724 ....DLB-39, 80
Mackey, William Wellington 1937- DLB-38	Maillu, David G. 1939- DLB-157	Mann, Abby 1927- DLB-44
Mackintosh, Elizabeth (see Tey, Josephine)	Main Selections of the Book-of-the-Month Club, 1926-1945 .... DLB-9	Mann, Heinrich 1871-1950 .....DLB-66, 118
Mackintosh, Sir James 1765-1832 ....DLB-158	Main Trends in Twentieth-Century Book Clubs DLB-46	Mann, Horace 1796-1859 ....DLB-1
Maclaren, Ian (see Watson, John)	Mainwaring, Daniel 1902-1977 .....DLB-44	Mann, Klaus 1906-1949 ....DLB-56
Macklin, Charles 1699-1797 ....DLB-89	Mair, Charles 1838-1927 ....DLB-99	Mann, Thomas 1875-1955 ....DLB-66
MacLean, Katherine Anne 1925- DLB-8	Mais, Roger 1905-1955 ....DLB-125	Mann, William D'Alton 1839-1920 ....DLB-137
MacLeish, Archibald 1892-1982 .... DLB-4, 7, 45; Y-82	Major, Andre 1942- DLB-60	Manning, Marie 1873?-1945 ....DLB-29
MacLennan, Hugh 1907-1990 ....DLB-68	Major, Clarence 1936- DLB-33	Manning and Loring DLB-49
Macleod, Fiona (see Sharp, William)	Major, Kevin 1949- DLB-60	Mannyng, Robert flourished 1303-1338 ....DLB-146
MacLeod, Alistair 1936- DLB-60	Major Books DLB-46	Mano, D. Keith 1942- DLB-6
Macleod, Norman 1906-1985 ....DLB-4	Makemie, Francis circa 1658-1708 ...DLB-24	Manor Books DLB-46
Macmillan and Company DLB-106	The Making of a People, by J. M. Ritchie DLB-66	Mansfield, Katherine 1888-1923 ....DLB-162
The Macmillan Company DLB-49	Maksimović, Desanka 1898-1993 ...DLB-147	Mapanje, Jack 1944- DLB-157
Macmillan's English Men of Letters, First Series (1878-1892) DLB-144	Malamud, Bernard 1914-1986 ......DLB-2, 28, 152; Y-80, 86	March, William 1893-1954 ....DLB-9, 86
MacNamara, Brinsley 1890-1963 ....DLB-10	Malet, Lucas 1852-1931 ....DLB-153	Marchand, Leslie A. 1900- DLB-103
MacNeice, Louis 1907-1963 ......DLB-10, 20	Malleson, Lucy Beatrice (see Gilbert, Anthony)	Marchant, Bessie 1862-1941 ....DLB-160
MacPhail, Andrew 1864-1938 ....DLB-92	Mallet-Joris, Françoise 1930- DLB-83	Marchessault, Jovette 1938- DLB-60
Macpherson, James 1736-1796 .....DLB-109	Mallock, W. H. 1849-1923 ......DLB-18, 57	Marcus, Frank 1928- DLB-13
Macpherson, Jay 1931- DLB-53	Malone, Dumas 1892-1986 ....DLB-17	Marden, Orison Swett 1850-1924 ....DLB-137
Macpherson, Jeanie 1884-1946 ......DLB-44	Malone, Edmond 1741-1812 ....DLB-142	Marechera, Dambudzo 1952-1987 ....DLB-157
Macrae Smith Company DLB-46	Malory, Sir Thomas circa 1400-1410 - 1471 ....DLB-146	Marek, Richard, Books DLB-46
Macrone, John [publishing house] DLB-106	Malraux, André 1901-1976 ....DLB-72	Mares, E. A. 1938- DLB-122
MacShane, Frank 1927- DLB-111	Malthus, Thomas Robert 1766-1834 ....DLB-107, 158	Mariani, Paul 1940- DLB-111
Macy-Masius DLB-46	Maltz, Albert 1908-1985 ....DLB-102	Marie-Victorin, Frère 1885-1944 .....DLB-92
Madden, David 1933- DLB-6	Malzberg, Barry N. 1939- DLB-8	Marin, Biagio 1891-1985 ....DLB-128
Maddow, Ben 1909-1992 ....DLB-44	Mamet, David 1947- DLB-7	Marinković, Ranko 1913- DLB-147
Maddux, Rachel 1912-1983 ....Y-93	Manaka, Matsemela 1956- DLB-157	Marinetti, Filippo Tommaso 1876-1944 ....DLB-114
Madgett, Naomi Long 1923- DLB-76	Manchester University Press DLB-112	Marion, Frances 1886-1973 ....DLB-44
Madhubuti, Haki R. 1942- DLB-5, 41; DS-8	Mandel, Eli 1922- DLB-53	Marius, Richard C. 1933- Y-85
Madison, James 1751-1836 ....DLB-37	Mandeville, Bernard 1670-1733 ....DLB-101	The Mark Taper Forum DLB-7
Maginn, William 1794-1842 ..DLB-110, 159		Mark Twain on Perpetual Copyright Y-92
Mahan, Alfred Thayer 1840-1914 ....DLB-47		

Markfield, Wallace 1926- DLB-2, 28

Markham, Edwin 1852-1940 .... DLB-54

Markle, Fletcher 1921-1991 ... DLB-68; Y-91

Marlatt, Daphne 1942- DLB-60

Marlitt, E. 1825-1887 .... DLB-129

Marlowe, Christopher 1564-1593 ... DLB-62

Marlyn, John 1912- DLB-88

Marmion, Shakerley 1603-1639 ..... DLB-58

Der Marner
    before 1230-circa 1287 .... DLB-138

The *Marprelate Tracts* 1588-1589 .... DLB-132

Marquand, John P. 1893-1960 ... DLB-9, 102

Marqués, René 1919-1979 .... DLB-113

Marquis, Don 1878-1937 .... DLB-11, 25

Marriott, Anne 1913- DLB-68

Marryat, Frederick 1792-1848 .. DLB-21, 163

Marsh, George Perkins
    1801-1882 .... DLB-1, 64

Marsh, James 1794-1842 .... DLB-1, 59

Marsh, Capen, Lyon and Webb DLB-49

Marsh, Ngaio 1899-1982 .... DLB-77

Marshall, Edison 1894-1967 .... DLB-102

Marshall, Edward 1932- DLB-16

Marshall, Emma 1828-1899 ..... DLB-163

Marshall, James 1942-1992 .... DLB-61

Marshall, Joyce 1913- DLB-88

Marshall, Paule 1929- DLB-33, 157

Marshall, Tom 1938- DLB-60

Marsilius of Padua
    circa 1275-circa 1342 .... DLB-115

Marson, Una 1905-1965 .... DLB-157

Marston, John 1576-1634 .... DLB-58

Marston, Philip Bourke 1850-1887 .. DLB-35

Martens, Kurt 1870-1945 .... DLB-66

Martien, William S.
    [publishing house] DLB-49

Martin, Abe (see Hubbard, Kin)

Martin, Charles 1942- DLB-120

Martin, Claire 1914- DLB-60

Martin, Jay 1935- DLB-111

Martin, Johann (see Laurentius von Schnüffis)

Martin, Violet Florence (see Ross, Martin)

Martin du Gard, Roger 1881-1958 .. DLB-65

Martineau, Harriet
    1802-1876 .... DLB-21, 55, 159, 163, 166

Martínez, Eliud 1935- DLB-122

Martínez, Max 1943- DLB-82

Martyn, Edward 1859-1923 .... DLB-10

Marvell, Andrew 1621-1678 ...... DLB-131

Marvin X 1944- DLB-38

Marx, Karl 1818-1883 .... DLB-129

Marzials, Theo 1850-1920 .... DLB-35

Masefield, John
    1878-1967 .... DLB-10, 19, 153, 160

Mason, A. E. W. 1865-1948 .... DLB-70

Mason, Bobbie Ann 1940- Y-87

Mason, William 1725-1797 .... DLB-142

Mason Brothers DLB-49

Massey, Gerald 1828-1907 .... DLB-32

Massinger, Philip 1583-1640 .... DLB-58

Masson, David 1822-1907 .... DLB-144

Masters, Edgar Lee 1868-1950 ...... DLB-54

Mather, Cotton
    1663-1728 .... DLB-24, 30, 140

Mather, Increase 1639-1723 .... DLB-24

Mather, Richard 1596-1669 .... DLB-24

Matheson, Richard 1926- DLB-8, 44

Matheus, John F. 1887- DLB-51

Mathews, Cornelius
    1817?-1889 .... DLB-3, 64

Mathews, Elkin
    [publishing house] DLB-112

Mathias, Roland 1915- DLB-27

Mathis, June 1892-1927 .... DLB-44

Mathis, Sharon Bell 1937- DLB-33

Matoš, Antun Gustav 1873-1914 ... DLB-147

The Matter of England
    1240-1400 .... DLB-146

The Matter of Rome
    early twelfth to late fifteenth
    century DLB-146

Matthews, Brander
    1852-1929 .... DLB-71, 78; DS-13

Matthews, Jack 1925- DLB-6

Matthews, William 1942- DLB-5

Matthiessen, F. O. 1902-1950 .... DLB-63

Matthiessen, Peter 1927- DLB-6

Maugham, W. Somerset
    1874-1965 ..... DLB-10, 36, 77, 100, 162

Maupassant, Guy de 1850-1893 .... DLB-123

Mauriac, Claude 1914- DLB-83

Mauriac, François 1885-1970 .... DLB-65

Maurice, Frederick Denison
    1805-1872 .... DLB-55

Maurois, André 1885-1967 ....DLB-65

Maury, James 1718-1769 ....DLB-31

Mavor, Elizabeth 1927- DLB-14

Mavor, Osborne Henry (see Bridie, James)

Maxwell, H. [publishing house] DLB-49

Maxwell, John [publishing house] ...DLB-106

Maxwell, William 1908- Y-80

May, Elaine 1932- DLB-44

May, Karl 1842-1912 ....DLB-129

May, Thomas 1595 or 1596-1650 ....DLB-58

Mayer, Bernadette 1945- DLB-165

Mayer, Mercer 1943- DLB-61

Mayer, O. B. 1818-1891 ....DLB-3

Mayes, Herbert R. 1900-1987 ......DLB-137

Mayes, Wendell 1919-1992 ....DLB-26

Mayfield, Julian 1928-1984 .... DLB-33; Y-84

Mayhew, Henry 1812-1887 ..... DLB-18, 55

Mayhew, Jonathan 1720-1766 .....DLB-31

Mayne, Jasper 1604-1672 ....DLB-126

Mayne, Seymour 1944- DLB-60

Mayor, Flora Macdonald
    1872-1932 ....DLB-36

Mayrocker, Friederike 1924- DLB-85

Mazrui, Ali A. 1933- DLB-125

Mažuranić, Ivan 1814-1890 ....DLB-147

Mazursky, Paul 1930- DLB-44

McAlmon, Robert 1896-1956 ..... DLB-4, 45

McArthur, Peter 1866-1924 ....DLB-92

McBride, Robert M., and
    Company DLB-46

McCaffrey, Anne 1926- DLB-8

McCarthy, Cormac 1933- DLB-6, 143

McCarthy, Mary 1912-1989 .... DLB-2; Y-81

McCay, Winsor 1871-1934 ....DLB-22

McClatchy, C. K. 1858-1936 ....DLB-25

McClellan, George Marion
    1860-1934 ....DLB-50

McCloskey, Robert 1914- DLB-22

McClung, Nellie Letitia 1873-1951 ...DLB-92

McClure, Joanna 1930- DLB-16

McClure, Michael 1932- DLB-16

McClure, Phillips and Company DLB-46

McClure, S. S. 1857-1949 ....DLB-91

McClurg, A. C., and Company DLB-49

McCluskey, John A., Jr. 1944- DLB-33

McCollum, Michael A. 1946- Y-87

McConnell, William C. 1917- DLB-88

McCord, David 1897- DLB-61

McCorkle, Jill 1958- Y-87

McCorkle, Samuel Eusebius
 1746-1811 ....DLB-37

McCormick, Anne O'Hare
 1880-1954 ....DLB-29

McCormick, Robert R. 1880-1955 ...DLB-29

McCourt, Edward 1907-1972 ....DLB-88

McCoy, Horace 1897-1955 ....DLB-9

McCrae, John 1872-1918 ....DLB-92

McCullagh, Joseph B. 1842-1896 ....DLB-23

McCullers, Carson 1917-1967 ......DLB-2, 7

McCulloch, Thomas 1776-1843 .....DLB-99

McDonald, Forrest 1927- DLB-17

McDonald, Walter
 1934- DLB-105, DS-9

McDonald, Walter, Getting Started:
 Accepting the Regions You Own—
 or Which Own You DLB-105

McDougall, Colin 1917-1984 ....DLB-68

McDowell, Obolensky DLB-46

McEwan, Ian 1948- DLB-14

McFadden, David 1940- DLB-60

McFall, Frances Elizabeth Clarke
 (see Grand, Sarah)

McFarlane, Leslie 1902-1977 ....DLB-88

McFee, William 1881-1966 ....DLB-153

McGahern, John 1934- DLB-14

McGee, Thomas D'Arcy
 1825-1868 ....DLB-99

McGeehan, W. O. 1879-1933 ....DLB-25

McGill, Ralph 1898-1969 ....DLB-29

McGinley, Phyllis 1905-1978 .....DLB-11, 48

McGirt, James E. 1874-1930 ....DLB-50

McGlashan and Gill DLB-106

McGough, Roger 1937- DLB-40

McGraw-Hill DLB-46

McGuane, Thomas 1939- DLB-2; Y-80

McGuckian, Medbh 1950- DLB-40

McGuffey, William Holmes
 1800-1873 ....DLB-42

McIlvanney, William 1936- DLB-14

McIlwraith, Jean Newton
 1859-1938 ....DLB-92

McIntyre, James 1827-1906 ....DLB-99

McIntyre, O. O. 1884-1938 ....DLB-25

McKay, Claude
 1889-1948 ....DLB-4, 45, 51, 117

The David McKay Company DLB-49

McKean, William V. 1820-1903 .....DLB-23

McKinley, Robin 1952- DLB-52

McLachlan, Alexander 1818-1896 ...DLB-99

McLaren, Floris Clark 1904-1978 ...DLB-68

McLaverty, Michael 1907- DLB-15

McLean, John R. 1848-1916 ....DLB-23

McLean, William L. 1852-1931 .....DLB-25

McLennan, William 1856-1904 .....DLB-92

McLoughlin Brothers DLB-49

McLuhan, Marshall 1911-1980 .....DLB-88

McMaster, John Bach 1852-1932 ....DLB-47

McMurtry, Larry
 1936- DLB-2, 143; Y-80, 87

McNally, Terrence 1939- DLB-7

McNeil, Florence 1937- DLB-60

McNeile, Herman Cyril
 1888-1937 ....DLB-77

McPherson, James Alan 1943- DLB-38

McPherson, Sandra 1943- Y-86

McWhirter, George 1939- DLB-60

McWilliams, Carey 1905-1980 .....DLB-137

Mead, L. T. 1844-1914 ....DLB-141

Mead, Matthew 1924- DLB-40

Mead, Taylor ?- DLB-16

Mechthild von Magdeburg
 circa 1207-circa 1282 ....DLB-138

Medill, Joseph 1823-1899 ....DLB-43

Medoff, Mark 1940- DLB-7

Meek, Alexander Beaufort
 1814-1865 ....DLB-3

Meeke, Mary ?-1816? DLB-116

Meinke, Peter 1932- DLB-5

Mejia Vallejo, Manuel 1923- DLB-113

Melançon, Robert 1947- DLB-60

Mell, Max 1882-1971 ....DLB-81, 124

Mellow, James R. 1926- DLB-111

Meltzer, David 1937- DLB-16

Meltzer, Milton 1915- DLB-61

Melville, Herman 1819-1891 .....DLB-3, 74

Memoirs of Life and Literature (1920),
 by W. H. Mallock [excerpt] DLB-57

Menantes (see Hunold, Christian Friedrich)

Mencke, Johann Burckhard
 1674-1732 ....DLB-168

Mencken, H. L.
 1880-1956 ....DLB-11, 29, 63, 137

Mencken and Nietzsche: An Unpublished
 Excerpt from H. L. Mencken's My Life
 as Author and Editor Y-93

Mendelssohn, Moses 1729-1786 .....DLB-97

Méndez M., Miguel 1930- DLB-82

Mercer, Cecil William (see Yates, Dornford)

Mercer, David 1928-1980 ....DLB-13

Mercer, John 1704-1768 ....DLB-31

Meredith, George
 1828-1909 ....DLB-18, 35, 57, 159

Meredith, Louisa Anne
 1812-1895 ....DLB-166

Meredith, Owen (see Lytton, Edward Robert
 Bulwer)

Meredith, William 1919- DLB-5

Mergerle, Johann Ulrich
 (see Abraham à Sancta Clara)

Mérimée, Prosper 1803-1870 ....DLB-119

Merivale, John Herman
 1779-1844 ....DLB-96

Meriwether, Louise 1923- DLB-33

Merlin Press DLB-112

Merriam, Eve 1916-1992 ....DLB-61

The Merriam Company DLB-49

Merrill, James
 1926-1995 ....DLB-5, 165; Y-85

Merrill and Baker DLB-49

The Mershon Company DLB-49

Merton, Thomas 1915-1968 ...DLB-48; Y-81

Merwin, W. S. 1927- DLB-5, 169

Messner, Julian [publishing house] ...DLB-46

Metcalf, J. [publishing house] DLB-49

Metcalf, John 1938- DLB-60

The Methodist Book Concern DLB-49

Methuen and Company DLB-112

Mew, Charlotte 1869-1928 .....DLB-19, 135

Mewshaw, Michael 1943- Y-80

Meyer, Conrad Ferdinand
 1825-1898 ....DLB-129

Meyer, E. Y. 1946- DLB-75

Meyer, Eugene 1875-1959 ....DLB-29

Meyer, Michael 1921- DLB-155

Meyers, Jeffrey 1939- DLB-111

Meynell, Alice
 1847-1922 ....DLB-19, 98

Meynell, Viola 1885-1956 .... DLB-153

Meyrink, Gustav 1868-1932 .... DLB-81

Michaels, Leonard 1933- DLB-130

Micheaux, Oscar 1884-1951 ...... DLB-50

Michel of Northgate, Dan
circa 1265-circa 1340 .... DLB-146

Micheline, Jack 1929- DLB-16

Michener, James A. 1907?- DLB-6

Micklejohn, George
circa 1717-1818 .... DLB-31

Middle English Literature:
An Introduction DLB-146

The Middle English Lyric DLB-146

Middle Hill Press DLB-106

Middleton, Christopher 1926- DLB-40

Middleton, Richard 1882-1911 ..... DLB-156

Middleton, Stanley 1919- DLB-14

Middleton, Thomas 1580-1627 ..... DLB-58

Miegel, Agnes 1879-1964 .... DLB-56

Miles, Josephine 1911-1985 .... DLB-48

Milius, John 1944- DLB-44

Mill, James 1773-1836 .... DLB-107, 158

Mill, John Stuart 1806-1873 .... DLB-55

Millar, Kenneth
1915-1983 .... DLB-2; Y-83; DS-6

Millar, Andrew
[publishing house] DLB-154

Millay, Edna St. Vincent
1892-1950 .... DLB-45

Miller, Arthur 1915- DLB-7

Miller, Caroline 1903-1992 ....DLB-9

Miller, Eugene Ethelbert 1950- DLB-41

Miller, Heather Ross 1939- DLB-120

Miller, Henry 1891-1980 ..... DLB-4, 9; Y-80

Miller, J. Hillis 1928- DLB-67

Miller, James [publishing house] DLB-49

Miller, Jason 1939- DLB-7

Miller, May 1899- DLB-41

Miller, Paul 1906-1991 .... DLB-127

Miller, Perry 1905-1963 .... DLB-17, 63

Miller, Sue 1943- DLB-143

Miller, Walter M., Jr. 1923- DLB-8

Miller, Webb 1892-1940 .... DLB-29

Millhauser, Steven 1943- DLB-2

Millican, Arthenia J. Bates
1920- DLB-38

Mills and Boon DLB-112

Milman, Henry Hart 1796-1868 ..... DLB-96

Milne, A. A.
1882-1956 .... DLB-10, 77, 100, 160

Milner, Ron 1938- DLB-38

Milner, William
[publishing house] DLB-106

Milnes, Richard Monckton (Lord Houghton)
1809-1885 .... DLB-32

Milton, John 1608-1674 .... DLB-131, 151

The Minerva Press DLB-154

Minnesang circa 1150-1280 .... DLB-138

Minns, Susan 1839-1938 .... DLB-140

Minor Illustrators, 1880-1914 ..... DLB-141

Minor Poets of the Earlier Seventeenth
Century DLB-121

Minton, Balch and Company DLB-46

Mirbeau, Octave 1848-1917 .... DLB-123

Mirk, John died after 1414? DLB-146

Miron, Gaston 1928- DLB-60

A Mirror for Magistrates DLB-167

Mitchel, Jonathan 1624-1668 .... DLB-24

Mitchell, Adrian 1932- DLB-40

Mitchell, Donald Grant
1822-1908 .... DLB-1; DS-13

Mitchell, Gladys 1901-1983 .... DLB-77

Mitchell, James Leslie 1901-1935 .... DLB-15

Mitchell, John (see Slater, Patrick)

Mitchell, John Ames 1845-1918 ..... DLB-79

Mitchell, Julian 1935- DLB-14

Mitchell, Ken 1940- DLB-60

Mitchell, Langdon 1862-1935 .... DLB-7

Mitchell, Loften 1919- DLB-38

Mitchell, Margaret 1900-1949 .... DLB-9

Mitchell, W. O. 1914- DLB-88

Mitchison, Naomi Margaret (Haldane)
1897- DLB-160

Mitford, Mary Russell
1787-1855 .... DLB-110, 116

Mittelholzer, Edgar 1909-1965 ..... DLB-117

Mitterer, Erika 1906- DLB-85

Mitterer, Felix 1948- DLB-124

Mitternacht, Johann Sebastian
1613-1679 .... DLB-168

Mizener, Arthur 1907-1988 .... DLB-103

Modern Age Books DLB-46

"Modern English Prose" (1876),
by George Saintsbury DLB-57

The Modern Language Association of America
Celebrates Its Centennial Y-84

The Modern Library DLB-46

"Modern Novelists – Great and Small" (1855),
by Margaret Oliphant DLB-21

"Modern Style" (1857), by Cockburn
Thomson [excerpt] DLB-57

The Modernists (1932), by Joseph Warren
Beach DLB-36

Modiano, Patrick 1945- DLB-83

Moffat, Yard and Company DLB-46

Moffet, Thomas 1553-1604 .... DLB-136

Mohr, Nicholasa 1938- DLB-145

Moix, Ana María 1947- DLB-134

Molesworth, Louisa 1839-1921 ..... DLB-135

Möllhausen, Balduin 1825-1905 ..... DLB-129

Momaday, N. Scott 1934- DLB-143

Monkhouse, Allan 1858-1936 .... DLB-10

Monro, Harold 1879-1932 .... DLB-19

Monroe, Harriet 1860-1936 ..... DLB-54, 91

Monsarrat, Nicholas 1910-1979 ..... DLB-15

Montagu, Lady Mary Wortley
1689-1762 .... DLB-95, 101

Montague, John 1929- DLB-40

Montale, Eugenio 1896-1981 .... DLB-114

Monterroso, Augusto 1921- DLB-145

Montgomerie, Alexander
circa 1550?-1598 .... DLB-167

Montgomery, James
1771-1854 .... DLB-93, 158

Montgomery, John 1919- DLB-16

Montgomery, Lucy Maud
1874-1942 .... DLB-92; DS-14

Montgomery, Marion 1925- DLB-6

Montgomery, Robert Bruce (see Crispin,
Edmund)

Montherlant, Henry de 1896-1972 ...DLB-72

The Monthly Review 1749-1844 .... DLB-110

Montigny, Louvigny de 1876-1955 ... DLB-92

Montoya, José 1932- DLB-122

Moodie, John Wedderburn Dunbar
1797-1869 .... DLB-99

Moodie, Susanna 1803-1885 .... DLB-99

Moody, Joshua circa 1633-1697 ...... DLB-24

Moody, William Vaughn
1869-1910 .... DLB-7, 54

Moorcock, Michael 1939- DLB-14

Moore, Catherine L. 1911- DLB-8

Moore, Clement Clarke 1779-1863 ... DLB-42

Moore, Dora Mavor 1888-1979 ...... DLB-92

Moore, George
1852-1933 .... DLB-10, 18, 57, 135

Moore, Marianne
1887-1972 .... DLB-45; DS-7

Moore, Mavor 1919- DLB-88

Moore, Richard 1927- DLB-105

Moore, Richard, The No Self, the Little Self,
and the Poets DLB-105

Moore, T. Sturge 1870-1944 .... DLB-19

Moore, Thomas 1779-1852 ..... DLB-96, 144

Moore, Ward 1903-1978 .... DLB-8

Moore, Wilstach, Keys and
Company DLB-49

The Moorland-Spingarn Research
Center DLB-76

Moorman, Mary C. 1905-1994 ..... DLB-155

Moraga, Cherríe 1952- DLB-82

Morales, Alejandro 1944- DLB-82

Morales, Mario Roberto 1947- DLB-145

Morales, Rafael 1919- DLB-108

Morality Plays: *Mankind* circa 1450-1500 and
Everyman circa 1500 .... DLB-146

More, Hannah
1745-1833 ...... DLB-107, 109, 116, 158

More, Henry 1614-1687 .... DLB-126

More, Sir Thomas
1477 or 1478-1535 .... DLB-136

Moreno, Dorinda 1939- DLB-122

Morency, Pierre 1942- DLB-60

Moretti, Marino 1885-1979 .... DLB-114

Morgan, Berry 1919- DLB-6

Morgan, Charles 1894-1958 ..... DLB-34, 100

Morgan, Edmund S. 1916- DLB-17

Morgan, Edwin 1920- DLB-27

Morgan, John Pierpont
1837-1913 .... DLB-140

Morgan, John Pierpont, Jr.
1867-1943 .... DLB-140

Morgan, Robert 1944- DLB-120

Morgan, Sydney Owenson, Lady
1776?-1859 .... DLB-116, 158

Morgner, Irmtraud 1933- DLB-75

Morhof, Daniel Georg
1639-1691 .... DLB-164

Morier, James Justinian
1782 or 1783?-1849 .... DLB-116

Mörike, Eduard 1804-1875 .... DLB-133

Morin, Paul 1889-1963 .... DLB-92

Morison, Richard 1514?-1556 ..... DLB-136

Morison, Samuel Eliot 1887-1976 ... DLB-17

Moritz, Karl Philipp 1756-1793 ..... DLB-94

Moriz von Craûn
circa 1220-1230 .... DLB-138

Morley, Christopher 1890-1957 ...... DLB-9

Morley, John 1838-1923 .... DLB-57, 144

Morris, George Pope 1802-1864 ..... DLB-73

Morris, Lewis 1833-1907 .... DLB-35

Morris, Richard B. 1904-1989 ...... DLB-17

Morris, William
1834-1896 .... DLB-18, 35, 57, 156

Morris, Willie 1934- Y-80

Morris, Wright 1910- DLB-2; Y-81

Morrison, Arthur 1863-1945 .... DLB-70, 135

Morrison, Charles Clayton
1874-1966 .... DLB-91

Morrison, Toni
1931- DLB-6, 33, 143; Y-81

Morrow, William, and Company DLB-46

Morse, James Herbert 1841-1923 .... DLB-71

Morse, Jedidiah 1761-1826 .... DLB-37

Morse, John T., Jr. 1840-1937 ...... DLB-47

Mortimer, Favell Lee 1802-1878 ... DLB-163

Mortimer, John 1923- DLB-13

Morton, Carlos 1942- DLB-122

Morton, John P., and Company DLB-49

Morton, Nathaniel 1613-1685 .... DLB-24

Morton, Sarah Wentworth
1759-1846 .... DLB-37

Morton, Thomas
circa 1579-circa 1647 .... DLB-24

Moscherosch, Johann Michael
1601-1669 .... DLB-164

Möser, Justus 1720-1794 .... DLB-97

Mosley, Nicholas 1923- DLB-14

Moss, Arthur 1889-1969 .... DLB-4

Moss, Howard 1922-1987 .... DLB-5

Moss, Thylias 1954- DLB-120

The Most Powerful Book Review in America
[*New York Times Book Review*] Y-82

Motion, Andrew 1952- DLB-40

Motley, John Lothrop
1814-1877 .... DLB-1, 30, 59

Motley, Willard 1909-1965 ..... DLB-76, 143

Motte, Benjamin Jr.
[publishing house] DLB-154

Motteux, Peter Anthony
1663-1718 .... DLB-80

Mottram, R. H. 1883-1971 .... DLB-36

Mouré, Erin 1955- DLB-60

Movies from Books, 1920-1974 .... DLB-9

Mowat, Farley 1921- DLB-68

Mowbray, A. R., and Company,
Limited DLB-106

Mowrer, Edgar Ansel 1892-1977 ..... DLB-29

Mowrer, Paul Scott 1887-1971 ...... DLB-29

Moxon, Edward
[publishing house] DLB-106

Mphahlele, Es'kia (Ezekiel)
1919- DLB-125

Mtshali, Oswald Mbuyiseni
1940- DLB-125

Mucedorus DLB-62

Mudford, William 1782-1848 ...... DLB-159

Mueller, Lisel 1924- DLB-105

Muhajir, El (see Marvin X)

Muhajir, Nazzam Al Fitnah (see Marvin X)

Mühlbach, Luise 1814-1873 .... DLB-133

Muir, Edwin 1887-1959 .... DLB-20, 100

Muir, Helen 1937- DLB-14

Mukherjee, Bharati 1940- DLB-60

Mulcaster, Richard
1531 or 1532-1611 .... DLB-167

Muldoon, Paul 1951- DLB-40

Müller, Friedrich (see Müller, Maler)

Müller, Heiner 1929- DLB-124

Müller, Maler 1749-1825 .... DLB-94

Müller, Wilhelm 1794-1827 .... DLB-90

Mumford, Lewis 1895-1990 .... DLB-63

Munby, Arthur Joseph 1828-1910 .... DLB-35

Munday, Anthony 1560-1633 .... DLB-62

Mundt, Clara (see Mühlbach, Luise)

Mundt, Theodore 1808-1861 ...... DLB-133

Munford, Robert circa 1737-1783 .... DLB-31

Mungoshi, Charles 1947- DLB-157

Munonye, John 1929- DLB-117

Munro, Alice 1931- DLB-53

Munro, H. H. 1870-1916 .... DLB-34, 162

Munro, Neil 1864-1930 .... DLB-156

Munro, George
[publishing house] DLB-49

Munro, Norman L.
[publishing house] DLB-49

Munroe, James, and Company DLB-49

Munroe, Kirk 1850-1930 .... DLB-42

Munroe and Francis DLB-49

Munsell, Joel [publishing house] DLB-49

Munsey, Frank A. 1854-1925 .... DLB-25, 91

Munsey, Frank A., and
  Company DLB-49

Murav'ev, Mikhail Nikitich
  1757-1807 .... DLB-150

Murdoch, Iris 1919- DLB-14

Murdoch, Rupert 1931- DLB-127

Murfree, Mary N. 1850-1922 .... DLB-12, 74

Murger, Henry 1822-1861 .... DLB-119

Murger, Louis-Henri (see Murger, Henry)

Muro, Amado 1915-1971 .... DLB-82

Murphy, Arthur 1727-1805 .... DLB-89, 142

Murphy, Beatrice M. 1908- DLB-76

Murphy, Emily 1868-1933 .... DLB-99

Murphy, John H., III 1916- DLB-127

Murphy, John, and Company DLB-49

Murphy, Richard 1927-1993 ...... DLB-40

Murray, Albert L. 1916- DLB-38

Murray, Gilbert 1866-1957 .... DLB-10

Murray, Judith Sargent 1751-1820 ... DLB-37

Murray, Pauli 1910-1985 .... DLB-41

Murray, John [publishing house] ... DLB-154

Murry, John Middleton
  1889-1957 .... DLB-149

Musäus, Johann Karl August
  1735-1787 .... DLB-97

Muschg, Adolf 1934- DLB-75

The Music of *Minnesang* DLB-138

Musil, Robert 1880-1942 .... DLB-81, 124

Muspilli circa 790-circa 850 .... DLB-148

Mussey, Benjamin B., and
  Company DLB-49

Mwangi, Meja 1948- DLB-125

Myers, Gustavus 1872-1942 .... DLB-47

Myers, L. H. 1881-1944 .... DLB-15

Myers, Walter Dean 1937- DLB-33

N

Nabbes, Thomas circa 1605-1641 ... DLB-58

Nabl, Franz 1883-1974 .... DLB-81

Nabokov, Vladimir
  1899-1977 ..... DLB-2; Y-80, Y-91; DS-3

Nabokov Festival at Cornell Y-83

The Vladimir Nabokov Archive
  in the Berg Collection Y-91

Nafis and Cornish DLB-49

Naipaul, Shiva 1945-1985 .... DLB-157; Y-85

Naipaul, V. S. 1932- DLB-125; Y-85

Nancrede, Joseph
  [publishing house] DLB-49

Naranjo, Carmen 1930- DLB-145

Narrache, Jean 1893-1970 .... DLB-92

Nasby, Petroleum Vesuvius (see Locke, David
  Ross)

Nash, Ogden 1902-1971 .... DLB-11

Nash, Eveleigh
  [publishing house] DLB-112

Nashe, Thomas 1567-1601? DLB-167

Nast, Conde 1873-1942 .... DLB-91

Nastasijević, Momčilo 1894-1938 ... DLB-147

Nathan, George Jean 1882-1958 .... DLB-137

Nathan, Robert 1894-1985 .... DLB-9

The National Jewish Book Awards Y-85

The National Theatre and the Royal
  Shakespeare Company: The
  National Companies DLB-13

Naughton, Bill 1910- DLB-13

Nazor, Vladimir 1876-1949 .... DLB-147

Ndebele, Njabulo 1948- DLB-157

Neagoe, Peter 1881-1960 .... DLB-4

Neal, John 1793-1876 .... DLB-1, 59

Neal, Joseph C. 1807-1847 .... DLB-11

Neal, Larry 1937-1981 .... DLB-38

The Neale Publishing Company DLB-49

Neely, F. Tennyson
  [publishing house] DLB-49

Negri, Ada 1870-1945 .... DLB-114

"The Negro as a Writer," by
  G. M. McClellan DLB-50

"Negro Poets and Their Poetry," by
  Wallace Thurman DLB-50

Neidhart von Reuental
  circa 1185-circa 1240 .... DLB-138

Neihardt, John G. 1881-1973 ...... DLB-9, 54

Neledinsky-Meletsky, Iurii Aleksandrovich
  1752-1828 .... DLB-150

Nelligan, Emile 1879-1941 .... DLB-92

Nelson, Alice Moore Dunbar
  1875-1935 .... DLB-50

Nelson, Thomas, and Sons [U.S.] DLB-49

Nelson, Thomas, and Sons [U.K.] ...DLB-106

Nelson, William 1908-1978 ....DLB-103

Nelson, William Rockhill
  1841-1915 ....DLB-23

Nemerov, Howard 1920-1991 ... DLB-5, 6; Y-83

Nesbit, E. 1858-1924 .... DLB-141, 153

Ness, Evaline 1911-1986 ....DLB-61

Nestroy, Johann 1801-1862 ....DLB-133

Neukirch, Benjamin 1655-1729 .....DLB-168

Neugeboren, Jay 1938- DLB-28

Neumann, Alfred 1895-1952 ....DLB-56

Neumark, Georg 1621-1681 ....DLB-164

Neumeister, Erdmann 1671-1756 ....DLB-168

Nevins, Allan 1890-1971 ....DLB-17

Nevinson, Henry Woodd
  1856-1941 ....DLB-135

The New American Library DLB-46

New Approaches to Biography: Challenges
  from Critical Theory, USC Conference
  on Literary Studies, 1990 .... Y-90

New Directions Publishing
  Corporation DLB-46

A New Edition of *Huck Finn* Y-85

New Forces at Work in the American Theatre:
  1915-1925 ....DLB-7

New Literary Periodicals:
  A Report for 1987 .... Y-87

New Literary Periodicals:
  A Report for 1988 .... Y-88

New Literary Periodicals:
  A Report for 1989 .... Y-89

New Literary Periodicals:
  A Report for 1990 .... Y-90

New Literary Periodicals:
  A Report for 1991 .... Y-91

New Literary Periodicals:
  A Report for 1992 .... Y-92

New Literary Periodicals:
  A Report for 1993 .... Y-93

The New Monthly Magazine
  1814-1884 ....DLB-110

The New *Ulysses* Y-84

The New Variorum Shakespeare Y-85

A New Voice: The Center for the Book's First
  Five Years Y-83

The New Wave [Science Fiction] DLB-8

New York City Bookshops in the 1930s and
  1940s: The Recollections of Walter
  Goldwater Y-93

Newbery, John [publishing house] DLB-154

Newbolt, Henry 1862-1938 ....DLB-19

Newbound, Bernard Slade (see Slade, Bernard)

Newby, P. H. 1918- DLB-15

Newby, Thomas Cautley [publishing house] DLB-106

Newcomb, Charles King 1820-1894 ...DLB-1

Newell, Peter 1862-1924 ....DLB-42

Newell, Robert Henry 1836-1901 ....DLB-11

Newhouse, Samuel I. 1895-1979 ....DLB-127

Newman, Cecil Earl 1903-1976 .....DLB-127

Newman, David (see Benton, Robert)

Newman, Frances 1883-1928 ....Y-80

Newman, John Henry 1801-1890 .... DLB-18, 32, 55

Newman, Mark [publishing house] ...DLB-49

Newnes, George, Limited DLB-112

Newsome, Effie Lee 1885-1979 ......DLB-76

Newspaper Syndication of American Humor DLB-11

Newton, A. Edward 1864-1940 .....DLB-140

Ngugi wa Thiong'o 1938- DLB-125

The *Nibelungenlied* and the *Klage* circa 1200 ....DLB-138

Nichol, B. P. 1944- DLB-53

Nicholas of Cusa 1401-1464 ....DLB-115

Nichols, Dudley 1895-1960 ....DLB-26

Nichols, Grace 1950- DLB-157

Nichols, John 1940- Y-82

Nichols, Mary Sargeant (Neal) Gove 1810-1884 ....DLB-1

Nichols, Peter 1927- DLB-13

Nichols, Roy F. 1896-1973 ....DLB-17

Nichols, Ruth 1948- DLB-60

Nicholson, Norman 1914- DLB-27

Nicholson, William 1872-1949 .....DLB-141

Ní Chuilleanáin, Eiléan 1942- DLB-40

Nicol, Eric 1919- DLB-68

Nicolai, Friedrich 1733-1811 ....DLB-97

Nicolay, John G. 1832-1901 and Hay, John 1838-1905 ....DLB-47

Nicolson, Harold 1886-1968 ...DLB-100, 149

Nicolson, Nigel 1917- DLB-155

Niebuhr, Reinhold 1892-1971 ....DLB-17

Niedecker, Lorine 1903-1970 ....DLB-48

Nieman, Lucius W. 1857-1935 ......DLB-25

Nietzsche, Friedrich 1844-1900 .....DLB-129

Niggli, Josefina 1910- Y-80

Nightingale, Florence 1820-1910 ...DLB-166

Nikolev, Nikolai Petrovich 1758-1815 ....DLB-150

Niles, Hezekiah 1777-1839 ....DLB-43

Nims, John Frederick 1913- DLB-5

Nin, Anaïs 1903-1977 ....DLB-2, 4, 152

1985: The Year of the Mystery: A Symposium Y-85

Nissenson, Hugh 1933- DLB-28

Niven, Frederick John 1878-1944 ....DLB-92

Niven, Larry 1938- DLB-8

Nizan, Paul 1905-1940 ....DLB-72

Njegoš, Petar II Petrović 1813-1851 ....DLB-147

Nkosi, Lewis 1936- DLB-157

Nobel Peace Prize
The 1986 Nobel Peace Prize Nobel Lecture 1986: Hope, Despair and Memory Tributes from Abraham Bernstein, Norman Lamm, and John R. Silber Y-86

The Nobel Prize and Literary Politics ...Y-86

Nobel Prize in Literature
The 1982 Nobel Prize in Literature Announcement by the Swedish Academy of the Nobel Prize Nobel Lecture 1982: The Solitude of Latin America Excerpt from *One Hundred Years of Solitude* The Magical World of Macondo A Tribute to Gabriel García Márquez Y-82

The 1983 Nobel Prize in Literature Announcement by the Swedish Academy Nobel Lecture 1983 The Stature of William Golding Y-83

The 1984 Nobel Prize in Literature Announcement by the Swedish Academy Jaroslav Seifert Through the Eyes of the English-Speaking Reader Three Poems by Jaroslav Seifert Y-84

The 1985 Nobel Prize in Literature Announcement by the Swedish Academy Nobel Lecture 1985 ....Y-85

The 1986 Nobel Prize in Literature Nobel Lecture 1986: This Past Must Address Its Present Y-86

The 1987 Nobel Prize in Literature Nobel Lecture 1987 ....Y-87

The 1988 Nobel Prize in Literature Nobel Lecture 1988 ....Y-88

The 1989 Nobel Prize in Literature Nobel Lecture 1989 ....Y-89

The 1990 Nobel Prize in Literature Nobel Lecture 1990 ....Y-90

The 1991 Nobel Prize in Literature Nobel Lecture 1991 ....Y-91

The 1992 Nobel Prize in Literature Nobel Lecture 1992 ....Y-92

The 1993 Nobel Prize in Literature Nobel Lecture 1993 ....Y-93

The 1994 Nobel Prize in Literature Nobel Lecture 1994 ....Y-94

The 1995 Nobel Prize in Literature Nobel Lecture 1995 ....Y-95

Nodier, Charles 1780-1844 ....DLB-119

Noel, Roden 1834-1894 ....DLB-35

Nolan, William F. 1928- DLB-8

Noland, C. F. M. 1810?-1858 ....DLB-11

Nonesuch Press DLB-112

Noonday Press DLB-46

Noone, John 1936- DLB-14

Nora, Eugenio de 1923- DLB-134

Nordhoff, Charles 1887-1947 ....DLB-9

Norman, Charles 1904- DLB-111

Norman, Marsha 1947- Y-84

Norris, Charles G. 1881-1945 ....DLB-9

Norris, Frank 1870-1902 ....DLB-12

Norris, Leslie 1921- DLB-27

Norse, Harold 1916- DLB-16

North Point Press DLB-46

Nortje, Arthur 1942-1970 ....DLB-125

Norton, Alice Mary (see Norton, Andre)

Norton, Andre 1912- DLB-8, 52

Norton, Andrews 1786-1853 ....DLB-1

Norton, Caroline 1808-1877 ....DLB-21, 159

Norton, Charles Eliot 1827-1908 ..DLB-1, 64

Norton, John 1606-1663 ....DLB-24

Norton, Mary 1903-1992 ....DLB-160

Norton, Thomas (see Sackville, Thomas)

Norton, W. W., and Company DLB-46

Norwood, Robert 1874-1932 ....DLB-92

Nossack, Hans Erich 1901-1977 .....DLB-69

Notker Balbulus circa 840-912 .....DLB-148

Notker III of Saint Gall circa 950-1022 ....DLB-148

Notker von Zweifalten ?-1095 ......DLB-148

A Note on Technique (1926), by Elizabeth A. Drew [excerpts] DLB-36

Nourse, Alan E. 1928- DLB-8

Novak, Vjenceslav 1859-1905 .....DLB-147

Novalis 1772-1801 ....DLB-90

Cumulative Index

Novaro, Mario 1868-1944 .... DLB-114

Novás Calvo, Lino 1903-1983 ..... DLB-145

"The Novel in [Robert Browning's] 'The Ring and the Book' " (1912), by Henry James DLB-32

The Novel of Impressionism, by Jethro Bithell DLB-66

Novel-Reading: *The Works of Charles Dickens, The Works of W. Makepeace Thackeray* (1879), by Anthony Trollope DLB-21

The Novels of Dorothy Richardson (1918), by May Sinclair DLB-36

Novels with a Purpose (1864), by Justin M'Carthy DLB-21

Noventa, Giacomo 1898-1960 ..... DLB-114

Novikov, Nikolai Ivanovich 1744-1818 .... DLB-150

Nowlan, Alden 1933-1983 .... DLB-53

Noyes, Alfred 1880-1958 .... DLB-20

Noyes, Crosby S. 1825-1908 ...... DLB-23

Noyes, Nicholas 1647-1717 .... DLB-24

Noyes, Theodore W. 1858-1946 .... DLB-29

N-Town Plays circa 1468 to early sixteenth century DLB-146

Nugent, Frank 1908-1965 .... DLB-44

Nusic, Branislav 1864-1938 .... DLB-147

Nutt, David [publishing house] DLB-106

Nwapa, Flora 1931- DLB-125

Nye, Edgar Wilson (Bill) 1850-1896 .... DLB-11, 23

Nye, Naomi Shihab 1952- DLB-120

Nye, Robert 1939- DLB-14

O

Oakes, Urian circa 1631-1681 ...... DLB-24

Oates, Joyce Carol 1938- DLB-2, 5, 130; Y-81

Ober, William 1920-1993 .... Y-93

Oberholtzer, Ellis Paxson 1868-1936 .... DLB-47

Obradović, Dositej 1740?-1811 .... DLB-147

O'Brien, Edna 1932- DLB-14

O'Brien, Fitz-James 1828-1862 ..... DLB-74

O'Brien, Kate 1897-1974 .... DLB-15

O'Brien, Tim 1946- DLB-152; Y-80; DS-9

O'Casey, Sean 1880-1964 .... DLB-10

Ochs, Adolph S. 1858-1935 .... DLB-25

Ochs-Oakes, George Washington 1861-1931 .... DLB-137

O'Connor, Flannery 1925-1964 ..... DLB-2, 152; Y-80; DS-12

O'Connor, Frank 1903-1966 .... DLB-162

Octopus Publishing Group DLB-112

Odell, Jonathan 1737-1818 ...... DLB-31, 99

O'Dell, Scott 1903-1989 .... DLB-52

Odets, Clifford 1906-1963 .... DLB-7, 26

Odhams Press Limited DLB-112

O'Donnell, Peter 1920- DLB-87

O'Donovan, Michael (see O'Connor, Frank)

O'Faolain, Julia 1932- DLB-14

O'Faolain, Sean 1900- DLB-15, 162

Off Broadway and Off-Off Broadway . DLB-7

Off-Loop Theatres DLB-7

Offord, Carl Ruthven 1910- DLB-76

O'Flaherty, Liam 1896-1984 .... DLB-36, 162; Y-84

Ogilvie, J. S., and Company DLB-49

Ogot, Grace 1930- DLB-125

O'Grady, Desmond 1935- DLB-40

Ogunyemi, Wale 1939- DLB-157

O'Hagan, Howard 1902-1982 .... DLB-68

O'Hara, Frank 1926-1966 .... DLB-5, 16

O'Hara, John 1905-1970 ... DLB-9, 86; DS-2

Okara, Gabriel 1921- DLB-125

O'Keeffe, John 1747-1833 .... DLB-89

Okigbo, Christopher 1930-1967 .... DLB-125

Okot p'Bitek 1931-1982 .... DLB-125

Okpewho, Isidore 1941- DLB-157

Okri, Ben 1959- DLB-157

Olaudah Equiano and Unfinished Journeys: The Slave-Narrative Tradition and Twentieth-Century Continuities, by Paul Edwards and Pauline T. Wangman DLB-117

Old English Literature: An Introduction DLB-146

Old English Riddles eighth to tenth centuries DLB-146

Old Franklin Publishing House DLB-49

Old German Genesis and *Old German Exodus* circa 1050-circa 1130 .... DLB-148

Old High German Charms and Blessings DLB-148

The *Old High German Isidor* circa 790-800 .... DLB-148

Older, Fremont 1856-1935 .... DLB-25

Oldham, John 1653-1683 .... DLB-131

Olds, Sharon 1942- DLB-120

Olearius, Adam 1599-1671 .... DLB-164

Oliphant, Laurence 1829?-1888 .... DLB-18, 166

Oliphant, Margaret 1828-1897 .... DLB-18

Oliver, Chad 1928- DLB-8

Oliver, Mary 1935- DLB-5

Ollier, Claude 1922- DLB-83

Olsen, Tillie 1913?- DLB-28; Y-80

Olson, Charles 1910-1970 .... DLB-5, 16

Olson, Elder 1909- DLB-48, 63

Omotoso, Kole 1943- DLB-125

"On Art in Fiction "(1838), by Edward Bulwer DLB-21

On Learning to Write Y-88

On Some of the Characteristics of Modern Poetry and On the Lyrical Poems of Alfred Tennyson (1831), by Arthur Henry Hallam DLB-32

"On Style in English Prose" (1898), by Frederic Harrison DLB-57

"On Style in Literature: Its Technical Elements" (1885), by Robert Louis Stevenson DLB-57

"On the Writing of Essays" (1862), by Alexander Smith DLB-57

Ondaatje, Michael 1943- DLB-60

O'Neill, Eugene 1888-1953 .... DLB-7

Onetti, Juan Carlos 1909-1994 ..... DLB-113

Onions, George Oliver 1872-1961 .... DLB-153

Onofri, Arturo 1885-1928 .... DLB-114

Opie, Amelia 1769-1853 ...... DLB-116, 159

Opitz, Martin 1597-1639 .... DLB-164

Oppen, George 1908-1984 ..... DLB-5, 165

Oppenheim, E. Phillips 1866-1946 ...DLB-70

Oppenheim, James 1882-1932 .... DLB-28

Oppenheimer, Joel 1930- DLB-5

Optic, Oliver (see Adams, William Taylor)

Orczy, Emma, Baroness 1865-1947 .... DLB-70

Origo, Iris 1902-1988 .... DLB-155

Orlovitz, Gil 1918-1973 .... DLB-2, 5

Orlovsky, Peter 1933- DLB-16

Ormond, John 1923- DLB-27

Ornitz, Samuel 1890-1957 .... DLB-28, 44

Ortiz, Simon 1941- DLB-120

Ortnit and *Wolfdietrich* circa 1225-1250 .... DLB-138

Orton, Joe 1933-1967 .... DLB-13

Orwell, George 1903-1950 .... DLB-15, 98

The Orwell Year Y-84

Ory, Carlos Edmundo de 1923- ... DLB-134

Osbey, Brenda Marie 1957- DLB-120

Osbon, B. S. 1827-1912 .... DLB-43

Osborne, John 1929-1994 .... DLB-13

Osgood, Herbert L. 1855-1918 ..... DLB-47

Osgood, James R., and Company DLB-49

Osgood, McIlvaine and Company DLB-112

O'Shaughnessy, Arthur 1844-1881 .... DLB-35

O'Shea, Patrick [publishing house] DLB-49

Osipov, Nikolai Petrovich 1751-1799 .... DLB-150

Osofisan, Femi 1946- DLB-125

Ostenso, Martha 1900-1963 .... DLB-92

Ostriker, Alicia 1937- DLB-120

Osundare, Niyi 1947- DLB-157

Oswald, Eleazer 1755-1795 .... DLB-43

Otero, Blas de 1916-1979 .... DLB-134

Otero, Miguel Antonio 1859-1944 .... DLB-82

Otero Silva, Miguel 1908-1985 ..... DLB-145

Otfried von Weißenburg circa 800-circa 875? DLB-148

Otis, James (see Kaler, James Otis)

Otis, James, Jr. 1725-1783 .... DLB-31

Otis, Broaders and Company DLB-49

Ottaway, James 1911- DLB-127

Ottendorfer, Oswald 1826-1900 ..... DLB-23

Otto-Peters, Louise 1819-1895 ...... DLB-129

Otway, Thomas 1652-1685 .... DLB-80

Ouellette, Fernand 1930- DLB-60

Ouida 1839-1908 .... DLB-18, 156

Outing Publishing Company DLB-46

Outlaw Days, by Joyce Johnson DLB-16

Overbury, Sir Thomas circa 1581-1613 .... DLB-151

The Overlook Press DLB-46

Overview of U.S. Book Publishing, 1910-1945 .... DLB-9

Owen, Guy 1925- DLB-5

Owen, John 1564-1622 .... DLB-121

Owen, John [publishing house] DLB-49

Owen, Robert 1771-1858 ..... DLB-107, 158

Owen, Wilfred 1893-1918 .... DLB-20

Owen, Peter, Limited DLB-112

The Owl and the Nightingale circa 1189-1199 .... DLB-146

Owsley, Frank L. 1890-1956 .... DLB-17

Ozerov, Vladislav Aleksandrovich 1769-1816 .... DLB-150

Ozick, Cynthia 1928- DLB-28, 152; Y-82

P

Pace, Richard 1482?-1536 .... DLB-167

Pacey, Desmond 1917-1975 .... DLB-88

Pack, Robert 1929- DLB-5

Packaging Papa: *The Garden of Eden* Y-86

Padell Publishing Company DLB-46

Padgett, Ron 1942- DLB-5

Padilla, Ernesto Chávez 1944- ... DLB-122

Page, L. C., and Company DLB-49

Page, P. K. 1916- DLB-68

Page, Thomas Nelson 1853-1922 .... DLB-12, 78; DS-13

Page, Walter Hines 1855-1918 ... DLB-71, 91

Paget, Francis Edward 1806-1882 .... DLB-163

Paget, Violet (see Lee, Vernon)

Pagliarani, Elio 1927- DLB-128

Pain, Barry 1864-1928 .... DLB-135

Pain, Philip ?-circa 1666 .... DLB-24

Paine, Robert Treat, Jr. 1773-1811 .. DLB-37

Paine, Thomas 1737-1809 .... DLB-31, 43, 73, 158

Painter, George D. 1914- DLB-155

Painter, William 1540?-1594 ...... DLB-136

Palazzeschi, Aldo 1885-1974 .... DLB-114

Paley, Grace 1922- DLB-28

Palfrey, John Gorham 1796-1881 .... DLB-1, 30

Palgrave, Francis Turner 1824-1897 .... DLB-35

Palmer, Michael 1943- DLB-169

Paltock, Robert 1697-1767 .... DLB-39

Pan Books Limited DLB-112

Panamaa, Norman 1914- and Frank, Melvin 1913-1988 .... DLB-26

Pancake, Breece D'J 1952-1979 ..... DLB-130

Panero, Leopoldo 1909-1962 .... DLB-108

Pangborn, Edgar 1909-1976 .... DLB-8

"Panic Among the Philistines": A Postscript, An Interview with Bryan Griffin Y-81

Panneton, Philippe (see Ringuet)

Panshin, Alexei 1940- DLB-8

Pansy (see Alden, Isabella)

Pantheon Books DLB-46

Paperback Library DLB-46

Paperback Science Fiction DLB-8

Paquet, Alfons 1881-1944 .... DLB-66

Paradis, Suzanne 1936- DLB-53

Pareja Diezcanseco, Alfredo 1908-1993 .... DLB-145

Pardoe, Julia 1804-1862 .... DLB-166

Parents' Magazine Press DLB-46

Parisian Theater, Fall 1984: Toward A New Baroque Y-85

Parizeau, Alice 1930- DLB-60

Parke, John 1754-1789 .... DLB-31

Parker, Dorothy 1893-1967 .... DLB-11, 45, 86

Parker, Gilbert 1860-1932 .... DLB-99

Parker, James 1714-1770 .... DLB-43

Parker, Theodore 1810-1860 .... DLB-1

Parker, William Riley 1906-1968 ... DLB-103

Parker, J. H. [publishing house] DLB-106

Parker, John [publishing house] DLB-106

Parkman, Francis, Jr. 1823-1893 .... DLB-1, 30

Parks, Gordon 1912- DLB-33

Parks, William 1698-1750 .... DLB-43

Parks, William [publishing house] ... DLB-49

Parley, Peter (see Goodrich, Samuel Griswold)

Parnell, Thomas 1679-1718 .... DLB-95

Parr, Catherine 1513?-1548 .... DLB-136

Parrington, Vernon L. 1871-1929 .... DLB-17, 63

Parronchi, Alessandro 1914- DLB-128

Partridge, S. W., and Company DLB-106

Parton, James 1822-1891 .... DLB-30

Parton, Sara Payson Willis 1811-1872 .... DLB-43, 74

Pasolini, Pier Paolo 1922- DLB-128

Pastan, Linda 1932- DLB-5

Paston, George 1860-1936 .... DLB-149

The *Paston Letters* 1422-1509 .... DLB-146

Pastorius, Francis Daniel
  1651-circa 1720 .... DLB-24

Patchen, Kenneth 1911-1972 .... DLB-16, 48

Pater, Walter 1839-1894 .... DLB-57, 156

Paterson, Katherine 1932- DLB-52

Patmore, Coventry 1823-1896 ... DLB-35, 98

Paton, Joseph Noel 1821-1901 ...... DLB-35

Paton Walsh, Jill 1937- DLB-161

Patrick, Edwin Hill ("Ted")
  1901-1964 .... DLB-137

Patrick, John 1906- DLB-7

Pattee, Fred Lewis 1863-1950 .... DLB-71

Pattern and Paradigm: History as
  Design, by Judith Ryan DLB-75

Patterson, Alicia 1906-1963 .... DLB-127

Patterson, Eleanor Medill
  1881-1948 .... DLB-29

Patterson, Eugene 1923- DLB-127

Patterson, Joseph Medill
  1879-1946 .... DLB-29

Pattillo, Henry 1726-1801 .... DLB-37

Paul, Elliot 1891-1958 ....DLB-4

Paul, Jean (see Richter, Johann Paul Friedrich)

Paul, Kegan, Trench, Trubner and Company
  Limited DLB-106

Paul, Peter, Book Company DLB-49

Paul, Stanley, and Company
  Limited DLB-112

Paulding, James Kirke
  1778-1860 .... DLB-3, 59, 74

Paulin, Tom 1949- DLB-40

Pauper, Peter, Press DLB-46

Pavese, Cesare 1908-1950 .... DLB-128

Paxton, John 1911-1985 .... DLB-44

Payn, James 1830-1898 .... DLB-18

Payne, John 1842-1916 .... DLB-35

Payne, John Howard 1791-1852 .... DLB-37

Payson and Clarke DLB-46

Peabody, Elizabeth Palmer
  1804-1894 ....DLB-1

Peabody, Elizabeth Palmer
  [publishing house] DLB-49

Peabody, Oliver William Bourn
  1799-1848 .... DLB-59

Peace, Roger 1899-1968 .... DLB-127

Peacham, Henry 1578-1644? DLB-151

Peachtree Publishers, Limited DLB-46

Peacock, Molly 1947- DLB-120

Peacock, Thomas Love
  1785-1866 ....DLB-96, 116

Pead, Deuel ?-1727 .... DLB-24

Peake, Mervyn 1911-1968 ......DLB-15, 160

Pear Tree Press DLB-112

Pearce, Philippa 1920- DLB-161

Pearson, H. B. [publishing house] DLB-49

Pearson, Hesketh 1887-1964 .... DLB-149

Peck, George W. 1840-1916 ..... DLB-23, 42

Peck, H. C., and Theo. Bliss
  [publishing house] DLB-49

Peck, Harry Thurston
  1856-1914 ....DLB-71, 91

Peele, George 1556-1596 ....DLB-62, 167

Pellegrini and Cudahy DLB-46

Pelletier, Aimé (see Vac, Bertrand)

Pemberton, Sir Max 1863-1950 ...... DLB-70

Penguin Books [U.S.] DLB-46

Penguin Books [U.K.] DLB-112

Penn Publishing Company DLB-49

Penn, William 1644-1718 .... DLB-24

Penna, Sandro 1906-1977 .... DLB-114

Penner, Jonathan 1940- Y-83

Pennington, Lee 1939- Y-82

Pepys, Samuel 1633-1703 .... DLB-101

Percy, Thomas 1729-1811 .... DLB-104

Percy, Walker 1916-1990 ....DLB-2; Y-80, 90

Perec, Georges 1936-1982 .... DLB-83

Perelman, S. J. 1904-1979 ....DLB-11, 44

Perez, Raymundo "Tigre"
  1946- DLB-122

Peri Rossi, Cristina 1941- DLB-145

Periodicals of the Beat Generation ... DLB-16

Perkins, Eugene 1932- DLB-41

Perkoff, Stuart Z. 1930-1974 .... DLB-16

Perley, Moses Henry 1804-1862 ..... DLB-99

Permabooks DLB-46

Perrin, Alice 1867-1934 .... DLB-156

Perry, Bliss 1860-1954 .... DLB-71

Perry, Eleanor 1915-1981 .... DLB-44

Perry, Sampson 1747-1823 .... DLB-158

"Personal Style" (1890), by John Addington
  Symonds DLB-57

Perutz, Leo 1882-1957 ....DLB-81

Pesetsky, Bette 1932- DLB-130

Pestalozzi, Johann Heinrich
  1746-1827 ....DLB-94

Peter, Laurence J. 1919-1990 ....DLB-53

Peter of Spain circa 1205-1277 ......DLB-115

Peterkin, Julia 1880-1961 ....DLB-9

Peters, Lenrie 1932- DLB-117

Peters, Robert 1924- DLB-105

Peters, Robert, Foreword to
  Ludwig of Bavaria DLB-105

Petersham, Maud 1889-1971 and
  Petersham, Miska 1888-1960 .....DLB-22

Peterson, Charles Jacobs
  1819-1887 ....DLB-79

Peterson, Len 1917- DLB-88

Peterson, Louis 1922- DLB-76

Peterson, T. B., and Brothers DLB-49

Petitclair, Pierre 1813-1860 ....DLB-99

Petrov, Gavriil 1730-1801 ....DLB-150

Petrov, Vasilii Petrovich
  1736-1799 ....DLB-150

Petrović, Rastko 1898-1949 ....DLB-147

Petruslied circa 854? DLB-148

Petry, Ann 1908- DLB-76

Pettie, George circa 1548-1589 .....DLB-136

Peyton, K. M. 1929- DLB-161

Pfaffe Konrad
  flourished circa 1172 ....DLB-148

Pfaffe Lamprecht
  flourished circa 1150 ....DLB-148

Pforzheimer, Carl H. 1879-1957 ....DLB-140

Phaer, Thomas 1510?-1560 ....DLB-167

Phaidon Press Limited DLB-112

Pharr, Robert Deane 1916-1992 ......DLB-33

Phelps, Elizabeth Stuart
  1844-1911 ....DLB-74

Philander von der Linde
  (see Mencke, Johann Burckhard)

Philip, Marlene Nourbese
  1947- DLB-157

Philippe, Charles-Louis
  1874-1909 ....DLB-65

Philips, John 1676-1708 ....DLB-95

Philips, Katherine 1632-1664 ....DLB-131

Phillips, Caryl 1958- DLB-157

Phillips, David Graham
  1867-1911 .... DLB-9, 12

Phillips, Jayne Anne 1952- Y-80

Phillips, Robert 1938- DLB-105

Phillips, Robert, Finding, Losing, Reclaiming: A Note on My Poems DLB-105

Phillips, Stephen 1864-1915 .... DLB-10

Phillips, Ulrich B. 1877-1934 .... DLB-17

Phillips, Willard 1784-1873 .... DLB-59

Phillips, William 1907- DLB-137

Phillips, Sampson and Company DLB-49

Phillpotts, Eden 1862-1960 .... DLB-10, 70, 135, 153

Philosophical Library DLB-46

"The Philosophy of Style" (1852), by Herbert Spencer DLB-57

Phinney, Elihu [publishing house] ...DLB-49

Phoenix, John (see Derby, George Horatio)

PHYLON (Fourth Quarter, 1950), The Negro in Literature: The Current Scene DLB-76

Physiologus circa 1070-circa 1150 .... DLB-148

Piccolo, Lucio 1903-1969 .... DLB-114

Pickard, Tom 1946- DLB-40

Pickering, William [publishing house] DLB-106

Pickthall, Marjorie 1883-1922 ......DLB-92

Pictorial Printing Company DLB-49

Piel, Gerard 1915- DLB-137

Piercy, Marge 1936- DLB-120

Pierro, Albino 1916- DLB-128

Pignotti, Lamberto 1926- DLB-128

Pike, Albert 1809-1891 .... DLB-74

Pilon, Jean-Guy 1930- DLB-60

Pinckney, Josephine 1895-1957 ...... DLB-6

Pindar, Peter (see Wolcot, John)

Pinero, Arthur Wing 1855-1934 ..... DLB-10

Pinget, Robert 1919- DLB-83

Pinnacle Books DLB-46

Piñon, Nélida 1935- DLB-145

Pinsky, Robert 1940- Y-82

Pinter, Harold 1930- DLB-13

Piontek, Heinz 1925- DLB-75

Piozzi, Hester Lynch [Thrale] 1741-1821 .... DLB-104, 142

Piper, H. Beam 1904-1964 ....DLB-8

Piper, Watty DLB-22

Pisar, Samuel 1929- Y-83

Pitkin, Timothy 1766-1847 .... DLB-30

The Pitt Poetry Series: Poetry Publishing Today Y-85

Pitter, Ruth 1897- DLB-20

Pix, Mary 1666-1709 .... DLB-80

Plaatje, Sol T. 1876-1932 .... DLB-125

The Place of Realism in Fiction (1895), by George Gissing DLB-18

Plante, David 1940- Y-83

Platen, August von 1796-1835 ...... DLB-90

Plath, Sylvia 1932-1963 .... DLB-5, 6, 152

Platon 1737-1812 .... DLB-150

Platt and Munk Company DLB-46

Playboy Press DLB-46

Plays, Playwrights, and Playgoers ... DLB-84

Playwrights and Professors, by Tom Stoppard DLB-13

Playwrights on the Theater DLB-80

Der Pleier flourished circa 1250 .... DLB-138

Plenzdorf, Ulrich 1934- DLB-75

Plessen, Elizabeth 1944- DLB-75

Plievier, Theodor 1892-1955 .... DLB-69

Plomer, William 1903-1973 .... DLB-20, 162

Plumly, Stanley 1939- DLB-5

Plumpp, Sterling D. 1940- DLB-41

Plunkett, James 1920- DLB-14

Plymell, Charles 1935- DLB-16

Pocket Books DLB-46

Poe, Edgar Allan 1809-1849 .... DLB-3, 59, 73, 74

Poe, James 1921-1980 .... DLB-44

The Poet Laureate of the United States Statements from Former Consultants in Poetry Y-86

Pohl, Frederik 1919- DLB-8

Poirier, Louis (see Gracq, Julien)

Polanyi, Michael 1891-1976 .... DLB-100

Pole, Reginald 1500-1558 .... DLB-132

Poliakoff, Stephen 1952- DLB-13

Polidori, John William 1795-1821 .... DLB-116

Polite, Carlene Hatcher 1932- DLB-33

Pollard, Edward A. 1832-1872 ...... DLB-30

Pollard, Percival 1869-1911 .... DLB-71

Pollard and Moss DLB-49

Pollock, Sharon 1936- DLB-60

Polonsky, Abraham 1910- DLB-26

Polotsky, Simeon 1629-1680 .... DLB-150

Ponce, Mary Helen 1938- DLB-122

Ponce-Montoya, Juanita 1949- DLB-122

Ponet, John 1516?-1556 .... DLB-132

Poniatowski, Elena 1933- DLB-113

Pony Stories DLB-160

Poole, Ernest 1880-1950 .... DLB-9

Poole, Sophia 1804-1891 .... DLB-166

Poore, Benjamin Perley 1820-1887 .... DLB-23

Pope, Abbie Hanscom 1858-1894 .... DLB-140

Pope, Alexander 1688-1744 ....DLB-95, 101

Popov, Mikhail Ivanovich 1742-circa 1790 .... DLB-150

Popular Library DLB-46

Porlock, Martin (see MacDonald, Philip)

Porpoise Press DLB-112

Porta, Antonio 1935-1989 .... DLB-128

Porter, Anna Maria 1780-1832 .... DLB-116, 159

Porter, Eleanor H. 1868-1920 .... DLB-9

Porter, Gene Stratton (see Stratton-Porter, Gene)

Porter, Henry ?-? DLB-62

Porter, Jane 1776-1850 .... DLB-116, 159

Porter, Katherine Anne 1890-1980 ... DLB-4, 9, 102; Y-80; DS-12

Porter, Peter 1929- DLB-40

Porter, William Sydney 1862-1910 .... DLB-12, 78, 79

Porter, William T. 1809-1858 ..... DLB-3, 43

Porter and Coates DLB-49

Portis, Charles 1933- DLB-6

Postans, Marianne circa 1810-1865 .... DLB-166

Postl, Carl (see Sealsfield, Carl)

Poston, Ted 1906-1974 .... DLB-51

Postscript to [the Third Edition of] *Clarissa* (1751), by Samuel Richardson ... DLB-39

Potok, Chaim 1929- ... DLB-28, 152; Y-84

Potter, Beatrix 1866-1943 .... DLB-141

Potter, David M. 1910-1971 .... DLB-17

Potter, John E., and Company ... DLB-49

Pottle, Frederick A. 1897-1987 .... DLB-103; Y-87

Poulin, Jacques 1937- DLB-60

Pound, Ezra 1885-1972 ...... DLB-4, 45, 63

Powell, Anthony 1905- DLB-15

Powers, J. F. 1917- DLB-130

Pownall, David 1938- DLB-14

Powys, John Cowper 1872-1963 .... DLB-15

Powys, Llewelyn 1884-1939 .... DLB-98

Powys, T. F. 1875-1953 .... DLB-36, 162

Poynter, Nelson 1903-1978 .... DLB-127

The Practice of Biography: An Interview with Stanley Weintraub Y-82

The Practice of Biography II: An Interview with B. L. Reid Y-83

The Practice of Biography III: An Interview with Humphrey Carpenter Y-84

The Practice of Biography IV: An Interview with William Manchester Y-85

The Practice of Biography V: An Interview with Justin Kaplan Y-86

The Practice of Biography VI: An Interview with David Herbert Donald Y-87

The Practice of Biography VII: An Interview with John Caldwell Guilds Y-92

The Practice of Biography VIII: An Interview with Joan Mellen Y-94

The Practice of Biography IX: An Interview with Michael Reynolds Y-95

Prados, Emilio 1899-1962 .... DLB-134

Praed, Winthrop Mackworth 1802-1839 .... DLB-96

Praeger Publishers DLB-46

Praetorius, Johannes 1630-1680 .... DLB-168

Pratt, E. J. 1882-1964 .... DLB-92

Pratt, Samuel Jackson 1749-1814 .... DLB-39

Preface to *Alwyn* (1780), by Thomas Holcroft DLB-39

Preface to *Colonel Jack* (1722), by Daniel Defoe DLB-39

Preface to *Evelina* (1778), by Fanny Burney DLB-39

Preface to *Ferdinand Count Fathom* (1753), by Tobias Smollett DLB-39

Preface to *Incognita* (1692), by William Congreve DLB-39

Preface to *Joseph Andrews* (1742), by Henry Fielding DLB-39

Preface to *Moll Flanders* (1722), by Daniel Defoe DLB-39

Preface to *Poems* (1853), by Matthew Arnold DLB-32

Preface to *Robinson Crusoe* (1719), by Daniel Defoe DLB-39

Preface to *Roderick Random* (1748), by Tobias Smollett DLB-39

Preface to *Roxana* (1724), by Daniel Defoe DLB-39

Preface to *St. Leon* (1799), by William Godwin DLB-39

Preface to Sarah Fielding's *Familiar Letters* (1747), by Henry Fielding [excerpt] DLB-39

Preface to Sarah Fielding's *The Adventures of David Simple* (1744), by Henry Fielding DLB-39

Preface to *The Cry* (1754), by Sarah Fielding DLB-39

Preface to *The Delicate Distress* (1769), by Elizabeth Griffin DLB-39

Preface to *The Disguis'd Prince* (1733), by Eliza Haywood [excerpt] DLB-39

Preface to *The Farther Adventures of Robinson Crusoe* (1719), by Daniel Defoe ... DLB-39

Preface to the First Edition of *Pamela* (1740), by Samuel Richardson DLB-39

Preface to the First Edition of *The Castle of Otranto* (1764), by Horace Walpole DLB-39

Preface to *The History of Romances* (1715), by Pierre Daniel Huet [excerpts] DLB-39

Preface to *The Life of Charlotta du Pont* (1723), by Penelope Aubin DLB-39

Preface to *The Old English Baron* (1778), by Clara Reeve DLB-39

Preface to the Second Edition of *The Castle of Otranto* (1765), by Horace Walpole DLB-39

Preface to *The Secret History, of Queen Zarah, and the Zarazians* (1705), by Delariviere Manley DLB-39

Preface to the Third Edition of *Clarissa* (1751), by Samuel Richardson [excerpt] DLB-39

Preface to *The Works of Mrs. Davys* (1725), by Mary Davys DLB-39

Preface to Volume 1 of *Clarissa* (1747), by Samuel Richardson DLB-39

Preface to Volume 3 of *Clarissa* (1748), by Samuel Richardson DLB-39

Préfontaine, Yves 1937- DLB-53

Prelutsky, Jack 1940- DLB-61

Premisses, by Michael Hamburger ... DLB-66

Prentice, George D. 1802-1870 ...... DLB-43

Prentice-Hall DLB-46

Prescott, William Hickling 1796-1859 .... DLB-1, 30, 59

The Present State of the English Novel (1892), by George Saintsbury DLB-18

Prešeren, Francè 1800-1849 .... DLB-147

Preston, Thomas 1537-1598 .... DLB-62

Price, Reynolds 1933- DLB-2

Price, Richard 1723-1791 .... DLB-158

Price, Richard 1949- Y-81

Priest, Christopher 1943- DLB-14

Priestley, J. B. 1894-1984 .... DLB-10, 34, 77, 100, 139; Y-84

Primary Bibliography: A Retrospective Y-95

Prime, Benjamin Young 1733-1791 ...DLB-31

Primrose, Diana floruit circa 1630 .... DLB-126

Prince, F. T. 1912- DLB-20

Prince, Thomas 1687-1758 ..... DLB-24, 140

The Principles of Success in Literature (1865), by George Henry Lewes [excerpt] ...DLB-57

Printz, Wolfgang Casper 1641-1717 .... DLB-168

Prior, Matthew 1664-1721 .... DLB-95

Pritchard, William H. 1932- DLB-111

Pritchett, V. S. 1900- DLB-15, 139

Procter, Adelaide Anne 1825-1864 ...DLB-32

Procter, Bryan Waller 1787-1874 .... DLB-96, 144

The Profession of Authorship: Scribblers for Bread Y-89

The Progress of Romance (1785), by Clara Reeve [excerpt] DLB-39

Prokopovich, Feofan 1681?-1736 ....DLB-150

Prokosch, Frederic 1906-1989 ....DLB-48

The Proletarian Novel DLB-9

Propper, Dan 1937- DLB-16

The Prospect of Peace (1778), by Joel Barlow DLB-37

Proud, Robert 1728-1813 .... DLB-30

Proust, Marcel 1871-1922 ....DLB-65

Prynne, J. H. 1936- DLB-40

Przybyszewski, Stanislaw 1868-1927 .... DLB-66

Pseudo-Dionysius the Areopagite floruit circa 500 .... DLB-115

The Public Lending Right in America Statement by Sen. Charles McC. Mathias, Jr. PLR and the Meaning of Literary Property Statements on PLR by American Writers Y-83

The Public Lending Right in the United Kingdom Public Lending Right: The First Year in the United Kingdom Y-83

The Publication of English Renaissance Plays DLB-62

Publications and Social Movements [Transcendentalism] DLB-1

Publishers and Agents: The Columbia Connection Y-87

A Publisher's Archives: G. P. Putnam ... Y-92

Publishing Fiction at LSU Press Y-87

Pückler-Muskau, Hermann von 1785-1871 .... DLB-133

Pufendorf, Samuel von 1632-1694 .... DLB-168

Pugh, Edwin William 1874-1930 .... DLB-135

Pugin, A. Welby 1812-1852 .... DLB-55

Puig, Manuel 1932-1990 .... DLB-113

Pulitzer, Joseph 1847-1911 .... DLB-23

Pulitzer, Joseph, Jr. 1885-1955 .... DLB-29

Pulitzer Prizes for the Novel, 1917-1945 .... DLB-9

Pulliam, Eugene 1889-1975 .... DLB-127

Purchas, Samuel 1577?-1626 .... DLB-151

Purdy, Al 1918- DLB-88

Purdy, James 1923- DLB-2

Purdy, Ken W. 1913-1972 .... DLB-137

Pusey, Edward Bouverie 1800-1882 .... DLB-55

Putnam, George Palmer 1814-1872 .... DLB-3, 79

Putnam, Samuel 1892-1950 .... DLB-4

G. P. Putnam's Sons [U.S.] DLB-49

G. P. Putnam's Sons [U.K.] DLB-106

Puzo, Mario 1920- DLB-6

Pyle, Ernie 1900-1945 .... DLB-29

Pyle, Howard 1853-1911 .... DLB-42; DS-13

Pym, Barbara 1913-1980 ...... DLB-14; Y-87

Pynchon, Thomas 1937- DLB-2

Pyramid Books DLB-46

Pyrnelle, Louise-Clarke 1850-1907 ... DLB-42

Q

Quad, M. (see Lewis, Charles B.)

Quarles, Francis 1592-1644 .... DLB-126

The Quarterly Review 1809-1967 .... DLB-110

Quasimodo, Salvatore 1901-1968 ... DLB-114

Queen, Ellery (see Dannay, Frederic, and Manfred B. Lee)

The Queen City Publishing House ... DLB-49

Queneau, Raymond 1903-1976 ...... DLB-72

Quennell, Sir Peter 1905-1993 ..... DLB-155

Quesnel, Joseph 1746-1809 .... DLB-99

The Question of American Copyright in the Nineteenth Century
  Headnote
  Preface, by George Haven Putnam
  The Evolution of Copyright, by Brander Matthews
  Summary of Copyright Legislation in the United States, by R. R. Bowker
  Analysis of the Provisions of the Copyright Law of 1891, by George Haven Putnam
  The Contest for International Copyright, by George Haven Putnam
  Cheap Books and Good Books, by Brander Matthews DLB-49

Quiller-Couch, Sir Arthur Thomas 1863-1944 .... DLB-135, 153

Quin, Ann 1936-1973 .... DLB-14

Quincy, Samuel, of Georgia ?-? DLB-31

Quincy, Samuel, of Massachusetts 1734-1789 .... DLB-31

Quinn, Anthony 1915- DLB-122

Quintana, Leroy V. 1944- DLB-82

Quintana, Miguel de 1671-1748
  A Forerunner of Chicano Literature DLB-122

Quist, Harlin, Books DLB-46

Quoirez, Françoise (see Sagan, Franççise)

R

Raabe, Wilhelm 1831-1910 .... DLB-129

Rabe, David 1940- DLB-7

Raboni, Giovanni 1932- DLB-128

Rachilde 1860-1953 .... DLB-123

Racin, Kočo 1908-1943 .... DLB-147

Rackham, Arthur 1867-1939 .... DLB-141

Radcliffe, Ann 1764-1823 .... DLB-39

Raddall, Thomas 1903- DLB-68

Radiguet, Raymond 1903-1923 ..... DLB-65

Radishchev, Aleksandr Nikolaevich 1749-1802 .... DLB-150

Radványi, Netty Reiling (see Seghers, Anna)

Rahv, Philip 1908-1973 .... DLB-137

Raimund, Ferdinand Jakob 1790-1836 .... DLB-90

Raine, Craig 1944- DLB-40

Raine, Kathleen 1908- DLB-20

Rainolde, Richard circa 1530-1606 .... DLB-136

Rakić, Milan 1876-1938 .... DLB-147

Ralph, Julian 1853-1903 .... DLB-23

Ralph Waldo Emerson in 1982 .... Y-82

Ramat, Silvio 1939- DLB-128

Rambler, no. 4 (1750), by Samuel Johnson [excerpt] DLB-39

Ramée, Marie Louise de la (see Ouida)

Ramírez, Sergío 1942- DLB-145

Ramke, Bin 1947- DLB-120

Ramler, Karl Wilhelm 1725-1798 .... DLB-97

Ramon Ribeyro, Julio 1929- DLB-145

Ramous, Mario 1924- DLB-128

Rampersad, Arnold 1941- DLB-111

Ramsay, Allan 1684 or 1685-1758 ... DLB-95

Ramsay, David 1749-1815 .... DLB-30

Ranck, Katherine Quintana 1942- DLB-122

Rand, Avery and Company DLB-49

Rand McNally and Company DLB-49

Randall, David Anton 1905-1975 .... DLB-140

Randall, Dudley 1914- DLB-41

Randall, Henry S. 1811-1876 .... DLB-30

Randall, James G. 1881-1953 .... DLB-17

The Randall Jarrell Symposium: A Small Collection of Randall Jarrells Excerpts From Papers Delivered at the Randall Jarrell Symposium Y-86

Randolph, A. Philip 1889-1979 ...... DLB-91

Randolph, Anson D. F. [publishing house] DLB-49

Randolph, Thomas 1605-1635 .. DLB-58, 126

Random House DLB-46

Ranlet, Henry [publishing house] DLB-49

Ransom, John Crowe 1888-1974 .... DLB-45, 63

Ransome, Arthur 1884-1967 .... DLB-160

Raphael, Frederic 1931- DLB-14

Raphaelson, Samson 1896-1983 .... DLB-44

Raskin, Ellen 1928-1984 .... DLB-52

Rastell, John 1475?-1536 .... DLB-136

Rattigan, Terence 1911-1977 .... DLB-13

Rawlings, Marjorie Kinnan 1896-1953 .... DLB-9, 22, 102

Raworth, Tom 1938- DLB-40

Ray, David 1932- DLB-5

Ray, Gordon Norton 1915-1986 .... DLB-103, 140

Ray, Henrietta Cordelia 1849-1916 .... DLB-50

Raymond, Henry J. 1820-1869 ... DLB-43, 79

Raymond Chandler Centenary Tributes from Michael Avallone, James Elroy, Joe Gores, and William F. Nolan Y-88

Reach, Angus 1821-1856 .... DLB-70

Read, Herbert 1893-1968 .... DLB-20, 149

Read, Herbert, "The Practice of Biography," in *The English Sense of Humour and Other Essays* DLB-149

Read, Opie 1852-1939 .... DLB-23

Read, Piers Paul 1941- DLB-14

Reade, Charles 1814-1884 .... DLB-21

Reader's Digest Condensed Books DLB-46

Reading, Peter 1946- DLB-40

Reaney, James 1926- DLB-68

Rèbora, Clemente 1885-1957 ...... DLB-114

Rechy, John 1934- DLB-122; Y-82

The Recovery of Literature: Criticism in the 1990s: A Symposium Y-91

Redding, J. Saunders 1906-1988 .... DLB-63, 76

Redfield, J. S. [publishing house] DLB-49

Redgrove, Peter 1932- DLB-40

Redmon, Anne 1943- Y-86

Redmond, Eugene B. 1937- DLB-41

Redpath, James [publishing house] .. DLB-49

Reed, Henry 1808-1854 .... DLB-59

Reed, Henry 1914- DLB-27

Reed, Ishmael 1938- DLB-2, 5, 33, 169; DS-8

Reed, Sampson 1800-1880 .... DLB-1

Reed, Talbot Baines 1852-1893 .... DLB-141

Reedy, William Marion 1862-1920 .. DLB-91

Reese, Lizette Woodworth 1856-1935 .... DLB-54

Reese, Thomas 1742-1796 .... DLB-37

Reeve, Clara 1729-1807 .... DLB-39

Reeves, James 1909-1978 .... DLB-161

Reeves, John 1926- DLB-88

Regnery, Henry, Company DLB-46

Rehberg, Hans 1901-1963 .... DLB-124

Rehfisch, Hans José 1891-1960 .... DLB-124

Reid, Alastair 1926- DLB-27

Reid, B. L. 1918-1990 .... DLB-111

Reid, Christopher 1949- DLB-40

Reid, Forrest 1875-1947 .... DLB-153

Reid, Helen Rogers 1882-1970 ...... DLB-29

Reid, James ?-? DLB-31

Reid, Mayne 1818-1883 .... DLB-21, 163

Reid, Thomas 1710-1796 .... DLB-31

Reid, V. S. (Vic) 1913-1987 .... DLB-125

Reid, Whitelaw 1837-1912 .... DLB-23

Reilly and Lee Publishing Company DLB-46

Reimann, Brigitte 1933-1973 .... DLB-75

Reinmar der Alte circa 1165-circa 1205 .... DLB-138

Reinmar von Zweter circa 1200-circa 1250 .... DLB-138

Reisch, Walter 1903-1983 .... DLB-44

Remarque, Erich Maria 1898-1970 ... DLB-56

"Re-meeting of Old Friends": The Jack Kerouac Conference Y-82

Remington, Frederic 1861-1909 ...... DLB-12

Renaud, Jacques 1943- DLB-60

Renault, Mary 1905-1983 .... Y-83

Rendell, Ruth 1930- DLB-87

Representative Men and Women: A Historical Perspective on the British Novel, 1930-1960 .... DLB-15

(Re-)Publishing Orwell Y-86

Rettenbacher, Simon 1634-1706 .... DLB-168

Reuter, Christian 1665-after 1712 ... DLB-168

Reuter, Fritz 1810-1874 .... DLB-129

Reuter, Gabriele 1859-1941 .... DLB-66

Revell, Fleming H., Company DLB-49

Reventlow, Franziska Gräfin zu 1871-1918 .... DLB-66

Review of Reviews Office DLB-112

Review of [Samuel Richardson's] *Clarissa* (1748), by Henry Fielding DLB-39

The Revolt (1937), by Mary Colum [excerpts] DLB-36

Rexroth, Kenneth 1905-1982 .... DLB-16, 48, 165; Y-82

Rey, H. A. 1898-1977 .... DLB-22

Reynal and Hitchcock DLB-46

Reynolds, G. W. M. 1814-1879 ..... DLB-21

Reynolds, John Hamilton 1794-1852 .... DLB-96

Reynolds, Mack 1917- DLB-8

Reynolds, Sir Joshua 1723-1792 ..... DLB-104

Reznikoff, Charles 1894-1976 ... DLB-28, 45

"Rhetoric" (1828; revised, 1859), by Thomas de Quincey [excerpt] ... DLB-57

Rhett, Robert Barnwell 1800-1876 ... DLB-43

Rhode, John 1884-1964 .... DLB-77

Rhodes, James Ford 1848-1927 ...... DLB-47

Rhys, Jean 1890-1979 ..... DLB-36, 117, 162

Ricardo, David 1772-1823 .... DLB-107, 158

Ricardou, Jean 1932- DLB-83

Rice, Elmer 1892-1967 .... DLB-4, 7

Rice, Grantland 1880-1954 .... DLB-29

Rich, Adrienne 1929- DLB-5, 67

Richards, David Adams 1950- DLB-53

Richards, George circa 1760-1814 .... DLB-37

Richards, I. A. 1893-1979 .... DLB-27

Richards, Laura E. 1850-1943 .... DLB-42

Richards, William Carey 1818-1892 .... DLB-73

Richards, Grant [publishing house] DLB-112

Richardson, Charles F. 1851-1913 .... DLB-71

Richardson, Dorothy M. 1873-1957 .... DLB-36

Richardson, Jack 1935- DLB-7

Richardson, John 1796-1852 .... DLB-99

Richardson, Samuel 1689-1761 .... DLB-39, 154

Richardson, Willis 1889-1977 .... DLB-51

Riche, Barnabe 1542-1617 .... DLB-136

Richler, Mordecai 1931- DLB-53

Richter, Conrad 1890-1968 .... DLB-9

Richter, Hans Werner 1908- DLB-69

Richter, Johann Paul Friedrich 1763-1825 .... DLB-94

Rickerby, Joseph [publishing house] DLB-106

Rickword, Edgell 1898-1982 .... DLB-20

Riddell, Charlotte 1832-1906 .... DLB-156

Riddell, John (see Ford, Corey)

Ridge, Lola 1873-1941 .... DLB-54

Ridge, William Pett 1859-1930 ..... DLB-135

Riding, Laura (see Jackson, Laura Riding)

Ridler, Anne 1912- DLB-27

Ridruego, Dionisio 1912-1975 ...... DLB-108

Riel, Louis 1844-1885 .... DLB-99

Riemer, Johannes 1648-1714 ...... DLB-168

Riffaterre, Michael 1924- DLB-67

Riis, Jacob 1849-1914 ....DLB-23

Riker, John C. [publishing house] DLB-49

Riley, John 1938-1978 ....DLB-40

Rilke, Rainer Maria 1875-1926 ......DLB-81

Rinehart and Company DLB-46

Ringuet 1895-1960 ....DLB-68

Ringwood, Gwen Pharis
   1910-1984 ....DLB-88

Rinser, Luise 1911- DLB-69

Ríos, Alberto 1952- DLB-122

Ríos, Isabella 1948- DLB-82

Ripley, Arthur 1895-1961 ....DLB-44

Ripley, George 1802-1880 .... DLB-1, 64, 73

The Rising Glory of America:
   Three Poems DLB-37

The Rising Glory of America: Written in 1771
   (1786), by Hugh Henry Brackenridge and
   Philip Freneau DLB-37

Riskin, Robert 1897-1955 ....DLB-26

Risse, Heinz 1898- DLB-69

Rist, Johann 1607-1667 ....DLB-164

Ritchie, Anna Mowatt 1819-1870 .....DLB-3

Ritchie, Anne Thackeray
   1837-1919 ....DLB-18

Ritchie, Thomas 1778-1854 ....DLB-43

Rites of Passage
   [on William Saroyan] Y-83

The Ritz Paris Hemingway Award Y-85

Rivard, Adjutor 1868-1945 ....DLB-92

Rive, Richard 1931-1989 ....DLB-125

Rivera, Marina 1942- DLB-122

Rivera, Tomás 1935-1984 ....DLB-82

Rivers, Conrad Kent 1933-1968 .....DLB-41

Riverside Press DLB-49

Rivington, James circa 1724-1802 ....DLB-43

Rivington, Charles
   [publishing house] DLB-154

Rivkin, Allen 1903-1990 ....DLB-26

Roa Bastos, Augusto 1917- DLB-113

Robbe-Grillet, Alain 1922- DLB-83

Robbins, Tom 1936- Y-80

Roberts, Charles G. D. 1860-1943 ...DLB-92

Roberts, Dorothy 1906-1993 ....DLB-88

Roberts, Elizabeth Madox
   1881-1941 .... DLB-9, 54, 102

Roberts, Kenneth 1885-1957 ....DLB-9

Roberts, William 1767-1849 ....DLB-142

Roberts Brothers DLB-49

Roberts, James [publishing house] ..DLB-154

Robertson, A. M., and Company DLB-49

Robertson, William 1721-1793 .....DLB-104

Robinson, Casey 1903-1979 ....DLB-44

Robinson, Edwin Arlington
   1869-1935 ....DLB-54

Robinson, Henry Crabb
   1775-1867 ....DLB-107

Robinson, James Harvey
   1863-1936 ....DLB-47

Robinson, Lennox 1886-1958 ....DLB-10

Robinson, Mabel Louise
   1874-1962 ....DLB-22

Robinson, Mary 1758-1800 ....DLB-158

Robinson, Richard
   circa 1545-1607 ....DLB-167

Robinson, Therese
   1797-1870 .... DLB-59, 133

Robison, Mary 1949- DLB-130

Roblès, Emmanuel 1914- DLB-83

Roccatagliata Ceccardi, Ceccardo
   1871-1919 ....DLB-114

Rochester, John Wilmot, Earl of
   1647-1680 ....DLB-131

Rock, Howard 1911-1976 ....DLB-127

Rodgers, Carolyn M. 1945- DLB-41

Rodgers, W. R. 1909-1969 ....DLB-20

Rodríguez, Claudio 1934- DLB-134

Rodriguez, Richard 1944- DLB-82

Rodríguez Julia, Edgardo
   1946- DLB-145

Roethke, Theodore 1908-1963 ....DLB-5

Rogers, Pattiann 1940- DLB-105

Rogers, Samuel 1763-1855 ....DLB-93

Rogers, Will 1879-1935 ....DLB-11

Rohmer, Sax 1883-1959 ....DLB-70

Roiphe, Anne 1935- Y-80

Rojas, Arnold R. 1896-1988 ....DLB-82

Rolfe, Frederick William
   1860-1913 .... DLB-34, 156

Rolland, Romain 1866-1944 ....DLB-65

Rolle, Richard
   circa 1290-1300 - 1340 ....DLB-146

Rolvaag, O. E. 1876-1931 ....DLB-9

Romains, Jules 1885-1972 ....DLB-65

Roman, A., and Company DLB-49

Romano, Octavio 1923- DLB-122

Romero, Leo 1950- DLB-122

Romero, Lin 1947- DLB-122

Romero, Orlando 1945- DLB-82

Rook, Clarence 1863-1915 ....DLB-135

Roosevelt, Theodore 1858-1919 .....DLB-47

Root, Waverley 1903-1982 ....DLB-4

Root, William Pitt 1941- DLB-120

Roquebrune, Robert de 1889-1978 ...DLB-68

Rosa, João Guimarães
   1908-1967 ....DLB-113

Rosales, Luis 1910-1992 ....DLB-134

Roscoe, William 1753-1831 ....DLB-163

Rose, Reginald 1920- DLB-26

Rosegger, Peter 1843-1918 ....DLB-129

Rosei, Peter 1946- DLB-85

Rosen, Norma 1925- DLB-28

Rosenbach, A. S. W. 1876-1952 ....DLB-140

Rosenberg, Isaac 1890-1918 ....DLB-20

Rosenfeld, Isaac 1918-1956 ....DLB-28

Rosenthal, M. L. 1917- DLB-5

Ross, Alexander 1591-1654 ....DLB-151

Ross, Harold 1892-1951 ....DLB-137

Ross, Leonard Q. (see Rosten, Leo)

Ross, Martin 1862-1915 ....DLB-135

Ross, Sinclair 1908- DLB-88

Ross, W. W. E. 1894-1966 ....DLB-88

Rosselli, Amelia 1930- DLB-128

Rossen, Robert 1908-1966 ....DLB-26

Rossetti, Christina Georgina
   1830-1894 ....DLB-35, 163

Rossetti, Dante Gabriel 1828-1882 ...DLB-35

Rossner, Judith 1935- DLB-6

Rosten, Leo 1908- DLB-11

Rostenberg, Leona 1908- DLB-140

Rostovsky, Dimitrii 1651-1709 .....DLB-150

Bertram Rota and His Bookshop Y-91

Roth, Gerhard 1942- DLB-85, 124

Roth, Henry 1906?- DLB-28

Roth, Joseph 1894-1939 ....DLB-85

Roth, Philip 1933- DLB-2, 28; Y-82

Rothenberg, Jerome 1931- DLB-5

Rotimi, Ola 1938- DLB-125

Routhier, Adolphe-Basile
   1839-1920 ....DLB-99

Routier, Simone 1901-1987 . . . . DLB-88

Routledge, George, and Sons DLB-106

Roversi, Roberto 1923- DLB-128

Rowe, Elizabeth Singer
 1674-1737 . . . . DLB-39, 95

Rowe, Nicholas 1674-1718 . . . . DLB-84

Rowlands, Samuel
 circa 1570-1630 . . . . DLB-121

Rowlandson, Mary
 circa 1635-circa 1678 . . . . DLB-24

Rowley, William circa 1585-1626 . . . DLB-58

Rowse, A. L. 1903- DLB-155

Rowson, Susanna Haswell
 circa 1762-1824 . . . . DLB-37

Roy, Camille 1870-1943 . . . . DLB-92

Roy, Gabrielle 1909-1983 . . . . DLB-68

Roy, Jules 1907- DLB-83

The Royal Court Theatre and the English
 Stage Company DLB-13

The Royal Court Theatre and the New
 Drama DLB-10

The Royal Shakespeare Company
 at the Swan Y-88

Royall, Anne 1769-1854 . . . . DLB-43

The Roycroft Printing Shop DLB-49

Royster, Vermont 1914- DLB-127

Ruark, Gibbons 1941- DLB-120

Ruban, Vasilii Grigorevich
 1742-1795 . . . . DLB-150

Rubens, Bernice 1928- DLB-14

Rudd and Carleton DLB-49

Rudkin, David 1936- DLB-13

Rudolf von Ems
 circa 1200-circa 1254 . . . . DLB-138

Ruffin, Josephine St. Pierre
 1842-1924 . . . . DLB-79

Ruganda, John 1941- DLB-157

Ruggles, Henry Joseph 1813-1906 . . . DLB-64

Rukeyser, Muriel 1913-1980 . . . . DLB-48

Rule, Jane 1931- DLB-60

Rulfo, Juan 1918-1986 . . . . DLB-113

Rumaker, Michael 1932- DLB-16

Rumens, Carol 1944- DLB-40

Runyon, Damon 1880-1946 . . . . . DLB-11, 86

Ruodlieb circa 1050-1075 . . . . DLB-148

Rush, Benjamin 1746-1813 . . . . DLB-37

Rusk, Ralph L. 1888-1962 . . . . DLB-103

Ruskin, John 1819-1900 . . . . DLB-55, 163

Russ, Joanna 1937- DLB-8

Russell, B. B., and Company DLB-49

Russell, Benjamin 1761-1845 . . . . DLB-43

Russell, Bertrand 1872-1970 . . . . . . DLB-100

Russell, Charles Edward
 1860-1941 . . . . DLB-25

Russell, George William (see AE)

Russell, R. H., and Son DLB-49

Rutherford, Mark 1831-1913 . . . . DLB-18

Ryan, Michael 1946- Y-82

Ryan, Oscar 1904- DLB-68

Ryga, George 1932- DLB-60

Rymer, Thomas 1643?-1713 . . . . DLB-101

Ryskind, Morrie 1895-1985 . . . . DLB-26

Rzhevsky, Aleksei Andreevich
 1737-1804 . . . . DLB-150

S

The Saalfield Publishing
 Company DLB-46

Saba, Umberto 1883-1957 . . . . DLB-114

Sábato, Ernesto 1911- DLB-145

Saberhagen, Fred 1930- DLB-8

Sacer, Gottfried Wilhelm
 1635-1699 . . . . DLB-168

Sackler, Howard 1929-1982 . . . . DLB-7

Sackville, Thomas 1536-1608 . . . . . . DLB-132

Sackville, Thomas 1536-1608
 and Norton, Thomas
 1532-1584 . . . . DLB-62

Sackville-West, V. 1892-1962 . . . . DLB-34

Sadlier, D. and J., and Company DLB-49

Sadlier, Mary Anne 1820-1903 . . . . . . DLB-99

Sadoff, Ira 1945- DLB-120

Saenz, Jaime 1921-1986 . . . . DLB-145

Saffin, John circa 1626-1710 . . . . DLB-24

Sagan, Françoise 1935- DLB-83

Sage, Robert 1899-1962 . . . . DLB-4

Sagel, Jim 1947- DLB-82

Sagendorph, Robb Hansell
 1900-1970 . . . . DLB-137

Sahagún, Carlos 1938- DLB-108

Sahkomaapii, Piitai (see Highwater, Jamake)

Sahl, Hans 1902- DLB-69

Said, Edward W. 1935- DLB-67

Saiko, George 1892-1962 . . . . DLB-85

St. Dominic's Press DLB-112

Saint-Exupéry, Antoine de
 1900-1944 . . . . DLB-72

St. Johns, Adela Rogers 1894-1988 . . . DLB-29

St. Martin's Press DLB-46

St. Omer, Garth 1931- DLB-117

Saint Pierre, Michel de 1916-1987 . . . DLB-83

Saintsbury, George
 1845-1933 . . . . DLB-57, 149

Saki (see Munro, H. H.)

Salaam, Kalamu ya 1947- DLB-38

Salas, Floyd 1931- DLB-82

Sálaz-Marquez, Rubén 1935- DLB-122

Salemson, Harold J. 1910-1988 . . . . . . DLB-4

Salinas, Luis Omar 1937- DLB-82

Salinas, Pedro 1891-1951 . . . . DLB-134

Salinger, J. D. 1919- DLB-2, 102

Salkey, Andrew 1928- DLB-125

Salt, Waldo 1914- DLB-44

Salter, James 1925- DLB-130

Salter, Mary Jo 1954- DLB-120

Salustri, Carlo Alberto (see Trilussa)

Salverson, Laura Goodman
 1890-1970 . . . . DLB-92

Sampson, Richard Henry (see Hull, Richard)

Samuels, Ernest 1903- DLB-111

Sanborn, Franklin Benjamin
 1831-1917 . . . . DLB-1

Sánchez, Luis Rafael 1936- DLB-145

Sánchez, Philomeno "Phil"
 1917- DLB-122

Sánchez, Ricardo 1941- DLB-82

Sanchez, Sonia 1934- DLB-41; DS-8

Sand, George 1804-1876 . . . . DLB-119

Sandburg, Carl 1878-1967 . . . . . . DLB-17, 54

Sanders, Ed 1939- DLB-16

Sandoz, Mari 1896-1966 . . . . DLB-9

Sandwell, B. K. 1876-1954 . . . . DLB-92

Sandy, Stephen 1934- DLB-165

Sandys, George 1578-1644 . . . . . DLB-24, 121

Sangster, Charles 1822-1893 . . . . DLB-99

Sanguineti, Edoardo 1930- DLB-128

Sansom, William 1912-1976 . . . . DLB-139

Santayana, George
 1863-1952 . . . . DLB-54, 71; DS-13

Santiago, Danny 1911-1988 . . . . DLB-122

Santmyer, Helen Hooven 1895-1986 .... Y-84

Sapir, Edward 1884-1939 .... DLB-92

Sapper (see McNeile, Herman Cyril)

Sarduy, Severo 1937- DLB-113

Sargent, Pamela 1948- DLB-8

Saro-Wiwa, Ken 1941- DLB-157

Saroyan, William
1908-1981 .... DLB-7, 9, 86; Y-81

Sarraute, Nathalie 1900- DLB-83

Sarrazin, Albertine 1937-1967 .... DLB-83

Sarton, May 1912- DLB-48; Y-81

Sartre, Jean-Paul 1905-1980 .... DLB-72

Sassoon, Siegfried 1886-1967 .... DLB-20

Saturday Review Press DLB-46

Saunders, James 1925- DLB-13

Saunders, John Monk 1897-1940 .... DLB-26

Saunders, Margaret Marshall
1861-1947 .... DLB-92

Saunders and Otley DLB-106

Savage, James 1784-1873 .... DLB-30

Savage, Marmion W. 1803?-1872 .... DLB-21

Savage, Richard 1697?-1743 .... DLB-95

Savard, Félix-Antoine 1896-1982 ..... DLB-68

Saville, (Leonard) Malcolm
1901-1982 .... DLB-160

Sawyer, Ruth 1880-1970 .... DLB-22

Sayers, Dorothy L.
1893-1957 .... DLB-10, 36, 77, 100

Sayles, John Thomas 1950- DLB-44

Sbarbaro, Camillo 1888-1967 ...... DLB-114

Scannell, Vernon 1922- DLB-27

Scarry, Richard 1919-1994 .... DLB-61

Schaeffer, Albrecht 1885-1950 .... DLB-66

Schaeffer, Susan Fromberg 1941- ... DLB-28

Schaff, Philip 1819-1893 .... DS-13

Schaper, Edzard 1908-1984 .... DLB-69

Scharf, J. Thomas 1843-1898 .... DLB-47

Scheffel, Joseph Viktor von
1826-1886 .... DLB-129

Scheffler, Johann 1624-1677 .... DLB-164

Schelling, Friedrich Wilhelm Joseph von
1775-1854 .... DLB-90

Scherer, Wilhelm 1841-1886 .... DLB-129

Schickele, René 1883-1940 .... DLB-66

Schiff, Dorothy 1903-1989 .... DLB-127

Schiller, Friedrich 1759-1805 .... DLB-94

Schirmer, David 1623-1687 .... DLB-164

Schlaf, Johannes 1862-1941 .... DLB-118

Schlegel, August Wilhelm
1767-1845 .... DLB-94

Schlegel, Dorothea 1763-1839 ...... DLB-90

Schlegel, Friedrich 1772-1829 .... DLB-90

Schleiermacher, Friedrich
1768-1834 .... DLB-90

Schlesinger, Arthur M., Jr. 1917- .. DLB-17

Schlumberger, Jean 1877-1968 ...... DLB-65

Schmid, Eduard Hermann Wilhelm (see
Edschmid, Kasimir)

Schmidt, Arno 1914-1979 .... DLB-69

Schmidt, Johann Kaspar (see Stirner, Max)

Schmidt, Michael 1947- DLB-40

Schmidtbonn, Wilhelm August
1876-1952 .... DLB-118

Schmitz, James H. 1911- DLB-8

Schnabel, Johann Gottfried
1692-1760 .... DLB-168

Schnackenberg, Gjertrud 1953- ... DLB-120

Schnitzler, Arthur 1862-1931 ... DLB-81, 118

Schnurre, Wolfdietrich 1920- DLB-69

Schocken Books DLB-46

Schönbeck, Virgilio (see Giotti, Virgilio)

School Stories, 1914-1960 .... DLB-160

Schönherr, Karl 1867-1943 .... DLB-118

Scholartis Press DLB-112

The Schomburg Center for Research
in Black Culture DLB-76

Schopenhauer, Arthur 1788-1860 .... DLB-90

Schopenhauer, Johanna 1766-1838 .. DLB-90

Schorer, Mark 1908-1977 .... DLB-103

Schottelius, Justus Georg
1612-1676 .... DLB-164

Schouler, James 1839-1920 .... DLB-47

Schrader, Paul 1946- DLB-44

Schreiner, Olive 1855-1920 ..... DLB-18, 156

Schroeder, Andreas 1946- DLB-53

Schubart, Christian Friedrich Daniel
1739-1791 .... DLB-97

Schubert, Gotthilf Heinrich
1780-1860 .... DLB-90

Schücking, Levin 1814-1883 .... DLB-133

Schulberg, Budd
1914- DLB-6, 26, 28; Y-81

Schulte, F. J., and Company DLB-49

Schulze, Hans (see Praetorius, Johannes)

Schupp, Johann Balthasar
1610-1661 .... DLB-164

Schurz, Carl 1829-1906 .... DLB-23

Schuyler, George S. 1895-1977 ...DLB-29, 51

Schuyler, James 1923-1991 .... DLB-5, 169

Schwartz, Delmore 1913-1966 ....DLB-28, 48

Schwartz, Jonathan 1938- Y-82

Schwarz, Sibylle 1621-1638 .... DLB-164

Schwerner, Armand 1927- DLB-165

Schwob, Marcel 1867-1905 .... DLB-123

Science Fantasy DLB-8

Science-Fiction Fandom and
Conventions DLB-8

Science-Fiction Fanzines: The Time
Binders DLB-8

Science-Fiction Films DLB-8

Science Fiction Writers of America and the
Nebula Awards DLB-8

Scot, Reginald circa 1538-1599 ..... DLB-136

Scotellaro, Rocco 1923-1953 .... DLB-128

Scott, Dennis 1939-1991 .... DLB-125

Scott, Dixon 1881-1915 .... DLB-98

Scott, Duncan Campbell
1862-1947 .... DLB-92

Scott, Evelyn 1893-1963 .... DLB-9, 48

Scott, F. R. 1899-1985 .... DLB-88

Scott, Frederick George
1861-1944 .... DLB-92

Scott, Geoffrey 1884-1929 .... DLB-149

Scott, Harvey W. 1838-1910 .... DLB-23

Scott, Paul 1920-1978 .... DLB-14

Scott, Sarah 1723-1795 .... DLB-39

Scott, Tom 1918- DLB-27

Scott, Sir Walter
1771-1832 ... DLB-93, 107, 116, 144, 159

Scott, William Bell 1811-1890 .... DLB-32

Scott, Walter, Publishing
Company Limited DLB-112

Scott, William R.
[publishing house] DLB-46

Scott-Heron, Gil 1949- DLB-41

Scribner, Charles, Jr. 1921-1995 .... Y-95

Charles Scribner's Sons DLB-49; DS-13

Scripps, E. W. 1854-1926 .... DLB-25

Scudder, Horace Elisha
1838-1902 .... DLB-42, 71

Scudder, Vida Dutton 1861-1954 .... DLB-71

Scupham, Peter 1933- DLB-40

Seabrook, William 1886-1945 ....DLB-4

Seabury, Samuel 1729-1796 ....DLB-31

Seacole, Mary Jane Grant
  1805-1881 ....DLB-166

The Seafarer circa 970 ....DLB-146

Sealsfield, Charles 1793-1864 ......DLB-133

Sears, Edward I. 1819?-1876 ....DLB-79

Sears Publishing Company DLB-46

Seaton, George 1911-1979 ....DLB-44

Seaton, William Winston
  1785-1866 ....DLB-43

Secker, Martin, and Warburg
  Limited DLB-112

Secker, Martin [publishing house] ..DLB-112

Second-Generation Minor Poets of the
  Seventeenth Century DLB-126

Sedgwick, Arthur George
  1844-1915 ....DLB-64

Sedgwick, Catharine Maria
  1789-1867 ....DLB-1, 74

Sedgwick, Ellery 1872-1930 ....DLB-91

Sedley, Sir Charles 1639-1701 .....DLB-131

Seeger, Alan 1888-1916 ....DLB-45

Seers, Eugene (see Dantin, Louis)

Segal, Erich 1937- Y-86

Seghers, Anna 1900-1983 ....DLB-69

Seid, Ruth (see Sinclair, Jo)

Seidel, Frederick Lewis 1936- Y-84

Seidel, Ina 1885-1974 ....DLB-56

Seigenthaler, John 1927- DLB-127

Seizin Press DLB-112

Séjour, Victor 1817-1874 ....DLB-50

Séjour Marcou et Ferrand, Juan Victor (see
  Séjour, Victor)

Selby, Hubert, Jr. 1928- DLB-2

Selden, George 1929-1989 ....DLB-52

Selected English-Language Little Magazines
  and Newspapers [France,
  1920-1939] DLB-4

Selected Humorous Magazines
  (1820-1950) DLB-11

Selected Science-Fiction Magazines and
  Anthologies DLB-8

Self, Edwin F. 1920- DLB-137

Seligman, Edwin R. A. 1861-1939 ...DLB-47

Seltzer, Chester E. (see Muro, Amado)

Seltzer, Thomas
  [publishing house] DLB-46

Selvon, Sam 1923-1994 ....DLB-125

Senancour, Etienne de 1770-1846 ...DLB-119

Sendak, Maurice 1928- DLB-61

Senécal, Eva 1905- DLB-92

Sengstacke, John 1912- DLB-127

Senior, Olive 1941- DLB-157

Šenoa, August 1838-1881 ....DLB-147

"Sensation Novels" (1863), by
  H. L. Manse DLB-21

Sepamla, Sipho 1932- DLB-157

Seredy, Kate 1899-1975 ....DLB-22

Sereni, Vittorio 1913-1983 ....DLB-128

Serling, Rod 1924-1975 ....DLB-26

Serote, Mongane Wally 1944- DLB-125

Serraillier, Ian 1912-1994 ....DLB-161

Serrano, Nina 1934- DLB-122

Service, Robert 1874-1958 ....DLB-92

Seth, Vikram 1952- DLB-120

Seton, Ernest Thompson
  1860-1942 ....DLB-92; DS-13

Settle, Mary Lee 1918- DLB-6

Seume, Johann Gottfried
  1763-1810 ....DLB-94

Seuss, Dr. (see Geisel, Theodor Seuss)

The Seventy-fifth Anniversary of the Armistice:
  The Wilfred Owen Centenary and the
  Great War Exhibit at the University of
  Virginia Y-93

Sewall, Joseph 1688-1769 ....DLB-24

Sewall, Richard B. 1908- DLB-111

Sewell, Anna 1820-1878 ....DLB-163

Sewell, Samuel 1652-1730 ....DLB-24

Sex, Class, Politics, and Religion [in the
  British Novel, 1930-1959] DLB-15

Sexton, Anne 1928-1974 ....DLB-5, 169

Seymour-Smith, Martin 1928- DLB-155

Shaara, Michael 1929-1988 ....Y-83

Shadwell, Thomas 1641?-1692 .....DLB-80

Shaffer, Anthony 1926- DLB-13

Shaffer, Peter 1926- DLB-13

Shaftesbury, Anthony Ashley Cooper,
  Third Earl of 1671-1713 .....DLB-101

Shairp, Mordaunt 1887-1939 ......DLB-10

Shakespeare, William 1564-1616 ...DLB-62

The Shakespeare Globe Trust Y-93

Shakespeare Head Press DLB-112

Shakhovskoi, Aleksandr Aleksandrovich
  1777-1846 ....DLB-150

Shange, Ntozake 1948- DLB-38

Shapiro, Karl 1913- DLB-48

Sharon Publications DLB-46

Sharp, Margery 1905-1991 ....DLB-161

Sharp, William 1855-1905 ....DLB-156

Sharpe, Tom 1928- DLB-14

Shaw, Albert 1857-1947 ....DLB-91

Shaw, Bernard 1856-1950 .....DLB-10, 57

Shaw, Henry Wheeler 1818-1885 ...DLB-11

Shaw, Joseph T. 1874-1952 ....DLB-137

Shaw, Irwin 1913-1984 ....DLB-6, 102; Y-84

Shaw, Robert 1927-1978 ....DLB-13, 14

Shaw, Robert B. 1947- DLB-120

Shawn, William 1907-1992 ....DLB-137

Shay, Frank [publishing house] DLB-46

Shea, John Gilmary 1824-1892 .....DLB-30

Sheaffer, Louis 1912-1993 ....DLB-103

Shearing, Joseph 1886-1952 ......DLB-70

Shebbeare, John 1709-1788 ....DLB-39

Sheckley, Robert 1928- DLB-8

Shedd, William G. T. 1820-1894 ...DLB-64

Sheed, Wilfred 1930- DLB-6

Sheed and Ward [U.S.] DLB-46

Sheed and Ward Limited [U.K.] ...DLB-112

Sheldon, Alice B. (see Tiptree, James, Jr.)

Sheldon, Edward 1886-1946 ....DLB-7

Sheldon and Company DLB-49

Shelley, Mary Wollstonecraft
  1797-1851 ....DLB-110, 116, 159

Shelley, Percy Bysshe
  1792-1822 ....DLB-96, 110, 158

Shelnutt, Eve 1941- DLB-130

Shenstone, William 1714-1763 .....DLB-95

Shepard, Ernest Howard
  1879-1976 ....DLB-160

Shepard, Sam 1943- DLB-7

Shepard, Thomas I,
  1604 or 1605-1649 ....DLB-24

Shepard, Thomas II, 1635-1677 ....DLB-24

Shepard, Clark and Brown DLB-49

Shepherd, Luke
  flourished 1547-1554 ....DLB-136

Sherburne, Edward 1616-1702 ....DLB-131

Sheridan, Frances 1724-1766 ....DLB-39, 84

Sheridan, Richard Brinsley
  1751-1816 ....DLB-89

Sherman, Francis 1871-1926 .... DLB-92

Sherriff, R. C. 1896-1975 .... DLB-10

Sherry, Norman 1935- DLB-155

Sherwood, Mary Martha
  1775-1851 .... DLB-163

Sherwood, Robert 1896-1955 ...... DLB-7, 26

Shiel, M. P. 1865-1947 .... DLB-153

Shiels, George 1886-1949 .... DLB-10

Shillaber, B.[enjamin] P.[enhallow]
  1814-1890 .... DLB-1, 11

Shine, Ted 1931- DLB-38

Ship, Reuben 1915-1975 .... DLB-88

Shirer, William L. 1904-1993 .... DLB-4

Shirinsky-Shikhmatov, Sergii Aleksandrovich
  1783-1837 .... DLB-150

Shirley, James 1596-1666 .... DLB-58

Shishkov, Aleksandr Semenovich
  1753-1841 .... DLB-150

Shockley, Ann Allen 1927- DLB-33

Shorthouse, Joseph Henry
  1834-1903 .... DLB-18

Showalter, Elaine 1941- DLB-67

Shulevitz, Uri 1935- DLB-61

Shulman, Max 1919-1988 .... DLB-11

Shute, Henry A. 1856-1943 .... DLB-9

Shuttle, Penelope 1947- DLB-14, 40

Sibbes, Richard 1577-1635 .... DLB-151

Sidgwick and Jackson Limited DLB-112

Sidney, Margaret (see Lothrop, Harriet M.)

Sidney, Mary 1561-1621 .... DLB-167

Sidney, Sir Philip 1554-1586 .... DLB-167

Sidney's Press DLB-49

Siegfried Loraine Sassoon: A Centenary Essay
  Tributes from Vivien F. Clarke and
  Michael Thorpe Y-86

Sierra, Rubén 1946- DLB-122

Sierra Club Books DLB-49

Siger of Brabant
  circa 1240-circa 1284 .... DLB-115

Sigourney, Lydia Howard (Huntley)
  1791-1865 .... DLB-1, 42, 73

Silkin, Jon 1930- DLB-27

Silko, Leslie Marmon 1948- DLB-143

Silliman, Ron 1946- DLB-169

Silliphant, Stirling 1918- DLB-26

Sillitoe, Alan 1928- DLB-14, 139

Silman, Roberta 1934- DLB-28

Silva, Beverly 1930- DLB-122

Silverberg, Robert 1935- DLB-8

Silverman, Kenneth 1936- DLB-111

Simak, Clifford D. 1904-1988 .... DLB-8

Simcoe, Elizabeth 1762-1850 .... DLB-99

Simcox, George Augustus
  1841-1905 .... DLB-35

Sime, Jessie Georgina 1868-1958 .... DLB-92

Simenon, Georges
  1903-1989 .... DLB-72; Y-89

Simic, Charles 1938- DLB-105

Simic, Charles,
  Images and "Images" DLB-105

Simmel, Johannes Mario 1924- DLB-69

Simmons, Ernest J. 1903-1972 ..... DLB-103

Simmons, Herbert Alfred 1930- ... DLB-33

Simmons, James 1933- DLB-40

Simms, William Gilmore
  1806-1870 .... DLB-3, 30, 59, 73

Simms and M'Intyre DLB-106

Simon, Claude 1913- DLB-83

Simon, Neil 1927- DLB-7

Simon and Schuster DLB-46

Simons, Katherine Drayton Mayrant
  1890-1969 .... Y-83

Simpkin and Marshall
  [publishing house] DLB-154

Simpson, Helen 1897-1940 .... DLB-77

Simpson, Louis 1923- DLB-5

Simpson, N. F. 1919- DLB-13

Sims, George 1923- DLB-87

Sims, George Robert
  1847-1922 .... DLB-35, 70, 135

Sinán, Rogelio 1904- DLB-145

Sinclair, Andrew 1935- DLB-14

Sinclair, Bertrand William
  1881-1972 .... DLB-92

Sinclair, Catherine
  1800-1864 .... DLB-163

Sinclair, Jo 1913- DLB-28

Sinclair Lewis Centennial
  Conference Y-85

Sinclair, Lister 1921- DLB-88

Sinclair, May 1863-1946 .... DLB-36, 135

Sinclair, Upton 1878-1968 .... DLB-9

Sinclair, Upton [publishing house] ... DLB-46

Singer, Isaac Bashevis
  1904-1991 .... DLB-6, 28, 52; Y-91

Singmaster, Elsie 1879-1958 .... DLB-9

Sinisgalli, Leonardo 1908-1981 ..... DLB-114

Siodmak, Curt 1902- DLB-44

Sissman, L. E. 1928-1976 .... DLB-5

Sisson, C. H. 1914- DLB-27

Sitwell, Edith 1887-1964 .... DLB-20

Sitwell, Osbert 1892-1969 .... DLB-100

Skármeta, Antonio 1940- DLB-145

Skeffington, William
  [publishing house] DLB-106

Skelton, John 1463-1529 .... DLB-136

Skelton, Robin 1925- DLB-27, 53

Skinner, Constance Lindsay
  1877-1939 .... DLB-92

Skinner, John Stuart 1788-1851 ..... DLB-73

Skipsey, Joseph 1832-1903 .... DLB-35

Slade, Bernard 1930- DLB-53

Slater, Patrick 1880-1951 .... DLB-68

Slaveykov, Pencho 1866-1912 ...... DLB-147

Slavitt, David 1935- DLB-5, 6

Sleigh, Burrows Willcocks Arthur
  1821-1869 .... DLB-99

A Slender Thread of Hope: The Kennedy
  Center Black Theatre Project DLB-38

Slesinger, Tess 1905-1945 .... DLB-102

Slick, Sam (see Haliburton, Thomas Chandler)

Sloane, William, Associates DLB-46

Small, Maynard and Company DLB-49

Small Presses in Great Britain and Ireland,
  1960-1985 .... DLB-40

Small Presses I: Jargon Society Y-84

Small Presses II: The Spirit That Moves Us
  Press Y-85

Small Presses III: Pushcart Press Y-87

Smart, Christopher 1722-1771 ...... DLB-109

Smart, David A. 1892-1957 .... DLB-137

Smart, Elizabeth 1913-1986 .... DLB-88

Smellie, William
  [publishing house] DLB-154

Smiles, Samuel 1812-1904 .... DLB-55

Smith, A. J. M. 1902-1980 .... DLB-88

Smith, Adam 1723-1790 .... DLB-104

Smith, Alexander 1829-1867 ..... DLB-32, 55

Smith, Betty 1896-1972 .... Y-82

Smith, Carol Sturm 1938- Y-81

Smith, Charles Henry 1826-1903 .... DLB-11

Smith, Charlotte 1749-1806 ..... DLB-39, 109

Smith, Cordwainer 1913-1966 . . . . .DLB-8

Smith, Dave 1942- DLB-5

Smith, Dodie 1896- DLB-10

Smith, Doris Buchanan 1934- DLB-52

Smith, E. E. 1890-1965 . . . . .DLB-8

Smith, Elihu Hubbard 1771-1798 . . . DLB-37

Smith, Elizabeth Oakes (Prince)
1806-1893 .DLB-1

Smith, F. Hopkinson 1838-1915 . . . . . . .DS-13

Smith, George D. 1870-1920 . . . . . . DLB-140

Smith, George O. 1911-1981 . . . . .DLB-8

Smith, Goldwin 1823-1910 . . . . DLB-99

Smith, H. Allen 1907-1976 . . . . . . DLB-11, 29

Smith, Hazel Brannon 1914- DLB-127

Smith, Horatio (Horace)
1779-1849 . . . . DLB-116

Smith, Horatio (Horace) 1779-1849 and
James Smith 1775-1839 . . . . DLB-96

Smith, Iain Crichton
1928- DLB-40, 139

Smith, J. Allen 1860-1924 . . . . DLB-47

Smith, John 1580-1631 . . . . DLB-24, 30

Smith, Josiah 1704-1781 . . . . DLB-24

Smith, Ken 1938- DLB-40

Smith, Lee 1944- DLB-143; Y-83

Smith, Logan Pearsall 1865-1946 . . . . DLB-98

Smith, Mark 1935- Y-82

Smith, Michael 1698-circa 1771 . . . . . DLB-31

Smith, Red 1905-1982 . . . . DLB-29

Smith, Roswell 1829-1892 . . . . DLB-79

Smith, Samuel Harrison
1772-1845 . . . . DLB-43

Smith, Samuel Stanhope
1751-1819 . . . . DLB-37

Smith, Sarah (see Stretton, Hesba)

Smith, Seba 1792-1868 . . . . DLB-1, 11

Smith, Sir Thomas 1513-1577 . . . . . DLB-132

Smith, Stevie 1902-1971 . . . . DLB-20

Smith, Sydney 1771-1845 . . . . DLB-107

Smith, Sydney Goodsir 1915-1975 . . . DLB-27

Smith, William
flourished 1595-1597 . . . . DLB-136

Smith, William 1727-1803 . . . . DLB-31

Smith, William 1728-1793 . . . . DLB-30

Smith, William Gardner
1927-1974 . . . . DLB-76

Smith, William Henry
1808-1872 . . . . DLB-159

Smith, William Jay 1918- DLB-5

Smith, Elder and Company DLB-154

Smith, Harrison, and Robert Haas
[publishing house] DLB-46

Smith, J. Stilman, and Company DLB-49

Smith, W. B., and Company DLB-49

Smith, W. H., and Son DLB-106

Smithers, Leonard
[publishing house] DLB-112

Smollett, Tobias 1721-1771 . . . . . DLB-39, 104

Snellings, Rolland (see Touré, Askia
Muhammad)

Snodgrass, W. D. 1926- DLB-5

Snow, C. P. 1905-1980 . . . . DLB-15, 77

Snyder, Gary 1930- DLB-5, 16, 165

Sobiloff, Hy 1912-1970 . . . . DLB-48

The Society for Textual Scholarship and
TEXT Y-87

The Society for the History of Authorship,
Reading and Publishing Y-92

Soffici, Ardengo 1879-1964 . . . . DLB-114

Sofola, 'Zulu 1938- DLB-157

Solano, Solita 1888-1975 . . . . DLB-4

Sollers, Philippe 1936- DLB-83

Solmi, Sergio 1899-1981 . . . . DLB-114

Solomon, Carl 1928- DLB-16

Solway, David 1941- DLB-53

Solzhenitsyn and America Y-85

Somerville, Edith Œnone
1858-1949 . . . . DLB-135

Song, Cathy 1955- DLB-169

Sontag, Susan 1933- DLB-2, 67

Sorrentino, Gilbert 1929- DLB-5; Y-80

Sorge, Reinhard Johannes
1892-1916 . . . . DLB-118

Sotheby, William 1757-1833 . . . . DLB-93

Soto, Gary 1952- DLB-82

Sources for the Study of Tudor and Stuart
Drama DLB-62

Souster, Raymond 1921- DLB-88

The *South English Legendary*
circa thirteenth-fifteenth
centuries DLB-146

Southerland, Ellease 1943- DLB-33

Southern Illinois University Press Y-95

Southern, Terry 1924- DLB-2

Southern Writers Between the
Wars DLB-9

Southerne, Thomas 1659-1746 . . . . . .DLB-80

Southey, Caroline Anne Bowles
1786-1854 . . . . DLB-116

Southey, Robert
1774-1843 . . . . DLB-93, 107, 142

Southwell, Robert 1561?-1595 . . . . . .DLB-167

Sowande, Bode 1948- DLB-157

Soyfer, Jura 1912-1939 . . . . .DLB-124

Soyinka, Wole 1934- . . . DLB-125; Y-86, 87

Spacks, Barry 1931- DLB-105

Spalding, Frances 1950- DLB-155

Spark, Muriel 1918- DLB-15, 139

Sparks, Jared 1789-1866 . . . . DLB-1, 30

Sparshott, Francis 1926- DLB-60

Späth, Gerold 1939- DLB-75

Spatola, Adriano 1941-1988 . . . . .DLB-128

Spaziani, Maria Luisa 1924- DLB-128

The Spectator 1828- DLB-110

Spedding, James 1808-1881 . . . . .DLB-144

Spee von Langenfeld, Friedrich
1591-1635 . . . . .DLB-164

Speght, Rachel 1597-after 1630 . . . . .DLB-126

Speke, John Hanning 1827-1864 . . . .DLB-166

Spellman, A. B. 1935- DLB-41

Spence, Thomas 1750-1814 . . . . .DLB-158

Spencer, Anne 1882-1975 . . . . . . DLB-51, 54

Spencer, Elizabeth 1921- DLB-6

Spencer, Herbert 1820-1903 . . . . .DLB-57

Spencer, Scott 1945- Y-86

Spender, J. A. 1862-1942 . . . . .DLB-98

Spender, Stephen 1909- DLB-20

Spener, Philipp Jakob 1635-1705 . . . .DLB-164

Spenser, Edmund circa 1552-1599 . . .DLB-167

Sperr, Martin 1944- DLB-124

Spicer, Jack 1925-1965 . . . . DLB-5, 16

Spielberg, Peter 1929- Y-81

Spielhagen, Friedrich 1829-1911 . . . .DLB-129

"*Spielmannsepen*"
(circa 1152-circa 1500) DLB-148

Spier, Peter 1927- DLB-61

Spinrad, Norman 1940- DLB-8

Spires, Elizabeth 1952- DLB-120

Spitteler, Carl 1845-1924 . . . . .DLB-129

Spivak, Lawrence E. 1900- DLB-137

Spofford, Harriet Prescott
  1835-1921 .... DLB-74

Squibob (see Derby, George Horatio)

Stacpoole, H. de Vere
  1863-1951 .... DLB-153

Staël, Germaine de 1766-1817 ...... DLB-119

Staël-Holstein, Anne-Louise Germaine de
  (see Staël, Germaine de)

Stafford, Jean 1915-1979 .... DLB-2

Stafford, William 1914- DLB-5

Stage Censorship: "The Rejected Statement"
  (1911), by Bernard Shaw
  [excerpts] DLB-10

Stallings, Laurence 1894-1968 ..... DLB-7, 44

Stallworthy, Jon 1935- DLB-40

Stampp, Kenneth M. 1912- DLB-17

Stanford, Ann 1916- DLB-5

Stanković, Borisav ("Bora")
  1876-1927 .... DLB-147

Stanley, Henry M. 1841-1904 .... DS-13

Stanley, Thomas 1625-1678 .... DLB-131

Stannard, Martin 1947- DLB-155

Stanton, Elizabeth Cady 1815-1902 ... DLB-79

Stanton, Frank L. 1857-1927 .... DLB-25

Stanton, Maura 1946- DLB-120

Stapledon, Olaf 1886-1950 .... DLB-15

Star Spangled Banner Office DLB-49

Starkey, Thomas circa 1499-1538 ... DLB-132

Starkweather, David 1935- DLB-7

Statements on the Art of Poetry DLB-54

Stead, Robert J. C. 1880-1959 ...... DLB-92

Steadman, Mark 1930- DLB-6

The Stealthy School of Criticism (1871), by
  Dante Gabriel Rossetti DLB-35

Stearns, Harold E. 1891-1943 ...... DLB-4

Stedman, Edmund Clarence
  1833-1908 .... DLB-64

Steegmuller, Francis 1906-1994 ..... DLB-111

Steel, Flora Annie
  1847-1929 .... DLB-153, 156

Steele, Max 1922- Y-80

Steele, Richard 1672-1729 ...... DLB-84, 101

Steele, Timothy 1948- DLB-120

Steele, Wilbur Daniel 1886-1970 .... DLB-86

Steere, Richard circa 1643-1721 .... DLB-24

Stegner, Wallace 1909-1993 ..... DLB-9; Y-93

Stehr, Hermann 1864-1940 .... DLB-66

Steig, William 1907- DLB-61

Stieler, Caspar 1632-1707 .... DLB-164

Stein, Gertrude 1874-1946 .... DLB-4, 54, 86

Stein, Leo 1872-1947 .... DLB-4

Stein and Day Publishers DLB-46

Steinbeck, John 1902-1968 ... DLB-7, 9; DS-2

Steiner, George 1929- DLB-67

Stendhal 1783-1842 .... DLB-119

Stephen Crane: A Revaluation Virginia
  Tech Conference, 1989 .... Y-89

Stephen, Leslie 1832-1904 ...... DLB-57, 144

Stephens, Alexander H. 1812-1883 .. DLB-47

Stephens, Ann 1810-1886 .... DLB-3, 73

Stephens, Charles Asbury
  1844?-1931 .... DLB-42

Stephens, James
  1882?-1950 .... DLB-19, 153, 162

Sterling, George 1869-1926 .... DLB-54

Sterling, James 1701-1763 .... DLB-24

Sterling, John 1806-1844 .... DLB-116

Stern, Gerald 1925- DLB-105

Stern, Madeleine B. 1912- ... DLB-111, 140

Stern, Gerald, Living in Ruin DLB-105

Stern, Richard 1928- Y-87

Stern, Stewart 1922- DLB-26

Sterne, Laurence 1713-1768 .... DLB-39

Sternheim, Carl 1878-1942 ..... DLB-56, 118

Sternhold, Thomas ?-1549 and
  John Hopkins ?-1570 .... DLB-132

Stevens, Henry 1819-1886 .... DLB-140

Stevens, Wallace 1879-1955 .... DLB-54

Stevenson, Anne 1933- DLB-40

Stevenson, Lionel 1902-1973 ...... DLB-155

Stevenson, Robert Louis
  1850-1894 ... DLB-18, 57, 141, 156; DS-13

Stewart, Donald Ogden
  1894-1980 .... DLB-4, 11, 26

Stewart, Dugald 1753-1828 .... DLB-31

Stewart, George, Jr. 1848-1906 ..... DLB-99

Stewart, George R. 1895-1980 .... DLB-8

Stewart and Kidd Company DLB-46

Stewart, Randall 1896-1964 .... DLB-103

Stickney, Trumbull 1874-1904 ..... DLB-54

Stifter, Adalbert 1805-1868 .... DLB-133

Stiles, Ezra 1727-1795 .... DLB-31

Still, James 1906- DLB-9

Stirner, Max 1806-1856 .... DLB-129

Stith, William 1707-1755 .... DLB-31

Stock, Elliot [publishing house] DLB-106

Stockton, Frank R.
  1834-1902 .... DLB-42, 74; DS-13

Stoddard, Ashbel
  [publishing house] DLB-49

Stoddard, Richard Henry
  1825-1903 .... DLB-3, 64; DS-13

Stoddard, Solomon 1643-1729 .... DLB-24

Stoker, Bram 1847-1912 .... DLB-36, 70

Stokes, Frederick A., Company DLB-49

Stokes, Thomas L. 1898-1958 .... DLB-29

Stokesbury, Leon 1945- DLB-120

Stolberg, Christian Graf zu
  1748-1821 .... DLB-94

Stolberg, Friedrich Leopold Graf zu
  1750-1819 .... DLB-94

Stone, Herbert S., and Company DLB-49

Stone, Lucy 1818-1893 .... DLB-79

Stone, Melville 1848-1929 .... DLB-25

Stone, Robert 1937- DLB-152

Stone, Ruth 1915- DLB-105

Stone, Samuel 1602-1663 .... DLB-24

Stone and Kimball DLB-49

Stoppard, Tom 1937- DLB-13; Y-85

Storey, Anthony 1928- DLB-14

Storey, David 1933- DLB-13, 14

Storm, Theodor 1817-1888 .... DLB-129

Story, Thomas circa 1670-1742 ..... DLB-31

Story, William Wetmore 1819-1895 ... DLB-1

Storytelling: A Contemporary
  Renaissance Y-84

Stoughton, William 1631-1701 ...... DLB-24

Stow, John 1525-1605 .... DLB-132

Stowe, Harriet Beecher
  1811-1896 .... DLB-1, 12, 42, 74

Stowe, Leland 1899- DLB-29

Stoyanov, Dimitŭr Ivanov (see Elin Pelin)

Strachey, Lytton
  1880-1932 .... DLB-149; DS-10

Strachey, Lytton, Preface to *Eminent
  Victorians* DLB-149

Strahan and Company DLB-106

Strahan, William
  [publishing house] DLB-154

Strand, Mark 1934- DLB-5

The Strasbourg Oaths 842 .... DLB-148

Stratemeyer, Edward 1862-1930 .... DLB-42

Stratton and Barnard DLB-49

Stratton-Porter, Gene 1863-1924 ...... DS-14

Straub, Peter 1943- Y-84

Strauß, Botho 1944- DLB-124

Strauß, David Friedrich
 1808-1874 .... DLB-133

The Strawberry Hill Press DLB-154

Streatfeild, Noel 1895-1986 .... DLB-160

Street, Cecil John Charles (see Rhode, John)

Street, G. S. 1867-1936 .... DLB-135

Street and Smith DLB-49

Streeter, Edward 1891-1976 .... DLB-11

Streeter, Thomas Winthrop
 1883-1965 .... DLB-140

Stretton, Hesba 1832-1911 .... DLB-163

Stribling, T. S. 1881-1965 .... DLB-9

Der Stricker circa 1190-circa 1250 .. DLB-138

Strickland, Samuel 1804-1867 ...... DLB-99

Stringer and Townsend DLB-49

Stringer, Arthur 1874-1950 .... DLB-92

Strittmatter, Erwin 1912- DLB-69

Strode, William 1630-1645 .... DLB-126

Strother, David Hunter 1816-1888 ....DLB-3

Strouse, Jean 1945- DLB-111

Stuart, Dabney 1937- DLB-105

Stuart, Dabney, Knots into Webs: Some Autobiographical Sources DLB-105

Stuart, Jesse
 1906-1984 .... DLB-9, 48, 102; Y-84

Stuart, Lyle [publishing house] DLB-46

Stubbs, Harry Clement (see Clement, Hal)

Stubenberg, Johann Wilhelm von
 1619-1663 .... DLB-164

Studio DLB-112

The Study of Poetry (1880), by
 Matthew Arnold DLB-35

Sturgeon, Theodore
 1918-1985 .... DLB-8; Y-85

Sturges, Preston 1898-1959 .... DLB-26

"Style" (1840; revised, 1859), by
 Thomas de Quincey [excerpt] ... DLB-57

"Style" (1888), by Walter Pater DLB-57

Style (1897), by Walter Raleigh
 [excerpt] DLB-57

"Style" (1877), by T. H. Wright
 [excerpt] DLB-57

"Le Style c'est l'homme" (1892), by
 W. H. Mallock DLB-57

Styron, William 1925- DLB-2, 143; Y-80

Suárez, Mario 1925- DLB-82

Such, Peter 1939- DLB-60

Suckling, Sir John 1609-1641? ...DLB-58, 126

Suckow, Ruth 1892-1960 .... DLB-9, 102

Sudermann, Hermann 1857-1928 ... DLB-118

Sue, Eugène 1804-1857 .... DLB-119

Sue, Marie-Joseph (see Sue, Eugène)

Suggs, Simon (see Hooper, Johnson Jones)

Sukenick, Ronald 1932- Y-81

Suknaski, Andrew 1942- DLB-53

Sullivan, Alan 1868-1947 .... DLB-92

Sullivan, C. Gardner 1886-1965 .... DLB-26

Sullivan, Frank 1892-1976 .... DLB-11

Sulte, Benjamin 1841-1923 .... DLB-99

Sulzberger, Arthur Hays
 1891-1968 .... DLB-127

Sulzberger, Arthur Ochs 1926- ... DLB-127

Sulzer, Johann Georg 1720-1779 ..... DLB-97

Sumarokov, Aleksandr Petrovich
 1717-1777 .... DLB-150

Summers, Hollis 1916- DLB-6

Sumner, Henry A.
 [publishing house] DLB-49

Surtees, Robert Smith 1803-1864 .... DLB-21

A Survey of Poetry Anthologies,
 1879-1960 .... DLB-54

Surveys of the Year's Biographies

A Transit of Poets and Others: American
 Biography in 1982 .... Y-82

The Year in Literary Biography ...Y-83–Y-95

Survey of the Year's Book Publishing

The Year in Book Publishing Y-86

Survey of the Year's Children's Books

The Year in Children's Books Y-92–Y-95

Surveys of the Year's Drama

The Year in Drama
 ....Y-82–Y-85, Y-87–Y-95

The Year in London Theatre Y-92

Surveys of the Year's Fiction

The Year's Work in Fiction:
 A Survey Y-82

The Year in Fiction: A Biased View Y-83

The Year in
 Fiction Y-84–Y-86, Y-89, Y-94, Y-95

The Year in the
 Novel Y-87, Y-88, Y-90–Y-93

The Year in Short Stories Y-87

The Year in the
 Short Story Y-88, Y-90–Y-93

Survey of the Year's Literary Theory

The Year in Literary Theory Y-92–Y-93

Surveys of the Year's Poetry

The Year's Work in American
 Poetry Y-82

The Year in Poetry ... Y-83–Y-92, Y-94, Y-95

Sutherland, Efua Theodora
 1924- DLB-117

Sutherland, John 1919-1956 ....DLB-68

Sutro, Alfred 1863-1933 ....DLB-10

Swados, Harvey 1920-1972 ....DLB-2

Swain, Charles 1801-1874 ....DLB-32

Swallow Press DLB-46

Swan Sonnenschein Limited DLB-106

Swanberg, W. A. 1907- DLB-103

Swenson, May 1919-1989 ....DLB-5

Swerling, Jo 1897- DLB-44

Swift, Jonathan
 1667-1745 .... DLB-39, 95, 101

Swinburne, A. C. 1837-1909 .... DLB-35, 57

Swineshead, Richard floruit
 circa 1350 ....DLB-115

Swinnerton, Frank 1884-1982 .... DLB-34

Swisshelm, Jane Grey 1815-1884 ..... DLB-43

Swope, Herbert Bayard 1882-1958 ...DLB-25

Swords, T. and J., and Company DLB-49

Swords, Thomas 1763-1843 and
 Swords, James ?-1844 ....DLB-73

Sylvester, Josuah
 1562 or 1563 - 1618 ....DLB-121

Symonds, Emily Morse (see Paston, George)

Symonds, John Addington
 1840-1893 .... DLB-57, 144

Symons, A. J. A. 1900-1941 ....DLB-149

Symons, Arthur
 1865-1945 .... DLB-19, 57, 149

Symons, Julian
 1912-1994 .... DLB-87, 155; Y-92

Symons, Scott 1933- DLB-53

A Symposium on The Columbia History of
 the Novel Y-92

Synge, John Millington
 1871-1909 .... DLB-10, 19

Synge Summer School: J. M. Synge and the Irish Theater, Rathdrum, County Wiclow, Ireland Y-93

Syrett, Netta 1865-1943 ....DLB-135

T

Taban lo Liyong 1939?- DLB-125

Taché, Joseph-Charles 1820-1894 ....DLB-99

Tafolla, Carmen 1951- DLB-82

Taggard, Genevieve 1894-1948 ......DLB-45

Tagger, Theodor (see Bruckner, Ferdinand)

Tait, J. Selwin, and Sons DLB-49

Tait's Edinburgh Magazine 1832-1861 ....DLB-110

The Takarazaka Revue Company Y-91

Talander (see Bohse, August)

Tallent, Elizabeth 1954- DLB-130

Talvj 1797-1870 ....DLB-59, 133

Taradash, Daniel 1913- DLB-44

Tarbell, Ida M. 1857-1944 ....DLB-47

Tardivel, Jules-Paul 1851-1905 ......DLB-99

Targan, Barry 1932- DLB-130

Tarkington, Booth 1869-1946 ....DLB-9, 102

Tashlin, Frank 1913-1972 ....DLB-44

Tate, Allen 1899-1979 .... DLB-4, 45, 63

Tate, James 1943- DLB-5, 169

Tate, Nahum circa 1652-1715 ....DLB-80

Tatian circa 830 ....DLB-148

Tavčar, Ivan 1851-1923 ....DLB-147

Taylor, Ann 1782-1866 ....DLB-163

Taylor, Bayard 1825-1878 ....DLB-3

Taylor, Bert Leston 1866-1921 ......DLB-25

Taylor, Charles H. 1846-1921 ......DLB-25

Taylor, Edward circa 1642-1729 .....DLB-24

Taylor, Elizabeth 1912-1975 ......DLB-139

Taylor, Henry 1942- DLB-5

Taylor, Sir Henry 1800-1886 ....DLB-32

Taylor, Jane 1783-1824 ....DLB-163

Taylor, Jeremy circa 1613-1667 .....DLB-151

Taylor, John 1577 or 1578 - 1653 ....DLB-121

Taylor, Mildred D. ?- DLB-52

Taylor, Peter 1917-1994 ....Y-81, Y-94

Taylor, William, and Company DLB-49

Taylor-Made Shakespeare? Or Is "Shall I Die?" the Long-Lost Text of Bottom's Dream? Y-85

Teasdale, Sara 1884-1933 ....DLB-45

The Tea-Table (1725), by Eliza Haywood [excerpt] DLB-39

Telles, Lygia Fagundes 1924- DLB-113

Temple, Sir William 1628-1699 ....DLB-101

Tenn, William 1919- DLB-8

Tennant, Emma 1937- DLB-14

Tenney, Tabitha Gilman 1762-1837 ....DLB-37

Tennyson, Alfred 1809-1892 ....DLB-32

Tennyson, Frederick 1807-1898 ....DLB-32

Terhune, Albert Payson 1872-1942 ...DLB-9

Terhune, Mary Virginia 1830-1922 ....DS-13

Terry, Megan 1932- DLB-7

Terson, Peter 1932- DLB-13

Tesich, Steve 1943- Y-83

Tessa, Delio 1886-1939 ....DLB-114

Testori, Giovanni 1923-1993 ......DLB-128

Tey, Josephine 1896?-1952 ....DLB-77

Thacher, James 1754-1844 ....DLB-37

Thackeray, William Makepeace 1811-1863 ....DLB-21, 55, 159, 163

Thames and Hudson Limited DLB-112

Thanet, Octave (see French, Alice)

The Theater in Shakespeare's Time DLB-62

The Theatre Guild DLB-7

Thegan and the Astronomer flourished circa 850 ....DLB-148

Thelwall, John 1764-1834 ...... DLB-93, 158

Theodulf circa 760-circa 821 ......DLB-148

Theriault, Yves 1915-1983 ....DLB-88

Thério, Adrien 1925- DLB-53

Theroux, Paul 1941- DLB-2

Thibaudeau, Colleen 1925- DLB-88

Thielen, Benedict 1903-1965 ......DLB-102

Thiong'o Ngugi wa (see Ngugi wa Thiong'o)

Third-Generation Minor Poets of the Seventeenth Century DLB-131

Thoma, Ludwig 1867-1921 .... DLB-66

Thoma, Richard 1902- DLB-4

Thomas, Audrey 1935- DLB-60

Thomas, D. M. 1935- DLB-40

Thomas, Dylan 1914-1953 .... DLB-13, 20, 139

Thomas, Edward 1878-1917 .... DLB-19, 98, 156

Thomas, Gwyn 1913-1981 ....DLB-15

Thomas, Isaiah 1750-1831 ....DLB-43, 73

Thomas, Isaiah [publishing house] ...DLB-49

Thomas, Johann 1624-1679 ....DLB-168

Thomas, John 1900-1932 ....DLB-4

Thomas, Joyce Carol 1938- DLB-33

Thomas, Lorenzo 1944- DLB-41

Thomas, R. S. 1915- DLB-27

Thomasîn von Zerclære circa 1186-circa 1259 ....DLB-138

Thomasius, Christian 1655-1728 ...DLB-168

Thompson, David 1770-1857 ....DLB-99

Thompson, Dorothy 1893-1961 .....DLB-29

Thompson, Francis 1859-1907 ......DLB-19

Thompson, George Selden (see Selden, George)

Thompson, John 1938-1976 ....DLB-60

Thompson, John R. 1823-1873 ....DLB-3, 73

Thompson, Lawrance 1906-1973 ...DLB-103

Thompson, Maurice 1844-1901 ....DLB-71, 74

Thompson, Ruth Plumly 1891-1976 ....DLB-22

Thompson, Thomas Phillips 1843-1933 ....DLB-99

Thompson, William 1775-1833 .....DLB-158

Thompson, William Tappan 1812-1882 ....DLB-3, 11

Thomson, Edward William 1849-1924 ....DLB-92

Thomson, James 1700-1748 ....DLB-95

Thomson, James 1834-1882 ....DLB-35

Thomson, Mortimer 1831-1875 .....DLB-11

Thoreau, Henry David 1817-1862 ....DLB-1

Thorpe, Thomas Bangs 1815-1878 ....DLB-3, 11

Thoughts on Poetry and Its Varieties (1833), by John Stuart Mill DLB-32

Thrale, Hester Lynch (see Piozzi, Hester Lynch [Thrale])

Thümmel, Moritz August von 1738-1817 ....DLB-97

Thurber, James 1894-1961 .... DLB-4, 11, 22, 102

Thurman, Wallace 1902-1934 ....DLB-51

Thwaite, Anthony 1930- DLB-40

397

Thwaites, Reuben Gold 1853-1913 ... DLB-47

Ticknor, George 1791-1871 ... DLB-1, 59, 140

Ticknor and Fields ... DLB-49

Ticknor and Fields (revived) ... DLB-46

Tieck, Ludwig 1773-1853 ... DLB-90

Tietjens, Eunice 1884-1944 ... DLB-54

Tilney, Edmund circa 1536-1610 ... DLB-136

Tilt, Charles [publishing house] ... DLB-106

Tilton, J. E., and Company ... DLB-49

Time and Western Man (1927), by Wyndham Lewis [excerpts] ... DLB-36

Time-Life Books ... DLB-46

Times Books ... DLB-46

Timothy, Peter circa 1725-1782 ... DLB-43

Timrod, Henry 1828-1867 ... DLB-3

Tinker, Chauncey Brewster 1876-1963 ... DLB-140

Tinsley Brothers ... DLB-106

Tiptree, James, Jr. 1915-1987 ... DLB-8

Titus, Edward William 1870-1952 ... DLB-4

Tlali, Miriam 1933- ... DLB-157

Todd, Barbara Euphan 1890-1976 ... DLB-160

Toklas, Alice B. 1877-1967 ... DLB-4

Tolkien, J. R. R. 1892-1973 ... DLB-15, 160

Toller, Ernst 1893-1939 ... DLB-124

Tollet, Elizabeth 1694-1754 ... DLB-95

Tolson, Melvin B. 1898-1966 ... DLB-48, 76

Tom Jones (1749), by Henry Fielding [excerpt] ... DLB-39

Tomalin, Claire 1933- ... DLB-155

Tomlinson, Charles 1927- ... DLB-40

Tomlinson, H. M. 1873-1958 ... DLB-36, 100

Tompkins, Abel [publishing house] ... DLB-49

Tompson, Benjamin 1642-1714 ... DLB-24

Tonks, Rosemary 1932- ... DLB-14

Tonna, Charlotte Elizabeth 1790-1846 ... DLB-163

Toole, John Kennedy 1937-1969 ... Y-81

Toomer, Jean 1894-1967 ... DLB-45, 51

Tor Books ... DLB-46

Torberg, Friedrich 1908-1979 ... DLB-85

Torrence, Ridgely 1874-1950 ... DLB-54

Torres-Metzger, Joseph V. 1933- ... DLB-122

Toth, Susan Allen 1940- ... Y-86

Tough-Guy Literature ... DLB-9

Touré, Askia Muhammad 1938- ... DLB-41

Tourgée, Albion W. 1838-1905 ... DLB-79

Tourneur, Cyril circa 1580-1626 ... DLB-58

Tournier, Michel 1924- ... DLB-83

Tousey, Frank [publishing house] ... DLB-49

Tower Publications ... DLB-46

Towne, Benjamin circa 1740-1793 ... DLB-43

Towne, Robert 1936- ... DLB-44

The Townely Plays fifteenth and sixteenth centuries ... DLB-146

Townshend, Aurelian by 1583 - circa 1651 ... DLB-121

Tracy, Honor 1913- ... DLB-15

Traherne, Thomas 1637?-1674 ... DLB-131

Traill, Catharine Parr 1802-1899 ... DLB-99

Train, Arthur 1875-1945 ... DLB-86

The Transatlantic Publishing Company ... DLB-49

Transcendentalists, American ... DS-5

Translators of the Twelfth Century: Literary Issues Raised and Impact Created ... DLB-115

Travel Writing, 1837-1875 ... DLB-166

Traven, B. 1882? or 1890?-1969? ... DLB-9, 56

Travers, Ben 1886-1980 ... DLB-10

Travers, P. L. (Pamela Lyndon) 1899- ... DLB-160

Trediakovsky, Vasilii Kirillovich 1703-1769 ... DLB-150

Treece, Henry 1911-1966 ... DLB-160

Trejo, Ernesto 1950- ... DLB-122

Trelawny, Edward John 1792-1881 ... DLB-110, 116, 144

Tremain, Rose 1943- ... DLB-14

Tremblay, Michel 1942- ... DLB-60

Trends in Twentieth-Century Mass Market Publishing ... DLB-46

Trent, William P. 1862-1939 ... DLB-47

Trescot, William Henry 1822-1898 ... DLB-30

Trevelyan, Sir George Otto 1838-1928 ... DLB-144

Trevisa, John circa 1342-circa 1402 ... DLB-146

Trevor, William 1928- ... DLB-14, 139

Trierer Floyris circa 1170-1180 ... DLB-138

Trilling, Lionel 1905-1975 ... DLB-28, 63

Trilussa 1871-1950 ... DLB-114

Trimmer, Sarah 1741-1810 ... DLB-158

Triolet, Elsa 1896-1970 ... DLB-72

Tripp, John 1927- ... DLB-40

Trocchi, Alexander 1925- ... DLB-15

Trollope, Anthony 1815-1882 ... DLB-21, 57, 159

Trollope, Frances 1779-1863 ... DLB-21, 166

Troop, Elizabeth 1931- ... DLB-14

Trotter, Catharine 1679-1749 ... DLB-84

Trotti, Lamar 1898-1952 ... DLB-44

Trottier, Pierre 1925- ... DLB-60

Troupe, Quincy Thomas, Jr. 1943- ... DLB-41

Trow, John F., and Company ... DLB-49

Truillier-Lacombe, Joseph-Patrice 1807-1863 ... DLB-99

Trumbo, Dalton 1905-1976 ... DLB-26

Trumbull, Benjamin 1735-1820 ... DLB-30

Trumbull, John 1750-1831 ... DLB-31

Tscherning, Andreas 1611-1659 ... DLB-164

T. S. Eliot Centennial ... Y-88

Tucholsky, Kurt 1890-1935 ... DLB-56

Tucker, Charlotte Maria 1821-1893 ... DLB-163

Tucker, George 1775-1861 ... DLB-3, 30

Tucker, Nathaniel Beverley 1784-1851 ... DLB-3

Tucker, St. George 1752-1827 ... DLB-37

Tuckerman, Henry Theodore 1813-1871 ... DLB-64

Tunis, John R. 1889-1975 ... DLB-22

Tunstall, Cuthbert 1474-1559 ... DLB-132

Tuohy, Frank 1925- ... DLB-14, 139

Tupper, Martin F. 1810-1889 ... DLB-32

Turbyfill, Mark 1896- ... DLB-45

Turco, Lewis 1934- ... Y-84

Turnbull, Andrew 1921-1970 ... DLB-103

Turnbull, Gael 1928- ... DLB-40

Turner, Arlin 1909-1980 ... DLB-103

Turner, Charles (Tennyson) 1808-1879 ... DLB-32

Turner, Frederick 1943- ... DLB-40

Turner, Frederick Jackson 1861-1932 ... DLB-17

Turner, Joseph Addison 1826-1868 .... DLB-79

Turpin, Waters Edward 1910-1968 .... DLB-51

Turrini, Peter 1944- DLB-124

Tutuola, Amos 1920- DLB-125

Twain, Mark (see Clemens, Samuel Langhorne)

The 'Twenties and Berlin, by Alex Natan DLB-66

Tyler, Anne 1941- DLB-6, 143; Y-82

Tyler, Moses Coit 1835-1900 ....DLB-47, 64

Tyler, Royall 1757-1826 .... DLB-37

Tylor, Edward Burnett 1832-1917 ...DLB-57

Tynan, Katharine 1861-1931 .... DLB-153

Tyndale, William circa 1494-1536 .... DLB-132

U

Udall, Nicholas 1504-1556 .... DLB-62

Uhland, Ludwig 1787-1862 .... DLB-90

Uhse, Bodo 1904-1963 .... DLB-69

Ujević, Augustin ("Tin") 1891-1955 .... DLB-147

Ulenhart, Niclas flourished circa 1600 .... DLB-164

Ulibarrí, Sabine R. 1919- DLB-82

Ulica, Jorge 1870-1926 .... DLB-82

Ulizio, B. George 1889-1969 .... DLB-140

Ulrich von Liechtenstein circa 1200-circa 1275 .... DLB-138

Ulrich von Zatzikhoven before 1194-after 1214 .... DLB-138

Unamuno, Miguel de 1864-1936 ....DLB-108

Under the Microscope (1872), by A. C. Swinburne DLB-35

Unger, Friederike Helene 1741-1813 .... DLB-94

Ungaretti, Giuseppe 1888-1970 ..... DLB-114

United States Book Company DLB-49

Universal Publishing and Distributing Corporation DLB-46

The University of Iowa Writers' Workshop Golden Jubilee Y-86

The University of South Carolina Press Y-94

University of Wales Press DLB-112

"The Unknown Public" (1858), by Wilkie Collins [excerpt] DLB-57

Unruh, Fritz von 1885-1970 ....DLB-56, 118

Unspeakable Practices II: The Festival of Vanguard Narrative at Brown University Y-93

Unwin, T. Fisher [publishing house] DLB-106

Upchurch, Boyd B. (see Boyd, John)

Updike, John 1932- ... DLB-2, 5, 143; Y-80, 82; DS-3

Upton, Bertha 1849-1912 .... DLB-141

Upton, Charles 1948- DLB-16

Upton, Florence K. 1873-1922 ..... DLB-141

Upward, Allen 1863-1926 .... DLB-36

Urista, Alberto Baltazar (see Alurista)

Urzidil, Johannes 1896-1976 .... DLB-85

Urquhart, Fred 1912- DLB-139

The Uses of Facsimile Y-90

Usk, Thomas died 1388 .... DLB-146

Uslar Pietri, Arturo 1906- DLB-113

Ustinov, Peter 1921- DLB-13

Uttley, Alison 1884-1976 .... DLB-160

Uz, Johann Peter 1720-1796 .... DLB-97

V

Vac, Bertrand 1914- DLB-88

Vail, Laurence 1891-1968 .... DLB-4

Vailland, Roger 1907-1965 .... DLB-83

Vajda, Ernest 1887-1954 .... DLB-44

Valdés, Gina 1943- DLB-122

Valdez, Luis Miguel 1940- DLB-122

Valduga, Patrizia 1953- DLB-128

Valente, José Angel 1929- DLB-108

Valenzuela, Luisa 1938- DLB-113

Valeri, Diego 1887-1976 .... DLB-128

Valgardson, W. D. 1939- DLB-60

Valle, Víctor Manuel 1950- DLB-122

Valle-Inclán, Ramón del 1866-1936 .... DLB-134

Vallejo, Armando 1949- DLB-122

Vallès, Jules 1832-1885 .... DLB-123

Vallette, Marguerite Eymery (see Rachilde)

Valverde, José María 1926- DLB-108

Van Allsburg, Chris 1949- DLB-61

Van Anda, Carr 1864-1945 .... DLB-25

Van Doren, Mark 1894-1972 ...... DLB-45

van Druten, John 1901-1957 ...... DLB-10

Van Duyn, Mona 1921- DLB-5

Van Dyke, Henry 1852-1933 .... DLB-71; DS-13

Van Dyke, Henry 1928- DLB-33

van Itallie, Jean-Claude 1936- DLB-7

Van Rensselaer, Mariana Griswold 1851-1934 .... DLB-47

Van Rensselaer, Mrs. Schuyler (see Van Rensselaer, Mariana Griswold)

Van Vechten, Carl 1880-1964 ......DLB-4, 9

van Vogt, A. E. 1912- DLB-8

Vanbrugh, Sir John 1664-1726 ... DLB-80

Vance, Jack 1916?- DLB-8

Vane, Sutton 1888-1963 .... DLB-10

Vanguard Press DLB-46

Vann, Robert L. 1879-1940 .... DLB-29

Vargas, Llosa, Mario 1936- DLB-145

Varley, John 1947- Y-81

Varnhagen von Ense, Karl August 1785-1858 .... DLB-90

Varnhagen von Ense, Rahel 1771-1833 .... DLB-90

Vásquez Montalbán, Manuel 1939- DLB-134

Vassa, Gustavus (see Equiano, Olaudah)

Vassalli, Sebastiano 1941- DLB-128

Vaughan, Henry 1621-1695 .... DLB-131

Vaughan, Thomas 1621-1666 ...... DLB-131

Vaux, Thomas, Lord 1509-1556 .... DLB-132

Vazov, Ivan 1850-1921 .... DLB-147

Vega, Janine Pommy 1942- DLB-16

Veiller, Anthony 1903-1965 .... DLB-44

Velásquez-Trevino, Gloria 1949- DLB-122

Veloz Maggiolo, Marcio 1936- DLB-145

Venegas, Daniel ?-? DLB-82

Vergil, Polydore circa 1470-1555 ...DLB-132

Veríssimo, Erico 1905-1975 .... DLB-145

Verne, Jules 1828-1905 .... DLB-123

Verplanck, Gulian C. 1786-1870 ..... DLB-59

Very, Jones 1813-1880 .... DLB-1

Vian, Boris 1920-1959 .... DLB-72

Vickers, Roy 1888?-1965 .... DLB-77

Victoria 1819-1901 .... DLB-55

Victoria Press DLB-106

Vidal, Gore 1925- DLB-6, 152

Viebig, Clara 1860-1952 .... DLB-66

Viereck, George Sylvester 1884-1962 .... DLB-54

Viereck, Peter 1916- DLB-5

Viets, Roger 1738-1811 .... DLB-99

Viewpoint: Politics and Performance, by David Edgar DLB-13

Vigil-Piñon, Evangelina 1949- ... DLB-122

Vigneault, Gilles 1928- DLB-60

Vigny, Alfred de 1797-1863 .... DLB-119

Vigolo, Giorgio 1894-1983 .... DLB-114

The Viking Press DLB-46

Villanueva, Alma Luz 1944- DLB-122

Villanueva, Tino 1941- DLB-82

Villard, Henry 1835-1900 .... DLB-23

Villard, Oswald Garrison 1872-1949 .... DLB-25, 91

Villarreal, José Antonio 1924- DLB-82

Villegas de Magnón, Leonor 1876-1955 .... DLB-122

Villemaire, Yolande 1949- DLB-60

Villena, Luis Antonio de 1951- ... DLB-134

Villiers de l'Isle-Adam, Jean-Marie Mathias Philippe-Auguste, Comte de 1838-1889 .... DLB-123

Villiers, George, Second Duke of Buckingham 1628-1687 ...... DLB-80

Vine Press DLB-112

Viorst, Judith ?- DLB-52

Vipont, Elfrida (Elfrida Vipont Foulds, Charles Vipont) 1902-1992 .... DLB-160

Viramontes, Helena María 1954- DLB-122

Vischer, Friedrich Theodor 1807-1887 .... DLB-133

Vivanco, Luis Felipe 1907-1975 .... DLB-108

Viviani, Cesare 1947- DLB-128

Vizetelly and Company DLB-106

Voaden, Herman 1903- DLB-88

Voigt, Ellen Bryant 1943- DLB-120

Vojnović, Ivo 1857-1929 .... DLB-147

Volkoff, Vladimir 1932- DLB-83

Volland, P. F., Company DLB-46

von der Grün, Max 1926- DLB-75

Vonnegut, Kurt 1922- DLB-2, 8, 152; Y-80; DS-3

Voranc, Prežihov 1893-1950 ..... DLB-147

Voß, Johann Heinrich 1751-1826 ... DLB-90

Vroman, Mary Elizabeth circa 1924-1967 .... DLB-33

W

Wace, Robert ("Maistre") circa 1100-circa 1175 .... DLB-146

Wackenroder, Wilhelm Heinrich 1773-1798 .... DLB-90

Wackernagel, Wilhelm 1806-1869 .... DLB-133

Waddington, Miriam 1917- DLB-68

Wade, Henry 1887-1969 .... DLB-77

Wagenknecht, Edward 1900- DLB-103

Wagner, Heinrich Leopold 1747-1779 .... DLB-94

Wagner, Henry R. 1862-1957 ..... DLB-140

Wagner, Richard 1813-1883 .... DLB-129

Wagoner, David 1926- DLB-5

Wah, Fred 1939- DLB-60

Waiblinger, Wilhelm 1804-1830 ..... DLB-90

Wain, John 1925-1994 .... DLB-15, 27, 139, 155

Wainwright, Jeffrey 1944- DLB-40

Waite, Peirce and Company DLB-49

Wakoski, Diane 1937- DLB-5

Walahfrid Strabo circa 808-849 .... DLB-148

Walck, Henry Z. DLB-46

Walcott, Derek 1930- DLB-117; Y-81, 92

Waldman, Anne 1945- DLB-16

Waldrop, Rosmarie 1935- DLB-169

Walker, Alice 1944- DLB-6, 33, 143

Walker, George F. 1947- DLB-60

Walker, Joseph A. 1935- DLB-38

Walker, Margaret 1915- DLB-76, 152

Walker, Ted 1934- DLB-40

Walker and Company DLB-49

Walker, Evans and Cogswell Company DLB-49

Walker, John Brisben 1847-1931 ... DLB-79

Wallace, Dewitt 1889-1981 and Lila Acheson Wallace 1889-1984 .... DLB-137

Wallace, Edgar 1875-1932 .... DLB-70

Wallace, Lila Acheson (see Wallace, Dewitt, and Lila Acheson Wallace)

Wallant, Edward Lewis 1926-1962 .... DLB-2, 28, 143

Waller, Edmund 1606-1687 .... DLB-126

Walpole, Horace 1717-1797 .... DLB-39, 104

Walpole, Hugh 1884-1941 .... DLB-34

Walrond, Eric 1898-1966 .... DLB-51

Walser, Martin 1927- DLB-75, 124

Walser, Robert 1878-1956 .... DLB-66

Walsh, Ernest 1895-1926 .... DLB-4, 45

Walsh, Robert 1784-1859 .... DLB-59

Waltharius circa 825 .... DLB-148

Walters, Henry 1848-1931 .... DLB-140

Walther von der Vogelweide circa 1170-circa 1230 .... DLB-138

Walton, Izaak 1593-1683 .... DLB-151

Wambaugh, Joseph 1937- DLB-6; Y-83

Waniek, Marilyn Nelson 1946- ...DLB-120

Warburton, William 1698-1779 .... DLB-104

Ward, Aileen 1919- DLB-111

Ward, Artemus (see Browne, Charles Farrar)

Ward, Arthur Henry Sarsfield (see Rohmer, Sax)

Ward, Douglas Turner 1930- ... DLB-7, 38

Ward, Lynd 1905-1985 .... DLB-22

Ward, Lock and Company DLB-106

Ward, Mrs. Humphry 1851-1920 .... DLB-18

Ward, Nathaniel circa 1578-1652 ....DLB-24

Ward, Theodore 1902-1983 .... DLB-76

Wardle, Ralph 1909-1988 .... DLB-103

Ware, William 1797-1852 .... DLB-1

Warne, Frederick, and Company [U.S.] DLB-49

Warne, Frederick, and Company [U.K.] DLB-106

Warner, Charles Dudley 1829-1900 .... DLB-64

Warner, Rex 1905- DLB-15

Warner, Susan Bogert 1819-1885 .... DLB-3, 42

Warner, Sylvia Townsend 1893-1978 .... DLB-34, 139

Warner Books DLB-46

Warr, Bertram 1917-1943 .... DLB-88

Warren, John Byrne Leicester (see De Tabley, Lord)

Warren, Lella 1899-1982 .... Y-83

Warren, Mercy Otis 1728-1814 ......DLB-31

Warren, Robert Penn 1905-1989 ......DLB-2, 48, 152; Y-80, 89

Die Wartburgkrieg circa 1230-circa 1280 .... DLB-138

Warton, Joseph 1722-1800 .... DLB-104, 109

Warton, Thomas 1728-1790 ... DLB-104, 109

Washington, George 1732-1799 ..... DLB-31
Wassermann, Jakob 1873-1934 ...... DLB-66
Wasson, David Atwood 1823-1887 .... DLB-1
Waterhouse, Keith 1929- DLB-13, 15
Waterman, Andrew 1940- DLB-40
Waters, Frank 1902- Y-86
Waters, Michael 1949- DLB-120
Watkins, Tobias 1780-1855 .... DLB-73
Watkins, Vernon 1906-1967 .... DLB-20
Watmough, David 1926- DLB-53
Watson, James Wreford (see Wreford, James)
Watson, John 1850-1907 .... DLB-156
Watson, Sheila 1909- DLB-60
Watson, Thomas 1545?-1592 ...... DLB-132
Watson, Wilfred 1911- DLB-60
Watt, W. J., and Company DLB-46
Watterson, Henry 1840-1921 ...... DLB-25
Watts, Alan 1915-1973 .... DLB-16
Watts, Franklin [publishing house] ... DLB-46
Watts, Isaac 1674-1748 .... DLB-95
Waugh, Auberon 1939- DLB-14
Waugh, Evelyn 1903-1966 ...... DLB-15, 162
Way and Williams DLB-49
Wayman, Tom 1945- DLB-53
Weatherly, Tom 1942- DLB-41
Weaver, Gordon 1937- DLB-130
Weaver, Robert 1921- DLB-88
Webb, Frank J. ?-? DLB-50
Webb, James Watson 1802-1884 ..... DLB-43
Webb, Mary 1881-1927 .... DLB-34
Webb, Phyllis 1927- DLB-53
Webb, Walter Prescott 1888-1963 ... DLB-17
Webbe, William ?-1591 .... DLB-132
Webster, Augusta 1837-1894 .... DLB-35
Webster, Charles L.,
    and Company DLB-49
Webster, John
    1579 or 1580-1634? DLB-58
Webster, Noah
    1758-1843 .... DLB-1, 37, 42, 43, 73
Weckherlin, Georg Rodolf
    1584-1653 .... DLB-164
Wedekind, Frank 1864-1918 .... DLB-118
Weeks, Edward Augustus, Jr.
    1898-1989 .... DLB-137

Weems, Mason Locke
    1759-1825 .... DLB-30, 37, 42
Weerth, Georg 1822-1856 .... DLB-129
Weidenfeld and Nicolson DLB-112
Weidman, Jerome 1913- DLB-28
Weigl, Bruce 1949- DLB-120
Weinbaum, Stanley Grauman
    1902-1935 .... DLB-8
Weintraub, Stanley 1929- DLB-111
Weise, Christian 1642-1708 .... DLB-168
Weisenborn, Gunther
    1902-1969 .... DLB-69, 124
Weiß, Ernst 1882-1940 .... DLB-81
Weiss, John 1818-1879 .... DLB-1
Weiss, Peter 1916-1982 .... DLB-69, 124
Weiss, Theodore 1916- DLB-5
Weisse, Christian Felix 1726-1804 ... DLB-97
Weitling, Wilhelm 1808-1871 ...... DLB-129
Welch, Lew 1926-1971? DLB-16
Weldon, Fay 1931- DLB-14
Wellek, René 1903- DLB-63
Wells, Carolyn 1862-1942 .... DLB-11
Wells, Charles Jeremiah
    circa 1800-1879 .... DLB-32
Wells, Gabriel 1862-1946 .... DLB-140
Wells, H. G. 1866-1946 ..... DLB-34, 70, 156
Wells, Robert 1947- DLB-40
Wells-Barnett, Ida B. 1862-1931 ..... DLB-23
Welty, Eudora
    1909- ... DLB-2, 102, 143; Y-87; DS-12
Wendell, Barrett 1855-1921 .... DLB-71
Wentworth, Patricia 1878-1961 ..... DLB-77
Werder, Diederich von dem
    1584-1657 .... DLB-164
Werfel, Franz 1890-1945 .... DLB-81, 124
The Werner Company DLB-49
Werner, Zacharias 1768-1823 .... DLB-94
Wersba, Barbara 1932- DLB-52
Wescott, Glenway 1901- DLB-4, 9, 102
Wesker, Arnold 1932- DLB-13
Wesley, Charles 1707-1788 .... DLB-95
Wesley, John 1703-1791 .... DLB-104
Wesley, Richard 1945- DLB-38
Wessels, A., and Company DLB-46
Wessobrunner Gebet
    circa 787-815 .... DLB-148
West, Anthony 1914-1988 .... DLB-15

West, Dorothy 1907- DLB-76
West, Jessamyn 1902-1984 ...... DLB-6; Y-84
West, Mae 1892-1980 .... DLB-44
West, Nathanael 1903-1940 ..... DLB-4, 9, 28
West, Paul 1930- DLB-14
West, Rebecca 1892-1983 ..... DLB-36; Y-83
West and Johnson DLB-49
Western Publishing Company DLB-46
The Westminster Review 1824-1914 ... DLB-110
Wetherald, Agnes Ethelwyn
    1857-1940 .... DLB-99
Wetherell, Elizabeth
    (see Warner, Susan Bogert)
Wetzel, Friedrich Gottlob
    1779-1819 .... DLB-90
Weyman, Stanley J.
    1855-1928 .... DLB-141, 156
Wezel, Johann Karl 1747-1819 ...... DLB-94
Whalen, Philip 1923- DLB-16
Whalley, George 1915-1983 .... DLB-88
Wharton, Edith
    1862-1937 ...... DLB-4, 9, 12, 78; DS-13
Wharton, William 1920s?- Y-80
Whately, Mary Louisa
    1824-1889 .... DLB-166
What's Really Wrong With Bestseller
    Lists Y-84
Wheatley, Dennis Yates
    1897-1977 .... DLB-77
Wheatley, Phillis
    circa 1754-1784 .... DLB-31, 50
Wheeler, Anna Doyle
    1785-1848? DLB-158
Wheeler, Charles Stearns
    1816-1843 .... DLB-1
Wheeler, Monroe 1900-1988 .... DLB-4
Wheelock, John Hall 1886-1978 ..... DLB-45
Wheelwright, John
    circa 1592-1679 .... DLB-24
Wheelwright, J. B. 1897-1940 .... DLB-45
Whetstone, Colonel Pete
    (see Noland, C. F. M.)
Whetstone, George 1550-1587 ..... DLB-136
Whicher, Stephen E. 1915-1961 .... DLB-111
Whipple, Edwin Percy
    1819-1886 .... DLB-1, 64
Whitaker, Alexander 1585-1617 ..... DLB-24
Whitaker, Daniel K. 1801-1881 ..... DLB-73
Whitcher, Frances Miriam
    1814-1852 .... DLB-11

Cumulative Index

White, Andrew 1579-1656 .... DLB-24

White, Andrew Dickson
1832-1918 .... DLB-47

White, E. B. 1899-1985 .... DLB-11, 22

White, Edgar B. 1947- DLB-38

White, Ethel Lina 1887-1944 .... DLB-77

White, Henry Kirke 1785-1806 ..... DLB-96

White, Horace 1834-1916 .... DLB-23

White, Phyllis Dorothy James
(see James, P. D.)

White, Richard Grant 1821-1885 .... DLB-64

White, T. H. 1906-1964 .... DLB-160

White, Walter 1893-1955 .... DLB-51

White, William, and Company DLB-49

White, William Allen
1868-1944 .... DLB-9, 25

White, William Anthony Parker (see Boucher, Anthony)

White, William Hale (see Rutherford, Mark)

Whitechurch, Victor L.
1868-1933 .... DLB-70

Whitehead, Alfred North
1861-1947 .... DLB-100

Whitehead, James 1936- Y-81

Whitehead, William
1715-1785 .... DLB-84, 109

Whitfield, James Monroe
1822-1871 .... DLB-50

Whitgift, John circa 1533-1604 ..... DLB-132

Whiting, John 1917-1963 .... DLB-13

Whiting, Samuel 1597-1679 .... DLB-24

Whitlock, Brand 1869-1934 .... DLB-12

Whitman, Albert, and Company DLB-46

Whitman, Albery Allson
1851-1901 .... DLB-50

Whitman, Alden 1913-1990 .... Y-91

Whitman, Sarah Helen (Power)
1803-1878 .... DLB-1

Whitman, Walt 1819-1892 .... DLB-3, 64

Whitman Publishing Company DLB-46

Whitney, Geoffrey
1548 or 1552?-1601 .... DLB-136

Whitney, Isabella
flourished 1566-1573 .... DLB-136

Whitney, John Hay 1904-1982 ..... DLB-127

Whittemore, Reed 1919- DLB-5

Whittier, John Greenleaf 1807-1892 ... DLB-1

Whittlesey House DLB-46

Who Runs American Literature? Y-94

Wideman, John Edgar
1941- DLB-33, 143

Widener, Harry Elkins 1885-1912 ... DLB-140

Wiebe, Rudy 1934- DLB-60

Wiechert, Ernst 1887-1950 .... DLB-56

Wied, Martina 1882-1957 .... DLB-85

Wieland, Christoph Martin
1733-1813 .... DLB-97

Wienbarg, Ludolf 1802-1872 .... DLB-133

Wieners, John 1934- DLB-16

Wier, Ester 1910- DLB-52

Wiesel, Elie 1928- DLB-83; Y-87

Wiggin, Kate Douglas 1856-1923 .... DLB-42

Wigglesworth, Michael 1631-1705 ... DLB-24

Wilberforce, William 1759-1833 .... DLB-158

Wilbrandt, Adolf 1837-1911 .... DLB-129

Wilbur, Richard 1921- DLB-5, 169

Wild, Peter 1940- DLB-5

Wilde, Oscar
1854-1900 ... DLB-10, 19, 34, 57, 141, 156

Wilde, Richard Henry
1789-1847 .... DLB-3, 59

Wilde, W. A., Company DLB-49

Wilder, Billy 1906- DLB-26

Wilder, Laura Ingalls 1867-1957 .... DLB-22

Wilder, Thornton 1897-1975 .... DLB-4, 7, 9

Wildgans, Anton 1881-1932 .... DLB-118

Wiley, Bell Irvin 1906-1980 .... DLB-17

Wiley, John, and Sons DLB-49

Wilhelm, Kate 1928- DLB-8

Wilkes, George 1817-1885 .... DLB-79

Wilkinson, Anne 1910-1961 .... DLB-88

Wilkinson, Sylvia 1940- Y-86

Wilkinson, William Cleaver
1833-1920 .... DLB-71

Willard, Barbara 1909-1994 .... DLB-161

Willard, L. [publishing house] DLB-49

Willard, Nancy 1936- DLB-5, 52

Willard, Samuel 1640-1707 .... DLB-24

William of Auvergne 1190-1249 .... DLB-115

William of Conches
circa 1090-circa 1154 .... DLB-115

William of Ockham
circa 1285-1347 .... DLB-115

William of Sherwood
1200/1205 - 1266/1271 .... DLB-115

The William Chavrat American Fiction
Collection at the Ohio State University
Libraries Y-92

Williams, A., and Company DLB-49

Williams, Ben Ames 1889-1953 ..... DLB-102

Williams, C. K. 1936- DLB-5

Williams, Chancellor 1905- DLB-76

Williams, Charles
1886-1945 .... DLB-100, 153

Williams, Denis 1923- DLB-117

Williams, Emlyn 1905- DLB-10, 77

Williams, Garth 1912- DLB-22

Williams, George Washington
1849-1891 .... DLB-47

Williams, Heathcote 1941- DLB-13

Williams, Helen Maria
1761-1827 .... DLB-158

Williams, Hugo 1942- DLB-40

Williams, Isaac 1802-1865 .... DLB-32

Williams, Joan 1928- DLB-6

Williams, John A. 1925- DLB-2, 33

Williams, John E. 1922-1994 .... DLB-6

Williams, Jonathan 1929- DLB-5

Williams, Miller 1930- DLB-105

Williams, Raymond 1921- DLB-14

Williams, Roger circa 1603-1683 ..... DLB-24

Williams, Samm-Art 1946- DLB-38

Williams, Sherley Anne 1944- DLB-41

Williams, T. Harry 1909-1979 .... DLB-17

Williams, Tennessee
1911-1983 .... DLB-7; Y-83; DS-4

Williams, Ursula Moray 1911- DLB-160

Williams, Valentine 1883-1946 ...... DLB-77

Williams, William Appleman
1921- DLB-17

Williams, William Carlos
1883-1963 .... DLB-4, 16, 54, 86

Williams, Wirt 1921- DLB-6

Williams Brothers DLB-49

Williamson, Jack 1908- DLB-8

Willingham, Calder Baynard, Jr.
1922- DLB-2, 44

Williram of Ebersberg
circa 1020-1085 .... DLB-148

Willis, Nathaniel Parker
1806-1867 ...... DLB-3, 59, 73, 74; DS-13

Willkomm, Ernst 1810-1886 .... DLB-133

Wilmer, Clive 1945- DLB-40

Wilson, A. N. 1950- DLB-14, 155

Wilson, Angus 1913-1991 ....DLB-15, 139, 155

Wilson, Arthur 1595-1652 ....DLB-58

Wilson, Augusta Jane Evans 1835-1909 ....DLB-42

Wilson, Colin 1931- DLB-14

Wilson, Edmund 1895-1972 ....DLB-63

Wilson, Ethel 1888-1980 ....DLB-68

Wilson, Harriet E. Adams 1828?-1863? DLB-50

Wilson, Harry Leon 1867-1939 ....DLB-9

Wilson, John 1588-1667 ....DLB-24

Wilson, John 1785-1854 ....DLB-110

Wilson, Lanford 1937- DLB-7

Wilson, Margaret 1882-1973 ....DLB-9

Wilson, Michael 1914-1978 ....DLB-44

Wilson, Mona 1872-1954 ....DLB-149

Wilson, Thomas 1523 or 1524-1581 ....DLB-132

Wilson, Woodrow 1856-1924 ....DLB-47

Wilson, Effingham [publishing house] DLB-154

Wimsatt, William K., Jr. 1907-1975 ....DLB-63

Winchell, Walter 1897-1972 ....DLB-29

Winchester, J. [publishing house] DLB-49

Winckelmann, Johann Joachim 1717-1768 ....DLB-97

Winckler, Paul 1630-1686 ....DLB-164

Windham, Donald 1920- DLB-6

Wingate, Allan [publishing house] ..DLB-112

Winnifrith, Tom 1938- DLB-155

Winsloe, Christa 1888-1944 ....DLB-124

Winsor, Justin 1831-1897 ....DLB-47

John C. Winston Company DLB-49

Winters, Yvor 1900-1968 ....DLB-48

Winthrop, John 1588-1649 ....DLB-24, 30

Winthrop, John, Jr. 1606-1676 ......DLB-24

Wirt, William 1772-1834 ....DLB-37

Wise, John 1652-1725 ....DLB-24

Wiseman, Adele 1928- DLB-88

Wishart and Company DLB-112

Wisner, George 1812-1849 ....DLB-43

Wister, Owen 1860-1938 ....DLB-9, 78

Wither, George 1588-1667 ....DLB-121

Witherspoon, John 1723-1794 ......DLB-31

Withrow, William Henry 1839-1908 ...DLB-99

Wittig, Monique 1935- DLB-83

Wodehouse, P. G. 1881-1975 ....DLB-34, 162

Wohmann, Gabriele 1932- DLB-75

Woiwode, Larry 1941- DLB-6

Wolcot, John 1738-1819 ....DLB-109

Wolcott, Roger 1679-1767 ....DLB-24

Wolf, Christa 1929- DLB-75

Wolf, Friedrich 1888-1953 ....DLB-124

Wolfe, Gene 1931- DLB-8

Wolfe, Thomas 1900-1938 ....DLB-9, 102; Y-85; DS-2

Wolfe, Tom 1931- DLB-152

Wolff, Helen 1906-1994 ....Y-94

Wolff, Tobias 1945- DLB-130

Wolfram von Eschenbach circa 1170-after 1220 ....DLB-138

Wolfram von Eschenbach's *Parzival*: Prologue and Book 3 ....DLB-138

Wollstonecraft, Mary 1759-1797 ....DLB-39, 104, 158

Wondratschek, Wolf 1943- DLB-75

Wood, Benjamin 1820-1900 ....DLB-23

Wood, Charles 1932- DLB-13

Wood, Mrs. Henry 1814-1887 ......DLB-18

Wood, Joanna E. 1867-1927 ....DLB-92

Wood, Samuel [publishing house] ...DLB-49

Wood, William ?-? DLB-24

Woodberry, George Edward 1855-1930 ....DLB-71, 103

Woodbridge, Benjamin 1622-1684 ...DLB-24

Woodcock, George 1912- DLB-88

Woodhull, Victoria C. 1838-1927 ...DLB-79

Woodmason, Charles circa 1720-? ...DLB-31

Woodress, Jr., James Leslie 1916- DLB-111

Woodson, Carter G. 1875-1950 .....DLB-17

Woodward, C. Vann 1908- DLB-17

Wooler, Thomas 1785 or 1786-1853 ....DLB-158

Woolf, David (see Maddow, Ben)

Woolf, Leonard 1880-1969 ....DLB-100; DS-10

Woolf, Virginia 1882-1941 ..... DLB-36, 100, 162; DS-10

Woolf, Virginia, "The New Biography," *New York Herald Tribune,* 30 October 1927 .... DLB-149

Woollcott, Alexander 1887-1943 .....DLB-29

Woolman, John 1720-1772 ....DLB-31

Woolner, Thomas 1825-1892 ....DLB-35

Woolsey, Sarah Chauncy 1835-1905 ....DLB-42

Woolson, Constance Fenimore 1840-1894 ....DLB-12, 74

Worcester, Joseph Emerson 1784-1865 ....DLB-1

Wordsworth, Christopher 1807-1885 ....DLB-166

Wordsworth, Dorothy 1771-1855 ....DLB-107

Wordsworth, Elizabeth 1840-1932 ....DLB-98

Wordsworth, William 1770-1850 ....DLB-93, 107

The Works of the Rev. John Witherspoon (1800-1801) [excerpts] DLB-31

A World Chronology of Important Science Fiction Works (1818-1979) DLB-8

World Publishing Company DLB-46

World War II Writers Symposium at the University of South Carolina, 12–14 April 1995 ....Y-95

Worthington, R., and Company DLB-49

Wotton, Sir Henry 1568-1639 ......DLB-121

Wouk, Herman 1915- Y-82

Wreford, James 1915- DLB-88

Wren, Percival Christopher 1885-1941 ....DLB-153

Wrenn, John Henry 1841-1911 .....DLB-140

Wright, C. D. 1949- DLB-120

Wright, Charles 1935- DLB-165; Y-82

Wright, Charles Stevenson 1932- ...DLB-33

Wright, Frances 1795-1852 ....DLB-73

Wright, Harold Bell 1872-1944 ....DLB-9

Wright, James 1927-1980 ....DLB-5, 169

Wright, Jay 1935- DLB-41

Wright, Louis B. 1899-1984 ....DLB-17

Wright, Richard 1908-1960 .... DLB-76, 102; DS-2

Wright, Richard B. 1937- DLB-53

Wright, Sarah Elizabeth 1928- DLB-33

Writers and Politics: 1871-1918, by Ronald Gray DLB-66

Writers and their Copyright Holders: the WATCH Project Y-94

Writers' Forum Y-85

Writing for the Theatre, by
  Harold Pinter DLB-13
Wroth, Lady Mary 1587-1653 ..... DLB-121
Wyatt, Sir Thomas
  circa 1503-1542 .... DLB-132
Wycherley, William 1641-1715 ..... DLB-80
Wyclif, John
  circa 1335-31 December 1384 .. DLB-146
Wylie, Elinor 1885-1928 .... DLB-9, 45
Wylie, Philip 1902-1971 .... DLB-9
Wyllie, John Cook 1908-1968 ..... DLB-140

Y

Yates, Dornford 1885-1960 ..... DLB-77, 153
Yates, J. Michael 1938- DLB-60
Yates, Richard 1926-1992 ... DLB-2; Y-81, 92
Yavorov, Peyo 1878-1914 .... DLB-147
Yearsley, Ann 1753-1806 .... DLB-109
Yeats, William Butler
  1865-1939 .... DLB-10, 19, 98, 156
Yep, Laurence 1948- DLB-52
Yerby, Frank 1916-1991 .... DLB-76
Yezierska, Anzia 1885-1970 .... DLB-28
Yolen, Jane 1939- DLB-52
Yonge, Charlotte Mary
  1823-1901 .... DLB-18, 163
The York Cycle
  circa 1376-circa 1569 .... DLB-146

A Yorkshire Tragedy DLB-58
Yoseloff, Thomas
  [publishing house] DLB-46
Young, Al 1939- DLB-33
Young, Arthur 1741-1820 .... DLB-158
Young, Edward 1683-1765 .... DLB-95
Young, Stark 1881-1963 .... DLB-9, 102
Young, Waldeman 1880-1938 .... DLB-26
Young, William [publishing house] .. DLB-49
Yourcenar, Marguerite
  1903-1987 .... DLB-72; Y-88
"You've Never Had It So Good," Gusted by
  "Winds of Change": British Fiction in the
  1950s, 1960s, and After DLB-14
Yovkov, Yordan 1880-1937 .... DLB-147

Z

Zachariä, Friedrich Wilhelm
  1726-1777 .... DLB-97
Zamora, Bernice 1938- DLB-82
Zand, Herbert 1923-1970 .... DLB-85
Zangwill, Israel 1864-1926 ...... DLB-10, 135
Zanzotto, Andrea 1921- DLB-128
Zapata Olivella, Manuel 1920- ... DLB-113
Zebra Books DLB-46
Zebrowski, George 1945- DLB-8
Zech, Paul 1881-1946 .... DLB-56

Zeidner, Lisa 1955- DLB-120
Zelazny, Roger 1937-1995 .... DLB-8
Zenger, John Peter 1697-1746 ... DLB-24, 43
Zesen, Philipp von 1619-1689 ..... DLB-164
Zieber, G. B., and Company DLB-49
Zieroth, Dale 1946- DLB-60
Zigler und Kliphausen, Heinrich Anshelm von
  1663-1697 .... DLB-168
Zimmer, Paul 1934- DLB-5
Zingref, Julius Wilhelm
  1591-1635 .... DLB-164
Zindel, Paul 1936- DLB-7, 52
Zinzendorf, Nikolaus Ludwig von
  1700-1760 .... DLB-168
Zola, Emile 1840-1902 .... DLB-123
Zolotow, Charlotte 1915- DLB-52
Zschokke, Heinrich 1771-1848 .... DLB-94
Zubly, John Joachim 1724-1781 ...... DLB-31
Zu-Bolton II, Ahmos 1936- DLB-41
Zuckmayer, Carl 1896-1977 .... DLB-56, 124
Zukofsky, Louis 1904-1978 ..... DLB-5, 165
Župančič, Oton 1878-1949 .... DLB-147
zur Mühlen, Hermynia 1883-1951 .... DLB-56
Zweig, Arnold 1887-1968 .... DLB-66
Zweig, Stefan 1881-1942 .... DLB-81, 118

ISBN 0-8103-9932-6

(Continued from front endsheets)

117 *Twentieth-Century Caribbean and Black African Writers, First Series*, edited by Bernth Lindfors and Reinhard Sander (1992)

118 *Twentieth-Century German Dramatists, 1889-1918*, edited by Wolfgang D. Elfe and James Hardin (1992)

119 *Nineteenth-Century French Fiction Writers: Romanticism and Realism, 1800-1860*, edited by Catharine Savage Brosman (1992)

120 *American Poets Since World War II, Third Series*, edited by R. S. Gwynn (1992)

121 *Seventeenth-Century British Nondramatic Poets, First Series*, edited by M. Thomas Hester (1992)

122 *Chicano Writers, Second Series*, edited by Francisco A. Lomelí and Carl R. Shirley (1992)

123 *Nineteenth-Century French Fiction Writers: Naturalism and Beyond, 1860-1900*, edited by Catharine Savage Brosman (1992)

124 *Twentieth-Century German Dramatists, 1919-1992*, edited by Wolfgang D. Elfe and James Hardin (1992)

125 *Twentieth-Century Caribbean and Black African Writers, Second Series*, edited by Bernth Lindfors and Reinhard Sander (1993)

126 *Seventeenth-Century British Nondramatic Poets, Second Series*, edited by M. Thomas Hester (1993)

127 *American Newspaper Publishers, 1950-1990*, edited by Perry J. Ashley (1993)

128 *Twentieth-Century Italian Poets, Second Series*, edited by Giovanna Wedel De Stasio, Glauco Cambon, and Antonio Illiano (1993)

129 *Nineteenth-Century German Writers, 1841-1900*, edited by James Hardin and Siegfried Mews (1993)

130 *American Short-Story Writers Since World War II*, edited by Patrick Meanor (1993)

131 *Seventeenth-Century British Nondramatic Poets, Third Series*, edited by M. Thomas Hester (1993)

132 *Sixteenth-Century British Nondramatic Writers, First Series*, edited by David A. Richardson (1993)

133 *Nineteenth-Century German Writers to 1840*, edited by James Hardin and Siegfried Mews (1993)

134 *Twentieth-Century Spanish Poets, Second Series*, edited by Jerry Phillips Winfield (1994)

135 *British Short-Fiction Writers, 1880-1914: The Realist Tradition*, edited by William B. Thesing (1994)

136 *Sixteenth-Century British Nondramatic Writers, Second Series*, edited by David A. Richardson (1994)

137 *American Magazine Journalists, 1900-1960, Second Series*, edited by Sam G. Riley (1994)

138 *German Writers and Works of the High Middle Ages: 1170-1280*, edited by James Hardin and Will Hasty (1994)

139 *British Short-Fiction Writers, 1945-1980*, edited by Dean Baldwin (1994)

140 *American Book-Collectors and Bibliographers, First Series*, edited by Joseph Rosenblum (1994)

141 *British Children's Writers, 1880-1914*, edited by Laura M. Zaidman (1994)

142 *Eighteenth-Century British Literary Biographers*, edited by Steven Serafin (1994)

143 *American Novelists Since World War II, Third Series*, edited by James R. Giles and Wanda H. Giles (1994)

144 *Nineteenth-Century British Literary Biographers*, edited by Steven Serafin (1994)

145 *Modern Latin-American Fiction Writers, Second Series*, edited by William Luis and Ann González (1994)

146 *Old and Middle English Literature*, edited by Jeffrey Helterman and Jerome Mitchell (1994)

147 *South Slavic Writers Before World War II*, edited by Vasa D. Mihailovich (1994)

148 *German Writers and Works of the Early Middle Ages: 800-1170*, edited by Will Hasty and James Hardin (1994)

149 *Late Nineteenth- and Early Twentieth-Century British Literary Biographers*, edited by Steven Serafin (1995)

150 *Early Modern Russian Writers, Late Seventeenth and Eighteenth Centuries*, edited by Marcus C. Levitt (1995)

151 *British Prose Writers of the Early Seventeenth Century*, edited by Clayton D. Lein (1995)

152 *American Novelists Since World War II, Fourth Series*, edited by James and Wanda Giles (1995)

153 *Late-Victorian and Edwardian British Novelists, First Series*, edited by George M. Johnson (1995)

154 *The British Literary Book Trade, 1700-1820*, edited by James K. Bracken and Joel Silver (1995)

155 *Twentieth-Century British Literary Biographers*, edited by Steven Serafin (1995)

156 *British Short-Fiction Writers, 1880-1914: The Romantic Tradition*, edited by William F. Naufftus (1995)

157 *Twentieth-Century Caribbean and Black African Writers, Third Series*, edited by Bernth Lindfors and Reinhard Sander (1995)

158 *British Reform Writers, 1789-1832*, edited by Gary Kelly and Edd Applegate (1995)

159 *British Short Fiction Writers, 1800-1880*, edited by John R. Greenfield (1996)

160 *British Children's Writers, 1914-1960*, edited by Donald R. Hettinga and Gary D. Schmidt (1996)

161 *British Children's Writers Since 1960, First Series*, edited by Caroline Hunt (1996)

162 *British Short-Fiction Writers, 1915-1945*, edited by John H. Rogers (1996)

163 *British Children's Writers, 1800-1880*, edited by Meena Khorana (1996)

164 *German Baroque Writers, 1580-1660*, edited by James Hardin (1996)

165 *American Poets Since World War II, Fourth Series*, edited by Joseph Conte (1996)

166 *British Travel Writers, 1837-1875*, edited by Barbara Brothers and Julia Gergits (1996)

167 *Sixteenth-Century British Nondramatic Writers, Third Series*, edited by David A. Richardson (1996)

168 *German Baroque Writers, 1661-1730*, edited by James Hardin (1996)

169 *American Poets Since World War II, Fifth Series*, edited by Joseph Conte (1996)

55 *Victorian Prose Writers Before 1867*, edited by William B. Thesing (1987)

56 *German Fiction Writers, 1914-1945*, edited by James Hardin (1987)

57 *Victorian Prose Writers After 1867*, edited by William B. Thesing (1987)

58 *Jacobean and Caroline Dramatists*, edited by Fredson Bowers (1987)

59 *American Literary Critics and Scholars, 1800-1850*, edited by John W. Rathbun and Monica M. Grecu (1987)

60 *Canadian Writers Since 1960, Second Series*, edited by W. H. New (1987)

61 *American Writers for Children Since 1960: Poets, Illustrators, and Nonfiction Authors*, edited by Glenn E. Estes (1987)

62 *Elizabethan Dramatists*, edited by Fredson Bowers (1987)

63 *Modern American Critics, 1920-1955*, edited by Gregory S. Jay (1988)

64 *American Literary Critics and Scholars, 1850-1880*, edited by John W. Rathbun and Monica M. Grecu (1988)

65 *French Novelists, 1900-1930*, edited by Catharine Savage Brosman (1988)

66 *German Fiction Writers, 1885-1913*, 2 parts, edited by James Hardin (1988)

67 *Modern American Critics Since 1955*, edited by Gregory S. Jay (1988)

68 *Canadian Writers, 1920-1959, First Series*, edited by W. H. New (1988)

69 *Contemporary German Fiction Writers, First Series*, edited by Wolfgang D. Elfe and James Hardin (1988)

70 *British Mystery Writers, 1860-1919*, edited by Bernard Benstock and Thomas F. Staley (1988)

71 *American Literary Critics and Scholars, 1880-1900*, edited by John W. Rathbun and Monica M. Grecu (1988)

72 *French Novelists, 1930-1960*, edited by Catharine Savage Brosman (1988)

73 *American Magazine Journalists, 1741-1850*, edited by Sam G. Riley (1988)

74 *American Short-Story Writers Before 1880*, edited by Bobby Ellen Kimbel, with the assistance of William E. Grant (1988)

75 *Contemporary German Fiction Writers, Second Series*, edited by Wolfgang D. Elfe and James Hardin (1988)

76 *Afro-American Writers, 1940-1955*, edited by Trudier Harris (1988)

77 *British Mystery Writers, 1920-1939*, edited by Bernard Benstock and Thomas F. Staley (1988)

78 *American Short-Story Writers, 1880-1910*, edited by Bobby Ellen Kimbel, with the assistance of William E. Grant (1988)

79 *American Magazine Journalists, 1850-1900*, edited by Sam G. Riley (1988)

80 *Restoration and Eighteenth-Century Dramatists, First Series*, edited by Paula R. Backscheider (1989)

81 *Austrian Fiction Writers, 1875-1913*, edited by James Hardin and Donald G. Daviau (1989)

82 *Chicano Writers, First Series*, edited by Francisco A. Lomelí and Carl R. Shirley (1989)

83 *French Novelists Since 1960*, edited by Catharine Savage Brosman (1989)

84 *Restoration and Eighteenth-Century Dramatists, Second Series*, edited by Paula R. Backscheider (1989)

85 *Austrian Fiction Writers After 1914*, edited by James Hardin and Donald G. Daviau (1989)

86 *American Short-Story Writers, 1910-1945, First Series*, edited by Bobby Ellen Kimbel (1989)

87 *British Mystery and Thriller Writers Since 1940, First Series*, edited by Bernard Benstock and Thomas F. Staley (1989)

88 *Canadian Writers, 1920-1959, Second Series*, edited by W. H. New (1989)

89 *Restoration and Eighteenth-Century Dramatists, Third Series*, edited by Paula R. Backscheider (1989)

90 *German Writers in the Age of Goethe, 1789-1832*, edited by James Hardin and Christoph E. Schweitzer (1989)

91 *American Magazine Journalists, 1900-1960, First Series*, edited by Sam G. Riley (1990)

92 *Canadian Writers, 1890-1920*, edited by W. H. New (1990)

93 *British Romantic Poets, 1789-1832, First Series*, edited by John R. Greenfield (1990)

94 *German Writers in the Age of Goethe: Sturm und Drang to Classicism*, edited by James Hardin and Christoph E. Schweitzer (1990)

95 *Eighteenth-Century British Poets, First Series*, edited by John Sitter (1990)

96 *British Romantic Poets, 1789-1832, Second Series*, edited by John R. Greenfield (1990)

97 *German Writers from the Enlightenment to Sturm und Drang, 1720-1764*, edited by James Hardin and Christoph E. Schweitzer (1990)

98 *Modern British Essayists, First Series*, edited by Robert Beum (1990)

99 *Canadian Writers Before 1890*, edited by W. H. New (1990)

100 *Modern British Essayists, Second Series*, edited by Robert Beum (1990)

101 *British Prose Writers, 1660-1800, First Series*, edited by Donald T. Siebert (1991)

102 *American Short-Story Writers, 1910-1945, Second Series*, edited by Bobby Ellen Kimbel (1991)

103 *American Literary Biographers, First Series*, edited by Steven Serafin (1991)

104 *British Prose Writers, 1660-1800, Second Series*, edited by Donald T. Siebert (1991)

105 *American Poets Since World War II, Second Series*, edited by R. S. Gwynn (1991)

106 *British Literary Publishing Houses, 1820-1880*, edited by Patricia J. Anderson and Jonathan Rose (1991)

107 *British Romantic Prose Writers, 1789-1832, First Series*, edited by John R. Greenfield (1991)

108 *Twentieth-Century Spanish Poets, First Series*, edited by Michael L. Perna (1991)

109 *Eighteenth-Century British Poets, Second Series*, edited by John Sitter (1991)

110 *British Romantic Prose Writers, 1789-1832, Second Series*, edited by John R. Greenfield (1991)

111 *American Literary Biographers, Second Series*, edited by Steven Serafin (1991)

112 *British Literary Publishing Houses, 1881-1965*, edited by Jonathan Rose and Patricia J. Anderson (1991)

113 *Modern Latin-American Fiction Writers, First Series*, edited by William Luis (1992)

114 *Twentieth-Century Italian Poets, First Series*, edited by Giovanna Wedel De Stasio, Glauco Cambon, and Antonio Illiano (1992)

115 *Medieval Philosophers*, edited by Jeremiah Hackett (1992)

116 *British Romantic Novelists, 1789-1832*, edited by Bradford K. Mudge (1992)

(Continued on back endsheets)